Studies in Logic
Logic and Argumentation
Volume 77

Argumentation and Inference
Proceedings of the 2nd European Conference on Argumentation
Volume II

Volume 66
Logical Consequences. Theory and Applications: An Introduction.
Luis M. Augusto

Volume 67
Many-Valued Logics: A Mathematical and Computational Introduction
Luis M. Augusto

Volume 68
Argument Technologies: Theory, Analysis, and Applications
Floris Bex, Floriana Grasso, Nancy Green, Fabio Paglieri and
Chris Reed, eds

Volume 69
Logic and Conditional Probability. A Synthesis
Philip Calabrese

Volume 70
Proceedings of the International Conference. Philosophy, Mathematics,
Linguistics: Aspects of Interaction, 2012 (PhML-2012)
Oleg Prosorov, ed.

Volume 71
Fathoming Formal Logic: Volume I. Theory and Decision Procedures for
Propositional Logic
Odysseus Makridis

Volume 72
Fathoming Formal Logic: Volume II. Semantics and Proof Theory for
Predicate Logic
Odysseus Makridis

Volume 73
Measuring Inconsistency in Information
John Grant and Maria Vanina Mrtinez, eds.

Volume 74
Dictionary of Argumentation. An Introduction to Argumentation Studies
Christian Plantin. With a Foreword by J. Anthony Blair

Volume 75
Theory of Effective Propositional Paraconsistent Logics
Arnon Avron, Ofer Arieli and Anna Zamansky

Volume 76
Argumentation and Inference. Proceedings of the 2nd European Conference
on Argumentation. Volume I
Steve Oswald and Didier Maillat, eds.

Volume 77
Argumentation and Inference. Proceedings of the 2nd European Conference
on Argumentation. Volume II
Steve Oswald and Didier Maillat, eds.

Studies in Logic Series Editor
Dov Gabbay   dov.gabbay@kcl.ac.uk

Argumentation and Inference
Proceedings of the 2nd European Conference on Argumentation
Volume II

Edited by

Steve Oswald

and

Didier Maillat

© Individual author and College Publications, 2018
All rights reserved.

ISBN 978-1-84890-284-8

College Publications
Scientific Director: Dov Gabbay
Managing Director: Jane Spurr

http://www.collegepublications.co.uk

Printed by Lightning Source, Milton Keynes, UK

All rights reserved. No part of this publication may be reproduced, stored in a retrieval system or transmitted in any form, or by any means, electronic, mechanical, photocopying, recording or otherwise without prior permission, in writing, from the publisher.

Table of contents

Regular Papers

1. Inference and Virtue 1
   Andrew Aberdein

2. Why the Dialectical Tier is an Epistemic Animal 11
   Scott F. Aikin

3. (Digital and) Material Representations of the European Green 23
   Belt: Juxtaposing Nature and Technology in Our Collective
   Memory of the Cold War
   Marcia Allison & Emma Frances Bloomfield

4. Reasoning Together: Fostering Rationality Through Group 33
   Deliberation
   Mark Battersby & Sharon Bailin

5. Argumentation Profiles vs. Argumentation Schemes 49
   Angelina Bobrova

6. Believing, Inferring, and Basing 63
   Patrick Bondy

7. Be Committed to Your Premises, or Face the Consequences: 79
   A Pragmatic Analysis of Commitment Inferences
   Kira Boulat & Didier Maillat

8. The Role of Argumentative Discussions in the Transmission............ 93
   of Implicit Norms and Values in Families with Young Children
   Antonio Bova

9. Virtue Argumentation Theory Reconsidered: Towards 107
   a Complete Account of Good Argument
   Tracy Bowell & Justine Kingsbury

10. A Conflict Index for Arguments in an Argumentation Graph 115
    Giulia Cesari, Francesca Fossati & Stefano Moretti

11. Argumentation Traits, Frames, and Dialogue Orientations 133
    Ioana A. Cionea, Dale Hample, Stacie Wilson Mumpower,
    Eryn N. Bostwick, Cameron W. Piercy & Candace L. Foutch

12. From Semi-Abstract Argumentation to Logical Consequence 151
    Esther Anna Corsi & Christian G. Fermüller

13. Reconstructing Multimodal Argument: The Dove vs. Nivea 165
    Advertisement
    Hédi Virág Csordás & Gábor Forrai

14. Rhetorical Inferences for Divine Authority: The Case of 179
    Classical Greek Divination
    Julie Dainville

15. Irresolvable Rational Disputes 191
    Istvan Danka

16. Processing Linguistic Persuasion: A Study in Message 205
    Discrepancy
    Kamila Dębowska-Kozłowska

17. Questioning the Explicit Cancellability of Scalar Implicatures 221
    Laura Devlesschouwer

18. Reasoning via Dialogue: An Illustrative Analysis 233
    of Deliberation
    Stéphane Dias & Jane Silveira

19. Are Humans Poor at Arguing? From 'the Argumentative 247
    Theory of Reasoning' back to a Rhetorical Theory of
    Argumentation
    Salvatore Di Piazza, Francesca Piazza & Mauro Serra

20. The Critical Question Model of Argument Assessment 263
    Ian J. Dove & E. Michael Nussbaum

21. Topoi and Refutations in Aristotle 281
    Iovan Drehe

22. Gender, Argumentation and Inference in Mexican Political 289
    and Media Discourse
    Olga Nelly Estrada & Griselda Zárate

23. "Conductive" Argumentation in the UK Fracking Debate 299
    Isabela Fairclough

24. The Development of Prototypical Patterns of Weighing 313
    and Balancing in the Justification of Judicial Decisions
    Eveline T. Feteris

25. "Metaphors are No Arguments, My Pretty Maiden" 327
    The Pragma-Dialectical Reconstruction of Figurative
    Analogy
    Bart Garssen

26. Against the Intentional Definition of Argument 339
    G. C. Goddu

27. Visual emotional argumentation: Analytical models 349
    Julieta Haidar

28. The Concept of an Argument 363
    David Hitchcock

29. Metaphors as Arguments: Perspectives from 377
    Psycholinguistics
    Curtis Hyra, Hamad Al-Azary, Lori Buchanan &
    Catherine Hundleby

30. Justifying a Bill Before Parliament: Beyond Instrumental 397
    Rationality
    Constanza Ihnen Jory

31. Ethos and Inference: Insights from a Multimodal Perspective 413
    Jérôme Jacquin

32. *Ad Populum* Arguments in a Political Context 425
    Henrike Jansen

33. View on Inference from Argumentation Tiers Perspective 439
    Iryna Khomenko

34. Studying the Process of Interpretation on a School Task: 453
    Crossing Perspectives
    Alaric Kohler & Teuta Mehmeti

35. Modes of Inference in Aristotle's Concept of the Enthymeme 479
    Manfred Kraus

36. "I Think Any Reasonable Person Will Agree...": A Corpus 493
    and Text Study of Keywords in Irish Political Argumentation
    Davide Mazzi

37. Arguing Inter-Issue in Public Political Arguments 511
    Dima Mohammed

38. Inference and Argument: Normative and Descriptive 527
    Dimensions
    Andrei Moldovan

39. Inferring Argumentative Patterns in Polylogues about 541
    Energy Issues
    Elena Musi & Mark Aakhus

40. Apologia in a Networked Society: The Case of Volkswagen's 561
    Polylogical Challenge
    Cassandra Oliveras-Moreno, Mark Aakhus & Marcin Lewiński

41. The Strategic Use of Examples in Supporting a Positive 581
    Evaluation of a Political Group
    Ahmed Omar

42. It Ought To Be Therefore It Is: On Fallaciousness of So-Called 597
    Moralistic Fallacy
    Tomáš Ondráček & Iva Svačinová

43. Pragmatic Inference and Argumentative Inference 615
    Steve Oswald

44. Challenging Judicial Impartiality: When Accusations of 631
    Derailments of Strategic Manoeuvring Derail
    H. José Plug

45. The "Neoliberal Agenda": How Portuguese Parties Use 647
    The "Neoliberalism" Concept to Argue Against Austerity
    Vera Ramalhete & Marco Lisi

46. Effect of Explicitness of Teachers' Arguments on Quality of 665
    Adolescent Students' Inferences in Science and History
    Chrysi Rapanta

47. On the Epistemic Basing Relation 683
    Juho Ritola

48. Dialogical Argumentation in Financial Conference Calls: 699
    the Request of Confirmation of Inference (ROCOI)
    Andrea Rocci & Carlo Raimondo

49. How to Create Rhetorical Exercises? 717
    Benoît Sans

50. Filling in the Gaps: The Role of Audience Inference in 731
    Exigence and Ethos
    Blake Scott

51. Multimodal Argumentation in Factual Television 741
    Andrea Sabine Sedlaczek

52. Bakhtin at the White House: The Argumentative Dimension 755
    of the Direct Address in the TV Series House of Cards
    *Carmen Spanò, Antonio Bova, Carlo Galimberti, Daniela Tacchi &
    Ilaria Vergine*

53. Arguments from Other Cases 769
    Katharina Stevens

54. The Attraction of the Ideal Has No Traction on the Real: 787
    On Adversariality and Roles in Argument
    Katharina Stevens & Daniel H. Cohen

55. Cultural Disagreements and Legal Argumentation: 805
    An Educational Program in Middle Schools
    Serena Tomasi

56. The Explicit/Implicit Distinction in Multimodal 821
    Argumentation: Comparing the Argumentative Use of
    Nano-Images in Scientific Journals and Science Magazines
    Assimakis Tseronis

57. Social Costs of Epistemic Vigilance and Premises 843
    in Arguments
    Christoph Unger

58. Sets of Situations, Topics, and Question Relevance 857
    Mariusz Urbański & Natalia Żyluk

59. Strategic Maneuvering with Presentational Devices: 873
    A Systematic Approach
    Ton Van Haaften & Maarten Van Leeuwen

60. Criticism and Justification of Negotiated Compromises 887
    Jan Albert van Laar & Erik C. W. Krabbe

61. Argumentative Functions of Metaphors: How can Metaphors 909
    Trigger Resistance?
    Lotte van Poppel

62. Argument Structures and Frame Theory as Tools of 925
    Linguistic Discourse Analysis
    Simon Varga

63. Straw Man as Misuse of Rephrase 941
    *Jacky Visser, Marcin Koszowy, Barbara Konat, Kasia Budzynska
    & Chris Reed*

64. Inferences Across Normative Domains 963
    Sheldon Wein

65. Conspiracy Ideations in Healthcare: A Rhetorical and 973
    Argumentative Analysis
    Roberta Martina Zagarella & Marco Annoni

66. With the Best Intentions, and the Worst Arguments. 989
    The "Fertility Day" Campaign in Italy
    Marta Zampa & Chiara Pollaroli

67. Of Inference and Argumentation in Financial Discourse: 1015
    The Crisis of 2007-2008
    Griselda Zárate & Homero Zambrano

68. What Warrants the Warrant? 1027
    David Zarefsky

1

Inference and Virtue

ANDREW ABERDEIN
Florida Institute of Technology
aberdein@fit.edu

What are the prospects (if any) for a virtue-theoretic account of inference? This paper compares three options. Firstly, assess each argument individually in terms of the virtues of the participants. Secondly, make the capacity for cogent inference itself a virtue. Thirdly, recapture a standard treatment of cogency by accounting for each of its components in terms of more familiar virtues. The three approaches are contrasted and their strengths and weaknesses assessed.

KEYWORDS: acceptability, cogency, epistemological approach, inference, relevance, reliabilism, responsibilism, sufficiency, virtue argumentation

1. INTRODUCTION

Virtue theories of argumentation (VTA) have recently attracted significant interest (Aberdein and Cohen, 2016). This paper addresses the possibility of analysing inference in terms of VTA. There aren't an enormous number of virtue argumentation theorists, but almost all of them, possibly all of them bar me, believe that virtue theory is not sufficient to describe inferences in a meaningful way (see, for example, Bowell and Kingsbury, 2013; Cohen, 2013; Gascón, 2015). The evaluation of inferences is seen as something that should be handed off to some other argumentation theory, of one of the many available flavours. The idea that it could be in any sense "virtues all the way down" is a position that I have advocated for, but so far I don't think I've succeeded in persuading anyone else. This paper is an attempt to do that—or at least to defend my obduracy.

There are at least three options for a virtue-theoretic account of inference. Firstly, we might adopt "virtue eliminativism", by analogy with similar positions in virtue epistemology which maintain that traditional epistemological concepts are incommensurable with—and should be replaced by—virtue theory. In the context of argument, this could take the form of rejecting altogether the existence of argument patterns that all virtuous arguers accept, and thereby assessing each argument individually in terms of the virtues of the participants. Secondly, we could seek to transpose cogency into virtue-theoretic terms by making the capacity to produce and recognize cogent inference a virtue. Thirdly, we could attempt to recapture a standard account of cogency in terms of more familiar virtues. This is perhaps the most ambitious of the three (and fittingly corresponds to what Fabio Paglieri has dubbed "ambitious moderate" VTA: Paglieri, 2015, p. 77). Such an approach would require accounts of all components of cogency in terms of virtues. I shall address each of these options in turn.

2. VIRTUE ELIMINATIVISM

In dubbing the first option "virtue eliminativism", I follow the virtue epistemologist Heather Battaly: "virtue-eliminativism … argues that epistemological projects other than explorations of the virtues should be eliminated: we should abandon discussions of knowledge and justification, and replace them with analyses of the virtues" (Battaly, 2008, p. 642). She is describing one strategy virtue epistemologists could take, and not what most of them do; most virtue epistemologists use virtues to recapture more traditional approaches to epistemology. By contrast, virtue eliminativists seek to rebuild everything from a fresh virtuistic foundation and if there are certain things which just don't get to be rebuilt on that foundation, bite that bullet and jettison those things. This is not a position Battaly is advocating, and its VTA counterpart is not a position I am advocating. I am not sure whether anyone actually is advocating either position. There is an undeniable anarchic thrill in dismissing all talk of cogency as a bad idea and concentrating on the virtues of arguments, first and last, even if that means we need to abandon any prospect of a shared structure between arguments and evaluate each argument individually. But although it is a position open to somebody, it is not one that I support.

3. COGENCY AS A VIRTUE

Another approach that is much more likely to prosper is what I shall refer to as "cogency as a virtue". We might also think of this as the easy road, or the low road. The key thought here is that being able to produce cogent arguments, and being able to recognize them when they are produced by other people, is a virtue. That solves the problem, at least to some extent. This would be continuous with the virtue reliablilism of Ernest Sosa, in which deductive inference is identified as a virtue:

> Whatever exactly the end may be, the virtue of a virtue derives not simply from leading us to it, perhaps accidentally, but from leading us to it reliably: e.g., 'in a way *bound* to maximize one's surplus of truth over error'. Rationalist intuition and deduction are thus prime candidates, since they would always lead us aright. But it is not so clearly virtuous to admit no other faculties, seeing the narrow limits beyond which intuition and deduction will never lead us. What other faculties might one admit? ... There are faculties of two broad sorts: those that lead to beliefs from beliefs already formed, and those that lead to beliefs but not from beliefs. The first of these we call 'transmission' faculties, the second 'generation' faculties. Rationalist deduction is hence a transmission faculty and rationalist intuition a 'generation' faculty. Supposing reason a single faculty with subfaculties of intuitive reason and inferential reason, reason itself is then both a transmission faculty and a generation faculty. (Sosa, 1985, p. 227).

One reason for the lengthy quotation is that Sosa can be somewhat skittish about actually using the word "virtue". He often prefers to speak of "faculties". I quote him here at length to remove any doubt that these faculties are indeed intended to be virtues. He acknowledges "rationalist intuition and deduction" as "prime candidates" for being virtues, and he talks about reason as "a single faculty", that is virtue, "with sub[virtues] of intuitive reason and inferential reason". Thus we have a faculty of reason that is cashed out in virtue terms. So, at least in some respects, that does what we need. Indeed, it would have the advantage that successful cogent inference is more readily observed than valid deductive inference, answering the charge that virtues should be attainable, and not a seldom achieved ideal. Nonetheless, it may seem a bit of a cop out: in essence, we've just said, "There's a virtue for that!" And that's it.

4. COGENCY RECAPTURE: RSA

A more interesting approach may be to take an off-the-shelf account of cogent argument, and then cash it out in virtue terms. The go-to off-the-shelf account of cogent argument is the RSA, aka ARG, approach to cogency. Here is Trudy Govier's version:

> The basic elements of a cogent argument, referred to here as the **ARG conditions**, are as follows:
>
> i. It has **acceptable premises**. That is, it is reasonable for those to whom the argument is addressed to believe these premises. There is good reason to accept the premises—even if, in some cases, they are not known to be true—and there is no good evidence indicating that the premises are false. When you are evaluating an argument, the person to whom the premises must be acceptable is you yourself. You have to think about whether you do accept them, or have good reason to accept them. ...
> ii. Its premises are relevant to its conclusion. By this we mean that the premises state evidence, offer reasons that support the conclusion, or can be arranged into a demonstration from which the conclusion can be derived. The **relevance of premises** is necessary for the cogency of an argument. ...
> iii. The premises provide adequate or **good grounds** for the conclusion. In other words, *considered together*, the premises give sufficient reason to make it rational to accept the conclusion. (Govier, 2010, p. 87).

What are our prospects for cashing all this out in terms of virtues? There are three sub-projects here, obviously: acceptability, relevance, and grounds (or sufficiency).

4.1 Acceptability

Acceptability is the easiest of the three. We already have some acknowledgement of its agent-relativity in Govier's definition: it is "reasonable for those to whom the argument is addressed". So, being the sort of person who has the right sense of what acceptable means is a plausible candidate for a virtue. Likewise, for Ralph Johnson, "acceptability will have to be understood in terms of a dialectical situation, of the interplay between arguer and Other" (Johnson, 2000, p. 195). Of course, "dialectical" is still quite a long way from virtue, but we're heading in the right direction. Specifically, it is not hard to see how that interplay could be cashed out in terms of the virtues of the respective parties, such as recognition of reliable authority or willingness to question the obvious.

4.2 Relevance

Relevance is going to be harder work. Johnson states that "relevance is a dialectical criterion" too (Johnson, 2000, p. 204). As noted above, this is at least a step in the right direction. More promisingly, Paglieri proposes that VTA take a lead from relevance theory in pragmatics. Specifically, he cites this definition of "Relevance of an input to an individual at a time" from the pioneering relevance theorists Deirdre Wilson and Dan Sperber:

> (a) Everything else being equal, the greater the positive cognitive effects achieved in an individual by processing an input at a given time, the greater the relevance of the input to that individual at that time.
> (b) Everything else being equal, the smaller the processing effort expended by the individual in achieving those effects, the greater the relevance of the input to that individual at that time. (Wilson and Sperber, 2002, p. 602).

As Paglieri observes,

> Here relevance is no longer a property of the argument per se, but rather a feature of the interaction between argument, context, and interpreter. While relevance theorists may leave it at that, virtue theorists will want to go a step further and add that also the ability to *be* argumentatively relevant (that is, to produce arguments that are relevant to one's intended audience within the appropriate context) is a virtue worth having—now for the producer of the argument, rather than its interpreter. (Paglieri, 2015, pp. 79-80).

However, at least as written, this won't work: we are looking for an explication of the relevance of premisses to conclusion, not of arguments to audiences. (Paglieri acknowledges as much: his purposes are not mine.) Indeed, audience relevance would seem to fall under acceptability, so we would appear to be no further forward. Nonetheless, we may reasonably ask whether premiss–conclusion relevance can be addressed in virtuistic terms. Certainly there are some virtues, such as recognition of salient facts, which seem suited to play a role here. But in order for them to do so, we must overcome the intuition that premiss–conclusion relevance must be agent-neutral.

4.3 Sufficiency

The trickiest of the RSA/ARG triple is sufficiency. Once again, we can take some comfort from Johnson: "It seems clear that the question to be asked requires that the critic look at all the evidence produced by the arguer and ask whether the premises, taken together, provide enough support for the conclusion" (Johnson, 2000, p. 205). At least we've got some sense of this being something which people do. So we have a process-based understanding of how this goes, that this isn't just a property of arguments as abstract objects, this is something that arises out of the interplay between the various parties concerned. Perhaps we can make a stronger case to cash this out in detail in virtue terms, for example by stressing the importance of intellectual empathy, in the form of insight into problems, and of intellectual perseverance.

But we always have a fall-back option, courtesy of Sosa and the low road. We have travelled some way up the high road, by noting that acceptability at least can plausibly be cashed out in virtue terms. Perhaps we have made less progress on relevance and sufficiency, but we can detour back onto the low road to finish the job. Thereby we can at least narrow down the task that we want the virtues to do from analysing cogency as a whole to analysing some of its essential components. So the route is not blocked, but perhaps we end up back on the low road after all. Nonetheless, I think we've learned something by going some distance on the high road.

5. COGENCY RECAPTURE: MERITS

Another high road strategy would be to observe that, although the RSA/ARG account has been very influential and widely adopted it is not the only way of thinking about cogency. A different analysis of cogency may do a better job of giving us a virtue account. Here, for example, is William Rehg:

> I have distinguished three types of merits of cogent arguments as products of argumentative practices. *Content merits* can be identified in the text of the argument itself by applying various analytic tools to an interpretation of that text. An argument has *transactional merits* to the extent that it wins acceptance in a local dialogue (an exchange in a small group and/or between an arguer and a text) conducted in a way that fosters reasonable judgment. The conditions for ascribing transactional merits vary according to the particular transactional context—the capacities of the participants to

process information, their background knowledge, local conventions of argument, and so on. An argument has *public merits* insofar as it can travel across different transactional locales whose macrosocial arrangement and aggregate conditioning sustain collective reasonableness (Rehg, 2005, p. 110).

He talks about "merits" rather than virtues, but the difference would seem to be primarily terminological. The major point of contrast is that these are not virtues of persons, they are the virtues of the arguments themselves. But this is akin to an alternative strategy in virtue epistemology, perhaps especially the use of virtue theory in the philosophy of science, of talking about virtues of theories rather than virtues of persons. So this is potentially a starting point for another way of scouting out the high road, and a rather different approach to VTA, to take a virtue approach to arguments as products, and then either satisfy yourself with that, or try and understand how an account of arguers might emerge from that project.

6. COGENCY RECAPTURE: THE EPISTEMOLOGICAL APPROACH

There is another approach to the high road which offers more immediate promise. In an earlier paper, I pointed out that, while different approaches to argumentation were more or less promising for VTA, if you take an epistemological approach to argumentation and you are also a virtue epistemologist, it is hard to avoid being a virtue argumentation theorist (Aberdein, 2014, p. 79). This does not work for every iteration of the epistemological approach to argumentation. But Christoph Lumer's helpful survey allows us to pinpoint where it does work (Lumer, 2005). Most obviously perhaps, we could employ Lumer's "responsibilist criteria" (RE), which place the emphasis squarely on the arguer:

> RE1 1 The arguer justifiedly believes in the reasons.
> 2 In case of uncertain arguments the arguer does not dispose of further information relevant to the implication.
> RE2 1 The arguer justifiedly believes that the reasons' acceptability, according to an effective epistemological principle, implies the thesis' acceptability.
> 2 Because of these beliefs the arguer believes in the thesis. (Lumer, 2005, p. 195).

Alternatively, we could prioritize the respondent, as Lumer does in his "gnostic or weak epistemic criteria" (G):

> G1 1 The argumentation's addressee justifiedly believes in the argument's reasons.
> 2 And he has no further information that would defeat that argument.
> G2 It is reasonable for the addressee to proceed from believing in the reasons to believing in the argument's thesis. (Lumer, 2005, p. 194).

Lumer has other criteria for the evaluation of arguments which are not as good a fit. But these two would fit a virtue-based approach very nicely. Of course, we are then faced with another question within the epistemological approach, and between it and its critics, of whether these criteria yield a good enough account of cogency. I shall not address that question here; for present purposes it suffices to observe that some flavours of the epistemological approach to argumentation, when combined with a virtue approach to epistemology, lead to a virtue approach to argument evaluation, that is, another path to the high road.

7. CONCLUSION

We have seen that there are a number of live options for a virtuistic account of inference. The eliminativist approach would be very radical indeed. This I remark upon as being a possibility which nobody has as yet seriously entertained, probably for very good reasons. But it is at least a position in the conceptual space. Perhaps it is worth someone, even more extreme than me, investigating exactly what it would entail. Another approach would be what I've been calling the low road: make the ability to produce cogent arguments and the ability to recognise them themselves virtues. The low road will have implications for what the virtues of argument will look like, since it seems to take VTA further in the direction of reliabilism than most accounts. Nonetheless, it offers a clear route to a virtue-based account of inference. As for the third option, piecemeal cogency recapture, what I've been calling the high road, we have seen that there are a number of different possible projects which one might pursue under that heading. Recapture of the RSA/ARG account remains a work in progress, but we have a fall-back position of diverting onto the low road to cover any gaps. Alternatively, a synthesis of VTA with the epistemological approach offers a promising alternative plan for constructing a high road. So my overall moral is that my lonely position is lonely for bad reasons—I would welcome company!

REFERENCES

Aberdein, A. (2014). In defence of virtue: The legitimacy of agent-based argument appraisal. *Informal Logic, 34*(1), 77–93.
Aberdein, A. & Cohen, D. H. (2016). Introduction: Virtues and arguments. *Topoi, 35*(2), 339–343.
Battaly, H. (2008). Virtue epistemology. *Philosophy Compass, 3*(4), 639–663.
Bowell, T. & Kingsbury, J. (2013). Virtue and argument: Taking character into account. *Informal Logic, 33*(1), 22–32.
Cohen, D. H. (2013). Virtue, in context. *Informal Logic, 33*(4), 471–485.
Gascón, J. Á. (2015). Arguing as a virtuous arguer would argue. *Informal Logic, 35*(4), 467–487.
Govier, T. (2010). *A practical study of argument* (7th ed.). Belmont, CA: Wadsworth.
Johnson, R. H. (2000). *Manifest rationality: A pragmatic theory of argument.* Mahwah, NJ: Lawrence Erlbaum Associates.
Lumer, C. (2005). The epistemological approach to argumentation: A map. *Informal Logic, 25*(3), 189–212.
Paglieri, F. (2015). Bogency and goodacies: On argument quality in virtue argumentation theory. *Informal Logic, 35*(1), 65–87.
Rehg, W. (2005). Assessing the cogency of arguments: Three kinds of merits. *Informal Logic, 25*(2), 95–115.
Sosa, E. (1985). Knowledge and intellectual virtue. *The Monist, 68*(2), 226–245.
Wilson, D. & Sperber, D. (2002). Truthfulness and relevance. *Mind, 111*(443), 583–632.

2

Why the Dialectical Tier is an Epistemic Animal

SCOTT F. AIKIN
Vanderbilt University
scott.f.aikin@vanderbilt.edu

Ralph Johnson has proposed a "two tiered" conception of argument, comprising of the illative core and the dialectical tier. This paper's two-part thesis is that (i) the dialectical tier is best understood as an epistemic requirement for argument, and (ii) once understood epistemically, the dialectical tier requirement can be defended against the leading objections.

KEYWORDS: dialectical tier, Ralph Johnson, epistemic theory of argument

1. INTRODUCTION

In *Manifest Rationality* (2000) and other work (1996 and 2003), Ralph Johnson defends what he calls a "two-tiered" conception of argument, comprising of an *illative core* (of premises and conclusions) and a *dialectical tier* (of replies to objections). Johnson holds that his two-tiered model for argument is necessary, because with argument, we are out to persuade an interlocutor with an act of manifest rationality. That is, an argument's objective is not only to rationally persuade, but for the rationality of that persuasion to be clear to the persuaded. It must not only *be* a rational inference from the reasons to the target claim, but that move must also be *clearly* rational. And so, "the practice of argumentation is best understood as an exercise in manifest rationality" (2000, p. 1).

Johnson's view that arguments must have a dialectical tier has been the target for a number of objections, three of which I will consider here. The first is that there is no obvious connection between the dialectical tier and the rationality of the illative core of argument (as argued by Ohler, 2003 and Liang & Xie, 2011). The second is that the dialectical tier is argumentatively supererogatory; it is not a dialectical obligation (as argued by Adler 2004). Third, and finally, it has been objected that the dialectical tier yields a vicious regress (as argued by Govier 1999).

This paper's two-part thesis is that with an account of epistemic defeat and defeater-defeat, it is possible to explain why arguments, if they are to provide hearers with justification for accepting a conclusion, must have a dialectical tier. Once this epistemic notion of dialectical considerations is in place, it is possible to answer the three pressing objections to Johnson's dialectical tier. As a consequence, it is best to take the dialectical tier to be one derived from the epistemic normativity of argument.

2. THE DIALECTICAL TIER AND ITS DISCONTENTS

Johnson's notion of the dialectical tier is dependent on his commitment that argument is an exercise in manifest rationality. An act is manifestly rational when it is not only rational, but it seems to all the relevant parties to the act that it is rational, too. Johnson explains with regard to argument, in particular:

> What is distinctive of argumentation is that it is an exercise in manifest rationality, by which I mean not only that a good argument is itself a rational product – a product of reasons, reasoning, and reasoners – but that it is part of the nature of the enterprise that this product appear as rational as well (2000, p. 144)

One may see this notion of argument arising from a requirement of respecting the dignity of others with whom one disagrees and thereby with whom one must argue – one must first see them, insofar as one takes it one can argue with them, as rational and movable by reason. Further, one must take it that if they are rational and moved by reasons, they, by their own lights, have been moved to their views (even if wrong) by reasons. Johnson makes this thought explicit by requiring of any theory of argument that it be able to recognize that for any view, there can be good arguments for and against it (2000, p. 53). What manifest rationality requires, then, is that those reasons be sorted in a way that allows them to be seen as reasons but also provides accessible reason to those who have disagreed for going one way and not another.

In light of the connection between this notion of manifest rationality and the view that argumentation must countenance conflicting reasons, the dialectical tier is a requirement of successful argument. Johnson's definition of argument bears this out:

> An argument is a type of discourse or text ... in which the arguer seeks to persuade the Other(s) of the truth of a thesis by producing the reasons that support it. In addition to this illative core, an argument possesses a dialectical tier in which the arguer discharges his dialectical obligations (2000, p. 168)

With the illative core, Johnson holds there must be "a premise-conclusion structure: a set of premises in support of some other proposition that is the conclusion" (2000, p. 150). With the dialectical tier, the arguer must "address standard objections" and "alternate positions" (2000, p. 125 and p. 328). Tying the dialectical tier to the notion of manifest rationality, Johnson explains:

> [I]f the arguer takes seriously the positions of others and in the course of his own argument addresses himself to them, the result is a display that is not only rational, but is one that appears to be rational (2000, p. 151).

In short, the dialectical tier is in the service of the *manifestness* of manifest rationality, the clarification and exemplification of the rationality of accepting the conclusion on the basis of the premises. Johnson's case for the dialectical tier, then, may be stated as follows:

> *Premise 1:* Arguments must be exercises of manifest rationality.
> *Premise 2*: Arguments are manifestly rational only if they have a dialectical tier.
> *Therefore,* arguments must have a dialectical tier.

Stated as such, the argument is valid, and Johnson's reasons for holding the premises are clear. For Premise 1, argument's objective is of rational persuasion, which requires the rationality of the terms of persuasion to themselves be clear to those persuaded. For Premise 2, clarity of this sort is possible only if one's worries, misgivings, objections, and clarifying questions are answered. The dialectical tier is the function of making the rationality of accepting an argument's conclusion manifest.

The three objections to the dialectical tier are best as targeting the premises of Johnson's case. First, one may object to Premise 1 on the grounds that good arguments need not be *manifestly rational*, but only *rational*. To have good arguments that one has a good argument is to ask more than what is required for argument. Alternately, one may object to Premise 2. One may do so in two ways. On the one hand, one may hold that there is no clear connection between the dialectical tier and manifesting rationality – that chasing down and answering objections actually impedes that goal. On the other hand, one may object that the dialectical tier sets before us a task that is impossible to complete – since for every objection answered, there must be an argument, which will need to answer yet further objections. With this insight, we have a rough taxonomy of objections to the dialectical tier.

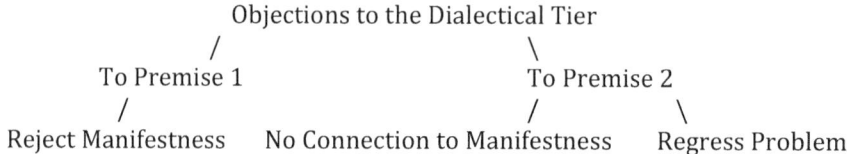

My plan is to review these objections in order. First, one may object to the dialectical tier because one sees manifestness of rationality as, instead of an obligation of argument, rather as something that is *supererogatory*. Johnathan Adler has argued that the dialectical tier is still an "imperfect duty" one has to oneself and one's view, but disagreements from others is not a defeater for arguments. The dialectical tier, then, "impos(es) excessively burdensome costs on arguers" (2004, p. 281). Human inquiry, Adler holds, must be bound by the demands for economizing, and this means that there must be a division of epistemic labour, so our arguments need only be appropriate for the time and resources we have at hand. Fulfilling the dialectical tier, then, would "diminish the vitality of argument and inquiry" (2004, p. 284).

The second objection to the dialectical tier is to Premise 2 along the lines that the dialectical tier has not obvious connection to the rationality of the support the conclusion has by the premises. Addressing objections that, by hypothesis, are unfounded or do not affect the argument's quality is not only a misuse of one's time and efforts, but it is actually contrary to the spirit of the manifestness of the rationality of the support. Amy Ohler captures this objection, noting that:

> To have to respond to criticism believed or known to be misguided is in one important sense of the word *irrational* (2003, p. 70).

Jonathan Adler, too, holds that this line of thought yields an absurd notion of what the argument must achieve.

> [A]n arguer is entitled to dismiss objections that he regards –
> and for good reason – as failing, without having to explain why
> it fails (2004, p. 289).

And further, Liang and Xie hold that the dialectical tier doesn't lead to improved or clarified rationality, since addressing bad, or low-quality, objections is "unlikely to affect the cogency of our argument" (2011, p. 233). As a consequence, the objection is that the dialectical tier adds nothing to the argument's quality, but it stands to obscure the connection between the premises and conclusions.

Third and finally, the dialectical tier, it is objected, sets arguers on the road to a vicious regress. The argument for regress runs as follows (standardized for presentation):

i. Every argument must have a dialectical tier.
ii. Dialectical tiers are arguments (or have arguments as components)
iii. *Therefore*, every dialectical tier must have a dialectical tier.
iv. *Therefore*, every argument has an infinite number of dialectical tiers.

Trudy Govier concludes that "Johnson's view seems to imply an infinite regress" (1999, pp. 232-3). On the assumption that giving a successful argument is a finite task, line 4 of the argument is absurd, and so we have a *reductio* of the dialectical tier. As Govier puts it, "surely it is not plausible to say that an arguer has an obligation to put forward an infinite number of arguments in order to build a good case for a single conclusion!" (1999, p. 233).

3. EPISTEMIC DEFEAT AND THE STRUCTURE OF OBJECTION AND REPLY

A successful case for an epistemic reading of the dialectical tier must address three issues. First, an account of what objections and replies are on an epistemic theory must be clear. So, an epistemic view of what the dialectical tier is must be manifest. Second, the epistemic theory must be shown to be pursuant of the broader objective of manifest rationality in argument. Third, the epistemic theory of the dialectical tier must have successful replies to the three standing objections.

The epistemic view of the dialectical tier depends on an account of the relationship between (a) inferential justification, (b) epistemic defeaters, and (c) defeat of defeaters. To begin, inferential epistemic justification is the status of a belief or commitment has when it is justified on the basis of some other commitment's inferential support. This, for argument, is the *illative core* – conclusions are supported by premises and the appropriate logical relation they have bearing on it.

Once a commitment is justified by the inferential support another justified commitment provides, new information can defeat that support. With this new information, the support that the premise provides for the conclusion can be eliminated. Here is a rough notion of what a *defeater* is:

> D is a *defeater* for a subject S's justification for holding that a proposition P is true on the basis of evidence E iff: (i) D is true or justified for S, and (ii) if D were added to S's evidence E, S would no longer be justified in holding that P.

Importantly, defeat can come in two forms, because we can no longer be justified in holding a proposition in two ways: either that justification is eliminated by new information but the proposition's truth value remains in question, or the proposition is shown to be false for reasons that are better than (or perhaps equal to) the support of the initial justification. And so, there are two types of defeat:

> D is an *undercutting* defeater iff D is a defeater that eliminates S's justification (but P's truth value remains indeterminate in light of D)

> D is a *rebutting* defeater iff D is a defeater that provides S with reasons to hold that P is false

An example will help keep the two kinds of defeater distinct. Let S hold that P on the basis of some attester A's say-so. S's evidence E, then, is this testimony. An *undercutting defeater* for E's support of P would be the information that A has some motive to lie about P. A's motive to lie about P isn't itself evidence that P is false, but it certainly eliminates our justification for holding that P on the basis of A's say-so. Alternately, let there be some powerful evidence, perhaps some undoctored photograph or video provided by a very reputable source that Q is true (and Q is a contrary of P). Now, S's justification for holding that P is defeated, because S now has reason to believe that P is false. This second form of defeat is *rebutting* defeat.

Objections can come along the lines of challenges to premises, illation, clarity, and to the conclusion. Insofar as reasons to consider defeat (of one of the two kinds) immanent, we can take the following complex to represent the possibilities for objection-types.

	Premise	Illation	Clarity	Conclusion
Undercutting				
Rebutting				

For example, one can provide undercutting reasons or rebutting objections against the conclusions of an argument (as we have seen above), or one could provide those objections against whether the premises are acceptable. Additionally, one can show that one does not have sufficient grasp on how clearly the premises support the conclusion or that it is positively clear that they do not. Every one of these sets of *objections* may be stated as a form of *defeater* for the justification a subject may have for holding the target proposition or conclusion as true.

Replies, then, must show that the target proposition (or some properly precisified version of it) is justified, even in light of the standing objection. These, too, may be rebutting or undercutting in form. And so:

> R is a *reply* in defence of S's holding that P on the basis of E (with defeater D) iff were R and D are added to E, S would have justification for holding that P is true.

Replies, given the way justification can arise from the coordination of defeat and new information, can come in two forms.

> R is a *restoring reply* iff were R and D are added to E, S's justification for holding P is true would solely be on E

> R is a *reestablishing reply* iff were R and D are added to E, S's justification for holding that P is true would arise with R's new information

To clarify these, let us return to the defeater case from before. Let S believe that P on the basis of some attestor A's say-so. If we had the defeating reason D to hold that A had motive to lie, we would call this an undercutting case of defeat. However, our justification for D itself could be undercut by R, perhaps showing that our source for the belief that A has a conflict is confused and had wrongly misnamed A as an unreliable source. Or we could find some rebutting evidence, perhaps showing that A has no conflict of interest in this case. In either way, we would *restore* our original justification. Alternately, we may find some other line of reasoning to P, perhaps along the lines of some more credible attesters to P, which would be a form of *reestablishing* justification.

The point of this taxonomy is to show that accounts of epistemic defeat and defeaters of defeat provide a model for how objections and replies may be aligned and assessed. Moreover, it provides a way of explaining not only the *rationality* of the process of objection and reply, but it explains what makes this exercise of rationality itself *manifestly rational*.

Manifestness arises from this nexus of objections and replies because if an arguer provides a hearer with reasons to accept P as true, but if the hearer has an objection, even if unjustified, the hearer is not rational in accepting p unless and until the hearer sees that the objection does not defeat the conclusion. And so, for the sake of manifest rationality, not only in the *arguing*, but in the hearer's being *rationally persuaded*, the arguer must address first the hearer's objections and, second, the objections the hearer would likely encounter in similar critical discussions. The activity and its product must not only be *rational*, in proportioning belief to evidence, but it should also *appear as rational* to those participating. So objections must be answered, and epistemic models of defeat and reply explain why this is so.

4. DEFENDING THE DIALECTICAL TIER

The final stage of the case for an epistemic reading of the dialectical tier is showing that the epistemic model of defeat and reply allows for successful defence against the three prominent objections surveyed earlier. To the objection to the dialectical tier not having a tight connection to manifest rationality (that is, to the second premise), the epistemic theory is that if we are to take ourselves to be justifying our commitments with argument and yielding rational change of view as we exchange reasons, then replies to objections must be part of the process. This is because a failure to reply to an objection, even if the objection is unfounded, does not defeat the defeater in the eyes of the person wielding the objection. This, for the objector, then nevertheless has justification defeated. Insofar as the objective of argument is *manifest* rationality, *showing* the rationality of a line of reasoning, then replies to objections is an intrinsic feature of the project.

For those who hold that the dialectical tier and manifestness is supererogatory (as does Adler) or that it is actually contrary to the exercise of rationality (as does Ohler and Xie & Liang), a distinction is in order. Let a speaker S be challenged by a hearer H about one of S's commitments, P. H knows enough about the domain of discourse concerning P to provide some concerns about P's truth, and H calls attention to them. S, in this case, is very knowledgeable about the domain and has been aware of the concerns H has raised for quite some time, and S is also aware of reasons not to take those too seriously, but rather to focus on the high-quality evidence E that supports P. Let's also stipulate that E, even despite H's concerns, supports P. In this case, from S's perspective, worrying about H's objections is pointless, and so it would be irrational to address them or give them more attention. But from H's perspective, these seem to be serious defeaters, and S's not addressing them seems intransigent dogmatism.

In one sense, S's argument from E to P is satisfactory, and is considerably more efficient than one that must engage in a dialectical tier with H's concerns. But in another sense, S would fail at achieving the interactive and communicative goal of argument, giving H reasons H can herself recognize as good reasons to accept P as true. Call the first sense of argumentative success *the absolute notion* – one does not need to look to an audience to assess whether the argument is good. Call the second sense of argumentative success *the relational notion* – one must look to audiences and the intellectual milieu to determine whether the argument is appropriate.

The absolute notion of argumentative quality is certainly *logically* prior to the absolute notion, as one cannot help but ask, when one evaluates an argument, whether one accepts the argument absolutely. However, the relational notion is *epistemically* prior, since when one evaluates an argument for absolute quality, it is one's own dialectical criteria (and then those of others) from which one evaluates an argument. The notion of relational quality depends on that of absolute quality (as one asks *do the premises support the conclusion?* internal to any evaluation). The notion of relational quality depends on that of absolute quality, but we cannot help but start from that of a relational perspective, namely, that of our own set of acceptable premises and what relevancies are manifest.

Insofar as we take the Gricean cooperative elements of communication in argument – that of producing reasons we think warranted, relevant, of an appropriate quantity, and of a manner accessible to our audience – are in the service of our collaborative weighing of reasons, the dialectical notion of argument seems an inescapable requirement. Arguments given without care for whether an audience accepts, understands, can keep track of the number, and can efficiently see the relevance of the premises are less *arguments*, and more instances of *browbeating* or *hectoring*. And, consequently, there are occasions where argumentative dialogues can become more sessions for information-sharing or inculcating a sense of what is relevant. Without these background conditions either in place or in the process of being established, argument is destined for communicative failure. And so, even though the absolute notion of argument quality is what we must evaluate arguments *in situ*, those deployments themselves depend on the fact of the relational conception – we deploy the absolute conception of argument against our own background of competencies and knowledge, without which, we would not be able to see the argument's quality. This is why argument must be an exercise of manifest rationality, and consequently, why there must be a dialectical tier.

The final challenge is to answer the problem of the regress. If arguments must have dialectical tiers, and if tiers themselves are arguments, then we face a vicious regress of reasons. There is a variety of views about the *epistemic* regress problem, and each may provide a method for addressing this *dialectical* regress. If I am right that the dialectical tier is an epistemic phenomenon, then any structural feature of it that has an epistemic problem will have a parallel epistemic solution. In this case, given that the dialectical tier and requirements of manifest rationality are highly demanding parallels to epistemic internalism, epistemic internalism's most significant problem will be a problem for argumentation theory – particularly, the problem of the regress of reasons.

Sextus Empiricus's Five Modes (PH 1.164) is the *locus classicus* for the regress challenge for justification (though it arguably is at work in Plato's *Meno* 75d and Aristotle's *Posterior Analytics* 72b5). For any commitment, Sextus reasons, one may challenge it with *disagreement* (someone who denies it) or with *relativity* (some information that defeats the evidence by showing it is not connected to truths). Those holding the commitment must then reply with a reason, and to that reason, one may challenge again with disagreement or relativity. And so another reason must be given. Sextus held that one could go in of three patterns for the reasons: (i) one could end with commitments without supporting reasons, (ii) one could keep giving reasons without end, or (iii) one could argue in a circle. Sextus held that none of these options, which he had termed *hypothesis, regress,* and *circularity,* yielded a structure of justifying reasons, and so he concluded that there seems to be no way for our commitments to be justified.

However, it seems there are many ways to solve Sextus' trilemma. The first strategy of reply is *foundationalism*, the view that some reasons do not need to come in the form of *arguments*, but in the form of *logical intuitions, experiences,* or *ethical impulses* (see Freeman 2003, p. 5 for an exemplary reply to Govier along these lines). Alternately, there may be *benign circularity*, as when commitments come in coherent theoretical systems. Within a theoretical program, one may explain and predict phenomena, one may revise and refine the details of the views in light of new information. And so, instead of an *argument*, at some point the *coherence of the system* itself becomes a reply to challenges.

Finally, there may be a more skeptical turn to our thoughts with the infinitist line with the regress – in particular, it may be the case that *every argument is incomplete*, all finite cases are necessarily leaving important critical questions unanswered. Perhaps it is not so absurd to think that arguments must often be opened up again, since there is always more to say. Surely, one must stop here or there, but that is for *pragmatic* reasons, not purely argumentative or epistemic purposes. We grow tired, bored, or hungry, and we let it go. Or sometimes, our audience finally comes around to agree, and so we leave off. But we always do so with a promise to keep talking if it arises that there are more questions later.

5. CONCLUSION

The takeaway from this discussion is that the dialectical tier is best conceived in terms of an epistemic feature of manifest rationality. We have the obligation to answer objections, because if we are to display the rationality of our commitments in a way that others can recognize, we must address concerns as rational. The nexus of epistemic defeat and varieties of justification-restoring and re-establishing reply provides a means of explaining why the dialectical tier is necessary for justification to be clear. Further, once the epistemic categories are in place, replies to the leading objections are not only possible, but are clear.

ACKNOWLEDGEMENTS: Thanks go to Gabriela Bašić, Ian Dove, Geoff Goddu, Hans Hansen, Michael Hoppmann, Catherine Hundleby, Harvey Siegel, and Robert Talisse for discussions of this paper.

REFERENCES

Adler, J. (2004). Shedding dialectical tiers. *Argumentation*, *18*(3), 279-293.
Freeman, J. (2003). Progress without regress on the dialectical tier. *OSSA Conference Archive*, 22. Retrieved from https://scholar.uwindsor.ca/ossaarchive/OSSA5/papersandcommentaries/22.

Govier, T. (1999). *The philosophy of argument.* Newport News: Vale Press.
Johnson, R. (1996). *The rise of informal logic.* Newport News: Vale Press.
Johnson, R. (2000). *Manifest rationality.* Mahwah, NJ: Lawrence Erlbaum.
Johnson, R. (2003). The dialectical tier revisited. In F. H. van Eemeren, J.A. Blair, C. A. Willard & F. A. Snoeck Henkemans (Eds.), *Anyone who has a view* (pp. 41-53). Berlin: Springer.
Liang, Q. and Xie, Y. (2011). How critical is the dialectical tier? *Argumentation,* 25(2), 229-42.
Ohler, A. (2003). A dialectical tier within reason. *Informal* Logic, 23(1), 65-75.

(Digital and) Material Representations of the European Green Belt: Juxtaposing Nature and Technology in Our Collective Memory of the Cold War

MARCIA ALLISON
University of Southern California
mcalliso@usc.edu

EMMA FRANCES BLOOMFIELD
University of Nevada, Las Vegas
emma.bloomfield@unlv.edu

We analyse the visual, verbal, and material arguments present at the European Green Belt (EGB), a contemporary conservation project built from the former Iron Curtain. The EGB argues for unity and presents itself as a "living memorial", that fuses together former warring countries. In this project, we compare digital representations and physical manifestations of the EGB's arguments about history and memory, nature and technology, peace and war, memorial and tourism, and preservation and restoration.

KEYWORDS: European Green Belt, environmental memory, material rhetoric, juxtaposition, transcendence, Cold War

1. INTRODUCTION

The European Green Belt (EGB) is a contemporary conservation and memory project that reclaims the forbidden zone of the former Iron Curtain that divided Eastern and Western Europe. Announced by Winston Churchill, the Iron Curtain was a metaphor for the linked borders of Soviet controlled territories after WWII. Over the 40 years that the Iron Curtain physically and ideologically divided Eastern and Western Europe, novel ecologies of flora, fauna, and animal species returned and regrew in the border's negative space inaccessible to ordinary citizens. Today, the EGB unites unique ecologies along with the material relics of the war – the abandoned border defences, military facilities and other such ruins. Through its online digital representation and marketing, the EGB thus presents itself as a pan-European conservation and biodiversity project that is also a natural "living memorial" to European history.

The EGB fuses together former warring countries, cultures, and people through a celebration of nature. We put these digital performances and representations in conversation with analysis from rhetorical criticism performed at the EGB. In this project, we address the EGB as a series of juxtapositions between history and memory, nature and technology, peace and war, memorial and tourism, and preservation and restoration. These problematic associations are mediated through the EGB's online argument that the site is a unifying force that transcends, and in the process potentially minimizes, issues of the past. The tensions and contradictions wrapped up in the EGB mirror the complicated relationship between humans and nature. To understand how the EGB functions as an environmental memory site, we will first outline the theoretical basis for this project, including material rhetoric, juxtaposition, and transcendence, before highlighting a few important tensions in the discursive and material reality of the EGB. Then, we will explicate our conclusions about what the EGB reflects about current environmental discourse and conservation efforts through exemplar sites along the EGB.

2. PUBLIC MEMORY AND THE ENVIRONMENT

Kenneth Burke (1966) argued that a defining characteristic of humans is that we are separated from our "natural condition by instruments of our own-making" (p. 13). Primarily, Burke argues that language and symbols are the tools we use to divide us from our identities as animals. It is our use of language, stories, and the tools and technologies of humanity that construct a hierarchy of human life over plant and animal life. This hierarchy is often present in our language when we discuss "conquering" nature and "taming" the wilderness, as if we are creatures separate and distinguishable from the world around us. Burke (1966) argued that there are no negatives in nature - "the negative is an idea, there can be no image of it" (p. 430), whilst humans are the inventors of the negative (p. 9). And yet, it is this lack of the negative in nature that allowed the EGB to flourish – the natural elements of the EGB refute the literal negative space of the Iron Curtain border through growth and presence.

Some groups attempt to level the status of humans and nature and sometimes reverse the hierarchy, such as Earth First! members living in trees scheduled to be cut down. In equalizing the relationship, humans and natures are juxtaposed linguistically, or connected as equal through language. Juxtaposition is also a visual rhetorical strategy, where images are placed together to create an argument through their comparison (Bloomfield & Sangalang, 2014). We argue that the visual theory of juxtaposition can also be applied to material reality. The EGB is a prime example of how the juxtaposition of contradictory elements produces tension and conflict in service of an argument. The EGB's natural elements represent life, new beginnings, and a physical world that has not been tampered with or altered. The EGB's war relics represent death, endings, and the destructive power of human intervention in the natural world. By placing them together, the EGB argues that current conservation efforts can transcend the evils of the past, creating a unified European identity. In place of war, we have peace; in place of technologies of death, we have the renewal of nature; in place of separation and borders, we have a cohesive European identity performed through the maintenance of this shared land and conservation projects.

Transcendence is often considered a linguistic strategy to overcome guilt by reimagining a transgression in a new light. For example, Brummett (1981) argued that Reagan used transcendence to get Americans spending in a down economy, transforming spending from a sin to an integral component of citizenship and patriotism. Transcendence can also overcome dichotomies by producing a narrative where competing binaries are united. Kaylor (2011) argued that John F. Kennedy transcended the conflict over his Catholic faith by noting his shared ideals of the American civil religion, which Mitt Romney failed to do. For the EGB, the Cold War serves as humanity's previous transgressions, for which we perform penance by enabling nature to reclaim the lost land. The EGB attempts an argument of transcendence by uniting the dichotomies of war and nature under a new vision of the future. The relics of war remain as a reminder of the past, but also as a foil to proclaim the new triumph of European unity. The EGB reveals an enduring push and pull between what Dunlap (1978) called the "Dominant Social Paradigm", which supports abundance, progress, private property, technology, and growth, and the "New Environmental Paradigm", which values a balance between nature and humanity, recognizes limits to growth, and rejects ideas that nature exists only for human use. The EGB attempts to value equally nature and humanity, but also is fuelled by tourism, the need to mark territory along the belt, and to insert relics of human presence in the natural space. To see how transcendence and juxtaposition function in material forms and how the EGB negotiates a united European identity, we analyse two primary tensions: between nature and technology, and between conservation and intervention.

3. NATURE AND TECHNOLOGY

Discoveries of these ecosystems along the Curtain began in the 1970s and nature conservation efforts soon followed. In 2003, local, state and regional conservation initiatives finally merged into a single Pan-European effort with the creation of the "European Green Belt" Initiative. Its associated NGO, the EGB Association, was established in 2014. The EGB is thus managed as a coherent whole but also split into four biogeographical regions: the FennoScandic, the Baltic, the Central European and the Balkan. All 24 countries through which the Iron Curtain existed now participate, providing a unified scheme for conservation, social reformation, improved diplomatic relations and economic prosperity. Under EU conservation laws, the EGB Association and Initiative acts as the supranational organization for local and state conservation actors in the differing biogeographical regions.

The EGB website captures the holistic nature of the project under the slogan "Borders Separate. Nature Unites!" (European Green Belt, 2017a). It proffers information on threats, projects and events to the site. Hyperlinks both literally and figuratively construct the EGB as an "ecological network" (European Green Belt, 2017b), linking to other state and local actor communication infrastructures. The banner of the landing page visually and rhetorically frames its position: "Vision! The European Green Belt, our shared natural heritage along the line of the former Iron Curtain, is to be conserved and restored as an ecological network connecting high-value natural and cultural landscapes while respecting the economic, social and cultural needs of local communities" (ibid.). Situated next to a photograph of the rare Balkan Lynx and with a map of the former Iron Curtain underneath, the border is now depicted in an symbolically ecological green to show which 24 contemporary countries it runs through. With the description "From Deathzone to Lifeline" (European Green Belt, 2017c), the Initiative portrays nature's prosperity as transforming the border from a place of annihilation into a site of anthropocentric transboundary cooperation. In a cyclical process, natural flourishing leading to transboundary cooperation thus becomes one of the central topoi of the website.

At the *Iron Curtain* tab, the website explicates the other aim of the Initiative: preservation of the border as a "memorial landscape" as "remains of border fortifications (watchtowers, patrol paths, ditches or border buildings) provide a vivid picture of the inhumanity of the border regime" (European Green Belt, 2017d). These relics of the past are thus repurposed for present consumption and performance, both for passive viewing and active engagement. The EGB Association website posits an alternate way for the traveller to visit this memorialized landscape through its transformation into a 6,000km long distance cycling route called the Iron Curtain Trail (ICT) or EuroVelo 13. This separate but cooperative scheme works in tandem with the EGB Association in a promotion of eco-tourism along former border pathways and roads as an EU Green Infrastructure Project (European Commission, 2013). The former border thus offers an eco-tourist gambit alongside conservation and memorialization.

The eco-tourist gaze upon the EGB memorial landscape finds tension presenting the border as both conservation and memory project at once: "The trail is not purely a scheme created for sustainable tourism. It preserves the memory of what the Iron Curtain once stood for [...] Along sections of the trail in Germany, plaques have been placed to mark where people, attempting to flee to the West, were shot dead by border police. By leaving these historical features in, it is as though one is riding through an open air museum [...] He who masters the past, masters the future" (Iron Curtain Trail, 2014). We see here how Dunlap's paradigms enter into conflict, as touristic progression and development of the ruins pushes against a harmonization with the EGB's natural environment. In a justification of memorialization and eco-tourism, nature here becomes characterized as a museum, placing a human name and container around it, defying Dunlap's Dominant Environmental Paradigm. Nature is not free and open – despite the flourishing that originated its existence – but instead becomes a collective object upon which to gaze, make meaningful and rhetorically argue. The nature makes sense of the *human* history to transcend and assuage collective guilt in the service of unity today. Thus, not only is the physical experience of the EGB juxtaposed to the adversarial, mediated rhetoric of the EGB, but the website itself juxtaposes two differing narratives: of biodiversity conservation and human historical memorialization.

4. CONSERVATION AND INTERVENTION

The EGB conserves and preserves existing flora, fauna and animal species, and encourages growth in the former border space. The Initiative, through local, state and EU actors, has set up various programs and laws to protect the space, prevent development on the land and remove interventions that threaten biodiversity. In response to Germany's decision to turn 62 former military bases into wildlife sanctuaries, Germany's Environment Minister Barbara Hendricks noted that "We are fortunate that we can now give these places back to nature" (Davies Boren, 2015). The atrocities of the past may have been responsible for taking ownership of the land and destroying it. But now, with time and with new people interested in conservation, ownership of the land is seen to be restored to the non-owner of natural habitation.

However, one of the tensions the EGB faces is how it resolves anthropocentric inclusivity as the Initiative makes the site approachable for humans in its conservation efforts. In this open-air museum, it is not just plaques or memorials of technology but the ruins that have become modified for modern-day consumption. Originally to keep humans out of the border, border patrol towers have now become an inviting presence, recontextualized in their new visual frame of the natural EGB setting. Some have been repurposed as observation stations. Others exist just as historical reminders, to be gazed upon by the visitor, approached by preserved German Democratic Republic (GDR) and other such roads. Watch towers thus become watching towers. Human visitors enact their tourist gaze upon the ruins and EGB landscape – the gaze of selected presence (Perelman & Olbrecht-Tyteca, 1971). Transcendence is literally achieved through coming towards, gazing upon or climbing inside these towers as an EGB tourist. By existing as ruins to a previous time, now transformed for a different rhetorical present, the towers transform this space into a sacred space of remembrance and natural flourishing in this juxtaposition of an eco-tourist memory project.

At the Pasvik-Inari Trilateral Park, co-owned and managed by Russia, Finland and Norway, an old border patrol tower has been transformed into a "peace promoting" (EUROPARC Federation, 2010, p. 25) bird watching tower. This technology of war has transcended its original intention and looks towards a peaceful future of trans-boundary cooperation, symbolized through the watchful gaze upon the birds. To watch from the tower is to view the EGB vista as *gestalt*. Nature and technology is not oppositionally presented as they would be in a discursive format but are fused together in their presentational form. Nature does not negate war but transcends it.

This contrasts with the town of Gosdorf in Austria which built The Murturm Nature Observation Tower on its stretch of the EGB. Rather than repurposing an old watch tower, this new structure is a spiralling metal framework resembling DNA, gazing out over the surrounding nature. While built to honour the EGB nature and to give visitors a better view, the tower itself is a representation of human intervention in the space. It denotes how that the EGB is preserved by human objectification of the wild frontier. The DNA structure nods towards the collapse of architectural technologies and organic material, without striving to blend in.

Observation Point Alpha (OPA) in Germany is a complex example of conservation and intervention as it intertwines repurposed ruins, a newly built museum, a memorial and nature trails to create an ultimate eco-tourist memorial site. OPA was a Cold War observation post between Rasdorf, Hesse in what was then West Germany, and Geisa, Thuringia in East Germany. It was a U.S. Army observation post that overlooked the Fulda Gap. This is an area between the border that contained two corridors of lowlands which were thought to be at risk of tanks driven in a surprise attack by the Soviets and their Warsaw Pact allies to gain crossing of the Rhine River and attack West Germany, starting World War Three. The Point Alpha memorial therefore presents "at this authentic historical site [...] the confrontation of the two world power blocs during the Cold War, as well as the painful time of division within Germany" (Point Alpha Foundation, n.d.a).

As a memorial landscape couched in the rhetoric of transboundary cooperation, the OPA barracks are transformed into a museum about American and Allied solidarity, whilst a modern sculpture to European unity stands amongst many European, U.S. and Russian flags outside the new museum. This museum houses both a history of the Fulda Gap alongside history of the EGB and current nature conservation efforts in the local area. As one learns about the division of Germany and its transcendence through the rewilding of the border, this is a symbolic and literal experience for the visitor as they traverse the memorial landscape of the OPA. The visitor takes a literal trail of transcendence as they traverse along a former GDR road from the museum, past the U.S. and German watch towers (witnessing border defences and Point Alpha barracks, to which they may take a detour), its end culminating in the flourishing nature of the EGB. Symbolizing European freedom, nature is both a literal and figurative peace broker. There are many nature trails that one can take, highlighting the different flora, fauna and other species that are flourishing in the region. The trails create a natural eco-tourist gaze through signs, icons and indexes.

Or one can gaze over the natural valley of the Fulda Gap, at nature's flourishing presence.

Despite leading from human intervention to conservation, there is also an additional anthropocentric feature: the Path of Hope. This is a 14 sculpture piece to celebrate the 20 year anniversary of the reunification of Germany along a 1.5 km stretch. The Path of Hope is "meant to inspire people to be reminded of the paths their own destinies took in times fraught with hardship and to reflect on them in similar sense of 'never again'" (Point Alpha Foundation, n.d.b.). Once again, a particular public memory is created: division and atrocity was overcome, but with the proselytizing that no European could want such a thing again. A common identity is thus sought after. War and division is transcended by nature, changing the polarity of the site to create a new visual frame (cf. McGeough, Palczewski & Lake, 2015). In the present, these war barracks no longer present solely a frame of war but become an epideictic monument to European transboundary cooperation that transcended the atrocities of the Cold War.

5. CONCLUSION

This project addresses the many tensions present at the EGB that complicate its identity as both a memory and conservation project. Two of the primary tensions are between technology and nature and conservation and intervention. The differences between portrayals of the EGB online compared to its physical and material reality represent a divide between the EGB's public presentation and how the site may be actually experienced. The technology and nature dichotomy is also present at the material site, where eco-tourists are confronted with technological remnants of war side-by-side with the flourishing of nature and life. The divide between conservation and intervention represents the inherent contradiction in taking action to preserve the environment and thereby intervening in its natural state. These juxtapositions are not static placements; they encourage movement and create dynamics between and within the binaries. They become the vehicles by which the transcendent arguments overcome these differences in search of a brighter future.

It is surely an important and worthwhile goal to preserve history and promote unity and harmony. But, it is equally, if not more, important to probe the implications of transcendent rhetoric when it potentially erases or complicates public memory of war. Furthermore, the EGB could be seen as promoting nature and appealing to the environment only as a tourist site for economic gain, or a co-opted symbol of unity for political gain. This inquiry engages theories of transcendence and juxtaposition to understand the layered arguments and meanings that occur in the negotiation between humans and the natural environment. In the space between life and death, peace and war, unity and division, lies the tumultuous task of protecting a natural world rife with human presence. In preserving the memory of a warring people and conserving the land on which the war was fought, the EGB is an exemplar of how contemporary conservation projects must rhetorically tackle humanity's ultimately destructive influence in nature's continued health and existence.

REFERENCES

Bloomfield, E. F., & Sangalang, A. (2014). Juxtaposition as visual argument: Health rhetoric in *Super Size Me* and *Fat Head*. *Argumentation and Advocacy*, *50*(3), 141-156.

Brummett, B. (1981). Burkean scapegoating, mortification, and transcendence in presidential campaign rhetoric. *Central States Speech Journal*, *32*(4), 254-264.

Burke, K. B. (1966). *Language as symbolic action.* Berkeley: University of California Press.

Davies Boren, Z. (2015, June 19). Germany is turning 62 military bases into wildlife sanctuaries. *Independent.* Retrieved from http://www.independent.co.uk/news/world/europe/germany-is-turning-62-military-bases-into-wildlife-sanctuaries-10332109.html

Dunlap, R. E. & Van Liere, K. D. (1978). The "new environmental paradigm". *The Journal of Environmental Education*, *9*(4), 10-19.

EUROPARC Federation. (2010). *Following nature's design. Promoting cross-border cooperation in nature conservation*. Retrieved from https://www.europarc.org/wp-content/uploads/2009/01/brochure_TransParcNet_final_low_resolution.pdf

European Commission. (2013, May 5). Green Infrastructure (GI) – Enhancing Europe's Natural Capital. *European Commission.* Retrieved from http://eur-lex.europa.eu/resource.html?uri=cellar:d41348f2-01d5-4abe-b817-4c73e6f1b2df.0014.03/DOC_1&format=PDF

European Green Belt. (2017a). Borders separate. Nature unites! *European Green Belt.* Retrieved from http://www.europeangreenbelt.org/initiative.html

European Green Belt. (2017b). European Green Belt. *European Green Belt.* Retrieved from http://www.europeangreenbelt.org/

European Green Belt. (2017c). From deathzone to lifeline. *European Green Belt.* Retrieved from http://www.europeangreenbelt.org/initiative/origin.html

European Green Belt. (2017d). Iron Curtain. *European Green Belt.* Retrieved from http://www.europeangreenbelt.org/iron-curtain.html

Iron Curtain Trail. (2014, August 27). Iron Curtain becomes 7,000-kilometer bike trail. *Iron Curtain Trail.* Retrieved from http://www.ironcurtaintrail.eu/en/ict_realisieren/realisierte_abschnitte/6663040.html

Kaylor, B. T. (2011). No Jack Kennedy: Mitt Romney's "Faith in America" speech and the changing religious-political environment. *Communication Studies, 62*(4), 491-507.

McGeough, R. E., Palczewski, C. H. & Lake, R. A. (2015). Oppositional memory practices: U.S. memorial spaces as arguments over public memory. *Argumentation and Advocacy, 51*(4), 231-254.

Perelman, C. & Olbrechts-Tyteca, L. (1971). *The new rhetoric: A treatise on argumentation.* Notre Dame: University of Notre Dame Press. (Original work published 1969)

Point Alpha Foundation (n.d.a.). The Point Alpha Memorial. *Point Alpha Foundation.* Retrieved from http://pointalpha.com/en/point-alpha-memorial

Point Alpha Foundation (n.d.b.) Path of hope ("Weg der Hoffnug"). *Point Alpha Foundation.* Retrieved from http://pointalpha.com/en/path-hope-weg-der-hoffnung

4

Reasoning Together: Fostering Rationality Through Group Deliberation

MARK BATTERSBY
Department of Philosophy, Capilano University, Vancouver Canada
mbattersby@criticalinquirygroup.com

SHARON BAILIN
Faculty of Education, Simon Fraser University, Vancouver Canada
bailin@sfu.ca

> This paper, which focuses on rational decision-making, has a threefold purpose: to argue for a view of rational decision making that includes the evaluation of ends as well as means; to argue that properly structured group deliberation can be an effective way to foster this kind of rationality; and to offer guidelines for achieving group decision-making rationality.
>
> KEYWORDS: deliberation guidelines, evaluative rationality, group deliberation, inquiry, procedural norms, rational decision-making

1. INTRODUCTION

This paper, which focuses on rational decision-making, has three purposes: 1) to argue that rational decision making includes the evaluation of ends as well as means, 2) to argue that group deliberation, properly structured, is an effective way to foster this kind of rationality, and 3) to offer guidelines for rational group decision-making that include the types of considerations and structures that promote this kind of rationality.

2. EVALUATIVE RATIONALITY

2.1 Problems with rational choice theory

One highly influential view of rational decision-making is an economic model called "rational choice theory". This view focuses exclusively on instrumental rationality usually aimed at maximizing rational self-interest.[1] Although not all work on decision-making takes such a narrow view of ends, the idea that rationality in decision-making is a matter of reasoning about the means to a given end has been pervasive. In the field of argumentation theory, for example, efforts to develop argumentation frameworks for practical reasoning by theorists such as Walton (1992) have, until recently, focused on instrumental reasoning.

We argue that rational decision-making goes beyond the limits of instrumental rationality and must include moral considerations and the evaluation of ends as well as means. Our view, "evaluative rationality", is an alternative view of rationality which includes such considerations (Battersby, 2016). Some more recent work on practical reasoning in argumentation has, in fact, begun to recognize the importance of the consideration of ends and values and has tried to incorporate this aspect into their models (Bench-Capon, 2003; Atkinson et al., 2006; Walton, 2013).

It is not clear the extent to which rational choice theory is meant to be descriptive of people's behaviour or normative. It has become increasingly well documented by cognitive psychologists and behavioural economists that people do not, perhaps even cannot, reliably make decisions which maximize their rational self-interest (or satisfy their preferences). But the descriptive failure of this theory has not affected its status as a normative theory. It is still assumed that the extent to which people's decisions and behaviour do not accord with this theory is a testimony to their irrationality.

A primary difficulty with this approach is that it treats the norms of rationality as purely instrumental, thus failing to allow for the rational assessment of ends and goals and the moral assessment of both ends and means. Real world decision-making, however, often involves (and usually should involve) the consideration of both. The question is not simply one of instrumentality: "Is action A the most likely to achieve the maximum realization of goal G?" Rather, even the means will be

---

[1] This is sometimes generalized to the goal of maximizing "preference satisfaction" but in most of the literature and experimental research, the assumption is that people will and should act to maximise their self-interest.

subject to evaluation, involving such questions as: "Will action A realize goal G with the least unintended consequences?" " . . . be most acceptable to the actors?" " . . . minimize downside risk?" And, crucially, "Is action A morally acceptable?" Importantly, the goal is also subject to evaluation, e.g., "Is goal G the correct goal?" And "Is goal G morally legitimate or commendable?"

The notorious decision of Ford to not fix the dangerous gas tank on the Pinto based on a cost-benefit analysis is a prime example of a decision which failed to take into account the moral aspect of both means (risking people's lives) and ends (unquestioned goal of profit versus public safety). It is a decision that would be judged rational according to rational choice theory but was both morally and strategically irrational given a broader view of rationality. The emphasis of rational choice theory on self-interest, or at best individual preferences, precludes acts motivated by moral considerations or concern for common goods.

Another inadequacy of rational choice theory is that it is product-oriented. Decisions are judged as rational or irrational simply on the basis of whether they are successful in achieving self-interest (or preference) maximization without regard to how the decisions were arrived at. A shift to a more evaluative and nuanced model of rationality also involves a shift from this type of "product" view of rationality to a "procedural" model. Such a view accords better with the ordinary view of rationality which involves judging a decision as rational on the basis of whether it was arrived at through a process of reasoned reflection. Such a procedural approach also allows for rational disagreement. People can follow rational procedural norms without coming to the same conclusion, either because they disagree about the appropriateness of the goals or because there is still a reasonable basis for uncertainty and disagreement about the facts and predictions. Of course, the uncertainly involved in judgment and decision-making means that a perfectly rational decision might turn out to be wrong, but this is not relevant to assessing the rationality of the judgment.

One important contribution of a theory of evaluative rationality can be to suggest procedural norms for rational deliberation. Rational choice theory, being a product-oriented view, is unable to provide such norms. The models of argumentation theorists such as Walton, Hitchcock, and Atkinson are procedural in nature, but their primary concern has been to articulate an abstract computational model of rationality. These models contain many useful insights into the criteria that are constitutive of rational deliberation, but they do not provide a procedural model that could be followed by humans engaging in rational deliberation. We also note that, while they include questions about goals and values as part of the procedural dialectic, they offer no guidelines for the moral evaluation of ends or means.

2.2 Evaluating goals

Most decisions to act involve relatively immediately achievable goals (e.g., seek an injunction), but these goals get their rationale from supervening goals (stop the pipeline) which in turn get their justification from higher level goals (reduced use of fossil fuels) and even higher level goals (prevent global climate change). Deliberation about an action will typically assume higher level goals that provide the frame for the immediate goals. But it is also, at times, rational to evaluate the supervening goals in light of moral or even practical reasonableness. Some argue, for example, that given the concentration of greenhouse gases in the environment, our goal should be adaptation rather than focusing on reducing carbon emissions. Such a change in the dominant goal would have obvious implications for more immediate goals. Procedural rationality allows for evaluation of any level of goal in light of new information or ethical implications.

2.3 Moral evaluation

By ignoring moral evaluation, rational choice theory implicitly treats moral evaluation as beyond the pale of rational deliberation. This view takes its support from the fact that people do disagree about moral judgments. But this fact does not provide a basis for ignoring moral concerns in rational deliberation. Clearly we do reason about morality and, indeed, even those researchers such as Haidt (2001), whose experiments demonstrate the significant role of "gut feeling" in moral decision-making, acknowledge that reason can influence moral decision-making (Haidt, 2001). We argue that the way to ensure rational deliberation about moral decisions is to require that relevant moral considerations are addressed in deliberation. We regard the long-standing debate in modern moral theorizing involving moral intuitionism, deontologism, and consequentialism as identifying three relevant considerations in moral deliberation: 1) the inherent or intuited moral quality of the act (in accord with our intuitions that some actions, e.g., lying, killing, being cruel to people, are inherently wrong), 2) relevant duties and obligations (related to our positions, relationships and more general moral duties such as treating people equally), and 3) consequences (are the actions harmful to others?

sufficiently beneficial to justify the action? can benefits be maximized and negative consequences minimized?).

In addition, because rationality and morality require judging similar acts similarly, a useful and traditional question is whether you would approve the decision if you were the recipient of the action rather than the actor.

What all this means is that, with moral issues as with other types of issues, coming to a reasoned judgment involves weighing a number of competing and in some cases conflicting considerations. An important criterion for evaluative rationality is that all these moral considerations are given due attention.

3. THE VALUE OF GROUP DELIBERATION

3.1 The nature of group deliberation

The next question, then, is how such evaluative rationality can best be fostered. We suggest that group deliberation, appropriately structured, can be a particularly effective means for developing this type of rationality.

Deliberation involves more than group discussion. Rather, it presupposes the appeal to public reasons (reasons that appeal to widely held values, credible sources, and accepted facts) and adherence to appropriate criteria and argumentative norms (Cohen, 1997).

The centrality of public reasons also distinguishes decision-making through deliberation from making decisions through negotiation. While the latter involves mediating among conflicting individual interests, the former involves appealing to reasons and values that go beyond individual interests and would be seen as reasons for all participants.

Deliberation can also be distinguished from advocacy. While advocacy entails "the exchange of arguments advocating for and against a proposed choice" (Blair, 2016, p. 57) and usually involves a fixed commitment to a claim and its justification, deliberation involves an open-ended inquiry into the "best and so choice-worthy alternative" (Blair, 2016, pp. 54–55).

3.2 The virtues of group deliberation

3.2.1 Epistemological

There are several ways in which groups can be helpful in rational decision-making. Individuals come to an inquiry with a limited store of information and, moreover, may be unaware of its limitations whereas a group can draw on a shared pool of information (Lunenburg, 2012). The variety of perspectives present in a group can provide different frames for evaluating the decision and more alternatives for consideration and can also create a synergy leading to new reasons and arguments not previously held by any of the individuals (Sunstein, 2006, p. 21; Druckman, 2004; Lunenburg, 2012). Individuals deliberating together in heterogeneous groups can often recognize and compensate for others' cognitive biases (Druckman, 2004; Schulz-Hardt et al., 2000; Tetlock & Kim, 1987). Finally, the requirement to publicly justify one's judgments in a group can move people to treat evidence objectively and to prioritize accuracy over buttressing their prior beliefs (Tetlock & Kim, 1987).

3.2.2 Ethical

In addition, individuals conducting a private inquiry may be limited by their own history and thus may neglect or be unaware of the ethical implications of various choices. In group deliberation, particularly in heterogeneous groups, such ethical concerns are more likely to be brought to light by the people affected or by others who are aware of the issues. It may not even be recognized that there are ethical considerations involved in what may seem like strictly instrumental questions. Consider, for example, a school board contemplating the move to an altered school day. Unless the point of view of parents is represented in the decision-making process, the hardship for working parents in terms of additional child-care may not be considered in the deliberation.

Group deliberation also provides a natural context for the critical comparison of goals and values. The encounter between individuals will likely bring to the table a variety of interests and goals which need to be taken into account.

Individual interests are not, however, the appropriate basis for group decision-making because the group must be concerned with the collective interest. Deliberation becomes a process of identifying not only the best means to achieve previously desired goals but also of evaluating which goals are worthwhile.

The confrontation of diverse interests and goals during collective deliberation can highlight issues regarding the tension between individual preference and collective well-being. As Habermas (1989) aptly points out, there is a difference between asking people "What do you as an individual prefer?" and "What should we as a group do?" While the former question will elicit personal preferences, the latter question appeals to what is best for the group, in other words, the common good. Dewey makes the point that through deliberation, conflicts grounded in individual interests "can be discussed and judged in the light of more inclusive interests than are represented by either of them separately" (Dewey, 1963, p. 56). It is the goals judged as worthwhile in light of these "more inclusive interests" that should be the basis for a rational group decision (Habermas, 1990).

Consider, for example, a community making a decision regarding zoning height restrictions. Although community members may initially oppose height restrictions because they want to build taller houses for a better view, through deliberation they should come to realize that an unlimited height allowance will thwart their desires because others will build even higher. Through this process, individuals may come to appreciate the need to have collective restrictions.

Doing what is best for all is not, of course, limited to doing what is best for those participating in the deliberative process. The common good will often extend beyond what is good for those present in the deliberating. So, for example, in a community deliberation over a proposed new housing development, the deliberations should take into account not only what will be beneficial to the deliberating group as a whole, but also broader considerations such as how the project will affect the environment, or whether the development will cause any social displacement.

Group deliberation is a particularly effective means for highlighting considerations of the common good because of the requirement to offer public reasons for one's suggestions or proposals. There is a need to advance reasons that will persuade others. And the reasoning of others may persuade us (Mercier, 2011). The need to offer justification thus moves the discussion toward considerations that others will accept, i.e., considerations of what is best for all (Cohen, 1997, p. 77).

3.2.3 Spirit of inquiry

Group deliberation has additional benefits which extend beyond the group context. Procedures and criteria for rational inquiry, such as considering alternatives or ends, which are readily fostered through group deliberation, are also vital for inquiries outside a group context. But one of the most far-reaching benefits may come in the fostering of what we have termed the "spirit of inquiry" (Bailin & Battersby, 2016a). Habits of mind such as open-mindedness and fair-mindedness are both required for, and can be further developed through, the critical interchange involved in group deliberation. Moreover, the very practice of giving reasons for one's views may, of itself, provide an important long-term benefit by contributing to the formation of a commitment to the resolution of issues through deliberation, and, more broadly, a commitment to rationality.

3.3 Does deliberation make a difference?

The claim that deliberation actually makes a difference in how people think about issues has considerable empirical support. A number of studies have found evidence that people engaging in deliberation arrive at different decisions than those thinking on their own or engaging in unstructured talk (Schneiderhan & Khan, 2008; Setälä et al., 2007; Neblo, 2007; Fishkin & Luskin, 1999; Gastil, 2000). This is particularly the case when the deliberations focus on reason-giving and are inclusive of all members' views (Schneiderhan & Khan, 2008).

The mere evidence of opinion change as a result of deliberation does not, of course, mean that the changes are normatively desirable. If such changes turned out to be a result of social power or conformity to the group, then the changes might indicate a decrease in rationality rather than an increase. In this regard, Neblo's 2007 study found evidence that the mechanisms of deliberative opinion change were, in fact, those specified by normative theories of deliberation. Habermas summarizes Neblo's findings thus:

> The process of group deliberation resulted in a unidirectional change and not in a polarization of opinions. Final decisions were quite different from the initial opinions expressed and opinions changed reflecting improved levels of information, and broader perspectives on a clearer and more specific definition of issues. Impersonal arguments tended to take priority over the influence of interpersonal relations, and

there was also an increasing trust expressed in the procedural legitimacy of fair argumentation (Habermas 2006, p. 414).

Particularly relevant to our argument is Neblo's finding that deliberation tended to enlarge subjects' perspectives, prompting a shift from a market frame ("what is it that *I*, as an individual, *prefer*?") to a forum frame ("what is it that *we*, as a group, *should do*?"). This is precisely the orientation toward the common good which we have argued characterizes evaluative rationality.

3.4 The challenges of group deliberation

Group deliberation is not, however, without its potential pitfalls which will need to be avoided or mitigated. The research surveyed thus far suggests some of the features of effective deliberation: a focus on reason-giving and justification, accountability for one's positions, inclusivity, and diversity of points of view.

One common obstacle to effective rational group decision-making is an adversarial stance which can result in a failure to consider opposing arguments and alternative views in a fair-minded and open-minded manner (Blair, 2016; Bowell & Kingsbury, 2016; Bailin & Battersby, 2016a). Adversariality can also result in an aggressive and confrontational stance which is counter-productive to rational inquiry (Bailin & Battersby, 2016a, 2016b).

A challenge emanating from the opposite direction is groupthink (Janis, 1972, 1982) – the common aversion to disagree with a growing group consensus based on the fear of group disapproval for dissent or lack of solidarity. This can result in biases in information-search, an avoidance of critical evaluation of proffered solutions, the lack of fair-minded appraisal of alternatives, and premature closure. Janis argues that some of the structural factors which contribute to groupthink are homogeneity of members' backgrounds and ideologies, the lack of impartial leadership, and the lack of norms requiring impartial procedures (Hardt, 1991).

The effect of group decision-making on cognitive biases has also been a source of criticism. There is evidence that groups that are homogeneous with respect to preferences or judgments are more predisposed toward confirmation bias, the underestimation of risks, and overconfidence than the statistical baseline (Schulz-Hardt et al., 2000; Schultz-Hardt et al., 2002). Group heterogeneity, on the other hand, tends to have a debiasing effect on deliberation (Druckman, 2004;

Schulz-Hardt et al., 2000; Tetlock & Kim, 1987; Mercier & Landemore, 2012).

Another hazard in group deliberation is the lack of inclusion of some voices and perspectives. This may be an artefact of group composition or of the failure to give encouragement or voice within the group to participants who feel excluded for a variety of personal, social, or structural reasons (Sanders, 1997).

3.5 Countering the challenges

There are, however, ways to structure and conduct group deliberation that can counter many of these potential problems. The problems of both adversariality and groupthink can be mitigated through the creation of a culture of inquiry. The focus of deliberation needs to be on reason-giving but the task needs to be framed in terms of the group arriving at the best decision as opposed to in terms of individuals winning or losing arguments. Rigorous but respectful critique should be the norm. This goal can be facilitated by introducing explicit expectations for group interaction in terms of respectful treatment, meaningful participation, and productive interaction (Bailin & Battersby, 2016a, pp. 278–280).

Particular attention needs to be paid to encouraging the meaningful participation of all group members. Everyone's contribution should be welcomed and care should be taken that no one is excluded, discouraged from, or hindered in their participation (Bailin & Battersby, 2016a). Strategies such as giving participants time to think about the issue or breaking into subgroups or pairs for initial discussion before the larger group discussion are useful because deliberation requires time and opportunity to exchange views (Neblo, 2007). This possibility is limited if the group is too large.

Creating groups which are heterogeneous which respect to perspectives, roles, and/or demographics appears to be an effective antidote to both groupthink and cognitive biases by encouraging critique and constructive conflict (Schweiger, 1986). Where this is not possible, the use of strategies such as devil's advocacy or structured dialectical inquiry (where group members are assigned positions to research and defend and then switch roles) can also be effective (Schweiger, 1986). Techniques such as randomly assigning and switching roles can be helpful in avoiding the problem of participants owning the positions they have taken on and becoming adversarial.

Certain heuristics can also be effective in promoting thorough and impartial evaluation, which is often missing in instances of groupthink and compromised by cognitive biases. Such heuristics as the use of guiding questions or a table for laying out arguments, objections and responses (Bailin & Battersby, 2016a, p. 61 & p. 194) can promote the critical evaluation of alternatives. A check-list of fallacies and cognitive biases to avoid can also be useful.

The necessity to structure deliberation so as to create a climate of respectful debate, ensure diversity, promote inclusivity, and institute appropriate deliberative procedures points to the importance of the role of the chair in setting up the deliberative context and guiding the process. One particular recommendation is for the chair to remain impartial and not state his or her view at the outset so as to avoid undue influence on the group's deliberation (Esser, 1998; Sunstein, 2006; Janis, 1982).

4. GUIDELINES FOR RATIONAL GROUP DELIBERATION

4.1 Introduction

The following guidelines draw on work by other theorists, in particular Blair (2016), Hitchcock et al. (2001), Walton (1996), Atkinson (2006), Habermas (1990), and Cohen (1989, p. 23) as well as our own previous work on inquiry (Bailin & Battersby, 2016a).

4.2 General guidelines

A. Groups should be as diverse as possible.
B. Ideally groups should be small enough to allow for extensive exchange of views.
C. Guidelines for the chair:
  i. Be impartial.
  ii. Ensure that all participants are encouraged to offer their views and that all voices are given a respectful hearing.
  iii. Ensure that deliberations are in accord with the guiding principles.
  Strategies:
  - The chair should avoid stating his/her views at the beginning.
  - Give participants time to think about and make notes about the issue before discussion.
  - Depending on group size, break into subgroups or pairs for initial discussion before the larger group discussion.

- Discussion of the various options should be encouraged before moving to a decision
  iv. Allow for the revisiting of the original framing of the issue as well as the revising of proposed options, arguments, and goals.
  v. Encourage constructive disagreement. If such diversity does not emerge spontaneously, these strategies should be tried:
  Strategies:
     - Devil's advocacy: chair assigns someone to provide a rigorous critique of options which are being favoured or which are not receiving sufficient critique; role should be rotated among group members
     - Structured dialectical inquiry: group members are assigned positions to research and defend and then switch roles (Johnson & Johnson, 1988).
D. Guidelines for participants
  i. Participants should understand the goal of the deliberation as arriving at the best decision, all views considered.
  ii. Participants should show respect for the contributions of other participants and charitably interpret others' contributions.
  iii. Criticisms should be directed at the arguments and not at the person who offers them.
  iv. Reasons should be given in support of any position taken.
  v. Everyone should consider arguments both for and against the various options.
  vi. Everyone should try to be aware of their biases and help identify those of others.

4.3 Guiding questions[2]

  i. What are we trying to achieve?
     - What are the different ways in which the issue could be framed?
     - Do any of the possible framings bias the direction of the inquiry?
     - Is the goal morally legitimate? Commendable? Feasible?

---

[2] Adapted from Bailin and Battersby's "Guiding questions for inquiry" (2016).

⇒ Consider issues of fairness, justice, minority rights, collective well-being.
ii. What actions might reasonably achieve our goal(s)?
iii. How morally acceptable or commendable is each of the proposed actions in terms of:
- the inherent moral quality of the action?
- the relevant duties and obligations of those making and executing the decision?
- the consequences of the action?

iv. What are the arguments for and against each option?
- Use a table of pro and con arguments to lay out options, arguments, objections, and responses.[3]

v. How strong are the arguments in support of each option?
- To what degree are the factual claims well supported and credible?
- Are the predictions based on reasonable inferences from the data?
- Is the context of the argument or action appropriately addressed?
- Have relevant ethical issues been adequately addressed?
- Has feasibility been adequately addressed?

vi. Given the evaluation of the arguments, which option has the most support?
- How much weight should we give to the pro and con arguments in light of their individual evaluation?[4]
- How much support do the arguments provide for each option?
- Is it possible and desirable to modify any of the options in order to address weaknesses revealed in the deliberation?
- Given the deliberation, are we confident we have identified the right goal?
- Given the deliberation, should we feel confident to decide to do X?

---

[3] For more on the structure and use of argument tables, see Bailin & Battersby (2016a, p. 61 & p. 19).

[4] For an account of factors in weighing arguments, see Bailin & Battersby (2016a, pp. 236 - 242).

REFERENCES

Atkinson, K., Bench-Capon, T. & McBurney, P. (2006). Computational representation of practical argument. *Synthese*, *152*(2), 157-206.

Bächtiger, A. & Steiner, J. (2005). Introduction, special issue: Empirical approaches to deliberation. *Acta Politica*, 41(2), 153-168.

Bailin, S., & Battersby, M. (2009). Inquiry: A dialectical approach to teaching critical thinking. *OSSA Conference Archive*. 9. Retrieved from https://scholar.uwindsor.ca/ossaarchive/OSSA8/papersandcommentaries/9

Bailin, S., & Battersby, M. (2016a). *Reason in the balance: An inquiry approach to critical thinking* (2nd ed.). Cambridge, MA: Hackett.

Bailin, S., & Battersby, M. (2016b). DAMed if you do; DAMed if you don't: Cohen's "Missed opportunities". *OSSA Conference Archive*. 90. Retrieved from https://scholar.uwindsor.ca/ossaarchive/OSSA11/papersandcommentaries/90.

Battersby, M. (2016). Enhancing rationality: Heuristics, biases, and the critical thinking project. *Informal Logic*, 36(2), 99-120.

Bench-Capon, T. (2003). Persuasion in practical argument using value-based argumentation frameworks. *Journal of Logic and Computation*, *13*(3), 429-448.

Blair, J. A. (2016). Advocacy vs. inquiry in small-group deliberations. In. D. Mohammed & M. Lewinski (Eds.), *Argumentation and Reasoned Action: Proceedings of the 1st European Conference on Argumentation, Lisbon 2015* (Vol. I , pp. 53-68). London: College Publications.

Bowell, T., & Kingsbury, J. (2016). Open mindedness. *OSSA Conference Archive*. 87. Retrieved from https://scholar.uwindsor.ca/ossaarchive/OSSA11/papersandcommentaries/87.

Cohen, J. (1986). An epistemic conception of democracy. *Ethics, 97*(1), 26–38.

Cohen, J. (1989). Deliberation and democratic legitimacy. In A. Hamlin & P. Pettit (Eds.), *The Good Polity* (pp. 67–92). Oxford: Blackwell.

Cohen, J. (1997). Deliberation and democratic legitimacy. In J. Bohman & W. Rehg (Eds.), *Deliberative democracy: Essays on reason and politics* (pp. 67 – 91). Cambridge, MA: MIT Press.

Dewey, J. (1963). *Liberalism and social action* (Vol. 74). New York: Capricorn books.

Druckman, J. N. (2004). Political preference formation. *American Political Science Review*, *98*(4), 671–686.

Esser, J. (1998). Alive and well after 25 years: A review of groupthink research. *Organizational Behavior and Human Decision Processes. 73*(2/3), 116-141.

Fishkin, J. S., & Luskin, R. C. (2005). Experimenting with a democratic ideal: Deliberative polling and public opinion. *Acta Politica, 40*(3), 284–298.

Gastil, J. (2000). *By popular demand: Revitalizing representative democracy through deliberative elections.* London: University of California Press.

Habermas, J. (1989). The structural transformation of the public sphere: An inquiry into a category of bourgeois society. Tr. T. Burger and F. G. Lawrence. Cambridge: The MIT Press.

Habermas, J. (1990) [1983]. *Moral consciousness and communicative action*, tr. C. Lenhardt & S. W. Nicholsen. Cambridge: Polity Press.

Habermas, J. (2006). Political communication in media society: Does democracy still enjoy an epistemic dimension? The impact of normative theory on empirical research. *Communication Theory, 16*(4), 411-426.

Haidt, J. (2001). The emotional dog and its rational tail: A social intuitionist approach to moral judgment. *Psychological Review, 108*(4), 814-834.

Hart, P. 't. (1991). Irving Janis' victims of groupthink. *Political Psychology, 12*(2), 247-278.

Hitchcock, D., McBurney, P. & Parsons, S. (2001). A framework for deliberation dialogues. *OSSA Conference Archive.* 57. Retrieved from https://scholar.uwindsor.ca/ossaarchive/OSSA4/papersandcommentaries/57.

Janis, I. L. (1972). *Victims of groupthink: Psychological studies of foreign policy decisions and fiascoes.* Boston: Houghton-Mifflin.

Janis, I. L. (1982a). *Groupthink.* Boston: Houghton Mifflin.

Johnson, D. W. & Johnson, R. (1988). Critical thinking through structured controversy. *Educational Leadership, 45*(8), 58-64.

Lunenburg, F. (2012). Devil's advocacy and dialectical inquiry: Antidotes to groupthink. *International Journal of Scholarly Academic Intellectual Diversity*, 14(1), 1–9.

Mercier, H. (2011). What good is moral reasoning? *Mind and Society*, 10(2), 131–148.

Neblo, M. A. (2007). Change for the better? Linking the mechanisms of deliberative opinion change to normative theory. *Common Voices: The Problems and Promise of a Deliberative Democracy.* (Book manuscript)

Sanders, L. M. (1997). Against deliberation. *Political Theory, 25*(3), 347-376.

Schulz-Hardt, S., Jochims, M. & Frey, D. (2002). Productive conflict in group decision making: Genuine and contrived dissent as strategies to counteract biased information seeking. *Organizational Behavior and Human Decision Processes, 88*(2), 563-586.

Schulz-Hardt, S., Frey, D., Luthgens, C., & Moscovici, S. (2000). Biased information search in group decision making. *Journal of Personality and Social Psychology, 78*(4), 655-669.

Schweiger, D., Sandberg, W., & Ragan, J. (1986). Group approaches for improving strategic decision-making: A comparative analysis of dialectical inquiry, devil's advocacy, and consensus. *Academy of Management Journal, 29*(1), 51-71.

Sunstein, C. R. (2006). *Infotopia: How many minds produce knowledge.* Oxford: Oxford University Press.

Tetlock, P. E. & Kim. I. T. (1987). Accountability and judgment processes in a personality prediction task. *Journal of Personality and Social Psychology 52*(4), 700-709.

Walton, D. N. (1996). *Argument schemes for presumptive reasoning*. Mahwah, NJ: Lawrence Erlbaum Associates.

Walton, D. N. (2013). Value-based practical reasoning. In K. Atkinson, H. Prakken and A. Wyner (Eds.). *Knowledge representation to argumentation in AI, law and policy making: A festschrift in honour of Trevor Bench-Capon* (pp. 259-282). London: College Publications.

5

Argumentation Profiles vs. Argumentation Schemes

ANGELINA BOBROVA
Russian State University for the Humanities
<u>angelinabobrova@gmail.com</u>

Argumentation profiles or profiles of dialog is a technique that might be useful in difficult cases of relevance evaluation. I base this on Walton and Macagno's approach and argue that it can be developed with Kant's critical method and his maxims. Besides this modification, I discuss Sperber & Wilson's relevance, credulous and sceptical reasoning (Mental model theory), and a dialogical core of logic (Dutilh Novaes). I claim that the modified version of the argumentation profiles technique can work without argumentation schemes.

KEYWORDS: argumentation profiles, profiles of dialog, Kant's critical method, maxims, relevance evaluation.

1. INTRODUCTION

The title of the paper is tricky. I will argue that argumentation profiles can be developed without argumentation schemes. The publication is devoted to argumentation profiles and says almost nothing about argumentation schemes. Argumentation profiles, or profiles of dialog, are mostly understood as graph or tree-shaped structures. They demonstrate actual and possible dialog movements in order to evaluate argumentation. "The method has been employed within various traditions and approaches to argumentation" (see: Koszowy & Walton, 2016). All approaches deal with real dialogs and imply correlations between these real dialogs and their normative ideals. At the same time, all approaches have similar problems. I would point to two major difficulties.

The first one concerns the concept of normativity. Its meaning is still vague. It is not clear what is a norm, who defines it, and how someone can be sure that a hypothetical path from the premises to an ultimate conclusion is not mistaken. A logically sound, even valid, reconstruction might be contextually wrong. Moreover, the same dialog can admit more than one of its normative modes. The second problem deals with the bounds of the dialog profiles method. There are no answers to (1) whether the technique is applicable to a certain list of cases, or it can be generally used, and (2) whether it deals with explicit dialogs (discourses with two clear sides) only, or it might be extended to analysis of some monologs, for instance, newspaper articles.

The problems are adjacent to each other, but the paper will be focused on the first one (concept of normativity). As for the second problem, it is hard to identify the application level. I will follow Walton and Macagno's (2016) position that the method can be fruitful in cases where the relevance is a problem. However, such cases are not necessary to be explicitly dialogic. The method can work with monologs too. Monologs should be treated as implicit dialogs as the gap between them and dialogs, from the structural point of view, is not wide. Logic has dialogic nature that is true even for deduction, which should be defined as a core of logic. As C. Dutilh Novaes states:

> the deductive method was obviously not created specifically to counter belief bias it does seem to have this effect in that it features a 'built-in opponent' who questions the unjustified assumptions that the agent may make, forcing her to look at all the models of the premises rather than just at the much smaller class of her own preferred models (Dutilh Novaes 2012, p. 159).

Since logic and argumentation theory have the same root, if logic is dialogic, argumentation theory is dialogic doubly so. That is all for the second problem, and I'm back to the first one.

The gap between actual dialogs and their abstract normative models is still too wide. The paper proposes a solution of how this gap can be reduced. I will argue that it can be partly done by Kantian critical method. This position will be illustrated through Walton and Macagno's (W&M's) approach. It proposes a fresh technique, which explicates normative ideal reconstructions.

The second section briefly introduces W&M's conception. The third section is devoted to Kantian method description. The way in which Kant's ideas specify W&M's model is presented in the fourth section. The fifth section gives an example of how the modified approach works. The sixth section points to correlations between Profiles of Dialog and Mental Model theory.

2. WALTON'S AND MACAGNO'S APPROACH

The profiles of dialogue method was introduced by Krabbe who defined a profile of dialogue as a tree-shaped structure displaying the various ways a reasonable dialogue could proceed (Krabbe, 1992). Van Eemeren (2010) used profiles to study argumentative strategic maneuvering in critical discussion. Walton applied dialog profiles to the fallacy of many questions (1989a). In 2016, he and Macagno proposed to extend the method to the cases where the relevance was a problem. Today, their method is used to model sequences of dialog moves to evaluate their relevance. For practical needs, the scholars distinguish topical and probative relevance. Their definitions are given below:

> A proposition P is topically relevant to a proposition Q if P shares subject-matter (discourse topic) overlap with Q, considering that the topic can be either explicitly stated or underlying the whole discourse. [...] [while] A proposition P is probative relevant to a proposition Q if there is a sequence of argumentation pro or con Q that starts at or contains P (Walton & Macagno 2016, p. 527).

These two types have been complemented with cognitive or Sperber and Wilson's relevance that is understood as follows:

> [It] consists in a ratio between cognitive effects and processing efforts. Cognitive effects can be described as improvements of a person's representation of the world, namely ways in which his previous assumptions are modified (Walton & Macagno 2016, p. 527).

The profile method is concentrated on probative relevance coupled with topical because cognitive relevance is applied in contexts of "information sharing, in which the goal of the interlocutors is to increase the shared knowledge base" (Walton & Macagno 2016, p. 528). Below, I will argue for the importance of "information sharing" and cognitive relevance.

Relevance is evaluated in three steps. A descriptive graph is constructed, then a normative one is built. After that the latter graph is incorporated into the former one, and the descriptive graph is evaluated. A descriptive graph is the graph of explicit argumentation, *i.e.* the graph of argumentation as it is done. A normative graph is a graph that

... is based on initial graph but it is more normative in nature because it is supposed to bring out some aspects missing in the original graph but necessary to judge features such as relevance[1]. The normative graph is mapped into the descriptive graph so that a comparison can be made to determine what is missing or otherwise problematic in the sequence displayed by the descriptive graph. The third part of the procedure goes the other way. The argument analyst compares the two graphs to examine and analyse what was missing in the descriptive graph. By this means the analyst is able to explain the problem that was displayed visually be the information in the descriptive graph, and by means of moving to the normative graph, diagnose the nature of the problem and build a recommendation on how to fix it (Walton & Macagno 2016, pp. 535-536).

This approach has some advantages. First, it explicates the idea of a normative graph that "represents the permissible sequence of speech acts (moves in a dialogue) for each party" (Koszowy & Walton 2016, p. 2). There is explicit correlation between a normative ideal and its actual realization. Besides that, the appeal to topical and probative relevance establishes a relationship between structural movements and their contexts. The term 'context' is understood "as the dialogue or discourse to which such a dialogue move contributes or is presumed to contribute" (Walton & Macagno 2016, p. 533).

On the other hand, the gap between normative models and real dialogs is still wide. Additionally, the treatment of normative reconstruction is not clear. This term usually points to logical 'tools', which are powerful but can lead astray if argumentation content is unknown.

3. KANTIAN CRITICAL METHOD AND MAXIMS

The problems with a real graph and its normative version can be clarified with Kantian critical method. The method has a certain potential, and it is still underestimated in argumentation theory. Actually, I. Kant did not write a single paper about the method. All we have is its contemporary reconstructions, one of which belongs to N. Hinske (2010). Hinske distinguishes three stages of this method

---

[1] Walton applies for that reason a Find Argument assistant of CAS but I leave this opportunity aside.

evolution. His first stage is quite profitable for the needs of profiles of dialog.

Kant claims that discussions are mostly based on vague or dubious presumptions, which are usually accepted by both sides. So, if we want to reach the truth, we should look not for the right position but for the vague presumption in order to "eliminate" the subject of the debates (Hinske, 2010). The truth in terms of argumentation theory can be understood as being synonymous with resulted or effective argumentation. Kant's method stresses a substantial argumentation peculiarity: the actual value of argumentation is information exchange. If this peculiarity is taken into account, a normative component loses its superiority. There is no more need to look for a normative ideal but there is a need to look for a way to have smooth information flow. In light of this, dubious presumptions identification is seen as balks. They demonstrate weak points of argumentation and point to the ways for it to improve.

The method declaration does not mean the way of its realization. What are the steps of critical method? Kant did not give any solution. However, it might be structured with Kantian maxims of how human understanding should be used. The term "understanding" is treated as understanding in general. The maxims, as they have been introduced in A. Krouglov (2014), are summarized as follows:

i. employ your own understanding;
ii. place yourself in your opponent's position;
iii. think consistently, compare your opponent's position
   with yours, and enrich your initial knowledge.

The maxims can be inscribed into argumentation analysis. At the first stage, an argument analyst employs her own understanding, which indicates that she has to scrutinize a given argumentation. At the next stage, she presumes her opponent's representations, knowledge and beliefs and reconstructs the same argumentation from that point of view. "To place yourself in your opponent's position" signifies a ban from your own ideas. Once an argumentation has been structured from both sides, the second structure is incorporated into the first one. This step is about consistent manners of enriching or correcting the initial position. If contradictions or inconsistencies appear, this position can be criticized or even abandoned.

It has to be mentioned that Walton almost repeats some of Kant's ideas in the paper of 1989: "as every argument has two sides to be considered, the pro and con of argument"; the study of dialog can help to develop "the ability to constructively understand the other side's point of view" and "the ability to detect bias" (Walton, 1989a, p. 169). However, the scholar did not correlate these ideas with Kant's method. As a result, he started working with normativity. At the same time, Kant's ideas look quite meaningful for Walton and Macagno's approach (2016), which I will demonstrate in the next section.

4. KANT'S IDEAS AND W&M'S MODEL

The core idea of Kant's critical method can essentially develop W&M's conception. The changes are presented in the comparison table 1.

Comments	W&M's conception	Kant's ideas and profiles of dialog
The aim	Build a normative graph to see the weaknesses in a descriptive version.	Find a vague presumption.
The 1st step	Reconstruction of argumentation as it is given (a descriptive graph).	Reconstruction of argumentation as an analyst sees it.
The 2st step	A normative graph generation (it fills in gaps in the first graph representing hypothetical paths leading from the premises to the ultimate conclusion).	Reconstruction of argumentation from the opponent's point of view.
The 3st step	The graphs combination and evaluation (it examines what is missing in the descriptive graph, and this graph is evaluated).	The graphs combination and evaluation (the second graph enhances initial representations).
The type of relevance	Probative relevance Topical relevance	Cognitive relevance Probative and topical

Table 1 – Models comparison

The first line of the table illustrates the aim differences. Kant's method stresses that all discussions are biased and their evaluation presupposes vague presumptions identification. Normative graph is not a "remedy" for its descriptive version, but normativity is, as it is shown below, a foundation for the technique in general.

The first step considers the current argumentation analysis. An analyst looks for the purpose of a dialog or for a thesis. When it is found, she builds a whole graph. According to W&M's conception, it is a descriptive graph as it reflects the explicit argumentation features that are given in the text. The first Kantian maxim (argument analyst employs her own understanding) specifies this position. In fact, the graph presents not the features as they are done, but the features as an analyst sees them. An argument analyst presents her version of how an ultimate conclusion is supported in a text. It means that even this graph is a reconstruction, which attaches a normative tint to it. Any reconstruction admits information insufficiency, but a lack of information leads to logic and its rules. They make our life easier and safer. Thus, each reconstruction is based on logic. Actually, the graph can be logically poor as it is too naïve to think that "being logical" automatically correlates with human's nature. As Dutilh Novaes's (2012) mentions, logical capability is something that can be taught. However, this situation does not influence on the very idea.

The next step in W&M's conception deals a normative graph, which prescribes the 'right' way of the discussion. With a help of implicit premises and known argument schemes an argument analyst builds a logically sound structure. The term 'implicit premises' can be treated as follows:

> An implicit premise is one that has not been explicitly stated, but that is necessary to insert as an assumption [...] that makes sense as a provisional interpretation of what an arguer might be trying to say (Walton and Macagno 2016, p. 551).

At the same time, this step is not trivial as there is no warranty that the normative graph is ideal, indeed. Usually there are several ways of logical reconstruction, as well as several sets of implicit premises. The second maxim (place yourself in your opponent's position) revises this problem. Instead of an abstract graph, it urges the development a graph that reflects the same argumentation from the opposite side. This alteration seems to be promising as it substitutes several ways of argumentation reconstruction (Krabbe's position) with the only structure; it justifies the choice of taken alternative and supports the idea of information development. The second reconstruction is also based on logic. Moreover, its role is even more sufficient as the means of reconstruction is defined with presumptions that increase the level of uncertainty. Presumptions don't give a definite information. Thus, the less information is known, the more weight logic gets.

The final step presumes graphs comparison and evaluation. As both positions have the same importance, an analyst has to decide whether the second graph consistently correlates with the first one. The process in both versions is quite similar technically, but there is a substantial difference. In the second case, the procedure is done not because of the evaluation itself but because of the opportunity to expand initial representations. The search for possible inconsistencies or contradictions is not primary but supporting.

The proposed graph evaluation implies knowledge sharing that sharpens the role of cognitive relevance. It goes first in the process of relevance evaluation. Last, but not least, this models divergence is presented in the last line of the comparison table.

5. EXAMPLE

Now it is time to test the method through an example. It is a piece titled "Online dissidents expose the Russian prime minister's material empire", an article that was published in *The Washington post* (the issue from 9.03.2017)[2]. This slippery material is devoted to Russian opposition activist Alexei Navalny and the Russian Prime Minister. The material is easy to analyse as there is no previous discussion in Mass Media. Here is the quotation:

> (1) THE STORY begins with a pair of Nike Air Max 95 shoes with neon-green soles. Russian Prime Minister Dmitry Medvedev was photographed wearing them; emails stolen from his iPhone by dissident Russian hackers show they, along with several other articles of clothing Mr. Medvedev has been seen in, were ordered online and delivered to an address linked to a web of companies and charities controlled by close confidants and relatives. After months spent probing the network, the Anti-Corruption Foundation, headed by opposition activist Alexei Navalny, has released a video and documentation it says show that Mr. Medvedev has accumulated more than $1 billion worth of property, including vast estates in Russia and Tuscany and two yachts.
> Mr. Medvedev's spokesperson shrugged off the story, as did the Kremlin. But as of Thursday, Mr. Navalny's biting, often humorous and slickly produced video (in Russian with English subtitles) had been viewed 7.4 million times on YouTube and attracted 40,000 comments. It's a testimony not only to the staggering corruption of the regime of Vladimir Putin, but also to the

---

[2] See: https://goo.gl/ytzWyu

power of the Internet and social media to expose it and inform Russians about it.

The ultimate proposition is a complex claim that the regime of Vladimir Putin is corrupt and the Internet is influential. The way arguments support the ultimate proposition demonstrates the 1st figure. Weak connections are marked with dotted lines. A sound connection is marked with a simple line. The ultimate conclusion is divided into two parts. The figure shows that the second part of the ultimate conclusion is supported with an argument, whereas the first one is not. It does not infer from the given arguments (the number of video viewers and the behaviour of Mr. Medvedev's spokesperson).

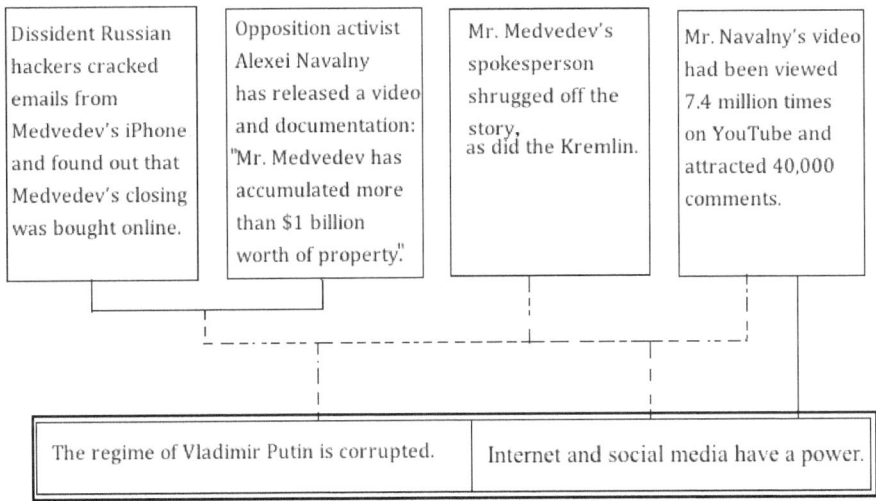

Figure 1 – The initial graph

To be honest, even the first stage presumes a common knowledge that annual salaries of top officials are high, but not too high, so the graph can be detailed. This is done in the 2nd figure where the presumption is put at the top of the graph. As it is not an argument, it has no certain connections, but governs argumentation and explains 'implicit' arguments (italicized). A supplementary conclusion appears, which is put in bold. At the same time, this specification does not alter the situation, *i.e.* the ultimate conclusion is not grounded.

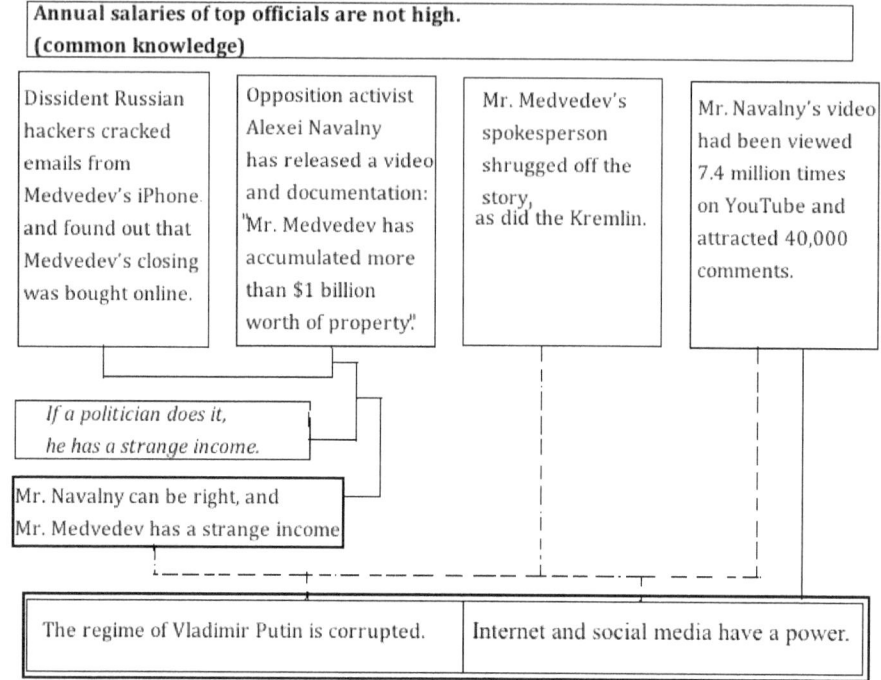

Figure 2 – The initial graph with a common knowledge presumption

The poor linkage should be explained at the next stage of profiles of dialog analysis, which is an opponent's position reconstruction. It starts with presumptions identification. In this case they are defined with the author's representations or newspaper orientations. *The Washington post* is the government translator. As the US–Russian relations have strained, the newspaper might use each opportunity to blame the Kremlin. Presuming that, some implicit premises should be added and another reconstruction is done. The 3rd figure introduces the way in which the arguments could support the ultimate proposition.

At the final stage, this graph is incorporated into the first one. Such incorporation gives an opportunity to evaluate argumentation. The evaluation starts with information sharing (an analyst tries to enrich her knowledge), *i.e.* with cognitive relevance.

The evaluation reveals that the proposition about Mr. Putin's regime contains in the presumption that Washington DC blames Moscow. There is a kind of begging the question (*petitio principii*) fallacy. To conclude, the second graph provides new information, but it

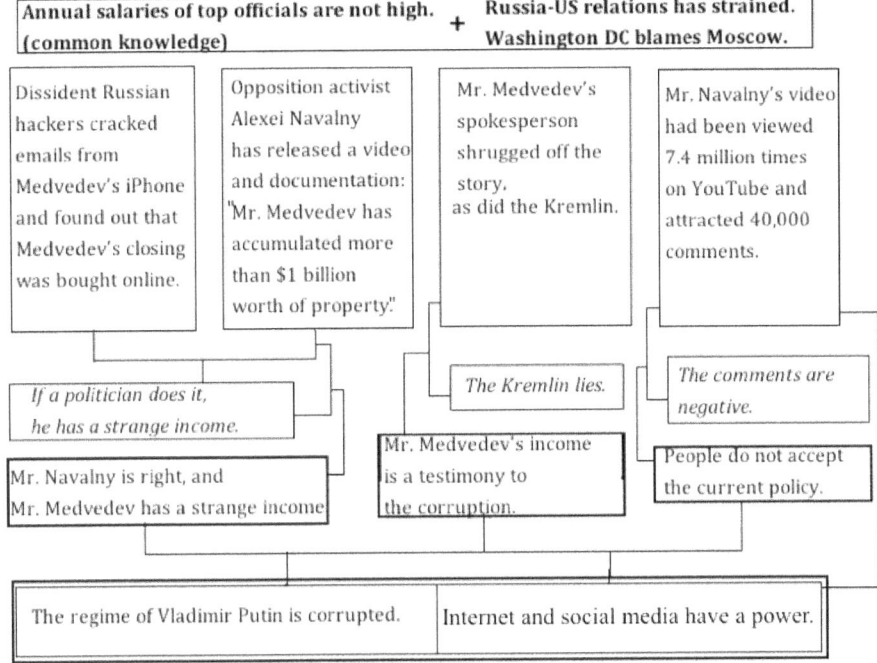

Figure 3 – The graph with opponent's presumptions

does not change the first conclusion: Navalny's video does not support the ultimate proposition (Putin's regime might be corrupted or not).

6. PROFILES OF DIALOG AND MENTAL MODEL THEORY

The altered Profiles of Dialog technique has much in common with mental model theory. It reinforces reasoning movements within the latter theory. To be more accurate, Mental Model theory has been realized in the concept of two types of reasoning, *i.e.* credulous processing of discourse and sceptical classical reasoning. The credulous reasoning is reasoning that

> ... is aimed at finding ideally a single interpretation which makes the speaker's utterance true, generally at the expense of importing all sorts of stuff from our assumed mutual general knowledge. Sceptical reasoning is aimed at finding only conclusions which are true in all interpretations of the premises. <...> In sceptical understanding, we consider ourselves as ... regards authority for inferences, and may well challenge what is said on the basis that a conclusion doesn't follow because we can find a single interpretation of the

premises in which that conclusion is false" (Stenning and van Lambalgen 2008, p. 26).

These reasoning types work within the source founding model, which "is designed to capitalise on the logical peculiarities of the domain which make credulous processing of discourse come so close to sceptical classical reasoning" (Stenning and van Lambalgen 200, p. 312). It is aimed at demonstration of credulous and sceptical reasoning cooperation.

Prototypes of credulous and sceptical reasoning and styles of their cooperation can be found in the modified version of W&M's profiles of dialogs. The profiles technique is aimed at argument identification and analysis. An analyst wants to understand an argument; she looks at it from two different points, and creates graphs. And here she predominantly operates with credulous reasoning. At the next stage, a comparison comes. If it cannot be done without contradictions, credulous reasoning is switched to sceptical one. The sceptical reasoning is aimed at inconsistencies identification, and once they are found, a counterexample appears. So, credulous reasoning goes hand in hand with sceptical reasoning.

7. CONCLUSION

In the paper, I've argued for the profiles of dialog, as for the technique that might be useful in difficult cases of relevance evaluation. It includes three steps those can be regulated with Kant's critical method and his maxims.

Kantian ideas modify the understanding of profiles of dialog, which appears to be built on the idea of information exchange. The vague presumptions orientation points to possible biases in argumentation foundations. The substitution of descriptive and normative graphs with graphs of (1) initial and (2) opposite positions respectively revises the role of logic and normativity. Normative tints are treated as basic characteristics of graph reconstruction. They become more apparent in the second graph as the opponent's position identification is accompanied with a higher level of information uncertainty. The further details can be found in the 4th section of the paper. As argumentation has to be understood before it is evaluated, and the understanding implies information sharing, the method "rehabilitates" the concept of cognitive relevance.

The modified W&M's Profiles of Dialog method reinforces some ideas of Mental Model Theory. Such correlations can be treated as evidence that various spheres of knowledge come to similar results.

Finally, presumptions coupled with logical basis give a nice alternative to argument schemes analysis. It may be too ambitious to say, but the profiles of dialog method can work without them. Argumentation schemes are profitable in simple cases, but they can be redundant in difficult cases. However, this question should be scrutinized in a separate paper.

REFERENCES

Dutilh Novaes, C. (2011). The different ways in which logic is (said to be) formal. *History and Philosophy of Logic*, *32*(4), 303-332.

Dutilh Novaes, C. (2012). *Formal languages in logic: A philosophical and cognitive analysis*. Cambridge: Cambridge University Press.

Eemeren, F. H. van (2010). *Strategic maneuvering in argumentative discourse: Extending the pragma-dialectical theory of argumentation*. Amsterdam: John Benjamins Publishing.

Hinske, N. (2010) Die Rolle des Methodenproblems im Denken Kants. Zum Zusammenhang von dogmatischer, polemischer, skeptischer und kritischer Methode, in Kants Grundlegung einer kritischen Metaphysik. In *Einführung in die Kritik der reinen Vernunft* (pp. 343-354). Hamburg: Norbert Fischer.

Krabbe, E. C. W. (2002). Profiles of dialogue as a dialectical tool. In F. H.van Eemeren (ed.), *Advances in Pragma-Dialectics* (Chapter 10), Amsterdam: Sic Sat /Newport News, VA: Vale Press.

Krabbe, E. C. W. (1992). So what? Profiles for relevance criticism in persuasion dialogues. *Argumentation*, *6*(2), 271-283.

Krouglov, A.N. (2014). Nesoversennoletie I zadacha istinnogo preobrazovaniya obraza mishleniya (Part 1). *Kantovskij sbornik*, *3*(49), 19-39. (In Russian).

Koszowy, M., & Walton, D. (2017). Profiles of dialogue for repairing faults in arguments from expert opinion. *Logic and Logical Philosophy*, *26*(1), 79-113.

Johnson-Laird, P.N. (1983). *Mental models*. Cambridge: Cambridge University Press.

Stenning, K., & van Lambalgen, M. (2008). *Human reasoning and cognitive science*. MIT Press, Cambridge.

Walton, D. (1989a). Dialogue theory for critical thinking. *Argumentation* 3(2), 169-184.

Walton, D. (1989b). *Question-reply argumentation*. Westport, Connecticut: Greenwood Press.

Walton, D., & Macagno, F. (2016). Profile of dialogue for relevance. *Informal logic*, *36*(4), 523-562.
Wilson, D., & Sperber, D. (2004). Relevance theory. In L. Horn & G. Ward (Eds.), *Handbook of pragmatics* (pp. 607-632). Oxford: Blackwell. Retrieved from http://doi.org/10.1016/j.pragma.2009.09.021

6

Believing, Inferring, and Basing

PATRICK BONDY
Department of Philosophy, Brandon University
patrbondy@gmail.com

This paper addresses inference and the epistemic basing relation. It articulates accounts of the basing relation that incorporate casual conditions, and meta-belief conditions, and mixtures of the two. It then explains the distinction between occurrent beliefs, dispositional beliefs, and dispositions to form beliefs, and it considers explicit and implicit inference, and how the meta-beliefs required by some accounts of the basing relation bear on these sorts of inferences.

KEYWORDS: belief, epistemic basing relation, inference, meta-belief, causal deviance

1. INTRODUCTION

This paper is about inference and the epistemic basing relation. The basing relation is the relation which obtains between beliefs and the reasons on the basis of which they are held. The term "inference" is used in various related ways, but the sense that I'm interested in here is the sense in which someone reasons consciously from premises to a conclusion. This is the sense of "inference" that various authors have recently been aiming to clarify.[1] In this sense, inference is more or less conscious, and at least somewhat voluntary. This is the sense of inference that is relevant for argumentation theory, because argumentation is exactly the kind of situation where premises are offered for inferentially accepting conclusions.

I begin, in section 2, by briefly laying out Robert Audi's way of distinguishing occurrent beliefs, dispositional beliefs, and dispositions

---

[1] In particular, Boghossian 2014; Broome 2014; Neta 2013; Wright 2014; Setiya 2013.

to form beliefs, and I'll illustrate that beliefs can be held on the basis of reasons either inferentially or non-inferentially.

Then, in section 3, I argue for several conclusions: (1) Inferring *p* from *q* is sufficient for basing (of one's belief B with content *p*, on reason R with content *q*). (2) Inference is not necessary for basing. (3) Causation is not necessary for inference (and therefore not necessary for basing). (4) Inferential basing is always partly explained by the subject's *appropriate meta-belief*, to the effect that the premises provide good reason for accepting the conclusion. And, (5) non-inferential basing always comes with a disposition to form an appropriate meta-belief, and the ground of the disposition to form the appropriate meta-belief is also the ground of the non-inferential basing relation.

2. OCCURRENT BELIEF, DISPOSITIONAL BELIEF, AND DISPOSITIONS TO FORM BELIEFS

At any given time, the overwhelming majority of our beliefs are stored in memory, more or less ready to be called into our conscious awareness when needed or when prompted. So there is a familiar distinction to be drawn, between *occurrent* beliefs—those that a subject S is currently entertaining—and *dispositional* beliefs—those that S is disposed to call to mind in appropriate conditions.

There are a few wrinkles we need to be careful about when thinking about dispositional belief, however. The wrinkle that is important for our purposes here is that dispositional beliefs need to be carefully distinguished from dispositions to form beliefs.[2] This distinction can be difficult to keep track of, because there are many propositions which we do not currently believe, but which we are disposed to come to believe immediately upon considering them. For example, consider the arithmetical proposition <15 + 112 = 127>. Immediately, or at least very quickly upon considering that proposition, we all are disposed to form the belief that it is true. But, I assume, it's not something that we already had in mind as a standing, dispositional belief.

Audi illustrates the distinction between occurrent beliefs, dispositional beliefs, and dispositions to form beliefs with a computer analogy: "Take the two relevant kinds of actual belief."—i.e., occurrent and dispositional. "What is dispositionally as opposed to occurrently

---

[2] Another wrinkle is that dispositions can be present in an object but fail to be manifested in characteristic manifestation conditions for the disposition, if other conditions are present which block it.

believed is analogous to what is in a computer's memory but not on its screen" (2015, pp. 12-13). This is like thinking about a proposition that one endorses, as opposed to something one believes but which needs to be called up by a memory-retrieval process. Both of these contrast with dispositions to believe: these are like what "a computer would display only upon doing a calculation, say addition: the raw materials...are present, but the proposition is not yet in the memory bank or on the screen" (ibid.)

Now, when we're wondering whether a subject S has a dispositional belief B, we will be thinking about whether there is a stored representation-as-true of B's content in S's mind (this is the ground of the dispositional belief), which would be called into S's awareness, or which would guide S's behaviour, in certain characteristic circumstances (this is what makes it a dispositional belief), and we will be looking for characteristic manifestations of that disposition in its characteristic manifestation conditions.

By contrast, when we're wondering whether S has a disposition to form a belief B with content p, we are not particularly concerned with whether there is a representation-as-true, or even a representation at all, of p in S's mind. S might have a representation of p as something possibly true, or to-be-feared, etc., without either believing or being disposed to believe p. Rather, we will be looking to see whether S possesses grounds, reasons, or perhaps some non-rational causes that would normally cause S to hold B; and we will be looking to see whether any of the characteristic manifestation conditions for the disposition to form B obtain; and we will be looking for characteristic manifestations of the disposition to form belief B, in its characteristic manifestation conditions.

One key difference between dispositional beliefs and dispositions to form beliefs is that, when one has a dispositional belief with content p, S is disposed to guide her behaviour in accord with p (perhaps even without thinking about p explicitly), whereas if S is only disposed to form the belief that p, S would guide her behaviour in accord with p only once she has thought of p or otherwise manifested her disposition to form the belief that p (cf. Audi, 2015, p. 17). Beliefs, but not dispositions to believe, are *presuppositionally available*.

3. INFERENCE

In this section, I will discuss some cases where a subject S infers p from some set of premises, and I will aim to establish five claims:

i. Inference is sufficient for basing.
ii. Inference is not necessary for basing.
iii. Causation is not necessary for inference.
iv. Inferential basing is partly explained by an appropriate meta-belief.
v. Non-inferential basing always comes with a disposition to form an appropriate meta-belief, and the ground of the non-inferential basing relation is exactly the ground that explains the disposition to form the appropriate meta-belief.

3.1 Inference is sufficient for basing

Consider an ordinary case of inference:

P1. I've just run out of milk!
(P2.) I like milk.
(P3.) I can get milk at the store.
C. I should go to the store!³

P2 and P3 would normally be left implicit; an ordinary reasoner's explicit inference would move directly from P1 to C. But the reconstruction and attribution of implicit premises isn't my point here; the point is just that this inference, or at least the move from P1 to C, is a perfectly ordinary case of inference. And in this case, it's because I *drew the inference* from P1 to C, that I have come to hold the belief that I should go to the store *on the basis* of my belief that I've run out of milk (and, plausibly, also on the basis of my beliefs that I like milk and that I can get milk at the store).

---

³ As it stands this is an instance of theoretical reasoning, resulting in a belief about what I should do. I am inclined to think that all inference can be appropriately modelled as theoretical inference in this way, though others (e.g. Boghossian 2012, Broome 2014) think that we should distinguish theoretical inference, which issues in beliefs, from practical inference, which issues in intentions. The distinction doesn't matter here; the suggestion that inference is sufficient and not necessary for basing can be extended to cover intentions as well as beliefs, for intentions can be had on the basis of reasons, just as beliefs can, and the relation between beliefs and reasons they're based on will presumably be the same as the relation between intentions and reasons they're based on. In any case, Boghossian (2012) and Broome (2014) are only concerned with giving an account theoretical inference. Neta (2013) gives a general account of inference meant to handle theoretical and practical cases.

We'll see a more complicated and more interesting case shortly (the superstitious lawyer). But for now, at least, it looks like typical cases of inference are also cases where the subject believes the conclusion of the inference on the basis of its premises.

3.2 Inference is not necessary for basing

There are cases where beliefs are formed non-inferentially but on the basis of a reason. Of course the term "inference" is used in various related ways, and there's a sense in which it's natural to say that any time S holds a belief B on the basis of R, S has at least implicitly inferred B from R. But there's a more involved sense of "inference", the one that people like Boghossian and Broome and Neta have been trying to explain, and in that sense, inference is at least somewhat reflective, conscious, and voluntary. In this sort of inference, the subject must think of at least one premise, and the conclusion, and the subject must take the premises to adequately support the conclusion.

Thinking of that sort of inference, it seems clear that there are cases of non-inferential basing. Here's an example:

(1) Suppose you had always believed that platypuses are mammals, but in some way you came to believe that platypuses are not mammals. Probably an automatic process caused you at the same time to stop believing platypuses are mammals; automatic processes normally prevent you from having contradictory beliefs. Then your non-belief in the proposition that platypuses are mammals (a negative attitude) is based on your belief in the proposition that platypuses are not mammals. Yet you did not acquire the non-belief by reasoning (Broome, 2013, p. 189).

Here's another, this time a case involving the formation of a dispositional belief in response to a perceptual stimulus:

(2) When absorbed in conversation, one might come to believe, through hearing a distinctive siren, that an ambulance went by, but without thinking of this proposition or considering the matter. This is the formation of a dispositional belief (Audi, 2015, p. 13).

In these cases, it seems, a belief is formed on the basis of a reason, but the subject does not consciously move from the reason to the belief. So these aren't cases of inference in the relevant sense.

3.3 Causation of belief in the conclusion is not necessary for inference

This is perhaps the most important, if only because so often overlooked, point I want to make in this paper.

First of all: inference need not issue in a new doxastic attitude. At a time t_1, S might hold B on the basis of an apparent reason R1, or on the basis of no reason at all, and then at a later time t_2, S might come to see that a different reason R2 supports B. Then at t_2, S can infer p (the content of B) from q (the content of R2), even if S doesn't change her attitudes at all with respect to p. S might retain a belief in p; S's feeling of confidence in the truth of p might not even change (S might already have been maximally confident in p, for instance). —What I'm describing here are ways a belief might be causally overdetermined by multiple independent reasons, or by both reasons and things other than reasons.

So, for example, when Boghossian characterizes the sort of inference he's interested in as the sort where "you start off with some beliefs and then, after a process of reasoning, end up either adding some new beliefs, or giving up some old beliefs, or both" (2014, p. 2), this characterization is too narrow: the sort of conscious inference, moving from premises to a conclusion, which he aims to analyze can also occur when there is no change in the subject's doxastic attitudes with respect to the conclusion of the inference.

I don't think that what I'm describing is controversial. Surely it is plausible to think that, just as a window's shattering might be causally overdetermined by multiple large rocks striking it at the same moment, so too a belief might be causally overdetermined. And so an inference might be performed, which concludes with a proposition which the subject already fully believes. Then her attitude would not change. I think that people simply ignore these sorts of cases when they're discussing inference because these seem like just special cases of ordinary inference, where a doxastic attitude does not change. Or maybe these seem like degenerate cases.

But recognizing that inference need not issue in a new doxastic attitude paves the way, I think, for us to understand that in inference, premises don't even need to causally contribute to the subject's hold a belief in the conclusion at all. It is possible for S to infer B's content p from R's content q even if R does not causally contribute to S's holding B. This is what we can see is going on in the case of the superstitious lawyer, from the literature on the epistemic basing relation. The case goes as follows:

> (3) The counterexample concerns a [sic. superstitious] lawyer who, like the rest of his contemporaries, takes his client to be guilty. However, because of his [superstitious] nature, the lawyer is inclined to trust what the tarot cards say, and upon learning that the tarot cards say that his client is innocent, comes to believe that his client is innocent. What the tarot cards say also prompts the lawyer to re-examine the evidence, which the lawyer comes to recognize conclusively establishes that his client is innocent. However, given his rather impressionable character, the lawyer also realizes that were the sustaining power of the tarot cards removed, the sway of public opinion would cause him to be unable to see that the evidence establishes his client's innocence. Nonetheless, the lawyer now justifiably believes that his client is innocent on the basis of his examination of the evidence. But this examination of the evidence neither prompts his belief that his client is innocent nor does it sustain his belief that his client is innocent—his belief in what the tarot cards say holds the dubious distinction of being responsible for both (Kvanvig 1985, pp. 153-154, summing up the case set out in Lehrer 1971).

Here, we have a reason R1, which is the evidence that the lawyer has available, which he has re-examined; we have a belief B, the belief that his client is innocent; and we have another (apparent, or motivating) reason R2 which the lawyer possesses, the Tarot card reading he has performed. In this case, R2 is causally sufficient in the context for the lawyer to hold B, and R2 is also causally necessary in the context for the lawyer to take R1 (the good evidence) to support B. If the lawyer were to lose R2—say, if he were to perform another Tarot reading which indicated that the first reading was mistaken—then he would also at the same time lose R1, and he would lose B.

But: in spite of the evidence's lack of causal efficacy, the lawyer does seem to draw an inference here from the evidence, which he has taken pains to re-examine, to the conclusion that his client is innocent. He now recognizes that R1 is a good reason for holding B, and if only he weren't so vulnerable to the emotional factors in the case, R1 would causally sustain B.

That is the central lesson of the case of the superstitious lawyer: inference is possible without the premises having any causal efficacy with respect to the doxastic attitude the subject has toward the conclusion of the inference. Of course, most cases of inference will be much more straightforward, and the subject's taking the premises to support the conclusion will in fact make a causal difference. Normally, in inference, the subject's doxastic attitude regarding the conclusion changes, or at least becomes causally overdetermined. Still, it's important to recognize that superstitious lawyer cases are possible, which means that causation is not necessary for inference.

And, if I was right to say (above, section 3.1) that inferring p from q is sufficient for basing belief in p on a reason with content q, that means that causation of belief in the conclusion is not necessary for basing a belief on a reason, either.

3.4 Meta-beliefs explain inferential basing

In any case where S draws an inference, S has an appropriate meta-belief M to the effect that R is a good reason for B. The belief need not be occurrent; it may be, and it typically is, dispositional. M guides S's behaviour—in particular, it guides his behaviour in drawing the inference in question—and under characteristic manifestation conditions for the dispositional meta-belief M, S would be disposed to occurrently hold M.

M need not exist prior to the moment when S draws the inference; the inference might be performed and the dispositional belief M might be formed at the same time, and they might both be justified by the very same thing. But S's appropriate meta-belief M is part of what *explains* why S draws the inference from (the content of) R to (the content of) B. I am proposing, in effect, a doxastic construal[4] of Boghossian's Taking Condition:

> **(Taking Condition):** Inferring necessarily involves the thinker taking his premises to support his conclusion and drawing his conclusion because of that fact. (2014, p. 5)[5]

---

[4] Setiya (2013) also proposes a doxastic construal of the Taking Condition, though he holds the specific view that S must take the conclusion to be *evidentially* supported by the premises of the inference. I propose the more general view that S need only take the conclusion to be normatively supported in some way or other by the premises.

[5] "because of that fact" seems to indicate that taking the premises to support the conclusion is part of what *causes* the subject to accept the conclusion. If that

It's very plausible that in order for S to infer a conclusion from some premises, S must take the premises to support the conclusion. Otherwise, even if S's acceptance of the premises somehow gives rise to an acceptance of the conclusion, the causal relation would surely be deviant; it would not be a case of inference.[6]

According to a doxastic construal of the Taking Condition, S takes the premises to support the conclusion when and only when S has an appropriate meta-belief to the effect that the premises provide good reason for accepting the conclusion.

Boghossian resists a doxastic construal of the Taking Condition, for, he argues, it over-intellectualizes the activity of drawing inferences, ruling out conceptually unsophisticated subjects such as animals and small children from counting as able to draw any inferences.[7] He provides the following example:

(4) A child, we are inclined to think, can reason. Luke and Drew are playing hide-and-seek. Seeing Drew's

---

is what's meant, then the Taking Condition would have to be modified, to handle cases of inference without causation, as we've seen already. But I'm not sure that the Taking Condition really needs to be modified, for it only says that the subject *draws his conclusion* because of that fact. The subject in the superstitious lawyer case plausibly does satisfy the Taking Condition so construed, for he draws his conclusion on the basis of the evidence, and he draws his conclusion *because* he takes the evidence to support that conclusion. It's just that his drawing that conclusion has no causal efficacy with respect to whether he holds his belief in the conclusion, in the circumstances he finds himself in.

[6] See Boghossian (2014) for further arguments for the Taking Condition.

[7] Boghossian raises several other worries as well, including especially the worry that a doxastic construal of the Taking Condition faces problems when it comes to explaining how an inference can *transmit justification* from its premises to its conclusion. I think that Boghossian's worries on this score can be handled, but they would take us too far from the main focus of this paper. I am only interested in discussing the *nature* of inference, not what makes it possible for inferences to justify belief in their conclusions. After all, the possibility of gaining justified belief in a conclusion on the basis of premises from which it is inferred must surely be explained by invoking our best account of the nature of inference. But we should not build our account of the nature of inference specifically so that it will yield the normative results that we are looking for; that good inferences transmit justification from their premises to their conclusions must be *shown* by invoking an account of the nature of inference, rather than presupposed as an adequacy condition for accounts of the nature of inference.

> bicycle leaning against the tree, Luke thinks: "If he were hiding behind that tree, he would not have left his bicycle there. So, he must be behind the hedge." That looks like reasoning. But do children have meta-beliefs about the relations between their premise judgments and their conclusions? Do children have the concepts of premises and conclusions? Do they have the normative concept of one belief justifying another? (Boghossian, 2014, pp. 6-7)

The doxastic construal of the Taking Condition is no doubt the most natural, and it is therefore the one Boghossian considers first, to rule it out before moving on to other less obvious construals of the condition. Seeing that the doxastic construal apparently runs into trouble, he moves on to try other construals, and he lands on his preferred rule-following construal.

In their commentaries on Boghossian's paper, Broome (2014) approves of the move to rule-following, while Wright (2014) instead argues that, given the difficulty in articulating a satisfactory non-circular account of how rule-following can be involved in inference, we should reject the Taking Condition after all. But all of this assumes that we should in fact reject the doxastic construal of the Taking Condition. And, it seems to me, we should accept the doxastic construal.

As we've seen, the most important objection to the doxastic construal is that it over-intellectualizes inference. This is a common enough worry, but it can be handled. For one thing, children *do* have some rudimentary understanding of the concept of a reason, even if they don't yet know the word "reason". Very early on in a child's linguistic development, it will learn how to employ and answer why-questions, such as "why did that happen?", "why did you do that?", "why do you say that?", "why do we have to go to gramma's?", and so on. Being able to say why something happened presupposes possessing at least a rudimentary concept of an explanatory reason; being able to say why she has done something, or why she thinks something, is sufficient for a child to possess at least a rudimentary concept of a motivating reason, i.e. a reason which explains from her perspective explains why she did something; and being able to say why someone should do something is to possess at least a rudimentary concept of a normative reason, i.e. a reason which favours, or supports, doing something. In Boghossian's example, Luke is able to wonder *why* Drew left his bicycle behind the tree. So he has at least a rudimentary concept of a motivating reason.

Second, there are of course very young children, and perhaps higher animals, that do not even have these rudimentary reason-concepts (i.e., they do not yet understand how to ask and answer why-questions), and yet they seem to base beliefs on reasons. (Say, a dog smells fresh food, and believes that his owner has just put food in his dish.) But even if that is correct, these kinds of cases do not pose any kind of problem for the doxastic construal of the Taking Condition. These unsophisticated subjects do not draw inferences at all—at least, they do not draw the sort of conscious, voluntary inferences that are at issue here. We can grant (for the moment; we'll come back to this at the end) that very small children and perhaps higher animals have beliefs, and that they even hold their beliefs on the basis of reasons. But subjects like these will hold beliefs non-inferentially on the basis of reasons. Precisely because they lack the concept of a reason, they are not in a position to formulate an inference from premises to a conclusion, which is the sort of inference that we, along with Boghossian, are interested in here.

So the over-intellectualization objection is no problem at all for a doxastic construal of the Taking Condition, because we're only intellectualizing the intellectual sort of inferences.

3.5 Non-inferential basing and the disposition to form an appropriate meta-belief

As we have seen (e.g., in example (2), the ambulance example), beliefs can be formed non-inferentially, but still be formed on the basis of reasons. In the ambulance case, one has formed a dispositional belief that an ambulance has gone by, and the belief that an ambulance went by is based on the sound of the siren. But what makes it the case that one has formed the dispositional belief that an ambulance has gone by is the sound of the siren, together with one's general belief that that kind of siren sound is produced by ambulances.

I've argued above that in cases of inferential basing, the subject has an appropriate meta-belief relating the premises and the conclusion of the inference. But what about in a case like this one with the ambulance? The appropriate meta-belief would have content along the following lines: <that sort of siren is indicative of ambulances>. Does the subject in a case like this *dispositionally believe* that content to be true? Or does she only have a *disposition to form* that belief? And, if the latter, then are the *grounds* of the disposition to form the appropriate meta-belief *responsible* for the actual basing of the belief that an ambulance went by on the reason consisting of the sound of the siren?

It seems to me that the appropriate meta-belief is *sometimes* an actual dispositional belief that the subject S has: S might be aware of the sound of the siren, and like most people, S of course also has the dispositional meta-belief that that sort of siren sound is indicative of ambulances. These grounds together could easily generate a dispositional belief to the effect that an ambulance has just gone by, which would be based on them. And importantly, as we've seen, if S has the dispositional belief that that sort of siren sound is indicative of ambulances, then S need not even explicitly draw to mind that belief; it is *presuppositionally available*: it is available as a presupposition for drawing inferences and guiding behaviour. It can guide belief-formation as a rule of inference without being explicitly invoked (Cf. Audi, 2015, p. 17).

But it can also happen that S is only disposed to form, rather than already dispositionally possessing, an appropriate meta-belief M about R and B. The subject possesses grounds that would generate M in the right circumstances, but those grounds might not yet generate an actual dispositional appropriate meta-belief. And if that is so, then before S is in position to employ M as a rule of inference in his belief-formation, some realization condition for the disposition to form the meta-belief M must be instantiated. For example, it might occur to S to wonder whether M is true, and therefore available to justify a rule permitting inferences from R to B. Perhaps immediately upon thinking about M, S will see that it is self-evident; or perhaps, S would have to determine whether M follows from other things he believes. The point is just that S needs to form the meta-belief M before S is in position to rely on M in inference.

This can happen, for example, if S lives in a crime-ridden neighbourhood, and has a belief that when one hears a loud bang, one should take cover (for loud bangs are indicative of gunshots). Then S goes to a neighbouring town, and hears a loud bang. S drops to the ground, but then it occurs to S that this is a safe neighbourhood. So S forms the belief that here, loud bangs are not indicative of gunshots. S therefore suspends his meta-belief that when one hears a loud bang, one ought to believe that a gun has just been fired. This doxastic process is very quick, and its realization condition was just that S consider the relatively crime-free nature of this neighbourhood while also thinking about loud banging sounds.

Again, as we've seen, it's possible to have beliefs based non-inferentially on reasons. I'm suggesting that the basing relation in these cases can, but need not, involve S's possessing an appropriate meta-belief; S might only possess a disposition to form an appropriate meta-belief. And the grounds of the disposition to form the appropriate meta-belief are exactly the grounds that bring S to form the belief B non-inferentially on the basis of R.

In example 2, the ambulance case, there is non-inferential basing, where the subject already possesses the relevant dispositional meta-belief. On the other hand, example 1, the platypus example, appears to be a case of non-inferential basing, where the subject (you) does not possess the relevant appropriate meta-belief, but where you are surely disposed to form that belief immediately upon thinking of it. An appropriate meta-belief M in that case would presumably be something to the effect that one shouldn't believe both that a platypus is a mammal and that it is not a mammal. And surely M is a belief that most of us would form, immediately upon thinking about it, for we recognize that it represents a prohibition on believing a particular contradiction. Most of us likely did not already hold the specific belief M, but most of us *do* already believe that one shouldn't believe contradictions. This more general meta-belief (call it M2) is a ground of a disposition to form M, immediately upon thinking about it.

But if M2 is the ground of the disposition to form M (the belief that one shouldn't believe both that a platypus is a mammal and that it's not a mammal), then the disposition to form M has the very same grounds as the ground of the basing relation between R (the belief that platypuses are not mammals) and B (the non-belief in the proposition that platypuses are mammals—non-beliefs are not beliefs, but they are still doxastic states, so I'll continue to use "B" to represent them).

If Broome is right, then B is based on R in the platypus case. And what explains that basing relation? S's possessing the relevant meta-belief can't be part of the explanation, because S doesn't possess the relevant meta-belief. We can perhaps aim to invoke the fact that R is a non-deviant cause of B: R, the belief that platypuses are not mammals, is taken into account by some subconscious process, and *together with the general prohibition on believing contradictions* which the subject in the platypus case endorses, caused the elimination of the belief that platypuses are mammals. (The elimination of the belief that platypuses are mammals is the formation of B, the non-belief that platypuses are mammals.) The general prohibition on believing contradictions is surely a key part of the explanation of the basing relation between R and B, for without it or something very much like it, the causal relation between R and B would likely be deviant. But the general prohibition on believing contradictions is also the ground of the disposition to form the appropriate meta-belief M, i.e., the belief that one shouldn't believe both that a platypus is a mammal and that a platypus is not a mammal.

So to say that S non-inferentially forms B on the basis of R implies that S has the disposition to form a meta-belief to the effect that R is a good reason for B, and the ground of that disposition is also what enables S's formation of B on the basis of R.

3.6 Objection: unsophisticated believers

There is one objection I want to address before closing, which rests on the assumption that small children and higher animals have beliefs, and that they hold beliefs non-inferentially on the basis of reasons. I granted this assumption above, in section 3.4, in order to show that even if it were true, it would not have any bearing on the account of the nature of conscious inference from premises to conclusions.

But this assumption does pose a problem for my claim that in every case of non-inferential basing of B upon R, the subject has at least a disposition to form a meta-belief M, to the effect that R supports B. For very small children and higher animals don't possess the concept of a reason, and they are therefore not disposed to form the belief that R supports (i.e., is a normative reason for holding) B. No one is disposed to form a belief which constitutively involves a concept the subject does not possess.

There are two replies to this objection. First: it's not clear that such unsophisticated subjects really do fail to possess the disposition to form the relevant meta-belief. They fail to possess the relevant *concept*, which prevents them from being able to *enter* characteristic realization conditions of the disposition to form the relevant meta-belief (such as the condition where a subject thinks of the proposition that R is a good reason for B). But being prevented from entering realization conditions for the manifestation of a disposition isn't automatically sufficient for lacking the disposition. For example, suppose that God takes a fragile glass, and encases it in an eternal, impermeable, indestructible soft bubble. The glass still has the disposition to shatter upon striking a hard surface; it's just that it's forever prevented from ever entering that realization condition for its fragile disposition.

Perhaps the case of the dispositions of very young children and higher animals to form a relevant meta-belief, when they hold a belief on the basis of a reason, is like that. A key realization condition for the disposition to form the relevant meta-belief regarding reason R and belief B is thinking of the proposition that R is a good reason for B. A higher animal presumably can't enter that realization condition, because it lacks the concept of a good reason, but the animal might still be such that, *if* it ever were to enter the relevant realization condition (which would presuppose that it had acquired the concept of belief), it would form that belief.

My preferred reply to the objection from very small children and higher animals, however, is simply to set them aside when we're theorizing about beliefs and reasons. What we should theorize about are the clear cases where subjects have beliefs, and base their beliefs on reasons. The clear cases are those of adult humans, as well as relatively conceptually sophisticated children. Once we have an adequate account of beliefs, reasons, inference, and basing, for the clear cases, we can apply that account to the unclear cases. So we should not begin by assuming that very small children and higher animals have genuine beliefs, or that their beliefs are held on the basis of reasons. That should be, at best, a result of our theories, not a data point to be taken as a starting-point in our theorizing. And surely, if it turns out in the end that our best theories of beliefs and reasons entail that very small children and higher animals do not instantiate basing relations of the kind which obtain between beliefs and the reasons for which they are held in older children and adult humans, that's not such an intolerable result.

REFERENCES

Audi, Robert (2015). Dispositional beliefs and dispositions to believe In *Rational belief: Structure, grounds, and intellectual virtue* (pp 11-26). Oxford: Oxford University Press. (Originally published 1994)

Bondy, Patrick (2016). Counterfactuals and epistemic basing relations. *Pacific Philosophical Quarterly*, *97*(4), 542-569.

Bondy, Patrick and Carter, J. Adam (forth.). The basing relation and the impossibility of the debasing demon. *American Philosophical Quarterly*.

Boghossian, Paul (2014). What is inference? *Philosophical Studies*, *169*(1), 1-18.

Broome, John (2013). *Rationality through reasoning*. Malden, MA: Blackwell.

Broome, John (2014). Comments on Boghossian. *Philosophical Studies*, *169*(1), 19-25.

Evans, Ian (2013). The problem of the basing relation. *Synthese*, *190*(14), 2943-2957.

Lehrer, Keith (1971). How reasons give us knowledge, or the case of the gypsy lawyer. *The Journal of Philosophy*, *68*(10), 311-313.

Leite, Adam (2004). On justifying and being justified. *Philosophical Issues*, *14*(1), 219-253.

Kvanvig, Jon (1985). Swain on the basing relation. *Analysis*, *45*(3), 153-158.

Moser (1989). *Knowledge and evidence*. Cambridge, UK: Cambridge University Press.

Plantinga, Alvin (1993). *Warrant: the current debate*. Oxford: Oxford University Press.

Neta, Ram (2013). What is an inference? *Philosophical Issues*, *23*(1), 388-407.
Setiya, Kieran (2013). Epistemic agency: Some doubts. *Philosophical Issues*, *23*(1), pp. 179-198.
Swain, Marshall (1981). *Reasons and knowledge*. Ithaca, NY: Cornell University Press.
Turri, John (2010). On the relation between propositional and doxastic justification. *Philosophy and Phenomenological Research*, *80*(2), 312-326.
Turri, John (2011). Believing for a reason. *Erkenntnis*, *74*(3), 383-397.
Wright, Crispin (2014). Comment on Paul Boghossian, "what is inference". *Philosophical Studies*, *169*(1), 27-37.

7

Be Committed to Your Premises, or Face the Consequences: A Pragmatic Analysis of Commitment Inferences

KIRA BOULAT
University of Fribourg
kira.boulat@unifr.ch

DIDIER MAILLAT
University of Fribourg
didier.maillat@unifr.ch

In this paper we look at a pragmatic account of commitment phenomena in argument processing. We develop a typology of commitment and propose a pragmatic account of the processes involved in the interpretation of commitment markers such as *I think that X, I was told that X* in terms of certainty and reliability. We show that the relevance-theoretic notion of cognitive strength can be used to capture the cognitive effect of commitment markers. We conclude by reviewing experimental evidence that supports our model.

KEYWORDS: commitment, epistemic strength, epistemic vigilance, evidentiality, experimental evidence, pragmatics, relevance theory

1. DEFINING COMMITMENT

From Speech Act Theory to the French tradition of *Théorie de l'énonciation*, from philosophy of language to cognitive pragmatics, commitment has been a feature of many different schools and approaches to linguistic phenomena (see the review in Boulat & Maillat 2017). In Argumentation Theory, it has appeared prominently in the work of Hamblin's (1970), who devised a commitment store, as well as, for instance, in the research by Walton & Krabbe (1995). This explains why to this day, commitment constitutes a central notion in models of argumentation.

More recently, commitment has been integrated to a pragmatic approach to argumentative phenomena. Oswald (2016) and Morency et al. (2008) have looked at how commitment interacts with the pragmatic processes. Boulat & Maillat (2017) establish a systematic proposal to account for the cognitive processes which underlie commitment, while offering a detailed typology that distinguishes between production and comprehension processes on the one hand, and a linguistic from a cognitive level on the other.

1.1 Commitment in Argumentation Theory

Overwhelmingly, argumentation theorists have argued in favour of a *binary* representation of commitment. Arguing for instance that an argument can be legitimately inferred or not, or that an arguer can be regarded as plausibly holding to a given premise or not, as can be seen in the pragma-dialectic reconstruction of premises in van Eemeren & Grootendorst (1992), or the relevance-theoretic procedure in Oswald (2016).

In this paper, we wish to contend that a more adequate account of commitment phenomena should construe commitment as a graded notion. In this view, a claim can be said to have been communicated by an arguer – and the hearer's commitment to it might be held – at *various degrees*. Specifically from a cognitive perspective, we propose that commitment be analysed as a cognitive function of the epistemic strength of a claim presented by an arguer in an exchange. In this sense an arguer can be regarded as being *more or less* committed to the components of a given argument.

Building on Paglieri's (2007) claim that cognitive processes bear on our assessment of arguments, we posit that cognitive processes governing commitment impact on the way we represent arguments in our cognitive environment.

In that respect, our proposal is cognate to Sperber et al.'s (2010) suggestions. They have thus argued that humans have evolved a series of mechanisms to ensure that arguers accept all and only reliable information. These mechanisms are grouped under the heading *Epistemic Vigilance*. According to these authors, two main components determine the strength of an argumentative claim.

   i. The credibility of the information, i.e. the degree of certainty of a CONTENT

ii. The quality of the source of information, i.e. the degree of reliability of its SOURCE

Crucially, commitment is a central aspect of this model as arguers will "follow the commitments of the different speakers and determine, at any given point of the argument, who has the burden of proof" (Mercier & Sperber 2009: 169).

1.2 Commitment in Pragmatics

According to Sperber et al. (2010) human communication involves both a comprehension and an acceptance stage, whereby, having understood a given utterance, the hearer must decide to integrate – or not – the information conveyed by the utterance. In this view, argumentation can be regarded as an attempt to warrant the acceptance of a claim by supporting it with other accepted claims.

> Claim:
> The degree of commitment to a claim marked by a speaker is a means to warrant its acceptance by the hearer.

This is expected because commitment is one of the 'trust calibration' mechanisms at work in epistemic vigilance: commitment to a piece of information that is later found to be unreliable imposes a social cost on the arguer as the hearer will lower his trust in a source known to be unreliable.

In what follows we are going to propose a cognitive pragmatic account of commitment that allows us to model these aspects of the acceptance process within epistemic vigilance. Commitment intuitively reflects the linguistic possibilities communicators have to endorse a given utterance at various degrees, but also to dissociate themselves from it. Basically, the notion of commitment in pragmatics captures the idea that communicators constantly assess and infer the degree to which the speaker and themselves are committed to the piece of information conveyed by a given argument. In this paper, we therefore argue that commitment to an utterance, and more specifically commitment to an argument, is cognitively determined by the strength of its corresponding assumptions in the hearer's cognitive environment (i.e. the set of contextual assumptions entertained by an individual).

2. A PRAGMATIC MODEL OF COMMITMENT

According to this alternative pragmatic model of commitment (see Boulat 2015, and Boulat & Maillat 2017), these different degrees of certainty and reliability communicated by the speaker's utterance are represented in the hearer's cognitive environment through the derivation of higher-level explicatures. A higher-level explicature is defined as the embedding of the proposition expressed under a higher-level description such as a speech act, a propositional attitude description or some other comments on the embedded proposition. The result is the speaker's intended epistemic stance or attitude towards she expressed (Wilson & Sperber 1993; Ifandidou 2001; Carston 2002).

In our view, strength is a function of both certainty and reliability. According to this alternative model of commitment, certainty is about the content (i.e. the communicated piece of information), whereas reliability is about the speaker and/or the reported speaker's reputation and her access to evidence.

Certainty generally refers to the speaker's appraisal of the epistemic status of the communicated state of affairs and is therefore related to the communicated content. Even though a communicated content can be intrinsically believable or unbelievable (for instance, in the case of tautologies or logical contractions, see Sperber et al. 2010), it is most of the time presented as more or less certain by the speaker. Indeed, she can express more or less certainty by using different linguistic markers, including categorical assertions (e.g. *Jane is a writer*), epistemic modals (e.g. *can, could, may, might*, etc.) or evidential expressions (e.g. propositional attitudes or parenthetical expressions such as *I think, I guess, I know* or adverbials like *probably, obviously*, and so on), among others. These linguistic markers of certainty are thought to have either a weakening or a strengthening function with respect to the speaker's commitment. As a result, they impact upon how the hearer interprets a given utterance and how he subsequently integrates this piece of information in his cognitive environment.

Reliability usually refers either to the speaker's competence and benevolence or to the soundness of the information (in terms of evidence or of lack thereof). It is thus linked to the speaker and/or reported speaker. The speaker's competence refers to the fact that she possesses genuine information, whereas her benevolence indicates that she wishes to share her genuine pieces of information with her addressees (Sperber et al. 2010; Mazzarella 2013). Besides, the type of evidence the speaker has when she communicates a piece of information is also thought to provide information with respect to the speaker and/or reported speaker's reliability. Indeed, what is communicated can have been acquired by direct or indirect perception, inference, memory, or hearsay. It is generally agreed that a communicated content based on direct evidence (that is, evidence acquired via direct perception) is more accurate and hence more likely to be accepted by a hearer than a content based on indirect evidence (Cornillie & Delbecque 2008: 39).

If it is theoretically useful to link certainty to the communicated content and reliability to the source of information, the overlap between these two categories cannot be denied, especially when it comes to the linguistic marking of certainty and reliability. For instance, and following Ifantidou (2001), some evidentials give indications about the certainty of the content (such as adverbials or propositional attitude markers) but they also convey information regarding the speaker's access to evidence (be it direct or indirect). Therefore, one must acknowledge that certainty and reliability are interrelated without being interchangeable.

3. A TYPOLOGY OF COMMITMENT

In previous works (Boulat 2015; Boulat & Maillat 2017), we proposed a commitment typology which takes into account the various (and often thought of as mutually exclusive) aspects related to commitment, namely: the speaker, the hearer, as well as its public and private sides. This typology distinguishes among four different kinds of commitment: speaker commitment, communicated commitment, attributed commitment and hearer commitment. Speaker commitment refers to what is generally called the private side of commitment, i.e. the degree of strength assigned by the speaker to assumptions in her cognitive environment. Communicated commitment refers to the public expression of what the speaker wants the hearer to infer as her commitment, that is, it refers to the speaker's ways of presenting her piece of information with more or less certainty, and herself or the reported speaker as more or less reliable. Obviously, speaker commitment and communicated commitment are not always aligned. On the hearer's side, attributed commitment corresponds to the result of his assessment of the certainty of the communicated information and of the speaker's and/or reported speaker's reliability, based on available linguistic cues and contextual assumptions. Finally, hearer commitment refers to the degree of strength assigned to this same piece of information as it is integrated in the hearer's own cognitive environment.

This typology of commitment takes into consideration both the speaker and the hearer and deals with mental representation (i.e. whether the notion captures commitment as a property of a cognitive representation) and linguistic marking (whether it captures commitment as a property of a linguistic form). We argue that the last two types of commitment, namely attributed commitment and hearer commitment, are influenced by two main factors which will be considered in turn. On the one hand, linguistic markers of certainty; and on the other hand, the hearer's appraisal of the speaker's and/or reported speaker's reliability (see Boulat 2015). These claims translate into two experimentally testable predictions which will be briefly presented in what follows.

4. PREDICTIONS: CERTAINTY AND RELIABILITY

The first prediction is based on the simple idea that linguistic markers of certainty like categorical assertions, epistemic modal and evidential expressions give an indication regarding the degree of certainty assigned by the speaker to her utterance (e.g. her argument). Consequently, the more the speaker presents her piece of information or argument as linguistically certain, the more likely the hearer is to attribute a strong commitment to the speaker (modulo the other prediction). The hearer will then be likely to integrate this piece of information or argument in his cognitive environment with a high degree of strength. Therefore, from an argumentative perspective, we argue that certainty markers increase the argumentative strength of an utterance.

The second prediction considers the hearer's appraisal of the speaker's and/or the reported speaker's reliability. We claim that, in order to decide whether or not to accept an incoming piece of information or an argument, the hearer needs to take into considerations his assumptions regarding the source of that piece of information. Indeed, we think that the hearer's hypotheses regarding the source's reliability are likely to influence attributed commitment and hearer commitment. Other things being equal, the hearer will be more likely to integrate with a high degree of strength a piece of information or an argument provided by a reliable speaker than the same piece of information or same argument conveyed by its unreliable counterpart. As a result, from an argumentative perspective, we argue that reliability markers also increase the argumentative strength of an utterance.

5. EXPERIMENTAL EVALUATION OF THE MODEL

In the following part of the paper we present some of our empirical findings and assess their relative fit with the proposed model.

5.1 Experiment 1: certainty

Following a body of experimental data (see, for instance, Moore & Davidge 1989; Dudley, Orita, Hacquart & Lidz 2015; Sabbagh & Baldwin 2001; Jaswal, Vikram & Malone 2007; Stock, Graham & Chambers 2009; Bernard, Mercier & Clément 2012 and Mercier, Bernard & Clément 2014), we posit that linguistic markers of certainty influence the acceptance or the non-acceptance of a piece of information or argument in an individual's cognitive environment. In other words, we think that these markers impact on attributed commitment and hearer commitment.

In order to test this hypothesis, 3 different groups of linguistic markers were created (see Boulat *forthcoming*). The influence of non-commitment markers (such as *I don't know if*, *I'm not sure*, *I hope*); weak commitment markers (like *I guess*, *I think*, *It seems*) and high commitment markers (e.g. *I am sure*, *I know*, *No doubt*) was tested on a yes-no recognition task, presenting 30 factual statements about a fictional narrative.[1] Within this recognition paradigm, participants were expected to perform better in the case of statements conveying high certainty (i.e. statements displaying a high commitment marker) than in the case of statements conveying low certainty (i.e. statements displaying a weak commitment marker) or uncertainty.

5.1.1 Participants

133 native English speaking Mturk workers (60 female; 73 male; aged 18-60) participated for monetary compensation to an online survey.[2]

---

[1] All the linguistic markers of certainty used in experiment 1 were tested and evaluated by 41 native English speaking Mechanical Turk workers (from the United States) aged 18 to 61 (23 female, 18 male), in a pre-test (see Boulat *forthcoming* for details).

[2] Mturk is described as a crowdsourcing internet market which enables its users to post HITS (Human Intelligence Tasks) in exchange for a monetary compensation.

5.1.2 Materials

30 short factual statements were created (e.g. *The old lady found a picture* or *Mr Black called is old mother*) and were controlled for number of words (*M*=6.03 words). Critical words (the last word of the statement, n=30), either old (i.e. previously studied, for instance *picture* and *mother* in the examples provided above) or new (e.g. *paper* and *father*), were selected on the basis of their length (between 1-2 syllables), part of speech (i.e. nouns), and frequency (50-600 occurrences for a million words). All the words in the 30 statements were obtained from Kucera and Francis's (1967) list providing the 2200 most frequent English words.[3] New words were matched to the critical words based on length, part of speech, frequency, and meaning.

5.1.3 Recognition test

The within-subject yes-no recognition test consisted of 30 trials of statements, 15 of which were old (i.e. previously studied) and 15 of which were new (i.e. previously unstudied). Of the 30 statements, 10 were presented with a high commitment marker, 10 with a weak commitment marker and 10 with a non-commitment marker. The statements were rotated through the different test conditions: the commitment levels (non-commitment, weak commitment, high commitment) and recognition (old vs new). Half of the trials were designed to elicit a "yes" response and the other half a "no" response. In the study phase, the statements were distributed in 6 lists combining linguistic markers and old-new critical words using a Latin Square technique.

5.1.4. Procedure

After agreeing to participate in the survey and after answering a few demographics questions, participants were told that they would read statements the police got from a witness, regarding a crime committed in Mr Black's house. They were instructed to carefully read the 30 statements provided by the witness. The format of the memory task was not specified. Participants were warned that during the study phase, the to-be-recalled statements would appear on the screen for 3 seconds. Then, participants performed 3 practice trials. During the study phase,

---

[3] This list can be found here: http://www.auburn.edu/~nunnath/engl6240/kucera67.html

each participant was presented 30 statements. In order to avoid primacy and recency effects, it was made sure that the linguistic markers were balanced in the 5 first and 5 last items of each. Statements were visually and individually presented for 3 seconds and appeared one at a time, before automatically disappearing. The experiment lasted 15 to 20 minutes. Following Ditman et al. (2010), a delay was placed between the study phase and the recognition test with a distractor task (i.e. 60 simple arithmetic questions), which approximatively took 10 minutes to perform.

5.1.5. Procedure for the recognition test

After the distractor task, participants were shown a message telling them that their memory of the statements provided by the witness would be tested. It also explained that they would be presented with the question "Did the witness say the following to the police?", which was followed by a statement such as *Mrs Lily loved dark chocolate*. Participants were instructed to indicate whether it was one of the statements they previously read in the study phase or not. They were also instructed to answer "yes" only if the statement was exactly the same (for instance, *Mrs Lily loved dark chocolate*) and that they should answer "no" only if the statement was not exactly the same (e.g. if they were presented with the statement *Mrs Lily loved white chocolate*).

Finally, the participants were presented with a yes-no memory recognition task where the final word of each statement was manipulated according to the old-new experimental condition (e.g. *I am unsure whether the old lady found a picture/paper* or *The butler clearly moved to the North/South*). They were instructed to tick the "yes" or "no" box for each trial. 30 trials were individually and randomly presented (see Ditman et al. 2010). If participants correctly ticked "yes" when the statement was old, it was scored as a correct answer whereas if they incorrectly pressed "yes" when it was a new statement, it was recorded as an incorrect answer.

5.1.6. Results

Our model uses commitment level as a fixed effect (i.e. non-commitment, weak commitment, and high commitment, based on the pre-test results) and as a categorical predictor for accuracy in the recognition task. The analysis shows that the two categories of non- and high commitment are good predictors for accuracy in the recognition task. Converting the log odds given in the model provides us with the probability of correct answers in the recognition task in each category of commitment markers: non-commitment category (0.61), weak commitment category (0.61) and high commitment category (0.65, $p < 0.05$). These results indicate that participants' performance is significantly affected by markers of certainty. Indeed, they show better retention of statements when participants were presented with a high commitment marker than with a non-commitment marker. Yet, weak commitment markers are not good predictors for accuracy in this task. These results suggest that one of the two parameters which determine commitment (i.e. certainty) impacts on the accessibility of assumptions and thus on their retrieval.

5.2. Experiment 2: Reliability

In line with a body of experimental data (see, for instance, Stewart, Haigh & Ferguson 2013; Koenig & Echols 2003; Clément, Koenig & Harris 2004; Koenig & Harris 2007; Jaswal et al. 2007; Southgate, Chevallier & Csibra 2010; Sobel, Sedivy, Buchanan & Hennessy 2012, *inter alia*), we posit that the hearer's appraisal of the speaker's reliability influences the acceptance or the non-acceptance of a piece of information or an argument in an individual's cognitive environment. In other words, these assessments also impact on attributed commitment and hearer commitment. In this second recognition task, participants were expected to perform better in the case of statements communicated by reliable sources than in the case of statements provided by unreliable ones (see Boulat *forthcoming* for discussion).

In order to test this second hypothesis, 2 different groups of sources (reliable vs unreliable) were manipulated and their influence was tested on a yes-no recognition task, displaying 20 factual statements about a fictional narrative.

5.2.1. Participants

128 native English speaking Mturk workers (61 female; 67 male; aged 18-60) participated to the online survey for monetary compensation.

5.2.2. Materials

20 simple statements were created, based on the list of stimuli used in experiment 1. These statements were followed by a percentage indicating the source of information's reliability (syllables per statements, M= 7,65). In the instructions, participants were told that

various witnesses provided the police with different pieces of information regarding a crime committed in Mr Black's house. They were told that the police had evaluated the source's reliability for each piece of information on a reliability scale and that information below 30% on that scale was regarded as unreliable, whereas information above 70% was reliable.

5.2.3. Recognition test

The within-subject yes-no recognition test contained 20 trials of statements, 10 of which were old (i.e. previously studied) and 10 of which were new (i.e. previously unstudied). Of the 20 statements, 10 were communicated by a reliable source, i.e. they were presented with a percentage above 70%, whereas the other 10 were provided by unreliable sources, i.e. they were presented with a percentage below 30%. The stimuli were individually displayed for 3 seconds and were directly followed by the next statement. The statements were rotated through the different test conditions: the source's reliability (unreliable vs reliable) and recognition (old vs new). Participants were asked "Did you read the following statement: *The French colonel broke his arm*"? Half of the trials were designed to elicit a "yes" response and the other half a "no" response. For instance, if the participants read *"The French colonel broke his arm* (9%)" in the study phase, they would be presented with either "Did you read the following statement: *The French colonel broke his arm ?*" (which should elicit a "yes" response), or "Did you read the following statement: *The French colonel broke his leg ?*" (which should elicit a "no" response).

5.2.4. Procedure for the study phase

Participants took 4 practice trials, explicitly specifying the source's reliability status. For instance, participants read the first practice trial statement *Mrs Lily went to the gym. (15%) = unreliable*. Participants were told that they would be asked about the identity of the murderer at the end of the survey. The distractor task was the same as in experiment 1.

5.2.5. Procedure for the recognition test

The procedure for the recognition test was the same as in experiments 1, except that it contained only 20 questions about the 20 statements.

5.2.6. Results

A mixed effect (logistic regression) model which used reliability as a categorical predictor of accuracy showed that the two categories "unreliable" and "reliable" are good predictors for accuracy. Converting the log odds given in the model provides us with the probability of correct answers in the recognition task in each category of source reliability: unreliability (0.67) and reliability (0.74, $p= 0.000237$).

6. CONCLUSION AND PERSPECTIVES

The results of these two experiments suggest that the two parameters which determine commitment (i.e. certainty and reliability) have an impact on participants' performance in a recognition task. That is, results show that when the participants are provided with certain and reliable statements, their recognition score is significantly higher than when they are presented with uncertain and unreliable statements.

We take these results to indicate that, as predicted by Relevance Theory, certainty and reliability impact on the strength of assumptions in an individual's cognitive environment. Consequently, the stronger the assumption, the more manifest it is and hence, the more relevant it is. Thus, at the level of Argumentation Theory, we claim that by increasing the manifestness of a certain claim, commitment determines a greater argumentative strength. In a nutshell, we posit that the higher the certainty of a claim and the higher the arguer's reliability, the stronger hearer commitment, the more manifest the corresponding assumption, and the better the recognition of that assumption. Finally, from a pragmatic perspective, we claim to have found evidence that the representations of utterances are stored at a different degree of cognitive strength in the individual's cognitive environment which is (in part) determined by the degree of hearer commitment.

ACKNOWLEDGEMENTS: This work benefitted from the expertise of, and many discussions with, our colleagues Napoleon Katsos for the experimental design and Davis Ozols for the statistical analysis. The first author carried out the research reported in this paper while on a Doc.Mobility grant of the Swiss National Science Foundation (SNF155140).

REFERENCES

Bernard, S., Mercier, H., & Clément, F. (2012). The power of well-connected arguments: Early sensitivity to the connective *because*. *Journal of Experimental Child Psychology*, *111*(1), 128–135.

Boulat, K. (2015). Hearer-oriented processes of strength assignment: a pragmatic model of commitment. In B. Cornillie & J. I. Marín Arrese (Eds.), *Evidentiality and the semantics pragmatics interface* (pp. 19-39). Amsterdam: John Benjamins publishing

Boulat, K. & Maillat, D. (2017). She said you said I saw it with my own eyes: a pragmatic account of commitment. In J. Blochowiak, S. Durrlemann-Tame, & C. Laenzlinger (Eds.), *Formal Models in the Study of Language: Applications in Interdisciplinary Contexts* (pp. 261-279). London: Springer.

Cornillie, B., & Delbecque, N. (2008). Speaker commitment: Back to the speaker: Evidence from Spanish alternations. In P. De Brabanter & P. Dendale (Eds), *Commitment* (pp. 37-62). Amsterdam: John Benjamins.

Carston R. (2002). *Thoughts and utterances: The pragmatics of explicit communication*. Oxford: Blackwell.

Clément, F., Koenig, M. A., & Harris, P. L. (2004). The ontogenesis of trust. *Mind and Language*, *19*(4), 360-379.

Ditman, T., Brunyé, T. T., Mahoney, C. R., & Taylor, H. A. (2010). Simulating an enactment effect: Pronouns guide action simulation during narrative comprehension. *Cognition*, *115*(1), 172-178.

Dudley, R., Orita, N., Hacquard; V., & Lidz, J. (2015). Three-year-olds' understanding of *know* and *think*. In F. Schwarz (Ed.), *Experimental perspectives on presuppositions* (pp. 241-262). New York: Springer.

Eemeren, F. H. van, & Grootendorst, R. (2004). *A Systematic Theory of Argumentation: the pragma-dialectical approach*. Cambridge: Cambridge University Press.

Hamblin, C. L., (1970). *Fallacies*. London: Methuen.

Ifantidou, E. (2001). *Evidentials and relevance*. Amsterdam: John Benjamins.

Jaswal, V. J., & Malone, L. S. (2007). Turning believers into skeptics: 3-years-olds' sensitivity to cues to speaker credibility. *Journal of Cognition and Development*, *8*(3), 263-283.

Koenig, M. A., Clément, F., & Harris, P. L. (2004). Trust in testimony: Children's use of true and false statements. *Psychological Science*, *15*(10), 694-698.

Koenig, M. A., & Harris, P. L. (2007). The basis of epistemic trust: Reliable testimony or reliable Sources? *Episteme*, *4*(3), 264-284.

Kucera, H., & Francis, N. W. (1967). *Computational analysis of present day American English*. Providence, RI: Brown University Press. Retrieved from: http://www.auburn.edu/~nunnath/engl6240/kucera67.html.

Mazzarella, D. (2013). "Optimal relevance" as a pragmatic criterion: The role of epistemic vigilance. *UCL Working Papers in Linguistics*, *25*, 20-45.

Mercier, H., Bernard, S., & Clément, F. (2014). Early sensitivity to arguments: How pre-schoolers weight circular arguments. *Journal of Experimental Child Psychology*, *125*, 102-109.

Mercier, H. & Sperber, D. (2009). Intuitive and reflective inferences. In J. St. B. T. Evans, & K. Frankish (Eds.), *In Two Minds: Dual processes and beyond*, (pp. 149–70). Oxford: Oxford University Press.

Moore, C., & Davidge, J. (1989). The development of mental terms: Pragmatics or semantics? *Journal of Child Language*, *16*(3), 633-641.

Morency, P., Oswald, S., & de Saussure, L. (2008). Explicitness, implicitness and commitment attribution: A cognitive pragmatic approach. In P. De Brabanter, and P. Dendale (Eds.), *Commitment* (pp. 197-219). Amsterdam: John Benjamins.

Oswald, S. (2016). Pragmatic constraints on argument processing. In M. Padilla Cruz (Ed.), *Relevance theory: Recent developments, current challenges and future directions* (pp. 261-285). Amsterdam: John Benjamins.

Sabbagh, M. A., & Baldwin, D. A. (2001). Learning words from knowledgeable versus ignorant speakers: Links between preschoolers' theory of mind and semantic development. *Child Development, 72*(4), 1054–1070.

Sobel, D. M., Sedivy, J., Buchanan, D. W., & Hennessy, R. (2012). Speaker reliability in preschoolers' inferences about the meanings of novel words. *Journal of Child Language*, *39*(1), 90-104.

Southgate, V., Chevallier, C., & Csibra, G. (2010). Seven-month-olds appeal to false beliefs to interpret others' referential communication. *Developmental Science*, *13*(6), 907-912.

Sperber, D., Clément, F., Heintz, C., Mascaro, O., Mercier, H., Origgi, C, & Wilson, D. (2010). Epistemic Vigilance. *Mind and Language*, *25*(4), 359-393.

Sperber, D., & Wilson, D. (1987). Précis of relevance: Communication and cognition. *Behavioral and Brain Sciences*, *10*(4), 697-710.

Sperber, D., & Wilson, D. (1995). *Relevance: Communication and cognition* (2nd ed.). Malden, MA: Blackwell Publications.

Stewart, A. J., Haigh, M., & Ferguson, H. J. (2013). Sensitivity to speaker control in the online comprehension of conditional tips and promises: An eye-tracking study. *Journal of Experimental Psychology*: *Learning, Memory and Cognition*, *39*(4), 1022-1036.

Stock, H. R., Graham, S. A.; & Chambers, C. G. (2009). Generic language and speaker confidence guide preschoolers' inferences about novel animate kinds. *Developmental Psychology*, *45*(3), 884-888.

Walton, D. N., & Krabbe, E. C. W. (1995). *Commitment in dialogue : basic concepts of interpersonal reasoning*. Albany: State University of New York Press.

Wilson, D., & Sperber, D. (1993). Linguistic form and relevance. *Lingua*, *90*(2), 1-25.

8

The Role of Argumentative Discussions in the Transmission of Implicit Norms and Values in Families with Young Children.

ANTONIO BOVA
Franklin University Switzerland, Switzerland
abova@fus.edu

In this study, I set out to show how the transmission of parental norms and values can lead parents and children to engage in argumentative discussions. The research design implies a corpus of 30 video-recorded separate meals of 10 middle to upper-middle-class Swiss and Italian families. The results of this study indicate that implicits in argumentation are particularly effective in transmitting what is taken for granted during family interactions at mealtime.

KEYWORDS: Family; Argumentation, Parent-child interaction, Implicit, Pragma-dialectical approach, Argumentum Model of Topics

1. INTRODUCTION

The family context is recently coming into light as an important context for the study of argumentation. Mealtime, in particular, can be considered a privileged moment to examine how parents and children interact and argue since it is one of the few activities that can bring all family members together daily (Ochs et al., 1989; Blum-Kulka, 1997; Bova & Arcidiacono, 2014a). It is more than a particular time of day at which to eat. Rather, it is an activity during which the parents can be "consulted" by their children about a large number of topics and, given the relative freedom of speech, there could be some dissensus about what is the right thing to do or to think (Arcidiacono & Bova, 2015; Bova & Arcidiacono, 2015). Although the degree of conversational freedom at mealtimes can vary from family to family and depends on various contextual and social factors (Beals, 1997; Fiese et al., 2006), a series of earlier studies have indicated that the argumentative interactions between parents and children during mealtimes contribute to the socialization of children toward the rules and behavioural models typical of their family and their own community (Pontecorvo et al., 2001; Bova & Arcidiacono, 2013a, 2013b, 2014b; Bova, 2015a, 2015b). As everyday life is full of taken for granted assumptions (Garfinkel, 1964), it follows that, in the family context, a part of the children's socialization is based on implicit aspects in the transmission of normative beliefs.

In the present study, I set out to show how the transmission of parental norms and values, which are often taken for granted within family mealtime conversations, can lead parents and children to engage in argumentative discussions. I will focus specifically on conversations where norms, values, or beliefs are implicitly embedded in communicated information. For this reason, I will consider the role of the social setting, of the actors, and of situational effects in contributing meaning to the conversation, with specific attention to the implicit in the transmission of normative beliefs or value. In this endeavour, I opted for an idiographic methodology based on the contemporary argumentation theory. The pragma-dialectical ideal model of critical discussion (van Eemeren & Grootendorst, 2004) represents the analytical approach used to identify the argumentatively relevant moves. In addition, I decided to integrate this model with the Argumentum Model of Topics (hereafter AMT) (Rigotti & Greco Morasso, 2010) to systematically reconstruct the inferential configuration of arguments. By doing this, we intend to move one step forward beyond the analytical reconstruction of argumentation, in order to consider how an argument is connected to its standpoint. The AMT model allows distinguishing premises of procedural (logical) nature from contextual (cultural) premises, and it is particularly important in our case to understand the procedural implicit and explicit premises used by parents in their argumentation.

2. METHODOLOGY

2.1 Data corpus

This study is part of a project devoted to the analysis of family argumentation. The research design involves a corpus of thirty video-recorded separate family meals (constituting about twenty hours of video data), constructed from two different sets of data, named sub-corpus 1 and sub-corpus 2. All participants are Italian-speaking. The length of each recording varies from 20 to 40 minutes.

Sub-corpus 1 consists of 15 video-recorded meals in five middle- to upper-middle-class Italian families in high socio-demographic group living in Rome. The criteria adopted in the selection of the Italian families were the following: the presence of both parents and at least two children, of whom the younger is of preschool age (three to six years). All families in sub-corpus 1 had two children. Sub-corpus 2 consists of 15 video-recorded meals in five middle- to upper-middle-class Swiss families in a high socio-demographic group, all residents in the Lugano area. The criteria adopted in the selection of the Swiss families mirror the criteria adopted in the creation of sub-corpus 1. Families had two or three children.

2.2 Identification of the conversational sequences

The analyses I present in this study are limited to and focused on the study of analytically relevant argumentative moves, i.e. "those speech acts that (at least potentially) play a role in the process of resolving a difference of opinion" (van Eemeren & Grootendorst, 2004, p. 73). In particular, the discussion is considered as argumentative if the following criteria are satisfied: (I) a difference of opinion among parents and children arises around a certain issue; (II) at least one standpoint advanced by one of the two parents is questioned by one or more children, or vice versa; (III) at least one family member puts forward at least one argument either in favour of or against the standpoint being questioned. For the present study, only the discussions that fulfill the three above-mentioned criteria were selected for analysis, while all non-argumentative conversations were excluded.

3. ANALYTICAL APPROACH

The analytical approach on which this study is based refers to the pragma-dialectical ideal model of critical discussion and to the AMT. I decided to integrate these approaches as two steps of the same process of analysis since they cover two relevant and complementary levels of the organization of the argumentative activity: the reconstruction of the structure of the argumentative discussions and the analysis of the procedural implicit and explicit premises used by parents in their argumentation.

In order to reconstruct the structure of the argumentative discussions, in a first phase of the analysis, I will refer to the ideal model of a critical discussion (van Eemeren & Grootendorst, 2004). This model does not describe reality, but how argumentative discourse would be structured were such discourse solely aimed at resolving differences of opinion. Confrontation, in which disagreement regarding a certain standpoint externalized in a discursive exchange or anticipated by the speaker, is, therefore, a necessary condition for an argumentative discussion to occur. This model is assumed, in the present study, as a grid for the analysis of argumentative discussions in the family context, since it provides the criteria for the identification and reconstruction (heuristic and analytic function) of the argumentative moves by parents and children.

After having reconstructed the structured of the argumentative discussions, for the analysis of the procedural implicit and explicit premises used by parents in their argumentation, I will refer to the AMT (Rigotti & Greco Morasso, 2010). This model is particularly important in our case to understand the procedural implicit and explicit premises used by parents in their argumentation since it allows distinguishing premises of procedural (logical) nature from contextual (cultural) premises. As stated by Rigotti and Greco Morasso (2010, p. 490), the AMT is an instrument that serves "to illustrate the structure of reasoning that underlies the connection between a standpoint and its supporting arguments". Despite its particular concern for the inferential aspects of argumentation, the AMT, de facto, accounts not only for the logical aspects of the argumentative exchange (topical component), but also for its embeddedness in the parties' relationship (endoxical component), and thus proves to be particularly suited for the argumentative analysis of ordinary conversations such as family mealtime discussions.

4. RESULTS

In this section, we will present and discuss some excerpts of family dinnertime interactions in order to show how norms and values are often taken for granted. In particular, the argumentative discussions between parents and children, occurring when the children do not respect taken for granted parental norms and values, have been organized into different sections according to the topics of the participants' discussions. More specifically, the first excerpt concerns a discussion around societal rules, gender representations, and adults'/children's attitude toward lifestyle. The second excerpt concerns a context-bound activity, namely the norms at the dinner table. For each excerpt, an analysis of the participants' discursive moves and an inferential reconstruction of the argument will be provided in order to build a collection of cases that will be discussed in the final part of the chapter.

>   Excerpt 1
>   Italian family. Participants: father (DAD, 39 years), mother (MOM, 36 years), Manuela (MAN, 9 years, 7 months), Adriano (ADR 5 years, 7 months), aunt (AUNT, 40 years). All family members are seated at the table. DAD sits at the head of the table, MOM sits on the right-hand side of DAD, whilst MAN, ADR and AUNT sit on the opposite side.

1.	*MAN:	when I become an adult, I would like to remain single/spinster ((not married))
2.	*AUNT:	not married? Don't you want a husband?
3.	*MAN:	no::
4.	%pau:	2.0
5.	*AUNT:	why?
6.	*MAN:	because I don't want to.
7.	*DAD:	because, men are no good, right? right, eh?
8.	%pau:	4.0
9.	*DAD:	because, men are beasts, do you think? ((MAN shakes her head from side to side))
→	*DAD:	eh, you know. this is a big question, eh.
10.	%pau:	5.0
11.	*DAD:	you women, you are the weak sex, right?
12.	*MAN:	how dare you!

In excerpt 1, Manuela presents a hypothetical future plan, opening a problematisation during the conversation. The question is about the possibility (or not) for Manuela to remain single (unmarried). The topic of marriage reveals a discrepancy between the adult expectations and the idea of the child. In fact, in her turn 1, Manuela expresses a plan for the future, confirming her own actual position (she is not yet adult). The reaction of the aunt seems to be surprised because Manuela's intention relates to an undervalued condition in society (to be an unmarried adult). However, it is important to specify that the term "zitella" corresponds to the English "spinster", that has a negative connotation. The Why-question asked by the aunt (who is married) in turn 5 is a good clue that the child's claim has been perceived as culturally unusual. This aspect is central and the father advances a series of arguments in order to "ridicule" the child and, ironically, to put Manuela in the obligation of finding arguments to defend her initial statement. In fact, in turn, 7 the father's intervention reveals an implicit negative perception of the judgment of male gender in Manuela's assertion. The father uses the tag-question "right" as an "extreme case formulation", that allows one to defend against challenges to the legitimacy of complaints and accusations (Pomerantz, 1986). In general, speakers tend to use extreme case formulations when they anticipate or expect the interlocutor to undermine their claims and when they are in adversarial situations. Thus, the father's statement produces an implicit reference to the general male and female positions. During the verbal exchange, Manuela doesn't provide arguments to reply to the father's interventions although his claim seems to ask for accounts (such as "Manuela, are you sure of what you are saying?").

Excerpt 1 illustrates how participants can exploit accusations as a discursive resource to build opposite points of view in conversation as expressions of different (implicit and/or explicit) positions. In addition, it shows that extreme case formulations are rhetorical devices that disputants can employ in defending against prospected challenges to the legitimacy of complaints. Therefore, as soon as Manuela expressed her choice, she is confronted (1) with the expression of surprise of her aunt and (2) with the irony of her father. It is interesting to notice that the father did not propose an explicit argumentation about the merits of marriage. By his extreme case formulation, he shows that such an idea can only be based on misconceptions; by default, the desirability of marriage is taken for granted.

The reconstruction of the argumentative discussion between Manuela and her father is summarized below:

Issue: Is Manuela right to not want to marry anyone (when she becomes an adult)?
Standpoint(s): (MAN) I am right to not want to marry anyone (when I become an adult)
(DAD) You are wrong to not want to marry anyone (when you become an adult)
Argument: (MAN) No argument in support of her standpoint
(DAD) Because men are good (they are not beasts)

In the analysis of the selected argumentative discussion, we will focus on the argument put forth by the father: "because men are good (they are not beasts)". The Y-structure (so-called because its form looks like the letter Y) in Figure 1 will be the graphical tool adopted for representing the AMT's reconstruction.

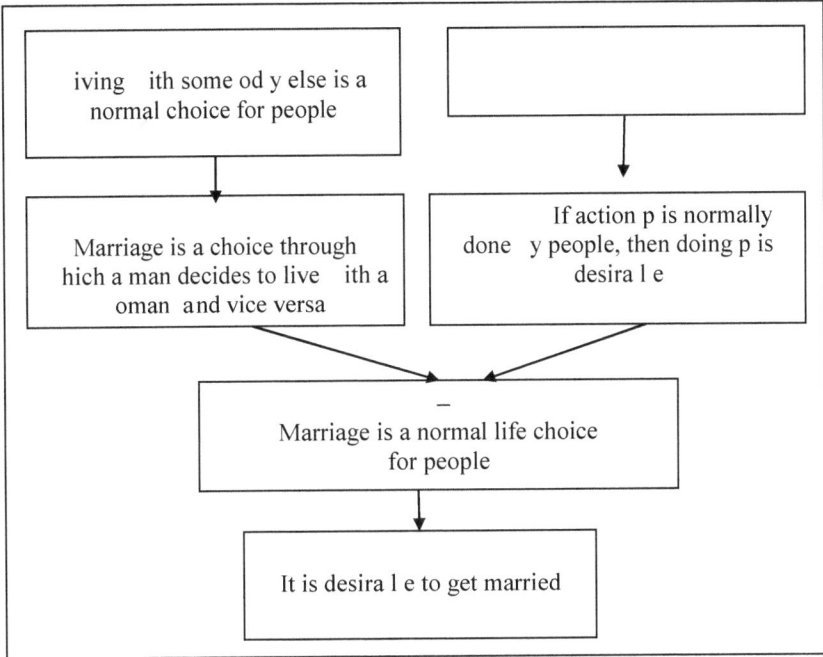

Figure 1 – AMT-based reconstruction of Dad's argument in excerpt 1.

The father's argument is based on a maxim that is engendered from the locus from termination and setting up: "If action p is normally done by people, then doing p is desirable." Greco Morasso (2011, p. 173), provides a clear definition of this locus:

> The locus from termination and setting up binds the acceptability of a state of affairs to the acceptability of one or more of its implications. For example, if a certain state of affairs is expected to bring positive consequences, one is led to conclude, on the basis of this locus, that it has to be accepted or even welcomed: for example, one might reason the positive value of going on a diet from the expected outcome to get fit, healthier, and so on. Contrariwise, a state of affairs is to be avoided if its consequences are negative.

The reasoning follows with a syllogistic (i.e. inferential) structure, "Marriage is a normal life choice for people" (minor premise), which leads to the conclusion that "It is desirable to get married". However, this is only one part of the argumentation. The fact that "Marriage is a normal life choice for people" needs further justification; unlike the maxim, this is not an inferential rule but a factual statement that must be backed by contextual knowledge. Looking at the endoxical syllogism of the diagram, the endoxon is the following: "Living with somebody else is a normal choice for people." The datum "Marriage is a choice through which a man decides to live with a woman (and vice versa)" combined with the endoxon lead to the first conclusion that "Marriage is a normal life choice for people". This example illustrates how some social norms or values are so deeply embedded in the social background that they do not necessitate any explanations. In a family, for instance, the fact that it could be preferable to live alone seems almost unthinkable (at least publicly...) and belongs to the common ground.

The second example is focused not on expectations about family life in general but in appropriate ways to behave in social settings.

Excerpt 2
Italian family. Participants: father (DAD, 38 years), mother (MOM, 37 years), Ugo (UGO, 9 years, 9 months), Luisa (LUI, 3 years, 10 months). All family members are seated at the table. DAD sits at the head of the table, MOM sits on the right-hand side of DAD, whilst UGO and LUI sit on the opposite side.

1.		%act:	Ugo tries to pour water into his glass by holding the bottle from the bottom with just one hand, similarly to the way in which waiters serve wine in a restaurant, risking dropping the bottle on the floor.
2.		*MOM:	Ugo, pour the water correctly, please.
→		*MOM:	but why do you have to do it in this way?
3.		*DAD:	Ugo! ((trying to strike the hand of Ugo))
→		*DAD:	you will force me to give you a slap some day!
4.		*LUI:	why all of it? ((referring to Ugo who takes the bottle and all the water into his glass))
5.		*MOM:	put the bottle behind you. ((talking to Ugo))
6.		%pau:	4.0
7.		*UGO:	what am I doing? ((talking to DAD, who has been looking at him for a few seconds))
8.		*DAD:	what you've done. not what you are doing.
9.		*UGO:	what have I done?
10.		%pau:	2.0
11.		*MOM:	you don't know, what have you done, Ugo?
12.		*UGO:	no.
13.		%pau:	1.5
14.		*UGO:	no: what have I done?
15.		*MOM:	you don't know?

→	*MOM:	well, next time we will explain it better to you through a big slap, right?
→	*MOM:	because otherwise, you have not understood.
16.	*DAD:	because you continue to behave stupidly.

Excerpt 2 concerns a sequence of conversation between the parents and the child around the contingent violation of a norm (how to pour the water correctly). The mother's intervention in turn 2 focuses attention on the inappropriate way used by Ugo: "why do you have to do it in this way?" implicitly assumes that "this way" is in contrast with another – correct – way to accomplish the action at stake. Immediately after, the father refers to a potential consequence of the inappropriate behaviour of Ugo; in fact, the reference to the possibility of "a big slap" is intended as the fact that a punishment is always the effect of a violation of a rule. After the initial statements and requests of the parents (also with the intervention of the daughter in turn 3, "why all of it?"), Ugo replies in turn 7, after a pause, trying to justify himself and asking for more explanations. Then, the adults focus their interventions on the child's violation of a norm (turn 11, "you don't know what you have done Ugo?"), while Ugo tries to use the negation as a reply to the parents' threats ("no...no what have I done?"). Finally, the parents close the sequence through a negative evaluation of the child (turn 16, "you continue to behave stupidly"), without providing an explanation of the child's conduct.

This example shows that not only social values can be transmitted implicitly but also behaviours that a social group considers as appropriate to a given situation. Ugo is manifestly making a "mistake" in the process of pouring some drinks but, interestingly, there is no attempt from either of his parents to explain or demonstrate the right way to do it. In other words, the adults take for granted that the young boy already masters the appropriate way to pour a drink from a bottle. Their irritation stems from the fact that Ugo, in their eyes, is doing it wrong not because of his lack of knowledge but because of his wish to act in a different way (probably observed in a restaurant).

The reconstruction of this argumentative discussion is summarized below:

Issue:	Did Ugo pour the water correctly (in an acceptable way)?
Standpoint:	(MOM and DAD) You are not pouring the water correctly
Argument:	(MOM and DAD) You know that you are doing it incorrectly

I now turn to the analysis of the inferential configuration of the argument put forward by the mother: "You know that you are doing it incorrectly". The reconstruction of the inferential configuration of this argument is illustrated below, in Figure 2:

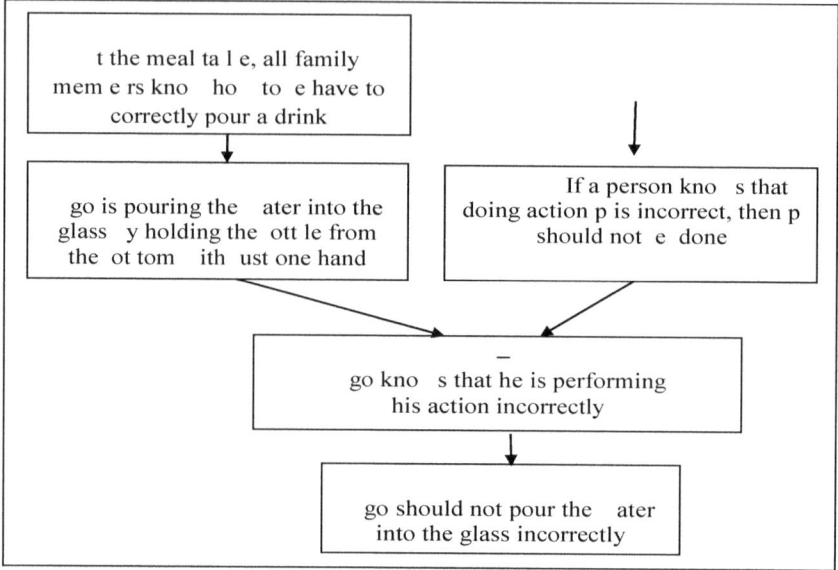

Figure 2 – AMT-based reconstruction of Mom's argument in excerpt 2.

On the right of the diagram the maxim on which the mother's argument is based is specified: "If a person knows that doing action p is incorrect, then p should not be done". This is another maxim engendered from the locus from termination and setting up. The minor premise of the topical syllogism is that "Ugo knows that he is performing his action incorrectly" which combined with the maxim leads to the following final conclusion: "Ugo should not pour the water into the glass holding the bottle from the bottom with just one hand". Looking at the endoxical dimension of the diagram, in this argument the endoxon is as follows: "At the meal table, all family members know when their behaviour is correct or incorrect, that is, in our case, that they know how to behave to correctly pour a drink". The minor premise, "Ugo is pouring the water into the glass in the wrong way," combined with the endoxon, produces the conclusion that "Ugo knows that he is performing his action incorrectly". This excerpt is interesting notably because it seems to contain a paradox. Indeed, when adults try to teach an appropriate behaviour to a child, they usually show or explain the correct use in an explicit way. In the literature, such learning processes have been described through the concept of scaffolding, i.e. the process of helping children's learning by controlling the elements of the task that are initially beyond the learners' ability and by giving them the opportunity to concentrate on the elements that are within their range of competence (Wood, Bruner & Ross, 1976, p. 90). In our example, such a scaffolding is strongly denied because the parents are convinced that their child is aware of the right way to proceed. In a way, we could speak here of a "negative scaffolding", with adults insisting on the fact that it is impossible for the child not to know what he was doing wrong. This emphasis shows implicitly how certain sorts of gestures are expected in social settings, like during mealtimes, and how parents take for granted that everyone in the family will submit to this way of behaving.

In family settings, it is actually not unusual that important norms are indicated, so to say, by the negative, either by insisting on the fact that it is impossible that someone does not know them already (as in excerpt 2), or by being extremely vague about an element that is crucial (but that one is not supposed to talk about). This latter example is typically encountered in the case of a subject marked by political correctness, as in the next excerpt.

6. DISCUSSION AND CONCLUSION

The two excerpts presented above illustrate specific ways in which what is left implicit in argumentative discussions plays an important role in socialization processes. The first excerpt highlights the role of sarcasm and irony in the depiction of a position that does not respect something that belongs to the cultural background (marriage is an essential component of a successful life). The second excerpt shows that parents can communicate implicitly not only certain abstract concepts but also ways of behaving. In conclusion, it is interesting to explicit in each case how what is left implicit in argumentation could play an important role in the shaping of a common ground between children and parents.

The example of Manuela, who expresses that she does not want to marry anyone when she becomes an adult, can be metaphorically seen as a reductio ad absurdum. The proposition "it is normal/desirable for a girl to get married" is never explicitly formulated. On the contrary, the father's argumentative effort aims to show that his daughter's desire can only be based on a distorted vision of reality. The lifestyle that is valued by the group is therefore indirectly emphasized as the normal life path for a young girl.

In the case of Ugo, we have seen that the way he is pouring the water is almost outrageous to his parents. By refusing to explicitly state or show what he is supposed to do and by insisting on the fact that he cannot be ignorant of his misbehaviour, they indicate how taken for granted the "normal" way to pour water is. The exchange takes place as if the boy was usurping a role, trying to be someone (a waiter) that he is not, especially not in the context of a family meal.

These excerpts also have the advantage of demonstrating an important consequence of the use of implicits in argumentation within socialization processes. First of all, they are based on presuppositions, i.e. background information not explicitly indicated as relevant. Information conveyed by presuppositions has the property, even if what is presupposed is new and relevant, of appearing as old, given and as not relevant in its own right (de Saussure, 2013). Moreover, when implicits and presuppositions are used in argumentation with children, such a complicity is even further accentuated. Firstly, children are largely dependent on adults for their well-being as well as for their knowledge acquisition. In such a dependent condition, it is hard to imagine how children could question what is presupposed by the persons who are taking care of them. Secondly, we have seen that the background necessary to understand an argument is often not present in younger minds. Children have therefore to figure out (initially vaguely) a certain context that enables them to make sense of the ongoing dialogue. This background will progressively be enriched thanks to other interactions. However, as it is "internally produced", it is very hard for the younger members of any social community to figure out that this common ground is not natural.

Finally, I would like to highlight the relevance of paying a particular attention to the context-based micro-level discourses that are framing and shaping argumentation as a dialogue-driven and context-specific communication process within family conversations. In fact, field-dependency of argumentative discourse relies primarily on the consideration of the social context (the family as a community, in our case) within which discourse is embedded. To understand how argumentative strategies are related to particular context-based activities pertaining to the family as a community of practice and to what extent argumentation practices are shaped by socio-cultural and interpersonal factors have been the objectives of this study. For this reason, the results of this investigation can contribute to the wider theme of argumentative practices and debates in societal and family issues: in particular, I have highlighted the interplay between different elements of argumentative practices, such as the social need to provide evidence for a particular assertion and the pragmatic functions of argumentation during a discussion. In this vein, talk and activities are viewed as the relevant units for the analysis of family dinnertime interactions, in order to shed light on situated frameworks that adults and children co-construct through their strategic maneuvering during everyday exchanges. Family interactions constitute a favourable discursive arena involving children and adults through different intersubjective positions that are shaped within the contingent context of the discussion.

ACKNOWLEDGEMENTS: This work was supported by the Swiss National Science Foundation (SNSF) [grant number PDFMP1-123093/1].

REFERENCES

Arcidiacono, F., & Bova, A. (2015). Activity-bound and activity-unbound arguments in response to parental eat-directives at mealtimes: Differences and similarities in children of 3-5 and 6-9 years old. *Learning, Culture and Social Interaction, 6*, 40-55.
Beals, D. E. (1997). Sources of support for learning words in conversation: Evidence from mealtimes. *Journal of Child Language, 24*(3), 673-694.
Blum-Kulka, S. (1997). *Dinner talk: Cultural patterns of sociability and socialization in family discourse.* Mahwah, NJ: Erlbaum.
Bova, A. (2015a). Children's responses in argumentative discussions relating to parental rules and prescriptions. *Ampersand, 2*, 109-121.
Bova, A. (2015b). "This is the cheese bought by Grandpa". A study of the arguments from authority used by parents with their children during mealtimes. *Journal of Argumentation in Context, 4*(2), 133-157.
Bova, A., & Arcidiacono, F. (2013a). Invoking the authority of feelings as a strategic maneuver in family mealtime conversations. *Journal of Community & Applied Social Psychology, 23*(3), 206-224.
Bova, A., & Arcidiacono, F. (2013b). Investigating children's Why-questions. A study comparing argumentative and explanatory function. *Discourse Studies, 15*(6), 713-734.
Bova, A., & Arcidiacono, F. (2014a). Types of arguments in parents-children discussions: An argumentative analysis. *Rivista di Psicolinguistica Applicata/Journal of Applied Psycholinguistics, 14*(1), 43-66.

Bova, A., & Arcidiacono, F. (2014b). "You must eat the salad because it is nutritious". Argumentative strategies adopted by parents and children in food-related discussions at mealtimes. *Appetite, 73*, 81-94.

Bova, A., & Arcidiacono, F. (2015). Beyond conflicts. Origin and types of issues leading to argumentative discussions during family mealtimes. *Journal of Language Aggression and Conflict, 3*(2), 263-288.

Eemeren, F. H. van, & Grootendorst, R. (2004). *A systematic theory of argumentation: The pragma-dialectical approach.* Cambridge: Cambridge University Press.

Fiese, B. H., Foley, K. P., & Spagnola, M. (2006). Routine and ritual elements in family mealtimes: Contexts for child well-being and family identity. *New Directions for Child and Adolescent Development, 111*, 67-89.

Garfinkel, H. (1964). Studies of the routine grounds of everyday activities. *Social Problems, 11*(3), 225-250.

Greco Morasso, S. (2011). *Argumentation in dispute mediation: A reasonable way to handle conflict.* Amsterdam: Benjamins.

Ochs, E., Smith, R., & Taylor, C. (1989). Detective stories at dinnertime: Problem solving through co-narration. *Cultural Dynamics, 2*(2), 238-257.

Pontecorvo, C., Fasulo, A., & Sterponi, L. (2001). Mutual apprentices: Making of parenthood and childhood in family dinner conversations. *Human Development, 44*(6), 340-361.

Rigotti, E., & Greco Morasso, S. (2010). Comparing the argumentum model of topics to other contemporary approaches to argument schemes: The procedural and material components. *Argumentation, 24*(4), 489-512.

Saussure (de), L. (2013). Background relevance. *Journal of Pragmatics, 59*(B), 178-189.

9

Virtue Argumentation Theory Reconsidered: Towards a Complete Account of Good Argument

TRACY BOWELL
University of Waikato
tracy.bowell@waikato.ac.nz

JUSTINE KINGSBURY
University of Waikato
justine.kingsbury@waikato.ac.nz

According to virtue argumentation theorists, virtues displayed by the arguer are constitutive of good argument. In earlier work we raise some problems for this approach, but as Paglieri points out, our objections presuppose a view of what argument is, and what *good* argument is, not accepted by virtue theorists. Here we first clarify our position. Then, prompted by Paglieri and Aberdein, we step back from this particular debate to consider more general questions it raises.

KEYWORDS: argument, character, virtue argumentation theory

1. INTRODUCTION

We seem to have got ourselves a reputation as people who beat up on virtue argumentation (hereafter VA) theorists - and to some extent we deserve this reputation.[1] But we think we actually have a lot in common with VA theorists. Indeed, if we can properly translate between our terminology - the terminology of what Fabio Paglieri calls 'cogency buffs' (Paglieri 2015, pp. 84-85) - those who think that the key factor in argument evaluation is how well an argument's premises support its conclusion) and the terminology of VA theorists, (those who think that

---

[1] See for example Bowell and Kingsbury 2013; Andrew Aberdein 2014; Fabio Paglieri 2015.

the key factor in argument evaluation is whether the arguer is virtuous in producing the argument) - in particular as regards what the word "argument" means and what's involved in argument evaluation, perhaps we won't turn out to disagree at all - or at least, only in emphasis. It might be that we all agree about what the facts of the matter are in this domain, but we disagree about which of them are important. This paper is a step in that direction. It doesn't get all the way: in the main what we do here is try to clarify our view, with occasional attempts at diagnosing where we're talking past the VA theorists.

2. VIRTUOUS ARGUMENT

To begin, here are some things we think about argumentative/epistemic virtues (we're not distinguishing between these here). They are important. We want to develop them, in ourselves, and our children, and our students, because we think everyone should have them. The standard of public debate on important issues would be higher if more people displayed more epistemic virtues: for example, if they were careful not to assert things they hadn't fact-checked, willing to seriously consider arguments against their own positions, and evaluated their own arguments with the same rigour with which they evaluate other people's. And in particular, if the people and groups who have the power to make policy displayed these virtues, the world would be a better place.

Intellectually virtuous people are more likely to produce good arguments, and more likely to accept other people's good arguments and argue well against bad arguments, and the more people are doing this, the more likely it is that the community to which they belong will move towards believing truths and rejecting falsehoods. There is common ground between us and some VA theorists, then: we think that, provided the virtues in question include reliabilist virtues such as being good at inductive and deductive reasoning, virtuous arguers tend to produce good arguments. But when either side goes on to make a definitional move, we run into disagreement: the VA theorist says that 'A good argument is one that's virtuously produced', whereas we are more inclined to think that 'a virtuous arguer is one who tends to produce good arguments'.

Does anything important turn on this? Paglieri seems to think not – he thinks it is a "which came first, the chicken and the egg" sort of problem (Paglieri 2015, pp. 68-69). By contrast, David Godden thinks of it as a matter of explanatory priority - which are basic, agent-based norms or product-based norms (Godden 2016, p.346)? And that is the way we tend to think about it too. In Paglieri's terms, we are mostly addressing here the ambitious moderate VA theorist, who considers cogency necessary, but not sufficient, for argument quality, and wants to define cogency in terms of the virtues of agents. (José Gascon (this volume) thinks this is a class of one, and that the theorist in question is Andrew Aberdein.) We wonder what motivates the latter move (defining cogency in virtue terms). If the driving force behind VA theory is, as Paglieri says, to restore the discussion of the character of arguers to a field from which it had been long absent, then, given what we've said about the importance of being a virtuous arguer, and the consistency of that with a non-virtue account of argument quality, we wonder whether the VA theorist actually needs to care about whether good argument is to be *defined* in terms of the exercise of argumentational virtues.

3. ARGUMENT AND GOOD ARGUMENT

On the face of it our position differs from that of the VA theorist with respect to our account of what a good argument is:

> A good argument is an argument that provides, via its premises, sufficient justification for believing its conclusion to be true or highly probable, or for accepting that the course of action it advises is one that certainly or highly probably should be taken. (Bowell and Kingsbury 2013, p. 23)

In Paglieri's terms, then, ours is a cogency account of what it is for an argument to be good. However, we do think there is more to say about what we do when we *give* arguments. The standard purpose of giving an argument is to move the listener from premises that they are already inclined to accept to a conclusion that they were not previously inclined to accept. When you're trying to persuade someone, you need to find some premises they'll be willing to accept and show them that if they accept those premises, they should also accept this conclusion. Note that we do not think this is the only purpose of argument – when you play devil's advocate, for example, you are doing something different. But we think it is the central or paradigm purpose of argument: the thing such that, if arguments didn't do it, we wouldn't have a practice of arguing at all. Note also (and this will be useful to us later) that this account of what arguments are for gives a tidy account of what's wrong with question-begging arguments: there is no listener who is in the target position (being inclined to believe the premises but not the conclusion), because the conclusion is also one of the premises, so question-begging arguments can't do the thing that arguments are meant to do. Given this, we can stick with our former definition of what it is for an argument to be a good argument (good=cogent), but add to it that not every use of such an argument is one that achieves the goal of rationally persuading.

Three different senses of "argument" appear to be in play in these debates: [2]

> Sense 1: an argument is a type, and when you try to persuade someone you utter a token of the type.
> Sense 2: an argument is a token of an argument in sense 1 uttered by a particular person at a particular time in a particular context.
> Sense 3: an argument is a whole collection of arguments in sense 2 – an 'argument' in the ordinary sense of a discussion in which people are disagreeing with each other.

We are using "argument" in Sense 1. On this view, an argument is an abstract thing, a type. This has its advantages – it captures the view that two different people at different times in different contexts can nevertheless present the same argument. Then an argument-presented-in-a-context can be an instance of a good argument and yet not succeed in rationally persuading. We could translate our view into sense 2: then we would say "a cogent argument can fail to be a good argument, because it fails to do the main thing arguments are supposed to do – it fails to rationally persuade." Now "argument" refers to a presentation of an argument (in the former sense) in a context. Given that we can translate in this way, we wonder whether the issue about evaluation is purely terminological. If it is, that might suggest that some of the apparent disagreements between us and the VA theorist aren't real.

4. SOME EXAMPLES

To illustrate and expand on our view, we consider some of the examples presented by Paglieri and Aberdein. Paglieri (2015, p. 69) labels these first two 'balid', that is, both bad and valid.

> Socrates is a man, therefore Socrates is mortal
> Pierre and Marie Curie are physicists, therefore Marie Curie is a physicist.

---

[2] Actually only the first two are in play in the present discussion, but we include the third for the sake of completeness.

If they really are bad as well as valid, on our view it's because, as with the question-begging argument, no-one is in the position of accepting the premise but not yet accepting the conclusion, so the argument can't be used for the main thing arguments are supposed to be used for; ie. rationally persuading someone of their conclusion. But, a couple of things seem important to note here: Firstly, simple valid arguments such as these can be used illustratively/pedagogically, and in fact we need them for that. It's hard enough to get students to understand validity without requiring your examples to be sufficiently complicated that one could easily believe the premises and not notice that they entail the conclusion. (We don't think rational persuasion is the *only* purpose of argument, though it is the paradigm purpose – both the usual purpose, and the purpose such that if no argument did ever rationally persuade, arguments would have no purpose.) Secondly, perhaps people actually often do need these kinds of simple consistency issues pointed out to them, especially on things they feel strongly about. They may not have ever put together the thoughts "All A's are B", "X is an A" and "X is not a B", and so pointing out the inconsistency, as the so-called "balid" argument does, might sometimes be worthwhile. The upshot then is that cogent arguments can be used well in ways that fail to achieve the usual aim of argument. Arguing well involves more than presenting good arguments. We don't think accepting this amounts to a move towards VAT.

In charitably reconstructing an incomplete argument, it helps to know things about the arguer. When we do this, we're trying to capture the arguer's intentions: where we don't have much evidence about their intentions, we try and do our best by them as regards reconstructing the argument as structurally strong and having plausible premises, but if we do know things about them, we should use that information. This may be one place where we have in the past sounded less interested in character than we actually are: we think of argument reconstruction as separate from and prior to argument evaluation, and our minimizing of the role that considerations of character play in argument evaluation was never intended to cover argument reconstruction as well. Aberdein (2014, p. 84), borrowing from Sorenson, presents the following example:

> Here it is hot.
> Here it is humid.
> Therefore here it is hot and humid.

We need to know the time and place of the utterances and use that information to make the argument precise before we can tell whether it's a good argument. For instance the following argument works:

> In Fribourg on June 21st 2017, it is hot.
> In Fribourg on June 21st 2017, it is humid.
> Therefore in Fribourg on June 21st 2017, it is hot and humid.
> But
> In Fribourg on June 21st 2017, it is hot.
> In Hamilton on June 21st 2017, it is humid.

are a pair of premises from which no such conclusion can be drawn. We take this to be a case in which facts about the utterer(s) change *what argument* is being given, but that once the argument is specified (that is to say, once we have reconstructed the incomplete argument given), facts about the arguer don't come into the evaluation of the argument.

Consider this further example reconstructed by Aberdein (2014, p. 85) from a politician's tweet:

> Two thirds of the cabinet – 18 out of 29 ministers – are millionaires. Tomorrow, unlike you, they'll get a 42,000 pound tax cut.

On the face of it, once we add the unstated premise "All millionaires will get a 42,000 pound tax cut tomorrow", the argument looks valid, but there is a problem: it equivocates on "millionaire". The stated premise uses it in the sense of "someone who has a net worth of at least a million". The unstated premise uses it in the sense of "someone who has an annual income of at least a million". Aberdein thinks the way to see what's wrong with the argument is to think about who is presenting it – probably a Labour Party person, given the content and the hashtag, and the fact that the (UK) Labour Party had taken to using "millionaire" in this idiosyncratic second sense. It is true that this might provide us with a pointer to a possible flaw in the argument. But we can surely find that flaw in other ways. The unstated premise is false, given the standard sense of "millionaire". The stated premise is false, given the new, Labour Party, sense of "millionaire". There is no way to charitably reconstruct this argument, since charity requires us not to attribute false premises to arguers as well as requiring us to do our best by the structure of the argument. Not only is it a bad argument, but the arguer is arguing badly. And we may well want to point that out – we might well want to negatively judge the character of the tweeter. But we think that's a separate thing from evaluating the argument.

We think you hardly ever need to rely on what you know about an arguer's character in order to evaluate an argument, and you don't in this case. However, facts about the arguer may prompt us to look more closely at an argument (as they indeed do in this case). Character considerations might be used as a short cut, or for triaging if you are short of time and need to decide which of a bunch of arguments to bother to inspect more closely. But character considerations are not what do the actual heavy lifting of the evaluative work. We don't deny that there are some legitimate ad hominem arguments. Sometimes an arguer's character is relevant to whether or not we should accept his premises, and therefore to whether we should be persuaded by his argument. But this is not the usual case. Usually we have independent means of checking whether the premises are true – we seldom have to accept a premise purely on an arguer's say-so. This is why it sounds odd to us to make *all* argument evaluation dependent on the character of the arguer.

5. CONCLUSION

What then are the remaining differences between us and the VA theorist? Let's return to our three different senses of argument (above). If you think cogency is all there is to good argument, you are using argument in sense 1. We don't think that cogency is the full story, but we are still using argument in sense 1, as well as distinguishing "a good argument" from "putting an argument to good use" or "arguing well". A particular use of a cogent argument may or may not succeed in achieving its goals. We think Aberdein is using "argument" in sense 2. Dan Cohen seems to use "argument" in sense 3. Our discussion of argument quality is narrower than his. We focus on particular moves within a passage of argumentation, rather than on the passage of argumentation as a whole. Cohen thinks we don't see the wood for the trees (Cohen 2013, p.478).

But how can you evaluate the whole without evaluating the parts? We need to be able to say, for example, that move A was a misstep that threatened to derail the whole discussion. It was good that Person X noticed it and pointed it out in a constructive way so that Person Y was unoffended and undid move A and replaced it with the better move B. If you don't talk about the rational persuasiveness of particular moves, the overall evaluation of the passage of argumentation ("satisfying to the virtuous participants") seems somewhat lacking in nuance. And it is difficult to see how we can tell that participants and their satisfaction are virtuous without saying something about the quality of the moves they are accepting and rejecting.

As well as a difference in the senses in which we use "argument", another difference between our position and that of the VA theorists is a difference in focus. We think that you can present a good argument without arguing well. Arguing well involves bringing to bear both reliabilist and responsibilist virtues. But if you argue well, then you do present good (cogent) arguments. Since we're focused on who should believe what, on the basis of what evidence, we think giving good arguments deserves more attention than arguing well. And we think the VA theorist thinks the opposite.

REFERENCES

Aberdein, A. (2014). In defence of virtue: The legitimacy of agent-based argument appraisal. *Informal Logic*, *34*(1), 77-93.
Bowell, T., & Kingsbury, J. (2013). Virtue and argument: Taking character into account. *Informal Logic*, *33*(1), 22-32.
Cohen, D. (2013). Virtue, in context. *Informal Logic*, *33*(4), 471-485.
Gascón, J. (2017). Virtuous arguers: Responsible and reliable. In S. Oswald & D. Maillat (Eds.), *Argumentation and inference. Proceedings of the 2nd European Conference on Argumentation, Fribourg 2017*. London: College Publications.
Godden, D. (2016). On the priority of agent-based argumentative norms. *Topoi*, *35*(2), 345-357.
Paglieri, F. (2015). Bogency and goodacies: On argument quality and virtue argumentation. *Informal Logic*, *35*(1), 65-87.

10

A Conflict Index for Arguments in an Argumentation Graph

GIULIA CESARI
Department of Mathematics, Politecnico di Milano, Italy
giulia.cesari@polimi.it

FRANCESCA FOSSATI
Sorbonne Université, CNRS, LIP6, F-75006 Paris, France
francesca.fossati@sorbonne-universite.fr

STEFANO MORETTI
Université Paris Dauphine, PSL Research University, CNRS, UMR [7243], LAMSADE, 75016 Paris, France
stefano.moretti@dauphine.fr

In this work, given a measure of the disagreement for argumentation graphs, we introduce a property driven approach aimed at defining a conflict index representing the controversy of arguments. The index can be interpreted as a ranking of arguments based on their potential of development inside a debate. Merging the abstract argumentation framework into a game theoretical cooperative setting, this index is reinterpreted as the Shapley value of a specific coalitional game.

KEYWORDS: abstract argumentation, gradual semantics, axiomatic approach, coalitional games, Shapley value

1. INTRODUCTION

Abstract argumentation (Dung, 1995) deals with the construction and the analysis of non-monotonic reasoning systems based on the complex interplay among distinct arguments. Basically, in this framework, arguments are represented as atomic entities (without any regard to their internal structure) whose interaction is modelled via a binary attack relation expressing a (possible) *disagreement* between pairs of arguments. In the literature, several *extension* semantics have been associated to the abstract argumentation framework with the objective to specify which arguments are accepted or not, and which are undecided (Dung, 1995; Caminada, & Gabbay, 2009].

Different from extension semantics, the aim of *gradual* semantics is to assign a degree of acceptability to each argument. An example of gradual semantic is the h-Categorizer introduced in Besnard & Hunter, 2001, which is intended to quantify the relative strength of an argument taking into account how much such an argument is challenged by other arguments, and by recursion, how much it challenges its counter-arguments. Another gradual interaction-based evaluation reflecting the way in which arguments weaken each other has been introduced in Cayrol & Lagasquie-Schiex, 2005, for a bipolar argumentation framework (i.e., supporting both attack and support relations between arguments). Other examples are the ranking-based semantics introduced in Amgoud & Ben-Naim, 2013, where a procedure to transform an argumentation graph (i.e., a digraph where the nodes are the arguments and the arrows represent the attack relation) is introduced following a property-driven approach. Still different examples are the probabilistic approaches studied in (Hunter, 2013), which interpret the probability of an argument as the degree to which the argument is believed to hold.

Game theory has also been used to define intermediate level of acceptability of arguments. Specifically, in Matt & Toni, 2008, a degree of acceptability is computed taking into account the minimax value of a zero-sum game between a 'proponent' and an 'opponent' and where the strategies and the payoffs of the players depend on the structure of an argumentation graph. More recently, coalitional games have been applied in Bonzon, Maudet & Moretti, 2014, to measure the relative importance of arguments taking into account both preferences of an agent over the arguments and the information provided by the attack relations.

In the aforementioned approaches, the weight attributed to each argument represents the strength of an argument to "force" its acceptability. On the other hand, acceptability is not the only arguments' attribute that has been studied in literature from a "gradual" perspective. In Thimm & Kern-Isberner, 2014, an index has been introduced to represent the *controversiality* of single arguments, where the most controversial arguments are those for which taking a decision on whether they are acceptable or not is difficult. In a similar direction, the problem of measuring the *disagreement* within an argumentation framework has been studied in Amgoud & Ben-Naim, 2017, where the authors provided an axiomatic analysis of different disagreement measures for argumentation graphs. Both definitions of controversial-based ranking and disagreement measure are strictly related to the notion of *enforcement* introduced in Baumann, 2012, and aimed at identifying the minimal changes needed to enforce the acceptability of a set of arguments (see Thimm & Kern-Isberner, 2014, for a discussion on the relation between controversiality and enforcement).

The objective of this paper is twofold. First, we want to show that the properties introduced in Amgoud & Ben-Naim, 2017, for argumentation graphs can be reformulated for single arguments, and may drive the definition of a *conflict-based* ranking, that can be seen as an alternative ranking for measuring the relevance of arguments to enforce or refute the acceptability of the other arguments. Our second goal, is to merge the abstract argumentation framework into a game theoretical coalitional framework similar to the one already proposed in Bonzon, Maudet & Moretti, 2014, and re-interpret our conflict-based ranking in terms of a classical solution for coalitional games, that is as the average marginal contribution of each argument to the enforcement or refutation of groups of other arguments. Considering persuasion scenarios, we argue that the conflict-based ranking introduced in this paper may drive agents to select those arguments that should be further developed in order to strengthen certain position in a debate, hence, responding to the question raised in Thimm & Kern-Isberner, 2014, about the definition of a ranking representing the *potential for development* of arguments.

We start in the next section with some preliminary notions on argumentation theory and on coalitional game theory. In Section 2 we focus on the properties for a conflict index and their interpretation with respect to problem of enforcing arguments. Section 3 and 4 are devoted to the property-driven analysis of conflict indices. Section 5 deals with the analysis of an associated coalitional framework and the reformulation of the conflict index introduced in Section 4 as a solution for these games. Section 6 concludes with some future research directions.

2. PRELIMINARIES

In this section we introduce some preliminary notations and definitions on argumentation graphs and coalitional games.

An argumentation framework is a directed graph in which nodes represent arguments and direct edges represent attack relations (Dung,

1995). Formally, an *argumentation framework* (or *argumentation graph*) **A** is a pair <A,R> where A is a non-empty finite set of arguments and $R \subseteq A \times A$ is an attack relation. Given two arguments a, b $\in$ A, $(a,b) \in R$ (or equivalently aRb) if a attacks b.

We denote by Ω the set of all argumentation graphs and by Ω^A the set of all argumentation graphs with A as the set of arguments.

The number of arguments in the graph, that is $|A|$, is called the *size* of the graph. A *path* in **A** is a sequence of arguments $(a_1,...a_k)$, where $a_i \in A$ for all i=1, ...,k such that $a_i R a_{i+1}$ for all $1 \le i < k$ and $a_i \ne a_j$ for all $i \ne j$. An *elementary cycle* is a path $(a_1,...a_k)$ such that $a_k R a_1$.

Let $i,j \in A$, $i \ne j$. The distance $d_{i,j}$ between i and j is defined as the length of the shortest path from i to j if such path exists, otherwise $d_{i,j} = |A|+1$. If i=j then $d_{i,j}$ is the length of the shortest elementary cycle in which i is involved, otherwise $d_{i,j} = |A|+1$.

An argumentation graph is by definition conflictual, since it describes the attack relations among arguments in an argumentation system. In order to quantify this inner conflict, it is possible to associate to each argumentation framework a disagreement measure.

A *disagreement measure* (Amgoud & Ben-Naim, 2017) is a function $k : \Omega \to [0,1]$ with the interpretation that, for every **A**, **A'** $\in \Omega$, **A** is more conflicting than **A'** if $k(\mathbf{A}) > k(\mathbf{A'})$. Note that $k=0$ corresponds to the absence of disagreement in a graph, while the maximum disagreement is set to $k=1$.

Let us now introduce some basic concepts of cooperative game theory. A *transferable utility (TU) game*, also referred to as *coalitional game*, is a pair (N,v) where N={1,...,n} is a finite set of players and $v: 2^N \to R$ is a real-valued function on the family of subsets of N that associate a value to each *coalition* $A \subseteq N$, such that $v(\emptyset)=0$. If the set N of players is fixed, we identify a coalitional game (N,v) with the characteristic function v. Let G be the set of all cooperative TU-games and G^N the class of all coalitional games with players set N. G is a vector space of dimension $2^n - 1$ and the family $\{u_A : A \subseteq N\}$ of the *unanimity games* is a basis for G^N. The game u_A is defined, for every $T \subseteq N$, $T \ne \emptyset$, as

$$u_A(T) = \begin{cases} 1, if\ A \subseteq T \\ 0, otherwise. \end{cases}$$

A *solution* is a map associating to each game v a set of vectors $(x_1,...,x_n)$ in which x_i is the amount assigned to player *i*. The solutions are classified into two groups: the ones which provide more than one solution vector (e.g. the core), called set-valued solutions, and the ones which provide only one solution vector (e.g. the nucleolus and the power indices), called one-point solutions. The *Shapley value* (Shapley, 1953) of a game is the one-point solution that satisfies the four following axioms: Efficiency, Symmetry, Dummy player property and Additivity. A formula for the Shapley value, for every $i \in N$, is the following:

$$\sigma_i(v) = \sum_{S \subseteq N: i \in S} \frac{(|S|-1)!(n-|S|)!}{n!}(v(S)-v(S-\{i\}))$$

It assigns to every player the average marginal contribution to all the coalitions he belongs to, with respect to a probability distribution that gives probability $p(S) = \frac{(|S|-1)!(n-|S|)!}{n!}$ to each coalition $S \subseteq N$.

3. A PROPERTY-DRIVEN APPROACH TO MEASURE CONFLICT IN AN ARGUMENTATION GRAPH

Different measures can be proposed in order to quantify the disagreement in a graph. A property-driven approach has been proposed in Amgoud & Ben-Naim, 2017, which leads to the introduction of a measure that is based on the concept of *global distance* among arguments in an argumentation framework.

Let **A** = <A,R> be an argumentation framework. The *global distance* D(**A**) is defined as

$$D(\mathbf{A}) = \sum_{i \in A} \sum_{j \in A} d_{ij}.$$

To compute D(**A**) it is possible to build a matrix of distances δ:

$$\delta = \begin{bmatrix} d_{11} & d_{12} & \cdots \\ d_{21} & d_{22} & \cdots \\ \cdots & \cdots & \cdots \end{bmatrix},$$

in which the element d_{ij} is the distance between argument i and argument j as defined in Section 2, and to sum all its elements.

The maximum value of $D(A)$ corresponds to the case where $R=\emptyset$ and it equals $n^2(n+1)$. On the other hand, when $R=A\times A$, $D(A)$ takes minimum value, that is n^2. The maximum value of $D(A)$ corresponds to the case in which the disagreement is minimal and the minimal value to the case in which the disagreement is maximal. For this reason it makes sense to define a disagreement measure that depends on the opposite value of $D(A)$.

Definition 1 (Distance-based measure [Amgoud & Ben-Naim, 2017]**)** Let $A = <A,R>$ be an argumentation graph. The distance-based measure $\kappa(A)$ is defined as:

$$\kappa(A) = \frac{\max - D(A)}{\max - \min}, \quad (1)$$

where $\max = n^2(n+1)$ and $\min = n^2$.

The distance-based measure is normalized by the value *max−min* in order to obtain a value between 0 and 1 that makes possible to easily compare different argumentation graphs. In Amgoud & Ben-Naim, 2017, the authors show that this measure satisfies the following set of axioms: it depends only on the structure of the graph and not on the label of the nodes with the consequence that it assigns the same value to isomorphic graphs (*abstraction*); it assigns 0 to graphs without attack relations and 1 to complete graphs (*coherence* and *maximality*); the disagreement value of a graph does not increase if an isolated argument is added, i.e. if an argument is added without modifying the set R, and it does not decrease if an attack between two arguments is added (*free independency* and *monotonicity*); it detects cycles by assigning them a higher disagreement with respect to acyclic graphs of the same size (*cycle sensitivity*) and, when comparing two cycles, it assigns a higher value to the cycle with less arguments, following the idea that cycles are seen as dilemmas or paradoxes and the less arguments are needed to produce a cycle, the stronger is the disagreement (*size sensitivity*).

Given a disagreement measure, it is interesting to assess which are the arguments that contribute the most to the total disagreement in an argumentation graph. To this purpose, we introduce, by using an axiomatic approach, a conflict index that evaluates the contribution of each argument to the total disagreement.

A *conflict index* $K: \Omega^N \to R^N$ is a function that assigns to every argumentation graph with $n=|N|$ nodes (arguments) a vector in R^n, representing the contributions of each argument to the conflict in the graph. The higher the value that such index assigns to an argument, the larger is the disagreement brought by that argument to the graph. In terms of enforcement, an argument with a large conflict index also plays a central role in the acceptability of the other arguments, acting as an important support for certain arguments (enforcement) or as a strong apposition for others (refutation).

Let $\mathbf{A} = <A,R>$ be an argumentation framework. For every argument $i \in A$, $K_i(\mathbf{A})$ measures the contribution of argument i to the total disagreement in the graph: an argument i brings (strictly) more disagreement than j if $K_i(\mathbf{A}) > K_j(\mathbf{A})$. Following the approach proposed in Amgoud & Ben-Naim, 2017, we introduce eight axioms that a conflict index should satisfy in order to describe the behaviour of arguments in a graph. The first axiom states that the contribution to the disagreement of each argument depends only on the structure of the graph. This means that the label of an argument does not add information about the contribution to the disagreement. Therefore if two argumentation graphs are isomorphic, the index assigns the same value to corresponding nodes.

> **Axiom 1 (Abstraction)** Let $\mathbf{A} = <A,R>$ and $\mathbf{A}' = <A',R'>$ be two argumentation graphs. If $\mathbf{A}$ and $\mathbf{A}'$ are isomorphic, then $K_i(\mathbf{A}) = K_{f(i)}(\mathbf{A}')$ for all $i \in A$, where $f(\cdot)$ is an isomorphism between $\mathbf{A}$ and $\mathbf{A}'$.

The second axiom sets to zero the conflict index of arguments in a graph without attack relations. This is a natural request that follows from the absence of disagreement in the graph.

> **Axiom 2 (Coherence)** Let $\mathbf{A} = <A,R>$ be an argumentation graph. If $R=\emptyset$, then $K_i(\mathbf{A}) = 0$ for all arguments $i \in A$.

In order to state the third axiom, the concept of *star* must be introduced. An argumentation graph is a star if an argument i exists that attacks all the other arguments and receives attacks from all the arguments, included itself, and there are no other attack relations between arguments. i is called centre of the star.

The third axiom states that the argument that brings the maximum of disagreement in a graph is the centre of the star.

Axiom 3 (Maximality) *Let A = <A,R> be a star and let i be the center of the star. Then $K_i(A) > K_j(A)$ for all $j \neq i \in A$, and $K_i(A) \geq K_k(A')$ for all $h \in A$, and all the argumentation graphs $A' = $ <A,R'> of the same size of A.*

The fourth axiom states that adding isolated arguments to an argumentation graph which contains attacks does not increase the contribution of each argument to the total disagreement.

Axiom 4 (Free independence) *Let Args be the universe of arguments. Let A = <A,R> be an argumentation graph with $R \neq \emptyset$ and let A' = <A ∪ {a},R> an argumentation graph with $a \in$ Args\A. Then $K_i(A) \geq K_i(A') \; \forall i \in A$.*

The next axiom states that if a new attack is added in an argumentation graph, no argument will decrease its contribution to the total disagreement.

Axiom 5 (Monotonicity) *Let A = <A,R> be an argumentation graph and let A' = <A,R ∪ R'> be an argumentation graph with $R' \subseteq (A \times A) \setminus R$. Then, for all $i \in A$, $K_i(A) \leq K_i(A')$.*

The next results for the centre of a star follow from the previous axioms.

Proposition 1 *If a conflict index satisfies Axioms 3 and 5, then adding attacks does not change the conflict index of the centre of the star.*
proof.
Let A = <A,R> be a star of size n and centre i and let A' be a graph in which one attack is added to a star of size n. Thanks to axiom 3, $K_i(A) \geq K_h(A') \; \forall h \in A'$, but axiom 5 states that $\forall j \in A$ and $\forall h \in A' \; K_i(A) \leq K_h(A')$, so the centre of the star does not change its value. □

Axiom 6 states that, among argumentation graphs of the same size, arguments that belong to an elementary cycle have larger contribution to the disagreement than arguments that belong to an acyclic graph.

Axiom 6 (Cycle sensitivity) Let $A = <A,R>$ be an acyclic argumentation graph and $A' = <A',R'>$ an elementary cycle. If $|A|=|A'|$ then, for all $i \in A$ and for all $j \in A'$, $K_i(A) < K_j(A')$.

The following axiom states that the larger is the size of an elementary cycle, the smaller is the contribution of each argument to the total disagreement.

Axiom 7 (Size sensitivity) Let $A = <A,R>$ and $A' = <A',R'>$ be two elementary cycles with $|A|<|A'|$. Then, for all $i \in A$ and for all $j \in A'$, $K_i(A) > K_j(A')$.

The last axiom states that the sum of the contributions of each node to the disagreement is equal to the total disagreement in the graph.

Axiom 8 (Efficiency) Let $A = <A,R>$ be an argumentation graph and let $\mathcal{K}$ be the distance-based disagreement measure defined by (1). Then $\sum_{i \in N} K_i(A) = \kappa(A)$.

4. A DISTANCE-BASED CONFLICT INDEX FOR ARGUMENTS IN AN ARGUMENTATION GRAPH

In this section we introduce a distance-based conflict index, which is linked to the disagreement measure introduced in Amgoud & Ben-Naim (2017), and we show that it satisfies all the axioms stated in the previous section. Moreover, we show how this index naturally induces a ranking based semantic which allows ordering arguments in an argumentation framework according to their acceptability.

Definition 2 (Distance-based conflict index) Let $A = <A,R>$ be an argumentation graph of size n. We define the distance-based conflict index as the conflict index that assigns to every $i \in A$ the following value:

$$K_i(A) = \frac{1}{\Delta}\left(\frac{\max}{n} - \phi_i\right), \quad (2)$$

where, $\Delta = \max - \min = n^2(n+1) - n^2 = n^3$ and

$$\phi_i = d_{ii} + \frac{1}{2}\sum_{j \in A: i \neq j} d_{ii} + \frac{1}{2}\sum_{j \in A: j \neq i} d_{ij} + \frac{1}{2}\sum_{j \in A: j \neq i} d_{ji}.$$

The value $K_i(\mathbf{A})$ depends on ϕ_i, which considers all the distances from i to the other arguments and vice versa.

Example 1 Let $\mathbf{A}=<A,R>$ be the argumentation graph where $A=\{1,2,3\}$ and $R=\{(1,1),(1,2),(2,3)\}$, as depicted in Figure 1.

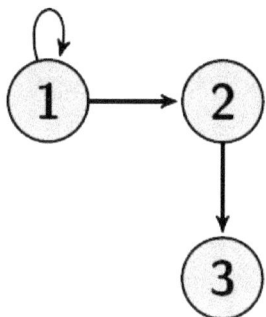

Figure 1 – An argumentation graph with three arguments.

The corresponding matrix of distances $\mathcal{D}$ is
$$\delta = \begin{bmatrix} 1 & 1 & 2 \\ 4 & 4 & 1 \\ 4 & 4 & 4 \end{bmatrix},$$
and the value of max and min are 36 and 9 respectively. Thus $\phi_1 = \frac{13}{2}$, $\phi_2 = 9$, $\phi_3 = \frac{19}{2}$ and the conflict index are $K_1 = 0.2, K_2 = 0.11, K_3 = 0.09$. The most conflictual argument according to our measure is argument 1, followed respectively by arguments 2 and 3.

Theorem 1 *The distance-based conflict index defined by (2) satisfies all the eight axioms introduced in Section 3.*
proof. Let $\mathbf{A} = <A,R>$ be an argumentation graph. In the following we prove that the index φ defined by relation (2) satisfies each axiom.

Abstraction: If $\mathbf{A}$ and $\mathbf{A}'$ are isomorphic, then $\phi_i = \phi_{f(i)}$ are $K_i(\mathbf{A}) = K_{f(i)}(\mathbf{A}')$.

Coherence: If $R=\emptyset$ then $\phi_i = n(n+1)$. It follows that

$$K_i(\mathbf{A}) = \frac{1}{\Delta}\left(\frac{\max}{n} - \phi_i\right) = \frac{1}{n^3}\left(\frac{n^2(n+1)}{n} - n(n+1)\right) = 0.$$

Maximality: Let $i \in A$ be an argument such that $\forall j \in A\ jRi$ and iRj and such that there are no other attack relations between arguments. Then $\phi_i = n$ and $\phi_j = 2n-1$ that implies

$$K_i(\mathbf{A}) = \frac{1}{n}$$

and

$$K_j(\mathbf{A}) = \frac{n^2 - n + 1}{n^3}.$$

It follows that $K_i(\mathbf{A}) > K_j(\mathbf{A})\ \forall j \in A$.

Furthermore n is the minimum possible value of ϕ for a node. This means that for all the argumentation graphs $\mathbf{A}'$ of the same size, $\forall h \in A', K_i(\mathbf{A}) \geq K_h(\mathbf{A}')$.

Free independence: Let $\mathbf{A}' = \langle A \cup \{a\}, R\rangle$ be an argumentation graph with $a \in Args \setminus A$. For all the arguments $i \in A$, $\phi_i(\mathbf{A}') \geq \phi_i(\mathbf{A})$. This implies $K_i(\mathbf{A}) \geq K_i(\mathbf{A}')\ \forall i \in A$.

Monotonicity: Let $\mathbf{A}' = \langle A, R'\rangle$ be an argumentation graph with $R' \subseteq (A \times A) \setminus R$. Then, for all the arguments $i \in A$, $\phi_i(\mathbf{A}) \geq \phi_i(\mathbf{A}')$ and it follows that $K_i(\mathbf{A}') \geq K_i(\mathbf{A})\ \forall i \in A$.

Cycle sensitivity: Let $\mathbf{A} = \langle A, R\rangle$ be an acyclic argumentation graph and $\mathbf{A}' = \langle A', R'\rangle$ an elementary cycle and let $|A| = |A'|$. In order to prove this axiom, first the acyclic configuration in which i gives the maximum possible disagreement is found and then it is shown that all the arguments in a cycle has a greater conflict index than i.

The acyclic graph in which i brings the largest disagreement is the one in which there is only one attack between i and all the other arguments $h \in A - \{i\}$ (i.e., $\forall h \in A\ iRh \vee hRi$). In this case ϕ_i has the minimum value possible for a node in an acyclic graph because for all i and j with $i \neq j$ (i.e., $[d_{ij} = 1$ and $d_{ji} = n+1] \vee [d_{ij} = n+1$ and $d_{ji} = 1]$). Therefore

$$\Phi_i(\mathbf{A}) = \frac{1}{2}(n+1)(n-1) + \frac{1}{2}(n-1) + (n+1) = \frac{1}{2}n^2 + \frac{3}{2}n$$

It follows that

$$K_i(\mathbf{A}) = \frac{n-1}{2n^2}.$$

For an elementary cycle, for all the arguments $j \in A'$ it holds that

$$\Phi_j(\mathbf{A'}) = 1 + 2 + \ldots + n = \frac{n(n+1)}{2}$$

and

$$K_j(\mathbf{A'}) = \frac{n+1}{2n^2}.$$

So, for all $i \in A$ and for all $j \in A'$, $K_i(\mathbf{A}) < K_j(\mathbf{A'})$.

> *Size sensitivity:* Let **A** be an elementary cycle. As it is proved in the cycle sensitive property, $\forall i \in A$

$$\Phi_i(\mathbf{A}) = \frac{n(n+1)}{2},$$

so it follows:

$$K_i(\mathbf{A}) = \frac{n+1}{2n^2}.$$

Let $\mathbf{A'} = <A', R'>$ be another elementary cycle with $|A| < |A'| = m$, then $\forall j \in A'$

$$K_j(\mathbf{A'}) = \frac{m+1}{2m^2}.$$

It's easy to check that if $n \geq 1$ and $n < m$, $K_i(\mathbf{A}) > K_j(\mathbf{A'})$ for all $i \in A$ and for all $j \in A'$.

Efficiency: Let $\kappa(\mathbf{A}) = \frac{\max - D(\mathbf{A})}{\max - \min} = 1 + \frac{1}{n} - \frac{D(\mathbf{A})}{n^3}$ be the disagreement measure of the graph **A**. Then

$$\sum_{i \in N} \frac{\max}{n} - \phi_i = \max - D(\mathbf{A})$$

implies

$$\sum_{i \in N} K_i(\mathbf{A}) = \frac{\max - D(\mathbf{A})}{\max - \min}.$$

The distance-based conflict index associates to each argument a value that depends on the distance from an argument to the others and vice versa. Every isolated argument has measure equal to zero which means that it does not bring any disagreement. Furthermore, this index detects arguments that belong to cycles giving them a high value. By assigning to each argument a real value, the index enables the comparison among arguments in contributing to the conflict in an argumentation framework, which naturally leads to consider the conflict index as a *ranking-based semantic*, that is, a function that transforms every argumentation graph into a ranking on the set of arguments Amgoud & Ben-Naim (2013). The distance-based conflict index generates a ranking between arguments since it assigns a real number to each argument and R is totally ordered. In particular, the higher is the conflict index of an argument, the least will be its position in the ranking, with the interpretation that an argument's controversiality is directly proportional to the amount of disagreement it produces.

5. A GAME-THEORETICAL INTERPRETATION

In the previous section, an index of conflict for arguments in an argumentation graph is proposed. The introduction of such an index has been justified by means of an axiomatic approach: we proved that the index satisfies a number of interesting properties. Moreover, in this section, we show that it coincides with the Shapley value of a properly defined game.

Let $\mathbf{A}$ = <A,R> be an argumentation graph, where A has cardinality n. We introduce a cooperative game (A,v), where the set of players coincides with the set of arguments A in the argumentation graph and the characteristic function is defined as follows for every $S \subseteq N$:

$$v(S) = \frac{\max - D(S)}{\max - \min}, \quad (3)$$

Where $D(S) = \sum_{i,j \in S} d_{ij}$, $max = n^2(n+1)$ is the maximal value that D can attain in an argumentation graph with n arguments and $min = n^2$ is the minimal one. $D(S)$ measures the global distance among arguments in a coalition S, taking into account the attack relations that exist among them on the whole graph: it is defined as the sum of distances among the nodes of the coalition S, where the distance between two nodes is computed on the entire graph.

The smaller is the global distance in a coalition, the higher is the value of that coalition in the game v, reflecting the fact that the overall conflict in a coalition of arguments conversely depends on the distance among arguments. It is possible to show that the conflict index introduced in the previous section coincides with the Shapley value of the game defined by relation (3). The following proposition holds.

Proposition 2 *The distance-based conflict index defined by (2) coincides with the Shapley value of the game (A,v) defined in (3).*
proof.
The expression in (3) can be written as follows:

$$v(S) = \frac{max}{\Delta} - \frac{D(S)}{\Delta}, \quad (4)$$

Where $\Delta = max - min = n^3$. Therefore, v can be written as the linear combination of two games: $v = \frac{1}{\Delta}(v^{max} + v^D)$, where v^{max} is a constant game that assumes value max for every coalition and $v^D(S) = D(S)$. Therefore, the Shapley value of v, for every $i \in A$, is given by

$$\sigma_i(v) = \frac{1}{\Delta}\left(\sigma_i(v^{max}) + \sigma_i(v^D)\right).$$

The value $\sigma_i(v^{max})$ is easy to calculate because v^{max} is a constant game and thus $\sigma_i(v^{max}) = \frac{max}{n}$.

Moreover, the game v^D can be written in terms of unanimity games as follows:

$$v^D = \sum_{i,j \in N: i \neq j} \alpha_{ij} u_{\{i,j\}} + \sum_{i \in N} \alpha_i u_{\{i\}},$$

Where $\alpha_{ij} = d_{ij} + d_{ji}$, and $\alpha_i = d_{ii}$.

It follows that the Shapley value of v^D for every $i \in A$ has the following expression:

$$\sigma_i(v^D) = d_{ii} + \frac{1}{2} \sum_{j \in A - \{i\}} (d_{ij} + d_{ji}),$$

Therefore, the Shapley value of v, for every $i \in A$, is given by

$$\sigma_i(v) = \frac{1}{\Delta}\left(\frac{\max}{n} - \phi_i\right),$$

which coincides with the expression in (2).

Example 3 Let **A** be the argumentation graph of Example 1. As an example, to compute the value of coalition {1,2}, we shall sum up all the terms in the following submatrix:

$$\begin{bmatrix} 1 & 1 \\ 4 & 4 \end{bmatrix},$$

which results in $v(\{1,2\}) = \dfrac{\max - 10}{\max - \min}$. Thus the game associated to this argumentation graph is such that $v(\emptyset) = 0$, $v(\{1\}) = \dfrac{35}{27}$, $v(\{2\}) = \dfrac{32}{27}$, $v(\{3\}) = \dfrac{32}{27}$, $v(\{1,2\}) = \dfrac{26}{27}$, $v(\{1,3\}) = \dfrac{25}{27}$, $v(\{2,3\}) = \dfrac{23}{27}$ and $v(A) = \dfrac{11}{27}$. The Shapley value of the game coincides with the disagreement measure found in Example 1.

Indeed, the distance-based measure we have introduced can be interpreted as an index of the contribution of each argument to the total disagreement in the graph, since it takes into account the marginal contribution, in terms of conflict, that each argument provides to any other coalition of arguments, weighting it according to a probabilistic coefficient that depends on the size of the coalition.

Note that it is possible to define other games on an argumentation graph, where the value of a coalition represents in some way the conflict among arguments in the coalition. The game we have considered takes into account the distance among arguments in the original graph. However, one can imagine restricting the attention to the only arguments in the coalition, and therefore compute the distances among them in the induced graph. This idea leads to the definition of the following game, for every $S \subseteq A$,

$$w(S) = \frac{\max_{|S|} - D_{|S|}(S)}{\max_{|S|} - \min_{|S|}}, \quad (5)$$

where $D_{|S|}(S) = \sum_{i \in S} \sum_{j \in S} d_{|S|,ij}$ is the global distance in the induced graph, $\max_{|S|} = |S|^2(|S|+1)$ is the maximum value of the global distance in a graph with $|S|$ nodes and $\min_{|S|} = |S|^2$ is the minimum one. We observe, however, that the Shapley value of game w, as defined by relation (5), does not satisfy all the axioms stated in the previous section. Indeed, the following proposition holds.

Proposition 3 *The Shapley value of game w, as defined by relation (5), does not satisfy Axioms 4 and 5.*
proof.
First we show that Axiom 4 is not satisfied. Let **A**=<A,R> the argumentation graph with A={1,2}, R={(1,1),(1,2)} and let **A'**=<A',R'>, with A'={1,2,3} and R'=R. On the graph **A**, w assumes the following values: $w(\{1\})=1$, $w(\{2\})=0$ and $w(\{1,2\})=0.5$. The Shapley value is $\sigma(w) = \left(\frac{3}{4}, -\frac{1}{4}\right)$. On the other hand, the game w on graph **A'** assumes the following values: $w(\{1\})=1$, $w(\{2\})=0$, $w(\{3\})=0$, $w(\{1,2\})=0.5$, $w(\{1,3\})=0.25$, $w(\{2,3\})=0$ and $w(A') = \frac{2}{9}$, and the Shapley value of argument 2 is $\sigma_2(w) = -\frac{10}{108} > -\frac{1}{4}$, so free independency is not satisfied.

In order to prove that Axiom 5 is not satisfied, consider for instance the argumentation graphs **A**=<A,R> such that A={1,2,3}, R={(1,2)}, and **A'**=<A',R'> such that $A'=A$, $R'=R \cup \{(2,3)\}$. It easy to check that the Shapley value of player 1 in the game defined over **A'** is smaller than the Shapley value of player 1 in the game defined over the graph **A**.

5. CONCLUSION

In this paper, we proposed a conflict index that quantifies the contribution of each argument to the total disagreement in an argumentation framework. Our index is introduced trough a property-driven approach, where the properties introduced in Amgoud & Ben-Naim (2017) are translated in terms of properties for arguments, and results in a ranking on the set of arguments, from the most controversial to the less controversial, therefore leading to the definition of a ranking-based semantic. Moreover, we show that the index coincides with the Shapley value of a suitable coalitional game built on argumentation graphs. Using a similar approach with different axioms (and alternative coalitional games), new measures could be proposed in order to define new ranking-based semantics, whose comparison with the one proposed here may help in providing better support to the analysis of the disagreement among arguments in complex decision processes.

ACKNOWLEDGEMENTS: S. Moretti acknowledges the support of the French National Research Agency (ANR) project AMANDE (grant no. ANR-13-BS02-0004).

REFERENCES

Amgoud, L., & Ben-Naim, J. (2013). Ranking-based semantics for argumentation frameworks. In W. Liu, V. S. Subrahmanian & J. Wijsen (Eds.), *International Conference on Scalable Uncertainty Management. SUM 2013.* (pp. 134-147). Berlin:Springer.

Amgoud, L., & Ben-Naim, J. (2017). Measuring disagreement in argumentation graphs. In S. Moral, O. Pivert, D. Sánchez & N. Marín (Eds.), *International Conference on Scalable Uncertainty Management. SUM 2017* (pp. 208-222). Cham, CH: Springer.

Baumann, R. (2012) What does it take to enforce an argument? Minimal change in abstract argumentation. In L. De Raedt, C. Bessiere, D. Dubois, P. Doherty, P. Frasconi, F. Heintz & P. Lucas. *Proceedings of the 20th*

European Conference on Artificial Intelligence, ECAI 2012, (Vol. 12, pp. 127-132). Amsterdam: IOS Press.

Besnard, P., & Hunter, A. (2001) A logic-based theory of deductive arguments. *Artificial Intelligence*, *128*(1), 203-235.

Bonzon, E., Maudet, N., & Moretti, S. (2014). Coalitional games for abstract argumentation. In S. Parson, N. Oren, C. Reed & F. Cerutti (Eds.) *Proceedings of the International Conference on Computational Models of Argument, COMMA 2014* (pp. 161-172). Amsterdam: IOS Press.

Caminada, M. W., & Gabbay, D.M. (2009) A logical account of formal argumentation. *Studia Logica*, *93*(2-3), 109-145.

Cayrol, C., & Lagasquie-Schiex, M.C. (2005) Graduality in argumentation. *Journal of Artificial Intelligence Research*, 23, 245–297.

Dung, P.M. (1995) On the acceptability of arguments and its fundamental role in nonmonotonic reasoning, logic programming and n-person games. *Artificial intelligence*, *77*(2), 321-357.

Hunter, A. (2013) A probabilistic approach to modelling uncertain logical arguments. *International Journal of Approximate Reasoning*, *54*(1), 47-81.

Matt, P.A., & Toni, F. (2008) A game-theoretic measure of argument strength for abstract argumentation. In L. Michael & A. Kakas (Eds.), *Logics in Artificial Intelligence, 15th European Conference, LNCS* (285-297). Cham, CH: Springer.

Thimm, M., & Kern-Isberner, G. (2014) On controversiality of arguments and stratified labelings. In S. Parson, N. Oren, C. Reed & F. Cerutti (Eds.),*Proceedings of the International Conference on Computational Models of Argument, COMMA 2014* (pp. 413-420). Amsterdam: IOS Press.

Shapley, L.S. (1953) A value for n-person games. In H. W. Kuhn & A. W. Tucker (Eds.), *Annals of Mathematics Studies: Vol. 28, Contributions to the theory of games* (Vol. II, pp. 307-317). Princeton: Princeton University Press.

11

Argumentation Traits, Frames, and Dialogue Orientations

IOANA A. CIONEA
Department of Communication, University of Oklahoma
icionea@ou.edu

DALE HAMPLE
Department of Communication, University of Maryland
dhample@ou.edu

STACIE WILSON MUMPOWER
Department of Communication, University of Oklahoma
stacie.wilsonmumpower@ou.edu

ERYN N. BOSTWICK
Department of Communication, University of Oklahoma
Eryn.N.Bostwick-1@ou.edu

CAMERON W. PIERCY
Department of Communication Studies, University of Kansas
cpiercy@ku.edu

CANDACE L. FOUTCH
Department of Communication, University of Oklahoma
Candace.L.Foutch-1@ou.edu

This paper investigates dialogue orientations in conjunction with argumentation traits and argument frames in dyads (friends or strangers). Respondents participated in a laboratory experiment and indicated the dialogue orientations they intended to use in an argument with another person. Findings revealed some correlations between friends' dialogue orientations and their traits and frames. Differences also emerged depending on whether the dyads were male or female. Some traits and frames modestly predicted planned dialogues.

KEYWORDS: dialogue orientations, argument frames, argumentativeness, verbal aggressiveness.

1. INTRODUCTION

Arguing occurs frequently in interpersonal relationships (Hample, 2005) as individuals try to negotiate their differences, persuade others of their point of view, or express their frustration. How individuals might go about arguing with others has been examined in a recent line of work proposed by Cionea and colleagues under the framework of dialogue orientations. Based on the theoretical conceptualizations of dialogues advanced by Walton (1998) and Walton and Krabbe (1995), Cionea (2011) proposed that dialogues can be applied empirically to map out people's argumentative exchanges in interpersonal relationships. Cionea, Hample, and Fink (2013), and Cionea and Hample (2014) provided initial empirical support for this social scientific approach to measuring people's preferred dialogues, which they called dialogue orientations. In this paper, we continue this line of research by examining whether individuals' dialogue orientations are related to their argumentation traits and frames in an experiment meant to examine people's arguing behaviours.

2. ARGUMENTATION TRAITS, FRAMES, AND DIALOGUES

The dialogue orientations framework is summarized graphically in Figure 1 below. Briefly, the type of situation arguers are in and the main goal of their argumentative exchange give rise to various dialogue types. Cionea and Hample (2014) added the information giving dialogue to the six already proposed by Walton.

Main Goal \ Initial Situation	Conflict	Open Problem	Unsatisfactory Spread of Information
Stable Agreement/Resolution	Persuasion	Inquiry	Information Seeking/Giving
Practical Settlement/ Decision (Not) to Act	Negotiation	Deliberation	--
Reaching a (Provisional) Accommodation	Eristic	--	--

Figure 1 – Dialogue orientations framework (adapted from Walton & Krabbe, 1995)

Argumentation traits (i.e., argumentativeness and verbal aggressiveness) have been examined extensively in previous research. In relation to dialogues, Hample and Cionea (2016) have proposed that people in long-term relationships (in their case, marriages) may match each other by using similar dialogue orientations. We extend this question by asking whether individuals in established relationships (in this study, friendships) may match on argumentation traits and frames. In other words, two individuals may become or stay friends partially due to their similar levels of argumentativeness and verbal aggressiveness, as well as due to the fact that they frame the role of arguing in similar ways. If this is the case, strangers should not exhibit more than chance connections between their traits and frames, whereas friends ought to exhibit some (stronger) correlations. We ask the following:

> RQ1: Is there a relationship between one's own argumentation traits, argument frames, or dialogue orientations and the other person's argumentation traits, argument frames, or dialogue orientations in the case of strangers or friends?

Another possibility we examine is that dialogue orientations are situational. Walton (1998) has explained that what type of dialogues arguers *ought to* engage in is determined partially by the situation they are in: conflict, open problem, or unsatisfactory spread of information. In everyday exchanges, however, other considerations also come into play, such as relational dynamics, power differences, topic of the argument, and gender differences (we examine the latter in this study). In respect to argumentation traits, differences between men and women have been documented consistently (e.g., Infante, Wall, Leap, & Danielson, 1984; Jordan-Jackson, Lin, Rancer, & Infante, 2008; Nicotera

& Rancer, 1994). In respect to dialogue orientations, though, only Hample and Cionea (2016) have examined differences between men and women, in marriages. This study extends the analysis of sex differences to same-sex friend and stranger dyads. The following research question is proposed:

> RQ2: Is there a difference in argumentation traits, argument frames, or dialogue orientations based on sex, argument partner, or topic of the argument?

Finally, Cionea and Hample (2014) have found that argumentation traits and frames predicted dialogue orientations with some success (explained variance ranged from 10% to 43%). They examined general dialogue orientations (i.e., dialogues individuals report using, in general). We seek to extend their research in a different setting by examining whether traits and frames can also explain the dialogues a person intends on using in an actual exchange with another person, as reported right before the argument. This study examines planned behaviours as opposed to general trait-like measures, which may be a more accurate representation of individuals' arguing behaviours. Therefore, we ask the following:

> RQ3: Are the dialogue orientations one intends to use during an argument predicted by a) one's argumentation traits, b) one's argument frames, c) the other person's argumentation traits, or d) the other person's argument frames, in the case of strangers and friends?

3. METHOD

3.1 Participants

Two hundred and eighty-two individuals participated in the study. They were mostly undergraduate students at a southwestern university in the United States, aged between 18 and 55 years old ($M = 20.18$, $SD = 2.96$). Of these, 140 participants identified as male, 141 identified as female, and one identified as "other". Most participants ($n = 114$) were freshmen, followed by sophomore ($n = 69$), juniors ($n = 57$), and seniors ($n = 38$), as well as four others (e.g., graduate students). The vast majority ($n = 204$) were White, with some who reported Asian ($n = 20$), Hispanic or Latino/Latina ($n = 16$), African-American ($n = 13$), and other ethnicities ($n = 29$). Approximately 20% of the participants were Communication majors, with the rest reporting a variety of majors, such as Finance, Sports Management, Public Relations, or Energy Management.

3.2 Procedures

Participants came to a laboratory to participate in an argumentation study. Roughly half (*n* = 142) of the respondents were paired with a stranger that signed up for the same timeslot; the other half (*n* = 140) argued with a friend they brought with them to the lab. First, participants completed a questionnaire containing the argumentativeness, the verbal aggressiveness, and the argument frames scales. Then, they were randomly assigned to one of two possible argument topics (i.e., "Texting while driving should be harshly penalized" or "Student athletes should be paid"), and then randomly assigned to be in favour or against that topic. Afterwards, participants were given some time to prepare what they wanted to say during the argument with the other person, and asked to indicate their dialogue orientations (i.e., what dialogues they intended to use while arguing with the other person).

3.3 Measures

All items were measured with a 1-7 Likert-type scale ranging from 1= *strongly disagree* to 7 = *strongly agree*. *Argumentativeness* was measured with 20 items from Infante and Rancer (1982). Ten items measured argument approach and ten items measured argument avoidance. *Verbal aggressiveness* was measured with 20 items as well, from Infante and Wigley (1986). Similarly, ten items measured constructive, pro-social tendencies and ten items measured destructive, anti-social tendencies. *Argument frames* were measured with Hample and Irions' (2015) scales. Eight items measured identity, six items measured dominance, four items measured play, ten items measured civility, eight items measured utility, and ten items measured blurting. *Dialogue orientations* were measured with Cionea and Hample's (2014) scales. The persuasive, negotiation, eristic, inquiry, and deliberation dialogue orientations were measured with six items each, whereas the information giving and information seeking dialogue orientations were measured with four items each. Means, standard deviations, and Cronbach's alpha reliabilities are included in Table 1.

Variable	Reliability	Strangers			Friends		
		N	M	SD	N	M	SD
Argum. approach	.88	142	4.41	1.02	140	4.42	1.03
Argum. avoid	.87	142	4.10	1.04	140	4.26	1.11
Verbal agg. pro[1]	.81	142	4.83	0.88	140	4.78	0.96
Verbal agg. anti	.85	142	3.04	1.00	140	3.25	1.00
Identity[2]	.71	142	4.91	0.91	140	4.85	0.71
Dominance	.78	142	3.32	1.14	140	3.40	1.19
Play	.77	142	3.55	1.40	140	3.65	1.37
Civility[3]	.74	142	4.86	0.79	140	4.81	0.77
Utility	.70	142	4.13	0.82	140	3.96	0.73
Blurting	.75	142	3.85	0.84	140	3.92	0.88
Persuasive dialogue	.70	142	5.27	0.74	140	5.31	0.77
Negotiation dialogue	.81	142	4.65	0.87	140	4.99	0.81
Info. seeking dialogue	.83	142	5.22	1.02	140	5.49	0.84
Info. giving dialogue	.77	142	5.44	0.93	140	5.54	0.84
Eristic dialogue	.71	142	3.01	0.89	140	3.26	0.99
Inquiry dialogue	.82	142	5.33	0.82	140	5.56	0.73
Delib. dialogue	.79	142	5.18	0.87	140	5.37	0.76

Table 1 – Descriptive Statistics and Reliability Estimates for Study Variables

4. RESULTS

To answer RQ1, we examined whether one's own argumentation traits, argument frames, or dialogue orientations were correlated with the other person's argumentation traits, argument frames, and dialogue orientations in the case of strangers and in the case of friends. For strangers, we found weak significant correlations between one's own argumentativeness avoidance and the other person's argumentativeness approach ($r = -.25$, $p < .05$) and avoidance ($r = .26$, $p < .05$), as well as between one's own constructive verbal aggressiveness and the other person's constructive verbal aggressiveness ($r = .24$, $p < .05$). In terms of frames, one's own utility frame was positively correlated to the other person's play frame ($r = .27$, $p < .05$). No significant correlations emerged for dialogue orientations.

For friends, we found more substantial correlations, as follows: One's own argumentativeness approach correlated negatively with the

---

[1] Item 10 dropped from this scale

[2] Item 6 dropped from this scale.

[3] Item 2 dropped from this scale

other person's constructive verbal aggressiveness ($r = -.34$, $p < .01$); one's own constructive verbal aggressiveness correlated moderately with the other person's constructive verbal aggressiveness ($r = .44$, $p < .01$) and with the other person's destructive verbal aggressiveness ($r = -.30$, $p < .05$); and one's destructive verbal aggressiveness correlated negatively with the other person's constructive verbal aggressiveness ($r = -.35$, $p < .01$). Furthermore, one's identity frame correlated positively with the other person's dominance frame ($r = .25$, $p < .05$) and with the other person's play frame ($r = .24$, $p < .05$), and the two arguers' play frames correlated positively ($r = .40$, $p < .01$). Finally, the other person's negotiation dialogue orientation correlated positively with one's own information seeking dialogue orientation ($r = .29$, $p < .05$), one's own deliberation dialogue orientation ($r = .26$, $p < .05$), and one's own inquiry dialogue orientation ($r = .23$, $p < .05$).

To answer RQ2, we conducted a multivariate analysis of variance (MANOVA) with argumentation traits, argument frames, and dialogue orientations as dependent variables, and argumentation partner (stranger or friend), the topic of the argument (texting or students athletes), and the sex of the dyad (male or female) as independent variables. Significant main effects occurred for partner in respect to destructive verbal aggressiveness, the negotiation dialogue orientation, the information seeking dialogue orientation, the eristic dialogue orientation, and the inquiry dialogue orientation (see Table 2). In all these cases, scores for friends were significantly higher than those for strangers.

Variable	F-test	p-value	Partial η^2	Friends		Strangers	
				M	SD	M	SD
Verbal agg. anti	4.63	< .05	.02	3.30	0.20	3.03	0.08
Negotiation dialogue	11.14	< .01	.04	4.85	0.18	4.65	0.07
Info seeking dialogue	5.64	< .05	.02	5.48	0.20	5.21	0.08
Eristic dialogue	6.28	< .05	.02	3.26	0.20	3.00	0.08
Inquiry dialogue	5.84	< .05	.02	5.61	0.16	5.33	0.07

Table 2 – RQ2 Main Effects for Argument Partner

Some significant main effects also occurred for the topic of the argument: constructive verbal aggressiveness and the information seeking dialogue orientation (see Table 3). In both cases, scores were significantly higher when the topic of the argument was "Student athletes should be paid" than when it was "Texting while driving should be harshly penalized".

Variable	F-test	p-value	Partial η^2	Texting M	SD	Athletes M	SD
Verbal agg. pro	6.53	< .01	.02	4.92	0.07	4.98	0.18
Info seeking dialogue	5.32	< .05	.02	5.22	0.08	5.48	0.20

Table 3 – RQ2 Main Effects for Argument Topic

Most main effects occurred for sex of the dyad, with significant differences in the following cases: both dimensions of argumentativeness and verbal aggressiveness, all frames except blurting, the persuasive dialogue orientation, and the eristic dialogue orientation (see Table 4).

Variable	F-test	p-value	Partial η^2	Male M	Male SD	Female M	Female SD
Argum. approach	12.02	<.001	.08	4.70	0.08	4.13	0.08
Argum. avoid	26.76	<.001	.16	3.77	0.08	4.59	0.08
Verbal agg. pro	27.23	<.001	.17	4.44	0.07	5.15	0.07
Verbal agg. anti	21.41	<.001	.14	3.52	0.08	2.79	0.08
Identity frame	6.55	<.01	.05	5.03	0.06	4.72	0.06
Dominance frame	12.72	<.001	.09	3.70	0.10	3.03	0.10
Play frame	28.75	<.001	.17	4.18	0.11	3.04	0.10
Civility frame	4.48	<.05	.03	4.94	0.07	4.74	0.07
Utility frame	3.33	<.05	.02	4.16	0.07	3.93	0.07
Persuasion dialogue	3.04	<.05	.02	5.41	0.06	5.19	0.06
Eristic dialogue	3.29	<.05	.02	3.23	0.08	3.00	0.08

Table 4 – *RQ2 Main Effects for Dyad Sex*

Finally, an interaction effect between one's argumentation partner and sex of the dyad also emerged, with significant differences in respect to: constructive verbal aggressiveness, the dominance frame, and the play frame (see Table 5 and Figures 2-4).

Variable	F-test	p-value	Partial η^2	Stranger				Friend			
				Male		Female		Male		Female	
				M	SD	M	SD	M	SD	M	SD
Verbal agg.	8.84	<.01	.03	4.65	0.10	5.06	0.10	4.24	0.10	5.24	0.10
Domin. pro	4.98	<.05	.02	3.50	0.13	3.12	0.14	3.91	0.14	2.94	0.13
Play	4.13	<.05	.02	3.94	0.15	3.11	0.15	4.42	0.16	2.97	0.15

Table 5 – RQ2 Interaction Effects for Argument Partner and Dyad Sex

Argumentation traits, frames, and dialogue orientations 143

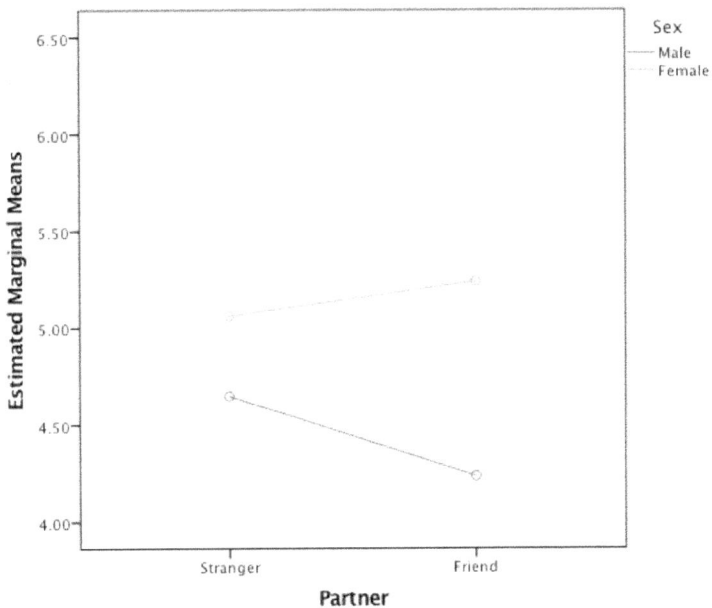

Figure 2 – Constructive verbal aggressiveness interaction plot

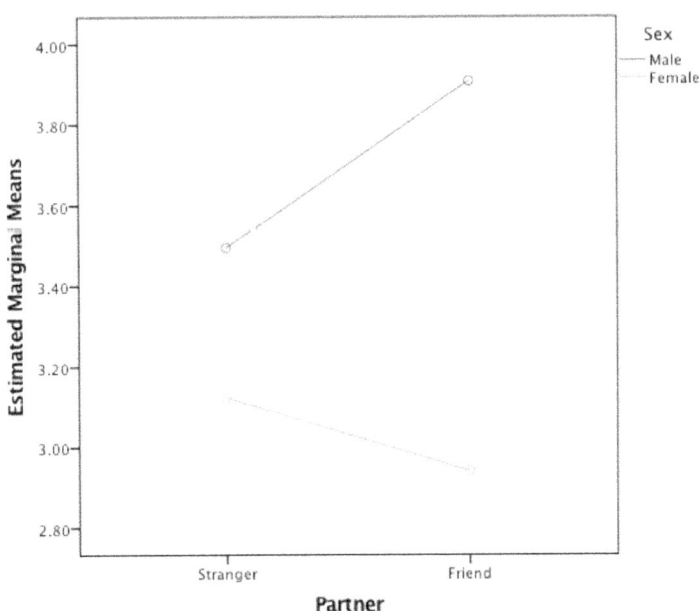

Figure 3 – Dominance frame interaction plot

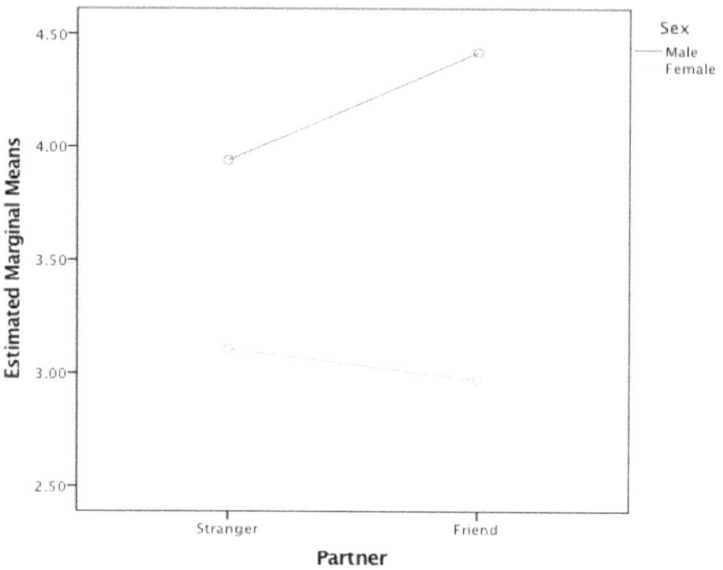

Figure 4 – Play frame interaction plot

To test RQ3, we conducted a series of linear regressions in which the dialogue orientations were entered as dependent variables, one at a time, and argumentation partner (stranger or friends, dummy-coded), one's own argumentation traits, and one's own argument frames were entered as independent variables in the first step. In the second step, we entered the other person's argumentation traits and argument frames.

The persuasive dialogue orientation was not significantly predicted by any of the independent variables. The negotiation dialogue orientation was predicted positively by one's own argumentativeness approach ($\beta = 0.25$, $p < .01$), and by one's own constructive verbal aggressiveness ($\beta = 0.33$, $p < .01$), $F(11, 130) = 3.74$, $p < .001$, adjusted $R^2 = .18$. Adding the other person's traits and frames did not significantly improve the model, and none of the added variables significantly predicted negotiation.

The information seeking dialogue orientation was predicted positively by one's own argumentativeness approach ($\beta = 0.35$, $p < .05$), and negatively by one's own destructive verbal aggressiveness ($\beta = -0.29$, $p < .05$), $F(11, 130) = 2.44$, $p < .01$, adjusted $R^2 = .10$. Adding the other person's traits and frames did not significantly improve the model nor did any of the added variables significantly predict information seeking.

The information giving dialogue orientation was predicted only marginally by one's own constructive verbal aggressiveness ($\beta = .22$, $p = .054$), $F(11, 130) = 2.10$, $p < .05$, adjusted $R^2 = .08$. No significant changes emerged once the other person's traits and frames were added.

The eristic dialogue orientation was predicted positively by two argument frames: play ($\beta = 0.24$, $p < .05$) and blurting ($\beta = .20$, $p < .05$), $F(11, 130) = 4.74$, $p < .001$, adjusted $R^2 = .23$. Adding the other person's argumentation traits and frames did not significantly change the model.

The deliberation dialogue orientation was predicted positively by one's argumentation partner ($\beta = .19$, $p < .05$) and by one's own constructive verbal aggressiveness ($\beta = .29$, $p < .01$), $F(11, 130) = 2.99$, $p < .01$, adjusted $R^2 = .13$. No significant changes occurred once the other person's traits and frames were added.

Finally, the inquiry dialogue was predicted positively by one's argumentation partner ($\beta = .20$, $p < .05$) and marginally by one's own constructive verbal aggressiveness ($\beta = .21$, $p = .06$), $F(11, 130) = 2.26$, $p < .05$, adjusted $R^2 = .09$. Adding the other person's traits and frames did not significantly change the model.

5. DISCUSSION

In this study, we examined possible relationships between a person's argumentation traits and frames and their dialogue orientations, in the case of strangers and friends who prepared to argue with each other about one of two topics. Our results, using actual (rather than recalled) arguments, have revealed several new findings that help refine our understanding of how dialogue orientations function and what influences them.

First, there is some weak evidence for the idea that friends may match each other on some traits and frames. Friends had stronger correlations than strangers between the constructive sub-dimension of verbal aggressiveness. Also, various combinations of the verbal aggressiveness sub-dimensions (e.g., one's constructive and the other person's destructive verbal aggressiveness) were significant. Additional evidence for this proposition comes from the finding that friends matched on the play frame – i.e., both believed that arguing was to be used for amusement or as a fun way to pass time. In a friendship, arguing for fun may be a shared activity in which friends engage as part of their relational maintenance. In addition, some dialogue orientations were also related in the case of friends – the more one sought to find out information from the other person, deliberate with the other person, or engage in inquiry, the more the other person was willing to negotiate. It appears that friends may be willing to accommodate when arguing with each other, especially in the case of dialogues that presuppose working together (i.e., deliberation and inquiry). In all, a similarity rather than complementary relationship seems to exist between friends in respect to arguing.

However, the findings regarding traits and frames should be interpreted with caution. Strangers, too, exhibited some of these correlations, albeit smaller in magnitude. For example, the argumentativeness sub-dimensions were correlated in the case of strangers (which was not the case for friends), as were strangers' constructive verbal aggressiveness scores. These significant correlations could potentially be due to the fact that, although strangers, participants were all students at the same university, in the same geographical area. Therefore, they may share a code of conduct and similar upbringing that has taught them to be polite and avoid arguments in interactions with strangers.

Second, we were also interested in differences in argumentation traits and frames as well as dialogue orientations depending on situational features (in this case, dyad sex, topic of the argument, and one's argumentation partner). In respect to traits and frames, most differences emerged when the sex of the dyad was taken into account as an explanatory variable. Traits differed by sex in that results indicated men were more argumentative and more verbally aggressive than women, which supports previous research findings. We add to these findings results pertaining to argument frames: men scored significantly higher than women on five of the six argument frames. Men believed arguing to be connected more closely to their identity [confirming Hample and Cionea's (2016) findings], to be more about dominance, having fun [also as Hample and Cionea (2016) found], and achieving instrumental needs than women did; men also believed arguing to be more civil than women did. Thus, these results indicate that, overall, the men in our study framed arguments in more positive, utilitarian terms than did women. Along the same lines, men also indicated that they intended to pursue the persuasion dialogue and the eristic dialogue in their subsequent argument in the lab more than women did. These dialogue orientations are consistent with the trait differences found – if men are more argumentative and verbally aggressive, it is reasonable they would capitalize on these tendencies in an argument by trying to convince the other person of their point of view (i.e., persuasion dialogue) or trying to quarrel more with the other person (i.e., eristic dialogue). The different socialization of men and women may explain these findings in that men may be encouraged to be more assertive and express themselves more than women are (Kosberg & Rancer, 1998).

Topic of the argument, one of the situational features we took into account for this study, yielded only two differences. In both cases, constructive verbal aggressiveness and the intent to pursue the information seeking dialogue were higher when the topic was "Student athletes should be paid" than when it was "Texting while driving should be harshly penalized". Sports have a long-standing tradition and important place in the culture of the university where data were collected. It may be the case that some participants felt the need to educate others about a topic that was important to them, and, in doing so, they relied on constructive verbal aggressive strategies. Other participants may know little about the status of student athletes at the university and may have wanted to gain more information (potentially from another person who was perceived as more knowledgeable, such as a student athlete himself/herself) from their argumentation partner.

Most differences in dialogue orientations emerged based on whether one's argumentation partner was a friend or a stranger. When participants expected to argue with a friend, they indicated they would pursue the negotiation dialogue, the information seeking dialogue, the inquiry dialogue, but also the eristic dialogue more than when they expected to argue with a stranger. Friends may feel more comfortable arguing with each other than strangers do; hence they feel more at ease with bargaining for a solution or examining critically the proposals that each side puts forth. Being more comfortable also means they can let out their feelings more freely during an argument, without being concerned about upsetting the other person or appearing aggressive or rude. Thus, one of the conclusions that we can draw based on these results is that one's dialogue orientations appear to change as the relationship with the other person becomes more intimate. This makes sense because as one gets to know the other person, one understands the affordances and boundaries of arguing in that relationship.

In addition, some interesting results emerged based on one's argumentation partner and sex of the dyad. Constructive verbal aggressiveness decreased in the case of men when they argued with friends instead of arguing with strangers, but it increased in the case of women who were arguing with friends instead of arguing with strangers. Thus, men appear to be more forceful and aggressive with strangers, which, again, may be explained by previous socialization (Kosberg & Rancer, 1998). These effects decrease in the case of friends, perhaps because greater care is taken to not criticize, insult, or attack a friend's self-concept during an argument. Women, on the other hand, appear to move from more restrained, polite interactions with strangers to more challenges and aggressive remarks with friends. Thus, in the case of women, familiarity with the other person functions to legitimize more aggression, potentially because the friendship is perceived as a safer environment in which to express such tendencies.

Furthermore, interaction effects occurred for two of the argument frames: dominance and play. Male friends believed arguing can be used to exhibit dominance more than male strangers, a result that was flipped in the case of women (i.e., women strangers scored higher than women friends). The same pattern occurred for play, meaning male friends viewed arguing as a fun activity more so than male strangers, with women strangers viewing arguing as play more so than women friends. Together, these results suggest that, compared with women friends, male friends are more competitive and focused on winning the exchange in an argument with the other person, but potentially interpret arguing as a playful endeavor or a way to re-enact fun, non-serious banter that characterizes their relational talk.

Finally, our investigations revealed that one's own argumentation traits and frames modestly predicted the dialogue orientation one was likely to resort to in an upcoming argument, whereas the other person's traits and frames exerted hardly any influence on one's planned dialogues. Specifically, argumentativeness approach predicted reliance on negotiation and information seeking positively; the more one enjoyed arguing, the more one intended to negotiate with the other person and ask the other person more questions. Constructive verbal aggressiveness predicted negotiation and deliberation, with marginal significant predictions of information seeking and inquiry. Individuals who took care when attacking the other person or who refrained from engaging in aggressive arguments were more likely to indicate they would negotiate and deliberate with the other person. Destructive verbal aggressiveness predicted information seeking negatively in that individuals who indicated they liked to attack the other person, to tell that person off, or to insult the other were not likely to plan on seeking out information from the other person during their argument. These results are relatively similar to Cionea and Hample (2014) who also found that constructive verbal aggressiveness predicted negotiation, information seeking, deliberation, and inquiry.

Argument frames worked to predict only the eristic dialogue. One was likely to report wanting to quarrel, vent, and let feelings out when indicating that arguing was a fun activity (in this case the eristic dialogue may be perceived as a friendly banter), and when one blurted during an argument, saying what was on their mind, without a filter (in this case, letting one's feelings or frustration about the topic of the argument out may be what motivated a person to choose this dialogue). Finally, arguing with a friend positively predicted intent to use inquiry and deliberation while arguing. Both of these two dialogues stem from an open problem; it appears individuals perceive they can tackle such an open situation better with friends than with strangers. This preference is likely due to their familiarity with the other person and potentially due to previous success throughout the friendship with resolving such situations.

Overall, the results of our study allow us to draw three important conclusions for how individuals argue in everyday, actual arguments. First, our study confirms a long line of research that has documented sex differences in arguing behaviours, adding findings regarding differences based on the relationship between arguing partners (stranger or friend). Second, the evidence that dialogues may be a trait-like characteristic is not supported in the current study. Instead, our results suggest one's dialogue orientations are a situational feature – individuals decide which orientation to rely on depending on whom the other person involved in the argument is and what the topic of the argument is. The sex differences found suggest men and women may rely on different dialogues, but it is unclear based only on our study whether this is a trait difference or whether other factors affect how men and women decide which dialogue to use. Finally, our study can offer only minimal support for the idea that friends match each other in their dialogue orientations, given that we found several matching effects in the case of strangers, too. Thus, the matching in dialogues idea proposed by Hample and Cionea (2016) needs further investigation in other studies.

REFERENCES

Cionea, I. A. (2011). Dialogue and interpersonal communication: How informal logic can enhance our understanding of the dynamics of close relationships. *Cogency, 3*(1), 93-105.

Cionea, I. A., & Hample, D. (2014). Dialogue types and argumentative behaviours. In B. J. Garssen, D. Godden, G. Mitchell, & A. F. Snoeck Henkemans (Eds.), *Proceedings of the 8th International Conference of the International Society for the Study of Argumentation* (pp. 245-256). Amsterdam: Sic Sat.

Cionea, I. A., Hample, D., & Fink, E. L. (2013). Dialogue types: A scale development study. In D. Mohammed & M. Lewiński (Eds.), *Virtues of*

Argumentation: Proceedings of the 10th international Conference of the Ontario Society for the Study of Argumentation (OSSA) (pp. 1-11). Windsor, ON: OSSA.

Hample, D. (2005). *Arguing: Exchanging reasons face to face*. Mahwah, NJ: Erlbaum.

Hample, D., & Cionea, I. A. (2016). Couple's dialogue orientations. In L. Benacquista & P. Bondy (Eds.), *Argumentation, Objectivity and Bias: Proceedings of the 11th International Conference of the Ontario Society for the Study of Argumentation*. Retrieved from http://scholar.uwindsor.ca/ossaarchive/OSSA11/papersandcommentaries/

Hample, D., & Irions, A. (2015). Arguing to display identity. *Argumentation, 29*(4), 389-416. doi: 10.1007/s10503-015-9351-9.

Infante, D. A., & Rancer, A. S. (1982). A conceptualization and measure of argumentativeness. *Journal of Personality Assessment, 46*(1), 72-80. doi: 10.1207/s15327752jpa4601_13.

Infante, D. A., Wall, C., Leap, C. J., & Danielson, K. (1984). Verbal aggression as a function of the receiver's argumentativeness. *Communication Research Reports, 1*(1), 33-37.

Infante, D. A., & Wigley, C. J. (1986). Verbal aggressiveness: An interpersonal model and measure. *Communication Monographs, 53*(1), 61-69. doi: 10.1080/03637758609376126.

Jordan-Jackson, F. F., Lin, Y., Rancer, A. S., & Infante, D. A. (2008). Perceptions of males and females' use of aggressive affirming and nonaffirming messages in an interpersonal dispute: You've come a long way baby? *Western Journal of Communication, 72*(3), 239-258.

Kosberg, R. L., & Rancer, A. S. (1998). Enhancing argumentativeness and argumentative behavior: The influence of gender and training. In L. Longmire & L. Merrill (Eds.), *Untying the tongue: Gender, power, and the word* (pp. 251-265). Westport, CT: Greenwood.

Nicotera, A. M., & Rancer, A. S. (1994). The influence of sex on self-perceptions and social stereotyping of aggressive communication. *Western Journal of Communication, 58*(4), 283-308. doi: 10.1080/10570319409374501.

Walton, D. N. (1998). *The new dialectic: Conversational contexts of argument*. Toronto, Canada: University of Toronto Press.

Walton, D. N., & Krabbe, E. C. W. (1995). *Commitment in dialogue: Basic concepts of interpersonal reasoning*. Albany: State University of New York Press.

12

From Semi-Abstract Argumentation to Logical Consequence

ESTHER ANNA CORSI
Institut für Computersprachen, TU Wien, Austria
esther@logic.at

CHRISTIAN G. FERMÜLLER
Institut für Computersprachen, TU Wien, Austria
chrisf@logic.at

The concept of "argumentative consequence" is introduced, involving only the attack relations in Dung-style abstract argumentation frames. Collections of attack principles of different strength, referring to the logical structure of claims of arguments, lead to new characterizations of classical and nonclassical consequence relations. In this manner systematic relations between structural constraints on abstract argumentation frames, sequent rules, and nondeterministic matrix semantics for corresponding calculi emerge.

KEYWORDS: abstract argumentation, sequent calculus

1. INTRODUCTION

Dung's landmark paper (Dung, 1995), introducing abstract argumentation to Artificial Intelligence, spawned a large body of research on computational aspects of argumentation. Part of its success can be explained by the simplicity of the main concept: an abstract argumentation frame (AF) is just a directed graph, where vertices are arguments and edges represent attacks between same. Various so-called 'semantics' intend to single out mutually attack-free sets of arguments, that are in some sense maximal and jointly defend attacking arguments. However, since only graph properties are used, but no attention is paid to the *logical structure* of the claims of the involved arguments, some oddities emerge. It may even happen that logically contradictory claims are not recognized as attacking each other. Also other forms of attacks that are logically implicit are not taken into consideration in Dung-style semantics. For example, one may want to see the following principle respected: if an argument with claim "A" is attacked, then this also entails an implicit attack on any argument which features a claim of the form "A and B". The purpose of this paper is to introduce "logical attack principles", like the just mentioned one in a simple formal setting, where the focus is solely on the (propositional) logical form of claims of - explicit or implicit - arguments. In particular, we are interested in a corresponding notion of logical consequence, that is not based on 'truth in a model', but rather treats attacking arguments like falsifying interpretations. As we will show, attack principles that are apt to recover classical logic in this manner, are unreasonably strong. More interestingly, we demonstrate that a collection of weaker attack principles, justifiable with respect to a simple modal interpretation of attack, leads to a notion of 'argumentative consequence' that matches a variant of classical sequent calculus (Gentzen, 1935), where some of the rules are omitted. Following recent work on canonical sequent systems (Avron & Lev, 2005; Avron & Zamansky, 2011) this in turn leads to a characterization of our 'logic of argumentation' in terms of nondeterministic matrix semantics. For the proofs of theorems and propositions please refer to (Corsi & Fermüller, 2017).

2. SEMI-ABSTRACT ARGUMENTATION

As already indicated, we look at arguments as logical propositions. Although one can find several examples in the literature, where arguments indeed just consist in single propositions (see, e.g., many articles in (Rahwan & Simari, 2009)), it is more common to distinguish explicitly between the *support* and the *claim* of an argument. Whereas the latter usually is indeed a (possibly logical complex) sentence, the support may consist of several statements, rules, or even of small theories, including default rules, see, e.g., (Besnard & Hunter, 2008). Here we focus on *semi-abstract argumentation*, which identifies claims of arguments with (interpreted) logical formulas, but ignores supports. More formally, let *PV* be an infinite set of propositional variables and define the set of propositional formulas $\mathcal{PL}$ over *PV* by

$$\mathcal{F} ::= \mathcal{F} \vee \mathcal{F} \mid \mathcal{F} \wedge \mathcal{F} \mid \mathcal{F} \supset \mathcal{F} \mid \neg \mathcal{F} \mid \mathcal{PV}$$

where $\mathcal{F}$ and $\mathcal{PV}$ are used as meta-variables for formulas and propositional variables, respectively.

Definition 1. *A semi-abstract argumentation frame (SAF) is a directed graph (A, R→), where each vertex a ∈ $\mathcal{A}$ is labelled by a formula of $\mathcal{PL}$, representing the claim of argument a, and the edges (R→) represent the attack relation between arguments.*

Note that an SAF is like an ordinary abstract argumentation frame as introduced by Dung (1995), except for attaching a formula (its claim) to each argument. We say that F *attacks* G and write $F \rightarrow G$ if there is an edge from an argument labelled by F to one labeled by G.[1] But the reader should be aware of the fact that, in general, $F \rightarrow G$ stands for "an argument with claim G is attacked by some argument with claim F". We abbreviate "not $F \rightarrow G$" by $F \nrightarrow G$.

Example (1). Consider the following statements:
(P) "The overall prosperity in society increases."
(W) "Warlike conflicts about energy resources arise."
(C) "The level of CO emissions is getting dangerously high."
(E) "Awareness regarding environmental problems increases."
An argumentation frame may consist of arguments, where the claims consist in these statements or in some simple logical compounds thereof. A concrete corresponding SAF S_E = (A, R→) is given by $\mathcal{A}$ = {P, E, W, P ⊃ C, E ∨ C, P ∧ C} and R→ = {E→P ⊃ C, W→E ∨ C, W→P ∧ C, E ∨ C→P}.

3. LOGICAL ATTACK PRINCIPLES

It seems reasonable to expect that an argument that attacks G also attacks claims that logically entail G. In our notation, this amounts to the following *general attack principle*:

(A) If $F \rightarrow G$ and $G' \models G$ then $F \rightarrow G'$.

Applied naively, principle **(A)** is problematic for at least two reasons. (1) We have not specified which logic the consequence relation ⊨ refers to. Classical logic may be a canonical choice, but we should not dismiss weaker logics, that are potentially more adequate in the context of defeasible reasoning, too quickly. (2) Even for classical propositional logic deciding logical consequence is computationally intractable, in

---

[1] The same formula may occur as claim of different arguments. Thus we (implicitly) refer to *occurrences* of formulas, rather than to formulas themselves when talking about attacks in a given SAF.

general. Arguably, a realistic model of argumentation might insist on constraining **(A)** to arguments G that *immediately* follow from G' in some appropriate sense. This motivates our focus on principles that follow already from *simple and transparent instances* of **(A)**:

(A. ∧) If $F \to A$ or $F \to B$ then $F \to A \wedge B$.
(A. ∨) If $F \to A \vee B$ then $F \to A$ and $F \to B$.
(A. ⊃) If $F \to A \supset B$ then $F \to B$.

These specific instances of **(A)** involve only very basic consequence claims, that are already valid in *minimal logic* (Troelstra & Schwichtenberg, 2000), i.e. in the positive fragment of intuitionistic logic (and in fact in even weaker logics).

> **Proposition 1.** *If "⊨" refers to minimal logic, then the general attack principle **(A)** entails the specific attack principles **(A. ∧)**, **(A. ∨)**, **(A.⊃)**.*

Principle **(A.⊃)** might be considered intuitively less obvious than **(A. ∧)** and **(A. ∨)**. Indeed, the above justification of **(A.⊃)** depends on the fact that $B \models A \supset B$ according to minimal logic and thus involves a logical principle that may be disputed, e.g., from the point of view of "relevant entailment" (see (Dunn & Restall, 2002)). Therefore we prefer to replace **(A.⊃)** by the following attack principle:

(B.⊃) If $F \to B$ and $F \not\to A$ then $F \to A \supset B$.

As we will see in Sections 5 and 6, **(B.⊃)** relates to a basic inference principle about implicative premises in logical consequence claims, that holds in a very wide range of logics. We have not yet specified any principle involving negation. Negation is often defined by $\neg F =_{df} F \supset \bot$, where $\bot$ is an atomic formula that signifies an elementary contradiction. But minimal logic treats $\bot$ just like an arbitrary propositional variable and thus does not give rise to any specific attack principle for negation. However the following principle seems intuitively plausible: If an argument attacks (an argument with claim) A then it does not simultaneously also attack the negation of A. In symbols:

(B.¬) If $F \to A$ then $F \not\to \neg A$.

We will show in Section 8 that the (weak) attack principles mentioned so far give raise to a logic that arises from dropping some logical rules from Gentzen's classical sequent calculus **LK**. But we are also interested in the question, which (stronger) attack principles have to be imposed on semi-abstract argumentation frames in order to recover ordinary classical logic. To this aim we introduce the following additional attack principles that are inverse to **(A. ∧)**, **(A. ∨)**, **(B.⊃)** and **(B.¬)**, respectively.

(C.∧) If $F \to A \wedge B$ then $F \to A$ or $F \to B$.
(C.∨) If $F \to A$ and $F \to B$ then $F \to A \vee B$.
(C.⊃) If $F \to A \supset B$ then $F \to B$ and $F \not\to A$.
(C.¬) If $F \not\to A$ then $F \to \neg A$.

Conditions like **(A. ∧)** seem to entail that the corresponding SAFs are infinite. However, we may *relativize* the attack principles to sets of formulas Γ (usually the set of claims of arguments of some finite argumentation frame). E.g.,

(A. ∧) For every $A, B, F \in \Gamma$: $F \to A$ or $F \to B$ implies $F \to A \wedge B$, if $A \wedge B \in \Gamma$.

In the following we will tacitly assume that attack principles are relativized to some (finite) set of formulas that will always be clear from the context.

4. ARGUMENTATIVE CONSEQUENCE

Viewing an argument attacking a certain claim F as a kind of counter-model to F suggests the following definition of consequence, that, in contrast to usual definitions of logical consequence, neither refers to truth values nor to interpretations in the usual (Tarskian) sense.

Definition 2. *F is an argumentative consequence of (the claims of) arguments $A_1, \ldots, A_n$ with respect to an SAF S ($A_1, \ldots, A_n \vDash_{arg}^{S} F$) if all arguments in S that attack F also attack A_i for some $i \in \{1, \ldots, n\}$.[2] For a set of SAFs $\mathcal{S}$ $A_1, \ldots, A_n \vDash_{arg}^{\mathcal{S}} F$ if $A_1, \ldots, A_n \vDash_{arg}^{S} F$ for all $S \in \mathcal{S}$*

To render this notion of consequence plausible from a logical perspective, the underlying SAFs should be rich enough to contain

---

[2] Note that, if we identify arguments with counter-models and if $\mathcal{S}$ contains all relevant counter-models, then argumentative consequence coincides with ordinary logical consequence: every counter-model of the conclusion must invalidate some premise.

(potential) arguments that feature also the subformulas of occurring formulas as claims. Moreover, we want these SAFs to satisfy at least some of our logical attack principles.

> **Definition 3**. *An SAF S is logically closed with respect to the set of formulas Γ if all subformulas of formulas in Γ occur as claims of some argument in S.*
> *Let $\mathcal{AP}$ be a set of attack principles, then F is an argumentative $\mathcal{AP}$- consequence of $A_1, \ldots, A_n$ ($A_1, \ldots, A_n \models_{arg}^{\mathcal{AP}} F$) if $A_1, \ldots, A_n \models_{arg}^{S} F$ for every SAF S that is logically closed with respect to $\{A_1, \ldots, A_n, F\}$ and moreover satisfies all (appropriately relativized) attack principles in $\mathcal{AP}$.*

We do *not* suggest that argumentation frames should always be logically closed. We rather view logical closure as an operation that augments a given SAF by "potential claims of arguments", which are *implicit* according to logical attack principles. Argumentative consequence thus refers to a "logical completion" of interpreted argumentation frames, rather than directly to arbitrarily given col- lections of arguments.

> Example (2). Continuing Example (1) of Section 2, we observe the SAF SE is almost, but not yet fully, logically closed with respect to the statements that appear as claims of arguments: we just have to add one more vertex with label C (for "Awareness about the need of environmental protection increases") to obtain logical closure.
> More interestingly, we may check which additional (implicit) attacks are induced by which of our attack principles: Since there is an argument with claim $E \vee C$ that attacks an argument with claim P, principle **(A. ∧)** stipulates that there is also an attack from $E \vee C$ to an argument with the stronger claim $P \wedge C$. In other words **(A. ∧)** induces the addition of the edge $E \vee C \to P \wedge C$. Note that this corresponds to the plausible assumption that an argument that attacks the claim that "the overall prosperity in society increases" also attacks the statement "The overall prosperity in society increases and (moreover) the level of CO_2 emissions is getting dangerously high". Similarly the principle **(A.∨)** stipulates that an argument with claim C (added for logical closure, as explained above) should be attacked by an argument with claim W, since such an argument already attacks the weaker claim $E \vee C$. The addition of $W \to C$, likewise induces the additional attack edge $W \to E$. Moreover, since we have $E \to P \supset C$, but $E \not\to P$, the principle **(B.⊃)** induces the addition of $E \to C$ to $R_\to$. The stronger

principle **(C.⊃)** would call for $E \to C$ even without $E \not\to P$. The strong conjunction principle **(C.∧)** demands that either $W \to P$ or $W \to C$. Both seem reasonable with respect to the intended interpretation of S_E. However, $W \to C$ is already present anyway if **(A.∨)** is imposed, as explained above. Likewise the strong disjunction principle **(C.∨)** is already satisfied.

In the next section we will investigate which collection $\mathcal{AP}$ of attack principles allows one to recover classical logical consequence as argumentative $\mathcal{AP}$-consequence. Gentzen's classical sequent calculus **LK** turns out to be a perfect tool for this task. Hence, we generalize the consequence relation to (disjunctive) sets of premises, as usual in proof theory. Moreover, we adopt the convention to identify finite lists and sets of formulas, and write, e.g., Γ, F for $\Gamma \cup \{F\}$.

Definition 4. *Let Γ and Δ be finite sets of formulas and let S be an SAF. Δ is an argumentative consequence of Γ with respect to an SAF S ($\Gamma \vDash_{arg}^{S} \Delta$) if all arguments in S that attack every $F \in \Delta$ attack at least some $G \in \Gamma$.*
The generalization to sets of SAFs and sets of attack principles is just as indicated above.

With respect to an SAF $S = (\mathcal{A}, R_{\to})$ we define:

$atts_S(F) =_{df} \{A \mid A \to F, A \in \mathbb{A}\}$,
$\overline{atts}_S(\Gamma) =_{df} \bigcup_{F \in \Gamma} atts_S(F)$,
$\underline{atts}_S(\Gamma) =_{df} \bigcap_{F \in \Gamma} atts_S(F)$.

The following simple facts will be useful below:

(a) $\Gamma \vDash_{arg}^{S} \Delta$ iff $\underline{atts}_S(\Delta) \subseteq \overline{atts}_S(\Gamma)$.
(b) $\overline{atts}_S(\Gamma, F) = \overline{atts}_S(\Gamma) \cup atts_S(F)$.
(c) $\underline{atts}_S(\Gamma, F) = \underline{atts}_S(\Gamma) \cap atts_S(F)$.

We will drop the index S if no ambiguity arises.

5. RELATING SEQUENT RULES AND ATTACK PRINCIPLES

In our version of Gentzen's classical sequent calculus **LK** (Gentzen, 1935), sequents are pairs of sets of formulas, written as $\Gamma \vdash \Delta$. Initial sequents (axioms) are of the form $A, \Gamma \vdash \Delta, A$. The logical rules (for introducing logical connectives) are as follows:

$$\frac{A,\Gamma \vdash \Delta}{\Gamma \vdash \Delta, \neg A} \; (\neg,r) \qquad \frac{\Gamma \vdash \Delta, A}{\neg A, \Gamma \vdash \Delta} \; (\neg,l)$$

$$\frac{\Gamma \vdash \Delta, A \quad \Gamma \vdash \Delta, B}{\Gamma \vdash \Delta, A \wedge B} \; (\wedge,r) \qquad \frac{A, B, \Gamma \vdash \Delta}{A \wedge B, \Gamma \vdash \Delta} \; (\wedge,l)$$

$$\frac{\Gamma \vdash \Delta, A, B}{\Gamma \vdash \Delta, A \vee B} \; (\vee,r) \qquad \frac{A, \Gamma \vdash \Delta \quad B, \Gamma \vdash \Delta}{A \vee B, \Gamma \vdash \Delta} \; (\vee,l)$$

$$\frac{A, \Gamma \vdash \Delta, B}{\Gamma \vdash \Delta, A \supset B} \; (\supset,r) \qquad \frac{\Gamma \vdash \Delta, A \quad B, \Gamma \vdash \Delta}{A \supset B, \Gamma \vdash \Delta} \; (\supset,l)$$

Note that we do not need to use structural rules: weakening is redundant because of the more general form of axioms compared to Gentzen's $A \vdash A$; contraction is eliminated because we treat sequents as pairs of *sets* of formulas. Moreover the calculus is cut-free complete with respect to the classical consequence relation $\vDash_{cl}$ (generalized to disjunctions of premises, as usual). We rely on the following well known facts (see, e.g., (Troelstra & Schwichtenberg, 2000).)

Proposition 2. $\Gamma \vdash \Delta$ *is derivable in* **LK** *iff* $\Gamma \vDash_{cl} \Delta$.

Proposition 3. (e.g., Troelstra & Schwichtenberg, 2000), Proposition 3.5.4) *The rules of* **LK** *are invertible; i.e., if a sequent* $\Gamma \vdash \Delta$ *is derivable and* $\Gamma \vdash \Delta$ *is an instance of a lower sequence of an* **LK***-rule then the corresponding instance(s) of the upper sequent(s) is (are) derivable, too.*

Let CAP consist of the attack principles **(A.∧)**, **(A.∨)**, **(B.⊃)**, **(B.¬)**, **(C.∧)**, **(C.∨)**, **(C.⊃)**, and **(C.¬)**. We first show that **LK** is sound with respect to argumentative CAP-consequence.

Theorem 1. *If* $\Gamma \vdash \Delta$ *is derivable in* **LK** *then* $\Gamma \vDash_{arg}^{CAP} \Delta$.

To show the completeness of **LK** with respect to argumentative CAP-consequence, we rely on the invertibility of the logical rules (Proposition 3).

Theorem 2. *If* $\Gamma \vDash_{arg}^{CAP} \Delta$ *then* $\Gamma \vdash \Delta$ *is derivable in* **LK**.

6. SORTING OUT ATTACK PRINCIPLES

We have seen that the collection CAP of attack principles leads to a characterization of classical logical consequence, that replaces model theoretic (Tarskian) semantics by a reference to specific structural properties of logically closed semi-abstract argumentation frames. Remember that classical logic is the top element in the lattice of all possible logics over the language $\mathcal{PL}$. (As usual, we identify a logic with a set of formulas that is closed under substitution and modus ponens, here.) Therefore it is hardly surprising that some of the principles in CAP might be considered too demanding to be adequate for models of logical argumentation. Consider for example **(C.¬)**: it says that any argument that does not attack a claim A can be understood as an argument that attacks $\neg A$. In other words every argument has to attack either A or $\neg A$. This is hardly plausible and should be contrasted with the inverse principle **(B.¬)**, which just stipulates that no argument attacks A and $\neg A$, simultaneously. We have justified **(A.∨)** and **(A.∧)** in Section 3 as immediate instances of the general principle that, if an argument attacks a claim A, then it (implicitly) also attacks claims from which A logically follows. But the plausibility of the inverse principles **(C.∨)** and **(C.∧)** remains in question. Intuitively, it seems justifiable to stipulate that an argument that attacks both, A and B, attacks also $A \vee B$ **(C.∨)**. However the requirement that any argument attacking a conjunction must attack also at least one of the conjuncts intuitively seems too strong: think of the instance $A \wedge \neg A$, against which an agent presumably may have a reasonable (general) argument, without knowing an argument that attacks either A or $\neg A$.

Rather than to simply appeal to pre-theoretic intuitions, as just outlined, we want to present a simple formal interpretation of attacks involving logically compound claims, that supports some, but not all of the attack principles in CAP. To this aim we employ standard modal logic and refer to *Kripke interpretations* $\langle W, R, V \rangle$, where W is a non-empty set of *states*, $R \subseteq W \times W$ the *accessibility relation*, and V a valuation $V: W \times PV \to \{\mathbf{t}, \mathbf{f}\}$ that assigns a truth value to each propositional variable in each state. The language $\mathcal{PL}$ is enriched by a modal operator $\Box$ and its dual $\Diamond = \neg \Box \neg$. The valuation V is extended from propositional variables to classical formulas (elements of $\mathcal{PL}$) as usual. For $\Box$ we have

$$V(w, \Box F) = \mathbf{t} \text{ iff } \forall v: wRv \text{ implies } V(v, F) = \mathbf{t}.$$

A Kripke interpretation $\langle W, R, V\rangle$ is a *model* of formula F if $V(w, F) = \mathbf{t}$ for all $w \in W$. F is a $\mathcal{K}$-consequence of $G_1, \ldots, G_n$ with respect to a class $\mathcal{K}$ of Kripke interpretations if every model of $G_1, \ldots, G_n$ is also a model of F.

We view the states W as possible states of affairs and interpret wRv as "v is a possible alternative from the viewpoint of w". An attack is considered to involve the claims of the involved arguments in two ways: (1) the claim of the attacking argument is asserted to hold in all alternatives; (2) the negation of the claim of the attacked argument is asserted to hold there. Accordingly, we define:

Definition 5. *For all $F, G \in \mathcal{PL}$:*
- $\iota(F \to G) =_{df} \Box\,(F \wedge \neg G)$;
- $\iota(F \nrightarrow G) =_{df} \Diamond(\neg F \vee G)$ *(or, equivalently, $\neg\Box(F \wedge \neg G)$).*

Recall that attack principles are implications between (disjunctions or conjunctions) of assertions of the form $F \to G$ or $F \nrightarrow G$. We call an attack principle $\mathcal{K}$-*justified* if the implication translates into a valid $\mathcal{K}$-consequence claim via ι.

We have not yet imposed any restriction on Kripke interpretations. The intended interpretation of the accessibility relation R as "possible alternative" might suggest that R is an equivalence relation, or at least reflective. However we will only impose the weaker condition of seriality. Let $\mathcal{D}$ be the corresponding class of Kripke interpretations, where for every $w \in W$ there is a $v \in W$ such that wRv.

Theorem 3. *The attack principles* **(A.∧)**, **(A.∨)**, **(C.∨)**, **(C.⊃)**, *and* **(B.¬)** *are all $\mathcal{D}$-justified.*

Theorem 4. *The attack principles* **(C.∧)**, **(C.¬)**, *and* **(B.⊃)** *are not $\mathcal{D}$-justified.*

To sum up, we have seen that our (admittedly rather unsophisticated and coarse) modal interpretation of the attack relation supports a formal justification of the collection of attack principles MAP = (**A.∧**), (**A.∨**), (**C.∨**), (**C.⊃**), (**B.¬**)}. Moreover, the modal interpretation allows us to reject the attack principles (**C.∧**) and (**C.¬**), thus suggesting that a corresponding "logic of argumentation" should be weaker than classical logic.

Of course, the interpretation of the attack relation using modalities is not unique. We briefly discuss three alternatives.

Definition 6. *For all $F, G \in \mathcal{PL}$:*
- $\iota_1(F \to G) =_{df} \Diamond (F \land \neg G);$
- $\iota_2(F \to G) =_{df} \Diamond (\neg F \lor \neg G);$
- $\iota_3(F \to G) =_{df} \Box (\neg F \lor \neg G);$
- $\iota_1(F \nrightarrow G) =_{df} \Box (\neg F \lor G).$
- $\iota_2(F \nrightarrow G) =_{df} \Box (F \land G).$
- $\iota_3(F \nrightarrow G) =_{df} \Diamond (F \land G).$

The interpretation ι_1 is similar to ι, but less demanding because (1) the claim of the attacking argument is asserted to hold in just one of the alternatives and (2) the negation of the claim of the attacked argument is asserted to hold only there. Interpretation ι_2 suggests that $F \to G$ means that there is at least one possible state in which $F \supset \neg G$ (equivalently: $\neg F \lor \neg G$) holds; whereas according to ι_3 all possible states must be of this form. These alternative interpretations of the attack relation (abstracted to the claims of the argument) are arguably more problematic than the interpretation ι suggested in Definition 5. This is also witnessed by the attack principles that are justified or rejected by the respective interpretations: The set of attack principles justified by ι_1 is MAP$_1$={**(A.∧)**, **(A.∨)**, **(B.⊃)**, **(C.∧)**}; ι_2 justifies MAP$_2$={**(A.∧)**, **(A.∨)**, **(B.⊃)**, **(C.∧)**, **(C.¬)**}; whereas ι_3 is extremely demanding and rejects all of our attack principles except **(A.∧)** and **(A.∨)**.

7. A 'LOGIC OF ARGUMENTATION'

We have seen that the attack principles in MAP are more plausible than the collection CAP that induces classical logical consequence. Thus the question arises, whether argumentative consequence relative to MAP can be characterized in a similar manner. We provide a positive by showing that $\vDash_{arg}^{CAP}$ matches the sequent calculus **LM**, that arises from dropping the rules $(\neg, l)$, $(\land, r)$, and $(\supset, l)$ from **LK**.

Theorem 5. $\Gamma \vdash \Delta$ *is derivable in* **LM** *iff* $\Gamma \vDash_{arg}^{CAP} \Delta$.

8. NONDETERMINISTIC MATRICES

We have identified a 'logic of argumentation' with a consequence relation arising from certain plausible principles about logically closed collections of (claims of) arguments and managed to characterize this logic in terms of a variant of Gentzen's classical sequent calculus **LK**, where some of the logical rules have been discarded. We finally ask whether our logic appears already in a different context, pointing to a different type of semantics. A positive answer is provided by the theory of *canonical signed calculi* and *nondeterministic matrix semantics* (see (Avron & Lev, 2005), (Avron & Zamansky, 2011)).

Definition 7. *A classical Nmatrix $\mathcal{N}$ consists in a function $\tilde{\neg}$: {t, f} $\to 2^{\{t,f\}} \setminus \emptyset$ and a function $\tilde{\odot}$: $\{t,f\}^2 \to 2^{\{t,f\}}$ for each $\odot \in \{\wedge, \vee, \supset\}$.*

A corresponding dynamic valuation *is a function $\tilde{v}_\mathcal{N}$: $\mathcal{PL} \to$ {t, f} such that $\tilde{v}_\mathcal{N}(\neg A) \in \tilde{\neg}(\tilde{v}_\mathcal{N}(A))$ and $\tilde{v}_\mathcal{N}(A \odot B) \in \tilde{\odot}(\tilde{v}_\mathcal{N}(A), \tilde{v}_\mathcal{N}(B))$ for $\odot \in \{\wedge, \vee, \supset\}$. $\tilde{v}_\mathcal{N}$ is a model of $\tilde{v}_\mathcal{N}(A) =$ t; it is a model of Γ if it is a model of every $A \in \Gamma$.*

Δ is a dynamical consequence *of Γ with respect to $\mathcal{N}$, written $\Gamma \vDash^{\mathcal{N}}_{dyn} \Delta$ if every model of Γ is a model of some $A \in \Delta$.*

Consider the following classical Nmatrix $\mathcal{M}_{\mathsf{MAP}}$:

	$\tilde{\wedge}$	$\tilde{\vee}$	$\tilde{\supset}$
t t	{t, f}	{t}	{t}
t f	{f}	{t}	{t, f}
f t	{f}	{t}	{t}
f f	{f}	{f}	{t}

	$\tilde{\neg}$
t	{t, f}
f	{t}

The following is just an instance of Theorem 62 of (Avron & Zamansky, 2011.)

Corollary 1. *$\Gamma \vdash \Delta$ is derivable in* **LM** *iff $\Gamma \vDash^{\mathcal{M}_{\mathsf{MAP}}}_{dyn} \Delta$.*

Combining this observation with Theorem 5, we have thus connected argumentative consequence with respect to the attack principles in **MAP** with logical consequence defined with respect to a specific nondeterministic valuation. (This type of analysis can straightforwardly be extended to, e.g., the collection **MFAP** of principles, Section 6.)

9. CONCLUSION

An impressive amount of recent literature focuses on Dung's abstract argumentation frames (Dung, 1995). The conceptual simplicity and flexibility of this notion is certainly instrumental for its success from a computational point of view. However certain logical aspects of Dung-style argumentation theory remain largely unexplored. Our 'logical attack principles' and the resulting notion of 'argumentative consequence' is intended to link Dung-style argumentation with the semantics and proof theory of certain types of non-classical logics.

The above results should by no means considered definitive or deflect attention from other logical aspects of argumentation. In particular, we have ignored here the important aspect of defeasibility of arguments that should induce a non-monotonic consequence relation, rather than our monotonic one. In a certain sense the 'logic of argumentation' suggested in this paper could be viewed as the monotonic core of a more general argumentation based consequence relation.

REFERENCES

Avron, A., & Lev, I. (2005). Non-deterministic multiple-valued structures. *Journal of Logic and Computation*, *15*(3), 241–261.

Avron, A., & Zamansky, A. (2011). Non-deterministic semantics for logical systems. In D. M. Gabbay & F. Guenthner (Eds.), *Handbook of philosophical logic* (2nd ed., Vol. 16, pp. 227-304). Dordrecht: Springer.

Besnard, P., & Hunter, A. (2008). *Elements of argumentation* (Vol. 47). Cambridge: MIT press.

Caminada, M. W., & Gabbay, D. M. (2009). A logical account of formal argumentation. *Studia Logica*, *93*(2), 109-145.

Corsi, E. A., & Fermüller, C. G. (2017, September). Logical Argumentation Principles, Sequents, and Nondeterministic Matrices. In A. Baltag, J. Seligman & T. Yamada (Eds.), *Proceedings of the International Workshop on Logic, Rationality and Interaction, LORI 2017* (pp. 422-437). Berlin: Springer.

Dung, P. M. (1995). On the acceptability of arguments and its fundamental role in nonmonotonic reasoning, logic programming and n-person games. *Artificial intelligence*, *77*(2), 321-357.

Dunn, J. M., & Restall, G. (2002). Relevance logic. In *Handbook of philosophical logic* (pp. 1-128). Springer Netherlands.

Gentzen, G. (1935). Untersuchungen über das Logische Schließen I & II. *Mathematische Zeitschrift*, *39*(1), 176–210.

Gorogiannis, N., & Hunter, A. (2011). Instantiating abstract argumentation with classical logic arguments: Postulates and properties. *Artificial Intelligence*, *175*(9-10), 1479-1497.

Grossi, D. (2010). Argumentation in the view of modal logic. In P. McBurney, I. Rahwan & S. Parsons (Eds.). *Arumentation in Multi-Agent Systems . ArgMAS 2010 (Vol. 6614*, pp. 190-208). Berlin: Springer.

Grossi, D. (2010, May). On the logic of argumentation theory. In *Proceedings of the 9th International Conference on Autonomous Agents and Multiagent Systems* (Vol. 1, pp. 409-416). Richland, SC: International Foundation for Autonomous Agents and Multiagent Systems.

Rahwan, I., & Simari, G. R. (Eds.). (2009). *Argumentation in artificial intelligence* (Vol. 47). Heidelberg: Springer.
Troelstra, A. S., & Schwichtenberg, H. (2000). *Basic proof theory* (Vol. 43). Cambridge University Press.

13

Reconstructing Multimodal Argument: The Dove vs. Nivea Advertisement

HÉDI VIRÁG CSORDÁS
Budapest University of Technology and Economics
hedi.csordas@filozofia.bme.hu

GÁBOR FORRAI
Eötvös Loránd University
forrai.gabor@filozofia.bme.hu

The paper argues that the reconstruction of visual and multimodal arguments, i.e. the explicit formulation of all the elements necessary for their evaluation, should follow the same pattern as that of verbal arguments. In particular, we cannot escape the verbal statement of elements presented by visual or other means. It is not just a matter of convenience: without verbalization we sometimes miss features of the argument which are crucially important for evaluation. This point is argued by an analysis of the Hungarian Competition Authority's ruling on a commercial by Unilever.

KEYWORDS: Visual argumentation, Comparative advertisement, Multi-modal visual arguments, KC table, Hungarian Competition Authority

1. INTRODUCTION

In the last quarter of a century informal logicians have devoted more and more attention to the analysis of arguments which essentially include visual elements. Most such arguments are not purely visual but consist of a mixture of visually and verbally presented elements. Blair named this type of argument 'hybrid' or 'multimodal', (Blair 2015, p. 2).

The thesis we will be arguing for is this: the reconstruction of visual and multimodal arguments should follow the same pattern as that of verbal arguments. By reconstruction we mean the transparent presentation of an argument, which displays everything that is relevant for its evaluation. So the reconstruction of an argument involves more than merely understanding it. Arguments are normally understood without being reconstructed. We will start by sketching the background, which is going to lead us to the Key Component analysis of multimodal arguments proposed by Leo Groarke in a recent paper, "Going to Multimodal: What is a Mode of Arguing and Why Does it Matter?" Then we will turn to a specific case, a Dove commercial which was challenged in front of the Hungarian Competition Authority, and show that the Hungarian Competition Authority's reasoning cannot be understood without a verbal reconstruction of the visual elements. This points to an important shortcoming of the method of KC – i. e. Key Component – Tables.

2. THE THEORETICAL BACKGROUND

One of the traditional objections against visual arguments – and also against multimodal arguments in so far as they are visual – runs as follows.

> P1: All arguments are propositional.
> P2: Pictures and moving images cannot express propositions.
> C: Therefore, there cannot be visual arguments.
> (Johnson, 2003, p. 6)

This argument may be attacked by challenging the second premise and saying that at least some pictures and moving images are propositional. (Roque, 2015, p. 178.) We will not follow this line though. The line of attack we are interested in consists in challenging the first premise instead, i.e. denying that all arguments are propositional.

One way to substantiate the idea that not all arguments are propositional, but not the only way, is to say that certain visual (or multimodal) arrangements constitute complete arguments without being explicitly rendered in propositional form. (Blair, 2004, pp. 48-49.) This response grants, at least for the sake of argument, that visual elements are not propositional or at least not propositional in the sense in which the sentences we find in logic books are. (Groarke, 2015, p. 135.) It does not deny that the content of a visual presentation can be expressed in the form of verbal propositions, but maintains that some visual presentations are arguments in themselves. They do not need to be translated into words in order to constitute arguments. To put it differently, they are not only arguments in the attenuated sense that they are not genuine arguments but their verbal counterparts are. It is not their expressibility in words which lends them argumentative force. This point is then justified by presenting examples which strike us as genuine arguments. (See for instance the examples in Roque, Birdsell and Groarke, Blair). (Roque, 2015, p. 189), (Birdsell & Groarke, 1996, p. 2), (Blair, 1996, p. 30)

This response can be understood in a weaker and a stronger sense. The weak sense is this: some visual presentations function as arguments without being verbalized. We look at a picture or a moving image and understand it as an argument and react to it in the way we react to arguments. Georges Roque puts this idea as follows:

> 'translation' or 'reconstruction' of an argument in a fully propositional form is not essential to understanding an argument (and, by extension, essential for one to accept it or reject it) (Roque, 2015, p. 183.);

or later:

> So my question is the following: is it absolutely necessary to "translate" the visual component of the poster into a verbal proposition? It is if we wish to analyse how the argument works. But the sober design of the poster is ample indication of the poster's meaning; we needn't verbalize the argument to understand it" (Roque, 2015, p. 188.).

We fully accept this weaker claim. The second quote, however, involves a qualification which points to a stronger one. Roque admits that it is "absolutely necessary" to translate the visual into verbal "if we wish to analyse how the argument works". We are not sure what Roque means by analysing how the argument works, but we take the stronger claim to be this: arguments can be reconstructed without being rendered in the form of verbal propositions. As we have already mentioned, we use 'reconstruction' to stand for a fully transparent representation of the argument which makes everything visible that we need to see so that can we can evaluate the argument. The box and arrow diagrams of informal logicians, the diagrams and numbering system provided by pragma-dialecticians, and the derivations consisting of numbered statements and cited rules are all reconstructions, and they all rely on fully verbal propositions. The stronger claim is then that we can capture everything (or nearly everything) these systems of representations capture without relying on fully-fledged verbal propositions.

It is this stronger claim that Leo Groarke seems to advocate in his recent paper just mentioned. To be sure, Groarke does not explicitly speak about 'reconstruction', he speaks about 'dressing' an argument. 'Dressing' an argument amounts to identifying premises and conclusions, supplying implicit premises, getting rid of diversions and merely rhetorical devices, and clarifying the structure of the argument (Groarke, 2015, p. 135.), and this is the same as reconstruction is usually taken to involve. The method he proposes for this purpose is the Key Component tables.

A KC table contains three columns and a diagram. The leftmost column contains the act of an arguing, such as: making a claim, directing the attention to something, asking something or displaying something. In the middle column we have the elements of argument, i.e. the premises and the conclusion. The rightmost column contains the modes of arguing, like verbal claims, arguing by non-verbal sounds, smells, tactile sensations, music and other non-verbal entities etc. We thus have in each row an act of arguing, an element of the argument and a mode. Below the table we have a diagram displaying how the premises support the conclusion. In one of Groarke's examples someone at the Amsterdam Airport Schiphol tries to persuade his companion that Professor Van Eemeren is in Amsterdam by showing him a picture of Van Eemeren and then pointing at Van Eemeren standing there. The KC table for this argument looks like this.

Act of arguing	Argument	Mode of arguing
Display of the photograph	Premise (p): photograph	Visual (photo)
Directing your attention to the man in front of us	Premise (l): what we see	Visual (facial expression, pointing, observation)
Asking 'Was I right?'	Conclusion (r): I was right (Van Eemeren is in Amsterdam).	Rhetorical question

p+l
↓
r

Figure 1 – Groarke's Van Eemeren example (Groarke, 2015, p. 137.)

KC tables are eminently suitable for rendering visual and all sorts of multimodal arguments: they state acts of arguing and argumentative elements in neutral terms then specify the modality in which they are expressed. The point is that the premises and the conclusions represented in the table do not have to be verbal propositions.

Groarke claims two advantages for this method over reformulating arguments verbally. First, it provides a more faithful account of what actually occurs in many multimodal acts of arguing: "an arguer uses non-verbal elements to establish a conclusion without some attempt to convert them into verbal counterparts" (Groarke, 2015, p. 138.). Second, it "eliminates many of the issues of interpretation" (Groarke, 2015, p. 139.), which is the most problematic part of the reconstruction, for two reasons. First, non-verbal presentations are hard to describe in words. (Try to describe Van Eemeren in words so that one can recognize him!) Second, there are often several different ways of expressing non-verbal presentations in words and it is very difficult to choose from them.

KC tables certainly have these advantages, but we are going to show that as reconstructions they are not quite satisfactory, and the problem is that at certain points we cannot do without propositions.

3. THE DOVE VS. NIVE COMPARATIVE ADVERTISEMENT

As an example we will use the comparative advertisement – also known as the "tulip commercial" – run by Unilever Hungary Ltd. between July and October 2005, which claimed that Dove Intensive Cream is a better moisturizer than Nivea's similar product. The Hungarian Competition Authority promptly initiated a proceeding against the advertisement for unfair manipulation of consumer decisions. Hungarian legislation at that time didn't prohibit comparative advertising if the comparative claims could be verified, i.e. backed up by scientific evidence. The Authority examined Dove's claim that it provides "better moisturization" and has eventually found it deceptive.

    The commercial runs as follows[1]: A female hand touches first the Dove then the Nivea product, placed left and right, respectively. Then we are shown two glasses with the names of the two brands. After that, the camera focuses on the containers and a dying tulip is placed first in the Dove cream, then another one in the Nivea moisturizer. The tulips are drooping in different directions; they obviously need water. The camera shows them from different angles to make their miserable condition evident. At the 12th second of the commercial the tulips are left alone to give them time to absorb the creams. After a while, the camera focuses on the tulip of Nivea and we see that its condition has not changed. The camera zooms out and both tulips are visible now. The ticking sound and the running clock in the top right corner demonstrate that the "race" between the tulips lasts ten hours. This is the point where the flower of Dove reaches its peak position. At the end of the commercial, the examiner makes her choice, opting, unsurprisingly, for the tulip moisturized by Dove. Nivea's humiliated tulip is left in container. The triumphant tulip is placed on the right beside the moisturizer. The text reads "New Dove Intensive Cream" and "Better moisturization, beautiful skin".

---

[1] tvspots.TV: Unilever Germany - tulip test. Retrieved from http://www.tvspots.tv/video/42773/unilever-germany--tulip-test

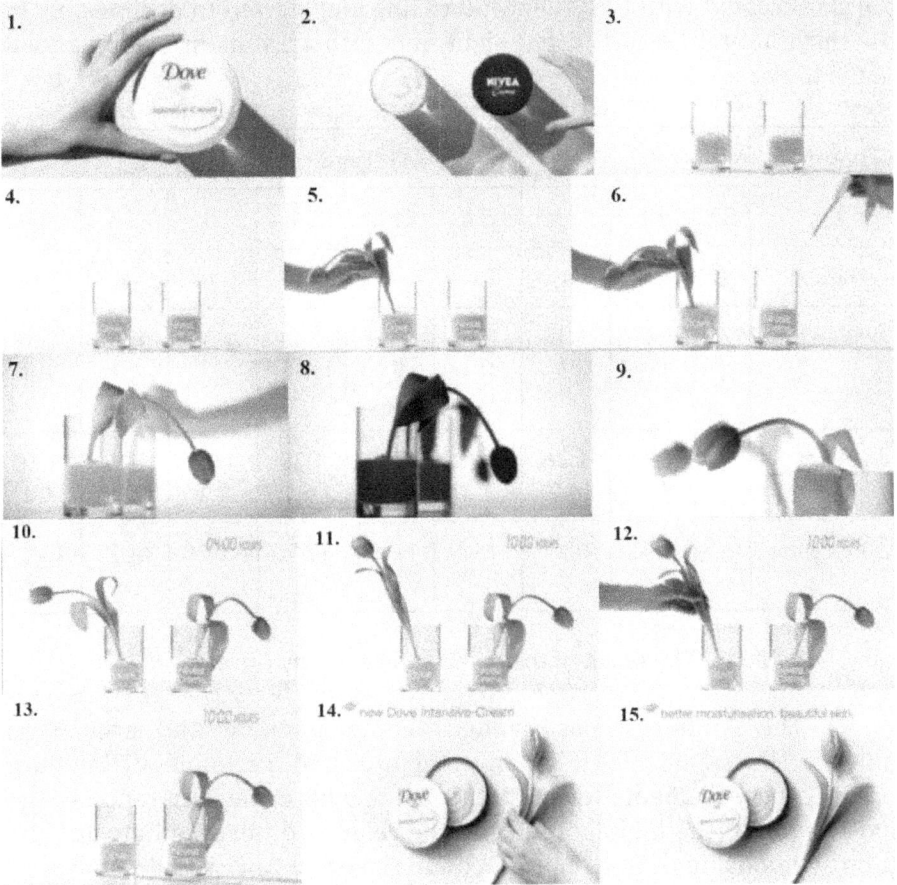

Figure 2 – The tulip test commercial in pictures

3.1 The reasoning of the Hungarian Competition Authority

Why has this commercial been found deceptive? The Authority ordered Dove and Nivea to furnish evidence for the moisturizing efficacy of their creams. The parties present questionnaires, rice paper test, spectrum analysis, and corneometry test. It was the last test the Authority based its decision on. In the corneometry test, which measures the hydration level of the skin, the effects of the Dove and the Nivea moisturizers on human skin were tested on thirty-member groups, comparing it with the untreated skin (baseline). In the first hour the hydration level of the skin was raised by both moisturizers, but Dove Intensive Cream increased it better. In the second hour the hydration level of the skin moisturized by Nivea started to decline, whereas the hydration level of the skin treated with Dove kept increasing and started to decline only in the third hour. The test lasted eight hours; the results are summarized in the table.

Treatment	Average								
	0h	1 h	2 h	3 h	4 h	5 h	6 h	7 h	8 h
Baseline	1	1	1	1	1	1	1	1	1
Dove	1	1.429 ↑	1.452 ↑	1.439 ↓	1.428 ↓	1.414 ↓	1.382 ↓	1.361 ↓	1.348 ↓
Nivea	1	1.380 ↑	1.364 ↓	1.333 ↓	1.280 ↓	1.254 ↓	1.231 ↓	1.192 ↓	1.190 ↓
Δ Percentage (%)		3.55	6.45	7.95	11.56	12.75	12.26	14.17	13.27

Table 1 – The result of the corneometry test by Dove

The Authority has banned the commercial and argued as follows: "Based on SIT [Skin Investigation and Technology Hamburg GmbH.] measurements, even at the point where the difference is the most marked [the highlighted cell], the relative difference between the tulip symbolising the Dove product and the other one would not exceed 15%, which, according to the Competition Board, is barely visible to the naked eye" (HCA, 2005, p. 16.). This reasoning may sound convincing, but it is probably not obvious at first sight which element of the visual argument is the one the court found false. We need a reconstruction to identify it.

3.2 Reconstruction by KC table

So how would a KC table for the commercial look like? It would be something like this.

Act of arguing	Argument	The mode of arguing
Presenting the tulip test that shows drooping tulips' condition change.	Premise (p1, ...px): Details of the commercial.	Visual moving image non-verbal sounds music
Claim that "Dove is a better moisturizer."	Conclusion (c): "New Dove Intensive Cream" and "Better moisturization, beautiful skin"	Verbal claim and visual static image

p1,...px
↓
c

Figure 3 – The reconstruction of the tulip test using by KC table

Now if this is indeed the complete reconstruction of the tulip test, then we are in trouble, because we just cannot identify what is wrong with the visual argument. It cannot be the premises presented visually, since those premises are concerned with the tulip, whereas the corneometry results were about the human skin. No test result about the skin can challenge what the commercial shows about tulips. It cannot be the conclusion, for the corneometry result indeed confirms that Dove moisturizes better, even if only by 15% at most. The verbal claim is just this: "New Dove Intensive Cream"; "Better moisturization, beautiful skin", not "20% better moisturization, beautiful skin". It cannot be the inferential move from the premises to the conclusion, for the Authority's reasoning, in effect, condemns Dove for making a scientifically unjustified, exaggerated claim rather than for fallacious inference. So, unless the Authority's reasoning is completely wrong, there must be more to the argument than its KC table shows.

3.3 Reconstruction by argumentation schemes

So let us use another method of reconstruction to find what is missing. We choose argumentation schemes, developed in most detail in Walton-Reed-Macagno's *Argumentation Schemes*. Like deductive inference schemes, an argumentation scheme is an abstract structure which can be filled in with various linguistic elements. Unlike deductive inference schemes, however, filling it in with linguistic elements which make the premises true does not necessarily make the conclusion true: it makes the conclusion only presumptively true; i.e. it entitles us to presume that the conclusion is true, which presumption can be defeated by a stronger argument. The argument in the commercial obviously relies on the similarity between the tulip and the skin, so it is an Argument from Analogy. Here is the scheme and the reconstruction of the argument in the commercial:

> Argument from Analogy:
> PI: Generally, case C1 is similar to case C2.
> PII: In case C1, A is true.
> C: A is true in case C2.
>
> P1: The skin is similar to the tulip.
> P2: Dove moisturizes the tulip much better than Nivea does.
> C: Dove moisturizes the skin much better than Nivea does.

If the conclusion is construed in this way, the Authority's reasoning makes perfect sense. 15% as the largest difference in a period of 8 hours certainly doesn't count as "much better"! So it is the conclusion which is false. (This is compatible with the truth of the premises, since we are talking about a presumptive inference.) But we can only arrive at this conclusion, if we construe P2 of the argument as including 'much better', otherwise the inference wouldn't be presumptively valid.[2]

The problem is that including 'much better' in the premise is not possible unless the visual presentation is translated into words. The reason is that the visual presentation in this case is ambiguous. Showing a big difference between the tulips is at the same time showing the mere existence of an unspecified degree of difference between them. This has to do with the intrinsic specificity of visual presentation: one cannot show the mere existence of a difference without showing a specific degree of difference. So the visual presentation in the commercial is ambiguous between 'better' and 'much better', and the difference between the two construals certainly matters for understanding the Authority's decision. The Authority's decision is justified only if the conclusion and, consequently, the second premise includes 'much

---

[2] Someone may object that construing C and P2 in this way is ruled out by the text of the commercial, which says 'better moisturization' and not 'much better moisturization'. However, 'better moisturization' is compatible with 'much better moisturization', and given the fact the commercials prefer slogans to precise statements nothing 'better moisturization' may equally mean 'much better moisturization'.

better'. But due the ambiguity of the visual presentation, the disambiguation can only be achieved by verbal means. In other words, the Authority couldn't have exposed the deceptive character of the commercial without translating the visual element of the commercial into words.

Groarke is certainly right that 'interpretation' -- to use his term for verbal rendering of nonverbally presented content -- is difficult: it is often hard and sometimes impossible to determine which verbal translation is better. But if we do not take the trouble of selecting one of the several possible verbal translations, we may sometimes miss something very important. If we rely solely on KC tables, which spares us the difficulty of verbalization, we only get an impoverished reconstruction which is unsuitable for the proper evaluation of at least certain arguments.[3]

As a final note, we would like to draw attention to a further troublesome point for reconstructing arguments by KC tables. Even though Groarke includes implicit premises in his KC tables, this does not quite fit the spirit of his approach (Fig. 4.).

Act of arguing	Argument	The mode of arguing
Claim that 'Socrates is a man'	Premise (s): Socrates is a man	Verbal claim
	Premise (a): All men is mortal	Enthymeme
Claim that 'Socrates is mortal'	Conclusion (m): Socrates is mortal	Verbal claim

s+a
↓
m

Figure 4 – Implicit premise in a KC table (Groarke, 2015, p. 136.)

First, there is no act of arguing corresponding to an implicit premise (see Figure 4). Second, by including them, Groarke loses something of the advantages he claims for his approach. The representation becomes less natural: implicit premises do not "actually

---

[3] An additional but related benefit of the verbal reconstruction in terms of argument schemes is that it helps us identify where the argument goes wrong. P1 is not acceptable, i.e. the similarity between the skin and the tulip is not sufficiently robust to support the analogical inference.

occur" in "acts of arguing"; the speaker does not "use" them in the sense in which he uses words or other modes of argumentation (Groarke, 2015, p. 138). Moreover, filling in the missing premises involves us in the problems of interpretation Groarke hopes to eliminate (Groarke, 2015, p. 139). At one point he signals agreement with Pinto who understands arguments as invitation to inferences. (Pinto, 2001, pp. 68-69) We think KC tables do a good job in representing what does the invitation, but they are unsuitable for representing the inferences invited.

4. CONCLUSION

Finally, let us sum up the argument. One way to resist the traditional argument against visual arguments is that they are not propositional in the sense in which verbal arguments are propositional. There are visual presentations which constitute full-fledged arguments without being verbalized. We granted that visual presentations may function as arguments without being translated into words, but we argued that they cannot be reconstructed, i. e. presented in a transparent form suitable for evaluation and criticism without being so translated. We have tried to make good on this claim by bringing an example showing that a reconstruction seeking to eschew verbalization – Groarke's KC tables – cannot expose all the significant elements needed for evaluation.

ACKNOWLEDGEMENTS: Special thanks go to Leo Groarke, János Tanács, István Danka and all of the colleagues at the Department of Philosophy and History of Science, Budapest University of Technology and Economics, who helped me with their advice during my research. This research progresses is carried out in the framework of the Integral Argumentation Studies (OTKA —K-109456) of the Doctoral School of Philosophy and History of Science, Budapest University of Technology and Economics.

REFERENCES

Birdsell, D. S., & L. Groarke (1996). Toward a theory of visual argument. *Argumentation and Advocacy*, *33*(1), 1-10.

Blair, J. A. (1996). The possibility and actuality of visual arguments. *Argumentation and Advocacy*, *33*(1), 23-39.

Blair, J. A. (2004). Defining the rhetoric of visual arguments. In C. A. Hill & M. Helmers (Eds.), *Visual rhetorics* (pp. 41-61). Mahwah, NJ: Lawrence Erlbaum Associates, Publishers.

Blair, J. A. (2015). Probative norms for multimodal visual arguments. *Argumentation, 29*(2), 217–233.

Georges, R. (2015). Should visual arguments be propositional in order to be arguments? *Argumentation, 29*(2), 177-195.

Hungarian Competition Authority (2005), Vj-145/2005/20, 1-17.

Johnson, R. H. (2003) Why "Visual Arguments" aren't Arguments (2003). OSSA Conference Archive. 49. Retrieved from https://scholar.uwindsor.ca/ossaarchive/OSSA5/papersandcommentaries/49.

Leo, Groarke (2015). Going multimodal: What is a mode of arguing and why does it matter? *Argumentation*, *29*(2), 135-155.

Pinto, Robert C. (2001). *Argument, inference and dialectic: Collected papers on informal logic*. Dordrecht: Kluwer.

Walton, Reed & Macagno (2008). *Argumentation Schemes.* Cambridge: Cambridge University Press.

14

Rhetorical Inferences for Divine Authority: The Case of Classical Greek Divination

JULIE DAINVILLE
Université libre de Bruxelles ; F.R.S.-FNRS
Julie.Dainville@ulb.ac.be

In classical Greece, concomitantly with the democratisation of the society, one can also observe a democratisation of the divinatory process. Concretely speaking, it means that the classical seers and prophets must prove their value to acquire authority. Their reliability relies on different kinds of clues allowing to *infer* their divine inspiration. The aim of this paper is to show how rhetoric can be useful to describe this status change of divination.

KEYWORDS: classical Greece, divination, Herodotus, prophets, Pythia, reliability, rhetorical inference, seers, Toulmin

1. INTRODUCTION

It is a well-known fact that divination is an important phenomenon in the Greek literature from the fifth century BC: the seer, for instance, is often a first plan character in classical tragedies (Teiresias in *Oedipous Rex* or *Antigone* or Calchas in *Agamemnon*, just to give the most famous examples). But divination is not limited to the mythological sphere: historians, especially Herodotus and Xenophon, also relate many stories involving oracles. Still, not in the same way. The main difference between both authors relies in the divinatory class they focus on: Xenophon mainly refers to seer technical divination. A hypothesis to explain this feature could be that the historian took an active part in some of the events he relates (especially in the *Anabasis*), and proposes a close-up on specific military expeditions. In this military context, divination is often restricted to the sacrifices required before starting a battle, and this function was assigned to the seers. Herodotus, on the other hand, gives his lectors a broader perspective, mentioning seers but also and most of all the Pythia at Delphi, who I will be focusing on in this paper. We leave here the technical sphere to approach inspired divination: the god speaks through the prophetess without her needing to master any specific skill.

I will begin this presentation with a brief historical note, to show how the way of talking about divination has changed from the epic times (around the 8th century BC) to the classical period. This will allow me to highlight the particularities of this period. Then, I will more specifically study an abstract from Herodotus' first book and show how Toulmin's model can be useful to describe the mechanisms of persuasion that will establish the Delphic authority. My aim will be to defend that there is not necessarily a contradiction between the spreading of rhetoric and democratic institutions and the use of divination.

2. DIVINATION IN LITERATURE: A TIMELESS INHERITANCE

2.1 Homeric poetry

In the archaic period, as Jean-Pierre Vernant (1974, p. 10-11) pointed out, the use of divination is obvious, and even highly recommended. The seer is often, if not always, questioned when a decision has to be made and advises kings and judges. He is a *demiourgos*, a craftsman, just like the physicians or the carpenters (see *Odyssey* XVII, 382-385). I would like to illustrate this statement with an example taken from Homer's *Iliad*. The first mention of a seer in the Homeric epic is at the beginning of the first book. The army is suffering from a plague, thought to be a divine sanction, and is in need for a seer or a priest to understand the reason of the gods' anger.

> [*Achilles is talking*]: Son of Atreus, now I think we shall return home, beaten back again, should we even escape death, if war and pestilence alike are to ravage the Achaeans. But come, let us ask some seer or priest, or some reader of dreams—for a dream too is from Zeus—who might say why Phoebus Apollo is so angry, whether he finds fault with a vow or a hecatomb; [...]. When he had thus spoken he sat down, and among them arose Calchas son of Thestor, far the best of bird-diviners, who knew the things that were, and that were to be, and that had been before, and who had guided the ships of the Achaeans to Ilios by his own prophetic powers which Phoebus Apollo had bestowed upon him. He with good intent addressed the

gathering, and spoke among them [...] (*Iliad* I, l. 59-73; translation: A. T. Murray).

Two points can be made after the reading of these verses. First, the use of divination is obvious, it needs no deliberation and is therefore beyond the argumentative sphere. The seer or priest[1] was likely to tell what to do to easy the angry god. The second point is the presentation of Calchas. This is the first time he is mentioned in the epic, but his name was probably familiar to the poet's listeners. Anyway, his authority is highlighted in the text by several elements: he got his divinatory skills from Apollo, god of divination, himself. He is said to be the best and to know the past, the present and the future and his value is not only theoretically assumed, it is also confirmed by a concrete example: he allowed the Greek ships to reach Troy[2].

Here again, Calchas was called and is able to explain the reason why Apollo is angry with the army: the plague was sent by the god as a vengeance for his priest Chryses whose daughter, Chryseis, had been captured by the Acheans (on this passage, see Di Sacco Franco, 2000, pp. 36-37). The girl was then king Agamemnon's captive. Calchas advises to give the girl back to her father to ease Apollo and end the plague. He knew that his answer would not please the king, and even ask for Achilles' support, to protect him in the eventuality of a physical attack. But still, even if this caused anger[3], the army gave the girl back to her father. The king Agamemnon himself took his decision according to the seer's answer. One could add at this point that Achilles, who called for a seer at first, is ready to protect Calchas because he knows his value. This shows that Calchas' legitimacy is well established in the minds of the Greek assembly: his prior *ethos* is without any doubt positive and his former deeds are even recalled or mentioned to the lectors. He, and the other epic seers who appear in the Homeric texts, have a good reputation: they are shown as wise, sincere and caring, even though

---

[1] Their respective functions were, according to what we know, close in some respects, especially regarding the practice of sacrifices (see Flower, 2009, pp. 27-30).

[2] The ships were not able to leave the harbour due to the absence of winds. Calchas explained that this situation ensued from Artemis' anger, because king Agamemnon had killed one of her deers. To conciliate the goddess, Agamemnon had to sacrifice his own daughter. And he followed the seer's recommendation.

[3] Agamemnon's anger towards Calchas could also be explained by the fact that he is, once again, the one who has to sacrifice something he cares about.

some characters occasionally doubt it, and these qualities are recalled by the poet throughout the text. From this point of view, the importance of the ethical criterion reminds of the working of arts, *technai*, in general. A physician, for example, will be preferred to another one due to his reputation; the seer was a craftsman, which means that his skills (whether a divine gift or the achievement of a training) had to be largely recognized before one called him. His prior *ethos* (as defined by Ruth Amossy, 2010, pp. 74-75) is the main factor of his authority.

2.2 Classical period

The situation is significantly different during the classical period. Three centuries passed, and the political and intellectual life changed. Deliberation and civic consultation got more and more important in the city organisation. Seers are still helping to interpret divine signs or to give advices, but the nature of their answers has significantly changed. Most of the time, the classical seer is attested in a military context, where he reads victims' liver to see if the gods allow or not the forthcoming battle. Their prestige faded, but this does not mean that their function was meaningless, and they also seem to have played a significant role in everyday life. Besides, another type of divination emerged after the epic period: oracular institutions, like the Apollo's sanctuary at Delphi, the most famous one[4], played an important part in the ancient political life. I would like to show that, contrary to a widespread opinion, the deliberation process development and the emergence of rhetorical theories and practices, which gave citizens another kind of tools to debate and achieve a decision, did not sweep divination away, even though it is not obvious anymore.

3. HERODOTUS AND THE PYTHIA: DEMONSTRATING THE DELPHIC AUTHORITY

I have chosen, to illustrate my point, to analyse an extract from Herodotus first book. This historian is our best source to approach the representation of divination in the classical period. He seems to have been very respectful of the divinatory institutions, and yet, as I will

---

[4] On the historical dimension of the Delphic oracle, see for instance Amandry 1950; Parke and Wormell, 1956, vol. 1; Fontenrose, 1978; or Bowden 2013. It is often assumed that the actual working of the sanctuary does not completely fit its literary representation. On this question with a particular focus on Herodotus, see Crahay 1956).

show, one can find marks of tensions in his writings, that may indicate that the oracular authority was no longer obvious in the society he lived in. The extract relates a test that Croesus, last king of Lydia (6th century BC; see *OCD*, 2003, p. 410 for more information) decided to give to the main oracular sanctuaries of his time. The king sent an embassy to the different temples and told them to ask the following question, on the hundredth day after they left: *what Croesus, king of Lydia, son of Alyattes, was doing then?* and to write down the different answers. Here is what follows:

> Now none relate what answer was given by the rest of the oracles. **But at Delphi**, no sooner had the Lydians entered the hall to inquire of the god and asked the question with which they were entrusted, than the Pythian priestess uttered the following hexameter verses:
> 'I know the number of the grains of sand and the extent of the sea,
> And understand the mute and hear the voiceless.
> The smell has come to my senses of a strong-shelled tortoise
> Boiling in a cauldron together with a lamb's flesh,
> Under which is bronze and over which is bronze'.
> Having written down this inspired utterance of the Pythian priestess, the Lydians went back to Sardis. When the others as well who had been sent to various places came bringing their oracles, Croesus then unfolded and examined all the writings. Some of them in no way satisfied him. **But when he read the Delphian message, he acknowledged it with worship and welcome, considering Delphi as the only true place of divination, because it had discovered what he himself had done.** For after sending his envoys to the oracles, he had thought up something which no conjecture could discover, and carried it out on the appointed day: namely, he had cut up a tortoise and a lamb, and then boiled them in a cauldron of bronze covered with a lid of the same. (Herodotus, *Histories* I, 47-48; translation: Godley)

From then on, Croesus will show a boundless devotion towards the Delphic sanctuary and consult Apollo on every matter. And the fact is that he will not be the only one to do so, as the Pythia, Apollo's prophetess at Delphi, is mentioned more than sixty times in the Herodotean corpus. This excerpt will allow me to illustrate different points

The first one is that consulting divination is no longer obvious. At this period, whether to consult or not the gods, and if one does, which sanctuary to choose, are deliberative and arguable subjects. And even if one chooses to do so, the sanctuaries had to deserve his respect and trust. On this matter, it is worth focusing on the words chosen by the historian. First the king *"acknowledged it with worship and welcome"*. The verbs used to describe his reaction are προσεύχομαι and προσδέχομαι, meaning respectively "offer prayers, worship" (*LSJ* 1996, p. 1511, *s. v.* προσεύχομαι) and "receive, admit, but also expect". In Herodotus text, we count only two attestations of προσεύχομαι, both in a religious context, after the recognition of a divinity[5]. Προσδέχομαι, on the contrary, occurs fifteen times, but the test is the only *religious* context. The word has different meanings in Greek, slightly depending on the object: a person will be "welcomed", a concrete object (like a present)[6] will be "accepted, received" (*LSJ* 1996, pp. 1505-1506, *s. v.* προσδέχομαι). But most of the time, it is used to talk about a fact, an event that occurred and was, usually, not expected: an ambush or an unexpected reaction, for example). Consequently, I would suggest that Croesus recognized the divinity of the answer thanks to its exactitude, but also that he did so with some surprise (either because of the global divine nature of the answers or because of the precision of the god's mastery). This is crucial, because, as I will show, destabilisation is one of the main elements of the divine authority construction.

Second, Croesus considers *Delphi as the only true place of divination because it had discovered what he himself had done*. Here, the verb is νομίζω, which means "think", "take into consideration" (*LSJ* 1996, pp. 1178-1179, *s. v.* νομάς). The verb thus includes a reflection, based on arguments. The case is slightly different from what we saw in Homer.

This is particularly clear in the Herodotus' narrative of the test, where the historian gives the details of the king's reasoning. For indeed, Croesus's decision to trust Apollo is the result of a scientific procedure, during which he decided to put the oracle to the test. It is not a dogmatic, obvious act; instead, its deliberative dimension is explicit. To advocate this point of view, I will focus on the argumentation details. For a clarity purpose, I decided to use Toulmin's model (Toulmin, 1958), which will allow me to provide a clearer view of the arguments. More specifically, it will be helpful to highlight the backing of the

---

[5] I, 48, 6 and I, 60, 29. The second attestation refers to the supposed recognition of the goddess Athena.

[6] I, 89, 8; II, 121ε, 6; III, 62, 14; 146, 14; V, 20, 8; 34, 2; VI, 100, 9; VII, 235, 20; VIII, 130, 15; IX, 6, 2; 45, 22; 48, 12; 116, 22.

argumentation. The authority establishment relies on a two steps deliberation, that can be mapped as follows:

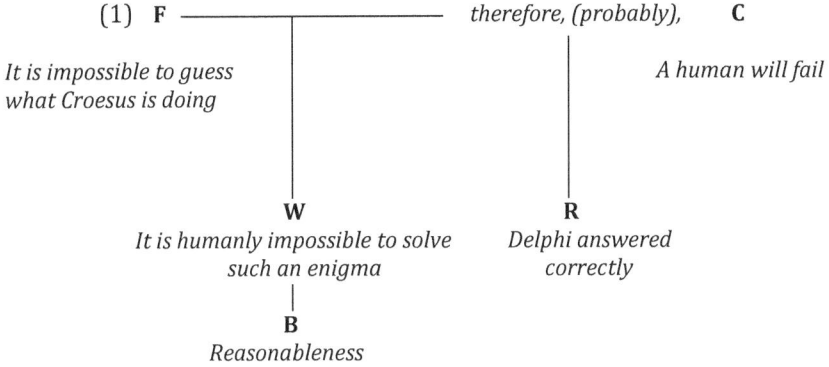

Figure 1 – first part of argumentation 1 (Toulmin's model)

Because of the difficulty of the task, a pure human would be expected to fail: it what impossible to guess, and impossible to know, due to the distance. Only an omniscient being, which means a divine being, in the Greek culture, could answer the question. That means that if the oracular sanctuary at Delphi was a fake, it would have failed. But it succeeded, and this leads us to the second part of the argumentation:

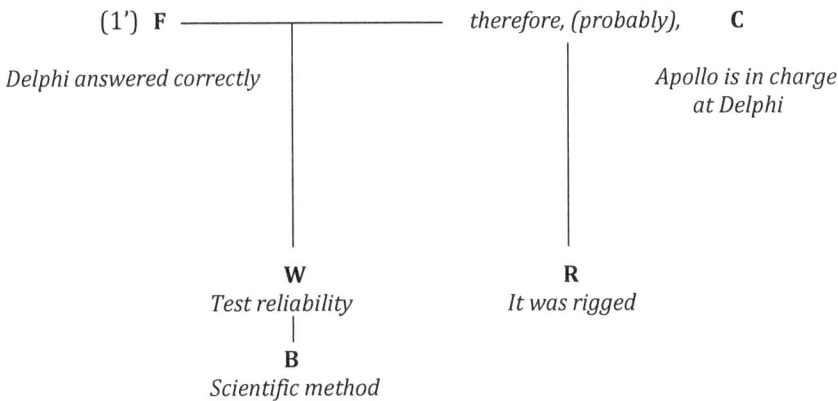

Figure 2 – second part of argumentation 1 (Toulmin's model)

By giving a correct answer, the sanctuary testified that Apollo was indeed in charge at Delphi. As the lexicon used in the text suggests, the king's surprise, destabilisation caused by the first restriction (*Delphi answered correctly*) is the first element. Human intelligence and knowledge is too restricted to be able to solve such an enigma. No one could expect that somebody could pass the test, and yet Delphi did. This fact is even more impressive that Croesus's doing was indeed impossible to figure out. The conclusion of the first argument has then to be modified, because Delphi and Apollo are worth trusting, and a new model needs to be shaped. Rhetoric here is without any doubt of a great help to understand the mechanisms at work, and the Toulmin's model allow to put into light that Croesus's trust does not rely on a cultural and irrational belief, but on a scientific method and the argumentative process is explicit[7].

Still the reflection has to be conducted one step further, because there is no direct communication between the god and the consultant. Indeed, Croesus, and the historian's readers[8], now knows that Apollo is worth trusting, but the actual person answering, in Herodotus' narrative, is Apollo's prophetess, the Pythia. The god answers through her. This implies that besides building the god's authority, it will also be necessary to let the consultant know that the Pythia speaks well in the name of the god, that she is *inspired* by him. Here again a reasoning may be shaped, and this consists the second step of the sanctuary's authority establishment.

Step 2

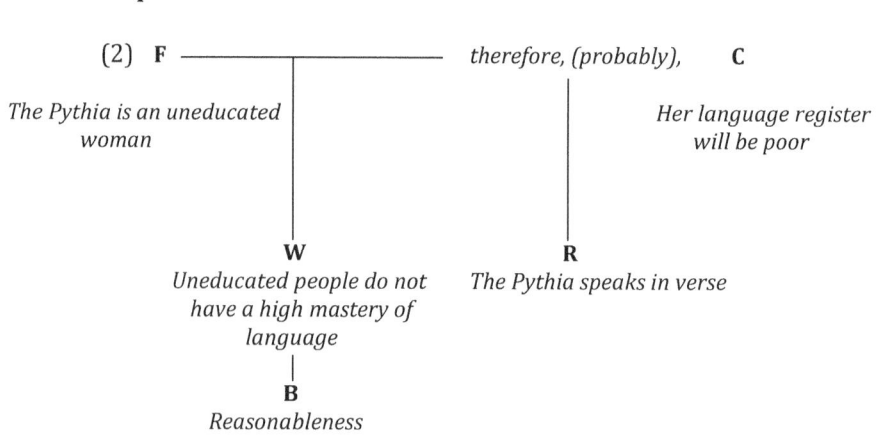

Figure 3 – first part of the second argumentation

---

[7] The warrant is implicit, as it is quite common, but it is quite easy to reconstruct. On this question, see Toulmin, 1958, p. 132 and Danblon, 2002, pp. 33-35.

[8] We may not forget that Ancient history had to have a purpose, to be useful for the readers (see for instance Devillers, 1994, pp.5-13).

and then:

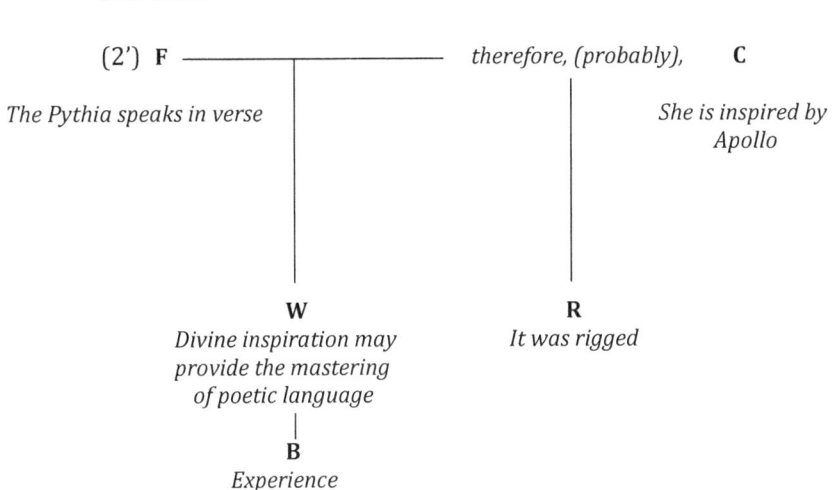

Figure 4 – second part of the second argumentation

The answer was given in verse, as it is usual when an oracle is quoted. And yet, the women who filled the Pythia function were known to be uneducated; some Ancient authors, like Plutarchus, even said that the less educated she was, the better (*De Pythiae oraculis*, 405c). In this second step, destabilisation also plays a part: people expect that an uneducated woman speaks simply, because, in a sense, this would fit her prior *ethos*, based on societal stereotypes but also on common sense: it is not possible to master a high language register, to master a challenging poetic verse, the dactylic hexameter, and archaisms if one were not trained to or divinely inspired. And yet the Pythia did, which leads us to the second argumentation scheme. The link with Apollo is easily established in the Ancient culture: he was the god of divination, but also of poetry. The Muses, who were under his control, inspired famous poets, like Homer and Hesiod, who then were also able to master a poetic language register. The point, then, will be to fit as less as possible to the stereotypes linked to her gender and social situation, in order to show that she is not talking as a particular woman, but as Apollo's prophetess. The legitimacy of what she says will not come from her personal reputation, but on the contrary, on the absence of personal *ethos*: it has to be put aside to let the god talk through her.

4. CONCLUSION

In the Homeric poems, Calchas was given his divinatory skills by Apollo, the divination god. The element that will grant his authority is the quality, the prior *ethos* of the seer. He already helped the Greeks in other contexts and has proved his value. To put it differently what I already said, his *prior ethos* is this of a competent, valuable and brave man, and this is the reason why people will listen to him. Besides, calling a seer to escape from a problematic situation seemed to be natural at the time. The situation at Delphi is quite different. The consultant questions a woman who fulfilled the prophetic function, the Pythia, and who was said to have no particular skills. She only had an authority when she was sitting on the tripod, in Apollo's sanctuary, letting the god speak through her. She did not have an authority of her own. The reputation that will be considered is that of the institution, of the god. And as I showed, even him has now to prove his value and is evaluated following the criteria of the scientific method.

As a conclusion of this paper, I would like to make clear that my point was not to defend that divination is indeed a pure *rational* process: however, I wanted to show how rhetoric and divination, belief in oracles and reasonableness did co-exist. The fact that Herodotus describes Croesus's reasoning may be seen as a clue of the decreasing importance of divination at the time. This hypothesis is grounded by our historical knowledge as well as by other historians' testimonies (Thucydides, for instance, has a totally different point of view on the matter and barely mentions divination in his work). And of course, the fact that he does not consider here the possibility of the Pythia being corrupted or a rigging of the test weakens the global argumentation, because corruption was one of the main argument against divination. Yet it would be a too simplistic view to say that democracy and argumentation skills sounded the death knell for oracles and divination. Those religious realities evolved, and fell from then on in the fields of arguable matters. Furthermore, I suggest that the ones in favour of divination used these new tools as well to prove the grounds of their opinions. For all these reasons, I think that adopting a rhetorical approach towards this religious question is not only interesting, but crucial.

REFERENCES

Amandry, P. (1950). *La mantique apollinienne à Delphes. Essai sur le fonctionnement de l'oracle*, Paris: E. de Boccard.
Amossy, R. (2010). *La présentation de soi. Ethos et identité verbale*, Paris: Presses universitaires de France.
Bowden, H. (2013). Seeking certainty and claiming authority: The consultation of Greek oracles from the Classical to the Roman Imperial periods. In V. Rosenberg (Ed.), *Divination in the Ancient World. Religious options and the individual* (pp. 41-60), Stuttgart: Franz Steiner Verlag.
Crahay, R. (1956). *La littérature oraculaire chez Hérodote*, Paris: Les Belles Lettres.
Danblon, E. (2002). *Rhétorique et rationalité. Essai sur l'émergence de la critique et de la persuasion*, Brussels: Editions de l'Université de Bruxelles.
Devillers, O. (1994). *L'art de la persuasion dans les Annales de Tacite*, Leuven: Peeters.
Di Sacco Franco, M. T. (2000). Les devins chez Homère. Essai d'analyse: *Kernos*, 13, 35-46.
Flower, M. (2009). *The seer in ancient Greece*, Berkley: University of California Press.
Fontenrose, J. (1978). *The Delphic Oracle: Its responses and operations with a catalogue of responses*, Berkeley: University of California Press.
Godley, A. D. (1920). *Herodotus, The Histories* (A. D. Godley, Trans.). Cambridge, MA: Harvard University Press.
Liddell, H. G., Scott, R., H. S. Jones & R. Mc Kenzie (1996). *A Greek English Lexicon*. Oxford: Clarendon Press.
Murray, A. T. (1924). *Homer. The Iliad*, (A. T. Murray, Trans.). Cambridge, MA-London: Harvard University Press.
Hornblower, S. & Spawforth, A., Eds., (2003). *The Oxford Classical Dictionary OCD* (3rd ed.). Oxford: Oxford University Press.
Parke, H. W. & Wormell, D. E. W. (1956). *The Delphic Oracle: Vol. 1. The history*. Oxford: Basil Blackwell.
Toulmin, S. (1958). *The uses of argument*, Cambridge: Cambridge University Press.
Vernant, J.-P. (1974). Paroles et signes muets. In J.-P. Vernant et al. (Eds.), *Divination et rationalité* (pp. 9-25), Paris: Seuil.

15

Irresolvable Rational Disputes

ISTVAN DANKA
Dept. of HPS, Budapest Univ. of Technology and Economics Hungary
danka.istvan@filozofia.bme.hu

Resolution of a dispute, a central aim of a rational debate, is irrational in cases when no decisive arguments can be given on either side. Distinguishing rhetorical and dialectical senses of rationality (see 'strategic manoeuvring'), the reason for that will be taken to be dialectical: if it is rational to be committed to both views in a conflict, it is dialectically irrational to resolve the dispute. Philosophical debates, in particular, are examples for that kind.

KEYWORDS: argumentative practices, dialectical, epistemic and rhetorical purposes, faultless disagreement, philosophical debates, pragma-dialectics, rational debates, strategic manoeuvring

1. INTRODUCTION

There is a widely held truism in argumentation theory (and esp. within the pragma-dialectical tradition) that rational debates aim at a resolution of a dispute. The resolution lies in some sort of agreement of the parties in the truth of one or the other opinion discussed. In other words, rational debates aim at an agreement of the parties about a claim (or set of claims) on which they formerly disagreed. A resolution is supposed to be rational, i.e., it is done in accordance with rules of logic and dialectics.

However, there are, as I will claim, debates the resolution of which is irrational in some sense. Obviously, a resolution of dispute e.g. by brute force is irrational; but sometimes the most rational possible way of resolving certain disputes is also irrational, and the reason why it is so is dialectical. In other words, the irrationality of resolution does not lie in contingent factors like the behaviour of debaters, nor does it lie in the interests of parties, not to speak of subject-specific factors like the nature of questions discussed. Although these can result in an irrational resolution in other cases, in the cases discussed they are irrelevant.

Debates e.g. on political, philosophical, religious or aesthetic questions are often endless. There seem to be no decisive, knockdown arguments for one side or the other. It is often claimed that the reason lies in the subjects themselves. I take the reason to be more general, arguing that the dialectical nature of certain debates is the ground for the phenomena called faultless disagreement that occur in debates where parties "agree to disagree" because no knockdown arguments can be found to reject either side.

Before turning to the question what exactly a 'faultless disagreement' is, I shall briefly sketch a methodological framework which is essential for developing my point. First, argumentative practices will be discussed as means-end structures; i.e., social activities in which parties aim at certain purposes for which they commit themselves to follow certain paths. In this means-end framework, strategic manoeuvring will be recalled, i.e., the idea that debating is normally done by balancing between the debaters' rhetorical and dialectical purposes. A third pole, namely epistemic purposes, will be also mentioned. Then a pluralistic account of rationality will be sketched in accordance with the approach that argumentative practices should be understood as means-end structures. This sort of pluralism will then allow me to identify different means-end structures in argumentative practices with different senses of rationality as defined by the triad of rhetorical, dialectical and epistemic ends of a debate. In this framework, faultless disagreement will be discussed as it occurs within (an extended version of) pragma-dialectics. Philosophical debates as illustrations for faultless disagreement will be mentioned. Finally, a (moderately) optimistic response will be developed to the problem what we should do with irresolvable disputes.

2. ARGUMENTATIVE PRACTICES AS MEANS-END STRUCTURES

A practice is a bunch of purposeful social/institutional activities that aims at some (necessary) purposes (MacIntyre, 1981). Consequently, an argumentative practice is a bunch of activities in which argumentative activities aim at certain necessary purposes (Kvernbekk, 2008) – typically a rational resolution of a dispute. The activities involved serve as means for the purposes, and the purposes may differ in different sorts of debate. In certain debates, rationality is less important than resolution; in other debates, resolution is less important than rationality.

Even within one and the same debate, more than one aim can be identified. The rhetorical tradition takes the purpose of a debate to be persuasion of the opponent or an audience; dialecticians take it to be consensus; finally, the epistemological tradition takes it to be accessing truth. These approaches are often seen in a conflict. I understand them, in accordance with Biro-Siegel (2006), rather as complementary: a pluralistic approach, based on the idea of balancing among rhetorical, dialectical and epistemic purposes of argumentative practices, helps us understand that these different purposes can be aimed in parallel, and an ideal resolution of dispute satisfies all requirements of being persuasive, consensual, and truth-seeking at the same time.

2.1 Strategic manoeuvring

The idea of strategic manoeuvring developed by van Eemeren and Houtlosser (1999, 2002) is a good starting point for harmonising different purposes of debates. Strategic manoeuvring understands rhetorical and dialectical aspects of a debate as two sides of the same coin: debaters – as rhetoricians would argue – naturally intend to win the debate by persuading their opponents and/or audience, but at the same time, in declaredly rational debates at least, they also must be fair enough to win the debate rationally – i.e., they cannot use any means for persuasion but only the ones allowed by rules of logic, dialectics, institutional requirements, etc. Balancing between doing whatever they can to win the debate on the one hand, and doing so in accordance with rules on the other is called as 'strategic manoeuvring'.

In terms of means and ends, a rhetorical purpose of a debater is winning the debate by persuasion; a dialectical purpose of her is resolving the dispute in a rational way (i.e., once again, in accordance with the rules of the debate). Her way of achieving her purposes is, accordingly, balancing between persuasion and rule-following argumentation. Ideally, she can win the debate by rationally resolving the dispute, namely in the case if her arguments are persuasive as well as rule-conforming at the same time. That makes her argumentative activity effective as well as reasonable, and if the opponent cannot respond equally well, it makes the debate resolved.

2.2 Epistemic purposes

A standard objection against pragma-dialectics is that they disregard epistemic aspects of a debate. Proponents of the epistemologist tradition argue that persuasion or consensus is a too weak criterion for a rational debate to be successful (Biro-Siegel, 1992). Persuading others is contingent upon the psychological conditions of the opponent(s) and audience. Consensus can be, and often factually is, reached on nonsensical points. Debates aiming at these purposes only are not rational enough. A different requirement should be introduced: rational disputes must aim at accessing truth, and their success should be evaluated on the ground whether they accessed truth or not.

The epistemological approach has its advantages as well as disadvantages in its relation to the pragma-dialectical tradition (Lumer, 2005). In my view, criticism from epistemologists helps pragma-dialectics developing towards a more complex theory: just as pragma-dialectics managed to include the rhetorical approach by introducing the idea of strategic manoeuvring, it can also learn from, and by that assimilate, most insights of epistemological approaches. Argumentative practices as means-end structures with multiple aims provide a suitable framework for doing so: just as dialectical and rhetorical aims can regulate the line of arguments in a debate at the same time, so can epistemic purposes add a further dimension to the picture.

Identifying epistemic purposes of an argumentative practice is essential for the line of thought below. Irresolvable disputes are not problematic insofar as they are not about issues that have some connection with truth. Debates on opinion or taste can be (dis)continued with no resolution because their irresolvability has no undesirable consequences. On subjective questions, no agreement is rationally expected. But agreement is required in objective matters, i.e., when truth is at issue. Debates on aesthetic issues, political or religious values are usually (though not necessarily) taken to be subjective. Truth-seeking enterprises like science are at the other extreme, and it is at least a widely held view that philosophy is also one of these enterprises. All the same, irresolvability seems to be prevalent in philosophical debates. Does that imply that philosophical (and other truth-seeking but irresolvable) debates are irrational to follow? It largely depends on what 'irrational' means.

2.3 Three senses of rationality

All possible senses of rationality cannot of course be mentioned here. Only those will be discussed that are relevant for argumentative practices analysed in means-end structures. Since means-end structures imply rationality to be instrumental, non-instrumental senses of rationality will not be discussed. Still, what makes argumentative practices rational can be understood in more than one sense, depending on which end their means serve primarily. Some are means for rhetorical, others for dialectical, still others for epistemic ends.

Accordingly, three notions of rationality can be introduced.

> (R-rationality): An argumentative move is rhetorically rational if it serves persuasion.

This is clearly a very weak sense of rationality, but still: it is irrational not to aim at persuasion if the arguer is committed to the standpoint she represents in the debate. Aiming at persuasion is not a pure self-interest either: for the best possible outcome of the debate, it is necessary that parties do whatever is in their power to defend their standpoint, as far as it can be defended 'rationally'.

This latter sense of rationality certainly differs from the former. Defending one's point 'rationally' does certainly not mean to act in a rhetorically rational way. If it did, there would be a circularity in the claim that persuasion should be done in a rational way. 'Rationality' in this sense is dialectical:

> (D-rationality): An argumentative move is dialectically rational if it is done in accordance with the rules of logic and dialectics.

These rules can be established implicitly or explicitly prior to the debate; in some cases, they can be even altered in a meta-debate running in parallel with the original. But insofar as they are (at least temporarily) fixed, it is irrational to break them, and if they are broken, it leads to formal errors, informal fallacies, derailments, etc.

D-rationality is rational in a stronger sense than R-rationality. It requires rigorous reasoning rather than psychological traps and tricks, and reasoning is a typical requirement for strong senses of rationality. But if the arguments of the epistemologist tradition are right, there is an even stronger sense of rationality: aiming at truth and knowledge rather than pure consensus. While a D-rational debate is, ideally at least, exempt from contingent psychological factors, it is nonetheless subject to no less contingent limitations of the arguers, e.g. in their argumentative skills, background knowledge, etc. Perhaps an agreement cannot be reached on obvious falsehoods, but it can probably be reached on sophisticated falsehoods if the opponent is not skilled or informed enough to refute them.

As a consequence, epistemic rationality as an even stronger sense of rationality should be introduced:

> (E-rationality): An argumentative movement is epistemically rational if it provides, or contributes to having, an access to truth.

The arguers, or even the analyst, should not necessarily be aware of this contribution; but for the sake of the greatest level of objectivity, E-rationality as an ideal is to be supposed as it is supposed to warrant that in the strongest sense, a debate is rational only if it reaches a true conclusion.

Three criteria of a rational resolution have been distinguished: a debate is rationally resolved if the concluding argument is winning (R-rationality); it is winning in an appropriate way (D-rationality); and it results in something more objective than just winning (E-rationality). Ideally, all these three criteria must be satisfied by a resolution. They can, however, sometimes be in a conflict, and as an effect, a resolution can be rational in some (weaker) sense and irrational in some other sense. It would be, however, an overly strong requirement for being rational that all three criteria be satisfied in every case as aiming at persuasion, consensus or truth, even if the other aims are not met, is rational – at least in some limited sense of the word.

3. FAULTLESS DISAGREEMENTS

A difference between truth and agreement is characteristic in the case of faultless disagreements (Kölbel, 2004). Faultless disagreements occur when participants consist of epistemic peers: they are equally well-skilled and well-trained, having the same empirical evidence, the same epistemic virtues and epistemic abilities, and they also acknowledge that they are epistemic peers of each other (Goldman, 2001), but they disagree in a matter relevant for their epistemic peerhood.

When epistemic peers disagree, they have equally good reasons and equally strong empirical evidence for their views which are nonetheless in a conflict. While their disagreement is R-rational for obvious reasons (all they want to win the debate), disagreement is also D-rational if no decisive evidence or argumentation is provided on either side. In that case, resolving the disagreement between them is D-irrational. Suppose that all possible evidence is uncovered, they provided all their arguments and still, there is no decisive argument for one side or the other. Hence, the dispute is irresolvable rationally (in a strong, dialectical sense).

It has been told that irresolvable disagreement is no problem for fields of subjective judgment. But objectivity requires agreement because it involves that a standpoint and its negation cannot be true at the same time. If both cannot be true but they can be held rationally at the same time, then one of the parties (D-)rationally holds false beliefs. A conflict arises between E- and D-rationality: aiming at holding true beliefs on the one hand, and giving up one's position only if decisive counter-arguments are provided on the other. Hence, one of these purposes cannot be achieved and the debate will be necessarily irrational in one sense or the other.

It seems therefore that D-rationally irresolvable disputes are E-irrational. This consequence is hard to accept: many debates that are normally taken to be E-rational (and philosophical debates in particular) seem to be D-rationally irresolvable. A prima facie promising solution would be to investigate whether that irresolvability is apparent or contingent only. If the debates in question, even if irresolved, were not irresolvable, E-rationality of this sort of debates could be saved. But as I will argue, this does not seem to be the case.

3.1 Disagreement in philosophy

A factual statement that disagreement in philosophy is apparent in many, if not all, questions is hardly questionable. According to a survey of Bourget and Chalmers (2014), even the greatest level of agreement is well below a consensus in philosophical debates. It is not a brave assumption to claim that disagreement in philosophy is factually apparent and general. But what about a normative level? Is disagreement in philosophy necessary? And if so, is it a bad thing?

The answer to these questions I shall develop is that disagreement in philosophy is necessary due to dialectical reasons, but it is not a bad thing because it helps us maintaining objectivity. The claim that disagreement is unavoidable and desirable at the same time is perhaps a seedbed for relativism in its most dangerous form as it allows that "anything goes". However, my reasons differ from a relativist reasoning. I find that disagreement must be held because it helps us being objective: if there is no decisive counterargument against a view, it would be D-irrational to reject it in favour of its rivals. Truth-seeking and objectivity are not opposed to dissensus but on the contrary: the latter is explained by pursuing the former(s). If an access to truth is not warranted, and hence E-rationality is an unrealistic requirement, I claim that the most reasonable thing to do is being dialectically rational; i.e., keeping committed to the position we have previously held as long as no decisive argument forces us to do otherwise.

My reasons are as follows. Regarding unavoidability, if there is no knock-down argument against a philosophical position (and at least in the case of sophisticated enough philosophical positions, this mostly stands), giving up that position is irrational. Obviously, the same holds for the opposite position (this is in fact the main reason why in epistemological debates on disagreement suspending judgment and/or conciliating views are taken as viable alternatives to the so-called 'steadfast' view that keeping committed to one's point is the best strategy – see esp. Lackey ed. 2014). But if my opponent has equally strong arguments for her point, that only means she also should not give up her views. Suspending judgments or harmonising the positions are R-irrational as well as D-irrational: what reasons we would have that make us revise a position that is well-argued (that is why we have accepted it) and no unforeseen counter-argument has been raised against it? From an epistemological point of view, there are reasons for suspension and revision: truth is unwarranted if rival views are equally reasonably held. But a dialectical approach implies that it is necessary to keep committed to one's unrefuted point in order to aim at a D-rational resolution of the dispute. If no rational resolution is possible then (in rational debates) rationality is a priority to resolution, and in those cases, D-rationality implies disagreement.

3.2 Knockdown arguments

The above line of thought takes it for granted what a 'decisive' or 'knockdown' argument is, and that participants of a debate can easily recognise them. A reason for a disagreement can be, however, that knockdown arguments are not easy to identify. In science, decisive experiments can often be identified only from a historical perspective. Similarly, it can be the case that a however good argument can be seen as decisive only after passing a great amount of time as the discussion progresses. A general account of what makes an argument decisive (or a general definition what 'decisive' is) can perhaps solve this problem.

I shall not develop such an account here. However, I think that all possible definitions for 'decisive argument' can be subsumed under one or another category of the following broad definitions:

> (R-decisive) Rhetorically speaking, an argument is decisive if it fully persuades the opponent/a neutral audience.
> (D-decisive) Dialectically speaking, an argument is decisive if it resolves a dispute.
> (E-decisive) Epistemically speaking, an argument is decisive if it results in a true/fully justified conclusion.

None of these broad definitions is suitable for classifying philosophical arguments as decisive, even though they seem to be fruitfully applicable outside philosophical debates. Persuasion is rarely found in philosophical disputes, and the case is similar with resolution. It is very rare that a philosopher finds an argumentation good enough to destroy her position. Similarly, even when relatively wide agreement is reached in what the fundamental questions and best methods in philosophy are, an agreement in the best answer is (perhaps) completely unprecedented. It is somehow in the very nature of philosophical debates that progress, if happens, lies in the garnish only: what questions to ask, what methods to choose, but never what to think or say.

Regarding truth or justification, what can be expected in debates unsupported by empirical evidence is coherence, consistency, clarity, etc. But these expectations can be satisfied by contradictory theories: e.g. if a set of assertions (p_1 & p_2 & p_3 & ... & p_n) is consistent, a set of their negations (non-p_1 & non-p_2 & non-p_3 & ... & non-p_n) is also consistent. Now all that can be done for justification is providing pro arguments and refuting counter-arguments; i.e., playing a game of dialectics. Even if assuming that truth is not the same as consensus, in philosophical debates, there is no access to truth other than debating, a dialectical aim of which is consensus.

All the same, if none of the broad definitions is suitable for taking philosophical arguments as knockdown, it is hard to imagine any narrow characterisation that is more suitable for that task. In one way or another, any definition of a knockdown argument must somehow relate to rhetorical, dialectical or epistemic categories. Insofar as these general characterisations do not seem to provide a good starting point, no further refinement would probably help either because refinement would make criteria more rather than less strict, and stricter criteria are less rather than more satisfiable.

4. REASONS FOR THE DISAGREEMENT

4.1 Internalist vs. externalist explanations

The most important difference between the point I defend and still existing accounts for philosophical disagreement is the reason why disagreement is prevalent in philosophy. These reasons are either internal or external to philosophy. Internal reasons can even support the sort of optimism apparent in Chalmers (2015), according to which the history of philosophy is indeed a progress towards gradually better answers to its questions, and if it seems to be to the contrary, it is so only because this progress is very slow.

Another internalist explanation implies a moderate sort of pessimism, claiming that due to the nature of philosophical questions, there is always more than one answer to them. This pluralist approach is held by Rescher (1978) or Goldman (2010). An advantage of these views to standard textbook relativism is that from the fact that more than one reasonable answer can be given to philosophical questions, they do not conclude that wrong answers cannot be rejected. There are equally or similarly good answers but from this, it does not follow that all answers are good.

A more radical pessimism implies radical scepticism regarding the issue. Associated with authors like van Inwagen (1996), this approach claims that due to the lack of knockdown arguments, philosophical questions cannot be reasonably answered in any way; we may ask the questions if we wish but we would never find reasons to choose among the options to respond them properly.

Externalist explanations follow a different route: they claim that the problem of undecidability can be attributed to factors external to philosophy: sociological, psychological and related reasons are behind this phenomenon. For example, philosophers are psychologically more immune to counter-arguments because they are personally committed to the views they hold; or they are sociologically more inclined to keep committed to a dissensus because philosophical debates constitute the very ground of the legitimacy of their profession – concluding a debate would literally mean that their jobs become worthless.

The opposition between internalism and externalism can be described by the two positions' commitment to different senses of rationality as described above. Internalism is biased towards E-rationality: it supposes that truth and an access to it are fundamental in dissolving disagreements. In contrast, externalism is biased towards R-rationality: its background supposition is that persuasion and hence human (psychological and sociological) factors can always enjoy priority to E-rationality and D-rationality. From the three-fold framework provided above, a third option arises as a viable alternative: rather than (over)emphasising epistemological or rhetorical aspects, at least some of the roots of disagreement should be found in the dialectical aspects of the debates in question.

4.2 A semi-internalist explanation: dialectics

A dialectical approach to faultless disagreement is semi-internal: dialectical reasons are internal to the *debates* in, but external to the *subjects* on which disagreement occurs. Faultless disagreement sets up a dialectical space so that the two (or more) opposing standpoints provide a potentially infinite number of moves in order to defend any of the options. In the lack of empirical evidence and/or decisive arguments, dialectical *expectations* (like resolving the dispute) do not meet dialectical *possibilities* (good enough arguments in the hands of one party or the other to come to a conclusion). That is the reason for the disagreement rather than the nature of subject-specific questions on the one hand, or the psychological/sociological nature of debaters on the other.

A dialectical approach does not take disagreement to be a *bad* thing like pessimist versions of internalism or externalism. But its reason to take it to be *good* differs from the reason of those who welcome relativism due to their commitment to diversity or other values. Disagreement is dialectically a necessary consequence of a lack of decisive arguments. Disagreement is good because and only if an agreement could be reached only on (D-)irrational grounds.

One may argue that disagreement is perhaps dialectically (and/or rhetorically) good but regarding E-rationality at least, disagreement is certainly a bad consequence. If it is impossible to come to a conclusion in philosophical debates then two consequences can follow: either multiple conflicting views must be taken to be true at the same time, or it must be acknowledged that we cannot have access to philosophical truths. The first implies relativism about philosophical truths, the latter implies scepticism (if there are philosophical truths but we cannot know them) or nihilism (if we do not have access to philosophical truths because there are no philosophical truths). None of these options is optimistic.

The semi-internalist can follow different strategies in responding this objection. First, she can join the camp of dialecticians who are anti-realists about truth and hence claim E-rationality to be unimportant (or even impossible) regarding the goals of a debate. Second, she can offer a characterisation of one of the above positions that makes that position attractive. (Relativism is probably the first candidate for attractive pessimism about truth.)

These options are perhaps not too desirable. Anti-realism about truth implies that the epistemic account does not fit into the background suppositions of semi-internalism. It is a too high price to re-open a new front that has just been closed by including E-rationality in the extended model of strategic manoeuvring. Promoting relativism is also a commitment that is unacceptable for many. While it is not a necessary consequence of semi-internalism that relativism should not be held, it is also not a necessary consequence that it should.

Rather than taking these unnecessary commitments to particular philosophical positions, the semi-internalist can argue that she is also a semi-optimist: she is optimistic about the dialectical but pessimistic about the epistemic goals of philosophical debates. Namely, insofar as (1) a debate is D-rational if no agreement is reached in the lack of decisive refutations on either side, and (2) in philosophical debates, no (sophisticated enough) position can be refuted, and (3) in philosophical debates, the debaters' *only access* to truth is (decisive) arguments, therefore, in philosophical debates, no access to truth is warranted. Pessimism about the epistemic purposes of philosophical debates is *a necessary consequence* of optimism about the dialectical purposes *and* the dialectical possibilities provided in philosophical debates. The semi-internalist loses truth in philosophy but at least rationality of the debate is maintained in some strong sense of rationality: philosophical debates are irresolvable *because* they are (D-)rational and philosophers do not accept compromise in arguments.

5. CONCLUSION

This paper has examined cases when rational debates are irresolvable rationally in a dialectical sense of rationality. Based on three senses of rationality relevant for argumentative practices understood in means-end structures and an extended version of strategic manoeuvring, a conflict between two strong senses of rationality has been demonstrated: dialectically rational irresolvability implies epistemic irrationality. Faultless disagreement has been investigated in debates with limited or no empirical evidence like philosophical debates. As argued, disagreement in these cases is not subject-specific but it is due to some dialectical characteristics of these debates. Disagreement has been found as necessary but desirable in these cases: necessary because a lack of decisive arguments disables consensus on dialectical grounds; and desirable because the only way to maintain objectivity in epistemically irrational situations is committing to dialectical rationality, and dialectical rationality requires dissensus in such situations.

ACKNOWLEDGEMENTS: Research has been supported by the Hungarian Scientific Research Fund (OTKA K-109456 and K-116191).

REFERENCES

Biro, J. & Siegel, H. (1992). Normativity, argumentation and an epistemic theory of fallacies. In F. H. van Eemeren, R. Grootendorst, J. A. Blair, & C. A. Willard (Eds.), *Argumentation illuminated* (pp. 85-103). Amsterdam: SicSat.

Biro, J., & Siegel, H. (2006) Pragma-dialectic versus epistemic theories of arguing and arguments: rivals or partners? In P. Houtlosser, & A. van Rees (Eds.), *Considering pragma-dialectics: A festschrift for Frans H. van Eemeren on the occasion of his 60th birthday* (pp. 1-11). New York: Routledge.

Bourget, D. & Chalmers, D. J. (2014). What do philosophers believe? *Philosophical Studies, 170*(3), 465–500.

Chalmers, D. J. (2015). Why isn't there more progress in philosophy? *Philosophy, 90*(1), 3-31.

Eemeren, F. H. van & Houtlosser, P. (1999). Strategic maneuvering in argumentative discourse. *Discourse Studies*, 1(4), 479-497.

Eemeren, F. H. van & Houtlosser, P. (2002). Strategic maneuvering: maintaining a delicate balance. In F. H. van Eemeren and P. Houtlosser (Eds.), *Dialectic and rhetoric: The warp and woof of argumentation analysis* (pp. 131-59). Dordrecht: Kluwer Academic Publishers.

Goldman, A. (2010). Epistemic relativism and reasonable disagreement. In R. Feldman, & T. Warfield (Eds.), *Disagreement* (pp. 187-215). Oxford: Oxford University Press.

Goldman, A. I. (2001). Experts: which ones should you trust? *Philosophy and Phenomenological Research, 63*(1), 85-110.

Inwagen, P. van (1996). It is wrong, everywhere, always, for anyone, to believe anything upon insufficient evidence. In J. Jordan, & D. Howard-Snyder (Eds.), *Faith, freedom and rationality* (pp. 137-154). Savage, Maryland: Rowman and Littlefield.

Kölbel, M. (2004). Faultless disagreement. *Proceedings of the Aristotelian Society, New Series, 104*(1), 53-73.

Kvernbekk, T. (2008). Johnson, MacIntyre, and the practice of argumentation. *Informal Logic*, *28*(3), 262-278.

Lackey, J. Ed. (2014). *Essays in collective epistemology.* Oxford: Oxford University Press.

Lumer, C. (2005). The epistemological approach to argumentation - a map. *Informal Logic*, *25*(3), 189-212.

MacIntyre, A. (1981). *After virtue. A study in moral theory.* Notre Dame: University of Notre Dame Press.

Rescher, N. (1978). Philosophical disagreement: an essay towards orientational pluralism in metaphilosophy. *The Review of Metaphysics*, *32*(2), 217-251.

16

Processing Linguistic Persuasion: A Study in Message Discrepancy

KAMILA DEBOWSKA-KOZLOWSKA
*Faculty of English, Department of Pragmatics of English,
Adam Mickiewicz University in Poznan, Poland*
kamila@wa.amu.edu.pl

A study using the Implicit Association Test programmed in E-Prime software is presented. The study seeks to explore the influence of a premessage counter-attitude and pro-attitude to an issue on the implicit processing of persuasive communication. The aim is to see how the results of the study on implicit evaluations might relate to the models and findings concerning explicit evaluations. The question whether implicit attitudes might mediate the effects of the unfavourable cognitive responses in explicit processing is explored.

KEYWORDS: cognitive responses, explicit, implicit, message discrepancy, persuasion

1. INTRODUCTION

The blossoming of experimental pragmatic research (Noveck & Sperber, 2004; Jończyk, 2016; Bąk, 2016) has opened the possibility to apply social and cognitive psychology measurement techniques to study the implicit evaluative reactions in relation to language use. The concept of implicitness in processing language has been thoroughly studied in linguistic pragmatics (e.g. Sperber and Wilson, 1995; Bromberek-Dyzman 2014). However, the influence of persuasiveness of linguistic material on spontaneous evaluations hasn't been a centre of its experimental investigations. This paper sets to explore the link between persuasive language use and implicit evaluations in experimental conditions.

The main interest of the present paper is the role of the processing of message discrepancy in implicit evaluations of persuasive language use. Message discrepancy is defined here as the difference between a position that a given message supports and a position of its recipient to the issue represented in the message that the recipient has held before being acquainted with the message (see also Fink & Cai, 2013). The concept of message discrepancy sets the ground for measuring a change in a recipient's position after being exposed to the message. The position of a recipient towards an issue is understood here as being represented by recipient's opinions, attitudes, belief and values with relation to the issue.

A study is presented that seeks to explore the influence of a premessage counter-attitude and pro-attitude to an issue on the implicit processing of a persuasive language content. A more specific aim is to see how the results of the study on implicit evaluations might relate to the models and findings concerning explicit evaluations. The paper focuses on the question whether implicit attitudes might induce the effects of unfavourable cognitive responses that occur while explicit processing. In other words, the question that has motivated the present paper centres on the idea whether implicit evaluations might be seen as uncontrolled mediators of final explicit opinion changes. Extremely variable correlations between implicit and explicit evaluations have been found in the literature on attitude change (see Devos, 2008). Some additional factors such as a motivation to hide a real preference not to expose prejudice have been reported to influence implicit-explicit dissociations. The unwillingness to truly self-report explicit attitudes is discussed in Devos (2008, pp. 69-70). This paper, however, pertains to those relationships between implicit and explicit evaluations that involved no internal or external motivation to misrepresent a preference. In such cases, strong correlations between explicit and implicit evaluations have been revealed (see Devos, 2008, p. 70).

Cognitive responses have long been explored in persuasion research, also in the message discrepancy research paradigm (Fink & Cai, 2013). Cognitive responses are characterised here as any kinds of thoughts that emerge in an individual's mind as a way of responding, elaborating on or predicting any kind of communication. In this paper, the focus is on a written type of persuasive communication that involves argumentative illocutionary force. According to the cognitive response approach (Greenwald, 1968), whenever an individual experiences a persuasive attempt, the incoming piece of communication is placed in the cognitive system of an individual. In other words, the knowledge the individual possesses is related to the new linguistic or non-linguistic material. The cognitions that pass through an individual's mind might be either 'supportive' or 'antagonistic' towards the persuasive communication. The supportive vs. antagonistic division of cognitive responses is often discussed in terms of favourable vs. unfavourable thoughts that arise in relation to communication (Petty et al., 2014, p. 13; Ciacioppo et al., 2014, pp. 41-42). Other categories than favourable or unfavourable thoughts have also been discussed in the research on persuasion to emphasise the 'polarity dimension' (Ciacioppo et al., 2014, pp. 41-42). This paper, however, employs the favourable/unfavourable categorization. Both favourable implicit cognitions that enhance processing of the message and unfavourable implicit cognitions that slow down the processing are of interest to the present paper (c.f. Greenwald et al., 2003). However, the hypothesis of the study is that only implicit favourable cognitions that are in agreement with a persuasive message are generated for both pro-attitude and counter-attitude participants.

Contrary to some mathematical models of the relation between opinion change and discrepancy (e.g. the push with pull back model), the experimental research on message discrepancy shows that unfavourable cognitive responses in the form of oscillations or counterarguing often do not hamper the influence of the persuasive language content on final explicit evaluations. This is the case if the source of a persuasive language use is highly credible. The question that arises is why some mathematical predictions are not confirmed in the results of experimental studies. Clearly, the existing mathematical models do not capture the attitude changes that result from automatic cognitive processing underlying people's spontaneous judgments and spontaneous social behaviours. Such implicit attitude changes are based on the past experience that cannot be accessed introspectively. It has been noted in the literature that direct experience precedes the stimulation of attitudes (see Regan & Fazio, 1977). The contact with a persuasive language use stimulates also those attitudes that function outside of conscious awareness and control. If a stimulated implicit attitude is strong then "crystallization, volatility, and stability, are consequences of [the] attitude strength" (Bassili, 2008, p. 238). Following Krosnick and Petty (1995) I define strong attitudes as those evaluations that are able to endure any type of unfavourable cognitive responses (e.g. counter-argumentation). I agree here with Smith et al. (2013) that direct contact with a persuasive language use might stimulate strong implicit attitudes. These implicit strong attitudes hinder or milden the processing of explicit/conscious counter-argumentation and lead to the conscious self-report of the adoption of a modified attitude if there is no need to conceal an individual's preference.

To verify the idea about the impact of implicit attitudes, a study is performed to check whether both the participants with a pro-message attitude and the participants with a counter-message attitude implicitly accept a persuasive message (through the production of favourable implicit cognitions towards the message) or whether the patterns for acceptance are different depending on message discrepancy. Convergent positive patterns for both groups in implicit evaluations could explain why some models of explicit evaluation are not confirmed in the explicit evaluation studies.

Section 2 of the present paper briefly discusses two lines of traditions for the use of the terms 'implicit' and 'explicit' that have been developed in the research on implicit cognition and have dominated the field of linguistic pragmatics (Bromberek-Dyzman, 2014). These two lines of tradition will allow me to better conceptualise the nature of the two types of attitudes that are of interest to the present paper, namely an implicit postmessage attitude and an explicit postmessage attitude. Section 3 presents two general trends for the relationship between opinion change and discrepancy in the explicit postmessage evaluation models and studies. Special attention is drawn to nonmonotonic, nonlinear and dynamic models. Section 4 presents the study that aims to show the immediate influence of a discrepant vs. non-discrepant text presented by the same qualified communicator on implicit evaluations of the same population. Section 5 discusses the results of the study and its implications for the final explicit postmessage evaluations.

2. IMPLICIT AND EXPLICIT COGNITION

Two lines of traditions for the use of the term 'implicit' that gave rise to distinct uses of the term have been distinguished. One of the traditions refers to the research on selective attention the other relates to the research on implicit memory.

The first line of the research (e.g. Broadbent, 1971; Treisman, 1969) inspired cognitive psychologists to distinguish between automatic and controlled modes of information processing. In this line of research, the term 'implicit' was used as synonymous with automatic and involuntary. Implicit/automatic/involuntary processing was described as involving just some attention and no individual's will for starting, stopping or modifying information processing. On the contrary, controlled processing was described as needing attention and individual's own will.

The second line of research, initiated by Greenwald and Banaji's (1995) review, introduced the conceptions of 'unconsciousness' and 'consciousness' into the discussion of implicitness and explicitness. Drawing some parallels between implicit memory and implicit attitudes, Greenwald and Banaji's (1995) point out to the unifying role of past experience. Implicit memory is described as the result of past experience that has been unconsciously acquired and is unconsciously realized through behaviour. Implicit attitude is perceived as fragments of past experience that that cannot be accessed consciously but generate cognitive responses that are either favourable or unfavourable. In contrast, explicit memory and explicit attitudes are identified as relying on conscious awareness and thus conscious processing of past experience.

Relying on the above research traditions, I define an implicit postmessage attitude as an unconscious evaluation that is influenced by traces of past experience that cannot be accessed introspectively at the moment of the evaluation. In contrast, an explicit postmessage attitude is understood here as a conscious evaluation that is influenced by past experience that is easily accessed introspectively at the moment of the evaluation.

In the next section, both mathematical models and studies of explicit postmessage beliefs with relation to discrepancy levels will be discussed. The influence of a discrepancy level on opinion change in the case of two credibility levels, i.e. a moderate and high credibility level, will be considered. Dynamic models' (e.g. the push with pull back model) predictions about the production and consequences of unfavourable cognitive responses towards a persuasive message will be juxtaposed with the results of experimental studies.

3. MESSAGE DISCREPANCY IN EXPLICIT EVALUATIONS

The relation between discrepancy and the successful outcome of a persuasive attempt needs to be considered in relation to some other variables to produce a more comprehensive picture (O'Keefe, 2002, p. 222). Only then can the differentiated, or even contradictory, results of some investigations be understood. Two exclusive trends have prevailed in the literature on discrepancy. One trend promotes a positive relationship between the size of discrepancy and the amount of attitude change obtained by a persuasive message. The other trend shows a negative relationship between attitude change and discrepancy (e.g. Cohen, 1959). In other words, the greater the discrepancy the greater or the lower the effectiveness of a persuasive message.

The linear discrepancy model (Anderson & Hovland, 1957), also known as the linear balance model, the distance proportional model and the proportional change model (e.g. Danes et al., 1978) represents the positive relationship. As Chung et al. (2008, p. 160) indicate, the model assumes, that "the amount of the belief change is a linear function of message discrepancy". Attention to and comprehension of the message are crucial according to the model since they facilitate a transition from subject's pre to post message attitude. The calculation of the transition is possible due to a fact that all variables, even if initially implicit, can be quantified by the application of the model's formulas (Fink & Cai, 2013, p. 86).

The linear model does not study the course of the belief change but only assumes that the time interval between reading or hearing a message and adoption of a modified position is enough for the message to sink in and become part of the conceptual system of an individual. The general assumptions of the model are similar to the assumptions of Anderson's (1974) information integration theory. Anderson's primary attention is paid to the close relation between incoming information and the final change, not to the way the belief change evolves over time. Still, the idea that time is the motor for change of beliefs is the centre of the theory's claims.

Many studies have confirmed that a linear model proposes an ideal view of the change that occurs in the belief system (see Fink & Cai, 2013). Some researchers (e.g. Chung et al., 2008) observe that linearity occurs only up to a certain point. For examples, it is noticed that belief change in a high credibility condition is most operative only when the discrepancy is small and moderate. In the cases of big discrepancy paired with high credibility the change that occurs is very subtle (see Aronson et al., 1963). This type of change from substantial to subtle has been described as nonlinear.

Fink and Cai (2013, p. 95) indicate that the nonlinear model by Fink et al. (1983) is characterized not only by monotonic but also nonmonotonic results and "was found to be statistically superior to the linear discrepancy model".

O'Keefe (2002, p. 222-223) indicates that nonmonotonicity (i.e. curvilinearity) describes a change that is increasing up to a certain point and then suddenly decreasing. When discrepancy increases, the attitude change increases as well but only for the cases of small levels of discrepancy. The peak of the curvilinear/nonmonotonic model is the moderate level of discrepancy. When discrepancy increases further (i.e. larger levels of discrepancy are considered), the attitude change starts to decrease. O'Keefe (2002) underlies that due to various circumstances in which a given persuasive message is processed, we should talk about a number of shapes of U inverted curves. The peak of the curve depends on other factors that accompany the discrepancy level. Credibility of the source and involvement of the receiver have been typically considered the most important factors. The general conclusion is that high-credibility sources are more effective with larger levels of discrepancy than low-credibility sources (Bochner & Insko, 1966). In contrast, the higher is the ego-involvement of the receiver, the smaller discrepancy levels are effective (see Kaplowitz & Fink, 1997, p. 78).

Laroche's (1977) nonlinear model presupposes that even for the largest discrepancy levels in the case of high credibility and low-ego involvement there will be no nonmonotonicity (c.f. Bochner & Insko, 1966). Laroche's (1977) model sets a formula that allows to see how source credibility, ego involvement and message discrepancy interact with each other. In Laroche's model, the main parameter of belief change, γ, is treated as the function of source credibility and non ego involvement. Chung et al. (2008, p. 161) explains that γ allows to specify "the degree to which the relationship between message discrepancy and belief change departs from linearity". γ_{Eq} is defined as "the belief position at equilibrium" (Fink & Cai, 2013, p. 94). The amount of belief change is measured from a pre-message position (i.e. an initial position). The following dependencies were proposed by Laroche (1977; see also Chung et al, 2008 p. 161) based on his model. If there is low credibility the relation between message discrepancy and belief change is nonmonotonic ($\gamma > 1$). If there is high credibility the relation is assumed to be monotonic (for $0 < \gamma < 1$).

Yet another way to approach the conception of discrepancy is to use the dynamic approach to belief change that has been inspired by McGuire's claims (1960). Dynamic models study belief change from one state of equilibrium to another (cf. Fink et al., 2002) and, in accordance with McGuire's claims (1960), indicate that after an individual is acquainted with a message the process of belief change is already initiated. The important aspect of the dynamic model is the idea of time. Although the importance of time interval for a position adoption is also mentioned in the standard nonlinear models, the dynamic models emphasize that the change might go back and forth in the process of belief modification. In this approach special attention is paid to unfavourable cognitive responses. Since the focus is on the non-straightforward way of the changing process itself, the dynamic models consider any fluctuations that happen on the way to the change. A Newtonian metaphor is often applied to explain the possible alternations before a new equilibrium belief value is obtained and the process of change is terminated (e.g. Kaplowitz et al., 1983). Using a Newtonian metaphor, Kaplowitz et al. (1983) argues that any concept in the mind of an individual can be perceived as having both location and mass. Any modification of a belief is represented as motion in the cognitive system of an individual. This motion is considered in terms of three forces. Except for a single push that induces motion leading to belief change, there are also two other forces. One of them initiates oscillation due to interattitude pressures, limitations and cognitive responses. The other one, "a frictional force", decelerates "the initial motion from the push and any oscillations" (Chung et al., 2008 p. 163) and puts it to an end.

Kaplowitz's et al. (1983)'s dynamic model has been referred to as the single-push with friction model by other scholars (e.g. Fink & Cai, 2013, p. 96). Kaplowitz et al. (1983)'s study with time used as a between subject variable and Kaplowitz and Fink's study with time used as a within subject variable (Kaplowitz & Fink, 1997, p. 96-99) were respectively designed to study cognitive responses that might lead to nonmonotonicity (i.e. to prove or disprove the assumptions of the single-push with friction model). In Kaplowitz et al. (1983)'s study with a between subject design, opinions of participants were measured after the message was presented. In this high involvement study, time and discrepancy were manipulated. The time of the measurement differed between participants from immediate to 10 minutes from the moment they were acquainted with the message. The study substantiated the claim of its designers that the belief trajectories of participants would display some opinion oscillation with a period of oscillation of about 13.5 seconds. The messages with extremely large discrepancy produced the greatest oscillation. Kaplowitz et al. (1983) final conclusion was that due to some oscillation message effectiveness "may be determined by the time interval from message to measurement" (Kaplowitz et al., 1983 p. 247). In the study with time used as between subject variable, the relationship between discrepancy and the position of the participant at three different points (10 sec., 20 sec. after reading the message and the final position) was investigated. Credibility and discrepancy were manipulated in a high involvement and low involvement scenario. When the source was highly credible there was a monotonic relation with a linear trend between discrepancy and final opinion for both high involvement and low involvement messages. However, when there was low credibility, discrepancy didn't have a significant effect on the final position for low involvement messages. Still, there was a significant linear relationship between discrepancy and final position for the high involvement messages. No nonmonotonic relationship was found. Thus, for both scenarios, after an initial change in the opinion, the cognitive responses (e.g. counter-arguments) didn't lead to the total rejection of the advocated position for all levels of discrepancy. Generally the same conclusions were reached in the study by Chung et al. (2008) that used a design similar to Kaplowitz and Fink's (although eleven time points at a participant's trajectory were studied). Although Chung et al. (2008) did not study any oscillations, the predictions of their push with pull back model about the production of counterarguments and rejection of the advocated position were generally not confirmed in their experiments. Nonmonotonicity between discrepancy and opinion change was shown only in a low credibility and low involvement condition.

The next sections will check whether monotonic, or even linear, trends for the relationship between high credibility and discrepancy level in the explicit attitude processing research might be motivated by the stimulation of strong implicit attitudes that are realized by the production of regular and repeating favourable cognitive responses towards a message. Indirect method of the Implicit Association Test will be applied to check whether the patterns for processing counterattitudinal and proattitudinal content are divergent or convergent.

4. THE STUDY: METHOD

4.1 Participants

The total number of 67 students of the Faculty of English at Adam Mickiewicz University (AMU), Poznan, Poland, who were highly proficient in English, took part in the study. All participants performed the Lexical Test for Advanced Learners of English (LexTale) and the Language History Questionnaire (the Language and Communication Laboratory custom version of LHQ 2.0 designed at the Brain, Language and Computation Lab at Penn State University). The average score in LexTale was 79%. In the LHQ, all participants indicated Polish as their first language and English as their second language. Their level of proficiency in English was between C1 and C2 according to the Common European Framework of Reference for Languages.

4.2 Hypothesis

The focus of this study was on implicit postmessage attitudes (i.e. unconscious evaluations) that were supposed to result from reading persuasive messages. A low involvement issue of junk food was applied as a persuasive scenario. The persuasive messages discouraged the consumption of junk food by describing its bad consequences (e.g.

"eating junk food contributes to high blood pressure"). The hypothesis of the study was that only implicit favourable cognitions that are in agreement with the persuasive messages will be generated for both pro-attitude and counter-attitude participants. In other words, the hypothesis was that the persuasive linguistic material will have the same effect on the pro-attitude and counter-attitude group. Implicit favourable cognitive responses towards the persuasive scenario were assumed to be realised through faster processing of evaluative stimuli in a message congruent experimental task than in a message incongruent experimental task.

4.3 Materials and procedure

On entering the Language and Communication Laboratory at the Faculty of English, AMU, participants were asked to characterise their attitude to junk food on a typical Likert scale. After subjective rating of their attitudes participants were divided into two groups: a counter-attitude group and a pro-attitude group. Subsequently, all participants read a note on a computer screen about bad consequences of eating junk food that was said to be prepared by both a doctor and an expert in a healthy diet. The high expertise condition and the same text about bad consequences were used for both groups. The counter-attitude group and the pro-attitude group performed the same Implicit Association Test afterwards. The standard IAT measures the strength of associations between pairs of target concepts and an attribute dimension and involves categorization of stimuli (Greenwald et al, 2003). The IAT, programmed in E-Prime software, was used to present stimuli and collect responses. Accuracy and response time were measured. The response time statistical analysis involved only correct responses. Blocks 1, 2, 3, 5, 6 of the study were training sessions. In the critical data collection experimental sessions (i.e. block 4 and block 7), concept attribute pairing was performed. In block 4, the categories *Health Food* and *I like* shared one response key (E) and the categories *Junk Food* and *I dislike* shared another response key (I). Block 4 was labelled message congruent. In block 7, the categories *Health Food* and *I dislike* shared one response key (E) and categories *Junk Food* and *I like* shared another response key (I). Block 7 was labelled message incongruent. Five adjectives with positive meaning and five adjectives with negative meaning were chosen for the evaluative stimuli that appeared in the middle of the screen and needed to be categorized in the super-ordinate categories.

4.4 Results

In the study, only the response times (RT) from the experimental sessions were taken for the analysis (RT in training sessions was not considered, cf. Greenwald et al., 2003). A repeated measures ANOVA 2x2 was performed for participants on response time (RT) and accuracy scores. RT latencies are of interest to the present paper. Two groups that differed in message discrepancy (a counter-attitude group and a pro-attitude group to the message) went through the same experimental conditions. They first read the information about bad consequences of eating junk food that was said to be prepared by a professional expert in a healthy diet and then performed the IAT. In Table 1, group 1 is represented by a pro-attitude group while group 2 stands for a counter-attitude group. The log-transformation on the RTs was used. LogRT_mean.4 refers to LogRT in the message congruent experimental session. LogRT_mean.7 relates to LogRT in the message incongruent experimental session.

Descriptive Statistics

	Group	Mean	Std. Deviation	N
LogRT_mean.4	1	6,4242	,11242	37
	2	6,3886	,12525	30
	Total	6,4082	,11876	67
LogRT_mean.7	1	6,5093	,17175	37
	2	6,5328	,18671	30
	Total	6,5198	,17762	67

Table 1 – LogRT, SD and number of participants in group 1 and group 2 for the congruent and incongruent experimental session

The overall ANOVA revealed a highly significant congruency effect: $F(1,65)=32.524$; $p<.0001$, with the experimental effect size $\eta^2=.333$. As shown in Figure 1, in both groups that differed in message discrepancy, message congruency was processed faster than message incongruence. Congruency 1 relates to message congruent session. Congruency 2 refers to message incongruent session.

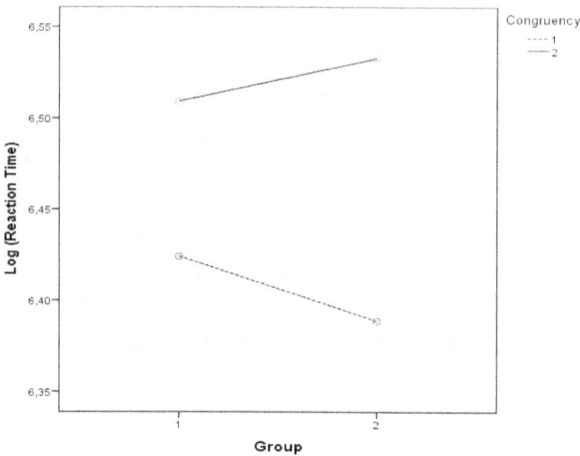

Figure 1 – LogRT for a pro-attitude group (group1) and a counter-attitude group (group 2)

The interaction between group and congruency showed no significant effect: F(1,65)=2.162, p.146. There was no group effect: F(1,65)=.037, p.849. There was no significant difference in LogRT between the counter-attitude and pro-attitude group.

5. CONCLUSIONS

The hypothesis of the study was confirmed. Implicit favourable cognitive responses towards the persuasive text were exposed through faster processing of message congruency than message incongruence. In both groups, the LogRT was significantly reduced when participants who were in favour of junk food processed the message congruent block (against junk food) than when they processed the message incongruent block (for junk food). The level of personal involvement was not manipulated. The study focused only on a low involvement issue that presented general bad consequences of eating junk food, but didn't focus on any immediate personal consequences for the lives of participants. Thus, high expertise had an effect on implicit evaluations in both groups in a specific low involvement scenario type. The results of the study indicate that for lower involvement issues implicit evaluations of persuasive language use might underlie further explicit decisions concerning the acceptance of persuasive language content. As the study showed, in a high expertise and low involvement condition, message congruent implicit cognitions are effectively stimulated and highly consistent. Taking into account the results of the present study and Bassili's (2008) and Devos' (2008) remarks on the implicit-explicit bond, it appears that the stimulation and consistency of message congruent implicit cognitions increases the accessibility of the prevailing implicit attitude that later shows up in an explicit evaluation which, therefore, might become resistant to counterarguing. Stimulation, repetition, consistency and durability of implicit message cognitions might thus explain why explicit belief change over time has a monotonic nature in high credibility conditions. This implication pertains only to the acceptance of the persuasive langue content that does not motivate internal or external concealment of a further explicit preference.

REFERENCES

Anderson, N. H. (1974). Cognitive algebra: Integration theory applied to social attribution. In L. Berkowitz (Ed.), *Advances in experimental social psychology* (Vol. 7, pp. 1–101). New York, NY: Academic Press.

Anderson, N. H., & Hovland, C. (1957). The representation of order effects in communication research. In C. Hovland, W. Mandell, E. H. Campbell, T. Brock, A. S. Luchins, A. R. Cohen, et al. (Eds.), *The order of presentation in persuasion* (pp. 158–169). New Haven, CT: Yale University Press.

Aronson, E., Turner, J. A., & Carlsmith, J. M. (1963). Communicator credibility and communication discrepancy as determinants of opinion change. *Journal of Abnormal and Social Psychology*, 67(1), 31–36.

Bassili, J.N. (2008) Attitude strength. In W. D. Crano, and R. Prislin (Eds.), *Attitudes and attitude change* (pp. 237-260). New York: Psychology Press.

Bąk H. 2016. *Emotional prosody processing for non-native English speakers: Towards an integrative emotion paradigm.* Switzerland: Springer.

Bochner, S., & Insko, C. A. (1966). Communicator discrepancy, source credibility, opinion change. *Journal of Personality and Social Psychology*, 4(6), 614–621.

Broadbent, D. (1971). *Decision and stress.* Oxford, UK: Academic Press.

Bromberek-Dyzman K. 2014. *Attitude and language: On explicit and implicit attitudinal meaning processing.* Poznań: Wydawnictwo Naukowe UAM.

Chung, S., Fink, E. L., & Kaplowitz, S. A. (2008). The comparative statics and dynamics of beliefs: The effect of message discrepancy and source credibility. *Communication Monographs*, 75(2), 158–189.

Ciacioppo, J. T., Harkins S. G., & Petty R. E. (2014). The nature of attitudes and cognitive responses and their relationships to behavior. In R. E. Petty, T. M. Ostrom, T. C. Brock (Eds.), *Cognitive responses in persuasion* (pp. 31-54). New York: Psychology Press.

Cohen, A. R. (1959). Some implications of self-esteem for social influence. In C. I. Hovland & I. L. Janis (Eds.), *Personality and persuasibility* (pp. 102-120). Oxford, England: Yale University Press.

Danes, J. E., Hunter, J. E., & Woelfel, J. (1978). Mass communication and belief change: A test of three mathematical models. *Human Communication Research*, 4(3), 243–252.

Devos, T. (2008). Implicit attitudes 101: Theoretical and empirical insights. In W. D. Crano & R. Prislin (Eds.), *Attitudes and attitude change* (pp. 61-84). New York: Psychology Press.

Fink, E. L., Kaplowitz, S. A., & Bauer, C. L. (1983). Positional discrepancy, psychological discrepancy, and attitude change: Experimental tests of some mathematical models. *Communication Monographs*, 50(4), 413–430.

Fink, E. L., Kaplowitz, S. A., & Hubbard, S. E. (2002). Oscillation in beliefs and decisions. In J. P. Dillard & M. Pfau (Eds.), *The persuasion handbook: Theory and practice* (pp. 17–37). Thousand Oaks, CA: Sage.

Fink, E. L., & Cai, D. A. (2013). Discrepancy models of belief change. In J. P. Dillard & L. Chen (Eds.), *The SAGE handbook of persuasion: Developments in theory and practice* (2nd ed., pp. 84-103). Los Angeles, CA: SAGE Publications, Inc.

Greenwald, A. G. (1968). Cognitive learning, cognitive response to persuasion and attitude change. In A. Greenwald, T. Brock, & T. Ostrom (Eds.), *Psychological foundations of attitudes* (pp. 147-170). New York: Academic Press.

Greenwald, A. G., & Banaji, M. R. (1995). Implicit social cognition: Attitudes, self-esteem, and stereotypes. *Psychological Review*, 102, 4-27.

Greenwald, A. G., Nosek, B.A. & Banaji, M. R. (2003). Understanding and using the Implicit Association Test: I. An improved scoring algorithm. *Journal of Personality and Social Psychology*, 85(2), 197-216.

Jończyk R. 2016. *Affect-language interactions in native and non-native English speakers: A neuropragmatic perspective.* Switzerland: Springer.

Kaplowitz, S. A., Fink, E. L., & Bauer, C. L. (1983). A dynamic model of the effect of discrepant information on unidimensional attitude change. *Behavioral Science*, 28(3), 233–250.

Kaplowitz, S. A., & Fink, E. L. (1997). Message discrepancy and persuasion. In G. A. Barnett & F. J. Boster (Eds.), *Progress in communication sciences* (Vol. 13, pp. 75–106). Greenwich, CT: Ablex.

Krosnick, J. A., & Petty, R. E. (1995). Attitude strength: An overview. In R. E. Petty & J. A. Krosnick (Eds.), *Attitude strength: Antecedents and consequences* (pp. 1–24). Mahwah, NJ: Erlbaum.

Laroche, M. (1977). A model of attitude change in groups following a persuasive communication: An attempt at formalizing research findings. *Behavioral Science*, 22(4), 246–257.

McGuire, W. J. (1960). Cognitive consistency and attitude change. *Journal of Abnormal and Social Psychology*, 60(3), 345-353.

Noveck, I., & Sperber, D. (Eds.). 2004. *Experimental pragmatics.* Basingstoke: Palgrave Macmillan.
O'Keefe, D.J. (2002). *Persuasion: Theory and research.* Thousand Oaks, CA: Sage.
Petty, R.E., Ostrom T.M., Brock T.C. (2014). Historical foundations of the cognitive responses approach to attitudes and persuasion. In R. E. Petty, T. M. Ostrom, T. C. Brock (Eds.), *Cognitive responses in persuasion* (pp 5-29). New York: Psychology Press.
Regan, D. T., & Fazio, R. (1977). On the consistency between attitudes and behavior: Look to the method of attitude formation. *Journal of Experimental Social Psychology*, 13(1), 28–45.
Smith C. T., De Houwer J, Nosek BA. (2013). Consider the source: Persuasion of implicit evaluations is moderated by source credibility. *Personality and Social Psychology Bulletin*, 39(2), 193–205.
Sperber, D. & Wilson, D. (1995). *Relevance theory: Communication and cognition.* Oxford: Blackwell.
Treisman, A. (1969). Strategies and models of selective attention. *Psychological Review*, 76(3), 282–299.

17

Questioning the Explicit Cancellability of Scalar Implicatures

LAURA DEVLESSCHOUWER
University of Antwerp & Université Libre de Bruxelles
(FWO fellowship)
devlessch@mail.com

The dominant view of scalar implicatures (e.g. the "Not all…"-implicature associated with "Some…") is that they are pragmatically inferred rather than conventional and encoded. One of the main arguments for this view is the fact that implicatures are explicitly 'cancellable' (e.g. by saying "Some, but not all, …"). However, when taking into account Anscombre and Ducrot's (1983) theory of argumentative scales, 'cancellability' is no longer an obstacle to a conventionalist theory of scalar implicatures.

argumentative orientation, implicature cancellation, scalar implicatures

1. INTRODUCTION

It is generally agreed that from the sentence "Some of my friends are Dutch", one can infer the information that *not all* of the speaker's friends are Dutch. The question is whether this information is *entailed* by the sentence, and whether 'not all' is part of the semantics of "some". The answer of logicians and Griceans to either of these questions is univocally "No". The reason is that "All of my friends are Dutch" entails "Some of my friends are Dutch". Indeed, what is true of the members of a superset ("all") is true of the members of the subset ("some"). Therefore, if a speaker says "Some of my friends are Dutch" while in fact *all* of her friends are Dutch, this is (strictly speaking) not a lie, but merely an *underinformative* utterance (cf. Carston, 2013, p. 9). The speaker has not said the whole truth, but she did say a part of the truth. However, it is generally expected that speakers do tell the whole truth.

This is part of being *cooperative* (Grice, 1989). Therefore, by saying "Some of my friends are Dutch", a speaker generally suggests or *implicates* that not all of her friends are Dutch. Thus, 'Not all of my friends are Dutch' is an implicature of the utterance "Some of my friends are Dutch". This implicature is in turn *inferred* by the hearer. The reasoning roughly proceeds as follows:

i. The speaker has said "Some...". She could have said "All...", which would have been more informative, but she did not.
ii. There is no reason to assume that the speaker is not being cooperative. So, if she did not say "All", she probably has a good reason for doing so.
iii. The reason is that she believes that "All..." is not the case.[1]

(freely from Geurts, 2010, p. 32).

One of the main arguments for regarding 'not all' as pragmatically inferred rather than part of the conventional, encoded meaning of "some", is that the inference is *cancellable* (cf. Geurts, 2010, p. 82; Levinson, 2000, p. 57; Recanati, 2010, p. 152). There are two types of cancellation: explicit and contextual (Grice, 1989, p. 44). In contextual cancellation, the context makes clear that the implicature is not intended. For instance, if I say "I hope that some of you will like my paper", it is clear that I do not hope that *not all* of you will like my paper, but rather, that *at least* some of you will like my paper.[2] In explicit cancellation, the speaker explicitly indicates that the opposite of the implicature is true, e.g. by saying "Some, indeed all of my friends like beer" or "Some of my friends like beer. In fact, all of them do". Contrary to implicatures, which are pragmatic, semantic meanings cannot be cancelled: one cannot say "Some, in fact none" (Geurts, 2010, p. 83). Therefore, it is argued that 'not all', since it is cancellable, cannot be part of the semantics of "some". In the present paper, I will criticize the explicit cancellability criterion and defend a semantic theory of scalar

---

[1] In fact, one can only infer that the speaker does not (necessarily) believe that "All..." is the case. This is compatible with the speaker knowing that "All..." is not the case, but also with the speaker *not knowing* whether or not "All..." is the case (cf. Geurts, 2010, p. 29). In the present paper, I will only consider the most widely investigated type of scalar implicature, namely the one whereby it is assumed that the speaker knows that the more informative statement is not the case.

[2] In future work, however, I contest the view that "some" means 'at least some' here. Also, I criticize the view that, in order to convey 'not all', "some" should be paraphrasable by "not all".

implicatures. Building on Anscombre and Ducrot's (1983) theory of 'argumentation in language', I will argue that explicit cancellations are attributable to argumentative phenomena, rather than to entailment logic.

The paper is structured as follows. First, I will introduce Anscombre and Ducrot's (1983) analysis of scalar terms, as well as a possible alternative account of explicit cancellation. Then, I will discuss five problems with explicit cancellability. When discussing the last problem, the uncancellability of some alleged scalar implicatures, it will transpire that it is argumentative scales rather than entailment scales that give the impression of scalarity. It will be concluded that explicit cancellation can no longer be used as an argument for the view that scalar implicatures are pragmatic and inferred rather than semantic and conventional.

2. SCALAR TERMS AND EXPLICIT CANCELLABILITY

2.1 Entailment scales and argumentative scales

In Gricean theory, scalar terms have *lower-bounded* semantics. This means that "some" literally means 'at least some', "warm" means 'at least warm', "two" means 'at least two', etc. This paraphrase may sound counterintuitive, but it reflects logical entailment: since a sentence such as "Some of my friends are Dutch" does not (on a literal reading) exclude the possibility that "All..." is the case, the meaning of "some" should be paraphrased as 'some (or many or most) or all' – in other words, 'at least some'.

Scalar implicatures owe their name to *entailment scales* (see Figure 1), on which they depend. What is higher on the scale *entails* what is lower on the scale, and the use of the term lower on the scale *implicates* that what is higher on the scale is not the case.

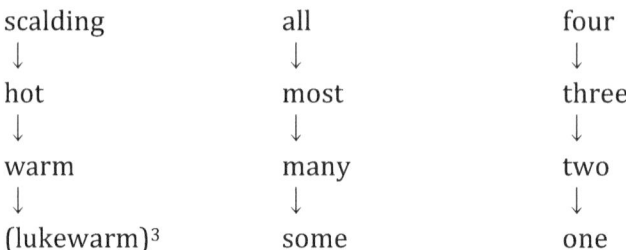

Figure 1 – Entailment Scales

In Anscombre and Ducrot's (1983) theory, scalar terms are ordered on *argumentative scales* (see Figure 2). On argumentative scales, terms that are higher on the scale give stronger arguments in favour of a certain conclusion than terms lower on the scale. Terms that are on the same scale all give arguments in favour of the same type of conclusions; they have the same *argumentative orientation*. For example, if "Some of my friends like beer" is used as an argument in favour of the conclusion "I'd better buy beer for my next party", "All of my friends like beer" will be a stronger argument in favour of that same conclusion. Because weaker arguments are used to support conclusions that would also be drawn from stronger arguments of the same type,[4] terms lower on the scale can be said to be oriented towards the terms higher on the scale (whence the upwards direction of the arrows in Figure 2).

Scalar terms on argumentative scales have *lower- and upper-bounded* semantics. That is to say, "some" means 'just some'/'some and not more'/'some, not all'. From an argumentative point of view, it makes sense to say that "Some of my friends like beer" is a weaker argument than "All of my friends like beer" because "some" denotes a *smaller quantity* than "all". If "some" had its lower-bounded meaning (as on the Gricean view), it would not necessarily denote a smaller quantity than "all" (since "Some..." does not exclude the possibility of "All..." being true).[5]

---

[3] It is not clear to everybody that "warm" should entail "lukewarm" (e.g. Hirschberg, 1985, p. 116), as on Horn's (1972, p. 59) account. If "lukewarm" is taken to mean '*neither warm* nor cold', it cannot be entailed by "warm".

[4] that is to say, arguments formed with terms from the same argumentative scale

[5] According to Philippe De Brabanter (personal communication), the fact that "some" *can* denote a smaller quantity than "all", suffices for the argumentative strength of "all" to be greater than that of "some". Also, he notes that "all" is more informative than "some", even on its 'some and not more' meaning,

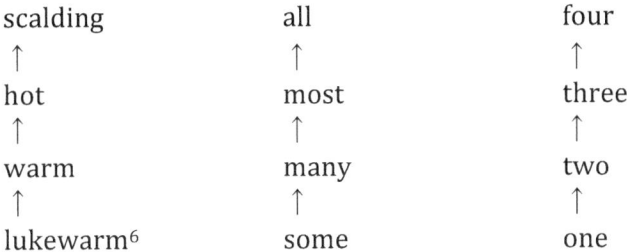

Figure 2 – Argumentative Scales

2.2 Explicit cancellability in argumentation theory

As mentioned in the introduction, cancellability is often used as an argument for the pragmatic (as opposed to semantic) nature of scalar implicatures. As for explicit cancellability, it is implicit on the Gricean view that the use of "Some, indeed all, ..." reflects language users' intuition about entailment scales: speakers are (unconsciously) aware that "all" entails "some", and it therefore makes sense for them to say "*indeed* all" (rather than "no, all").

In Anscombre and Ducrot's (1983) theory, on the other hand, the reason why an element higher on the scale can be mentioned after an element lower on the scale is that stronger arguments always come *after* weaker arguments.[7] Indeed, Ducrot's (1980, p. 23) "et même"-test helps to determine whether two elements are on the same scale (and whether the first element is weaker than the second), e.g. "beaucoup, et

---

because, if the total number of members of the set is known, "all" denotes a precise quantity whereas "some" does not. In other words, "all" then excludes more possible worlds than "some". However, I doubt that this fact is relevant in daily communication. For example, if you tell me, pointing to a bunch of leaves, that "Some of them are brown", I do not care about the exact number of brown leaves. Most of the time, what is relevant is whether "all" is the case, "most", "some", or "none". All four possibilities are then equally informative to ordinary language users.

[6] „Lukewarm" is also on the scale of "cold", as in "I don't want to go into the pool: the water is lukewarm, if not cold". An example where "lukewarm" is on the scale of "hot" is: "I don't want to drink this beer: it is lukewarm, if not warm". The argumentative orientation of "lukewarm" thus depends on whether the thing in question ought to be warm (e.g. a pool) or cold (e.g. a beer, cf. Ducrot, 1980, p. 69).

[7] Of course, strong arguments can be given without any preceding weaker arguments.

même tous" ("many, and even all"). Arguably, "indeed" (as it is found in explicit cancellations) is best translated by "et même" or by "voire" ("or even"). However, these translations are not appropriate for "in fact", another formula that is used in English cancellations (e.g. "Some, in fact all, of my friends are Dutch" or "Some of my friends are Dutch. In fact, all of them are"). These cancellations will be discussed below (see 3.2, 3.4, 3.5).

3. PROBLEMS WITH CANCELLABILITY

3.1 Cancellation with "or" vs. with "and"

As mentioned above, "Some, indeed all, ..." might be equivalent to French "Certains, et même tous", that is, to a formula containing the argumentatively loaded word "même" ("even"). It must be noted however that "ou même" and "voire" (which also means "or even") are more common than "et même" ("and even"): whereas "certains et même tous" has only 7 Google hits,[8] "certains ou même tous" has 2580 and "certains voire tous" 1460.[9] In English, the same trend can be observed: in the American Google Books corpus[10] containing 155 billion words, no results were found for "some and even all" versus 1331 for "some or even all". This fact is highly relevant, since the preferred use of "or" over "and" suggests that "some" and "all" are viewed as mutually exclusive terms. Indeed, it is unlikely that the inclusive "or" is intended here, since it would make little sense to say that the number of elements of a total is "some and all".[11]

In sum, the "or even"/"if not" test (cf. Horn, 1972, p. 49) does not identify entailment scales, but rather argumentative scales. Second, English "indeed" (as used in "Some, indeed all") may be equivalent to

---

[8] on 29/07/2017

[9] Note that these numbers from Google are not entirely reliable (on 07/06/2017, "certains voire tous" had 4070 hits). However, the very low number of hits for "certains et même tous" as compared to the other queries makes the comparison relevant.

[10] created by Mark Davies (http://googlebooks.byu.edu/x.asp, consulted on 29/07/2017)

[11] Pragmaticists may argue that the infelicity of "some and all" is due to the fact that "all" entails "some". After all, it is also odd to say "It is an animal and a monkey". However, it is equally odd to say "It is an animal or a monkey", whereas "some or all" is perfectly fine.

French "voire" (which in turn is more akin to "ou même" than to "et même").

3.2 The translation problem

As mentioned above, it is unclear how to translate "some, in fact all" in French. "Certains, en fait tous" or "certains, en effet tous" are infelicitous, just as Dutch "sommige, eigenlijk alle" and "sommige, in feite alle", as testified by the small or inexistent number of Google results for these queries. Also, in Dutch even "some, indeed all" seems untranslatable. "Sommige, inderdaad alle" is impossible. The only possible expression which comes close to "some, indeed all" is "sommige of zelfs alle" –again, an expression with "or" (see 3.1). "Sommige en zelfs alle" ("some and even all") is infelicitous (0 Google hits on 29/07/2017).

In sum, it could be asked whether "some, indeed all" and "some, in fact all" are idiosyncrasies of English. If so, this would be problematic for Gricean theory, which ought to be universal and uses cancellability as a main argument for viewing scalar implicatures as pragmatic.

3.3 The 'ability' in 'cancellability'

An implicature is cancellable if it *can* be cancelled, but what does this 'ability' consist in? Three different answers to this question can be found in the literature.

To some, an implicature can be cancelled if it can be denied without *logical contradiction* (e.g. Van Kuppevelt, 1996). This view is probably the closest to Grice's (1975) original definition. The problem with this view is that it is circular (cf. also Åkerman, 2014, p. 473) and therefore cannot help to determine whether the meaning of "some" is 'at least some' (as in Grice's theory) or 'just some' (as in Anscombre and Ducrot's theory). Indeed, the question to answer is whether "some, indeed all" is *semantically* contradictory (in other words, whether or not it is semantically felicitous). If one uses the absence of *logical* contradiction as an argument for viewing the logical meaning of "some" ('at least some') as its semantic meaning, one already presupposes that the logical meaning of scalar words is their semantic meaning –which is precisely what Anscombre & Ducrot (1983) contest.

To other theorists, such as Levinson (2000, p. 57), an implicature is explicitly cancellable if it can be denied without any 'feeling' of contradiction. In other words, the explicit cancellation must be pragmatically felicitous. The problem with this view is that even semantic contradictions can be pragmatically felicitous, e.g. "Everybody liked it. Well, except Fred". In other words, the absence of *pragmatic* contradiction still does not reveal anything about *semantic* contradiction or semantic felicity, and therefore nothing can be concluded about the semantics of "some".

Finally, a singular view is held by Åkerman (2015), according to whom an implicature is explicitly cancellable if it can be denied *tout court*: the resulting utterance need not be felicitous (Åkerman, 2015, p. 473). In his view, an implicature is cancellable if it can be "successfully" defeated (Åkerman, 2015, p. 469). However, how can one determine whether or not an implicature is successfully defeated? Arguably, totally infelicitous contradictions such as "some, in fact none" also succeed in defeating a meaning (which could then, falsely, be identified as an implicature). Yet it is precisely such infelicitous 'cancellations' of entailments which are contrasted with the felicity of implicature cancellations when cancellability is evoked as one of the defining characteristics of implicatures (e.g. Geurts, 2010, 83). Hence, *pace* Åkerman, it does not seem reasonable to dismiss the felicity criterion.

In sum, all existing views on the definition of explicit cancellability suffer from the same flaw: since they offer no account of semantic felicity, no conclusions can be drawn about the semantics of scalar terms. In fact, the cancellability test is bound to be circular: to know whether a string is semantically contradictory, one needs to know the semantics of the words involved, but this is precisely what we are trying to find out with the cancellability test.

3.4 Cancellation vs. contradiction

Some authors have pointed out that it is important to distinguish implicature cancellation from self-correction (e.g. Mayol & Castroviejo, 2013, p. 88f.). However, they provide no method for doing so.[12] A suggestion that could be made is that "in fact" and "indeed", as well as "and actually" (cf. Mayol & Castroviejo's example, 2013, p. 94), are typical markers for implicature cancellation, whereas "I mean" and

---

[12]Mayol and Castroviejo (2013) did experimental testing to compare the acceptability of (what they regarded as) implicature cancelations and (what they regarded as) self-repairs. Since they found differences in acceptability (repairs being more felicitous in some of the conditions created by the authors, cancelations in others), they concluded that this supports their distinction between repairs and cancelations.

"actually" would indicate self-repairs.[13] The problem is that strings such as "Some, I mean, all" and "Some, actually all" are perfectly natural. Hence, what is theoretically regarded as scalar implicature cancellation may, in the mind of a language user, in fact be a self-repair.[14]

Levinson (2000, p. 51) suggests that implicature cancellations are cognitively distinct from contradictions: implicatures would be mentally marked as 'cancellable', so that they can be cancelled at any time later. This implies that if a speaker cancels an implicature, this should cause less surprise to the hearer than when the speaker would contradict herself (as in self-corrections or changes of mind). This should be experimentally verifiable. However, to my knowledge, no such experiment has been conducted yet.

What is surprising, however, is that one of Levinson's (2000) examples of implicature cancellation seems to suggest that the speaker changes her mind: "John has two children and in fact, now I come to think of it, maybe more." (Levinson 2000, p. 56). Here, the "now I come to think of it" suggests that a new thought occurs to the speaker. Anscombre and Ducrot (1983, p. 65), after giving an example similar to Levinson's examples of implicature cancellation, comment that "le locuteur se reprend, au vu d'un fait nouveau". Hence, they do not seem to distinguish cancellation from self-repair or change of mind.

It could be argued that the new information which comes to the speaker's mind in these examples is not *contradictory* with the given information, but merely forms an *addition* to it. Although this distinction may be theoretically made by logicians (and Griceans), it is not obvious that language users do the same (whether consciously or unconsciously). It seems more likely that language users distinguish between "small contradictions" and "major contradictions", whereby what counts as "small" or "major" depends of the importance of the mistake in the context. Scalar implicature cancellations would then tend to be small contradictions, not because an implicature is being cancelled, but because the reparans ("some") and the reparandum ("all") have the same argumentative orientation. Inversely, there are

---

[13] Indeed, Mayol and Castroviejo (2013, p. 89) claim that "I mean" can only be used for self-repairs (self-corrections, as the meaning of the phrase "I mean" indicates). Also, there are examples of self-repairs given in Ginzburg et al. (2014, p. 38, 39) which involve "actually" (but not "and actually").

[14] It is important to note that I am only thinking of "in fact" here. "Indeed all" seems more likely to be a 'planned' correction. This is not to say that "in fact all" can never be used as a planned correction. These are matters for further investigation.

also repairs where no implicature is cancelled (and which would be regarded as real contradictions by logicians) which would tend to be considered small contradictions by language users. For instance, in "I love all fruits. Well, except for pears" the speaker contradicts her first statement "I love all fruits". Nevertheless, it is unlikely to be considered a real contradiction by language users. Hence, the intuitive distinction 'small contradiction/ major contradiction' does not coincide with the theoretical distinction 'implicature cancellation/real contradiction'.

3.5 The felicity of explicit cancellations

Until now, the felicity of explicit cancellations has very much been taken for granted. Mayol and Castroviejo (2013) constitute an exception: they asked participants to rate cancellations for felicity on a scale by using the graphical technique of 'magnitude estimation' (cf. Bard et al. 1996). However, the only factor that was manipulated was the presence or absence of a presuppositional verb in the correction clause (e.g. "in fact, all" vs. "in fact, it is amazing that all"). For the purposes of my research, it would be more interesting to manipulate other structural factors, such as the position of the "indeed/ in fact". In a questionnaire conducted in August 2016 with 21 non-native (but proficient) speakers of English (Dutch speakers), I have found that immediate cancellations with "indeed", as in "It was cold, indeed ice cold" were found more felicitous that later cancellations such as "It was cold. Indeed, it was ice-cold". Also, corpus studies should be conducted to find out whether short-distance cancellations with "in fact" ("Some, in fact all, ...") and longer-distance scalar implicature cancellations with "indeed" and "in fact" ("It was cold. Indeed/in fact, it was ice-cold.") can be found with any regularity.

Many of the scalar implicatures mentioned by Levinson (2000) and Horn (1989) do not seem to pass the explicit cancellability test. Take the following scales: <taller than, as tall as> (cf. Horn 1972, p. 51-52), <square, squarish> (cf. Levinson 2000, p. 34), <complete, partial> (Levinson 2000, p. 87), <long-term, short-term>, <at, near> (Levinson 2000, p. 96). Is it not semantically problematic to say "I'm as tall as you, indeed, I'm taller than you", "This object is squarish, indeed square", "I'm near the station, in fact, I'm at the station", "This is a partial solution, indeed a complete one", "This is a short-term solution, indeed a long-term one"? In the last two examples, the two scalar terms even seem to be opposites. Nevertheless, they do seem to be scalar, since it is possible to say "Give me a solution, at least a partial one", or "Give me a solution, at least a short-term one". All the examples given pass the "at least"-test. Moreover, most of them pass the "or even"-test: "I'm as tall as you, or even taller", "This object is squarish, or even square", "She must be near the station by now, or even at the station" sound perfectly fine. Thus, it must be concluded that the existence of scales does not imply the existence of cancellable scalar implicatures.

Another test which is sometimes used to check for explicit cancellability is the "Yes"-answer test (e.g. Horn 1992, p. 175, cited in Carston 1998, p. 1999). For example, in answer to the question "Are some of your friends Dutch?" it is more natural to answer "Yes, in fact all of them" than "No. All of them".[15] Similarly, if someone asks you "Are you near the station yet?" it is more likely that you will answer "Yes, in fact I'm at the station" than that you would say "No, I'm at the station". This is comparable to Anscombre and Ducrot's (1983, p. 65) example "Is dinner almost ready?" – "Yes, in fact, it is ready". In Anscombre and Ducrot's (ibid.) theory, there is an argumentative scale {almost x, x}, whereby "almost x" is oriented towards "x". This may seem surprising, since "almost x" presupposes "not-x" (ibid.). Thus, the question-answer pattern is "[presupp.] Not-X?" – "Yes, X". Therefore, it appears that argumentative orientation can 'override' presupposition. As for the examples with scalar implicature, there is a choice between two conclusions: 1. argumentative orientation can also override other types of semantic information (such as so-called 'scalar implicatures'), not just presuppositions; 2. scalar implicatures are in fact presuppositions.

4. CONCLUSION

The present paper has identified several problems with the explicit cancellability test for scalar implicatures. It was suggested that the "indeed", "or even", "at least", and "Yes"-answer tests most probably identify *argumentative scales* rather than entailment scales. While explicit cancellability might still be used as a defining criterion for implicatures, I have argued that it can no longer be used as an argument for the pragmatic (as opposed to semantic) nature of scalar implicatures. This insight opens up new ways for the thesis that scalar implicatures are conventional and encoded rather than pragmatic and inferred.

---

[15] This is not to say that this answer is impossible, but it would be less *cooperative*.

ACKNOWLEDGEMENTS: I would like to thank the people who attended my presentation at ECA 2017 for their interest and insightful remarks. I also thank Philippe De Brabanter and Walter De Mulder for their useful comments.

REFERENCES

Åkerman, J. (2015). Infelicitous cancellation: The explicit cancellability test for conversational implicature revisited. *Australian Journal of Philosophy, 93*(3), 465-474.
Anscombre, J.-C., & Ducrot, O. (1983). L'argumentation dans la langue (P. Mardaga, Ed.). *Dialogue: Canadian Philosophical Review, 23*(3), 514-517.
Bard, E., Robertson, D., & Sorace, A. (1996). Magnitude estimation of linguistic acceptability. *Language, 72*(1), 32-68.
Carston, R. (1998). Informativeness, relevance and scalar implicature'. In R. Carston & S. Uchida (Eds.), *Relevance theory: Applications and implications* (pp. 179-236). Amsterdam: John Benjamins.
Carston, R. (2013). Legal texts and canons of construction: A view from current pragmatic theory. In M. Freeman & F. Smith, *Law and language: current legal* issues (Vol. 15, pp. 8-33). Oxford: Oxford University Press.
Ducrot, O. (1980). *Les échelles argumentatives.* Paris: Minuit.
Geurts, B. (2010). *Quantity implicatures.* Cambridge: Cambridge University Press.
Ginzburg, J., Fernandez, R., & Schlangen, D. (2014). Disfluencies as intra-utterance dialogue moves. *Semantics & Pragmatics, 7*(article 9), 1-64.
Grice, H. P. (1989). *Studies in the way of words.* Cambridge, MA: Harvard University Press.
Hirschberg, J. (1985). *A theory of scalar implicature.* University of Pennsylvania PhD thesis. (Reprinted as Hirschberg, J. (1991). *A theory of scalar implicature.* New York: Garland)
Horn, L. R. (1972). *On the semantic properties of logical operators in English.* University of Indiana PhD thesis.
Horn, L. R. (1992). The said and the unsaid. In C. Baker & D. Dowty (Eds.), *SALT II: Proceedings of the second conference on semantics and linguistic theory* (pp. 163-192). Columbus, OH: Ohio State University Linguistics Department.
Mayol, L., & Castroviejo, E. (2013). How to cancel an implicature. *Journal of Pragmatics, 50*(1), 84-104.
Levinson, S. (2000). *Presumptive meanings: The theory of generalized conversational implicature.* Massachusetts: The MIT Press.
Recanati, F. (2010). *Truth-conditional pragmatics.* Oxford: Oxford University Press.
van Kuppevelt, J. (1996). Inferring from topics: Scalar implicatures as topic dependent inferences. *Linguistics and Philosophy, 19*(4), 393-443.

18

Reasoning via Dialogue: An Illustrative Analysis of Deliberation[1]

STÉPHANE DIAS
Federal Institute of Education, Science and Technology Farroupilha
Campus *Santo Augusto, Brazil*
stephanerdias@gmail.com

JANE SILVEIRA
The Pontifical Catholic University of Rio Grande do Sul, Brazil
jane.silveira@pucrs.br

> We will explore the relation between reasoning aiming at a practical goal (decision-making and ultimately action) and reasoning aiming at a proper evaluation of the evidence to reach a consensual truth. This will be accomplished by an illustrative analysis of 12 Angry Men, considering the reasoning process at the levels of (1) the institutional group itself (we, the jury), (2) internal to the group (as members, jurors), and (3) the individuals.
>
> KEYWORDS: agency, agents, reasoning, dialogue, argumentation, deliberation

1. INTRODUCTION

We have two illustrative goals and a central argumentative goal with this paper. Concerning the illustrative goals, we aim to address dialogue moves in a deliberative dialogue (McBurney, Hitchcock & Parsons, 2007; Walton, Atkinson, Bench-capon, Wyner & Cartwright, 2010; Bigi & Macagno, 2017). By means of these moves, reasons are presented to

---

[1] This is a substantially revised and extended version of Dias, S. R., & Silveira, J. R. C. (2015). Reasoning via dialogue: an illustrative analysis of deliberation. *Dialogue Under Occupation Proceedings.* EDIPUCRS: Porto Alegre.

support individual or collective positions[2]. Moreover, we want to exemplify the claim that humans reason better when they need to argue with one another (Mercier & Sperber, 2011). Mercier & Sperber (2017) challenged the traditional panorama of reasoning when they replaced the centrality of processes and goals of the individuals with the centrality of social interaction and its effects. For them, the main function of reasoning is to produce arguments, reasons to convince others, rather than being a means for someone to think better, achieve greater knowledge or make the best decision possible. More precisely, they suggest we use reason to justify our cognitive and social performances – i.e. thoughts, claims, actions, and to convince our audience to think and act accordingly, while, at the same time, we evaluate their performance using the same mechanism (Mercier & Sperber, 2017, p. 12).

For the second goal, we want to defend that a more coherent approach of human reasoning would also require us to replace the loose theoretical notion of *individual* with the one of *agent*. As we are in the social realm of reasons and communicative acts, we need to address the agency that performs such acts (Dias, 2016; Dias & Müller, 2017). For that aim, the notions of *levels of agency* and *agent-types* will be helpful when addressing the reasoning process established by dialogues, once they are tools to explain constraints on the set of reasons and commitments of an agent.

For illustrative purposes, we take a deliberation dialogue case (in an institutional decision-making scenario) played by those who are required to evaluate an evidential body *as members of a jury*. Since every man must act as a juror and help the collective body, the jury, to produce a final statement, different types of goals interact in their reasoning processes.

We hope the analysis of *12 Angry Men*[3] will help us to clarify the above-mentioned claims. Consequently, our illustrative goals are subordinated to our argumentative goal.

2. REASONING: WHAT SHOULD WE ACCOMPLISH TOGETHER?

Aristotle explored a Socratic question-answer model as the way of reaching the truth. By asking and answering questions, a set of men

---

[2] The arguments for this option can be found in Dias (2016).

[3] There are many versions of the text written by Reginald Rose. We will make use of the one written for television and staged in 1954. Retrieved from https://fischersoph.files.wordpress.com/2010/09/12-angry-men-script.pdf.

could reason together, watching each other's steps and pointing out reasoning weaknesses. In this dialectic scenario, if no error was identified, it was taken as evidence that the proposition corresponded to the truth.

Similarly, in many dialogue scenarios reasons are brought up by the participants to base a good decision. In the *Nicomachean Ethics*, Aristotle invokes three things which would control action and truth – sensation, reason, and desire (1999, p. 92 [orig. 350 B.C.E]). How each one guides action and reasoning is still disputed. In the context of deliberation, the philosopher states,

> We deliberate about things that are in our power and can be done [...] Now every class of men deliberates about the things that can be done by their own efforts. (Aristotle, 1999, book III, p. 38 [orig. 350 B.C.E])

This idea is twice relevant here, as it invokes the category of "every class of men" and their consequent "power." Thus, it addresses agents and their agency. By "agents" we refer to the sets of social functions[4] the same human entity performs. A person, when deliberating about a course of action, will consider her social function regarding that communicative situation. The result may be different courses of action given this person's commitments as a *human being*, as a *group member* and as a *group representative*.

Particularly, the teleplay *12 Angry Men* shows agents exchanging reasons based on their practical and epistemic goals.[5] They need to judge the case of, an eighteen-year-old man from the ghetto who is accused of killing his abusive father, thus being on trial. The jurors are asked to deliberate about a final position: *either* the boy is "guilty" (which means there is 'no reasonable doubt' that the accused committed the crime) *or* he is "not guilty" (which means that the members agree that there is a 'reasonable doubt' concerning the guilt of the accused). As the judge states, "it's now your duty to sit down and try to separate the facts from the fancy."

Let us now turn to the deliberation scenario, which comprises the task of evaluating the evidence for reaching a unanimous position.

---

[4] By social function, we mean the roles and their deontic consequences a person X collectively assumes in a specific context - in Searle's simple notation, X as Y in C. "To describe the basic structure of social-institutional reality, we need exactly three primitives, collective intentionality, the assignment of function, and constitutive rules and procedures" (Searle, 2008, p. 31).

[5] And the agents may have different, common, and joint goals.

2.1 Deliberation: reasoning and action

The jury adopts a summative method of deliberation (judgement aggregation) such that the decision is reached by counting the vote of each group member, where each vote has the same weight[6]; consensus is then obtained when all members vote accordingly (thus reaching the verdict).

The members must consider that if the defendant is found guilty, the death sentence is mandatory (there is a possibility of a recommendation for mercy on the part of the court, though). On the other hand, if they assume, as a group, that there is reasonable doubt on their minds, he is free. In the "jury room", one of the jurors conducts the deliberation process.

Clearly, the epistemic goals involved in the deliberation embrace an adequate evaluation of the evidence; presentation of counter-evidence against a conclusion, argument, or inferential step by others; use and acceptance of trustable data and of good reasons in the process of evaluation, and mutual manifestability or information-sharing.

Those agents learn some of the goals and expectations involved by means of an initial dialogue among themselves. For instance, they stated practical reasons for not extending the jury's meeting 'more than necessary' – "We've probably all got things to do here", "We can all get out of here pretty quick. I have tickets to that ball game tonight"; "I called the Weather Bureau this morning. This is gonna be the hottest day of the year". It is worth noting that they were in a small room with no ventilation system. This gives us a realistic scenario of sets of commitments and reasons for each person, who operates as a single individual, and as a member of the jury.

Once this scenario has been considered, our next step is to address dialogical and linguistic structures that reveal these cognitive and practical moves. To accomplish the task, let us contextualize some features of deliberative dialogues and why their analysis is worth-considering.

3. DIALOGUE-BASED APPROACHES

A dialogue-based approach of argumentation, or a 'pragmatic conception of argument', is a powerful theoretical tool. It will generally

---

[6] There are other procedures for group decisions, such as majority voting, just as pointed out by McBurney, Hitchcock & Parsons (2007).

assume that: a) argumentation is a dialogical process (Johnson & Blair, 1980); b) as such, arguments are dialogue-moves, thus being evaluated as part of these structures and not as independent objects; c) arguments are composed by or used inside speech-acts or dialogue-moves; d) there is a structure of interaction inside which agents assume argumentative-dialogue roles, such as questioners (Blair & Johnson, 1987); e) the process starts and evolves by the challenge of a position or assumption; f) each agent has a dialogue-goal concerning the process of argumentation, given the attitudes agents are committed to, by weakening or strengthening the plausibility of an assumption or position (see Walton & Godden, 2007).

Argumentation is therefore understood as a dialogue-oriented activity being part of dialogue-oriented games, such as deliberations, or rather as its main goal. Via argumentation, a position, or an initial assumption presented as is a valuable option, and via deliberation, agents will choose among the options made available.

The argumentative process is, thus, crucial because it starts with a question, or any expression of doubt (Blair & Johnson, 1987), towards reasons for believing something. For Walton, Atkinson, Bench-capon, Wyner & Cartwright (2010), along with the Amsterdam view, a dialogue-based argumentation has a simple structure: an opening, a body, and a concluding stage. Accordingly, in virtue of being inside this structure, an argument is itself a process that goes through all stages created by agents facing doubt or disagreement. Particularly, single dialogical sequences are taken as "dialogue moves", which are functional to achieving specific sub-goals in the dialogue towards a higher dialogical goal (Macagno & Bigi, 2017: 149). As higher communicative goals, we can think of making a deal or a joint decision. As sub-goals, we can think of dialogical steps that help the agent to achieve higher goals or her "global dialogical intentions." A move, then, besides being our minimal unit for the analysis of dialogues, is expressed by a turn-taking dialogical sequence (Weigand 2009, 2010 also addresses the action-reaction game).

Let us now focus on these structures of reasoning and action.

3.1 Deliberation dialogues

Taking into account the typology of dialogues proposed by Walton & Krabbe (1995), information seeking/inquiry dialogues involve precisely "a search for the true answer to some factual question" (McBurney, Hitchcock & Parsons, 2007, p. 3). It is claimed that, in such dialogues, the appeal to individual values, like goals and preferences, would be inappropriate. On the other hand, deliberations focus on collective action, on what is to be done (McBurney, Hitchcock & Parsons, 2007, p. 3). They may also help individuals to coordinate their individual goals. Nonetheless, our jury faces a deliberation involving factual questions. Thus, interconnecting both communicative structures.

In deliberation dialogues, besides the general aim of decision-making, we can have *internal goals* of the dialogue, as considered by McBurney, Hitchcock & Parsons (2007). However, in all these communicative scenarios, agents have a demand for using reasons to choose one option rather than the other. Argumentation, then, takes place, since agents may or may not agree with each other. In the end, the target is decision-making about what (is the best for them) to do.

In the plot invoked here, the jurors face a yes-no deliberation setting, where they need to decide over presenting a position that there is *or* a position that there is no reasonable doubt about the boy's guilt. In other words, to decide their course of action, they first need to reach a consensus about states-of-the-world (what may or may not have happened), and about their propositional attitude regarding the evidential body (as sufficient or not for their purposes). Since they do not have the condition to access decisive evidence to reach consensus concerning a proposition such as 'the boy killed the man,' they can only reach consensus on a propositional attitude of the type 'there is (no) reasonable doubt *that the boy killed the man*'.

Like the *deliberation dialogue framework* (McBurney, Hitchcock & Parsons, 2007), we assume some fixed dialogical norms. For us here, when an agent publicly asserts a locution in this game he is explicitly committed with it as an agent; and their actions can be explained by those commitments, which are explained by their agency, recursively.

With these considerations in mind, we can assume the existence of dialogical moves[7] as means to a set of ends. In the next section, we will address the agents in interaction and their dialogical moves motivated by their goals.

---

[7] These moves are taken to be calibrated by a Principle of Relevance of the type proposed by Sperber & Wilson (1986, 1995). According to them, our cognition processes information oriented by optimization, taken as a cost-benefit calculation of cognitive effects and cognitive costs; this proposal predicts that human cognition reacts based on innate expectations over human behavior, such that linguistic input or utterances "create expectations of optimal relevance" (Wilson & Sperber, 1995). Since this aspect is not our focus here, we will only be committed with it, given its explanatory importance.

4. AGENTS REASONING VIA DIALOGUE

After having contextualized the structure inside which practical and cognitive goals interact in the reasoning process evoked by dialogues, let us consider how the process takes place.

Human agents, as claimed, would stand for three main types: individuals, group members and groups.[8] Let us then assume that it is valuable for individual agents to improve their cognitive environment and decision power over actions that will fulfill their interests, values, desires, judgments, or that are important to that agent's welfare or maintenance. Collective agents also have goals to fulfill, practical ones. We already assumed that the same human being can stand for different *agents* in different contexts of interaction, or institutional contexts, such as at home, at work, at the parliament, and that this 'status-function' (Searle, 1995, 2010) will regulate their behavior, via a cost-benefit calculation (Dias, 2014, 2016). That means human beings have a social cognitive capacity to stand for roles, which is a crucial mechanism to regulate decision-making and interactions, not to say conceptualization itself.

For the collective agency to work, information sharing is required. And dialogue is our specialized form of information sharing. By questions, answers, comments, replies, members of a group construct a common ground of assumptions and evaluate their individual ones.

Thus, during the deliberative process presented in the plot of *12 Angry Man*, the jurors' individual inclinations and values, and their collective assumptions and goals interacted. As a collective entity, *the jury*, they must deliver a verdict. As jurors, they must reach consensus via deliberation.

During the process, by means of a preliminary vote, an assumption is made explicit: they need to reach an agreement by changing a member's mind or eleven members' mind, since one of them disagrees about the certainty of the defendant's guilt. The collective epistemic goal, then – taken simply as a practical goal imposed by institutional pressures – turns out to be the evaluation of data for finding either decisive evidence of the boy's culpability (represented by the position of the majority) or a reasonable doubt (represented by the dissonant opinion of juror number 8).

---

[8] The same division can be found in Wodak (2000).

Therefore, the members will reason aiming ultimately at consensus, via reasoning to find a *reasonable doubt*[9]. For all jurors but one, it was not necessary to evaluate the evidence, just to rely on it and accept the conclusion that the boy was guilty. Once the jurors learned that one among them was not certain of the conclusion, they had to use argumentative means to reach their joint goal (consensus). Consequently, many agents were asked to revise their own cognitive environment, coming to realize they had no good reasons to be certain that the boy was guilty.

5. THE ANALYSIS

In this section, we will analyze and illustrate the above-mentioned points considering the treatment of the evidential body. Again, interactions among 12 jurors were a means to achieve a set of ends. Our next step, then, will be to explore part of this dialogue structure.

The evidence is composed, among other pieces, by witness number 1 (an old man who testified to having heard a discussion between the boy and his father) and witness number 2 (the neighbor who claimed to have seen the crime). These two pieces appear at different times in the narrative. Let us now consider their dialogue moves.

Testimony (1)

N$^{o.}$3[10]: Okay, let's get to the facts. Number one (...).
FOREMAN: And the coroner fixed the time of death at around midnight.
N$^{o.}$3: Right. Now what else do you want?
N$^{o.}$ 8: An el takes ten seconds to pass a given point or two seconds per car. That el had been going by the old man's window for at least six seconds and maybe more, before the body fell, according to the woman. The old man would have had to hear the boy say "I'm going to kill you", while the front of the el was roaring past his nose. It's not possible that he could have heard it.
N$^{o.}$ 3: What d'ya mean! Sure he could have heard it.
N$^{o.}$ 8: Could he?
N$^{o.}$ 3: He said the boy yelled it out. That's enough for me.

---

[9] Once again, we will connect the goals of building agreement on some course of action and searching for truth. Truth has intrinsic practical value.

[10] Each juror is identified by a number.

Nº· 9: I don't think he could have heard it.
Nº· 2: Maybe he didn't hear it. I mean with the el noise....
Nº· 3: What are you people talking about? Are you calling the old man a liar?
Nº· 5: Well, it stands to reason.
[...]
Nº·5: I'd like to change my vote to not guilty.

This sub stage of the dialogue starts with the presentation of the evidence available to the group; we can count this as the first move of that stage. ("Okay, let's get to the facts. Number one (...)").

The next move is a follow-up comment about the evidence, adding information to the common ground ("And the coroner fixed the time of death at around midnight").

In the sequence, juror #3 makes clear that each contribution is directed towards the decision-making process, where they need to decide in favor or against the view that the boy was guilty; the juror's position rests on the certainty of the boy's guilt, which is implied by his dialogue move ("Now what else do you want?").

Juror #8, then, analyses the evidence; he does that by means of considerations and predictions over the testimony's content, ending with a conclusion ("An el takes ten seconds to pass a given point or two seconds per car. That el had been going by the old man's window for at least six seconds and maybe more, before the body fell, according to the woman. The old man would have had to hear the boy say 'I'm going to kill you', while the front of the el was roaring past his nose. It's not possible that he could have heard it").

As a reaction, juror #3 questions juror #8's conclusion ("What d'ya mean! Sure he could have heard it"). Juror #8 then poses a doubt to juror #3 and the others ("Could he?") instead of reasserting the conclusion. This gives room to other agents' manifestation ("I don't think he could have heard it", "Maybe he didn't hear it. I mean with the noise"). Given these moves, juror #3 reacts to a shared conclusion and its possible consequences ("What are you people talking about? Are you calling the old man a liar?").

After that, juror #5 observes that the conclusion reached follows from the premises ("Well, it stands to reason"). Because of these cognitive and dialogical effects, juror #5 changes his position to 'not guilty' ("I'd like to change my vote to not guilty").

From the collective evaluation of the evidence, some of them agreed upon the conclusion that there were insufficient time and conditions for the testimony to have heard the boy yelling at his father. Given the new interpretation of the evidence, the analysis weakened the conclusion that the boy was guilty; consequently, one juror changed his position.

Testimony (2)

N⁰. 10: Look, what about the woman across the street? If her testimony doesn't prove it, then nothing does.
[...]
N⁰. 12: (...) She looked into the open window and saw the boy stab his father. She saw it. Now if that's not enough for you....
[...]
N⁰. 8: All right. Let's go over her testimony. What exactly did she say?
N⁰. 4: I believe I can recount it accurately. (...) As far as I can see, this is unshakable testimony.
N⁰. 6: Well, I was thinking. You know the woman who testified that she saw the killing wears glasses.
N⁰. 3: So does my grandmother. So what?
N⁰. 8: Your grandmother isn't a murder witness.
N⁰. 6: Look, stop me if I'm wrong. This woman wouldn't wear her eyeglasses to bed, would she?
FOREMAN: Wait a minute! Did she wear glasses at all? I don't remember.
N⁰. 11: (excited). Of course she did! The woman wore bifocals. I remember this very clearly. They looked quite strong.
N⁰. 9: That's right. Bifocals. She never took them off.
N⁰. 4: She did wear glasses. Funny. I never thought of it. [...]
N⁰. 8: Does anyone think there still is not a reasonable doubt?
[...]
N⁰. 4: (quietly). No. I'm convinced.
N⁰. 3: Well, I told you I think the kid's guilty. What else do you want?
N⁰. 8: Your arguments. [They all look at N⁰. 3:] [...]
N⁰. 3: (thundering). All right!

The above dialogue sequence started with a crucial move, in which juror #10 manifests certainty about a witness' testimony ("If her testimony doesn't prove it, then nothing does"). This epistemic attitude is endorsed by another juror, juror #12, who explicitly addresses the other jurors aiming to convince them of a position to be chosen ("She saw it. Now if that's not enough for you").

Juror #8 is the one who conducts the collective evaluation of the evidence ("All right. Let's go over her testimony. What exactly did she say?"), implying that they should not start from the presumed conclusion, but from the premises' evaluation. It opens space for a 'thinking out loud' process, where each one contributes to the collective process ("You know the woman who testified that she saw the killing wears glasses"). The reaction to this move is a request for clarification regarding the value of that information to the process and their consequent goals ("So does my grandmother. So what?").

The follow-up moves are adequate reactions to this argumentative structure ("Look, stop me if I'm wrong (...)", "That's right (...)"). Moreover, cognitive effects followed from that ("(...) I never thought of it"). Given the cognitive effects of revision and addition of new assumptions, the agents were able to derive, jointly, the desirable conclusion implied by juror#8's question ("Does anyone think there still is not a reasonable doubt?"). At the end of the process, another juror openly recognized the argumentative game and properly reacted by changing his position ("No. I'm convinced").

On the other hand, one of the agents, juror #3, chose to reassert his position at the game by claiming that the boy was guilty ("Well, I told you I think the kid's guilty. What else do you want?"). Since juror #3 also asks them a question as part of his move, juror #8 replies to that by answering the question in such a way to request juror #3's arguments ("Your arguments").

With no arguments to provide and in view of the reasons already put forward in the dialogue by the other jurors, juror #3 accepts the fact he has no evidence justifying his position and then accepts the conclusion reached so far ("All right!"), culminating in consensus and the consequent verdict. The main epistemic and practical goals of the agents were, thus, reached.

6. CONCLUSION

First, by means of this analysis, we aimed to exemplify the claim that humans reason better when they need to argue with one another. We hope to have succeeded in that. Regarding the dialogical moves at stake, juror #8 was a central player, not only for positing doubt where there was certainty, and bringing about good reasons to the evaluation table, but also for his ability in the conduction of the dialogue, by asking key questions and providing answers such as, "I don't know" for questions like "Do you think he's guilty?"; or even comments, such as "I don't want to change your mind. I just want to talk for a while".

Moreover, through the collective evaluation of the evidence, each member had the chance to update his cognitive environment. Via argumentation, by questions, answers and claims, they tried to justify their positions, and, at the same time, they evaluated each other's performance. The verdict, then, was formed based on a conclusion and premises evaluated collectively. The reasoning path, dialogically constructed, led them to a unanimous decision.

We could observe clear constraints on the set of reasons and commitments of the agents. At the individual level, two jurors were committed to a position given their personal bias. As members, they restricted their set of reasons according to the set of evidence available to them, considering their collective goals, and practical pressures of time and space. It is worth noting that a knife found at the crime scene was another decisive piece of evidence that we haven't discussed. In view of counter-evidence to their conclusions, some of the jurors changed their initial position to not guilty. This reveals a shift in their understanding of the evidence during that dialogue exchange, considering cognitive changes made possible through communication.

The jury, then, concluded that there was reasonable doubt as to the guilt of the accused. And, once consensus was reached, the group was committed to that claim. Still, these agents composed a picture no one could privately see.

Inside this scope of deliberative dialogue, they constructed the required argumentative structure they needed to convince other members. Consequently, 11 members of the group changed their initial position, meaning the agents reached their epistemic goals (they performed better, both as individuals and as members) and succeeded in their collective practical goal (producing a consensual verdict). The move made by a juror, suggesting that it was up to them to convince juror #8 about the fact that they were right and he was wrong, can be read as an illustration of a claim of the *argumentative theory of reasoning*. The claim is that reasoning has an argumentative function, where people, taken here as agents, reason better when trying to persuade others. In fact, they reached their goals precisely by reasoning in an interactive context. Nonetheless, members acted inside a restricted scope of possible moves, given collective commitments to the group.

Via argumentation, agents persuaded others to change their course of action as members of a jury (their vote). Their collective cognitive goal (a collective evaluation of the evidence) was a means for their practical, collective, institutional goal (the production of a verdict). As we could observe, the **argumentative framework improved both outcomes** – better evaluation of the evidence and agreement – since the conclusion could not be proven, but only confirmed by premises.

Finally, after having recognized that practical and cognitive goals interact in the reasoning process evoked by dialogues, and given those goals are the goals of an agent, we claim that agency is a central feature of any account of dialogue or reasoning.

ACKNOWLEDGEMENTS: A special thanks to CNPq and Fulbright/Capes.

REFERENCES

Aristotle. (1999 [orig. 350 B.C.E]). *Nicomachean Ethics* (W. D. Ross, Trans.). Kitchener, Ontario: Batoche Books.
Blair, J. A., & Johnson, R. H. (1987). Argumentation as dialectical. *Argumentation*, 1(1), 41-56.
Dias, S. R. (2016). *Agency via dialogue*: A pragmatic, dialogue-based approach to agents (doctoral dissertation). Porto Alegre, PUCRS.
Dias, S. R. (2014). ToM e o aparato comunicativo da linguagem. In *Anais do XVII Congreso Internacional de la Asociación de Lingüística y Filología de América Latina* (pp. 1124-1136). Gran Canaria: Universidad de Las Palmas de Gran Canaria.
Dias, S. R., & Müller, F. de M. (2017). Speaker or agent: implications and applications. *Gate to pragmatics* [mbook]: uma introdução a abordagens, conceitos e teorias da pragmática / orgs. V. Wannmacher Pereira et al. (dados eletrônicos). Porto Alegre: EDIPUCRS.
Johnson, R. H., & Blair, J. A. (1980). *Informal logic*: The first international symposium. Inverness, CA: Edge Press.
Macagno, F.; Bigi, S. (2017). Analyzing the pragmatic structure of dialogues. *Discourse Studies*, 19(2), 148-168.
McBurney, P.; Hitchcock, D.; Parsons, S. (2007). The eightfold way of deliberation dialogue. *International Journal of Intelligent Systems*, 22(1), 95-132.
Mercier, H., & Sperber, D. (2011). Why do humans reason? Arguments for an argumentative theory. *Behavioral and brain sciences*, 34(2), 57-111.
Mercier, H., & Sperber, D. (2017). *The enigma of reason*: A new theory of human understanding. Harvard: Harvard University Press.
Rose, R. (writer), Schaffner, F (director), (1954). *Twelve angry men* [Teleplay](F. Jackson, Producer). New York: CBS Television Network
Searle, J. (1995). *The construction of social reality*. New York: Free Press.
Searle, J. (2008). *Philosophy in a new century: selected essays*. Cambridge: Cambridge University Press.
Searle, J. (2010). *Making the social world*: The structure of human civilization. Oxford: Oxford University Press.
Sperber, D., & Wilson, D. (1986). *Relevance*: Communication and cognition. Oxford: Wiley-Blackwell.
Sperber, D., & Wilson, D. (1995). *Relevance*: Communication and cognition (2nd ed.). Oxford: Blackwell.

Walton, D., Atkinson, K., Bench-capon, T., Wyner, A., Cartwright, D. (2010). Argumentation in the framework of deliberation dialogue. In C. Bjola & M. Kornprobst (Eds.), *Arguing global governance* (pp. 210-230). London: Routledge.
Walton, D., & Godden, D. M. (2007). Informal logic and the dialectical approach to argument. In H. V. Hansen & R. C. Pinto (Eds.), *Reason Reclaimed* (pp. 3-17). Newport News, VA: Vale Press.
Walton, D. N., & Krabbe, E. C. W. (1995). *Commitment in dialogue*: *Basic concepts of interpersonal reasoning*. Albany, NY: State University of New York Press.
Weigand, E. (2009). *Language as dialogue*. S. Feller (Ed.). Amsterdam/Philadelphia: John Benjamins Publishing.
Weigand, E. (2010). *Dialogue: The mixed game*. Amsterdam/Philadelphia: John Benjamins Publishing.
Wodak, R. (2000). *Does sociolinguistics need social theory? New perspectives in critical discourse analysis* (Keynote address, SS2000). Bristol, Bristol University.

19

Are Humans Poor at Arguing? From the 'Argumentative Theory of Reasoning' back to a Rhetorical Theory of Argumentation[1]

SALVATORE DI PIAZZA
University of Palermo
salvatore.dipiazza@unipa.it

FRANCESCA PIAZZA
University of Palermo
francesca.piazza@unipa.it

MAURO SERRA
University of Salerno
maserra@unisa.it

Starting from Sperber and Mercier's theory (2011) on the relationship between reasoning and arguing, we will try to rethink the link between rhetoric and argumentation. Using Aristotelian rhetoric as a theoretical framework, we will focus on two related features: 1) the nature and the role of argumentation inferences in classical models of rhetoric; 2) the role of normativity in assessing a naturalistic description of what we make when we argue.

KEYWORDS: argumentation, Aristotle, normativism, rhetoric

---

[1] Although all the authors collaborated in the conception of the article's general framework, Mauro Serra wrote sections 1, 2 and 3, Salvatore Di Piazza wrote section 4, Francesca Piazza wrote section 5 and Salvatore Di Piazza and Francesca Piazza wrote together section 6.

1. INTRODUCTION

According to a long philosophical tradition, reasoning is an individual activity aimed at improving our knowledge of reality and maximizing our personal utility. Recently, however, some scholars have supported the idea that rational activity is intrinsically dialogical.
From this perspective, they

> have argued that reasoning and argumentation are so strictly related that the former cannot be studied detached from its place of occurrence, that is dialogical, argumentative situations, regardless of whether reasoning is considered to be cognitive or social (Labinaz, 2014, p. 578).

These scholars include Dan Sperber and Hugo Mercier, who in 2011 presented a new and challenging theory of the relationship between argumentation and reasoning. The theory, which then assumed a wider and more fully articulated form in their volume *The Enigma of Reason* (2017), ultimately argues that the emergence of reasoning can be better understood in the evolutionary framework of human communication. Reasoning has, therefore, not been developed to help humans think better or make better and more effective decisions but to produce arguments in support of their claims and to evaluate the arguments provided by their interlocutors in a dialogical context. For this reason, the theory has become known as 'the argumentative theory of reasoning'. This summary explains the rationale behind such a name.

The theory moves on at least three different levels: a thesis of evolutionary nature related to the development of reasoning; a hypothesis about the primary function of reasoning (and why it developed); and, finally, a thesis concerning the relationship between argumentation and reasoning. We will not say anything about the first two aspects. In fact, we are unable to evaluate the plausibility of the evolutionary part of their work. And with regard to the question of the primary function of reasoning (if there is one), it is too complex to be addressed here. One can only observe that Sperber and Mercier's theory has been criticized for failing to take account of the variety of forms that fall within reason (Groarke, 2012) and because, while maintaining the dialogical perspective emphasized by scholars, it is possible to assume that the primary function of reasoning is different from what they think they have identified (Labinaz, 2014). Instead, we will focus on the third point and try to show how this theory allows us to rethink the relationship between rhetoric and argumentation, particularly the role of normativity in delineating a naturalistic description of what we do when we argue.

Our thesis is that the value of a rhetorical model of arguing (Tindale, 1999) lies in its attempt to combine success in argumentation with weak normativity. To support this thesis, we will use an Aristotelian theoretical framework and in particular the notion of *enthymeme* described and analysed by Aristotle in his *Rhetoric*.

2. SPERBER AND MERCIER'S THEORY

Let us now go back to Sperber and Mercier to look at their theory in greater detail.

As has been said, according to their hypothesis, reasoning has developed to make human communication more profitable (Sperber and Mercier, 2011, p. 60). Its fundamental function is, therefore, on the one hand, to produce arguments to convince our interlocutors, and, on the other hand, to acquire reliable information, and avoid being deceived (Labinaz, 2014, p. 579). The reasoning activity is characterized by two different types of inferences, which they call respectively intuitive and reflective inferences. The former are unconscious and uncontrollable; they occur at a sub-personal level and characterize much of our mental activity. The latter, however, occur when

> [we] accept a conclusion because of an argument in its favour that is intuitive enough [...] construct a complex argument by linking argumentative steps, each of which we see as having enough intuitive strength [...] and verbally produce the argument so that others will see its intuitive force and will accept its conclusion [...]. (Mercier and Sperber, 2011, p. 59)

In this case, we can say that we are at least partially aware of the reasons for drawing some conclusions and of their relationship. In Sperber and Mercier's view, humans tend to use arguments to confirm their unconscious inferences only when they are stimulated in a dialogical context into trying to improve these arguments.

The improvement of knowledge is, in this perspective, a by-product of argumentative reasoning and not its main function. Reasoning has an argumentative function because it has been developed to allow humans to argue in support of their claims and to evaluate the arguments of their interlocutors. It is in fact the result of the interaction of two different cognitive mechanisms: one related to the production of arguments and the other to their evaluation. The latter would have a function of epistemic vigilance, from which it would follow that, although not its primary function, reasoning still results in an improvement of our knowledge (at least under 'normal' conditions). Precisely this feature would allow us to accurately account for the empirical data about our reasoning abilities.

Many studies have shown that, on the one hand, even individuals with a high educational attainment perform poorly when asked to reason in non-dialogical or argumentative contexts, while, on the other hand, performance levels grow tremendously (for everyone who participates in experiments) when the same task is proposed in a dialogical context. This would follow the idea that

> there is an asymmetry between the production of arguments, which involves an intrinsic bias in favour of the opinions or decisions of the arguer whether they are sound or not, and the evaluation of arguments, which aims to distinguish good arguments from bad ones (Sperber and Mercier, 2011, p. 72).

In other words, while trying to convince our interlocutors, it is rational to search for arguments confirming our starting point; when it comes to deciding whether to accept the opinions of others, it is just as rational to try to evaluate adequately whether they are correct. This explains why

> interaction with an adversary will prompt the production of stronger reasons than will solitary thought, and also that the prospect of argumentative success is a stronger motivator than the search for truth or high-quality decisions (Zarefsky, 2012, p. 176).

In this perspective, it can then be argued that reasoning works best in argumentative contexts because it has developed to satisfy argumentative aims. It serves to produce convincing arguments for changing the opinion of our interlocutors, and thus plays an eminently persuasive function.

This is where the question of normativity comes into play. As Sperber and Mercier recognize, the standard for evaluating the argument is only efficacy; reasons that are judged strong according to this standard can be very weak according to critical or normative standards. This is one of the most delicate and controversial points in their theory, which has attracted much criticism. Thus, for example, Santibáñez Yáñez states that they abolish in one shot the distinction between rhetoric and argumentation, whereas the latter field studies precisely the criteria why and how good arguments can ultimately convince and persuade different audiences (2012, p. 151).

This is not surprising, because, particularly for the epistemic tendency (Siegel and Biro, 2008), persuasion and argumentation remain quite distinct and, even if it is allowed that persuasion may sometimes be the aim of argumentation, proponents of this point of view nevertheless consider that the validity of an argument must be evaluated on epistemic criteria alone.

But, one may ask, is this objection really convincing? Or does Sperber and Mercier's theory give us the opportunity to consider the features of our argumentative practice in a more naturalistic way, while at the same time implicitly providing indications on what theoretical and pedagogical perspectives can contribute to improving it? (Zarefsky, 2012). Here, rhetoric, and what it has been theorizing for more than two thousand years, comes into play. Meanwhile, it may be useful to remember that Ian Hacking (2013) and Catarina Dutilh Novaes (2013) have proposed that we consider deductive reasoning, commonly considered to be the reference standard of a correct form of reasoning, to be not so much a cognitive universal, but rather a cultural product, whose 'invention' can be traced back to Aristotelian syllogistics. Taking a brief look at this cultural evolution will prove helpful in rethinking the question of normativity.

3. THE GREEK 'INVENTION' OF DEDUCTIVE REASONING.

Recent research has shown that the appearance of deductive reasoning in ancient Greece – an event unique in human history – must be understood against the backdrop of ancient Greek culture, in which a fundamental role was played by debating practices that – as G. E. R. Lloyd has repeatedly stressed (e.g. 1979) – were essentially open in terms of both participants and audience and radical in their willingness to challenge everything (Netz, 1999, p. 292). Three main features should be emphasized in this context: 1) the fundamental role of rhetoric and persuasion; 2) the centrality of a polemical stance; 3) the topics of these debating practices. Some short comment is in order, starting from the last point.

In most cases, debated issues have two different but complementary characteristics: they are not subject to direct verification and do not allow a conclusive and definitive answer. As

Castelnérac and Marion (2009, p. 72) write referring to the dialectical games in the platonic Academy:

> For some classes of theses, e.g., cosmology, mathematical principles, ethical principles, statements in forensic, medical, or political contexts, etc. there is no possibility of a direct verification, and the Ancient Greeks appear to have considered the test of consistency offered by dialectical games as providing the best available warrant for a given thesis from one of these classes.

From this point of view, it is only natural that these practices have a polemical stance. In a sense there is not an effective difference between debates in the public sphere (political and legal debates) and scientific debates. In both cases, we have an instantiation of an adversarial model of communication that is in contrast with the majority of ordinary conversational situations, which are typically cooperative. As K. Stenning has pointed out:

> logic originated as a model of what might call adversarial communication at least in a technical sense of adversarial [...] and this is not accidentally related to the fact that logic arose as a model of legal and political debate. (2002, p. 138)

This polemical background, finally, explains the role of persuasion and rhetoric. A long tradition has accustomed us to a diminutive and negative image of rhetoric as deceptive practice. On the contrary, in our view, it is born in the context of the Greek culture of the fifth century by seeking a winning strategy within debate practices characterized by cognitive uncertainty and polemical nature. From this point of view, the impossibility of distinguishing between truth and success, or between truth and persuasion, is not really a fault; it is even possible to argue that logical (and mathematical) *apodeixis* "was, partly, a development of rhetorical *epideixis*, a public, oral presentation, akin in a sense to a political speech" (Netz, 1999, p. 292). The 'invention' of deductive reasoning was, in fact, favoured by the dialogical and polemical nature of the argumentative practice in which the identification of an indefeasible argument "corresponds to a winning strategy for the proponent" (Dutilh Novaes, 2013, p. 463). A possible confirmation of this reconstruction comes from the Aristotelian corpus. Although the chronology of the books comprising the *Organon* is a matter of debate, it is almost unanimously accepted that the discovery of the syllogism was a late one and that *Topics* and *Rhetoric* are thus earlier treatises than

Prior Analytics. As Ian Hacking says: "Aristotle had not yet discovered the syllogism at the time he lectured on rhetoric and dialectic" (2013, p. 426). It is worth highlighting that – as we will see more clearly in the next paragraph – this statement applies only to the deductive syllogism and that it is probable that Aristotle worked with a broad notion of syllogism by rethinking the rhetorical syllogism or enthymeme in the light of deductive one. In any case, once the procedure of the deductive syllogism was found it was subsequently considered (though not by Aristotle!) to be a kind of 'golden standard' of reasoning, contributing to devaluing completely the persuasive function (Hacking, 2013).

A last question remains: what prompted Aristotle to discover deductive syllogistics? The answer must be tentative but we think that Castelnérac and Marion (2009, p. 75) are on the right track when they hypothesize "a distinction, between the level of *matches*, where anyone who has mastered the rules can go on playing, and the level of *strategies*, which involves the handling of some procedure", because "it is an entirely different thing to look at strategies, i.e., not only at the ability to play through simple mastery of the rules, but through reflection on games, i.e., study of tricks that will ensure a win". This could correspond to the move from cram-books, as for example *Dissoi logoi*, to the *Rhetoric* and *Topics* at first, and the *Analytics* later, a move to which Aristotle seems to be alluding in the conclusion of his *Sophistical Refutations* (183b36-184a8). In other words, we are dealing here with a metalinguistic knowledge, which helps one to win dialectical games, via deduction. As Hintikka writes:

> [...]. He [Aristotle] was *as competitive as* the next Greek, and hence was *keenly interested in winning his questioning games*. Now any competent trial lawyer knows what the most important feature of successful cross-examination is: being able to predict witnesses' answers. Aristotle quickly discovered that certain answers were indeed perfectly predictable. In our terminology, they are the answers that are logically implied by the witness' earlier responses. By studying such predictable answers in their own right in relation to their antecedents, Aristotle became the founder of deductive logic. (2007, 3-4, our italics)

What is the suggestion that we can draw from this reconstruction? Deductive reasoning seems to be not only the result of an historical process but also a skill, which can arise by means of sophisticated and specific training, schooling in particular (Dutilh Novaes, 2013). But if that is right, perhaps it is not what we need as a standard in discursive practices in which we are usually engaged. We need instead a theoretical framework, which tries to combine effectiveness and weak normativity. This is the case made by Aristotle's *Rhetoric*.

4. ARISTOTLE'S THEORETICAL FRAMEWORK

In the attempt to combine effectiveness and weak normativity, the analysis of the notion of syllogism or rather *sylloghismos*, as Aristotle understood it, offers a good starting point. More precisely, we refer to a particular type of *sylloghismos*: the enthymeme. We prefer to use the Greek term *sylloghismos* in order to highlight the original Aristotelian understanding. Indeed, it is semantically broader than the syllogism of post-Aristotelian logic: it does not exactly coincide with deductive reasoning, and it is not independent of the purposes for which it is deployed.

We shall start by saying that the notion of *sylloghismos* is an Aristotelian invention, as written explicitly by Aristotle in the concluding parts of the *Sophistical Refutations*:

> Moreover, on the subject of rhetoric there exists much that has been said long ago, whereas on the subject of *sylloghizesthai* we had *absolutely nothing else of an earlier date to mention* (184a8-b2).

Aristotle is probably the first to use the word *sylloghismos* to refer to a particular form of reasoning, a form characterized by certain internal guarantees of consistency and correctness – aspects we shall discuss later on.

From an etymological perspective, the noun derives from the verb *sylloghizesthai*, which already existed with the broader and non-technical meaning of 'calculating', 'computing' and, by extension, 'reasoning' or 'concluding'.

To understand the deeper theoretical productivity of *sylloghismos*, it is necessary to note that the Aristotelian reflection on this notion has its roots – even before than in the *Analytics* – in dialectics (*Topics* and *Sophistical Refutations*) and rhetoric (*Rhetoric*) – fields akin to the modern theory of argumentation (Hacking, 2013). Furthermore, as is often the case in Aristotle, the technical use of the concept alternates with broader and less specific uses. This alternation can be identified within the same text, and this is the case with the *Analytics*. In our understanding, it is precisely this semantical amplitude that makes the notion even more interesting and challenging.

There are at least three meanings of the term *sylloghismos* (Smith, 1995):

i. A first meaning, the narrowest one, is *sylloghimos* as a deductive reasoning in which the connection between premises and conclusion is necessary, although, unlike modern logic, this is not an analytic process (*PrA* 24b 18-20);
ii. According to a second meaning, the most widespread diffused one, *sylloghimos* can be seen as deductive reasoning in which it is not specified whether the connection between premises and conclusion is necessary or not (*Rh.* 1356b 16-18);
iii. According to the third meaning, the broadest one, *sylloghismos* refers to reasoning in general so that it can include non-deductive forms of reasoning, as for instance *epagoghē* (induction, *PrA* 68b 15 ff.).

For the purpose of this paper, the meaning we mainly refer to is the second. This choice is directly linked to the fact that it is precisely this meaning that allows Aristotle to consider enthymeme as a *rhetorical syllogism* (*Rh.* 1356b 4-5) (see Grimaldi, 1972; Burnyeat, 1996; Braet, 1999; Piazza, 2000). For reasons of space, we cannot go more into depth into the different interpretations of the notion of *sylloghismos* and the relations with enthymeme. We limit ourselves to saying simply that we fully agree with scholars such as Grimaldi, Burnyeat and Braet, who show the reasons why the enthymeme can be included among syllogisms, thanks to the semantic amplitude of the term *sylloghismos* in Aristotle.

What makes the Aristotelian enthymeme particularly interesting for the topic of this article is, therefore, the fact that enthymeme is simultaneously *sylloghismos* and rhetorical, an association that allows us to keep together several features that have become separated in the post-Aristotelian tradition (Gerritsen, 1999).

As it is rhetorical, the Aristotelian enthymeme is a *sylloghismos* that has a practical purpose: more precisely, deliberation. Being oriented to a practical purpose, the enthymeme must necessarily aim to be effective. This aspect affects all the other features of the enthymeme, including the logical-argumentative ones.

5. THE FEATURES OF THE ENTHYMEME

All these features — intertwined and supporting each other— can schematically be summarized as such:

> i. The enthymeme has a practical and deliberative aim:

> The use of persuasive speech is to lead to decisions. (When we know a thing, and have decided about it, there is no further use in speaking about it). This is so even if one is addressing a single person and urging him to do or not to do something, as when we advise a man about his conduct or try to change his views: the single person is as much your judge as if he were one of many; we may say, without qualification, that anyone is your judge whom you have to persuade (*Rh.*1391 b18-21).

As is clear in the quotation above, all the rhetorical practice is oriented to persuade the hearer to make a decision and therefore the enthymeme (2007, 3-4 our italics) – which is the rhetorical reasoning *par excellence* — cannot disregard this persuasive goal and forget its deliberative aim. In a sense, all the other features we shall see can be viewed as ultimately due to this practical nature of the rhetorical discourse.

> ii. The content of the enthymeme belongs to the domain of so-called "for the most part":

> The enthymeme and the example must, then, deal with what is *for the most part capable of being otherwise*, the example being an induction (*epagoghē*), and the enthymeme a deduction (*sylloghismos*) (*Rh.*1357a 13-16).

This characteristic is a direct consequence of the practical nature of the enthymeme. Indeed, according to Aristotle, "deliberation is concerned with things that happen in a certain way for the most part, but in which the event is obscure, and with things in which it is indeterminate" (*NE*, 112b8-9). Therefore, due to this indeterminate nature of the domain of deliberation, the content of the enthymeme is, always and in principle, potentially questionable.

> iii. The conclusions of an enthymeme are mostly refutable:

> There are *few facts of the necessary type that can form the basis of rhetorical sylloghismoi*. Most of the things about which we

make decisions, and into which we inquire, present us with alternative possibilities (...). Again, conclusions that state what holds for the most part and is possible must be drawn from premises that do the same, just as necessary conclusions must be drawn from necessary premises; this too is clear to us from the *Analytics*. It is evident, therefore, that *the propositions forming the basis of enthymemes*, though some of them may be necessary, *will in the main hold for the most part* (*Rh.* 1357a 22-32).

This passage shows well how the practical aim and the indeterminate nature of the content affect the degree of certainty of the enthymeme's conclusions. Indeed,

the whole Aristotelian theoretical framework revolves around a fundamental methodological principle, that is to say the strong relationship between the object that has to be investigated and the method with which this object is investigated (*NE*, 11 1094b 11-28) (Di Piazza, 2012, 78-79).

This is the reason why the enthymeme, whose content is questionable, is a kind of reasoning that is almost always refutable and, therefore, the only normativity it can aim at is a weak one.

iv. An enthymeme can leave a premise implicit or, in some cases, even the conclusion:

It has already been pointed out that the enthymeme is a *sylloghismos*, and in what sense it is so. We have also noted the differences between it and the deduction of dialectic. Thus we must not carry its reasoning too far back, or the length of our argument will cause obscurity; nor must we put in all the steps that lead to our conclusion, or we shall waste words in saying what is manifest (*Rh.* 1395b 23-27).

This is probably the most famous feature of the enthymeme: so much so that it becomes the only definitory trait of the enthymeme itself in the post-Aristotelian tradition. Let us recall, for example, the traditional definition of the enthymeme as "truncated syllogism" or "syllogism with an unstated premise". However, it is worth mentioning that in the Aristotelian perspective this feature is not a flaw but an important rhetorical quality. Indeed, the opportunity to leave a part of the reasoning implicit increases the efficacy of the enthymeme thanks to the involvement of the hearer, who plays a crucial role in the persuasive process.

v. The listener of an enthymeme must be actively involved:

> The refutative enthymeme has a greater reputation than the demonstrative, because within a small space it works out two opposing arguments, and arguments put side by side are clearer to the audience. But of all *sylloghismoi*, whether refutative or demonstrative, those are most applauded of which we foresee the conclusions from the beginning, so long as they are not obvious at first sight—*for part of the pleasure we feel is at our own intelligent anticipation*; or those which we follow well enough to see the point of them as soon as the last word has been uttered (*Rh.* 1400b29-33).

The crucial role the hearer plays in persuasive process — which we have just discussed — is well described and explained here. According to Aristotle, the hearer of a deliberative discourse is always a judge (*Rh,* 1391b 7-21) and therefore he is never a passive target. Since, for Aristotle, the human being is at the same time reason (*logos*) and desire (*orexis*) (*NE* 1139b 4-5) the only way to persuade him to act is involving him both emotively and cognitively. As Aristotle says in the quotation above, the hearer can *feel pleasure* in taking part in the building of reasoning and this confirms that, from Aristotle's perspective, emotional and cognitive sphere are strictly intertwined. This is also the reason why the best enthymeme is the one able to include the hearer actively in the persuasive process. This need to involve the hearer necessarily affects the degree of cogency of the rhetorical reasoning.

6. WEAK NORMATIVISM

In general, all these features, schematically examined here, cannot but affect the type of normativity of the enthymeme. Indeed, as *sylloghismos*, the enthymeme guarantees a certain normative standard: so much so that Aristotle, for example, admits the existence of 'apparent' enthymemes. That is, enthymemes that do not comply with normative standards, in the sense that they seem to be a *sylloghismos* without being a real *sylloghismos* (*Rh.* 1356a 38-b 4, 1400b 34-1401a 1). On the other hand, however, as the enthymeme is *rhetorical* reasoning, its normative standard cannot be but weak and flexible.

These standards of weak normativity become clearer if we compare the various definitions of *sylloghismos* that Aristotle provides in his works:

i. A *sylloghismos* is a *logos* in which, certain things being stated, something other than what is stated follows of necessity from their being so (*PrA* 24b 18-20);

ii. A *sylloghismos* is a *logos* in which, certain things being laid down, something other than these necessarily comes about through them (*Top.* 100a 25-26);

iii. For a *sylloghismos* rests on certain statements such that they involve necessarily the assertion of something other than what has been stated, through what has been stated (*SR* 164b 27-165a 2);

iv. When it is shown that, certain propositions being true, a further and quite distinct proposition must also be true in consequence, whether universally or for the most part this is called *sylloghismos* in dialectic, enthymeme in rhetoric (*Rh.* 1356b 16-18).

Based on this comparison, it is possible to identify some common features among these different definitions. These features are a sort of minimal criteria for considering a certain chain of *logoi* as a *sylloghismos*. We can schematically summarize these features in this way: the conclusion has to originate semantically and syntactically from the premises and must be different from them. This allows the enthymeme to maintain a heuristic potential that does not affect its practical purpose but somehow depends on it. Although the persuasive purpose of the enthymeme remains its main feature, it still allows the acquisition of new knowledge, although the knowledge is uncertain and refutable. In this way, Aristotle does not separate the strictly logical component from the eminently persuasive one, but he adapts these components in order to keep together weak normativity and effectiveness.

In this way, to conclude our paper, the enthymeme can play a hinging role between argumentation, with its demand to normative standards, and persuasion, with its aim at efficacy. Therefore, Aristotle paves the way towards a naturalistic description of arguing, which is different but, perhaps, complementary to Sperber and Mercier's theory.

REFERENCES

Braet, A.C. (1999). The enthymeme in Aristotle's rhetoric: From argumentation theory to logic. *Informal Logic*, *19*(2-3), 101-117.
Burnyeat, M. F. (1996). Enthymeme: Aristotle on the rationality of rhetoric. In A. Oksenberg Rorty (Ed.), *Essays on Aristotle's rhetoric* (pp. 88-115). Berkeley: University of California Press.
Castelnérac, B. and Marion, M. (2009). Arguing for inconsistency: Dialectical games in the academy. In G. Primiero & S. Rahman (Eds.), *Acts of knowledge: History, philosophy, and logic*, (pp. 45-84). London: College Publications.
Di Piazza, S. (2012). Stochastic knowledge: For the most part and conjecture in Aristotle. In P. Olmos (Ed.), *Greek science in the long run: Essays on the Greek scientific tradition* (pp. 78-95). Newcastle: Cambridge Scholars Publishing.
Dutilh Novaes, C. (2013). A dialogical account of deductive reasoning as a case study for how culture shapes cognition. *Journal of Cognition and Culture*, *13*(5), 459-482.
Gerritsen, S. (1999). The history of the enthymeme. *Sic Sat*, *7*, 228-230.
Grimaldi, W. M. A. (1972). *Studies in the philosophy of Aristotle's rhetoric*. Wiesbaden: Steiner.
Groarke, L. (2012). Should Mercier and Sperber change the way we teach and study reasoning? *Argumentation and Advocacy*, *48*(3), 188-190.
Hacking, I. (2013). What logic did to rhetoric. *Journal of Cognition and Culture*, *13*(5), 419-436.
Hintikka, J. (2007). *Socratic epistemology: Explorations of knowledge-seeking by questioning*. Cambridge: Cambridge University Press.
Labinaz, P. (2014). Reasoning, argumentation and rationality. *Etica & Politica / Ethics & Politics*, *16*(2), 576-594.
Lloyd, G. E. R. (1979). *Magic, reason and experience*. Cambridge: Cambridge University Press.
Mercier, H. & Sperber, D. (2011). Why do humans reason? Arguments for an argumentative theory. *Behavioral and Brain Sciences*, *34*(2), 57-74.
Mercier, H. & Sperber, D. (2017). *The enigma of reason*. Cambridge: Harvard University Press.
Netz, R. (1999). *The shaping of deduction in Greek mathematic: A cognitive history*. Cambridge: Cambridge University Press.
Piazza, F. (2000). *Il corpo della persuasion: L'entimema nella retorica greca*. Palermo: Novecento.
Santibáñez Yáñez, C. (2012). Mercier and Sperber's argumentative theory of reasoning: From the psychology of reasoning to argumentation studies. *Informal Logic*, *32*(1), 132-159.
Siegel, H. & Biro, J. (2008). Rationality, reasonableness, and critical rationalism: Problems with the pragma-dialectical view. *Argumentation*, *22*(2), 91-103.
Stenning, K. (2002). *Seeing reason*. Oxford: Oxford University Press.

Tindale, C. (1999). *Acts of arguing. A rhetorical model of argument*. Albany: State University of New York Press.
Zarefski, D. (2012). A challenge and an opportunity for argumentation studies. *Argumentation and Advocacy*, *48*(3), 175-178.

20

The Critical Question Model of Argument Assessment

IAN J. DOVE
University of Nevada, Las Vegas
ian.dove@unlv.edu

E. MICHAEL NUSSBAUM
University of Nevada, Las Vegas
michael.nussbaum@unlv.edu

We propose a Critical Question Model of Argument (CQMA). It developed from research with argumentation schemes for teaching critical thinking. We combine critical questions with a graphical organization device to make apply this model. We discuss our research findings: The CQMA approach has many virtues of full argumentation schemes, without the vices. As CQMA is easy to apply in practice, it can be a preliminary, if not a competitor, to argumentation schemes.

KEYWORDS: argumentation schemes, critical questions, RSA conditions, VEE diagrams, Walton

1. INTRODUCTION: WHY ARGUMENT ASSESSMENT? WHY CRITICAL QUESTIONS?

Teaching argumentation can be difficult. Producing and critiquing arguments doesn't reduce to a single skill or set of skills. Instead, it is a manifold of skills, procedures, and tasks, any one of which can be daunting for students and teachers alike. Against this impediment, at least in the case of the teaching of composition in American universities, the so-called Toulmin Model or Toulmin Scheme, is the standard tool (Fulkerson, 1996). However, there are good reasons for thinking that this tool is inadequate to the task.

> There is no evidence that studying logic [via the Toulmin Model] had a positive effect on students' written arguments. In

fact, most evidence, though seldom statistically significant, points in the opposite direction. (McCleary, 1979, p. 196, cited in Fulkerson, 1988, p. 441; see also Nussbaum, 2011)

Perhaps just as telling, we have within the introduction to a recent collection of papers on the Toulmin Model, the following claim by the editors of the collection. "In *The Uses of Argument*, Toulmin gave no specific direction on how to evaluate arguments laid out according to his model" (Hitchcock & Verheij, 2006, p. 10). This is problematic because research suggests that argument assessment is a central activity for critical thinking (Halpern, 1988). Hence, we need a different model for argument assessment.

One alternative is argumentation schemes. Although argumentation schemes may trace their origins to Aristotle, recent work by Douglas Walton (Walton, Reed & Macagno, 2008, pp. 3ff.) locates the contemporary source of research into argumentation schemes with Arthur Hastings' dissertation in 1963. Personal experience teaching argumentation assessment using schemes suggests that students find even small numbers of schemes quite difficult to apply to actual, as opposed to textbook, examples. Perhaps, even with a tiny subset of schemes, students would need much more scaffolding than is generally available in a single-semester course to be able to apply schemes generally. Moreover, even when students have some grasp of the schemes, identifying THE scheme relevant to an actual argument can be controversial at best. Thus we need either to revise the teaching of schemes, or find another way to get at the important skill of assessing arguments.

Our suggestion is to use critical questions to guide the assessment process. Although critical questions, hereafter abbreviated "CQs", have generally been associated with particular schemes, we formulate a more generic approach that generalizes from the considerations of CQs from particular schemes. Indeed as students seem naturally to ask questions as part of the process of understanding and assessing arguments, it makes sense to refine and direct this ability in fruitful directions.

2. CRITICAL QUESTIONS AND ARGUMENT ASSESSMENT

Let argument assessment be the process that takes one from some initial confrontation with a text or speech, wherein one must first decide whether there is an argument at all, through to an overall evaluation, though perhaps of a preliminary sort, for the identified argument. This process will include analytical components, such as identifying conclusions, along with evaluative components, e.g., determining whether the premises are acceptable as stated. Our method would guide one through this process by asking critical questions as a cue to what is required to perform such an assessment.

As noted above, generally CQs have been associated with particular schemes. For example, Walton (1996), Walton & Gordon (2011 & 2005), Walton, Reed, & Macagno (2008), Walton & Godden (2007 & 2005), Verheij (2003), Bex & Verheij (2012), Song & Ferretti (2013), Nussbaum (2011), and Nussbaum & Edwards (2011) all identify CQs with particular schemes. The CQs thereby apply normative pressure to elements of individualized patterns of reasoning. This means that the particular CQs of one scheme are not generally applicable to other schemes. To see this, consider two schemes that would usually be considered closely related—argument from authority and argument from position to know. These schemes have a very similar structure. Both rely on features of some kind of witness as evidence for the truth of what is claimed by the witness. But, whereas the argument from authority puts pressure on the *expertise* of the witness, the argument from position to know puts pressure on the *position* of the witness. These are not the same thing. The lack of inter-applicability becomes even more apparent when you consider unrelated schemes, such as comparing CQs of an argument from authority to those of an analogical argument. Hence, to learn a scheme is to learn a specialized set of critical questions as well.

This may explain some of the difficulties associated with learning schemes. Yet, Song and Ferretti (2013) have shown the schemes are less effective when they are taught without the use of CQs compared to when the CQs are explicitly included as part of instruction regarding argument assessment.

> Compared to students in the contrasting conditions [learning argumentation schemes without also learning the associated critical questions and learning neither argumentation schemes nor critical questions], those who were taught to ask and answer critical questions wrote essays that were of higher quality, and included more counterarguments, alternative standpoints, and rebuttals. (Song & Ferretti, 2013, p. 67)

Similarly, Nussbaum (2008, 2011) and Nussbaum and Edwards (2011) found that the use of explicit critical questions enhanced students' abilities[1] regarding argument assessment. In these studies, however, rather than being taught schemes, students were introduced to Argumentation Vee Diagrams (AVDs). In one form, an AVD simply separates a student's perceived explication of the arguments and counterarguments. The only *critical questions* associated with this generic AVD are whether one side—the argument or the counterargument—is stronger, and whether it is possible to construct either a compromise solution or a novel solution to the tension between the arguments and counterarguments (See Figure 1).

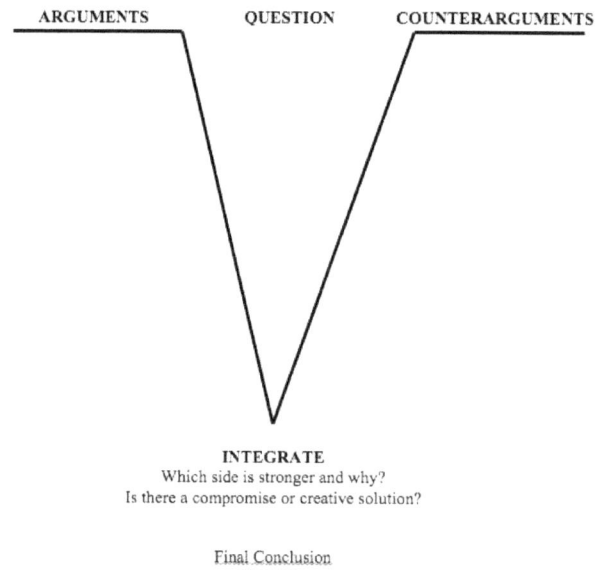

Figure 1 – Blank AVD without critical questions (Nussbaum, 2008, p. 552)

In response to discussions with the students, Nussbaum determined which schemes best captured the students' reasoning. From this he devised a lesson in which the students were taught, first, a generic account of CQs, and next, a particular account of the CQs associated with the schemes they used in assessing the argument. The CQs were questions such as "Why?" and "Who will pay?" Next, the blank AVD was edited to include an explicit set of critical questions. These AVDs also included the arguments and counterarguments that emerged from previous discussions with the students (See Figure 2).

---

[1] In this case, the students were 7th graders.

This approach focused more on teaching students to ask a few key critical questions and to generate their own critical questions based on the generic ones specified by Walton et al. (2008). In contrast to directly teaching schemes, this approach had the advantage of being easier to fit into a curriculum and involved less cognitive load, because students did not have to learn to recognize specific schemes (Nussbaum, 2011, p. 93).

Hence, though these CQs were derived from CQs associated with specific schemes, the teaching of the CQs was independent of the schemes. Yet, students, because of the CQs, were better able to assess arguments.

This all suggests that CQs are important whether or not one presents them within the context of explicit argumentation schemes. Moreover, insofar as some of the elements of argument assessment, for example whether a text contains an argument, are not handled by scheme-specific CQs, it is possible to construct a CQ-centred approach to assessment.

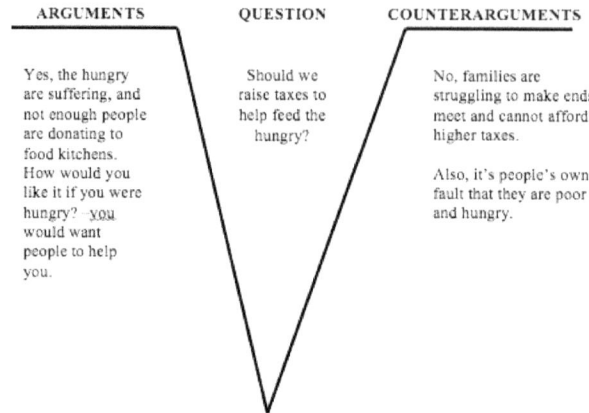

Figure 2 – Sample AVD with Critical Questions included. (Nussbaum, 2011, p. 92)

3. IS THERE AN EXTANT THEORY/ACCOUNT OF CRITICAL QUESTIONS?

To begin to construct such an approach, we want to avoid *reinventing the wheel*. Hence, if there is a theory for critical questions, then we can appeal to that theory to develop general critical questions. Unfortunately, there doesn't appear to be a single explanation for the construction of critical questions. Instead, there are three or four partial accounts that seem to have informed the CQs along with what we might term *historical practice*. Historical practice, for example, might explain the particular ways that arguments from authority have put pressure on the expertise of the asserting authority. The partial accounts include RSA/ARG conditions, an *implicit premise* account, and a general argumentative quality account. Insofar as these accounts can generate critical questions, we can collect and summarize the kinds of questions each account would generalize as part of our overall method for argument assessment.

3.1 Historical practice

Perhaps the most straightforward account of critical questions is an appeal to historical practice. Why, for example, do textbooks generally ask the same questions regarding appeal to authority? The answer is because this practice has long been part of the community of critical argumentation. Had some wording other than, for example, "do other experts agree?" been common, then that alternative would be reproduced in textbook after textbook.

However, to make this account work for our method, we need to look at the more general historical practice of argument assessment. Two particular practices stand out in this regard. First, it seems obvious that the question, "Is there an argument in this text?" is a natural starting point for any assessment. For, if there is no argument, then there is no reason to continue with the assessment at all. Second, there ought to be an explicit endpoint for the assessment. Something like the question, "Given the answers to all of the previous questions, do you think it is thereby rational to accept the conclusion?" seems to be typically invoked in critical thinking texts, though perhaps not as obviously as an explicit question. Since we will return to the question of general argument quality in subsection §3.4, we'll pick up the discussion of alternative general CQs there.

3.2 An RSA/ARG account

Walton (1996) and Walton and Godden (2007, 2005), in particular, focus on the so-called RSA-conditions for good argument as the[2] theory of critical questions. The general RSA conditions ask: (1) Are the premises relevant (in this context)? (2) Are the premises sufficient (for this conclusion)? and (3) Are the premises acceptable (or true or at least plausible)? Although the RSA-version of these questions is due to Blair and Johnson (1994), an equivalent account is given by Govier (2010) using the abbreviation ARG. The "R" and the "A" of Govier's ARG conditions are the same as Blair and Johnson's "R" and "A". Govier's "G" condition is for *grounding*: Is the conclusion grounded by the premises? This is equivalent to the "S" of RSA. Hence, these are equivalent accounts of the conditions for acceptable argumentation. Therefore, these conditions can serve as generic critical questions for any argument whatsoever.

In Walton's earlier text on argumentation schemes, for example, the critical questions associated with the argument from consequences scheme suss out the RSA/ARG conditions.

   i. How strong is the probability or plausibility that these cited consequences will (may, might, must) occur?
   ii. What evidence, if any, supports the claim that these consequences will (may, might, must) occur given the action?
   iii. Are there any consequences of the opposite value that ought to be taken into account? (Walton, 1996, pp. 76-77)

Question 1 probes the relevance of a premise for the context as well as the acceptability of a premise and the sufficiency of the premise for a given conclusion. This first question carries quite a bit of the assessment task. Question 2 probes the acceptability of the consequence premise. However, Question 3 doesn't really fit perfectly with the RSA account. It does, for example, probe the sufficiency of the premise for the given conclusion. However, it does so by probing the dialectical situation (i.e., counterarguments, rebuttals, etc.) rather than through the illative strength the premise provides for the conclusion. Later, we will add this dialectical element to our generic account of critical questions. Still, RSA/ARG do provide a good source for critical questions in the assessment process. Moreover, probing the RSA/ARG conditions would

---

[2] "The" is perhaps too strong here. RSA-conditions are *an* account of how critical questions work according to Walton and Godden.

not require appeal to particular schemes. And that is one of our goals with the method.

3.3 An implicit premise account and dialectical considerations

Walton and Gordon (2011, 2005), in the context of discussing how to include critical questions in argument diagramming software that doesn't explicitly include CQs already, treated some CQs as implicit premises, that is as assumptions presumed in the statement of the argument. The notion of *presumption* on offer here is (legal) burden of proof, since the paper was addressing legal argumentation specifically. However, we can take the notion of *onus* or *burden of proof* more generically to represent how to deal with (some) critical questions as appeals for implicit premises.

Notice that by appealing to presumption or onus or burden of proof, we are explicitly addressing the dialectical situation of argument. Hence, this discussion naturally continues where we left off our discussion of the RSA/ARG as a source of critical questions. Recall that we found it difficult to shoehorn all of the existing critical questions for the argument from consequences scheme into the RSA/ARG account. It was the appeal to dialectical (or perhaps even dialogical) elements that was left unexplained by RSA/ARG. Walton and Gordon, of course, can appeal to the legal notion of *burden of proof* that, although controversial among scholars, is an explicit element of most legal situations. Within the context of argument outside of legal domains, it is difficult to determine whether there really exist *burdens of proof*. We will, therefore, turn to more general dialectical features as possibly creating onus.

Nussbaum and Edwards take critical questions to have only dialectical import.

> Critical questions reflect the dialectical nature of argumentation; asking these questions creates a burden of proof on those advancing the argument being critiqued to answer those questions. [...] In fact, critical questions suggest possible refutations. (Nussbaum & Edwards, 2011, p. 450)

On this account the critical questions create the burden, but the source of the questions is in the possibility of refutation. We can see this kind of understanding of the place of critical questions in other research as well.

> The opponent might ask a series of critical questions (i.e., potential counterarguments) about the acceptability and relevance of these argumentative strategies. [...] In other words, critical questions can help establish the relevance of an argumentation scheme by encouraging consideration of alternative perspective through the generation of counterarguments and rebuttals. (Song & Ferretti, 2013, p. 69)

By locating a source of critical questions in the possibility of counterargument and rebuttal, one would surely consider RSA/ARG conditions. However, there are counterarguments that don't make use of any part of the argument they counter. Such arguments are independent except that their conclusion has a logical tension[3] with the target argument's conclusion. The dialectical pressure that such a counterargument puts on an argument cannot be captured completely by RSA/ARG conditions. Yet, the question, "Is there a (good/strong) counterargument?" would be important for any argument assessment to answer.

3.4 A general argumentative quality account

Though RSA/ARG conditions do probe the general quality of an argument, they are somewhat limited, as noted in the previous subsection. Verheij (2003) and Bex and Verheij (2012) have developed a very general account of critical questions that takes critical questions to be relevant to assessing the general quality of arguments. Both of these papers operate in legal or quasi-legal domains, and hence can appeal to the legal account of burden of proof, for example; yet the overall account is nevertheless useful for extra-legal argumentative situations.

Verheij claims that critical questions serve four roles. First, CQs can be used to *criticize* a schematic argument's[4] premises (Verheij, 2003, p. 182). Since this role makes explicit contact with argumentation schemes, Verheij's overall explanation will not serve our purpose without modifying Verheij's questions to eliminate the reliance on schemes. This may fit the RSA/ARG account as the criticism could take the form of a question regarding the relevance, acceptability, or

---

[3] The strongest tension would be contradictory. Weaker tensions are nevertheless tensions: an argument for a contrary conclusion—where the conclusions can't both be true—would put dialectical pressure on an argument.

[4] By "schematic argument", we just mean that the argument is being analyzed in the terms of a specific scheme.

sufficiency of the premise. Second, CQs explicate exceptions to schemes (Verheij, 2003, p. 182). As an example of an exception to a scheme, he notes that if an expert has made a mistake regarding something for which he or she is a bona fide expert, one would be justified in disregarding the expert testimony. A third role for CQs is as conditions for use (Verheij, 2003, p. 183). It is tempting to understand Verheij's explanation of this role in terms of premise relevance. But this would be narrow. Relevance is surely the right concept. However, as opposed to merely the relevance of the premise for the given argument, we take Verheij to mean that CQs can help an argument assessor to determine whether and to what extent an argument scheme is relevant. As Verheij is focused upon legal reasoning, the context is somewhat more static than that of extra-legal reasoning. Still, the idea here is that simply reflecting upon a critical question associated with a scheme can help you determine whether you are using the correct scheme for a case. Again, insofar as this role requires schemes, it doesn't really fit our needs.

The fourth role, though, recapitulates the dialectical considerations of the previous subsection. CQs aid in discovering possible and relevant arguments and counterarguments for an argument's conclusion (Verheij, 2003, p. 183).[5] As a gloss, then, consider the question, "What other arguments, counterarguments, and rebuttals should be considered in assessing this argument's overall strength?" This would lead to an obvious follow-up question, "Are any of these competing arguments or counterarguments stronger than the argument here assessed?"

Bex and Verheij (2012) consider critical questions in the context of so-called *narrative* arguments or stories. They consider story-specific schemes. However, we can leave the schemes to one side as we consider the particular kinds of critical questions they think are important for assessing arguments given as narratives. What follows isn't the complete list of questions, but the numbers correspond to Bex and Verheij's numbering of the questions.

---

[5] We here extend Verheij's actual explanation of this fourth role. Verheij only considers alternative possible *arguments* for the conclusion. This isn't dialectical. Instead, it is a request for alternative illation. But, in the very next section of his paper, he makes a strong case for dialectical considerations that seem to us to be related to these elements of critical questioning (cf. Verheij, 2003, pp. 183ff.).

1. Are the facts of the case made sufficiently explicit in the story? [...]
5. Have alternative stories been sufficiently taken into account? [...]
6. Have all opposing reasons been weighed? (Bex & Verheij, 2012, pp. 340-349)

The first question, regarding the explicit facts, points to an element of RSA/ARG which is perhaps tacit. When you ask whether the premises of an argument are sufficient for a conclusion, the notion of "the premises of the argument" can be ambiguous. For example, if there is an obvious assumption that would clarify, strengthen, or otherwise aid an argument, but that assumption is, as assumptions usually are, unstated, then is it really a part of the argument? To deny it would be uncharitable? To include it would be unfaithful to the text or speech. Yet we can probe the apparent but unstated elements of an argument through critical questioning. For example, "Are there any assumptions, principles, laws, or presuppositions, on which this argument relies and which should therefore be made explicit?"

Question 5, though obviously a part of the dialectical elements considered above, gets at these elements in an interesting way. Rather than simply querying the existence of counter and competing arguments, it questions whether the statement of the argument pays enough attention to such things. We take it that there are at least three elements to consider when looking at the dialectical situation of an argument. On the one hand, there are existential considerations—are there explicit counterarguments, in the literature, say, to this position? Another consideration is the strength of the counterarguments. But more than that, there is also the question of the amount of attention one must give to competing and counter arguments. The amount of attention that such dialectical considerations require will surely differ in different situations. Hence, a question that allows for this kind of appropriate ambiguity is apt for a project like ours that seeks to ground assessment in critical questions.

Question 6, then, rounds out the dialectical elements by considering their strength. This question could be obviated by the previous one (or at least our retelling of it).

The overall result of these considerations, though, is that critical questions need to probe the general strength of the arguments you consider. We should generalize, for example, Question 6 to make it explicit that in weighing the opposing reasons, these reasons are weighed against the positive support that the argument provides for the conclusion.

4. PUTTING IT TOGETHER: THE START OF A METHOD

In a graduate reasoning course in the College of Education at the University of Nevada, Las Vegas, Nussbaum had students develop critical questions as part of a discussion of policy issues. Students created a wide variety of question types. His classification of these types is useful here insofar as we want to enhance the critical questioning behaviour of argument assessors.

A first class of questions concerns *logical clarity*. These are questions that probe whether one can follow or reconstruct the reasoning as given, whether the terms are appropriately defined, and whether there is an argument at all. These questions, as with the other classes of questions given below, can occur at any stage of the assessment process. Thus, even though, for example, the question, "Is there an argument here?" would seem antecedent to other questions, it is possible that an initially positive answer to this question, which would direct further questioning, might be overturned by subsequent considerations. Hence, even the obvious starting point isn't completely concluded until the overall query completes.

A second class of question are *epistemological*. These concern justification and expertise, for example, and are probably well covered by RSA/ARG considerations. Still, it is worthwhile to note that at least part of the justification for considering RSA/ARG conditions is generally epistemological.

A third category is termed *practical*. Practical considerations don't always fit with any of the other sources of critical questions considered above. An example of a *practical* critical question that occurs in the context of a policy debate is, "Will it solve the problem?" A positive answer, though, doesn't warrant accepting the policy. Rather, it spurs a further *practical* question: "What will it cost?" as well as "What are the risks?" Notice that these practical considerations are in some sense external to the explicitly logical and epistemological questions. Practical considerations do have an obvious home in policy discussion where externalities are sure to intrude on otherwise good solutions, but they aren't foreign to other argumentative contexts. And, by explicitly probing the practicality of an argument's premises or assumptions or even the conclusions, one can make argument assessment more obviously relevant.

A fourth category of questions concerns systemic issues. In policy discussions, these include questions about alternative, but internal factors. Outside of policy discussion, such questions could implicate conceptual issues. Will alternative accounts of, causation, say, lead to different predictions? Such systemic considerations probe the depth and breadth of the context.

A final category includes *normative* considerations. Are some values in a policy debate, more important than others? Again, though this category is clearly insinuated in policy discussions, it is relevant outside of such discussions as well.

It would be nice if we could put all of these considerations into something of a flowchart for argument assessment. However, we've come to believe that such a flowchart would need to be reflexive insofar as the obvious starting point might need to be reconsidered after each phase of questioning. Still, there is a sense in which we start with a general question regarding whether a text contains an argument and we end, if there is an argument, with a general question regarding how we ought to judge the overall quality as a function of answers to previous questions. That we might need to revisit any question on the basis of answers to later questions isn't detrimental to the method.

In each of the following subsections, then, we will layout a general method of assessing arguments by asking and answering critical questions. This preliminary set of questions is meant to capture the many important question types discussed above.

4.1 Argument vs. non-argument recognition

To assess the argument contained within a text or speech, we need to initially consider whether there is an argument. Although some textbooks treat this as if it were an easy question to answer, this question requires having some definition of argument in place. Thus, to answer the question already requires some conceptual work. Here is our first critical question:

    i.   Does the passage/text/speech contain an argument?

4.2 Structural and pattern-focused questions

If you answer the first question affirmatively, there are many possible follow-ups. We don't think you necessarily need to ask structural and pattern focused questions immediately after the recognition question. But, if you are going to assess an argument, one will eventually need to do analytical work. The structural and pattern focused questions will direct this analysis. These questions are meant to probe both the explicit gross structure of the reasoning as presented as well as the, perhaps, unstated principles, assumption, and presuppositions along with the fine-structure of particular kinds of claims, e.g., appeals to testimony. Here is a preliminary set of structural/pattern questions:

    ii. What are the explicit premises, conclusions, sub-conclusions, assumptions, presuppositions, and/or principles in the argument?
    iii. What premises, conclusions, sub-conclusions, assumptions, presuppositions, and/or principles should be made explicit in assessing this argument?
    iv. What is the structure of this argument? Does this structure conform to any pattern of reasoning you recognize? (These patterns could include schemes but also other patterns such as convergent or coordinate arguments.)

4.3 Premise quality and sufficiency questions

Once you have identified the components of an argument, both the explicit and implicit elements, you can then probe the quality of these elements. Again, one isn't required to follow these questions as a necessary sequence. However, an assessment should probe the quality and relevance of an argument's premises.

    v. Are the premises true (or acceptable or plausible) as stated?
    vi. Is there an obvious paraphrase of any premise that would make it more plausible?
    vii. If you added a premise to the argument, what is the justification for adding this premise?
    viii. What makes these premises relevant (if anything)? (If the premises are irrelevant, one should explain this.)

4.4 Individual inference related questions

The steps in an argument could be recast as individual inference manoeuvres. Hence, it makes sense to probe the sufficiency or grounding of these inference moves.

    ix. Are the premises sufficient to ground the conclusion?
    x. Are there obvious or important exceptions to the inference?

4.5 Dialectical questions

For dialectical considerations, recall that we think there are at least three roles for dialectical considerations: existential, strength of counter, and proper attention to alternatives. With this in mind, we put forward the following dialectical critical questions.

> xi. Are there important or obvious, counter or competing arguments?
> xii. Is the conclusion or any sub-conclusion rebutted by some important or obvious counterargument?
> xiii. Does the argument pay proper attention to alternative considerations?

4.6 Practical or external considerations

We noted that, at least for policy discussions, externalities are likely to intrude upon otherwise useful considerations. We think that this is also likely in other argumentative contexts.

> xiv. Are there external factors that will (could, might) undermine the considerations of the argument?

4.7 Overall quality questions

Finally, we want to have an obvious stopping point, though we note that this may be arbitrary insofar as argumentation could continue beyond whatever judgment one makes regarding the overall quality of a particular argument. Moreover, we want this question to reference all of the other questions. In this way, the method is holistic in that it makes the overall judgment a function of the individual answers one gives to the prior critical questions.

> xv. Given the answers to the previous questions, is it thereby rational to accept the conclusion of this argument?

5. CONCLUSION

A method centred on critical questioning makes a satisfying alternative to either the so-called Toulmin-model or argumentation schemes. The method we sketched has the advantage of simplicity, as is found with Toulmin. One needn't learn a long list of schemes and scheme types in order to get on with the business of assessing arguments. And it is in the assessing that this method improves upon the Toulmin-model, which doesn't give a method of assessment. Rather, the Toulmin-model is a method for reconstructing arguments, i.e., it is an analytical method, without also giving tools for evaluation. Our CQMA method improves upon argumentation schemes by leaving the large list of schemes behind, or at least not centred on them. Importantly, we think this method better captures the actual methods that are taught, for example, in critical thinking courses. In such courses, except those that use argumentation schemes, some more general critical questions, such as, "Is there an argument here at all?" guide the assessment process. Thus, we think there will be little cost to most critical thinking instructors in adopting our method. Indeed, it would fit with any method that isn't centred on argument schemes.

ACKNOWLEDGEMENTS: We would like to thank the members of our research group, CarolAnne Kardash and David Vallett, for many helpful discussions as we formulated our account of argument assessment. Dove would also like to thank Steven Berghel and David Godden for numerous conversations regarding critical questions and argument assessment.

REFERENCES

Bex, F., & Verheij, B. (2012). Solving a murder case by asking critical questions. *Argumentation*, 26(3), 325-353.
Blair, A., & Johnson, R. (1994). *Logical self-defense* (1st ed.). New York, NY: McGraw-Hill.
Fulkerson, R. (1988). Technical logic, comp-logic, and the teaching of writing. *College Composition and Communication*, 39(4), 436-452.
Fulkerson, R. (1996). The Toulmin model of argument and the teaching of composition. In B. Emmel, P. Resch, & D. Tenney (Eds.), *Argument revisited, argument redefined: Negotiating meaning in the composition classroom* (pp. 45-72). Thousand Oaks, CA: Sage.
Govier, T. (2010). *A practical study of argument.* (7th ed.). Belmont CA: Wadsworth Publishing.
Halpern, D. (1998). Teaching critical thinking for transfer across domains. *American Psychologist*, 53(4), April, 449-455.
Hitchcock, D., & Verheij, B. (2006). Introduction. In D. Hitchcock & B. Verheij (Eds.), *Arguing on the Toulmin model* (pp. 1-25), New York, NY: Springer.

McCleary, W. (1979). *Teaching deductive logic: A test of the Toulmin and Aristotelian models for critical thinking and college composition.* Dissertation. University of Texas at Austin.

Nussbaum, E. M. (2008). Using argumentation vee diagrams (AVDs) for promoting argument/counterargument integration in reflective writing. *Journal of Educational Psychology*, 100(3), 549–565.

Nussbaum, E. M. (2011). Argumentation, dialogue theory, and probability modeling. *Educational Psychologist*, 46(2), 84-106.

Nussbaum, E. M., & Edwards, O. (2011). Critical questions and argument stratagems. *Journal of the Learning Sciences*, 20(3), 443-488.

Song, Y., & Ferretti, R. (2013). Teaching critical questions about argumentation through the revising process. *Reading and Writing: An Interdisciplinary Journal*, 26(1), 67-90.

Verheij, B. (2003). Dialectical argumentation with argumentation schemes. *Artificial Intelligence and Law*, 11(2-3), 167-195.

Walton, D. (1996), *Argumentation schemes for presumptive reasoning*, Mahwah: Lawrence Erlbaum Associates.

Walton, D., & Godden, D. (2005). The nature and status of critical questions. In D. Hitchcock (Ed.), *The uses of argument: Proceedings of a conference at McMaster University* (pp. 476-484). Hamilton, Ontario: Ontario Society for the Study of Argumentation.

Walton, D., & Godden, D. (2007). Advances in the theory of argumentation schemes and critical questions. *Informal Logic*, 27(3), 276-292.

Walton, D., & Gordon, T. (2005). Critical questions in computational models of legal argument. In P. Dunno & T. Bench-Capon (Eds.) *IALL, Workshop Series, International Workshop on Argumentation in Artificial Intelligence and Law* (pp. 103-111). Nijmegen, Wolf Legal Publishers.

Walton, D., & Gordon, T., (2011). Modeling critical questions as additional premises. OSSA Conference Archive. 51. Retrieved from https://scholar.uwindsor.ca/ossaarchive/OSSA9/papersandcommentaries/51.

Walton, D., Reed, C., & Macagno, F. (2008). *Argumentation schemes*. Cambridge: Cambridge University Press.

21

Topoi and Refutations in Aristotle

IOVAN DREHE
"Babeș-Bolyai" University of Cluj-Napoca
drehe_iovan@yahoo.com

The main presupposition of the present paper is that a discussion on the relation between refutation (elenchus) and common-place (topos) in Aristotle can bring about relevant clarifications in relation to the usage mechanics of the topoi. For this, I will discuss the way in which a Questioner should make use of topoi in order to obtain a refutation in a dialectical encounter and provide illustrations from Plato's earlier dialogues.

1. INTRODUCTION

One of the most important concepts in contemporary argumentation theory is that of "argumentation scheme". Argumentation theorists generally agree that at the historical roots of this concept we can find what Aristotle wrote about the concept of *topos*. This is a "crucial" concept in Aristotle's dialectic and its usage should be understood in the context of the dialectical argumentative dispute specific to the times Aristotle lived in. In short, a dialectical dispute consisted of the following: 2 agents engaged in argument around a dialectical problem which has the form: "Is X the definition of Y or not?", "Is A the genus of B or not?" etc. One of the agents is called Questioner and the other Answerer. The Answerer will choose one of the two possible answers to the dialectical problem: "X is the definition of Y". This is the thesis. From this point, the Answerer has a defensive role, and the Questioner needs to ask questions in order to secure concessions from the Answerer, concessions which are called dialectical premises. Based on these concessions, the Questioner needs to lead the Answerer to accept at some point a proposition that contradicts the initial accepted thesis. If this happens, it means that the Questioner was successful in building a refutation. This can be a refutation of the Answerer (because he will appear inconsistent) or of the thesis, because the conclusion of the refutation contradicts the initial thesis.

So, the purpose of the dispute being a refutation, the usage of the *topoi* should make sense in relation to this goal. The present paper will tackle some issues about the relation between these two concepts: 1. I will start with a concise presentation of the concept of refutation and the way the Question can strategize given the specific features of the thesis he has to refute; 2. I will then continue with a short presentation of the way a *topos* is considered to work; 4. Finally, for illustrative purposes, I will try to exemplify from Plato's earlier dialogues.

2. REFUTATION AND STRATEGY

Refutation is defined by Aristotle thus:

> Fallacies that depend on accident are clear once deduction has been defined. For the same definition ought to hold good of refutation too, except that a mention of the contradictory is here added; for a refutation is a deduction of the contradictory (*Sophistical Refutations* 6, 168a34-37; see also *Sophistical Refutations* 1, 165a3-4; *Sophistical Refutations* 5, 167a23-27 cf. *Prior Analytics* II, 20, 66b14-16).

I should mention here that the terminology used in the *Sophistical refutations* is different from that used in the *Topics* and I thank one of the anonymous reviewers for reminding me of this point: while in the *Sophistical Refutations* we have *elenchus*, in the Topics we find *anaskeue/anaskeuazein* – for theses in affirmative form; and for establishing, i.e. against theses in negative form - *kataskeuazein*.

When the Answerer has chosen his thesis at the beginning of the dialectical dispute, the Questioner should be able to conceive an argumentative strategy that will allow him to construct a refutation. This can be based on 3 levels: 1. On the endoxal character of the thesis (see *Topics* VIII, 5); 2. On the predicational form of the thesis (see *Topics* VII, 5); 3. On the propositional form of the thesis (see *Prior Analytics* I, 26). In short:

1. The endoxal character of the thesis: A reputable or plausible thesis (*endoxon*) should be refuted by a refutation with an implausible (*adoxon*) conclusion. An implausible one, the other way around, should be refuted with a plausible conclusion.

2. In relation to the predicational form of the thesis, the dialectical theses can express one of the 4 possible relations between the subject and the predicate, i.e. the predicables: 1. the predicate can be

the definition of the subject; 2. the predicate can be the property of the subject; 3. the predicate can be the genus of the subject; 4. the predicate can be the accident of the subject. Considering this classification, Aristotle says that the easiest to refute is the definition and the hardest to refute is the particular accident. The account to be found in *Topics* VII, 5, presents propositions where the predicate is the definition, property, genus or universal accident of the subject as refutable by SeP or SoP, while the propositions where the predicate is the particular accident of the subject only with SeP.

3. Finally, in relation to the logical form of the thesis, Aristotle presents us in *Prior Analytics* I, 26 with the following: If the thesis is a SaP, then it can be refuted with SeP and SoP (9 valid forms with conclusions of this type); if the thesis is a SeP, then it can be refuted with SaP and SiP (5 valid forms); if the thesis is a SiP, then it can be refuted with SeP (3 valid forms); finally, if the thesis is a SoP, then it can be refuted with SaP (1 valid form).

This strategic planning on the part of the Questioner is done before the actual dialectical dispute starts. This way the Questioner can identify the strategic objectives that should be reached in order for a refutation to come about.

I detailed this initial phase in a paper from 2015, "Dialectical Strategic Planning in Aristotle". When I first conceived this paper, I borrowed a distinction between strategy and tactics from military terminology, in the sense that in the case of a dialectical dispute we can find both strategy and tactics. Strategy is about planning ahead in order to determine which particular objectives need to be reached in order to achieve the final objective (this being the refutation), and tactics being about how to reach each particular intermediary objective. In short, in terms of tactics, one needs an arsenal of *topoi*.

3. WHAT IS A *TOPOS*

It is hard to understand the exact mechanics of the way a *topos* should be used in argumentation. In the scholarly work a *topos* has been described in a variety of ways:

- topoi = points of view (Hambruch, Viehweg, Prantl, Wieland)
- topoi = middle terms in syllogisms (Prantl)
- topoi = principles to solve the four types of dialectical problems (Gardeil)
- topoi = common genera for many arguments (Thionville)

- topoi = non-analytical premises (Plebe)
- topoi = research formulas (Lausberg)
- topoi = "a general principle out of which arguments must be drawn for concrete cases" (Bochenski)
- topoi = "pigeon-holes from which dialectical reasoning is to draw its arguments" (Ross)

(De Pater, 1965; Slomkowski, 1997; Rubinelli, 2009)

At this point, the most plausible interpretation seems to be that of J. Brunschwig, found in the first volume of his edition of the *Topics* (1967, XXXVIII - XLV). Brunschwig describes the *topos* as "a machinery to make premises starting from a given conclusion". The *topos* is like a rule of inference, for example Modus Ponens or Modus Tollens, where the major is the rule, the minor is a premise that it is already accepted by the Answerer in the dialectical debate, and the conclusion is the newly made premise, to be used for furthering the dialectical discussion towards a refutation.

So, I take the topoi to be rules that warrant (explicitly or implicitly) the passage from one proposition to another in terms of concession. In other words, they help making the conceded proposition more acceptable from an endoxal perspective. Or, they warrant the acceptability of the conceded proposition based on the greater acceptability of the topos rule and of another premise conceded before.

4. USE OF TOPOI

Before I continue, I have to underline the following: it should not be expected that everything that was said about strategy in Aristotle fits perfectly to Plato's earlier dialogues, because this would imply an anachronistic reading at least. In many cases we can happily cherry-pick some examples of syllogisms employed by Plato, but this should not be considered enough. However, until a more systematic presentation of the *topoi* present in the central books of the *Topics* emerges, we are condemned to use only examples that seem suitable for our purposes. Anyway, certain features, such as the endoxical character of the premises, do appear in Plato; or cases where particular instances undermine general statements, or inductive moves. But we should not expect to find illustrations or applications of a fully developed argumentative dialectic. With this in mind, we can observe that there are certain places where it seems possible for a *topos* to be applicable (N.B. the scholarly hypothesis that at least in part Aristotle collected the common places while in the Academy). In what follows, I will present one dialectical sequence (or *agon*) from one of Plato's earlier dialogues, the *Lesser Hippias*[1] which could be understood in light of certain common-places, which seem to work in the background of the argument.

The thesis defended there by Hippias, that "Achilles is better than Odysseus", from the perspective of its endoxal character, seems to be considered reputable or plausible in the highest degree in the Greek world. Now, to argue about the superiority of the Iliad over the Odyssey based on the fact that Achilles is better than Odysseus or the other way around can be based on the following *topos*:

T1: „if one thing is without qualification better than another, then also the best of the members of the former is better than the best of the members of the latter; e.g. if man is better than horse, then also the best man is better than the best horse. Also if the best is better than the best, then also the former is better than the latter without qualification; e.g. if the best man is better than the best horse, then also man is better than horse without qualification" (*Topics* III, 2, 117b33-38, tr. Pickard-Cambridge).

Here, „without qualification" means that is an *endoxon* of the type that is not contested (I follow here (Brunschwig 2007, p. 118): „absolument conforme ou absolument contraire à des idées admises". Now, is this the status of the proposition that "Achilles is better than Odysseus?" It seems so. Because in the time of Aristotle the superiority of Achilles over Odysseus was accepted as illustrated by an example given by Aristotle in the 3rd book of the *Topics*, where he says that "what is nearer to the good is better and more desirable. (...) Also, the one which is more like something better than them both, as. e.g. some say that Ajax was a better man than Odysseus because he was more like Achilles" (*Topics* II, 2, 117b10-16). Other arguments from the tradition or from more recent scholarship can be brought to show that the Greeks considered the Iliad as better. For example, the *scholia* on the Odyssey are much fewer; also, the Odyssey being only one return home (or *nostos*) among many others; or, Pseudo-Longinus in the treatise *On the sublime* 9 writes that Homer wrote the Iliad when he was at the zenith of his creativity, while the Odyssey was written much later, when his creative powers were already in decline. But this is the later tradition, in

---

[1] Context: There are two principal agones in the dialogue: 1. The thesis "Achilles is better than Odysseus" is refuted by "Odysseus is better than Achilles" (364c-371e); 2. The thesis "Those who do wrong willingly are worse than those who do not do wrong willingly" is refuted by "Those who do wrong willingly are better..." (375d-376b).

the time of Plato the general opinion seems to be the one attributed by Socrates to Apemantus at the beginning of the dialogue: "Apemantus used to say that the Iliad of Homer is a finer poem than the Odyssey, to just the extent that Achilles is a better man than Odysseus; for, he said, one of these poems is about Odysseus and the other about Achilles." (Hippias Minor, 363b). The character of Hippias agrees to this and this is the first thesis in the dialogue: "Achilles is better than Odysseus."

So, at this point, Socrates knows that he needs to reach the opposed conclusion, i.e. that „Odysseus is actually better than Achilles". This, of course, is implausible without qualification, i.e. generally considered not to be the case. Consider what Aristotle has to say in the *Topics* (VIII, 5, 159b13-23) about how should the conclusion of the refutation look like from the perspective of the endoxal character of the thesis: if the thesis is reputable, then the conclusion of the refutation needs to be implausible or less reputable, and the premises of the refutation need to be more reputable or plausible than the conclusion (cf. this with the premises which are better known than the conclusion in the case of a demonstration). It is clear that the conclusion Socrates wants to reach is implausible „Odysseus is better than Achilles". But he will do it nevertheless, using some plausible premises. In order to do this, then, Socrates needs to show that Odysseus is better in a certain way than Achilles. And he does this by bringing into discussion concepts like capacity and knowledge.

The argument goes on in the following manner:
Thesis: Achilles is better than Odysseus.

---

1. Liars = capable of doing something (c = concession)
2. Liars = possess intelligence (c)
3. Liars = know what they do (c)
4. Liars = capable, intelligent, knowledgeable (from 1, 2, 3)
5. The ignorant man is different from the one who knows. (c)
6. The one who knows is capable of lying (from 4, 5)
7. The one who knows is capable of telling the truth (induction from sciences)
8. The one who tells the truth is the same with the one who lies (from 6, 7).
9. The expert in certain matters can tell truths and lies voluntarily about those matters (induction).
10. Voluntary liars are better than involuntary liars (from 8, 9)
11. Achilles lies involuntarily; Odysseus lies voluntarily (c)

Refutation: Odysseus is better than Achilles (from 10, 11)

1. When Socrates introduces the idea that the liar is a person who has the power or is capable to do something, it is obvious that he determines Hippias in this way to take this as something acceptable. In the 4th book of the *Topics* Aristotle says the following:

> T2: 'a capacity is always a desirable thing; for even the capacities for doing bad things are desirable, and that is why we say that even God and the good man possess them; for they are capable (we say) of doing evil. So then capacity can never be the genus of anything blameworthy. Otherwise, the result

> will be that some blameworthy thing is desirable; for there will be a capacity that is blameworthy' (*Topics* IV, 5, 126a37-b3).

When Socrates makes this move, he changes the focus of the discussion from something blameworthy to something that is desirable: to have the power to do something. Then the discussion goes on about the capacity to tell lies, which, in itself can be something desirable.

So, the reasoning would look like this:

1. T2 – A capacity is always a desirable thing.
2. To lie is a capacity
----
3. This capacity is a desirable thing.

2. In the following steps of the argument, Socrates links this capacity to knowledge, which makes it "even more desirable" (cf. Aristotle in the *Eudemian Ethics* VII, 12, says that perception and knowledge are most desirable). This engages the discussion about lying in positive terms and this will eventually allow Socrates to lead Hippias to the improbable conclusion that Odysseus is better than Achilles. Linking knowledge and capacity is possible via another *topos* from the 3rd book of the *Topics*, a *topos* of "like degree" (as contrasted with *topoi* of lesser or greater degree):

> T3: 'For if a certain capacity is good in a like degree to knowledge, and a certain capacity is good, then also is knowledge; while if no capacity is good, then neither is knowledge' (*Topics* III, 6, 119b24-26).

So, in this manner Socrates can introduce "knowledge" in the equation, which, as can be seen in the *Ethics*, is a prerequisite for voluntary action, involuntary action being caused by ignorance (or by force in certain cases). These two things being preferable, and voluntary action being preferable as well, based on capacity and knowledge, it results that Odysseus is better than Achilles.

5. CONCLUDING REMARKS

Please keep in mind that these *topoi* need not to be stated explicitly in a discussion, especially between experts, who do not seem to need explanation that will warrant each argumentative step. So, they can be considered as being tacitly accepted by the discussants.

ACKNOWLEDGEMENTS: This work was supported by a grant of the Romanian National Authority for Scientific Research and Innovation, CNCS – UEFISCDI, project number PN-II-RU-TE-2014-4-1207.

REFERENCES:

Jonathan Barnes (Ed.), *The complete works of Aristotle* (Vol. II). Princeton: Princeton University Press.
Plato (2009). *Complete works* (J. M. Cooper, Ed.). Indianapolis: Hackett Publishing Company.
Brunschwig, J. (Ed. and Trans.) (1967). *Aristote, Topiques (livres I-IV)*. Les Belles Lettres: Paris.
Brunschwig , J. (Ed. and Trans.) (2007). *Aristote, Topiques (tome 2, livres V-VIII)*, Les Belles Lettres: Paris.
De Pater, W. A. (1965) W. *Les Topiques d'Aristote et la dialectique platonicienne*. Fribourg, CH:Editions St. Paul.
Drehe, I. (2015). Dialectical strategic planning in aristotle. *Symposon. Theoretical and Applied Inquiries in Philosophy and Social Sciences*, *2*(3), 287-309.
Rubinelli, S. (2009) *Ars topica: The classical technique of constructing arguments from Aristotle to Cicero*. Dordrecht: Springer.
Slomkowski, P. (1997). *Aristotle's Topics*. Leiden: Brill Academic Publishers.

22

Gender, Argumentation and Inference in Mexican Political and Media Discourse

OLGA NELLY ESTRADA
Department of Political Sciences, UANL, Mexico
olganellye@yahoo.com

GRISELDA ZÁRATE
Department of Humanities, Universidad de Monterrey, Mexico
griselda.zarate@udem.edu

> This paper aims to identify the inferential processes in argument production in Mexican media and political discourse from a theoretical interdisciplinary perspective including gender studies, argumentation theory. According to gender studies theorist, Gayle Rubin (1986) the concept of being a woman and to act in political scenarios is more criticized in a sex-gender culture. The study also incorporates an integrated operative model of argumentation (Zárate 2012; 2015), which contains logical, emotional, visceral and kisceral modes, drawing from Toulmin, Rieke and Janik (1979) and Gilbert (1997).
>
> KEYWORDS: Inference, gender, argumentation, political discourse, media discourse, conceptual metaphor

1. INTRODUCTION

This article studies the inferences in gender discourse used by Mexican politicians when they refer to women legislators in media, through the case of pre-candidate Giselle Arellano to a seat in the national congress for the state of Zacatecas, Mexico, in 2013. Special attention is given to the arguments that were made with a sexist and misogynist position against the feminine gender to infer in the Mexican collective imagination that women do not have the intellectual capacities to occupy a legislative position. From this space is argued and inferred a critique of political discourse to denote and denaturalize the androcentric language that harms and subordinates the feminine gender through political discourse (Estrada and Flores, 2016). Faced with such a situation of evident inequality, the reflection that is built on this work, pays for awareness, for non-discrimination, to reflect on equality between women and men to eradicate – reduce – the processes of naturalization of violence (Bourdieu, 2003). It also shows how women in politics are at a disadvantage and do not have the cultural and social privileges of their male counterparts. The representations of women imposed by the sex-gender system in Mexican society remain sexist, discriminatory and exclusive. The above is observed in some media narratives when they refer to the actions of women in politics, such is the case of several legislators who broke with the traditional roles of women to enter politics and / or protest against corruption (Estrada and Flores, 2016).

The current democracies and political parties are immersed in different changes and one of them is the incursion of women into politics, a consequence of the feminist movements in the world. At the beginning of the twentieth century in Mexico, thanks to the struggle of women, their demands for education and, particularly, for the right to vote were met. In 1953, Mexican women obtained citizenship, it is to say, the right to vote and the right for women to enter politics. In spite of having obtained the vote and the right to be voted to occupy a legislative position, the majority of the women returned to the domestic life, since it was the duty to be woman of that time in spite of the new law.

In the Mexican political system it was normal not to see women participating in Congress or to see them on popular election ballots, despite the fact that the law already contemplated that right, nevertheless in the collective social imaginary it was believed that The role of women was to be in the private and the care of the children only (Estrada and Ochoa, 2015). The system did not provide the necessary strategies to include them after the right to vote for women to enter politics or make policies for inclusion in the public world, but kept them private until a group of feminists in the 1990s questioned the absence of women in the seats of the Mexican congress is when the percentages of women in representation positions were analysed for the first time (Estrada, 2012). It is from these unequal results that strategies and inclusion mechanisms are proposed, giving as a reference the proposals of the first initiatives of the so-called positive discrimination or gender quotas in 2012. It was also reflected on the importance of women to be inserted in political participation and establish mechanisms for a parity democracy and that they could influence in this field of decision making for the benefit of women. These public policies paved the way for women in the twenty-first century to see their names on the ballots to be candidates and could win some popular election. Based on these findings and the fact that these inequalities became visible, some academics such as Graciela Hierro (1999) opened the feminist reflection within the Mexican academy and decided to give congresses, colloquiums and seminars in universities all over the country so that thought and inclusive language spread in many areas of everyday life and the importance of women entering Mexican politics. A society that pretends to be democratic, cannot leave women excluded, because for reasons of representation they are a little more than 50 percent of the national population and are citizens with the same rights and opportunities in accordance with the Mexican Constitution, in its article 4 3. There are also several ethical, political and efficiency fallacies that lead to the belief that there is less capacity in government or performance for public office. The figure of the feminine remains exposed to a continuous vulnerability and therefore is a political issue projected with different speeches that mask its reality. The prejudice of seclusion in the private space continues to subject women to a subjectivity that confuses them to a state of emergency (Agamben, 2000) and, on the other hand, the discourse of public institutions, which generate practices of disapproval, with a load of meaning in which cultural components of various kinds, whether religious, moral, conventional or traditional, are combined (Alencar Rodrigues & Cantera, 2013, UN Women, 2012). For these reasons, this research intends to identify how political discourse shapes practices of naturalized violence within the public space and how, in some way, the media legitimizes it and the political substratum endorses it.

2. METHODOLOGY

The theoretical framework is based on an integrated Operative Model of Argumentation (OMA) (Zárate, 2012; 2015), which draws from Toulmin, Rieke and Janik (1979) and includes logical, emotional, visceral and kisceral modes of argumentation Gilbert (1997), as shown in Figure 1.

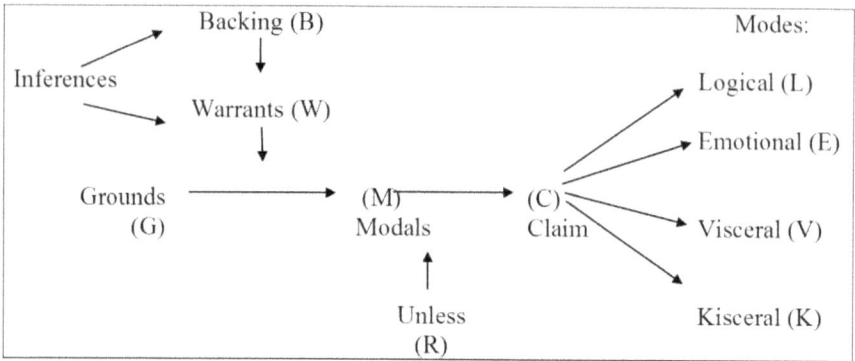

Figure 1 – Operative model of argumentation (OMA)

The study also incorporates a gender perspective. For gender studies theorist, Gayle Rubin (1986) the concept of being a woman and to act in political scenarios is more criticized in a sex-gender culture. In particular, it is important to identify the naturalization of violence in discourse (Bourdieu, 2003).

3. ANDROCENTRIC AND SEXIST INFERENCES IN MEXICAN POLITICAL AND MEDIA DISCOURSE

The twentieth century was called the women's century, because it was when women won the right to be citizens, but the present century has been characterized by the struggle to include women in politics and in different fields that were believed to be only for the male gender, for examples science or politics. On the other hand, the debate about the importance of using an inclusive gender language is a subject little discussed in Mexican politics and has generated diverse positions for and against that has generally fallen into misogynist jokes or mockery in the lexicon when they refer to the female gender.

However, for some years the gender language has been accepted by the Mexican government, which has permeated the different strata of society. The most obvious example is that of former President Vicente Fox, in his speeches: "Mexicanos y Mexicanas" [Spanish male and female linguistic constructions respectively], even if it was taken seriously or in a mocking tone, people of different socioeconomic levels had the opportunity to realize the role that women played in language and reflect on it. However, it is necessary to study inclusive language in the media and the academy since in the contemporary world communication through mass media has become the essential form of contact between human beings.

The volume and intensity with which these media reach the average citizen deserves not only a deep study, but also a particular approach to each form and scope of mass communication. Such is the case of political communication, it involves both the recipients, who are the citizens, and the issuers who are the political agents; thus, it constitutes a space of agreement or disagreement between those who form an organized society.

Our interest is to show how the patriarchal discourse has not changed and therefore permeates androcentrism in political discourse when it refers to women. Androcentrism is when men are given more value and hegemonic terms are expressed and has the vision of the world that places man at the centre of all things. Ana de Miguel (2016) and "political discourse" is used in allusion to any manifestation, message or expression directed intentionally to an audience, with the purpose of influencing, persuading or persuading it to adopt the position that the source of the communication sustains with reference to some public and / or political issue. We consider that this expression can take shape both through the word and the image, since we understand that the image can also be used as a discursive resource (Page, 1996). It is also the main means of a candidate and his political organization to get his message to the population and the electorate.

Sexism is the idea that men are superior to women physically, morally and intellectually, and the status of women is subordinated to (Ana de Miguel 2016). Theorist Michel Foucault has stated that power is exerted through discourse (2008). As Estela Serret says, she mentions that the subordination of women is seen regularly as the result of a set of discursive practices that are organized and intentionally reproduced by the beneficiaries as part of a project to preserve a system of domination (2006, p. 68).

Androcentric and sexist political discourse against women who participate in Mexican politics are reproduced in the press in their headlines and texts. Political narratives presented by the press about women in the political context, show specific generalizations and inferences of androcentric and sexist dimensions, which unfortunately reproduce the prejudices or gender stereotypes prevailing in Mexican society. Such ideas are based on gender roles differently for men and women, because of the criteria determined by the sex-gender system that establishes behaviours, attitudes, behaviours, tasks or activities for one or the other. These norms of social coexistence are established in a natural way and one of the ways to do so is through the political hierarchies of those who make decisions from power, from the church and one of the most important today is the media, as Guerrero mentions that

> The main agents of socialization through which our identity is developed and developed are the family, the school, the media and the language. Thanks to the power of language, human beings can communicate, transmit thoughts, feelings, knowledge, ideas. Language is at the same time a reflection and model of society (2012, p. 8).

In the same vein, Vilches points out that the process of socialization is a part of the process by which men and women adapt to their peers through the whole range of economic, social, technological, religious, aesthetic and linguistic traditions that have inherited. At present, the media in many cases replaces the family or school institution, now receives it through the media, mainly television and social networks (Celaya, 2008, p. 13). Socialization promotion has been observed thanks to popular networks such as Facebook, Twitter and Tuenti. The creation of networks has emerged at a time when the pace of life is so fast that it prevents us from maintaining contact as we would with friends and acquaintances, and thus, with a few minutes a day, we can keep in touch in a virtual way, but gifts.

3.1 Case – Modelo Giselle Arellano

A few days before election, aspiring congresswoman Giselle Arellano received a notification from the conservative political party, PAN (*Partido Acción Nacional* in Spanish) refusing her candidacy to run for a congress seat for the state of Zacatecas at the national congress in Mexico, on the grounds that she lacked an honest way of living. The issue was covered by media, for example, in the article "PAN refuses registration to migrant model" ["Niega PAN registro a modelo migrante"], written by Gerardo Romo and published in *Reforma* newspaper on March 15, 2013.

(1) The PAN's national committee led by Gustavo Madero and Cecilia Romero denied the registration on Friday to Giselle Arellano, as a pre-candidate for a migrant congress seat from Zacatecas, arguing that she has no honest way of living, so the young woman of 33 years, who is a professional model, said she had been denigrated. ["La dirigencia nacional del PAN encabezada por Gustavo Madero y Cecilia Romero

negaron el registro este viernes como precandidata a diputada migrante de Zacatecas, a Giselle Arellano, argumentando que no tiene un modo honesto de vivir, por lo que la joven de 33 años quien es modelo profesional se dijo denigrada".]

Following the operative model of argumentation (OMA), one can identify the (C) claim of the political discourse, reproduced in the media text, the refusal to grant the candidate registration to former model Giselle Arellano, on the (G) grounds "that she has no honest way of living". Thus, one can assume that behind those grounds, in the form of (W) warrants and (B) backing, is the inference that models are not permitted to contend for a political position. Nevertheless, one can mention the case of Marlene Benvenutti, a former lingerie model, who was a local congresswoman in the state of Nuevo Leon for the same conservative party PAN. One can deduct that party rules vary according to the state in question. In Benvenutti's case her protest for a state case of corruption, triggered the publication of photos of her former job a model (Estrada and Zárate, 2017). The general public knew of Giselle Arellano's situation through the social widespread of a video:

(2) The pre-candidate caused controversy by a video that appeared in social networks in which she appears modeling in underwear with angel wings for a company of Las Vegas. ["La precandidata causó polémica por un video que apareció en redes sociales en el que aparece modelando en ropa interior y con unas alas de ángel para una empresa de Las Vegas".]

Through discursive markers such as "controversy", it is inferred in the text that the appearance in a video modeling in underwear is not appropriate for a Mexican political candidate. As mentioned before, a woman who acts in a political scenario is more criticized in a sex-gender culture (Rubin, 1986). This is particularly acute when the political woman, such as Giselle Arellano has posed as a model before, because of the woman stereotype taken into consideration as housewife, honest worker, in Mexican society (Lagarde, 2006). It is important to indicate that Mexican collective imagination sees women as a reigning queen in her home – with moral virtues, modesty, and a physical and emotional caregiver; on the opposing side, as a woman of no moral virtues or qualities. A middle category is the one of a woman who exercises power (Ortner, 1979), or wants to exercises it such as Giselle Arellano's case. One can even identify the naturalization of violence (Bourdieu, 2003), as

Arellano was discriminated against because of her previous job. Sexist comments repeatedly appear in public life as this is a symptom of the poor democratic regime and normalization of violence against women from political power.

On the other hand, a misogynistic inferential dimension in the media text is expressed in the following quote:

> (3) The document with the resolution of the National Executive Committee, with the refusal to the pre-candidature that was shown by Giselle Arellano, appears with the signature of Cecilia Romero, Executive Secretary of the CEN of the PAN. ["El documento con la resolución del Comité Ejecutivo Nacional con la negativa a su precandidatura que fue mostrado por Giselle Arellano aparece con la firma de Cecilia Romero, Secretaria Ejecutiva del CEN del PAN"].

In the sex-gender system, men and women reproduce misogyny, since the only thing that counts in the system is male hegemony. This is shown by the approval of Cecilia Romero, Executive Secretary of the National Committee of the PAN. the leader and executive secretary of the PAN CEN, also in his political speech with clearly androcentric and misogynistic dimensions.

4. CONCLUSION

This study incorporates theoretical gender perspectives (Rubin, Lamas, Ortner), and an integrated operative model of argumentation (Zárate, 2012), which contains logical, emotional, visceral and kisceral modes, drawing from Toulmin, Rieke and Janik (1979) and Gilbert (1997) to analyse inferences in political discourse. As the previous pages show, political discourse shapes practices of naturalized violence within the public space and how, in some way, the media legitimizes it and the political substratum endorses it.

As it has been stated, Mexican collective imagination sees women as a reigning queen in her home – with moral virtues, modesty, and a physical and emotional caregiver; on the opposing side, as a woman of no moral virtues or qualities. A middle category is the one of a woman who exercises power (Ortner, 1979), or wants to exercises it such as Giselle Arellano's case. What are the inferential processes regarding gender construction in Mexican politics? According to

Mexican feminist theorist (Lamas, 2007), collective memory in Mexico has changed with new generations they assume government corruption is far worse than being a lingerie photo model. This theorist also mentions that being a model is a job as the next one. There is a strong inference of politics as a masculine field in Mexico in the general population, spite of constitutional law. Behind those inferences there is the assumption that women in Mexican politics are still subjected to the private dimension of taking care of others. XVII century scholastic and patristic discourse in education (Estrada y Ochoa, 2015) of moral excellence in regards to women is still valid in today's Mexican society.

ACKNOWLEDGEMENTS: Beca Matías Romero 2017, UT in Austin, Texas; and CONACYT.

REFERENCES

Agamben, G. (2000). *Estado de excepción. Homo sacer, I.* Buenos Aires, Argentina: Adriana Hidalgo editora.
Alencar Rodrigues, R. de, & Cantera, L. M. (2013). Intervención en violencia de género en la pareja: el papel de los recursos institucionales. *Athenea Digital, 13*(3), 75-100.
Álvarez, A de M. (2015). *El Neoliberalismo sexual.* Cathedra: Madrid.
Bourdieu, P. (2003). *La dominación masculina.* Barcelona: Anagrama.
Celaya, J., (2008). *La empresa en la web 2.0.* (3rd ed.).Madrid: Gestión 2000.
"De las garantías individuales", Constitución Política de los Estados Unidos Mexicanos, Anaya, México, 2009, pp. 123-140.
Estrada, O.N., y Ochoa, I. A. (2015). Argumentos y refutaciones de la supuesta inferioridad femenina: Un repaso histórico. In *Pasado, presente y porvenir de las humanidad y las artes* VI. Zacatecas: Textere Editores.
Estrada, O.N., y Flores, M.E. (2016). Género, imagen y discurso en el ámbito político a través de la tecnología de poder, en la prensa y las redes sociales. XI Congreso Iberoamericano de Ciencia Tecnología y Género
Estrada, O. N. (2012). *Vivencias, realidades y utopías. Mujeres, género y feminismo.* Monterrey: UANL.
Estrada, O.N., y Zárate, G. (2017). "Atenuación, género e intensificación en el discurso político: Las reacciones mediáticas del caso de una legisladora de Nuevo León, México, (2015)". Universitat de València. *Revista Normas* 7 (2), 125-138.
Foucault, M. (2008). *El orden del discurso* (A. González Troyano, Trans.) México, D. F.: Tusquets.
Gilbert, M. A. (1997). Coalescent argumentation. Mahwah: Lawrence Erlbaum Associates, Publishers.

Guerrero Salazar, S. (2012). Guía para un uso igualitario y no sexista del lenguaje y de la imagen en la Universidad de Jaén. Jaén: Universidad de Jaén. Retrieved from https://www10.ujaen.es/sites/default/files/users/facexp/TFG/Guia_lenguaje_no_sexista.pdf

Hierro, G. (1999). *De la domesticación a la educación de las mexicanas.* 4ª.edición. México, D.F.:Edición Torres Asociados.

Lagarde, M. (2006) *Los cautiverios de las mujeres: Madresposas, monjas, putas, presas y locas.* México: UNAM.

Lamas, M. (2007) *Miradas feministas sobre las mexicanas del siglo XX*, FCE, México.

Ley de Acceso de las Mujeres a una Vida Libre de Violencia, Colección Mujeres y Poder, IEMNL, Nuevo León, 2007.

Ortner, S. (1979): ¿Es la mujer con respecto al hombre lo que la naturaleza con respecto a la cultura?. In Harris y Young, comp. *Antropología y feminismo.* Barcelona: Anagrama.

Page, B. (1996). *Who deliberates? Mass media in modern democracy.* Chicago: University of Chicago Press.

Rubin, G. (1986). *El tráfico de mujeres: Notas sobre la economía política del sexo* (Vol VIII, , Núm. 30). Ciudad de México: Nueva Antropología.

Serret, E. (2006). *El género y lo simbólico: La constitución imaginaria de la identidad femenina.* Instituto de la mujer oaxaqueña, Oaxaca, IMO.

Toulmin, S., Rieke, R., & Janik, A. (1979). *An introduction to reasoning.* New York: Macmillan Publishing.

Vilches, L. (1992). La lectura de la imagen: Prensa, cine y televisión. Barcelona: Paidós comunicación.

Zárate, G. (2015). Argumentación en los textos de Andrea Villarreal (1907-1910). *Lenguas en contexto*, *12*, 173-184.

Zárate, G. 2012. El exilio del ningún lugar. Las voces utópicas de la familia Villarreal González (tesis doctoral). Monterrey, Mexico: ITESM (220-231)

23

"Conductive" Argumentation in the UK Fracking Debate

ISABELA FAIRCLOUGH
School of Humanities and Social Sciences
University of Central Lancashire, UK
ifairclough@uclan.ac.uk

From a critical rationalist perspective, I look at a fragment of the debate on shale gas exploration in the UK in order to make a proposal on the nature and representation of "conductive" argumentation, arguing it should not be viewed as a single argument, but in relation to deliberation as genre. There is no "conductive argumentation", only various possible outcomes of deliberation, seen as critical testing of (alternative) proposals.

KEYWORDS: argument scheme, conductive argumentation, critical rationalism, decision-making, deliberation, fracking, practical reasoning, shale gas

1. INTRODUCTION

The literature on "conductive argumentation" is by now fairly extensive, including Wellman's (1971) original statement of the problem, Govier's (1999; 2010; 2011) crucial contributions, Blair and Johnson's (2011) edited collection, as well as other significant developments (Hitchcock, 2013) and useful critical reviews (van Laar, 2013; Paglieri, 2013). According to these sources, a typical illustration of "conductive" argumentation is a "pro/con" or "balance-of-considerations" argument, in which both reasons in favour and reasons against are (convergently and defeasibly) supporting a conclusion, for example a practical conclusion – the agent ought to do A. Conductive arguments are taken to be single arguments (in favour of one conclusion) with two kinds of premises, positively and negatively relevant to the conclusion, with the reasons in favour outweighing the reasons against (called "counterconsiderations"). It has been suggested that, in addition to the pro and con sets of reasons, there should also be a specific premise that expresses the result of the process of weighing reasons, an OB (on-balance) premise (Hansen, 2011).

My proposal is to define "conductive" (pro/con) argumentation in favour of a practical-normative conclusion (the Agent ought to do A) in relation to deliberation as genre, and represent it as a possible configuration of the "deliberation scheme" I have developed in other publications (Fairclough, 2016; Fairclough, 2017). I suggest that there is no such thing as "conductive argument", and that speaking about pro/con argumentation as a type of argument is possibly a category mistake. In order to understand what is involved in pro/con argumentation, a change of perspective or level is needed, from the level of simple argument schemes to the level of genre, the level at which various kinds of argument schemes are interrelated in pursuit of a higher-order function, e.g. rational decision-making. By focusing on deliberation, what is usually called "conductive" argumentation appears to be one of two main possible configurations or outcomes of a deliberative process, the one where a pro-conclusion can still be maintained, in spite of the existence of reasons against, because the reasons against are not strong enough to refute it, and the reasons in favour "outweigh" the reasons against.

I advance this proposal from a critical rationalist logic of inquiry (Miller, 1994; 2006; 2013; 2014), seeing deliberation as the critical testing of alternative proposals for action, designed to enable rational decision-making. Critical testing of alternative proposals, resulting in the normative judgment that proposal A_n is not recommended (and ought to be discarded), but other proposals can be provisionally maintained, may be followed by choice of a "better" alternative among those proposals that have survived criticism and a decision to adopt that alternative. I propose two crucial distinctions: (a) between *counterconsiderations* (CCs) and *critical objections* (COs), arguing that, unlike CCs, COs can rebut a proposal; (b) between the concepts of *outweighing* and *overriding* reasons, which I see as occurring at different temporal stages in a deliberative process. Only if there are no *overriding* reasons against doing A, does it make sense for deliberating agents to move on to *weighing* the pros and the cons.

The deliberation scheme I am suggesting (Figures 1-3) basically involves an argument from goals, circumstances and means-goal relations (tentatively supporting the conclusion in favour of proposal A); an argument from positive consequences (also tentatively supporting that conclusion), and an argument from negative consequences, which can conclusively rebut the hypothesis that A is the right course of action when the potential undesirable consequences are unacceptable. Deliberation typically starts with one or more agents having a stated goal G (or several) in a set of circumstances C (including "problems"), and an open question (what should be done?), in response to which agents will propose a course of action A (or several), intended to transform their current circumstances into the future state-of-affairs corresponding to their goals (Fairclough & Fairclough, 2012). Based on all the knowledge available, the agents might conjecture that they ought to do A_1 (or A_2 or A_3...) to achieve G, in the circumstances. In order to decide rationally, the agents should subject each of these alternatives (hypotheses, conjectures) to critical testing, i.e. should try to expose potential negative consequences of each, and evaluate them as to their acceptability. Deciding to adopt proposal A_n will be reasonable if the conjecture (hypothesis) that A_n is the right course of action has been subjected to thorough critical testing in light of all the knowledge available and has withstood all attempts to find critical objections against it. A critical objection is an overriding reason why the action should not be performed. *Unacceptable* consequences (e.g. unacceptable impacts or risks of a course of action) are critical objections against a proposal and can conclusively rebut it. The purpose of critical testing is (1) to eliminate unreasonable proposals by examining their potential consequences; (2) to enable non-arbitrary choice of a better proposal, if several reasonable proposals have withstood criticism.

Normatively speaking, the underlying logic of deliberation (and of practical pro/con argumentation) is a logic of inquiry, not advocacy or justification: arguers do not know in advance which proposal is recommended, but should endeavour to find out, by trying to find reasons against each, discarding some on this basis and then comparing the remaining ones against each other. The fact that no critical objections may have been uncovered does not mean that there are no counterconsiderations, no reasons against that proposal at all, nor does it mean that no critical objections are likely to be uncovered in the future. Counterconsiderations (I suggest) are reasons against the proposal that can be outweighed by the reasons in favour. Critical objections, by contrast, are reasons against the proposal that cannot be outweighed in this way, but in fact override the reasons in favour. What is a mere CC to someone may be a CO to someone else. The inconvenience of a very early start to catch an early flight may be a CC to one person, for whom the cheaper cost of that flight outweighs the disadvantages, but may be a CO to someone else, for whom the inconvenience overrides whatever arguments in favour there may be.

2. ARGUING FOR AND AGAINST SHALE GAS EXPLORATION

I will test my conception of conductive argumentation against a few examples taken from the controversy on shale gas exploration in the UK, specifically from the debate that took place in the Lancashire County Council (LCC) on whether to approve the applications for hydraulic fracturing submitted by oil-and-gas drilling company Cuadrilla.[1] Following extensive deliberation over several days (23-29 June 2015) by the LCC Development Control Committee, including speeches against and in favour of the proposal by members of the public, the applications were rejected on account of *unacceptable impacts*. In rejecting the applications, councillors voted against the views of their own planning officers, who had recommended approval. Here is the Planning Officer's presentation – in an abbreviated form combining summary and direct quotation:

> According to all the evidence and expert opinion, the objections raised by the opponents are "not sustainable" ("cannot be supported"). It would therefore be "unreasonable" to delay making a decision. "Whilst there would be some negative impacts" (traffic, noise, dust, visual impact, loss of agricultural land), "most particularly for those living in the closest proximity to the site, they would be for a temporary period" and "could be minimized by the use of conditions". "It is therefore concluded that there would not be any unacceptable impacts associated with the proposal" on traffic, air quality, visual and noise grounds, and to refuse the application on such grounds "would be unsustainable". Furthermore, the Environment Agency has concluded that the risks of water and soil contamination are "very low". Consequently, "refusing the application in view of the risks to surface or ground water contamination … would be unsustainable". All risks can be controlled by the "permitting process" and "regulatory regime" in place. If approved, the development would achieve important goals ("would establish the presence and viability of exploiting an indigenous resource … which could contribute to the national energy needs, maintaining a diverse energy supply, and would bring some local benefits to the area in terms of employment and contributions to the local economy"). "It would not be acceptable to dismiss such exploration where it would not

---

[1] This is a corpus of approximately 130,000 words, transcribed from the video recordings of the 4 days of deliberations by Phillip Norris (UCLAN).

have an unacceptable impact that could not be adequately controlled and meet the policies of national guidance and the development plan." "It is considered that the proposal complies with the national guidance and the policy of the development plan" and all other relevant legislation, except SP2 & EP11 of the Fylde Local Plan (seeing as it constitutes "industrial development in the countryside"). However, "there is sufficient justification to override these two policies" in this case; "little weight should be attached to [them]... and more weight should be attached to the policies of minerals and waste". "So, in conclusion, overall, after extensive consultations and assessment of the proposals in light of the responses, representations, and most particularly and importantly against the policies of the development plan for the area, I am of the view that the principle of the development is acceptable or can be made acceptable by the use of conditions...". "I therefore recommend that... planning permission be granted subject to the conditions set out in the report...".

This is a conductive argument in which objections, mainly in the form of "impacts", are acknowledged (using the same argumentative indicator to express a concessive relation: "*whilst* there would be emissions..."/ "*whilst* [the drill] would still be visually apparent..."/ "*whilst* it is acknowledged that these operations would be noisy..."/ "*whilst* there would be some negative impacts..."), but are not considered serious enough to rebut the proposal, seeing as they can be "minimized" or "made acceptable by the use of conditions". As for risks, they are said to be very low, by implication manageable, therefore not unacceptable either. Regarding conflict with existing legislation, it is considered that the application complies with all relevant national and local legislation, save two local policies, which can be "overridden". In other words, although, in principle, laws provide non-overridable reasons against proposals which go against them (e.g. local people's *rights* or government's *obligations* must not be violated), this particular conflict is not unacceptable and does not suggest the proposal should be discarded. (For another perspective on pro/con argumentation over shale gas extraction, see Lewinski 2016).

I have elsewhere argued that the most significant perspective in light of which proposals are to be tested is a consequentialist one (Fairclough & Fairclough, 2012; Fairclough, 2016): would the consequences of a proposal, if adopted, be acceptable or not? The term "consequence" is used here broadly to refer to several types of states-of-affairs:

- the *goals* of the proposed action (as intended results or end-states): for example, Cuadrilla's immediate goal is to have their application approved, in order to move on to exploration and commercial exploitation, for the long-term stated goal of achieving energy security for the UK.
- the *risks* involved, as potential unintended and undesirable consequences, e.g. water and soil contamination;
- *impacts* on the natural environment, known to occur in the process of achieving the goals, e.g. the coming into existence of a drilling rig of a certain height, situated on a fracking pad of a certain size;
- *impacts* on the institutional, social world, e.g. the coming into existence of a situation in which the rights of the local population are being infringed.

Impacts are different from risks. Impacts are *known* (not merely probable or possible) consequences: if agents want to achieve their goals, certain impacts will be unavoidable. By contrast, risks may or may not materialize. A visual impact on the landscape is an unavoidable impact if the goal is to exploit shale gas; causing an earthquake by drilling is a risk. In the case of risk, proposals with potentially unacceptable consequences may nevertheless be allowed to stand, if the risks can be "managed" or "controlled". One way of managing risks is by being able to avoid them via a Plan B, an alternative course of action that agents can switch to if necessary; another is by transferring them to another party (e.g. by insuring against them in an acceptable way) (Miller, 2013). Agents may also choose to take the risk, if it is not possible or desirable to abandon the proposal. If the risk is accepted, then it is rational to try to minimize or optimize it, so as to reduce the probability and/or severity of the potential loss.

Overall, in the LCC debate, the argument scheme underlying argumentation against the proposal was mainly argumentation from negative consequence, where the negative consequences were deemed to be unacceptable (critical objections). By contrast, the supporters of the proposal tended to argue from desirable goals (e.g. energy security) and other alleged positive consequences. Since the undesirable consequences could not be overlooked altogether, the supporters' arguments tended to be of the pro/con type: impacts and risks were acknowledged, but were not considered serious enough to challenge the proposal, being allegedly mitigated and controlled in an acceptable way.

3. DELIBERATION AS CRITICAL TESTING OF PROPOSALS: POSSIBLE OUTCOMES

According to van Eemeren (2010, pp. 138-143), deliberation is a genre, at a higher level of abstraction than activity types. I suggest that argumentation in deliberative activity types can be succinctly represented as follows (Figure 1), where the conclusion of the practical argument from goals and circumstances (centre) is tested by a pragmatic argument from negative consequence (left). The pragmatic argument from negative consequence can potentially rebut the practical proposal (conclusion) itself if the consequences are deemed to be critical objections. To say that the conclusion "Agent ought to do A" is rebutted means that the opposite conclusion follows instead. For example, from the critic's perspective, the (unintended) consequences of proposal A can be such that A had better not be performed, even if the goal can be achieved by doing A. If this is the case, a critical objection to A has been exposed, and the hypothesis that the agent ought to do A has been rebutted (refuted, falsified). However, if the negative consequences, while undesirable, are not unacceptable, and do not therefore constitute critical objections against A – this could be because there is some "Plan B" or mitigating strategy in place, or because they can be traded off against positive consequences (outweighed by them) – then the conclusion in favour of A may still stand. Practical claims can also be tentatively supported by arguments from positive consequence (right-hand side). Positive consequences include desirable side effects that are not explicitly intended (are not goals that agents start from), but can be predicted to occur. Figure 1 is a development of the scheme proposed in Fairclough & Fairclough (2012), connecting two argument schemes, the practical argument from goals and the pragmatic arguments from positive and negative consequence. For convenience, only one goal, only one negative and only positive consequence are represented; there may be several of each, convergently supporting one conclusion or another.

Let us assume that three alternative proposals are tested, A_1, A_2, A_3, that can all presumably deliver a set of goals and possibly other positive consequences. Let us also assume that these goals have withstood critical questioning (they are not unacceptable) and the agent wants to achieve them. There is, potentially, a defeasible inference (from each of these sets of premises) to the conclusions *Proposal A_1, or A_2, or A_3 is recommended*. Let us also assume that, by testing A_1 (Figure 1), it is found that, in addition to various positive consequences and possibly some counterconsiderations, A_1 has a range of unacceptable consequences, e.g. unacceptable risks or impacts. As critical objections, these will conclusively rebut A_1 (so that the argument on the left-hand side of Figure 1 can be advanced), overriding whatever reasons in favour there may be (achievement of goals and other positive consequences). Thus, it can no longer follow, not even tentatively, that A_1 is recommended, because it conclusively follows that A_1 is not recommended, seeing as there are COs to A_1.

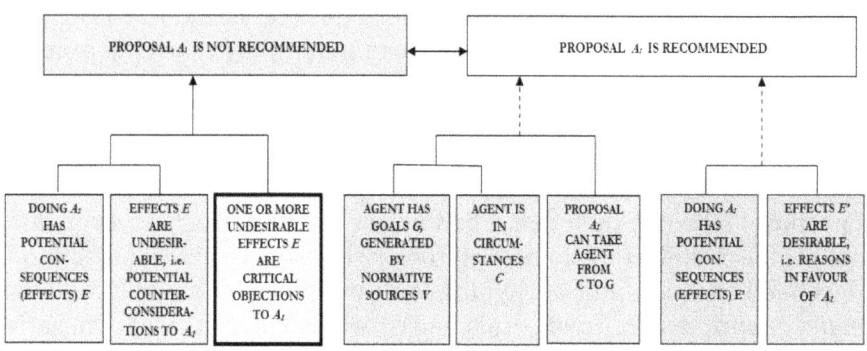

Figure 1 – Proposal A_1 is rebutted in light of critical objections

Let us now suppose that A_2 is being tested (Figure 2). A_2 can also deliver the goals and other positive consequences, and no critical objections come to light, which means it does not follow that A_2 is not recommended (i.e. that potential inference, left-hand side, is now being undercut). Proposal A_2 has therefore passed the critical test and can be provisionally maintained. If A_1 and A_2 were the only alternatives, then A_2 would be chosen at this stage, because, unlike A_1, A_2 has no critical objections against it. If the potential conclusion on the left does not obtain (if it is not the case that A_2 should not be performed – I have left the conclusion box blank, to suggest this), *then what remains of the deliberation scheme is what is commonly called a "conductive" argument which says*: in spite of various counterconsiderations (in principle, reasons against A_2), and seeing that there are no overriding reasons against A_2, but a number of reasons in favour (e.g. it will achieve the goals and other desirable effects), and also seeing that there is no better alternative, A_2 is the right course of action to achieve the goals.

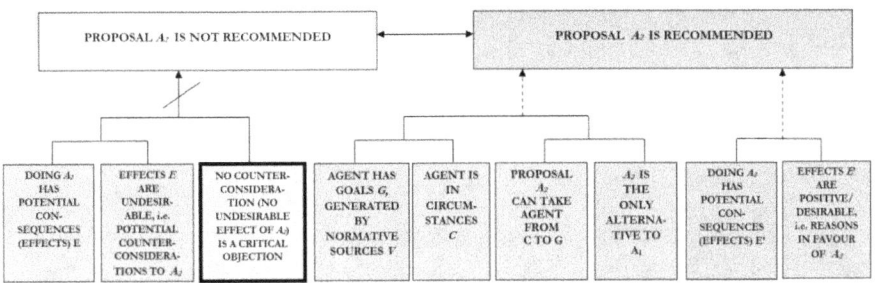

Figure 2 – Proposal A_2 has survived criticism (no COs, only reasons in favour and CCs): a "conductive" argument in favour of A_2 is possible, if A_2 is the only reasonable alternative.

The negatively relevant reasons (undesirable consequences), which – had they been COs – would have supported the conclusion that *A_2 is not recommended*, can thus be incorporated into the arguer's case as counterconsiderations (Figure 2). Unlike CCs, COs cannot be integrated into a conductive argument. Whenever there are COs (Figure 1), the potential conductive argument in favour of A_1 collapses into a deductive argument conclusively supporting the conclusion that *A_1 is not recommended*.

If more than one proposal passes the test, it is rational to choose one that is preferable from whatever perspective is important to the agents in the context. For example, for the opponents of fracking, the proposal to drill for shale gas in Lancashire was unreasonable, yet other proposals – for renewable energy sources – passed the critical test. Among renewables, all considered reasonable, some people might prefer solar, others wind or tidal energy, based on various criteria (cost, predictability of supply, etc.), weighed against each other in various ways. Essentially, in my view, the metaphor of *weighing* applies only at a later stage in the deliberation process, only in relation to proposals that have been found to be reasonable (i.e. without unacceptable consequences). Unacceptable consequences do not outweigh, but *override* the reasons in favour.[2]

The situation where more than one alternative has survived criticism is represented in Figure 3, where A_2 is finally chosen because it

---

[2] If the critical testing process yields no reasonable alternatives, and all available courses of action have unacceptable consequences, yet a decision has to be made, the metaphor of weighing may also apply to the choice among *unreasonable* proposals, in order to choose not a *better* but a *less bad* alternative.

has been found preferable to A_3, though both are reasonable courses of action.

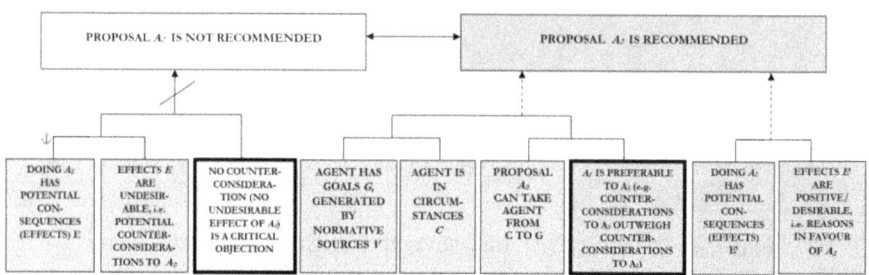

Figure 3 – Proposal A_2 has survived criticism (no COs, only reasons in favour and CCs): a "conductive" argument in favour of A_2 is possible if A_2 is preferable to other reasonable alternatives.

The premises highlighted in bold in each figure are the equivalent of the OB (on-balance) premise, and come in two kinds: a premise expressing the absence of overriding reasons against doing A_2 (no CC to A_2 is a CO), and a premise expressing the preferability of A_2 over A_3.

To recapitulate, all the alternative proposals put forward in response to the question "what should we do?" need to first pass the critical test of the deductive argument from negative consequence (left side of diagram). If one or more critical objections come to light, then it follows that the proposal is not recommended. Some proposals will be discarded at that stage, but some will pass the test and be allowed to stand, thus ending up on the right-hand side, tentatively supported (in light of the goals). To say that a proposal is recommended – i.e. asserting the conclusion on the right-hand side of the deliberation scheme – is to say that there are reasons in favour but, essentially, that *there are no serious reasons against (i.e. no critical objections)*. In addition, it is to say either that the proposal in question is *the only reasonable alternative*, or that, among reasonable alternatives, it is *preferable to other reasonable alternatives* (e.g. it has fewer counterconsiderations, or more benefits, or both). (It is assumed here that the goals are acceptable and the agent wants to achieve them).

There can be conclusive (deductive) arguments against a proposal but no conclusive arguments in favour of a proposal: any argument in favour can be defeated by new information, emerging feedback, etc. (Fairclough & Fairclough, 2012; Fairclough, 2015; 2016). A proposal that, for an arguer, has withstood critical testing, together with the set of pro/con reasons that are positively/negatively relevant to it, will take the form of a so-called "conductive" argument in favour of that proposal. Figure 2 represents the situation when a proposal has passed the critical test, seeing as *no counterconsideration is a critical objection*. If this is indeed the case, then the conclusion in favour will be (defeasible) supported, in light of the goals to be achieved, *because the conclusion against is not supported*, and the counterconsiderations are outweighed by the reasons in favour. If, however, this is not the case, i.e. if one or more reasons against are critical objections (Figure 1, left), then the potential "conductive" argument in favour of that proposal disintegrates, collapsing into a deductive argument against it.

As reasons against a practical conclusion, critical objections must be kept distinct from counterconsiderations. For every proposal there will be reasons against, however minor, but not every reason against a proposal is a critical objection. Unacceptable consequences, as critical objections, can rebut a proposal, conclusively indicating that it would be unreasonable to go ahead with it. Counterconsiderations can incline the balance towards one reasonable proposal or another, once the unreasonable ones have been weeded out. The point at which reasons against, taken to be counterconsiderations, may (singly or collectively) turn into critical objections, or at which merely undesirable (negative) consequences become *unacceptable* consequences, is a matter for deliberating agents to decide for themselves.[3]

4. CONCLUSION

On a critical rationalist view, practical reasoning can be modelled as a critical procedure that filters out those conclusions (and corresponding decisions) that do not pass the critical test of whether the intended or unintended consequences are acceptable. Proposals are tested in light of their consequences. For any alternative A, it is in principle possible to find not only counterconsiderations but critical objections: this is why *the two opposite conclusions should always be represented*, and

---

[3] I have retained both the words "practical" and "pragmatic" because this is how the two schemes are discussed in the literature: Walton (2006; 2007) calls the argument from goals a "practical" argument, while van Eemeren (2010) calls the argument from consequence a "pragmatic" argument. These schemes cannot be conflated and are both operative in deliberative practice. In my account, the argument from consequence is used to test the conclusion of the argument from goals. In other words, the argument from consequence has a critical function, while the argument from goals has a motivational function.

conductive argumentation is not a single argument, in my view, but one of two possible main outcomes of a deliberative process, understood as a process of critical testing of alternative proposals for achieving a goal.

What appears to be a "conductive" argument is a particular argumentative configuration that may (or may not) appear in the temporal unfolding of a deliberative process. A proposal that has withstood critical testing (i.e. the con reasons are not COs but CCs), together with the sets reasons that are positively/negatively relevant to it, will take the form of a so-called "conductive" argument in favour of that proposal. Whenever critical objections do come to light in the course of deliberation, the potential "conductive" argument tentatively supporting A will collapse into a deductive argument in support of not doing A. Thus, "conductive" argumentation attempting to justify doing A materializes, or emerges as an actual pro/con argument, only in those situations when the reasons against doing A are taken to be mere conterconsiderations, not critical objections. As Figure 2 suggests, "conductive" argumentation is that particular situation in which the conclusion that A should not be performed, *always possible in principle, does not follow (the inference to that conclusion is undercut), because none of the con reasons are strong enough to warrant that conclusion*. In other words, there are no overriding reasons against A, only reasons in favour and counterconsiderations, the latter are outweighed by the former, so an argument in favour of doing A can be tentatively put forward.

The two main configurations are presented in Figures 1 and 2, with Figure 3 being a variation of Figure 2, when more than one reasonable proposal has passed the test and they are now weighed against each other in order to choose one. The premises that fulfil the role of the OB premise (Hansen, 2011) are represented in bold, in each figure. Whenever there are no critical objections against a course of action, it does not follow that the agent should not do A_n (the inference to *Proposal A_n is not recommended* is undercut, as there are no overriding reasons why A_n should not be performed). Nor does it, nevertheless, follow (however defeasibly) that the agent should do A_n, unless, in addition to not having any unacceptable consequences, A_n is the only course of action that will deliver the goals or is preferable to other reasonable alternatives. This is where "weighing" comes into the picture: to say that A_2 is preferable to A_3, among reasonable alternatives (and therefore recommended), is to say that the reasons (CCs) against A_2 have been outweighed by the reasons (CCs) against A_3, or that the reasons in favour of A_2 outweigh those in favour of A_3 (or both: A_2 has comparatively fewer disadvantages and more advantages than A_3, though neither would be an unreasonable course of action). Or, to say that A_2 is recommended means that these two premises (in bold, in Figure 3) obtain: no CC to A_2 is a CO, and A_2 is in some sense preferable to its alternatives. These premises (in addition to whatever reasons in favour there are) are spread out across the various simple interrelated argument schemes that model deliberative activity types, which supports my claim that "conductive" argumentation must be analyzed at the "higher" level of genre, not as a type of argument.

To conclude, a conductive argument is *not* a single argument with pro and con reasons, but a particular argumentative configuration or outcome of a process of critical testing, where two opposite conclusions ("Do A" and "Do not do A") are always possible, and whichever becomes actualized for any given alternative depends on how that alternative survives critical testing. "Do not do A" follows conclusively when there are critical objections, in spite of there being reasons in favour (i.e. in such cases, "Do A" is rebutted, and the pro reasons are overridden). "Do A" follows defeasibly when there are no critical objections, only reasons in favour and counterconsiderations (e.g. merely undesirable, but not unacceptable consequences), and that particular alternative has fared comparatively better than others in the process of weighing reasons (e.g. its disadvantages are comparatively smaller, or its gains are comparatively superior to those of others, including doing nothing, etc.). If there are no overriding reasons against doing A, the potential inference to "Do not do A" is undercut, the "Do A" alternative can stand (however tentatively), and the arguer's argument may take the form of "conductive" argument saying: in spite of such-and-such counterconsiderations, doing A is recommended.

REFERENCES

Blair, J. A., & Johnson, R. H. (Eds.). (2011). *Conductive arguments, an overlooked type of defeasible reasoning*. London: College Publications.
Eemeren, F. H., van (2010). *Strategic maneuvering in argumentative discourse*, Amsterdam: John Benjamins.
Fairclough, I. & Fairclough, N. (2012). *Political discourse analysis*. London: Routledge.
Fairclough, I. (2015). A dialectical profile for the evaluation of practical arguments. In B. Garssen, D. Godden, G. Mitchell & A. F. Snoeck Henkemans (Eds.), *Proceedings of the 8th International Conference of the International Society for the Study of Argumentation*. Rozenberg Quarterly, Amsterdam: SicSat.

Fairclough, I. (2016). Evaluating policy as argument: the public debate over the first UK Austerity Budget. *Critical Discourse Studies*, *13*(1), 57-77. DOI: 10.1080/17405904.2015.1074595.

Fairclough, I. (2017). Deliberative discourse. In J. Richardson & J. Flowerdew (Eds.), *The Routledge handbook of critical discourse analysis* (pp. 242-256). London: Routledge.

Govier, T. (1999). *The philosophy of argument*. Newport News, VA: Vale Press.

Govier, T. (2010). *A practical study of argument* (7th ed.). Belmont, CA: Wadsworth.

Govier, T. (2011). Conductive arguments: Overview of the symposium. In J. A. Blair & R. H. Johnson (Eds.) *Conductive arguments, an overlooked type of defeasible reasoning* (pp. 262-276). London: College Publications.

Hansen, H. V. (2011). Notes on balance-of-consideration arguments. In J. A. Blair & R. H. Johnson (Eds.). *Conductive arguments, an overlooked type of defeasible reasoning* (pp. 31-51). London: College Publications.

Hitchcock, D. (2013). Appeals to considerations, *Informal Logic*, *33*(2), 195-237.

Laar, J. A. van (2014). Arguments that take counterconsiderations into account, *Informal Logic*, *34*(3), 240-272.

Lewiński, M. (2016). Shale gas debate in Europe: pro-and-con dialectics and argumentative polylogues. *Discourse & Communication*, *10*(6), 553-575.

Miller, D. (1994). *Critical rationalism: A restatement and defence*. Chicago: Open Court.

Miller, D. (2006). *Out of error. Further essays on critical Rationalism*. London: Routledge.

Miller, D. (2013). Deduktivistische Entscheidungsfindung (C. Kopetzky, Trans.). In R. Neck & H. Stelzer (Eds.) *Kritischer Rationalismus heute. Zur Aktuaklität de Philosophie Karl Poppers*, Schriftenreihe der Karl Popper Foundation, (pp. 45-78). Frankfurt am Main: Peter Lang.

Miller, D. (2014) Some hard questions for critical rationalism. *Discusiones filosoficas*, *15*(24), 15-40.

Paglieri, F. (2013). Critical review: Conductive argument, an overlooked type of defeasible reasoning. Edited by J. A. Blair and R.H. Johnson. *Informal Logic*, *33*(3), 438-461.

Walton, D. (2006). *Fundamentals of critical argumentation*. New York: Cambridge University Press.

Walton, D. (2007). *Media argumentation*. New York: Cambridge University Press.

Wellman, C. (1971). *Challenge and response: Justification in ethics*. Carbondale, IL: Southern Illinois University Press.

24

The Development of Prototypical Patterns of Weighing and Balancing in the Justification of Judicial Decisions

EVELINE T. FETERIS
University of Amsterdam
e.t.feteris@uva.nl

In this paper I describe the pattern underlying legal justification in the context of weighing and balancing as an inference process that is based on rational justification. On the basis of an integration of insights from legal theory and the pragma-dialectical theory I specify the burden of proof of a judge who takes a decision on the basis of weighing and balancing. The aim is to specify the elements of the justification that must be made explicit so that they can be submitted to rational critique.

KEYWORDS: argumentative pattern, decision rule, hard case, legal argumentation, legal decision-making, legal interpretation, rational justification, teleological-evaluative argumentation, weighing and balancing

1. INTRODUCTION

Weighing and balancing forms the basis of the justification of judicial decisions in cases in which there are reasons for and against the applicability of a legal rule in a specific case. Two alternative decisions are possible on the basis of different complexes of reasons that are both acceptable but incompatible. In such cases it is the task of the court to establish a priority among these reasons and it must decide how to apply the rule in the specific case on the basis of certain weighing criteria that are relevant in a particular field of law. The types of argument that constitute the weighing criteria can differ, depending on the field of law. For example, in constitutional law a court must establish a priority among different fundamental rights on the basis of a weighing rule that is not an existing rule of law. In its justification, the court must account for the decisions it has taken in establishing the priority from the perspective of the way in which it has used its discretionary space.

From the perspective of a rational justification of legal decisions it is important to specify the requirements a justification that is based on weighing and balancing should meet. What arguments have a function on different levels of the justification? How can the burden of proof of a court be established in light of the requirements of a rational justification of judicial decisions? For certain fields of law, such as the field of constitutional law, various authors have formulated decision rules or weighing rules that are to be taken into account in weighing and balancing. In the field of AI and Law, authors have specified general argumentation structures for arguing with conflicting conclusions in the context of defeasible reasoning. However, a general characterization of weighing and balancing as a form of rational justification in terms of an argumentative pattern that specifies the components on the different levels of the justification that must be made explicit so that they can be submitted to critique is not available yet.

The aim of my paper is to show that weighing and balancing can be reconstructed as a more general argumentative pattern that underlies a justification in which different forms of argumentation play a role that support different outcomes of the dispute. In my paper I will start with describing the process of weighing and balancing in legal decision-making as a particular inference process on the basis of insights from legal theory regarding the different types of considerations that are to be taken into account. Then I translate these insights in terms of an argumentative pattern. The function of the pattern is to give a characterization of the burden of proof of a judge who takes a decision that is based on weighing and balancing. The aim is to specify the elements of the justification that must be made explicit so that they can be submitted to rational critique.

2. WEIGHING AND BALANCING AS A FORM OF LEGAL JUSTIFICATION IN HARD CASES

In legal theory, weighing and balancing is considered as a form of justification that has to take place in accordance with certain norms for rational legal justification. Weighing and balancing has been studied in the context of different fields of law and authors have concentrated on different aspects of the weighing and balancing.

For the field of constitutional law a prominent theory about weighing and balancing is developed by the legal theorist Alexy (2003a and 2003b). Alexy has formulated various laws of balancing and weight formulas to be applied in the balancing. These laws and weight formulas have been elaborated by others such as Borowski (2013), Klatt and Meister (2012), Klatt and Schmidt (2012), and Sieckmann (2010, 2013). The authors concentrate on disputes about the infringement of a constitutional right and the way in which a decision about the collision of principles should be decided on the basis of balancing, applying a certain weight formula. For the use of precedents, Bernal (2012) has indicated how such a weight formula can be implemented and how courts can account for the discretionary space they have in applying the weight formula.

For the field of EU law and the field of human rights authors have also developed insights about the weighing of conflicting principles. For the field of EU law authors such as Pontier and Burg (2004) have analysed the weighing and balancing of conflicting principles. Pontier and Burg concentrate on the weighing of principles in light of the goals of EU law by analysing the way in which the European Court of Justice proceeds in weighing conflicting principles in specific cases. For the field of human rights Gerards (2013) concentrates on balancing in the context of interferences with human rights where the Convention right is balanced against the aims pursued by the interference.

For the field of AI and Law authors have developed insights about the justification of legal decisions in the context of conflicting conclusions. For the context of statutory interpretation incorporating teleological argumentation, Araszkiewicz (2015, p. 138) has formulated an informal model for the descriptive analysis for balancing. He develops a model for the representation of the structure and argumentation patterns in actual cases with the aim of implementing this model in legal knowledge systems in AI research.

The insights on weighing and balancing developed in legal theory are restricted to certain fields of law (constitutional law, EU law, human rights) or to certain types of argumentation in the context of statutory interpretation (teleological argumentation). The authors do not describe the characteristics of a general pattern that underlies the justification. They do not specify the different types of argument that are relevant and they do not go into their interrelations from the perspective of the requirements for the burden of proof of a judge in such hard cases. In what follows, by way of summary, on the basis of the insights formulated by the authors in legal theory, I describe the characteristics of weighing and balancing as form of rational justification. I take these characteristics as starting points for the translation in terms of a general argumentative pattern for a rational justification of decisions that are based on weighing and balancing.

On the basis of the insights discussed above, weighing and balancing can be characterized as a form of justification in hard cases. The decision about the interpretation and/or applicability of a particular legal rule is based on a priority among various conflicting considerations that are put forward in support of different, conflicting decisions. Weighing and balancing can be considered as a form of 'external justification' or 'second-order justification' in which the choice for the applicability (or non-applicability) of a particular legal rule in a particular interpretation for a specific case is defended. The justification consists of different types of arguments on different levels.

First, the justification consists of two or more arguments consisting of considerations that conflict with each other and that are balanced against each other. In the field of constitutional law these arguments consist of the legal rights that conflict in light of certain constitutional rules and principles. In the context of balancing in light of the interpretation of statute law, these arguments consist of the legal rules as they are formulated on the basis of different interpretation methods that conflict in light of certain value hierarchies.

Second, the justification consists of two or more arguments that specify the weighing and balancing. One argument constitutes the decision rule or weighing rule and the other argument the result of the application of this rule to the facts and circumstances of the specific case in terms of a priority on the basis of the weight attached to the conflicting arguments.

In their turn, all these arguments must be supported. In line with the requirements of a theory of rational legal discourse all arguments that form part of the burden of proof of a judge in terms of his argumentative obligations should be articulated. A court must account for its discretionary space in applying the weighing criteria of the decision rule in the circumstances of the specific case. This support consists of a justification of the decision/weighing rule as well as the justification of the application of the weighing rule to the different arguments in light of the facts of the specific case.

This characterization of weighing an balancing clarifies the central requirements with respect to the underlying assumptions that must be made explicit, but also leave open certain questions that have to be answered in order to be able to identify the argumentative patterns that are prototypical for weighing and balancing in terms of a general theory of legal justification. The different types of arguments that form part of the weighing must be specified in terms of their function in the justification, such as the arguments that constitute the considerations based on the interpretation of the applicable legal rule, the arguments that constitute the weighing criteria and the decision/weighing rule, the arguments that constitute the decision about the relative weight of the different considerations, etcetera. Furthermore, the different levels in the justification and the relations between the arguments on the various levels need to be specified. The development of a general pattern that takes into account the various parts of a rational justification makes clear which elements of the justification form part of general argumentative obligations in hard cases that must be submitted to rational critique.

3. A PROTOTYPICAL ARGUMENTATIVE PATTERN OF WEIGHING AND BALANCING AS A FORM OF RATIONAL JUSTIFICATION OF JUDICIAL DECISIONS

Weighing and balancing as form of argumentation is subjected to the rules and standards of rational legal discourse. This implies that all arguments on the different levels of the argumentation that form part of the argumentative obligations of the court should be made explicit so that they can be submitted to rational critique. From a pragma-dialectical perspective the forms of rational critique that are relevant can be formulated on the basis of the critical questions that belong to the argumentation schemes used in the argumentation.[1])

The first step in the description of the argumentative pattern is to establish the argumentative obligations in terms of the arguments and argumentation schemes that underlie the justification. To this end, it must be established which levels of argumentation can be distinguished, which arguments have a function on these levels and what argumentation schemes are underlying these arguments. Then it must be established how the relation between the arguments on the different levels can be explained in light of the critical questions that are relevant for the various types of argumentation schemes.

The argumentative pattern that is prototypical for weighing and balancing can be developed as an extension of the general argumentative pattern for hard cases that I have specified in earlier publications (Feteris, 2017)[2]. The extension consists of two levels of

---

[1] For a discussion of the concept of prototypical argumentative patterns and the role of argumentation schemes in specifying relevant forms of critique see van Eemeren (2016, 2017).

subordinate argumentation that are prototypical for weighing and balancing that are represented in Figure 1.

1	Legal decision: Legal consequence Y must (not) follow
1.1	Legal qualification of the facts X1, X2
1.1'	Reformulation of the applicable legal rule as R": If facts X1, X2, then legal consequence Y must (not) follow
1.1'.1a	Justification of formulation R' on the basis of C1
1.1'.1b	Justification of formulation R" on the basis of C2
1.1'.1c	Decision rule D: Specification of the criteria on the basis of which C2 weighs heavier than C1
1.1'.1d	Result of the application of the decision rule D
1.1'.1a.1	Justification of C1
1.1'.1b.1	Justification of C2
1.1'.1c.1	Justification of the decision rule D
1.1'.1d.1	Justification of the result of the application of the decision rule D

Figure 1 – Prototypical pattern of a justification in a hard case based on weighing and balancing

In the general argumentative pattern of a justification in a hard case the *argumentation of the first order* consists of a justification in which it is said that a particular legal rule R" is to be applied to the facts X1 and X2.[3] This argumentation is put forward as a justification for the decision that legal consequence Y follows. This argumentation of the first order constitutes *symptomatic* argumentation, a form of argumentation in which it is argued that in light of certain characteristics (certain legal facts) a particular legal predicate (a particular legal consequence) is applicable.[4] This legal rule R", that forms a specification of the rule R for the specific case, forms the *ratio decidendi* for the fact that the legal consequence Y is to be applied to the facts X1 and X2 in the specific case.

The formulation of the legal rule R" can be any legal rule in any field of law such as civil, criminal, constitutional, European or international law. It can concern a formulation of a rule that is a specification of a rule of a code of law in a continental law system, a formulation of a rule that constitutes a precedent or a principle in a

---

[3] This argumentation is also called internal argumentation (Alexy 1989; Wróblewski 1974).

[4] For a description of this prototypical pattern and the implementation of argumentation schemes see Feteris (2017).

common law system, or a specification of a rule in the context of EU law or international law.

In the *argumentation of the second order* the (re)formulation of the legal rule R" is justified.[5] This justification is necessary because the legal rule R" is not an existing legal rule so that the formulation (interpretation) must be justified in light of certain legal standards of acceptability such as coherence, stability and equality. In what follows, I will go deeper into this argumentation of the second order that is prototypical for weighing and balancing.

In weighing and balancing, the *argumentation of the second order* is a complex argumentation that consists of various arguments that are necessary to justify the choice for the formulation of rule R as R". Characteristic for weighing and balancing is that this argumentation consists of arguments of a different nature:

i. the arguments containing the considerations C1 and C2 that justify the different formulations of R as R' and R"
ii. the decision rule/weighing rule D
iii. the result of the application of the decision rule D, resulting in a preference for R".

Ad(1) The first type of argument are the arguments 1.1'.1a (in support of R') and the argument 1.1'.1b (in support of R") that constitute the considerations C1 and C2.

Ad (2) The second type of argument constitutes the specification of the decision rule/weighing rule D (argument 1.1'.1c in Figure 1) (that has to be distinguished from the formulation of the legal rules R' and R"). In order to justify the preference for R", a decision rule D has to be provided that justifies this choice. Such a decision rule must specify the weighing factors, the criteria to be applied in the weighing, as well as the way in which the priority among the considerations C1 and C2 is to be established.

This rule can be considered as a *decision rule* in line with the requirement of universalization that always a rule must be put forward to justify the choice for a particular alternative.[6] In the context of constitutional law this decision rule can be the

---

[5] This justification is also called external justification (Alexy 1989; Wróblewki 1974).

[6] For the formulation of the decision rules in the context of constitutional law see Alexy (2003a and 2003b). For a critique of the characterization of the application of weighing criteria in terms of rule application see Sieckmann (2013).

weighing rule formulated by Alexy (2003a and 2003b).[7]) In the context of the interpretation of statutes this weighing rule can be a rule specifying the weight attached to certain interpretation methods. In the context of human rights this decision rule can be formulated in terms of the proportionality of the interference and its legitimate objectives. In the context of EU law this weighing rule can be a rule specifying the weight to be attached to the principles and goals of the EU.

Ad (3) The third type of argument constitutes the specification of the result of the application of the decision rule D in the circumstances of the concrete case (the argument 1.1'.1d in Figure 1). It contains a specification of the priority among the considerations implying that application of the decision rule leads to the conclusion that the considerations C2 in support of R" weigh heavier than the considerations C1 in support of R'.

The argumentative pattern also consists of a level of *argumentation of the third order* in which the different components of the weighing and balancing, that constitute the argumentation of the second order, must be justified.

First, the arguments 1.1'.1a and 1.1'.1b must be justified. This justification consists of specifying the application of a particular interpretation method (such as the linguistic, systematic, teleological method) or a form of legal reasoning (such as analogy, a contrario, a fortiori, etcetera).

Second, the decision rule D, argument 1.1'.1c, must be justified. Depending on the field of law, this justification will have a different nature. The justification of the decision/weighing rule in the context of constitutional law consists of a justification of the weighing factors such as the importance and the abstract weight of the factor. The justification of the decision/weighing rule in the context of the interpretation of statutes consists of a specification of the importance and weight of a particular method/canon of interpretation in terms of a hierarchy of interpretation methods in light of certain values. An example of such a specification could be the hierarchy of interpretation methods as described by MacCormick and Summers (1991) in which certain general principles for the hierarchy of interpretation methods and the relative weight of various arguments that are based on these methods are formulated.

---

[7] See Sieckmann (2013) for another view on the rule-based character of weighing.

Third, the result of the application of the decision rule D with a preference for R" in light of the circumstances and facts of the specific case (argument 1.1'.1d) must be justified. The court must account for its discretionary space by explaining how the degree of importance and the abstract weight of the criteria, the weighing factors, of the decision rule have been established and what considerations have played a role in this decision process.[8])

4. THE DEVELOPMENT OF A PROTOTYPICAL PATTERN FOR TELEOGICAL-EVALUATIVE ARGUMENTATION IN THE CONTEXT OF WEIGHING AND BALANCING

For different types of dispute, different developments of the argumentative pattern of the second and third order of the argumentative pattern of Figure 1 must be developed. In this way, it can be made clear which argumentation schemes are underlying the different parts of the justification and how the extensions of the argumentation can be explained in light of the critical questions for the argumentation schemes that are anticipated or reacted to.

An example of such an implementation and extension is the justification of argument 1.1', the formulation of rule R as R", on the basis of teleological-evaluative considerations. Such a form of argumentation concerns the consequences of applying rule R in the formulation R" in light of the goal and values the rule is intended to realize. In different contexts in which weighing and balancing is applied, a justification on the basis of teleological-evaluative considerations plays an important role. As has been indicated by Araszkiewicz (2015) and Feteris (2008) for general forms of legal argumentation, by Bernal (2012) for balancing with precedents, by Gerards (2013) for human rights, and by Pontier and Burg (2004) for EU law, on different levels of the balancing process reference to consequences in light of the goals of the law plays an important role in the justification.

For teleological-evaluative argumentation, the general pattern of Figure 1 can be implemented as is indicated in Figure 2, in which the considerations C1 and C2 constitute the different interpretations I1 and I2. In the arguments 1.1'1c and 1.1'.1d it is specified why the interpretation I2 on the basis of teleological-evaluative considerations is given more weight in the specific case.

---

[8] For examples of the justification of decisions about the degree of importance and the abstract weight of competing principles by referring to factual and normative premises see Bernal (2012).

1	Legal decision: Legal consequence Y must (not) follow
1.1	Legal qualification of the facts X1, X2
1.1'	Reformulation of the applicable legal rule as R": If facts X1, X2, then legal consequence Y must (not) follow
1.1'.1a	Justification of formulation R' on the basis of interpretation method I1
1.1'.1b	Justification of formulation R" on the basis of interpretation method I2
1.1'.1c	Decision rule D: Specification of the criteria on the basis of which I2 weighs heavier than I1
1.1'.1d	Result of the application of the decision rule D
1.1'.1a.1	Justification of I1
1.1'.1b.1	Justification of I2
1.1'.1c.1	Justification of the decision rule D
1.1'.1d.1	Justification of the result of the application of the decision rule D

Figure 2 – Prototypical pattern of a justification of weighing and balancing involving teleological-evaluative considerations

The development of this pattern is based on the critical questions that are relevant for teleological-evaluative argumentation. As has been explained in Feteris (2008 and 2016b), the prototypical argumentative pattern for teleological-evaluative argumentation consists of a particular implementation of pragmatic argumentation. The extensions of this scheme constitute the justification that consists of the answers to relevant critical questions regarding the desirability of the consequences of the interpretation in light of the goals and values the rule is intended to realize. Characteristic for the role of teleological-evaluative arguments in the context of weighing and balancing is that they are located in specific 'slots' in the argumentation and take on a specific form.

The first argument that is prototypical for this form of argumentation is argument 1.1'.1c, the decision rule D. In the decision rule D, it must be specified what the importance and the weight of the teleological-evaluative considerations, the goal of the rule and the consequences in light of the goal of the rule, are. In the justification of this decision rule (the weighing rule), the underlying values regarding the weight attached to the teleological-evaluative considerations must be made explicit. It must be specified why certain teleological-evaluative considerations carry more weight than other criteria from the perspective of the purpose of the legal rule.

The second type of argument that is prototypical for this form of argumentation is argument 1.1'.1d, the result of application of the decision rule. In the description of the result of application of the decision rule it must be specified that application of the legal rule R" would lead to certain consequences in the given case and that those consequences are desirable from the perspective of the goal of the rule. In the justification of this argument it must be specified what the consequences of applying rule R in interpretation R' and R" are, what the circumstances of the given case are that are relevant for applying the criteria of the decision/weighing rule, what the goal of the rule is (whether it is a goal formulated by the legislator or a rational goal that underlies the legal system), and why the consequences of R" are more desirable in light of the goal of the rule than the consequences of R'.

5. CONCLUSION

In my contribution I have described the general characteristics of weighing and balancing in terms of an argumentative pattern that is prototypical for legal justification in hard cases. In this way it can be clarified which general assumptions are underlying the decision process from the perspective of the obligation for courts to account for their discretionary space in applying and interpreting the law and how these assumptions must be made explicit in terms of arguments that can be submitted to rational critique. I have specified how this pattern can be implemented further for a particular type of hard cases in which different interpretation methods and forms of argumentation have to be weighed on the basis of teleological-evaluative considerations. On the basis of this description I have explained that teleological-evaluative argumentation has a function in a specific part of the argumentation and I have explained how extensions of the argumentation can be explained on the basis of answers to relevant critical questions.

The importance of specifying the burden of proof of a court who applies weighing and balancing is that it can be specified what the margins are for courts to manoeuvre strategically. The argumentative patterns specify the normative and descriptive arguments that must be put forward on the basis of the commitments courts undertake when they decide to make a decision on the basis of balancing.

For the different types of argumentation that form part of the patterns, a further description of the inferential steps must be specified. For the various forms of argumentation that are based on certain interpretation methods and on forms of argumentation (analogy, *a contrario*) etc. the different components of the complete argument form must be formulated.

In my translation, I restricted myself to the formal argumentation structure, the types of arguments and their interrelations. For different fields of law and legal systems, the various components of the pattern must be implemented further on the basis of the criteria of correctness that are applicable in various legal cultures and fields of law. By carrying out analyses of actual examples of weighing and balancing it can be demonstrated how the different patterns are implemented in legal practice.

REFERENCES

Alexy, R. (2003a). On Balancing and Subsumption. A structural comparison. *Ratio Juris*, 16(4), 433-449.
Alexy, R. (2003b). Constitutional rights, balancing, and rationality. *Ratio Juris*, 16(2), 131-140.
Araszkiewicz, M. (2015). Argumentation structures in legal interpretation: Balancing and thresholds. In: T. Bustamante & C. Dahlman (Eds.). *Argument types and fallacies in legal argumentation* (pp. 129-150). Dordrecht: Springer.
Bernal Pulido, C. (2012). Precedents and balancing. In C. Bernal & T. Bustamante (Eds.), *On the philosophy of precedent: Proceedings of the 24th World Congress of the International Association for Philosophy of Law and Social Philosophy, Beijing, 2009* (pp. 51-58). Wiesbaden: Franz Steiner Verlag.
Borowski, M. (2013). Formelle Prinzipien und Gewichtsformel (Formal principles and weighing formulas). In: M. Klatt (Ed.). *Prinzipientheorie und Theorie der Abwägung* (pp. 151-199).Tübingen: Mohr Siebeck.
Eemeren, F. H., van (2016). Identifying argumentative patterns: A vital step in the development of pragma-dialectics. *Argumentation*, 30(1), 1-23.
Eemeren, F. H. van (2017). Context-dependency of argumentative patterns in discourse. *Journal of Argumentation in Context*, 6(1), 3-26.
Feteris, E. T. (2004). Rational reconstruction of legal argumentation and the role of arguments from consequences'. In A. Soeteman (Ed.), *Pluralism and law. Proceedings of the 20th IVR World Congress, Amsterdam, 2001.* (Vol. 4, pp. 69-78). *Archiv für Rechts-und Sozialphilosophie, ARSP Beiheft Nr. 91*. Stuttgard: Franz Steiner Verlag.
Feteris, E. T. (2008). The rational reconstruction of weighing and balancing on the basis of teleological-evaluative considerations in the justification of judicial decisions. *Ratio Juris*, 21(4), 481-495.
Feteris, E. T. (2016a). Argumentative patterns in the justification of judicial decisions: A translation of Robert Alexy's concept of weighing and balancing in terms of a general argumentative pattern of legal justification. *Analisi e diritto*, 223-240.

Feteris, E. T (2016b). Prototypical argumentative patterns in a legal context: The role of pragmatic argumentation in the justification of legal decisions. *Argumentation*, *30*(1), 61-79.

Feteris, E. T. (2017). The identification of prototypical argumentative patterns in the justification of judicial decisions. *Journal of argumentation in context*, *6*(1), 44-58.

Gerards, J. (2013). How to improve the necessity test of the European Court of Human Rights. *International Journal of Constitutional Law*, *11*(2), 466-490.

Klatt, M., & Meister, M. (2012). Proportionality – A benefit to human rights? Remarks on the I.CON controversy. *Journal of Constitutional Law*, *10*(3), 687-708.

Klatt, M., & Schmidt, J. (2012). Epistemic discretion in constitutional Law. *Journal of Constitutional Law*, 10(1), 69-105. (German version: Klatt, M., Schmidt, J. (2013). Abwägung unter Unsicherheit. In: Klatt, M. (Ed.), *Prinzipientheorie und Theorie der Abwägung* (pp. 105-150). Tübingen: Mohr).

MacCormick, N., & Summers, R. (1991). *Interpreting statutes. A comparative study*. Aldershot etc.: Dartmouth.

Pontier, J.A. & Burg, E. (2004). *EU Principles on jurisdiction and recognition and enforcement of judgments in civil and commercial matters according to the case law of the European Court of Justice*. The Hague: TMC Asser Press.

Sieckmann, J. (Ed.). (2010). *Legal reasoning: The methods of balancing. Proceedings of the Special Workshop held at the 24th World Congress of the International Association for Philosophy of Law and Social Philosophy (IVR), Beijing, 2009. ARSP*, Beiheft. Wiesbaden: Franz Steiner Verlag.

Sieckmann, J. (2013). Is balancing a method of rational justification *suigeneris*? On the structure of autonomous balancing. In C. Dahlman & E. T. Feteris (Eds.). *Legal argumentation theory: Cross-disciplinaryperspectives* (pp. 189-206). Dordrecht etc.: Springer.

Wróblewski, J. (1974). Legal syllogism and rationality of judicial decision. *Rechtstheorie*, *14*(5), 33-46.

25

"Metaphors are No Arguments, My Pretty Maiden" The Pragma-Dialectical Reconstruction of Figurative Analogy

BART GARSSEN
University of Amsterdam
b.j.garssen@uva.nl

> While in literal analogies the concrete characteristics of two items from the same domain are compared, in figurative analogies a comparison is made between items from very different domains. It is argued that figurative analogy should not be seen as analogy argumentation but as a way of expressing a different type of argument scheme.
>
> KEYWORDS: analogy, figurative language, metaphor, pragma-dialectical reconstruction

1. INTRODUCTION

The title of this paper refers to one of Sir Walter Scott's lesser-read novels, *The fortunes of Nigel*.[1] Margaret (the pretty maiden and the heroin of the story) is in a discussion with Lady Hermione and she claims that an advantageous life is favourable to a calm life. When she compares these life styles to the lark and to the weathercock respectively Lady Hermione replies that "metaphors are no arguments". Margaret is sorry for that [...] "for they are such a pretty indirect way of telling one's mind when it differs from one's betters—besides, on this subject there is no end of them, and they are so civil and becoming withal". The question is of course: is Lady Hermione right when she claims that metaphors are no arguments?

To answer this question we first need to know in what way metaphors are supposed to be arguments or what role they can play in argumentation. In the following argumentation the use of a

---

[1] Max Black (1955) uses this phrase as a motto in his paper on metaphorical meaning.

metaphorical expression in the explicit premise seems to be unproblematic:

> (1) Robert is a good accountant because he is a computer.

The metaphorical expression *Robert is a computer* serves as an argument because in this case, given the standpoint, we may take it that the arguer means that Robert, like a computer, "can process large amounts of numerical information and never makes mistakes, and so on". (Wilson 2011, p. 190). These are all characteristics of good accountants and can therefore serve as parts of the defence of the standpoint that Robert is a good accountant.

Metaphorical expressions can also be used in a different way. In the following argumentation the metaphorical expression is part of the premise that links the explicit premise with the standpoint:

> (2) The best time to meet the threat is in the beginning. It is easier to put out a fire in the beginning when it is small than after it has become a roaring blaze.
> (President Harry Truman in his address about US intervention in Korea)

This argumentation is special because a comparison is made between two very different things or situations (fire and war). The type of argument that is used here is commonly known as a figurative analogy. While there is general agreement amongst argumentation theorists that we can make a distinction between literal and figurative analogies there is not much agreement when it comes to the argumentative or probative force of figurative analogies (Garssen, 2009). For some theorists analogy argumentation that features a metaphorical comparison has no probative force at all and should not be regarded as serious argumentation. Freeley and Steinberg, for instance, are rather outspoken:

> Carefully developed literal analogies may be used to establish a high degree of probability. Figurative analogies on the other hand, have no value in establishing *logical* proof. If well chosen, however, they may have considerable value in establishing *ethical* or *emotional* proof, in illustrating a point, and in making a vivid impression on the audience (Freeley and Steinberg 2005, p. 162).

Waller (1991) has a similar view: "Figurative analogies do not argue, though they may elucidate" (1991, p. 200). On the other hand, theorists like Hastings (1962) believe that the figurative analogy can lend support to a standpoint although most of them urge the analyst to be careful when evaluating the figurative analogy.

Garssen (2009) and Garssen and van Eemeren (2014) take a different stance. Viewed from a pragma-dialectical perspective figurative analogies should be seen as argumentation but not as comparison argumentation. In order to evaluate figurative analogies a reconstruction of the argumentation is necessary. In this paper I continue this line of thought. First, I will go into the pragma-dialectical view of comparison argumentation or literal analogy. Next, I will explain why, in this view, the figurative analogy cannot be seen as comparison argumentation. Finally, I will go into the reconstruction of figurative analogies.

2. LITERAL AND FIGURATIVE ANALOGIES[2]

According to van Eemeren and Grootendorst, in analogy argumentation the argumentation is presented "as if there were a resemblance, an agreement, a likeness, a parallel, a correspondence or some other kind of similarity between that which is stated in the argument and that which is stated in the standpoint" (1992, p. 97). This type of argumentation can be generally characterized in the following way:

	Y is true of X
because:	Y is true of Z,
and:	Z is comparable to X.

In analogy argumentation, a comparison is made between the actual characteristics of one thing, person or situation and the actual characteristics of another thing, person or situation.[3] This happens, for instance, in the following example:

> (3) Camera surveillance in the centre of Amsterdam will be effective, because in London camera surveillance proved to be effective.

---

[2] The first part of this account is based on Van Eemeren and Garssen (2014).

[3] In pragma-dialectics a distinction is made between two forms of analogy argumentation: descriptive analogy and normative analogy. For the sake of brevity, I restrict myself here to descriptive analogies.

In this argumentation two things are compared which belong to the same class (London and Amsterdam are both cities). It is claimed that something will be the case in Amsterdam because it is already the case in London and Amsterdam is comparable to London.

It is assumed in this argumentation that there are a number of similarities directly relevant to safety between Amsterdam and London. Because of these similarities we have to take it that the property of London mentioned in the standpoint that camera surveillance is effective will also be shared by Amsterdam. In other words, the presence of the property that is discussed is "extrapolated" from the properties the two cities already share.

It is important to realize that these similarities are often not mentioned explicitly in the argumentation. In claiming that what is mentioned in the standpoint is similar to what is mentioned in the premise, the similarities are only assumed. However, as we shall see, the similarities often come into play in the evaluation of the analogy argumentation.

In analogy argumentation the pragmatic principle of analogy is used in extrapolating a property from a list of commonalities. Because the two things, persons or situations compared have a series of properties in common, they are assumed to share also another property that is mentioned in the standpoint.

Figure 1 provides a general characterization of this kind of analogy argumentation.[4]

---

[4] Most accounts of analogy argumentation follow this model. However, there are other models of analogy argumentation that do not involve this extrapolation of characteristics. Whately (1963), for instance, believes that analogy is a form of argument that actually comprises two separate arguments "for it is evident that there can be no reasoning from one individual to another, unless they come under some common genus" (86). In the first part a general rule is defended by means of an argument by example, while in the second part this general rule is used as the basis for defending the standpoint. Beardsley has a similar view: "What makes an analogical argument plausible is always a hidden generalisation" (1975, 114). According to Beardsley every analogy argumentation should be reduced to an argument based on such a generalisation.

Situation referred to in the premise	Situation referred to in the standpoint
Relevant similarity 1	Relevant similarity 1
Relevant similarity 2	Relevant similarity 2
Similarity 3	[Similarity 3 - extrapolated]

Figure 1 – Analogy argumentation as an extrapolation of properties

The first step in the pragma-dialectical testing procedure for analogy argumentation is to ask whether the things, persons or situations compared in the argumentation are, in principle, comparable. If asked to do so, the protagonist is obliged to show that the things, persons or situations compared indeed belong to the same class. From then onwards the testing procedure can take different routes. Since the relevant similarities on which the extrapolation is based remain implicit, the antagonist may ask the protagonist to add to what has been said by mentioning the relevant similarities. The protagonist is then forced to provide additional argumentation in which one or more relevant similarities are mentioned. The similarities to be mentioned are the properties that allow for the extrapolation step. Mentioning similarities can again lead to criticism from the antagonist because the similarities that are mentioned may not be recognized as actual similarities, or they may be seen as similarities which are not relevant to the issue at hand.

The antagonist can also criticize the argumentation by pointing at differences between the things, persons or situations that are compared. The protagonist is then forced to show either that these differences are not relevant or that the similarities outweigh the differences. In this way it is established in the testing procedure whether or not the intended extrapolation of characteristics is acceptable to both parties. The more similarities there are, the more likely it is in principle that an extrapolation is successful.

It is important to emphasize that it is not enough to just point to similarities and differences: these similarities and differences should also be *relevant* to the standpoint that is defended by putting forward analogy argumentation. Crucial in this regard is the claim that is made in the standpoint that is defended. In our example the (descriptive) standpoint is that in Amsterdam camera surveillance will lead to more safety. Relevant factors in this case are all conditions necessary for upholding this causal claim. The commonality between Amsterdam and

London that they are both capitals is not relevant here, since that is a fact that is not directly related to safety in the streets. Instead, in the case of this claim it may be an important fact that Amsterdam and London are both big cities.

These are the kind of issues that are pertinent to a continuation of the evaluation of argumentation that is part of the critical testing procedure for analogy argumentation. The question vital to the testing procedure for this type of analogy argumentation is whether the step of extrapolating properties is indeed acceptable to both parties.

In their linguistic realisation figurative analogies resemble literal analogies. At first sight figurative analogy seems to have the same general argument scheme of analogy argumentation, but argumentation in which a figurative analogy is used should be treated differently (see Garssen, 2009, pp. 137-139). The reason for this is that it is a distinctive feature of a figurative analogy that the elements that are being compared are situated on different levels of experience, stem from different "spheres", or are of a completely different kind. This is illustrated by the example (already mentioned in the introduction) of a figurative analogy, in which President Truman put forward at the beginning of the Korea conflict in defence of his claim that the United States should strike immediately:

> (4) The best time to meet the threat is in the beginning. It is easier to put out a fire in the beginning when it is small than after it has become a roaring blaze.

Fire and war belong to different classes of events, which makes it impossible to compare them directly. In this case however the similarities we have to look for are not so much between the concrete features of fire and war, but between the abstract relationships between what is said in the argument and what is said in the standpoint. It is predicted that the war in Korea will become unmanageable if we do not act now and this is the reason why we should act now. Because, viewed literally, President Truman does not make a direct comparison between war and fire, the standard critical questions going with analogy argumentation (*Are there relevant similarities? Are there relevant differences?*) do not really apply. In fact, if we would require asking these questions in this form, no figurative analogy could stand the test.

Perelman and Olbrechts-Tyteca (1969) characterize the "resemblance of structures" that is at issue in figurative analogy as follows: "*A* and *B* together, the terms to which the conclusion relates [...], we shall call the *theme*, and *C* and *D* together, the terms that serve to buttress the argument [...], we shall call the *phoros*" (p. 373). In figurative analogy there is a striking similarity between what is stated in the standpoint and what is stated in the argument and this similarity is normally expressed as A is to B (standpoint) is as C is to D (argument).

This similarity is the *only* similarity between what is stated in the standpoint and what is stated in the argument. That is the reason the model for the application of the pragmatic principle of analogy does not apply here, because this models assumes that there are many separate similarities between the argument and the standpoint.

One might reply that, in practice, there may be more similarities. However these similarities are on the same abstract level and they do not allow for the same kind of extrapolation from known similarities to the unknown controversial characteristic claimed in the standpoint. As we shall see, the similarities on this abstract level are all part of one and the same structural resemblance. This means that a different model for the figurative analogy is needed if we want to maintain that it is a specific form of argumentation. No such model is currently available.

What then, one may ask, if we change the critical questions that go with literal analogies so that this special kind of comparison (of ratios) is taken into account: *are there similarities (or differences) when it comes to the abstract relations between the argument and the standpoint?* This reformulation is of no use, since there is only one similarity to be observed.

3. THE RECONSTRUCTION OF FIGURATIVE ANALOGIES

A pragma-dialectical analysis of argumentation amounts to the reconstruction of the process of resolving a difference of opinion occurring in an argumentative discourse or text (Van Eemeren and Grootendorst, 2004, p. 95). Before an argumentative discourse or text can be analysed and evaluated systematically we need to reconstruct the speech event as part of a critical discussion. This reconstruction amounts to four different reconstruction transformation: deletion, addition, substitution and permutation. The deletion transformation involves taking out parts of the discourse that are not relevant for the resolution of the difference of opinion. The addition transformation comes down to adding elements that are implicit in the original text, but are nonetheless relevant. The substitution transformation entails replacing of formulations that are otherwise not relevant for the resolution process. The permutation transformation comes down to rearranging parts of the text so that their relevance becomes clear (Van Eemeren and Grootendorst 2004, 103).

In case of a figurative analogy the substitution transformation needs to be performed. We need to look for a relevant expression of the argument that is in line with the intentions of the speaker and enables a critical test of the argumentation. Because the stock critical questions for comparison argumentation do not fit figurative analogies, a reconstruction is needed so that it becomes possible to ask critical questions that are relevant.

For the reconstruction of figurative analogy one observation is vital: in a figurative analogy what is communicated is that the *phoros* and the *theme* are very different except for one thing: they are similar in respect to the relation(s) of elements (A is to B is as C is to D). There is a similarity between what is said in the argument and what is said in the standpoint but this is all there is. In other words: the analyst needs to trace what exactly is this similarity. Because the *phoros* and the *theme* are very different, tracing the similarity is generally not that difficult. In fact, it is this difference that makes it possible for the analyst to find out what is a shared element in the *phoros* and *theme*: the differences make the similarities stand out in a structural way. This understanding of the analogy is in line with Perelman and Olbrechts-Tyteca's remark that the situation mentioned in the *phoros* (argument) is supposed to be better known than the situation mentioned in the *theme* (standpoint): the *phoros* should clarify the structure of the *theme*. In other words, the principle underlying the situation that is mentioned in the standpoint and the situation mentioned in the premise stands out in the *phoros* (1969, p. 374).

The reconstruction of figurative analogy starts with finding what is similar in the *phoros* and *theme*. When the similarity is found it can be expressed in terms of a general principle. Next, this general principle should be applied to the case at hand (the situation in the standpoint). The analogy disappears from this point on.

This reconstruction can be illustrated by means of a much-quoted argument put forward by President Lincoln in his reply to the National Union League on 9 June, 1864:

> (5) I have not permitted myself, gentleman, to conclude that I am the best man in the country: but I am reminded, in this connection, of a story of an old Dutch farmer who remarked to a companion once that "it was not best to swap horses while crossing a stream".

The figurative analogy in Lincoln's argument (I should not resign at this moment because one should not swap horses while crossing a stream) can be schematised as follows:

> Standpoint: Theme (A – B): changing president – being in the midst of war
> Premise: Phoros (C – D): swapping horses – crossing a stream

What becomes clear in this schematization is that the two situations share only one similarity. In both cases the practical principle "it is risky to make important changes when there is hardly time for realizing these changes while being in a difficult process" applies.[5] Once this principle is reconstructed it makes no sense to look for more similarities. The analogy as such disappears after the reconstruction. The analogy provides the grounds for the reconstruction, but is not part of the argumentation as such.

These considerations lead to the following reconstruction of Lincoln's argument:

1. Lincoln should not resign at this point.
1.1 Changing the residing president in mid-war, when there is no time for a major change, is risky.

Figurative analogy argumentation is based on a *metaphorical* relation that serves as an indirect way of expressing a certain principle that forms the basis of the argument. After the reconstruction of the argument has taken place it becomes clear that in fact no analogy argumentation was used, but another type of argumentation: argumentation based on a causal relation or symptomatic argumentation.

The principle that is shared by the *phoros* and *theme* is a general statement that is usually acceptable right away. If one agrees with this principle we would immediately agree with the statement that is made in the *phoros*. That is why the *phoros* does not lend support to the general principle. The *phoros* can therefore not be seen as argument by

---

[5] Branham (1991, 178) uses the same example to explain his view of analogy as a strategy for moral argument. According to Branham, Lincoln used the analogy to highlight the risk associated with such change. "The easily visualized risk of the former is used to make clear the less apparent risks of the latter" (178). Analogy in his view involves a moral principle that goes for both the situation in the argument and the standpoint. His formulation of the principle that is at stake is less informative than the one I propose here

example supporting some sort of general statement. Interpreting the *phoros* like that would lead to circular argumentation. The *phoros* illustrates the principle by way of a more or less well-chosen image.

The general principle that is shared by the *phoros* and the *theme* often comes very close to a maxim, in many cases but not always a moral maxim (Branham, 1991, p. 178). Just as in the Lincoln example the maxim frequently functions as a warning against a certain type of behaviour.

What is interesting about the Lincoln example is that over the years *changing horses in midstream* has become an idiomatic expression.[6] It is immediately clear that in idiomatic expressions or comparisons like "that is like throwing out the baby with the bathwater" no real argumentative comparison is made. In evaluating argumentation in which such an idiomatic expression is used it does not make sense to ask the standard critical questions going with analogy argumentation, since it is immediately clear that the proverb should not be taken literally and that it has a conventional meaning: "You should be careful not to lose the good parts of something when you get rid of the bad parts".

It is clear that when such a proverbial expression is used the listener recognizes the intention of the speaker and realizes that it should not be treated as analogy argumentation. The similarity with figurative analogies is striking: the situation mentioned in the standpoint and the idiomatic expression mentioned in the argument are very different. They are so different that the similarity stands out. And this similarity is exactly what is communicated.

Idiomatic expressions containing a metaphorical structural comparison are in fact figurative analogies that have become conventional. This means that when it comes to the reconstruction of figurative analogies, we can follow the same rules. What can be said about the meaning and use of proverbial expressions used as arguments can also be said of figurative analogies.

An objection to the kind of reconstruction that is proposed in this paper is that by losing the analogy we also loose the vivid image provided by the analogy and all its associations. The images of danger, stupidity, war etc. are not taken up in the argumentation. That is to say they do not get a place in the analytic overview of the argumentative discourse. They will however be taken into account when the analyst

---

[6] Fogelin and Sinnott-Armstrong include this expression in their list of examples of idiomatic expressions they call "maxims of common sense" (2005, 33).

turns to the strategic manoeuvring that takes place in the argumentative discourse. The main question in analysing the strategic manoeuvring inherent in figurative analogies is why the arguer chooses to use a figurative analogy instead of a direct presentation of the argumentation.

In fact, our reconstruction of the figurative analogy is helpful not only for the reconstruction of the analytic overview containing all the relevant parts in the dialectical process of resolving the difference of opinion, but also for the analysis of the rhetorical aspects pertaining to the effectiveness of the argumentation. In the analysis the dialectical and the rhetorical aspects of figurative analogy are separated and therefore we can analyse both aspects more systematically.[7]

In conclusion, figurative analogies can be arguments and they are not necessarily weak forms of arguments. However, seen from a pragma-dialectical perspective they need to be reconstructed before we can evaluate them. However, they will prove to be very weak forms of argumentation if one treats them as normal (literal) analogy argumentation.

Who is right, Margaret or Lady Hermione? I am inclined to say that Margaret has the right view on using metaphor in argumentation, especially because "metaphors are such a pretty indirect way of telling one's mind when it differs from one's betters".

REFERENCES

Beardsley, M. C. (1975). *Thinking straight : principles of reasoning for readers and writers.* Englewood Cliffs: Prentice-Hall.
Branham, R. J. (1991). *Debate and critical analysis: The harmony of conflict.* Hillsdale, NJ: Lawrence Erlbaum.
Black, M. (1955). Metaphor. *Proceedings of the Aristotelian Society, New Series,* 55(*1954 - 1955*), 273-294.
Eemeren, F. H. van, & Garssen, B. J. (2014). Argumentation by analogy in stereotypical argumentative patterns. In H. J. Ribeiro (Ed.), *Systematic Approaches to Argument by Analogy* (pp. 41-56). Dordrecht: Springer.

---

[7] In a rhetorical analysis no such a division is made. The reason for this is that the main interest of a rhetorical analysis is the effectiveness of the argumentation without systematically caring for the dialectical aspects of reasonableness of arguments. This explains why Perelman and Olbrechts-Tyteca and their followers treat figurative analogy as a separate argument scheme without taken into account the worries concerning its lack of probative force expressed by others.

Eemeren, F. H. van, & Grootendorst, R. (2004). *A systematic theory of argumentation: The pragma-dialectical approach.* Cambridge: Cambridge University Press.

Fogelin, R. J. & Sinnott-Armstrong, W. (2005). *Understanding arguments: an introduction to informal logic.* Belmont, CA : Wadsworth.

Freeley, A. J. & Steinberg, D. L. (2005). *Argumentation and debate: Critical thinking for reasoned decision making.* Belmont, CA: Wadsworth.

Garssen, B. (2009). Comparing the incomparable: Figurative analogies in a dialectical testing procedure. In F. H. van Eemeren & B. Garssen (Eds.), *Pondering on problems of argumentation: Twenty essays on theoretical issues* (pp. 133-140). Dordrecht etc.: Springer.

Hastings, A. C. (1962). *A reformulation of the modes of reasoning in argumentation* (unpublished doctoral dissertation). Northwestern University, Evanston, IL.

Perelman, C. & Olbrechts-Tyteca, L. (1969). *The new rhetoric: A treatise on argumentation.* Notre Dame/London: University of Notre Dame Press.

Waller, B. N. (1991). Classifying and analyzing analogies. *Informal Logic, 21*(3), 199-218.

Whately. R. (1846/1963). *Elements of rhetoric.* Carbondale and Edwardsville: Southern Illinois University Press.

Wilson, D. (2011). Parallels and differences in the treatment of metaphor in relevance theory and cognitive linguistics. *Intercultural Pragmatics 8-2*, 177-196.

26

Against the Intentional Definition of Argument

G. C. GODDU
University of Richmond
ggoddu@richmond.edu

Intentional definitions of argument, i.e. the conclusion being intended to follow from the premises, abound. Yet, there are numerous problem cases in which we appear to have arguments, but no intention. One way to try to avoid these problem cases is to appeal to acts, in which case one has to give up on the repeatability of arguments. One can keep repeatability and intentions if one resorts to act types, but then it appears that the problem cases re-emerge.

KEYWORDS: acts, act-types, argument, definition, example, intention

1. INTRODUCTION

Peruse various logic and critical thinking textbooks and one will encounter definitions of 'argument' such as the following:

> In logic, argument refers strictly to any group of propositions of which one is claimed to follow from the others, which are regarded as providing support for the truth of that one (Copi and Cohen, 2009, pp. 6-7).

> the term argument … will be used to connote any set of assertions that is intended to support some conclusion or influence a person's belief (Nickerson, 1986, p. 68).

> An argument, in its most basic form, is a group of statements, one or more of which (the premises) are claimed to provide support for, or reasons to believe, one of the others (the conclusion (Hurley and Watson, 2018, p. 2).

> As used in the study of logic, an argument is any group of propositions (truth claims), one of which is claimed to follow logically from the others. The key phrase here is 'follows logically from.' For a group of propositions to be an argument, one of them must be claimed to follow logically from the others (Soccio and Barry, 1992, p. 5).

> One or more statements (premises) offered in support of another statement (a conclusion) (Kahane and Cavender, 2002, p. 378).

Additional examples abound.[1]
Nor is this sort of definition restricted to textbooks—it shows up in theoretical discussions as well.

> An argument may be described as [a] set of propositions, one of which is designated as the conclusion and the remainder as premises, whereby the conclusion is claimed to be based upon (e.g., derived from, supported by the premises (Ben-Ze'ev, 1995, p. 189).

> I take an argument or inference to be a collection of claims, one of which, the conclusion is put forth as following from the others, the premises (Berg, 1987, p. 13).

> a set of statements or propositions that one person offers to another in the attempt to induce that other person to accept some conclusion (Pinto, 2001, p. 32).

> From the pragmatic point of view, then, an argument is discourse directed toward rational persuasion. By rational persuasion, I mean that the arguer wishes to persuade the Other to accept the conclusion on the basis of the reasons and considerations cited, and those alone (R. H. Johnson, 2000, p. 150).

All of these examples are instances of what I call 'intentional' definitions of argument. They are intentional because, to have an argument, we need more than just sentences or statements or propositions, but also the intention, sometimes expressed in terms of claiming, affirming, or supposing that the statements are related in the correct way.

---

[1]For example, see also: (Stratton, 1999, p. 135); (R.M. Johnson, 2007 p. 2); (Layman, 1999, p.2); (Kelley, 1998, p. 89); (Klenk, 2002, p. 4.)

Contrast such definitions with what we might call 'minimalist' definitions of argument: An argument is a set of propositions, one of which is the conclusion.[2] Here, no intention is required to have an argument. But with intentional accounts you can have the propositions, or whatever one takes the constituents to be, and still not have an argument until there is the intention that the constituents be related in the proper way. As Berg (1987, p. 13) puts it:

> An argument is not merely a collection of claims, nor even a collection of claims bearing a certain logical relation to each other, but rather a collection of claims intended, by an arguer, to bear a certain logical relation to each other.

Van Eemeren and Grootendorst (2004, p. 3) write:

> It is important to realize right away that verbal expressions are not 'by nature' standpoints, arguments, or other kinds of units of language use that are interesting to argumentation theorists. They only become so when they occur in a context where they fulfil a specific function in the communication process. ... a series of utterances constitutes an argumentation only if these expressions are jointly used in an attempt to justify or refute a proposition ...

Despite the ubiquity of intentional accounts of argument, I shall argue that such accounts are seriously flawed.

2. PROBLEMS FOR INTENTIONAL ACCOUNTS

Why might intentional accounts arise? What at first appears important in identifying arguments is the relationship between the constituents of arguments, between the premises or reasons and the conclusion—this after all is what should distinguish arguments from mere lists of sentences or sonnets, say. But of course, bad arguments might be bad exactly because the relationship between premises and conclusion is lacking, so it cannot be the actual presence of the relationship that makes the constituents an argument—what's left? Presumably, some sort of intention that the relationship holds (even if in fact it doesn't).

---

[2] See for example: (Kalish and Montague, 1964, p. 13); (Skyrms, 2000, p. 13); (Godden, 2003, p. 1); (Bergmann, Moor and Nelson, 1998, p. 7); (Tomoczko and Henle, 1999, p. 1).

But intentional accounts raise puzzles of their own. Firstly, there is the problem of dealing with conflicting intentions. I intend or claim statement X follows from others, but you do not—do we have an argument or not? If we do, then why does the intention that X follows trump either the failure to intend X follows or the active intention that X does not follow? If not, then why does the failure to intend or the active intention that X does not follow trump the intention that X follows. Either way, the intentional accounts owes us an explanation of why, especially in the case of two conflicting active intentions, one intention has priority over the other without also undermining the need to appeal to intentions in identifying arguments in the first place.

One could avoid this problem by relativizing arguments to agents. Since I intend X follows, the group of statements is an argument for me, but since you do not so intend, either by failing to intend or by actively intending X not follow, the group of statements is not an argument for you. But such a solution makes substantive debate about whether someone is giving an argument or not impossible and yet argumentation theorists argue and debate about whether a particular passage of text is or is not an argument all the time.

Secondly, intending or claiming something to follow seems too easy. I hereby intend every sentence to follow from every possible set of sentences. Did I just make every set of sentences, i.e. mere lists, sonnets, etc., an argument? If so, then we have not solved the alleged problem the intentional account was supposed to solve., i.e. demarcating arguments from mere lists or other groups of sentences. But if not, why was this intention not enough to make all sets of sentences arguments?

In general, intentional accounts face the challenge of trying to specify the sort of intention that makes sets of propositions or sentences or statements arguments without somehow letting all sets of statements in as arguments. For example, I might consider several candidate 'arguments' for inclusion in this paper—but, prior to inclusion, I certainly do not intend or claim that any of the conclusions follow— indeed, some of the candidates may eventually be rejected precisely because I judge that the conclusion does not sufficiently follow from the premises given in the candidate 'arguments'. But if these candidate arguments are arguments, what is the intention that is making them arguments—my mere wondering if the conclusion follows? Or hypothesizing the conclusion follows? I can wonder or hypothesize about one sentence or statement or claim or proposition following from others, for any set of such things, in which case the intention again appears to be doing no distinguishing work—any set of statements say, be they a random list sentences or a sonnet or an instruction manual, and so on, can be an argument.

3. A SOLUTION?

There is a fairly straightforward solution to the debate between minimalist accounts and intentional accounts — they are actually accounts of two different kinds of things — objects, such as groups of propositions on the one hand, and actions, such as acts of arguing on the other. The minimalists are trying to identify the *thing* that is composed of propositions or sentences or whatever, whereas the intentionalists are trying to identify the *acts of arguing* (as opposed to acts of explaining or prophesying, etc.) It is not uncommon to try to distinguish acts in terms of intentions—the difference between murder and manslaughter, for example, hinges on the presence or absence of certain sorts of intentions. The minimalists certainly do not deny that there are acts of arguing; nor do the intentionalists deny that there are sets of propositions or statements. They might try to dispute which entity is properly labelled 'argument', but this would be a pointless terminological dispute—clearly, we use the term 'argument' to refer sometimes to sets of statements, such as Anselm's Ontological Argument and sometimes to acts of arguing such as in "their argument over the morality of capital punishment was sometimes loud, and certainly sustained, but always respectful".

If we solve the dispute by distinguishing two sorts of entities of concern, we can now make sense of my considering various arguments for this paper—my considering which arguments to include is not itself an act of arguing, since after all, there is no relevant intention of "claiming the conclusion follows" or "intending to convince anyone of the truth of the conclusion" or whatever the arguing making intention might be. But the relevant sets of propositions I consider are arguments—there is just no arguing going on yet.

I have no problem with this solution. There are arguments in the sense of sets of propositions and there are arguments in the sense of acts of arguing. I just ask that theorists (and textbook writers) make clear which entity they are talking about—the group of propositions or statements on the one hand or the acts of arguing on the other. Unfortunately, many definitions do not make clear what the target entity type is, with the result that some theorists mix the intentions of acts with abstract objects such as sets of propositions with the puzzling results we saw in section 2.

In some cases the mixing seems deliberate. For example, van Eemeren and Grootendorst (2004, p. 1) write of their definition of argument that a virtue of their definition is that it maintains the "process-product" ambiguity of the word "argumentation". Elsewhere I have argued (Goddu, 2011) that the process/product ambiguity is a confused version of the act/object distinction I have used above to make sense of intentionalist accounts of argument. If van Eemeran and Grootendorst are interpreted to have a definition that makes arguments both acts and objects, then I say the result is not a virtue, but rather a vice, since any definition that puts an object in two distinct ontological categories simultaneously is problematic.[3]

Even if the mixing is not deliberate, I suspect the mixing is an attempt to get the intentionality of arguing while keeping the generality afforded by objects such as sets of propositions.[4] After all, the theorists and textbook writers often go on to talk about assessing various properties of the arguments such as truth of the premises or the validity or support strength of the argument. But acts happen — they are not true or false or valid or strong. Acts are not composed of propositions or sentences. Perhaps some acts can contain statements or claims, but only in the sense of claimings or statings, and not in the sense of the content of those claimings. But it is the content that is being appealed to when we talk of truth or relevance or inferential strength.

Defenders of intentional accounts might grant that it is the content that is true or false or valid or whatever, but still maintain that the acts of arguing that express that content, derivatively at least, have the relevant properties, as in "she argued validly". Again, I do not have a problem with this solution — I just ask that theorists make clear that the arguments they are talking about are acts of arguing, where some of

---

[3] Their definition is, in part, as follows: **Argumentation** is a verbal, social, and rational activity aimed at convincing a reasonable critic of ... Given the crucial word 'activity', despite their own claim of respecting the process/product ambiguity, the best interpretation of their definition is likely to be that they are trying to define acts of arguing.

[4] For example, David Hitchcock (2007) tries to utilize the generality of sets, but the intentionality of acts by defining arguments in terms of sets of acts. In the face of criticism (Goddu, 2009; Freeman, 2009) Hitchcock (2009), and in conversation, briefly reverts to sets of propositions. But in the paper Hitchcock (2018) gave at this conference he returns to sets of acts (or perhaps act types) but within a two-tiered categorization of 'arguments in general' and 'actually used arguments.' Whether Hitchcock is trying to define arguments as objects or acts of arguing remains unclear to this author.

the properties of those acts may be derived from the properties of the content expressed in the act.

Suppose we grant the intentionalists that they are trying to define acts of arguing and have granted them a way of talking about the validity of those acts or the truth of some of the sub-acts. A significant problem still remains. Most argumentation theorists take arguments to be repeatable—it makes sense to ask our students or ourselves to reconstruct the arguments of others, i.e. to repeat them. But acts, which happen at specific space-time regions, are not repeatable. Hence, one can keep an intentional account of arguments, it seems, only by dropping the repeatability of arguments. To date, few, if any, theorists have pursued the non-repeatability option.

4. ACT TYPES

I conclude with one final attempt to salvage the intentional account of argument. Perhaps arguments are not acts, but act types. Act types are not spatio-temporal particulars, but rather are instantiated by spatio-temporal particulars. Two different spatio-temporal particulars might instantiate the same act type, and so act types are repeatable. Act types clearly involve intentions since that is how, at least in part, we distinguish something as a type of action rather than as a mere behaviour. So perhaps we could define an argument as follows:

> For any set, possibly empty, of propositions, P and for any proposition C, an argument, A is the act type that is the expressing of P and C with the intention (or claim) that P supports C.

Even with this definition, one might wonder how weak the intention that P supports C can be. Is wondering or hypothesizing whether P supports C enough to make an act instantiating that type an argument. If so, then my reciting a Shakespearean sonnet while wondering whether the last line follows from the previous lines is enough to make an act of arguing happen. But if the intention must be stronger than mere hypothesizing or wondering, then I cannot consider (and reject) arguments for inclusion in this presentation that do not have this stronger intention, since without it they would not be arguments. But beyond hypothesizing for as long as it takes me to realize that the conclusion does not follow, for at least some of the candidate 'arguments', I have no stronger intention. But then, despite appearances, on the current proposal these candidate 'arguments' are not arguments at all.

According to the current proposal it also seems impossible to give an example of an argument without also arguing. After all, for the example giving to be an example of an argument, the example giving must instantiate the expressing of P and C with the intention that P supports C, and the expressing of P and C with the intention that P supports C is just arguing for C on the basis of P. But we consider examples of arguments all the time without also arguing. For example, here is an argument I do not want to make (regardless of my attitude toward the conclusion):

> A: There are fewer than a million people in this room, so all my arguments are good ones.

I am certainly not arguing for that conclusion based on that premise and in fact, nothing I say in this paper I take to be an arguing for the goodness of all my arguments. And yet A still seems to be an example of an argument I am not making, i.e. one I am not actually arguing.

Finally, according to the current proposal it is impossible for me to program a computer (assuming computers have no intentions) to generate new arguments of which I am not aware. The computer could spit out millions of examples of the form 'P, so C' and none would be arguments since none would instantiate an act type that is the expressing of P and C with the intention that P sufficiently supports C. The computer has no intentions and I am aware of none of these outputs, so I certainly do not intend any of the C's to be supported by any of the P's. And yet, for any of the given examples, there is a fact, regardless of whether we know it or not, about whether the P's are all true in a given case, or whether C follows from P—the very properties we are often interested in with regards to arguments.

5. CONCLUSION

Being able to distinguish acts of arguing from other sorts of acts such as explaining or holding an incoherent press conference is certainly an important task for argumentation theorists. If intentional accounts of argument restrict themselves to this important task, then they can avoid the odd results of seemingly talking about some special 'intentionalised' kind of set of propositions. Understanding the properties of the content of actual and potential acts of arguing seems relevant to grasping the rationality or goodness of the inferences made in such acts of arguing and so is an important task for argumentation theorists. But the 'logical' properties of the content is independent of the intentions and so intentions should be kept out of any definition trying to capture the object that is the content of an arguing. If what I have argued above is correct, then trying to mix attempts to capture the content and the intentionality that makes some act an act of arguing into a single definition of argument is problematic. Even appeal to act types which are both repeatable abstract object and involve intentions fail to capture all the desired cases, most especially examples of arguments that are not arguings. The upshot: we should not mix and match the intentional language of acts with the abstract objects that may be the content of those acts.

In other words, stop trying to make ice sculptures out of the rolling waves.

REFERENCES

Ben-Ze'ev, A. (1995). Emotions and argumentation. *Informal Logic*, 17(2), 189-200.
Berg, J. (1987). Interpreting arguments. *Informal Logic*, 9(1), 13-21.
Bergmann, M., Moor, J., & Nelson, J. (1998). *The logic book* (3rd ed.). New York: McGraw-Hill.
Copi, I., & Cohen, C. (2009). *Introduction to logic* (13th ed.). Upper Saddle River, NJ: Pearson Education, Inc.
Freeman, J. (2009). Commentary on Goddu. OSSA Conference Archive. 56. Rerieved from: https://scholar.uwindsor.ca/ossaarchive/OSSA8/papersandcommentaries/56.
Godden, D. (2003). Reconstruction and representation: Deductivism as an interpretative strategy. OSSA Conference Archive. 26. Retrieved from https://scholar.uwindsor.ca/ossaarchive/OSSA5/papersandcommentaries/26.
Goddu, G. C. (2011). Is 'argument' subject to the process/product ambiguity? *Informal Logic*, 31(2), 75-88.
Goddu, G. C. (2009). Refining Hitchcock's definition of 'argument'. OSSA Conference Archive. 55. Retrieved from https://scholar.uwindsor.ca/ossaarchive/OSSA8/papersandcommentaries/55.
Hitchcock, D. (2018). The concept of argument. In S. Oswald & D. Maillat (Eds.), *Argumentation and inference. Proceedings of the 2nd European Conference on Argumentation, Fribourg 2017*. London: College Publications.

Hitchcock, D (2009). Commentary on Goddu. OSSA Conference Archive. 57. Retrieved from https://scholar.uwindsor.ca/ossaarchive/OSSA8/papersandcommentaries/57.

Hitchcock, D. (2007). Informal logic and the concept of argument. In D. Jaquette (Ed.), *Philosophy of logic* (pp. 101-129). Amsterdam: Elsevier.

Hurley, P., & Watson, L. (2018). *A concise introduction to logic* (13th ed.). Boston: Cenage Learning.

Johnson, R. H. (2000). *Manifest rationality: A pragmatic theory of argument.* Mahwah, NJ: Lawrence Erlbaum Associates.

Johnson, R. M. (2007). *A logic book* (5th ed.). Belmont: Wadsworth.

Kelley, D. (1998). *The art of reasoning* (3rd ed.). New York: W. W. Norton and Company.

Kahane, H. & Cavender, N. (2002). *Logic and contemporary rhetoric* (9th ed.). Belmont: Wadsworth.

Kalish, D. & Montague, R. (1964). *Logic, techniques of formal reasoning*. New York: Harcourt, Brace & World, Inc.

Klenk, V. (2002). *Understanding symbolic logic* (4th ed.). Upper Saddle River, NJ: Prentice Hall Publishing.

Layman, C. S. (1999). *The power of logic*. Mountain View, CA: Mayfield Publishing Company.

Nickerson, R. S. (1986). *Reflections on reasoning*. Hillsdale, NJ: Lawrence Erlbaum Associates.

Pinto, R. C. (2001). *Argument, inference, and dialectic.* Dordrecht: Kluwer Academic.

Skyrms, B. (2000). *Choice & chance* (4th ed.). Belmont: Wadsworth.

Soccio, D., & Barry, V. (1992). *Practical logic: An antidote for uncritical thinking.* Fort Worth, TX. Harcourt Brace Jovanovich College Publishers.

Stratton, J. (1999). *Critical thinking for college students*. Oxford: Rowman & Littlefield Publishers.

Tomoczko, T., & Henle, J. (1999). *Sweet reason: A field guide to modern logic*. New York: Springer-Verlag.

Van Eemeren, F. H. & Grottendorst, R. (2004). *A Systematic Theory of Argumentation*. Cambridge: Cambridge University Press.

27

Visual emotional argumentation: Analytical models

JULIETA HAIDAR
Escuela Nacional de Antropología e Historia
jurucuyu@gmail.com

We have several objectives in this article. First, we address the typology of inferences, and highlight some types, such as the deductive and the inductive to identify emotional inference. Next, we approach the emotional dimension from complexity and transdisciplinarity, highlighting the recursion between reason <> emotion. Third, we present different proposals to approach emotional argumentation in discourse, and in semiosis. We conclude, with an analysis of visual emotional argumentation.

KEYWORDS: emotional dimension, complexity, inference, visual / post-visual emotional argumentation, transdisciplinarity

1. INTRODUCTION

In this paper, we aim to present the problems related to emotional discursive argumentation and semiotic visual / post-visual argumentation, of which there are several analytical models, such as those of Edwards', Gilbert's, Plantin's, Charaudeau's, and Groarke's, among others. One can identify interesting analytical paths in these models, which complement each other, in order to study how emotion is materialized in semiotic-discursive practices, a category that encompasses verbal and visual as well as post-visual aspects. From a second theoretical perspective, we analyse how discursive argumentation becomes visual argumentation, which implies several theoretical-methodological adjustments, given the difficulties of interdiscursive processes, to which inter-semiotic translations are added; a step from the verbal dimension to the visual / post visual dimension.

It is important to consider in semiotic terms, what we understand by the visual and the post-visual. The visual is the semiotic production before the digital culture, before the digital images that involve complex constructions with tools and software created to produce other visual texts, like 3 D and 4 D paintings, which break the normal perceptual codes. Following these changes, comes the need to understand the new perspectives and to create different analytical models. In this respect, one can mention a revolution of the sensorial-perceptual codes, a revolution of eye perception, which are essential changes in all visual semiotic production, in the search of new analytical models. In these terms, we think it is important to introduce ourselves to the new visuals, which we call the post-visual, also called digital visual production, inserted in the complex digital culture of cyberspace, cyber-time, cyber-anthrop.

2. TYPES OF INFERENCE

In this section, we briefly review the broad field of inference and discuss in general terms what is an inference in classical and current logic. In classical logic, inference is related to deduction, although there are also immediate inferences, while in present-day logic other types of inference appear, for example, inference by evidence (inductive and enumerative), statistical inference, among others. Next, a typology lists some inferences, but some types are missing, such as emotional inference:

- Inference by classical logic: Inference that only admits two values: true or false.
- Three-valued inference: An inference of this style gives as possible results three values.
- Multi-valued inference: An inference of this style gives as possible multiple values.
- Diffuse Inference: An inference of this style describes all multivalued cases with accuracy and precision.
- Probabilistic inference: In the sense of an induction that allows establishing a truth with a greater probability index than the others.[1]

There are several studies that approach emotional inference, such as Barreyro, Molinari *et al.* (2008), who analyse emotional inferences derived from the reading of narrative texts, using the

---

[1] https://es.wikipedia.org/wiki/Inferencia#Tipos_de_inferencia)

contributions of cognitive psychology. Dávalos and León (2017) approach emotional inferences in the process of comprehension of discourse and provide several interesting analytical elements. In the research of Molinari, Barreyro *et al.* (2011), emotional inferences have been analysed in natural and offline readers of natural texts, using the Landscape Model, of a computational type in which interconnected networks are established between the propositions to establish which are nuclear, and are nodes that condense the emotional. The results are quite significant and some conclusions are reached.

Inferences are linked to cognitive processes, which must be articulated with multiple intelligences (Gardner 2001) and with Edgar Morin's thought (2006) on multiversity intelligences (or the multiversity of intelligence). From these contemporary perspectives that question the classic conception of a single intelligence, we want to return to the approaches on emotional intelligence proposed by Goleman (1996).

Emotional intelligence lies in a set of skills, such as self-control, enthusiasm, empathy, perseverance, self-motivation. These skills, which navigate between the genetic and the acquired, can be learned and perfected. The two types of intelligence coexist in all people, both cognitive and emotional are important, and must be analysed in a recursive way. From complexity, this type of intelligence involves complex operations of the human brain, conditioned to the physical-chemical-biological laws, as well as the sociohistorical-cultural environment. This complex recursion, therefore, leads to the fact that human behaviour oscillates between the predictable and unpredictable, which makes the impact of emotional intelligence relevant in all human practices (Lotman 2013).

Subjects' practices are conditioned by these two organic dimensions. One important thing from this, it is that human behaviour can be inferred from patterns and from models generated by the various sciences. But it is too difficult to predict them because subjects are complex and they are always oscillating between the rational and the emotional.

3. THE EMOTIONAL FROM COMPLEXITY AND TRANSDISCIPLINARITY

As a way to broaden the studies on emotional argumentation, in this third section we explore some significant aspects of emotional dimension not considered meaningful in such works. Authors working with emotional argumentation, mainly discursive, do not deepen the issues regarding the emotional dimension. They start from a priori premises that do not allow examining the emotional with rigor.

For Cosnier (2015), a great researcher in the field of emotions, basic emotions and social interactions convey the following premises, which nuanced are:

- Deep emotions are not frequent.
- Deep emotions always unfold in social contexts and peculiar social interactions.
- Different emotions have variable durations and intensities.
- Non-verbal manifestations also materialize with emotions.
- There are explicit and implicit emotions.
- Emotional regulation and control are intimately mixed with the types of expression.
- There are subjective, sexual and generic differences in relation to emotions.
- Emotions are of genetic, socio-cultural, educational origin.

From complexity and transdisciplinarity (Morin 1999; Nicolescu 1996), cognitive sciences (located in the *rational epistemological*) and emotional sciences (located in the *emotional epistemological*) are constituted and emerge almost simultaneously, but in a polarized direction at the beginning, which later converge to configure the "epistemological continuum reason <-> emotion" (Haidar 2006; 2013). The recursion reason <-> emotion / emotion <-> reason is inescapable. Piaget, the great transdisciplinary psychologist, said that there is no intelligent dimension, without passing through the affect (Piaget: 1991). This position is assumed by many scholars against Cartesian positivism, which separated reason from emotion, placing negative characteristics solely in the latter, and in particular, defending a rational, objective human being, what constitutes the great fallacy of positivist subjectivity.

Emphasis in this work is on emotional inferences, derived also from the reason-emotion recursive continuum as well as logical inferences. It is necessary to underline that emotional argumentation produces emotional inferences and that it is always connected to persuasion. Thus, in this way the definition of the emotional from complexity and transdisciplinarity implies a recursive relation to several cognitive fields, as can be seen in Figure 1, below:

Figure 1 – Cognitive fields in complexity and transdisciplinarity

There are multiple definitions of the emotion, affect and feeling categories, as well as many explanations, analyses, and applications in these cognitive fields and subfields, of which we do not intend to cover all possibilities and which are separated in the profuse and extensive bibliography consulted. It is important to regroup them in order to achieve a transdisciplinary definition of the emotional.

4. DISCURSIVE EMOTIONAL ARGUMENTATION

Assuming that in today's world the emotional is always present in intersubjective interactions, in this section we return to some reflections on discursive emotional argumentation. The theoretical approaches can be applied to visual emotional arguments, which are linked to inferences and persuasion. In the field of discourse analysis, there are several models studying the emotional component, for example, such as those of Edwards (1999), Plantin (1997), Charaudeau (2000), and Gilbert (1997), among others.

In Edwards (1999: 288), we find the following premises that link emotion with cognition, crucial for our objectives. Although it seems there is a certain predominance of the emotional compared to the logical, it is important to think of this relationship in a recursive way, that is, concurrent, complementary, antagonistic:

i. Emotion vs. cognition: considered as concurrent discursive resources, because actions and mental states are described and formulated as thoughts, opinions, emotions.
ii. Emotion vs. cognition, as the irrational vs the rational: for the author emotions are not irrational, since there is an integral part of rational responsibility.
iii. Emotion and cognition: there are many cognitive consequences derived from emotional experiences.
iv. Emotional behaviour as a controllable action or a passive reaction: this proposal refers to the emotions that may arise in human practices.
v. Honest (spontaneous) vs false; emotional reactions, particularly if they are immediate, provide an honest narrative and rhetoric, in contrast to the cognitive calculation that is considered false, insincere. The category of emotion as honest and spontaneous corresponds not only to the popular conception, but to quite many results of experimental research.

The last approach seems thought-provoking, because it relates cognitive-emotional processes present in common sense, with those of science. In short, according to this author, the conceptual repertoire of emotions provides an extraordinary tool for analysing actions, reactions, dispositions, motivations and other psychological characteristics that can be found in subjective semiotic-discursive practices.

Other analytical angles to approach the emotional dimension in argumentation are expressed in Plantin (1997: 83-88). Emotional inferences are produced in the following:

i. The psychological place is marked by a set of terms linked to emotion: nouns, adjectives, verbs, adverbs. This lexical approach is also found among psychologists, who in order to analyse "basic emotion" propose classes of emotion nouns.
ii. The statement of emotion also contains verbs of feeling or psychological verbs, which are considered in three classes: 1. Love, despise, 2. Impress, etc., 3. Pleasure, dislike, etc.
iii. Direct or indirect designation of emotions: emotions can be designated directly or indirectly.

iv. Emotions can be implicit or explicit. For example, in the statement, "children die of hunger and thirst in the desert," there are implicit emotions to develop the argument "ad misericordiam".
v. There are two modes of detection of the functioning of emotions in discourse: 1. Emotions that are reconstructed on the basis of linguistic descriptions of conventional emotional states, marked in the lexicon and in the grammar; and 2. Emotions that are not explicit, and are generated implicitly.
vi. The relationship topoi and emotion: this author states that there is a topic of emotion based on common ordinary topics such as: what, who, how, when, where, etc.
vii. Discursive strategies to express emotions are infinite, as are the games of language, which is why they are linked to rhetorical functioning in general, tropes and topoi.

In search of a synthesis, we propose the following strategies of emotional argumentation (Haidar 2006; 2013), linked to emotional inferences:

- Communicative-pragmatic device that enables the emotional component.
- Thematic fields and discursive objects conducive to the emotional component.
- The use of a lexicon linked to emotion that covers all morphological classes from nouns, adjectives, adverbs, to verbs.
- The use of intrinsic and extrinsic emotion statements.
- Explicit or implicit emotions.
- The use of emotional arguments (very close to fallacies).
- The use of rhetoric of emotions.
- The use of the topic of emotions, of passions.
- The functioning of beliefs and knowledge.
- Semiotic-discursive sociocultural representations of emotions.
- Emotions in the verbal, in the para-verbal and in the non-verbal

5. VISUAL AND POSTVISUAL EMOTIONAL ARGUMENTATION

It is essential to consider everything explained previously in relation to discursive argumentation as a way to further continue our theoretical approach in this section. In particular, because there are recursive relations of similarities and differences between verbal discursive argumentation and visual and post-visual semiotics, of which we consider some aspects.

Extensive production in this field has been done by Groarke (1999, 2016), who has developed significant contributions in several chapters and articles on visual argumentation. As a philosopher, his approaches are based mainly on two important trends in argumentative theory: Informal Logic and Pragma-dialectics. The reflections of this author on the subject are abundant. Firstly, he questions other argumentation perspectives that do not accept the existence of visual argumentation, and if they do so, it is with a great difficulty. Secondly, Groarke proposes theoretical-methodological approaches to analyse visual arguments focusing on visual premises and visual fallacies.

From our perspective in order to address the visual and post-visual emotional argumentation, and considering previous achievements in this direction, we propose to resort to two fields that can provide many elements and analytical tools to analyse these argumentative productions: the first one is Visual / Post-visual Semiotics, and the second one is Visual / Post-visual Rhetoric (Haidar 2013). It is critical to review the following problems in relation to Visual / Post-visual Semiotics:

- Define what is visual / post-visual and how to work with this semiotic dimension: shapes, colours, volumes, composition, scenarios.
- Establish the relationships between the visual and the discursive, which have been different in different cultures and historical epochs.
- The differences between the static visual, the kinetic visual, the visual / sign function.
- Analyse problems related to iconism, the different degrees of iconicity.
- Address the emergence of digital Semiotics, with different types of post-visual production that go through the simulations, by digital chronotopes that involve very different space-times, produced in hyper-reality.

In relation to Visual / Post-visual Rhetoric, we must address the problems derived from a position of Discursive Rhetoric to Visual / Post-Visual Rhetoric, which implies specifically the passage from verbal tropes to visual tropes, among other aspects. From Complexity and Transdisciplinarity, we redefine Rhetoric with the following premises (Haidar 2013):

- The transdisciplinary nature of Rhetoric is organic, because rhetorical operations are present in all semiotic-discursive productions.

- Rhetoric is not a peculiar art, as it was understood among the classics, because it is found in all semiotics and discourses, as Perelman and Olbrecht Tyteca (1989), Ducrot and Anscombre (1983) have noted, among others.
- The main parts of Rhetoric, such as inventio, dipositio, elocutio, actio, are present in all semiotic-discursive practices, in different degrees.
- The new Rhetoric is not oriented only to persuasion, but also to rebuttal.
- The Rhetoric from complexity and transdiscipline allows different developments for the analysis of the visual / post-visual: Rhetoric of image, plastic rhetoric, rhetoric of objects, of spaces, etc.

With these considerations, we resume the rhetorical operations both in the discursive as in the semiotic visual / post-visual, which allow to analyse tropes of thought, of diction, enthymemes, fallacies in discourse and in semiosis, as well as to orient the analysis of emotional inferences. The two theoretical-methodological perspectives, of Semiotics and of Rhetoric, can be very well articulated to the visual emotional argumentation, but starting from a redefinition of argumentation (Haidar 2006):

- Argumentation is a semiotic-discursive macro-operation that is oriented as much to persuasion as to rebuttal.
- This operation is present both in verbal discourses, as in semiotic visual / post-visual productions.
- The main problem, with several analytical challenges, is to go from discursive to visual / post-visual argumentation, to analyse premises, enthymemes, fallacies, visual tropes.
- In visual / post-visual argumentation it is fundamental to highlight the rational and emotional dimensions in a recursive way.
- It is necessary to provide a redefinition of the rational versus different logics (natural, informal, everyday, which are opposed to formal logic) and also, a transdisciplinary redefinition of the emotional (as already proposed).

6. VISUAL EMOTIONAL ARGUMENTATION EXAMPLES

In the selected examples, we analyse visual emotional argumentation from the dialectic between negative emotions / positive emotions. In addition, it is important to dwell on the type of visual text we are using. From complexity, they are post-visual texts in which the tools of digital semiotics, or cyber-semiotics, are applied. These two texts are artistic because there is a very detailed work on the form. Moreover, from these perspectives, art is trans-dimensional because it contains emotion, cognition, culture, history, and the social. The aesthetic-artistic can produce euphoric effects in the case of Image 1, although these are easily changed when the gaze stops; while in Image 2, it is impossible to avoid dysphoric effects.

The two visual emotional texts trigger continual cognitive <> emotional inferences, and from the relationship between art and complexity they oscillate between the photo and the painting, with the rupture between the arts that occurs in avant-garde and post-avant-garde movements. Rupture of borders in artistic sciences is impossible to avoid since the field of art is supposedly that of the greatest freedom, albeit the always present phantom of the canon. Between the epistemological continuum from complexity among natural sciences <> social sciences <> quantitative sciences <> artistic sciences, the latter has the most tendency to break the boundaries, which happened with the avant-garde and post-avant-garde movements. Consequently, in the two visual texts we find a convergence of photography techniques, together with those of painting, producing complex texts, with the heterogeneity of art, as Iuri Lotman suggests in many of his works.

For the analysis, we use a transdisciplinary operating model, in which the following dimensions are combined: emotional, argumentative, visual / post-visual, semiotic, rhetoric. Undoubtedly, complex but fascinating paths open up to address the production and reproduction of the senses in these powerful semiotic-discursive productions, with great impact for any type of observer. The use of semiotic visual / post-visual strategies produces many reactions, impacts on viewers. Memory of culture, historical memory, cognitive memory, emotional memory, they all condense and move spectators, traumatized recursively.

Image 1 – "Certain things hang on forever. Set the kids free from abuse and violence", Campaign against child abuse. Casa do Menor - Hands, Brasil – (2010)

This is a visual text produced by an institution dedicated to prevent the abuse and violence of minors. The post-visual construction is shocking from the start, and so traumatic that it produces intense emotional inferences, anchored in visual emotional argumentation. It is a complex visual text, composed of several semiosis: the dark colours, the space full of non-children furniture, the illumination in the girl and in the lamp in the corner, the abused child in the centre of the tragic scene, in which the only child object, the doll, is in her hand.

This construction conveys a visual emotional argumentation, which has a discursive base, which contains an indisputable enthymeme. Other enthymemes are derived from the central figure of the brutalized child, a post-visual object constructed with visual tropes of violence. Violence materializes in a child's body invaded by hands that destroy it, which is complemented by the girl's distorted face. The hands are metaphors-symbols of current violence of all kinds. Another enthymeme: violence and abuse are destructive. Actually, the aesthetic resources of this visual emotional argument are impressive. The gaze runs through the text, triggering multiple senses around the fact that is denounced and which stays in the eye. The visual emotional argument is one of refutation and contains emotional inferences such as: children must be defended from these attacks; we must fight with all means against child violence and abuse.

Image 2 – "If you don´t help feed them, who will?" Source: PIGLETS by Y&R Philippines for Concordia Children's Services, 2006 (Orphanage in Manila, Filipinas)

This is another example of a complex visual text, which uses a dialectical of negative / positive emotions, produced by the visual emotional argumentation and the emotional inferences that are derived. As in the previous example, it is a refutation argument to which emotional inferences are linked. The semiosis are several: the dark colour of despair, of death; the degeneration of the human being, since children are fed by a sow, which produces the visual enthymeme of the animalization of human beings by extreme poverty, a situation produced by the global crisis generated by the concentration of wealth.

Among many visual tropes, there is the oxymoron which is the irresolvable paradox: the feeding of human beings denigrated by a sow, as the only form of subsistence. Emotional dysphoric inferences that seek to impact to produce humanitarian actions that achieve struggles with these tremendous injustices, exploitations generated by this world full of uncertainties, of gloomy horizons due to the absence of minimum ethical values. The gaze produces mnemonic processes that allow the permanence of the traumatic, articulated to emotional inferences.

7. CONCLUSION

We can reach some conclusions with the material elaborated in the previous sections. First, it is crucial to recognize the importance of the epistemological position of complexity and transdisciplinarity in order to build theoretical and analytical perspectives of greater scope. Secondly, from these positions many cognitive fields can converge, as have been pointed out, to reach a greater depth and analytical rigor that are necessary to grasp the complex senses of visual, post-visual semiotic production related to emotional argumentation. Third, we cannot fail to point out that recursion is the complex relationship that exists between logical and emotional inferences in semiotic-discursive productions, the two visual / post-visual texts. In the analysis we attempt to briefly illustrate the transdisciplinary approaches to address the production and reproduction of the complex senses in these visual / post-visual texts.

In short, in the two visual texts there is the visual emotional argumentation that generates negative emotional inferences, since they are not pleasant, but traumatic, producing rejections, anxieties, anguish in front of two brutal phenomena of the current world: violence and sexually abused children, and the extreme poverty of a large part of humanity.

REFERENCES

Barreyro, J.P., Molinari Marotto, C., Yomha Cevasco, J., Duarte, Dionisio A. (2008). Generación de Inferencias Emocionales durante la lectura de textos narrativos. Anuario de Investigaciones en línea (pp. 19-24). Retrieved from
<http://www.redalyc.org/articulo.oa?id=369139944033>

Charaudeau, P. (2000). Une problématisation discursive de l'émotion - A propos des effets de pathémisation à la télévision. In Plantin, C.& Doury, M. & Traverso, V. (Comp.), *Les émotions dans les interactions* (pp. 125-155). Lyon: Presses Universitaires de Lyon.

Cosnier, J. J. (2015). *Psychologie des emotions et des sentiments*. Retrieved from <http://icar-lyon2.fr/membres/jcosnier>

Dávalos, M.T., & León, J.A. (2017). Inferencias emocionales en los procesos de comprensión de los discursos. *Revista Digital Internacional de Psicología y Ciencia social*, Vol 3, 1-13. Retrieved from <htpp://dx.doi.org/10.22402/j.rdipycs.unam.3.0.2017.125.53-63>

Ducrot, O., & Anscombre, J.C. (1983). *L'argumentation dans la langue*. Bruxelles: Pierre Mardaga Editeur.

Edwards, D. (1999). Emotion Discourse. *Revue Culture & Psychology, Vol. 5, No 3.* September, 271-291.
Edwards, D. (1997). *Discourse and Cognition.* London/ New Delhi: Sage Publications.
Gilbert, M. A. (1997). Prolegomenon to a Pragmatics of Emotion. In *Proceedings of the Ontario Society for the Study of Argumentation.* Canada: Brock University, St. Catherine's.
Goleman, D. (1995). *La Inteligencia Emocional.* México: Ediciones B México.
Groarke, L. (2016). Informal Logic. Stanford Encyclopedia of Philosophy. First published in 1997. Stanford: Stanford University. Retrieved from <https://plato.stanford.edu/entries/logic-informal/>
Groarke, L. (2002). Hacia una pragma-dialéctica de la argumentación visual. In Frans H. van Eemeren (Ed.), *Advances in Pregma-Dialectics* (pp. 137-151). Amsterdam: Sic Sat / Virginia, Vale Press / Newport News.
Haidar, J. (2013). De la argumentación verbal a la visual: lo emocional y la refutación en la argumentación visual. In Adrián Gimate-Welsh y Julieta Haidar (Coord.), *La Argumentación. Ensayos de análisis de textos verbales y visuales* (pp. 201-223). México: Universidad Autónoma Metropolitana (UAM).
Haidar, J. (2006). *Debate CEU-Rectoría. Torbellino pasional de los argumentos.* México: Universidad Nacional Autónoma de México.
Howard, G. (2001). *Estructuras de la mente. La teoría de las inteligencias múltiples.* Colombia: Fondo de Cultura Económica.
Lotman, J. M (2013). *The unpredictable workings of culture.* Tallinn: Tallinn University Press.
Molinari Marotto, C. & Barreyro, J.P.& Cevasco, J. & van den Broek, P. (2011). Generation of Emotional Inferences during Text Comprehension: Behavioral Data and Implementation through the Landscape Model. *Escritos de Psicología*, Vol. 4, nº 1, Enero-Abril, 9-17.
Morin, E. (2006). Complejidad Restringida, Complejidad General. In: Edgar Morin, Jean Louis Le Moigne (Edts.), *Inteligencia de la Complejidad. Epistemología y Pragmática* (Coloquio de Ceresy) (pp. 20-47). Paris: Editions de l´Aube.
Morin, E. (1999). *El conocimiento del conocimiento.* Madrid: Ediciones Cátedra.
Nicolescu, B. (1996). *La transdisciplinarité.* Paris: Editions du Rocher.
Perelman, Ch. & Olbrechts-Tyteca, L. (1989). *La nueva retórica. Tratado de la argumentación.* Madrid: Editorial Gredos.
Plantin, C. (1998). Les raisons des émotions (pp 1-29). Retrieved from <https://es.scribd.com/document/162760999/Plantin-1998-Les-Raisons-Des-Emotions>
Plantin, C. (1997). L'argumentation dans l'émotion. In *Revue Pratique – Enseigner l'Argumentation, No 96,* 81-100.
Piaget, J. (1991). *Seis estudios de psicología.* Barcelona: Editorial Labor.

27

The Concept of an Argument[1]

DAVID HITCHCOCK
Department of Philosophy, McMaster University, Hamilton, Canada
hitchckd@mcmaster.ca

I revise my (2006) definition of an argument in response to criticisms by G. C. Goddu and James Freeman. I retain the claims that the ultimate constituents of arguments are illocutionary acts and that complex arguments may be formed either by chaining or by embedding. But I now count attacks as well as supports as arguments, and I rest the unity of an expressed argument on its author's second-order illocutionary act of adducing.

KEYWORDS: argument, chaining, complex argument, concepts, definition, embedding, illocutionary acts, recursive definition, simple argument

1. INTRODUCTION

The present paper revises the definition of an argument in (Hitchcock, 2017), which responds to objections by Geoff Goddu (2010) and Jim Freeman (2010) to the definition in (Hitchcock, 2006). The concept of an argument for which I propose an analysis is the reason-giving sense in which one speaks for example about Daniel Kahneman's argument (2011, pp. 334-35) that the tendency of most people to be risk averse about gains but risk seeking about losses is irrational. This sense of the word 'argument' should be sharply distinguished from the disputational sense in which one speaks about two people having an argument. That these are two different senses is clear from the fact that languages other than English use two different words for the two senses.

---

[1] The present paper uses material from "The concept of argument" (Hitchcock 2017, pp. 518-529; © Springer International Publishing AG 2017), with permission of Springer.

We could give a rough lexical definition of the word 'argument' in this sense by quoting the definition by the Hellenistic Stoics of an argument as "a system composed of premisses and a conclusion" (*systêma ek lêmmatôn kai epiphoras*, Diogenes Laertius 1925/ca. 210-240, 7.45). Aside from its idiosyncratic failure to recognize one-premiss arguments, this definition is an acceptable starting point for a conceptual analysis. That analysis would need to answer a number of questions raised by the lexical definition. What is a premiss? What is a conclusion? What sorts of entities can function as a premiss? What sorts can function as a conclusion? How do premisses and a conclusion form a unified system? Does such a system have an intrinsic function or purpose, or on the contrary can it be used for various purposes? What about complex arguments?

We should recognize that arguments are not necessarily the content or product of argumentation. One can consider an argument for a certain position or policy even if nobody has ever used that argument. One can imagine crazy arguments that no sane person would ever put forward. One can apply the term 'argument' to unified stretches of solo reasoning where some conclusion is reached on the basis of reasons, as when one considers mentally various aspects of a situation and then describes "the argument that finally convinced me". Indeed, it seems consistent with our ordinary use of the term 'argument' in its reason-giving sense to say that there are arguments that nobody has ever thought of and nobody ever will. The most we can demand is that an argument must be thinkable and expressible. Arguments as a class thus have no common function or purpose. They are not necessarily used to justify or establish something. Nor are they necessarily used to persuade anybody.

2. SIMPLE ARGUMENTS

I begin by considering simple arguments, in the sense of single-inference systems that consists of one or more premisses and one conclusion. Of what sorts of entities are such arguments composed? That is, what kind of object can function as a premiss, and what kind can function as a conclusion? A common answer is that these components are propositions, in the sense of postulated timeless and non-located entities that can be expressed linguistically (or in some other way) and that can be objects of belief or knowledge. Propositions however are not the right candidates to be premisses or conclusions.

As to premisses, consider the following two arguments:

(1) Suppose that there is life on other planets in the universe. Then it makes sense to look for it.
(2) There is life on other planets in the universe. So it makes sense to look for it.

These arguments have the same conclusion, but different premisses. But both premisses express the same proposition, that there is life on other planets in the universe. The difference is in the illocutionary act performed by someone who utters the premiss in standard contexts. In uttering argument (1), the author *hypothesizes* the propositional content of the premiss. In uttering argument (2), on the other hand, the author *asserts* the proposition. The difference makes a difference to the evaluation of the two arguments: argument (2) requires a stricter condition of premiss adequacy than argument (1).

If arguments need not be expressed but their premisses are illocutionary acts, the premisses must be illocutionary act-types rather than illocutionary act-tokens. Sometimes the type may have no actual tokens. A premiss can be a supposition or an assertive or any other member of the class of illocutionary acts that Searle (1976) grouped under the label 'representatives', which he defined (p. 10) as acts whose point is to commit the speaker, perhaps hypothetically or guardedly, to something's being the case. It cannot be any other kind of illocutionary act, as we can see by noting the peculiarity of putting examples (taken from Hitchcock, 2006, pp. 103-104) of the other kinds in premissary position before an inferential 'so':

(3) * What time is it? So you must go home.

(4) ? I promise to pick up some milk on the way home. So you don't need to get it.

(5) * Congratulations on your anniversary. So you are married.

(6) * I hereby sentence you to two years less a day. So the guards will now take you to prison.

In standard contexts, the utterances in premissary position are respectively a directive (in particular, a request for information), a commissive (in particular, a promise), an expressive (in particular, a congratulation), and a declarative (in particular, a judicial verdict). They reveal the general inability of illocutionary acts other than representatives to function as premisses. The premiss in (4) is a borderline case, because a promise can be taken to imply a prediction (that the promise will be kept) and a prediction is a kind of assertion. It is the implicit prediction rather than the commitment that makes it possible to construe example (4) as an argument.

Among conclusions, we find a greater variety of illocutionary acts than among premisses. We have already seen two examples (1 and 2 above) in which the conclusion is a representative. Conclusions (and premisses too) can be hedged by such qualifiers as 'probably', 'presumably', 'possibly', and the like, which are best given a speech-act interpretation, as Toulmin (1958, pp. 47-62) and Ennis (2006, pp. 145-164) have argued. Thus a wide range of representatives can be conclusions of arguments. But the other main kinds of illocutionary acts can also be conclusions, as we can see from the following examples (taken from Hitchcock, 2006, p. 105):

(7) There is a forecast of thundershowers, so let's cancel the picnic.

(8) I know how difficult it will be for you to get the milk, so I promise you that I will pick it up on the way home.

(9) My conduct was inexcusable, so I apologize most sincerely.

(10) The evidence establishes beyond a reasonable doubt that you committed the crime of which you are accused, so I hereby find you guilty as charged.

When these arguments are expressed in standard contexts, their conclusions are respectively a directive, a commissive, an expressive, and a declarative. Thus the conclusion of an argument can be an illocutionary act type of any kind: a representative, a directive, a commissive, an expressive, or a declarative.

A reason can be advanced as a reason against some claim as well as a reason for it. Consider the following example:

(11) Proponent: We need a concerted global reduction of greenhouse gas emissions to mitigate future climate disruption.
Opponent: A cost-benefit analysis needs to be done first to determine whether it makes more sense to adapt to future climate disruption rather than to mitigate it.

Here the opponent's reason is put forward as a reason against the proponent's claim. The exchange has an inferential structure that is quite parallel to that in which a reason is put forward in support of a claim. In general, objections and criticisms seem to have just the same inferential structure as supports. In other words, there can be arguments against something as well as arguments for something. Consider an example. In the 11th century the monk Gaunilon objected to Anselm's ontological argument for the existence of God that by the same reasoning one could prove the existence of a perfect island (Anselm 1903/1077-78). Gaunilon's objection is an argument against the cogency of Anselm's argument. It would involve needless and misleading subtlety to recast his objection as an argument in support of some claim. It is better to follow a number of authors who have recognized that there can be arguments against as well as arguments for (Johnson, 2000; Rahwan et al., 2009; Freeman, 2010; Wohlrapp, 2014/2008).

In fact, the same reason can be adduced for a claim by one person and against the same claim by another person. For example, some people use the principle of self-determination as a reason for legalizing voluntary euthanasia, whereas others use it as a reason against legalizing voluntary euthanasia (Wohlrapp, 2014/2008, pp. 264-265). Each group advances an argument, and the two arguments clearly differ from each other. Thus it makes sense to add to the conception of an argument as a premiss-conclusion or claim-reason complex a third component indicating whether the reasons are to count for or against the claim. Further, to accommodate the existence of arguments against as well as arguments for, I shall hereafter use the term 'target' rather than 'conclusion' or 'claim' for the part of an argument to which its reasons are directed.

We are now in a position to say what a simple argument is:

> A simple argument is a triple whose first member is a set of one or more representative illocutionary act types (called 'the reasons'), whose third member (called 'the target') is an illocutionary act type of any kind, and whose second member is an indicator of whether the reasons count for or against the target.

There may be no tokens of the component illocutionary act types. In other words, simple arguments as just defined are abstract structures that are not necessarily actually realized. We need impose no further restrictions on what can count as a simple argument, thus widening the class of arguments to the craziest combinations that one can imagine.

If a simple argument is expressed or mentally entertained, something must constitute its unity as a single argument. I propose to found this unity on a second-order illocutionary act of adducing the argument's reasons as supporting or opposing its target. Echoing Searle (1969), I define this act by its content, essential, preparatory and sincerity conditions. The content of an act of adducing is a triple consisting of the reasons (a set of one or more representatives), an indicator of whether the set counts for or against the target, and the target (a first-order illocutionary act of any kind). The essence of adducing is that the utterance of the adducer counts as a claim that the reasons if true or otherwise acceptable[2] would provide the indicated epistemic support for or opposition to the target. Both the context of adducing and adducers' intentions vary widely. Hence there are few preparatory conditions common to all adducing. Perhaps one preparatory condition is that the addressee (who in the case of solo reasoning will be identical with the adducer) did not previously think that the reasons provide the claimed epistemic support for or opposition to the target. The sincerity condition for adducing is that the adducer believes that the reasons if true or otherwise acceptable would provide the claimed epistemic support for or opposition to the target.

Goddu (2018) has objected to this way of securing the unity of an actualized argument that a person can entertain an argument mentally without supposing that its reasons support (or oppose) its target. One can for example wonder whether the reasons in an argument one is considering actually support the target. In response to this objection, I propose to rest the unity of such merely considered arguments on a hypothetically possible act of adducing. If one is considering a complex of reasons, indicator and target as a whole that could be used to adduce the reasons as supporting or opposing the target, then one is considering an argument.

Someone who adduces reasons as counting for or against a target must express the indicator, and must be the author of either the reasons or the target. But such an adducer need not be the author of both reasons and target. One can draw a conclusion from something someone else has said, in which case the person who draws the conclusion adduces what the other person said as supporting the

---

[2] I add the phrase 'or otherwise acceptable' to accommodate representatives whose propositional content is not a description. It might for example be an evaluation or a prescription. Some people balk at calling such propositions 'true'. They can use instead whatever word they accept as an appropriate analogue of 'true' for such propositions.

conclusion drawn. One can provide reasons against someone else's claim, in which case the person who provides the reasons adduces them as opposing the other person's claim. In any situation where an argument is expressed, the adducer is the person who articulates the inferential claim conveyed by the indicator that points to the argument's target.

3. COMPLEX ARGUMENTS

So far my conception of an argument accommodates only simple arguments, i.e. single-inference arguments with a set of one or more reasons, a target, and an indication of whether the reasons count for or against the target. We need to allow as well for complex arguments that involve a chain of reasoning or embedded suppositional reasoning.

In chained arguments, a reason of one argument (which I will call 'the superordinate argument') is the target of another (which I will call the subordinate argument or 'sub-argument'). Since only representatives can be reasons, the target of any subordinate argument must be a representative. There is no limit to the depth of chaining. The ultimate target in a chain of reasoning is supported or opposed by one or more reasons, each of which may be the target of one or more reasons in a sub-argument, each of which in turn may be the target of one or more reasons in a sub-sub-argument, and so on indefinitely. In an expressed sub-argument at any level, the reasons must be adduced in support of the target. Otherwise the target would have to be the complement of the reason in the superordinate argument to which it was linked. But it is hard to define the complement of a representative illocutionary act. What, for example, is the complement of a hedged assertion of the form 'probably p'? Representatives incompatible with 'probably p' include 'definitely p', 'probably p'', and 'definitely p'', where p' is a contradictory of p. An argument against some target that was used to support a reason of the form 'probably p' would therefore need an unwieldy disjunction as its target, for which it would be difficult if not impossible to formulate a set of reasons that successfully opposed each disjunct simultaneously. Since subordinate simple arguments in an expressed complex argument make sense only if their reasons are presented in support of their target, it makes sense to limit unexpressed subordinate simple arguments in the same way.

The natural way to accommodate the indefinite complexity of chained arguments is to use a recursion clause that can be applied again and again so as to build up arguments of increasing complexity. The process is analogous to that by which one defines what a person's ancestor is by saying that a parent of a person is an ancestor of that person and that a parent of any ancestor of that person is also an ancestor of that person. This definition allows one to construct the class of a person's ancestors, starting with the person's parents, then adding the grandparents, then the great-grandparents, and so on without end. In defining recursively what an argument is, one needs to take some care in constructing the recursion clause for chaining arguments. The most sensible way to do so seems to be to add one at a time a simple argument for a reason in an already constructed argument. One can conceive of a simple argument, which is a triple, as a unit set, a set with one member. One can combine it with a simple argument whose target is a reason in the first argument by taking the union of the two sets, i.e. the set whose members are all the triples that are members of either set. And then one can combine this set of two triples with a third simple argument whose target is a reason in one of the first two triples. And so on. Let us call a reason in any triple in a set of such triples a 'reason in the argument'. The recursion clause might then read as follows:

> If in an argument something is a reason but is not a target, then the union of that argument with a simple argument whose target is that reason and whose indicator is positive is also an argument.

As with the definition of simple arguments, this clause allows that the most fantastic and crazy combinations are arguments in the abstract sense. The condition that the reason is not already a target is meant to exclude from being a single argument structures in which a reason is the target of more than one simple argument. Just as multiple arguments for or against the same ultimate target do not constitute a single argument, so too multiple arguments for the same intermediate target cannot be components of a single complex argument.

Chaining is one of two ways to construct complex arguments. The other is embedding, where suppositional reasoning is used to support or oppose a target, with one or more of its suppositions being "discharged" in the process. Any line of suppositional reasoning is an argument according to the recursive definition of argument developed so far. To allow for embedding one or more such lines of suppositional reasoning, we need to allow that a line of suppositional reasoning can count as something like a reason.[3] One way to do so is to take the line of

---

[3] It won't do however, to count it as a reason in the same sense as that in which a representative is a reason. Otherwise the recursion clause for chaining would

suppositional reasoning as an implicit assertion that the ultimate target of the line of suppositional reasoning follows from its ultimate supposition in combination with any other ultimate reasons used in derivation of the ultimate target. (The assertion may be qualified, if either an inference or an ultimate reason in the suppositional reasoning is qualified.) When the suppositional reasoning is embedded in a larger context, the target external to this line of reasoning is a representative, which in the case of nested suppositional reasoning may itself have a suppositional status. The recursion clause allowing embedding of suppositional reasoning thus needs to allow for the dual complexity of chains of reasoning from suppositions and nesting of suppositional reasoning inside suppositional reasoning. It also needs to allow that a line of suppositional reasoning can be used in opposition to a target as well in support of one. If the target of a line of suppositional reasoning is a reason in a sub-argument, however, then the expression of such a complex argument makes sense only if the suppositional reasoning is adduced in support of the target, for the same reason that an expressed chained simple sub-argument makes sense only if its reasons are adduced in support of its target: an opposed target would have to be too complex to count as the complement of the reason in the superordinate argument that is being indirectly supported by opposition to the target of the suppositional reasoning. In general, too, it makes sense to use a line of suppositional reasoning only if its internal ultimate target is argued for rather than against, since all the recognized legitimate ways of discharging a supposition assume that the supposition is used to support the ultimate target. Similar restrictions to arguments with a positive indicator are therefore appropriate for abstract arguments that need not be expressed. It seems an unnecessary further complication, however, to incorporate in a clause allowing embedding the specific ways in which a supposition in a piece of suppositional reasoning may legitimately be discharged.[4] The abstract definition of an argument will

---

allow for arguments that are subordinate to a line of suppositional reasoning, which makes no sense.

[4] One way of legitimately discharging a supposition is conditional proof, in which one derives a conditional from a line of suppositional reasoning that starts from the supposition of the conditional's antecedent and ends with the conditional's consequent. A variant form of conditional proof starts from the supposition of a contradictory of a conditional's consequent and ends with a contradictory of its antecedent. Another way to legitimately discharge a supposition is *reductio ad absurdum*, in which one uses a line of reasoning from a supposition to some absurdity as a reason for denying the supposed proposition. Another is argument by cases, in which one considers an allegedly

thus allow for embedding pieces of suppositional reasoning in totally illegitimate ways.

The following is a possible recursion clause allowing embedding;

> A triple is an argument if its first member is a set whose members include at least one argument with a suppositional ultimate reason, whose third member (the target) is an illocutionary act of any kind, and whose second member is an indicator of whether the members of the set count for or against the target.

As is usual with recursive definitions, there needs to be a final closure clause to the effect that nothing is an argument unless it is an argument according to the base clause and the recursion clauses.[5] One can illustrate and test the resulting definition by using its clauses to construct complex arguments as they appear in argumentative texts. Space limitations preclude inclusion of such illustrations in the present paper.

As with the abstract concept of a single argument, we need a basis for the unity of an expressed complex argument. An expressed chaining of two arguments is a complex illocutionary act of adducing the resulting chain of reasoning as supporting or opposing the ultimate target of the superordinate argument in the chain (i.e. the argument that has a reason which the subordinate simple argument targets). The

---

exhaustive set of possible cases, deriving the same ultimate target from the supposition of each case, and then drawing this ultimate target as a conclusion. Another is to argue for a proposition in a proof by mathematical induction by supposing at the inductive step that the proposition holds for the number n (or for every number up to and including n) and deriving from this supposition that then it also holds for the number $n + 1$. Another is universal generalization, in which one derives a universal generalization about a kind by reasoning from the supposition that some individual is of that kind to the conclusion that the generalization holds for this individual, without using any other assumption about the individual.

[5] One can express the recursive definition in the customary form of a statement in which there appears in the first part the term to be defined, in the last part the defining part of the definition, and in between these two parts an indicator (such as 'means', '=df', 'if and only if', or 'is a') that the defining part states the meaning of the defined term. For example, one could say that something is an argument if and only if it belongs to every set that includes everything that satisfies the base clause, as well as everything that can be constructed from its members using the recursion clauses for chaining and embedding.

essence of adducing in this case is that the utterance of the adducer counts as a claim that in each link of the chain the reasons if true or otherwise acceptable would provide epistemic support for the target or as a claim that the reasons if true or otherwise acceptable would provide epistemic opposition to the target. An embedding of an argument is a complex act of adducing the embedded suppositional reasoning, possibly along with one or more reasons, as support for or opposition to the target of the argument in which the suppositional reasoning is embedded. The essence of adducing in this case is that the utterance of the adducer counts as a claim that the suppositional reasoning would if the additional reasons (if any) were true or otherwise acceptable provide epistemic support for the target or as a claim that the suppositional reasoning would if the additional reasons (if any) were true or otherwise acceptable provide epistemic opposition to the target. The content conditions, preparatory conditions and sincerity conditions for these more complex acts of adducing are a function of the content, preparatory and sincerity conditions for the simple acts of adducing from which they are constituted.

As with simple expressed arguments, we can accommodate complex arguments that are merely considered by a hypothetically possible act of adducing. If one is considering a complex abstract argument as a whole that could be used to adduce the reasons as supporting or opposing its ultimate target, then one is considering an argument.

4. SUMMARY

The conception of an argument that I propose has the following distinctive features:
- It takes the ultimate constituents of arguments to be illocutionary act types rather than propositions, statements, utterances, and the like.
- It allows for arguments against something as well as arguments for something.
- It allows the reasons in an argument to be any kind of representative illocutionary act.
- It allows arguments to have as their target any kind of illocutionary act.
- It distinguishes arguments as abstract structures that may never be expressed or even thought of from expressed arguments.
- It locates the unity of an expressed or mentally entertained argument in a second-order illocutionary act

- of adducing, which may be actual or merely hypothetically entertained.
- It allows for a variety of uses of arguments, since neither the abstract conception of an argument nor the act of adducing that constitutes a complex of illocutionary act types as a single argument includes any conception of the purpose or function of an argument.
- It provides explicitly for complex arguments to be constructed recursively by steps of chaining and embedding.

REFERENCES

Anselm, Saint (1903). *Proslogium; Monologium: An appendix in behalf of the fool by Gaunilo; and Cur deus homo* (S. Norton Deane, Trans.). Chicago: Open Court. (Latin original written in 1077-78).

Diogenes Laertius (1925). *Lives of eminent philosophers* (R. D. Hicks, Trans). Cambridge, MA: Harvard University Press. (Originally published ca. 210-240 CE)

Ennis, Robert H. (2006). 'Probably'. In D. Hitchcock & B. Verheij (Eds.), *Arguing on the Toulmin model: New essays on argument analysis and evaluation* (pp. 145-164). Dordrecht: Springer.

Freeman, James B. (2009). Commentary on Geoffrey C. Goddu's "Refining Hitchcock's definition of 'argument'". *OSSA Conference Archive*. 56. Retrieved from https://scholar.uwindsor.ca/ossaarchive/OSSA8/papersandcommentaries/56

Goddu, G. C. (2009). Refining Hitchcock's definition of 'argument'. Goddu, G C., "Refining Hitchcock's Definition of 'Argument'" (2009). *OSSA Conference Archive*. 55. Retrieved from https://scholar.uwindsor.ca/ossaarchive/OSSA8/papersandcommentaries/55

Goddu, G. C. (2018). Against the intentional definition of argument. In S. Oswald & D. Maillat (Eds.), *Argumentation and inference. Proceedings of the 2nd European Conference on Argumentation, Fribourg 2017*. London: College Publications.

Hitchcock, D. (2006). Informal logic and the concept of argument. In D. Jacquette (Ed.), *Philosophy of logic*, Vol. 5 of Dov M. Gabbay, P. Thagard & J. Woods (Eds.), *Handbook of the philosophy of science* (pp. 101-129). Amsterdam: North Holland.

Hitchcock, D. (2017). The concept of argument. In David Hitchcock, *On reasoning and argument: Essays in informal logic and on critical thinking* (pp. 518-529). Dordrecht: Springer.

Johnson, R. H. (2000). *Manifest rationality: A pragmatic theory of argument*. Mahwah, NJ: Lawrence Erlbaum Associates.

Kahneman, D. (2011). *Thinking, fast and slow*. New York: Farrar, Straus and Giroux.
Rahwan, I., & C. Reed (2009). The argument interchange format. In G. R. Simari (Ed.), *Argumentation in artificial intelligence* (pp. 383-402). Boston: Springer.
Searle, J. R. (1969). *Speech acts: An essay in the philosophy of language*. Cambridge: Cambridge University Press.
Searle, J. R. (1976). A classification of illocutionary acts. *Language in Society*, 5(1), 1-23.
Toulmin, S. E. (1958). *The uses of argument*. Cambridge: Cambridge University Press.
Wohlrapp, H. R. (2014). *The concept of argument: A philosophical foundation*. Dordrecht: Springer. (German original first published in 2008)

29

Metaphors as Arguments: Perspectives from Psycholinguistics

CURTIS HYRA
Argumentation Studies, University of Windsor
hyra2@uwindsor.ca

HAMAD AL-AZARY
Department of Psychology, University of Western Ontario
halazary@uwo.ca

LORI BUCHANAN
Department of Psychology, University of Windsor
buchanan@uwindsor.ca

CATHERINE HUNDLEBY
Department of Philosophy, University of Windsor
hundleby@uwindsor.ca

Given the resemblance of metaphor to literal language, we argue that metaphor can be viewed as an argument; an attempt to persuade that one thing is another. Certainly, this is a compressed argument, and how that compression might be unpacked and related to more explicit arguments raises a number of questions. To consider how metaphors might scale up and unpack as fully fledged arguments, we consider empirical psychological work about the processing of metaphors.

KEYWORDS: argument, metaphor, psycholinguistics, rhetoric, semantic neighbourhood density

1. INTRODUCTION

The roots of metaphor in argumentation go back to Aristotle. For Aristotle, metaphors facilitate the acceptance of a fact or value in such a way that the audience is more likely to accept than in its blunt factual form[1] (*Poetics* Chapter 21, 1457b1-30, *Rhetoric* Book 3, 1410b). This is a rhetorical approach to metaphor that focuses on audience in its formulation. Indeed, metaphor and rhetoric have maintained close allegiance through the development of the theoretical issues concerning metaphor. Unfortunately, as brothers in arms, rhetoric and metaphor came to be seen as mere adornment. The argumentative and epistemic importance were downplayed in metaphor theory (Leff, 1984).

The tide is turning on the importance of metaphor, and it is thereby gaining increasing momentum in argumentation theory. An important event in this regard is the treatment of metaphor by Chaim Perelman and Lucie Olbrechts-Tyteca in *The New Rhetoric* (1969). Since then, the attention to metaphor has been swelling. For instance, Michael Leff emphasizes metaphor's argumentative function (1984), Christopher Tindale gives a treatment of metaphor in his book *Acts of Arguing* (1999), and various pragma-dialectical approaches give a treatment of metaphor (cf. van Eemeren and Gootendoorst, 2004). Even though he is not an argumentation theorist, George Lakoff has also made the case for argumentative function of metaphor (2006). Here, Lakoff provides the link between argumentation and cognitive linguistics that is further developed by Christian Santibanez (2007, 2010), and Steve Oswald and Alain Rihs (2014). It is at this intersection of cognitive linguistics and argumentation that our approach departs. Whereas Oswald and Rihs treat metaphors as arguments from the cognitive linguistic perspective, we show that metaphors are arguments from a psycholinguistic perspective.

In this paper we begin by outlining the conception of argument we endorse from Christopher Tindale (2015). Next, we conduct a literature review of metaphor is psycholinguistics. We then outline the research of Hamad Al-Azary and Lori Buchanan to show our conception of metaphor as argument and how this new conception deviates from the cognitive linguistics approaches. We then move to our case that metaphors are arguments as they exhibit propositional content and

---

[1] We might have said "literal form" rather than "blunt factual form". In this paper, we will argue that metaphor processing and literal language processing are similar, such that the distinction between literal and metaphorical is blurred. In light of this, we avoid using the term literal here.

persuasive force. We will draw from Perelman and Leff, however, we focus on the work of Santibanez and Oswald and Rihs. We will focus on Santibanez's association between conceptual metaphor and the Toulminian model of argument, and Oswald and Rihs' treatment of extended metaphor as argument. Our approach deviates in that it considers empirical work from psycholinguistics and draws on established phenomenon such as semantic neighbourhood density to show what insights from the psychology of metaphor processing add to the theory of metaphor in argumentation. The unique contribution of semantic neighbourhood density is twofold. First, the empirical fact of semantic neighbourhood density is itself support that a nominal metaphor unpacks into an argument. Secondly, empirical research on SND and the processing of metaphor provides data that shows that the cognitive processing of a metaphor plays a role in audience reception. We show what this effect is, and what SND tells us about effective metaphors as arguments with the added benefit of gaining predictive power for effective metaphors/arguments.

2. WHAT IS AN ARGUMENT?

Before making the claim that a metaphor is an argument, we must be clear about what we take an argument to be. We will follow a rhetorical approach to argument from Tindale (2015). He calls this the dynamic account of argument.

Traditional approaches treated arguments as abstract objects to be examined independently of the environment from which they arise, and/or in terms of their goals, yet independent of the audiences to which they are addressed. For instance, Irving Copi's definition of argument stresses the logical relation between premises (Copi, 1967). What is important for Copi is that an argument is an abstract object where some objective inferential link is maintained between premises, and between premises and conclusions. Informal Logicians deviate from this definition by emphasising the dialectic features of argument. Ralph Johnson, for instance, considers arguments (products) to be the "distillate" of the process of arguing, consisting of an "illative core" and a "dialectical tier" (Johnson, 2000, p. 168). The pragma-dialectic approach of Frans van Eeemeren and Rob Grootendorst differs slightly from Johnson in that they focus on the goal or aim of argumentation to resolve a difference of opinion, and, argumentation refers to both the product and process sense of the term (van Eemeren and Grootendorst, 2004, p. 2). In these characterizations of argument, we find little mention of the audience to which an argument is addressed.

Tindale forwards a dynamic conception of argument which includes both the abstract objects seen above, and the audience to which they are addressed. He conceives of arguments as having two movements. The first movement is between premises and conclusion. This first movement may include any of the conceptions of argument defined above. In this way, Tindale preserves the work of these argumentation theorists. However, essential to his account of argument is the second movement from argument to audience. This movement is recognition of the inherently social aspect of argumentation, as well as the fact that audiences shape arguments put forward by arguers (Tindale, 2015, p. 22). Insofar as the audience is taken into account, the rich set of concepts from rhetorical theory become fundamental to argumentation. Given the important role that the audience plays, we adopt an audience based conception of argument.

Therefore, in order to be an argument, a metaphor must meet two conditions. First, there must be a sense in which the metaphor aligns with the traditional definitions of argument above. For the sake of simplicity, we will take this as propositional content, or reasons, in support of a conclusion. Second, the metaphor must be delivered in such a way that it is considered an attempt to persuade an audience[2] of some conclusion or standpoint. We take our conception of metaphor as argument to be successful if it satisfies these two conditions.

3. METAPHOR IN PSYCHOLINGUISTICS

In psycholinguistics, metaphor comprehension is treated, as Black (1955) argued, as a result of topic-vehicle interaction (e.g., in a metaphor like lawyers are sharks, the topic is 'lawyers' and the vehicle is 'sharks'), where semantic properties relevant for understanding are enhanced and irrelevant properties are suppressed. Semantics can be modelled by representing words as vectors in a high-dimensional space, and their location in the space is determined by their co-occurrence in discourse (e.g., HAL, LSA, WINDSORS[3]). For example, words that appear together in discourse are closer in the space, and are therefore considered semantic neighbours. Kintsch's (2008) computational

---

[2] There are other interesting relationships between argument and audience that can be explored by psycholinguistics. For instance, the very nature of the psychological studies conducted in the field are audience-centric, since empirical results are measurements of phenomena within audiences.

[3] C.f. HAL - Lund and Burgess (1996), LSA - Kintsch (2001), WINDSORS - Durda and Buchanan (2008)

model, the predication algorithm, shows that metaphor comprehension depends on finding the appropriate semantic neighbours of the topic and vehicle. For example, computing the meaning of the metaphor *my lawyer is a shark* involves searching the semantic neighbourhood of *shark* for words which are related to *lawyer*. Such words, such as *vicious*, are strengthened whereas the unrelated words, such as *swim*, are inhibited. A vector situating the metaphor's meaning in the semantic space is computed based on the topic, vehicle, and the strengthened and inhibited semantic neighbours. The simulations of the model are intuitive and consistent with human interpretations of metaphors (Kintsch & Bowles, 2002).

Recent work has considered how the nature of the topic and vehicle's semantic representations may affect comprehension. For example, a word's semantic neighbours can be densely or sparsely distributed, a variable known as semantic neighbourhood density (SND). High SND words have many near semantic neighbours whereas low SND words have few near semantic neighbours. Moreover, words can be concrete, such as *dove* or abstract, such as *peace*. Al-Azary and Buchanan (2017) constructed metaphors where the topics and vehicles were both high or low SND words. In addition, the topic words were either abstract or concrete whereas the vehicles were all concrete. This resulted in four types of metaphors; (1) abstract high SND (e.g., *language is a bridge*); abstract low SND (e.g., *responsibility is a chain*); concrete high SND (e.g., *a mosquito is a vampire*); concrete low SND (e.g., *a pond is a mirror*). Participants rated the metaphors for comprehensibility in self-paced and speeded tasks. The result of both tasks was an interaction where the most comprehensible metaphors were the low SND, with no difference between the abstract and concrete counterparts. The next most comprehensible were the abstract high SND metaphors, and the least comprehensible were the concrete high SND metaphors.

Al-Azary and Buchanan (2017) reasoned that metaphor comprehension depends on fitting the topic word in the semantic neighbourhood of the vehicle. For effective comprehension, the semantic neighbourhood of the vehicle must have adequate space to accommodate the topic. When the semantic space is sparse, as is the case for low SND metaphors, there is enough room to assimilate the topic word, and fewer words need to be inhibited. In these cases, abstract or concrete words can both equally fit. This explains why low SND metaphors are equally comprehensible despite topic concreteness. However, when the semantic space is dense, there is less room to accommodate new words, and abstract words can fit better because they lack concrete features which may cause additional interference, and need to be inhibited. Conversely, concrete words have features that would make assimilation in a dense semantic neighbourhood the most difficult. Such metaphors are the least comprehensible, and require the most inhibition of unrelated semantic neighbours and features.

4. METAPHOR IN ARGUMENTATION – PROPOSITIONAL CONTENT

This section will outline the relationship between a single metaphor such as *my lawyer is a shark* and the propositional content associated with a metaphor. We look at treatments of metaphor and argumentation theory while keeping an eye towards psycholinguistics. We argue that a single metaphor implies propositional content beyond the written word such that it can be construed as an argument. To make this case we first show how metaphor has been treated in literature in argumentation. One prominent feature is the tie to analogy. We bridge the gap between metaphor and analogical argument, while at the same time showing that there is no necessary connection between the two. Next, we show the importance of psycholinguistic approach to metaphor to provide empirical study of the relations that metaphors highlight and diminish.

We began the paper with the rhetorical nature of Aristotle's conception of metaphor. A further point from Aristotle that has been carried through the literature in argumentation is metaphor's relation to analogy. From the Poetics, we find that Aristotle defines metaphor as a type of noun, one that, "is the application of a strange term either transferred from the genus and applied to the species or from the species and applied to the genus, or from one species to another, or else by analogy" (*Poetics* Chapter 21, 1457b7[4]). These are the four types of metaphor, according to Aristotle. As this section will focus on analogy,

---

[4] Loeb Translation. The Bywater translation is somewhat different: "Metaphor consists in giving the thing a name that belongs to something else; the transference being either from genus to species, or from species to genus, or from species to species, on the grounds of analogy" (*Poetics* Chapter 21 1457 b.7). Rather than four kinds of metaphor, this translation presents as if there are three types of metaphor, all grounded in analogy. Later translations are thought to more accurately capture Aristotle's theory of metaphor, especially since the idea that there are four kinds of metaphor squares with the Kennedy translation of the Rhetoric, where metaphor by analogy is the most important of the four kinds (*Rhetoric* 1410 b.).

we will only deal with that kind.[5] For Aristotle, metaphor by analogy is the most well liked (*Rhetoric* 1410 b). The following passage illustrates what he means by metaphor by analogy,

> Metaphor by analogy means this: when B is to A as D is to C, then instead of B the poet will say D and B instead of D. And sometimes they add that to which the term supplanted by the metaphor is relative. For instance, a cup is to Dionysus what a shield is to Ares; so he will call the cup 'Dionysus's shield' and the shield 'Ares' cup.' Or old age is to life as evening is to day; so he will call the evening 'day's old age' ... and old age he will call 'the evening of life' or 'life's setting sun.' (*Poetics* Chapter 21 1457 b1-30).

We see here that a metaphorical reference such as calling old age 'the evening of life', has an inherently analogical structure. That is, for Aristotle, the metaphor implies the analogy; its form is an enthymematic version of the scheme for analogy.

Chaim Perelman and Lucie Olbrechts-Tyteca have a similar conception of metaphor in *The New Rhetoric*. They follow Aristotle's conception of metaphor by analogy, "In the context of argumentation, at least, we cannot better describe a metaphor than by conceiving it as a condensed analogy, resulting from the fusion of an element from the phoros with an element from the theme" (Perelman and Olbrechts-Tyteca, 2006, p. 399). The idea of a metaphor as a condensed analogy is, clearly, not unique. However, Perelman and Olbrechts-Tyteca uniquely situate their view of metaphor within argumentation. In their conception, as the theory of analogy is developed, so too will the theory of metaphor be developed.

The link between metaphor and analogy is also taken up by Dedre Gentner et al. They represent an approach in the application of structure-mapping theory of analogy in the case of metaphor. However, the similarity between metaphor and analogy is controversial in psycholinguistics. In his paper "The psycholinguistics of metaphor", Sam Glucksberg (2003) challenges some of the main assumptions that lead theorists such as Gentner to relate metaphor and analogy. Gentner says that metaphoric meaning is derived from a literal representation of the topic and vehicle and meaning can be directly computed from structure-mapping. She assumes that the literal representation of the vehicle gives

---

[5] The paper "Aristotle's Theory of Metaphor" by Samuel R. Levin (1984) advocates for a more substantive view of the first three types than had traditionally been given.

rise to relations which are projected to the topic (Gentner et. al., 2001). Glucksberg thinks that dual reference is at play and the vehicle is not a literal representation but a metaphorical one. However, metaphorical processing is understood as a categorical assertion by the same process that a literal categorical assertion is understood. Indeed,

> When I say that '*my job is a jail*', in a sense I mean it literally. I do not mean that my job is merely like a jail, but that it actually is a member of the category of situations that are extremely unpleasant, confining and difficult to escape from (Glucksberg, 2003, p. 96).

Here we see that Gentner's theory assumes that the categorical comparative is 'jail', whereas Glucksberg re-frames the category to 'situations that are extremely unpleasant, confining, and difficult to escape from'. In this frame, the metaphor *my job is a jail* is processed the same way as a literal class inclusion, or, categorization statement. This constitutes a quantitative difference between other literal statements and the metaphorical statement rather than a qualitative difference between the two.

So, does this conception of metaphorical processing challenge the idea that metaphors are like analogies at the level of argumentation? Not necessarily. Just because we may not process metaphors analogically, it does not follow that metaphors are not compressed analogical arguments. The nature of metaphor in argumentation may still be analogical, as Aristotle and Perelman and Olbrechts-Tyteca suggest. This research in psycholinguistics also provides with evidence for another possibility, that metaphors are not necessarily compressed analogical arguments, but, compressed arguments in general. That is, a metaphor need not be compressed as an analogy, a metaphor can be compressed in different schematizations, or along different lines of argument, such as a class inclusion statement (i.e., X is a Y). Remember it is our main goal here to show that a single metaphor gives rise to propositional content such that it might be considered an argument. To do so we will first turn to Michael Leff, and then to Christian Santibanez to show that metaphors do unpack in this way.

In Michael Leff's essay "Topical Invention and Metaphorical Interaction", he challenges established conceptions of metaphor that dismiss its role in argument production (Leff, 2016, pp. 115 – 123). Leff argues that the assumption that metaphors are merely adornment, breaks down when we deal with metaphor in practice. To argue his point, Leff turns to an argument made by Loren Eisley. Eisley is arguing against the notion that artificial intelligence is significantly similar to human intelligence. Leff outlines the argument as follows:

A_1 – Machine intelligence is entirely rational
A_2 – Passion is the contrary of reason
A_3 – Therefore, machine intelligence lacks passion
B_1 – But human intelligence involves passion
B_2 – Therefore, human intelligence is not the same as machine intelligence (Leff, 2016, pp. 124-125).

This reconstruction is not the same, however, as the written argument, which depends largely on the following story to "fill in" proposition (partly) A_2 and all of B_2.

Eisley presents this point as a part of a narrative, where, in his youth, he had a job that required him to catch wild birds. One time, he caught a male and female hawk, but the male hawk attacked Eisley, and the female got away. The next day, feeling guilty that he had taken the hawk away from its partner, Eisley decided to release the hawk and both the male and female hawks exhibited behaviour that was remarkably similar to human behaviour, which is exhibited metaphorically (i.e., patiently waiting for the other) (Leff, 2016, p. 124).

This story involves an extended metaphor "by inviting us to view the hawks as human." (125). The whole case that Eisley makes against the similarities between human and machine intelligence relies on the metaphor involving human and animal passion. It is materially irrational for the female hawk to wait overnight and into the next day for her partner to return, but, she held out hope that he would, and knew right away when he had been freed. We see ourselves in the hawks, and realize that there is more to intelligence than rationality, we also rely on passion to guide us in our decision-making when it comes to our loved ones. There is no support in this particular argument for the proposition (B_1) that human reason involves passion without the metaphor of the hawk.

The point that we emphasize in this paper echoes Leff's point: metaphors fill in, or provide propositional content (missing premises) in an enthymematic way. That is, the propositions necessary to make an argument are implied or entailed by the metaphor. The metaphor itself is able to transfer literal propositional content in the form of a missing premise in an argument. Leff's example comes in the form of metaphorical language, or extended metaphor. So, we see how a metaphor can provide propositional content in this case, we make the further case that a nominal metaphor such as *my lawyer is a shark* implies or entails propositional content such that it can be considered an argument.

In his 2010 paper, "Metaphors and Argumentation: The Case of Chilean parliamentarian media participation", Christian Santibanez's 2010 account of metaphor in argumentation moves towards an account of metaphors as arguments. The example we will deal with comes from an example form Chilean politics:

> In all barrels of apples there are bad apples, but the point is, whether we think that all businessmen are shameless, or, on the contrary, we should worry about legislation so that those abuses are no longer committed (Santibanez, 2010, p. 985).

And his reconstruction:

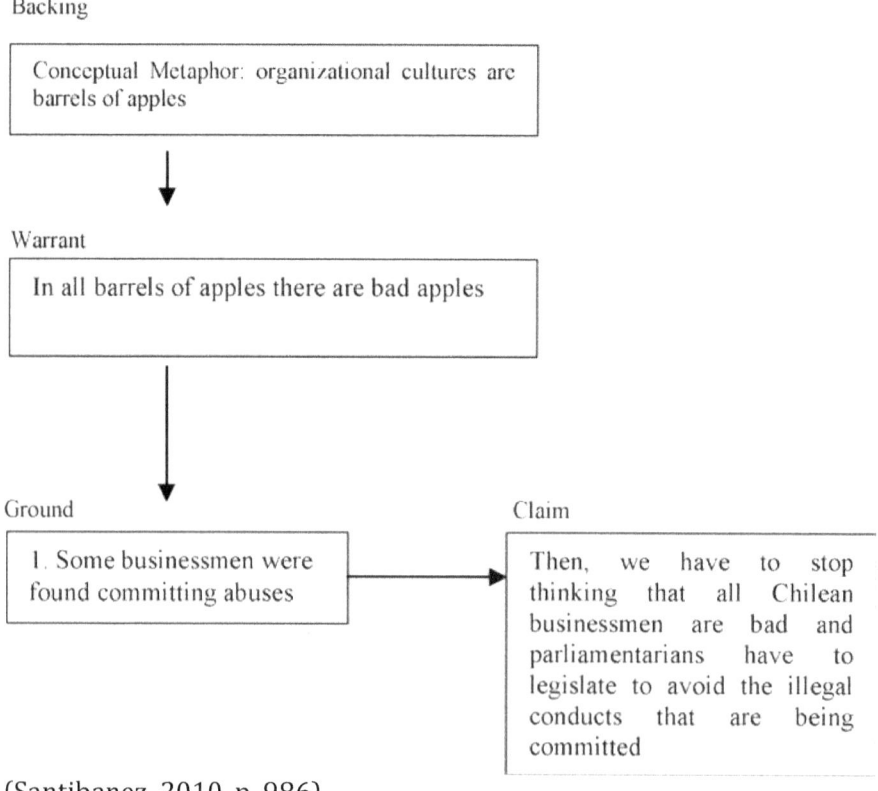

(Santibanez, 2010, p. 986)

Santibanez draws on research in cognitive linguistics to show how conceptual metaphors serve as backing in arguments. He uses a Toulmin model of argumentation to show how a conceptual metaphor such as *organizational structures are barrels of apples* is implied in a statement and serves as a backing to a more specific claim about the conduct or action of parliamentarians. The warrant for this claim is explicit, "in all barrels of apples, there are bad apples" (Santibanez, 2010, p. 985). As he notes,

> Because metaphors work as a cultural and social consistency (Lakoff and Johnson, 1980), actualized by mental procedures (Lakoff, 2006a), it could be suggested that, in terms of the metaphorical model previously described, the conceptual metaphor could be conceived as 'backing' when the warrant is reconstructed (Santibanez, 2010, p. 980).

When reconstructed in this way, the conceptual metaphor provides no further information, rather is implied as further support for the warrant. This reconstruction follows research in cognitive linguistics that assumes that metaphorical processing is not literal. That is, the literal meaning of the metaphor that links bad apples to businesses is a derivation of the metaphorical phrase. When viewed in this way, the reconstruction shows the role of metaphor in this case. Santibanez shows that there is implied or implicit propositional content in the use of a metaphor.

The reconstruction might differ, however, if we view the metaphor according to the literal processing theory we found in Glucksberg earlier.[6] The metaphor of companies and barrels of apples is seen as a class inclusion statement, rather than a metaphorical conceptualization. When viewed in this way, the metaphor *organizational structures are barrels of apples* serves an implicit warrant, and the explicit statement "in some barrels of apples there are bad apples" serves as the backing. The reconstruction is switched in this case because the metaphor itself provides the grounds for the relevance of the claim. In the reconstruction above we see that without the conceptual metaphor as backing, there is a disconnect between the warrant and the claim. When the metaphor is a class inclusion, the statement "in some barrels of apples there are bad apples" provides further support for the warrant *organizational structures are barrels of apples* by clarifying the class to which the topic and vehicle belong.

What the statement "in some barrels of apples there are bad apples" is doing, is clarifying the sense in which we should take the metaphor *organizational structures are barrels of apples*. As Walter

---

[6] For more of Glucksberg's account see Glucksberg and Keysar 1990).

Kintsch shows through his method of Latent Semantic Analysis (LSA) simple words such as *run* can have 30 or more senses in which they are used (Kintsch, 2001, p. 173-174). When we scale up to the case of metaphor, the possible senses which is intended by the use of a metaphor may be well served by clarification. The fact that there are different senses in which we use words is one of the motivating factors behind Glucksberg's defence of the literal theory of metaphor. He draws on American Sign Language (ASL) and a variety of Asian and Native American languages to show how these languages navigate meaning transfer without the use of metaphor. Furthermore, when we view metaphor as class inclusion statements, it explains the phenomenon of creating new, non-existent classes via the metaphor. For instance, in the case of *my job is a jail* we are pointing out that the topic and vehicle are members of the same class, one which is either 1) not straightforward, or 2) does not exist. The word "jail" loosely conveys this class, and the statement includes "my job" as part of that class as well. It is by linguistic features of languages such as English that compress large amounts of information into a single metaphor to represent class inclusion. While much of this information is implied, or given by the context, in natural language examples such as Santibanez, we see that different parts of a metaphor might be used to emphasize a point, or clarify the sense in which a metaphor is being used, without explicitly uttering the metaphor. What the example from Santibanez emphasizes is the condensed propositional nature of metaphor, and that different approaches to metaphor will have different explanations of that metaphor.

Another paper that makes the case that metaphors are arguments by Steve Oswald and Alain Rihs (2013) deals with the propositional content of extended metaphors. They draw on the concept of weak implicature to show the illocutionary and perlocutionary aspects of extended metaphors. The paper focuses on the perlocutionary aspect of extended metaphor, so, will figure prominently in the next section of this paper. For now, we want to focus on one of the claims made by Oswald and Rihs,

> Extended metaphors may therefore provide the grounds for rich inferential work *geared towards the derivation of specific conclusions.* We owe this possibility to the discursive nature of extended metaphors: the conclusions we draw from them "are cumulative, and, crucially, achieved by way of text and discourse processes, rather than sentence processes" (ibid.) (Oswald and Rihs, 2013, p. 140).

Whereas in Leff and Santibanez we saw extended metaphors give rise to propositions, here we see Oswald and Rihs consider the possibility that extended metaphors give rise to "specific conclusions". That is, in their construal of metaphor, the conceptual metaphor serves as the standpoint, and each extension of the metaphor in discourse is support for that standpoint (Oswald and Rihs, p. 143). The support they offer for this claim is grounded in the persuasive function of the metaphor, which we will see in the next section. Some combinations of linguistic features and context cause the hearer to draw particular inferences that lead to specific conclusions being drawn. Context plays an important part because it determines which inferences will be made, because, as Oswald and Rihs point out, "metaphors hardly ever have to be interpreted in neutral contexts" (2013, p. 137).

We contend that just as conceptual metaphors give rise to premises, or propositions, they also give rise to conclusions. That is, the use of a nominal metaphor can have the same effect in the case of a single conceptual metaphor as it does in the case of an extended metaphor, however, the hearer matches the metaphor to aspects of their cognitive environment in different ways than they would in an extended metaphor. There is much greater license on the part of the hearer in the case of a single conceptual metaphor that could cause the hearer to inferentially invoke premises and conclusions that relate to the metaphor in such a way that constitutes an argument. In this case, the premises and conclusions also follow from the linguistic features and context of metaphor, however they are less specific than in the case of extended metaphors. We call these "weak conclusions". We do not hear metaphors in isolation, but, we also do not process metaphors in isolation, we draw on our lived experiences, our cognitive environments, to process them.

We turn now to psycholinguistics, and a model to show how words relate to other words. Of course, any given word relates in a way to every other word. Some words relate more closely to others, or, are more likely to be associated with each other. Semantic neighbourhood density (SND) provides an algorithmic model of the relatedness of words. On Kintsch's (2001) model, the relatedness of words in calculated by computationally analysing a corpus of natural language ctext, and deriving a vector in "semantic space"[7]. The relatedness of

---

[7] This method draws on similar algorithms to the ones used in artificial neural networks, where representations are 'n-dimensional", that is, rather than merely (x,y,z) axes, any number are mathematically possible. Therefore, the representation is "muti-" or "n-" dimensional. Generally, we would say "n-

words is derived from the distance relations in this space. For instance, the most closely related words to "eye" are: "cornea", "retina" and "eyeball". The cluster of words that group around each other in semantic space, such as these, are called semantic neighbourhoods. There are other methods to calculate semantic neighbourhood density as well. For instance, the WINDSORS method (Durda and Buchanan, 2008). We need not get too far into the empirical metrics behind each method, we simply note that there are multiple methods for deriving semantic neighbourhood density, and these are highly valuable empirical tools for studying literal and metaphorical language processing.

For our purposes in this section, we point to the existence of SND as empirical evidence that we do, in practice, make many associations between words. Semantic neighbourhood densities have been used by psycholinguists to study the cognitive processing of metaphors, and predict which metaphors will be more or less likely to be accepted by study participants. We will discuss the predictive/prescriptive value of these studies in a later section. For now, we show that there are databases of empirically derived similarity vectors between words. To the extent that the topic and vehicle of metaphors have overlapping SND's, this metric shows which associations are reasonable for a person to make in the case of any given metaphor. So, while it is logically true that any word relates to any other word, sematic neighbourhood density provides a weighted average of how related words are, given the training corpus. In other terms, semantic neighbourhood density is a rough approximation of our shared cognitive environments.

That we share SND's, and that those SND's suggest what words we are most likely to associate, shows that we do bring to mind other sematic content in our everyday thought processes. Since we have such constrained associations in our mind, combined with contextual cues, we take it to be no stretch of the imagination to say that we do more than simple word association. We associate, construct, design, fill-in, full propositional content when processing a metaphor.

Ultimately, what is important to considering that metaphors imply the appropriate content such that they can be considered arguments is what is not included in a metaphor. That is, metaphors are effective by highlighting certain and particular relations between two things, while diminishing others. There are many ways in which lawyers

---

dimensional state-space", here, the "state" is determined, so it is called "sematic space".

are like sharks: They both eat, they both have a sense of touch, there are relatively few lawyers compared to people, there are relatively few sharks compared to fish, and many more. However, the metaphor diminishes these similarities because the semantic neighbourhood densities are conducive to some comparisons over others. Indeed, where two individuals do not share the relevant cognitive environments, a metaphor becomes completely ineffective. So, it is the background knowledge of the individuals, and the assumed shared cognitive environments that contribute to the success of metaphors.

Considering the examples from Leff, Santibanez and Oswald and Rihs together, along with work on semantic neighbourhood density, we find two different ways that metaphors unpack or decompress propositional content. In the second example, we find that switching our theory of metaphor from a cognitive approach, to a view from psycholinguistics, changes the reconstruction and importance of the statements in argumentation. What is also clear from these examples is that metaphors need not be condensed analogies. A metaphor can be a different kind of scheme altogether. Therefore, we find that metaphors do unpack, or decompress in an appropriate way such that we can say that they provide the propositional content necessary to be considered an argument.

5. METAPHOR IN ARGUMENTATION – PERSUASIVE FORCE AND PREDICTIVE POWER

This next section is aimed at fulfilling the second aspect of Tindale's conception of argument. The second criteria is the goal of persuading an audience. By way of research in argumentation and psycholinguistics, we show that this is indeed a goal that can be attributed to use of metaphor.[8] To do so we first draw on Oswald and Rihs to show that metaphors have an argumentative function, then show how this function can be exploited by drawing form research in psycholinguistics.

As we saw in the previous section, Oswald and Rihs support the idea that metaphors are arguments by grounding their point in the persuasional function of arguments. They make the claim that we can consider extended metaphors as arguments (where the metaphor serves as the standpoint) because each instantiation of the metaphor

---

[8] There will be an inherently circular, or tautological, aspect to this section. Something like: When a metaphor is uttered with the goal of persuading and audience, then metaphors have a goal of persuading an audience. However, to avoid this we will focus on the function of the metaphor.

strengthens the audience adherence to the standpoint. When viewed in this way, the argumentative function of the metaphor is revealed in the use of metaphor in text (Oswald and Rihs, p. 141-142). Cognitively, they claim, these extended metaphors also have rhetorical functions of framing the ethos of the speaker in the minds of the audience, and reinforces the concepts being carried by the metaphor, increasing their epistemic strength (Oswald and Rihs, p. 143).

So, insofar as arguments are put forward by an agent with the intention to convince an audience of a standpoint, we can view extended metaphors as arguments. Oswald and Rihs focus on the cognitive dimension of metaphor to make this claim, drawing on relevance theory to reinforce Perelman and Olbrechts-Tyteca's insight that multiple instantiations of extended metaphor increase audience adherence. This approach also outlines metaphors as arguments in terms of the aim or goal of the arguer, that is, the goal of persuasion. We take a similar approach to metaphors as arguments. The arguer utters with the intent to persuade, so this is an appropriate indicator that an argument is being made. If we begin with the attempt to persuade, then we can take a broad scope of what counts as an argument.

Oswald and Rihs, and Santibanez make very interesting and insightful cases for considering an extended metaphor as an argument. In the previous section, we argued that a nominal metaphor contains unstated or implied content, such that we can attribute to it a set of premises and conclusion, thereby making the metaphor itself an argument. We do not contend that metaphors are always arguments, or that their only function is argumentative. In the right context and uttered *with the intent to persuade*, metaphors can be considered arguments. This is true of a single conceptual metaphor, and in reinforced by work in psycholinguistics that shows that people are more likely to accept metaphors where the topic and vehicle have low semantic neighbourhood densities.

To make their case, Oswald and Rihs rely on research in cognitive linguistics that shows that processing conventional and novel metaphors is of the same kind. They use the example "Jefferey is a clown". They point out that, as a conventional metaphor, we might take *Jeffery is a clown* to be mean that he is a class clown, a joker, someone who is a funny person. In a different context, say, a shoe store, the same metaphor might be taken to mean that Jeffery has big feet. Since the context can change the nature of the metaphor, from conventional to novel, they argue that there is no difference in processing between the two types of metaphor (2014). We go a step further and follow Glucksberg (1993, 2001) and Kintsch (2001) in saying that metaphorical processing is not different in kind than literal processing. Oswald and Rihs point out that, if "Jeffery is a clown" were uttered in a shoe store, it would imply different meaning. It is also true that if the context was changed to a circus, the utterance would lose its metaphoricity; the statement would be literal. The point of approaches to metaphor such as Glucksberg, Kintsch, and Al-Azary and Buchanan is that the processing is the same in each case. All of the associations made in each different case are possible in any one of the given cases. This example shows that the difference between literal and metaphorical is a matter of degree, not of kind. A speaker may exploit this feature of metaphor to a persuasional end, thereby making an argument.

The work by Al-Azary and Buchanan shows the propositional nature, or possible decompression, of metaphor, but, also shows different strategies for an argumentative illocution of a metaphor to persuade. While Oswald and Rihs clearly agree with the persuasive function of metaphor, their argument deals with extended metaphors. We show how single nominal metaphors can unpack and have the same argumentative force, to be aimed at persuasion.

We now see that nominal metaphors can be considered arguments. Not only do they imply the propositional content needed to be considered an argument, they also have a persuasional aspect to them. These two features together fulfil the definition of argument that we outlined at the beginning of the paper. Based on a rhetorical model of argument taken from Tindale, we that metaphors can indeed be considered arguments.

Not only does a psycholinguistic approach to metaphor in argumentation provide for a way to show how and what features of the topic and vehicle metaphors highlight, it also provides a basis for empirical study of metaphors in argumentation. Given a conception of argument that takes audience into account and to the extent that a persuasive argument is a function of the cognitive processing of individuals, SND helps to identify good arguments. And, insofar as we can identify good arguments, SND provides a prescriptive model for constructing good metaphor as argument. A good metaphor is one that is more likely to be accepted by the audience, therefore, good metaphorical argument is one that makes use of a topic/vehicle with low SND, rather than high SND. We also discuss the role of abstractness vs. concreteness in this regard. This is an added benefit of meshing work on semantic neighbourhood density with argumentation theory, and could be a fruitful source for further empirical studies at the intersection of argumentation and psycholinguistics.

6. CONCLUSION

So, assuming the literal processing of metaphor, and taking the fact that the relation between words has a calculable measure in semantic neighbourhood density, we find that metaphors are invitation to inference of the kind consistent with arguments. We deliberately do not take a stand on which would be the proper ontology of argument, instead, we claim that different reconstructions are possible, depending on the situation. Furthermore, research in Semantic Neighborhood Density shows that metaphors come in degrees of acceptability, depending on certain specific features (i.e., abstract, concrete, high-SND, low-SND etc.). This suggests that metaphors serve a persuasive function as well. Therefore, metaphors meet the conditions necessary to be considered arguments in their own right. What is new on this theory is that research on semantic neighbourhood density provides us with an empirically verifiable study on how persuasive a metaphor is. Based on a view of arguments that takes audience into account, and to the extent that the effectiveness of an argument is a function of the cognitive processing of that argument, SND provides data that shows how we might guide the construction of an argument to be more persuasive. That is, the cognitive processing of an argument plays a role in accepting an argument, and how effective a metaphor is quantifiable.

ACKNOWLEDGEMENTS: Partially supported by SSHRC Partnership Grant.

REFERENCES

Al-Azary, H., & Buchanan, L. (2017). Novel metaphor comprehension: Semantic neighbourhood density interacts with concreteness. *Memory & Cognition*, *41*(2), 1-12.
Black, M. (1955). XII.—Metaphor. *Proceedings of the Aristotelian Society*, 55(1), pp. 273-294.
Copi, I. M. (1967). *Symbolic logic*. New York: The Macmillan Company.
Durda, K., & Buchanan, L. (2008). WINDSOR: Windsor improved norms of distance and similarity of representations of semantics. *Behavior Research Methods*, *40*(3), 705-712.
Eemeren, F., & Grootendorst, R. (2004). *A systematic theory of argumentation* (1st ed.). Cambridge: Cambridge University Press.
Gentner, D., Wolff, B. & Boronat, C. (2001).Metaphor is like analogy. In D. Gentner, K. J. Holyoak, & B. N. Kokinov (Eds.), *The analogical mind:*

Perspectives from cognitive science, (pp. 199-253). Cambridge, MA: MIT Press.
Glucksberg, S., & Keysar, B. (1990). Understanding metaphorical comparisons: Beyond similarity. *Psychological review,97*(1), 3-18.
Glucksberg, S. (2003). The psycholinguistics of metaphor. *Trends in cognitive sciences*, 7(2), 92-96.
Johnson, R. (2012) *Manifest rationality: A pragmatic theory of argument.* New York: Routledge.
Kennedy, G. A. (2006). On rhetoric: A theory of civic discourse. Oxford: Oxford University Press.
Kintsch, W. (2001). Predication. *Cognitive science*, 25(2), 173-202.
Kintsch, W., & Bowles, A. R. (2002). Metaphor comprehension: What makes a metaphor difficult to understand?.*Metaphor and symbol*, 17(4), 249-262.
Kintsch, W. (2008). How the mind computes the meaning of metaphor. In R. W. Gibbs Jr. (Ed.), *The Cambridge handbook of metaphor and thought*, (pp. 129-142). Cambridge: Cambridge University Press.
Halliwell, S. (1987). *The poetics of Aristotle: Translation and commentary.* Chapel Hill, NC: University of North Carolina Press Books.
Lakoff, G. (2006). *Whose freedom? The battle over America's most important idea.* New York: Farrar, Straus and Giroux.
Leff, M. (1983). Topical invention and metaphoric interaction. *Southern Journal of Communication*, 48(3), 214-229.
Levin, S. R. (1982). Aristotle's theory of metaphor. *Philosophy & Rhetoric*, 24-46.
Lund, K., & Burgess, C. (1996). Producing high-dimensional semantic spaces from lexical co-occurrence. *Behavior Research Methods, Instruments, & Computers*, 28(2), 203-208.
Oswald, S., & Rihs, A. (2014). Metaphor as argument: Rhetorical and epistemic advantages of extended metaphors. *Argumentation*, 28(2), 133-159.
Perelman, C., & Olbrechts-Tyteca, L. (2006). *The New Rhetoric: A treatise on argumentation* (J. Wilkinson & P. Weaver, Trans.). Notre Dame, IN: University of Notre Dame Press. (Originally published 1971)
Santibáñez, C. (2010). Metaphors and argumentation: The case of Chilean parliamentarian media participation. *Journal of Pragmatics*, 42(4), 973-989.
Velasco, A., Campbell, J., Henry, D. (Eds.) (2016). *Rethinking rhetorical theory, criticism, and pedagogy: The living art of Michael C. Leff.* East Lansing, MI: Michigan State University Press.
Tindale, C. W. (1999). *Acts of arguing: A rhetorical model of argument.* Albany, NY: State University of New York Press.
Tindale, C. W. (2015). *The philosophy of argument and audience reception.* Cambridge: Cambridge University Press.

30

Justifying a Bill Before Parliament: Beyond Instrumental Rationality

CONSTANZA IHNEN JORY
Institute of Argumentation, Law Faculty, University of Chile
cihnen@derecho.uchile.cl

How can legislators rationally justify or criticize the ends pursued by a bill? By means of which argument scheme? The aim of the paper *is to begin an inventory of goal-(de)legitimising arguments schemes used in law-making practices.* The schemes included in the inventory thus far are: from consequences, from model and anti-model, value-based, and from social demand.

KEYWORDS: instrumental rationality, legislative debates, maximalism, minimalism, argument from consequences, argument from model, value-based argument, argument from social demand

1. INTRODUCTION

A critical and long-standing debate in legal theory concerns the possibility of rationally justifying legislation. The debate is critical because scepticism on the capacity of the legislator to decide rationally can have an effect not only on the theory and practice of adjudication,[1] but also on the theory and practice of legislation: the higher the levels of

---

[1] According to García Amado (2000, p. 305), historically, the demystification of the rational legislator is generally correlated to a Herculean view of the judge, who must decide the case at hand by amending the mistakes (i.e., the irrationality) of the legislator. He draws this conclusion from the history of legal theory, but from a logical perspective, the demystification of the legislator does not entail the mystification of the judge. In fact, as García Amado himself points out, it would be perfectly consistent to hold simultaneously the irrationality of the legislator and the irrationality of the judge. For a more detailed exposition of this view, see García Amado, 1988.

scepticism, the weaker the mechanisms developed to control the law-making activity.

Positions within this debate are quite varied. Those that recognise at least some form of rationality in legislation fall under two main approaches: the "minimalist" and the "maximalist" approach (Marcilla Córdoba, 2013, p. 58).[2] Minimalism is premised on the meta-ethical view that evaluative and normative statements cannot be rationally justified. This premise is usually accompanied by an optimistic view of our capacity to determine the truth or falsity of analytic and factual statements. Accordingly, for minimalists, the rationality of a bill is based on the degree to which the bill optimises the means to achieve the ends defined by the legislator, whichever those ends are. The legislator's selection of social ends is a subjective matter, not open to rational enquiry. Minimalists will therefore strive to solve technical-juridical problems, such as the adequacy of the linguistic means used, the adequate integration of the new law into the legal system, the likeliness that the new law will be complied with in practice, and the probability that it will indeed achieve the social ends for which it was created. For minimalists, then, legislative rationality is simply instrumental rationality. Kelsen (1945/2007), Wroblèwski (1979) and Bulygin (1991), exemplify this view (García Amado, 2000, p. 301; Marcilla Córdoba, 2013, p. 58).

By contrast, maximalists start from the meta-ethical view that evaluative and normative statements can be rationally justified. They accordingly see legislative rationality as depending not only on the effectiveness and efficiency of a bill, but also on the validity of the legislative purposes it seeks to achieve. The conception of rationality underlying this view is broader. Karpen (1986), Habermas (1992a/1996), Atienza (1997), García Amado (2000) and Marcilla (2013) exemplify this view (García Amado, 2000, p. 309-17; Marcilla Córdoba, 2013).

How can normative statements, legal or otherwise, be rationally justified? A prominent answer to this question is found in Habermas' discourse principle.[3] According to this principle, a normative statement

---

[2] García Amado (2000) distinguishes between "non-normative (or weakly normative)" and "normative" theories of the rationality of legislation, and Atienza (1997) between "instrumental, technical or weak rationality" and "rationality of ends, ethical, or strong rationality". These distinctions broadly correspond to the "minimalist" and "maximalist" distinction.

[3] Habermas offers a procedural answer to the question on how to determine the rationality of normative statements. This answer contrasts with a material

is justified if all possibly affected persons could agree on its validity as participants in a rational discourse (Habermas, 1992a/1996, p. 107). The notion of rational discourse stands for an argumentative discussion that develops in accordance with the requirements of the ideal speech situation. Those requirements demand that participants have equal rights to participate in the discussion and that they are free from coercion and manipulation. To use Habermas' famous formulation, in an ideal speech situation, deliberation should be characterized by the absence of interfering pressures except for "the forceless force of the better argument" (Habermas, 2005, p. 384). Habermas himself (1992a/1996), and legal scholars such as García Amado (2000), see in the ideal speech situation a counterfactual model that can be used to compare and critically assess real law-making practices and institutions.

Thus, the central question to be answered by a maximalist theory inspired by Habermas' discourse principle is: *In what way should institutions and legislative discourse be articulated to achieve rational outcomes*? One way of answering this question is to identify the constitutional and legal mechanisms that can promote high-quality discussions in the law-making process and thereby improve the rationality of legislation. Habermas (1992a/1996); García Amado (2000); and Steiner et al (2004) exemplify this approach.[4] A complementary strategy is to move from the articulation of *institutional preconditions* for rational legislative discourse to the articulation of *discursive conditions* for rational legislative discourse. Atienza (1997, 2005), Oliver-Lalana (2008, 2013), and Fairclough & Fairclough (2012), among others, have made significant contributions to the study of law-making and political discourse.[5] Yet despite providing a general framework to analyse and critically assess the argumentative quality of

---

approach to this question. Such an approach would establish direct limits to the possible contents of (legislative) norms, limits which would be claimed to emanate from reason. From this point of view, there would be certain legal norms that are prohibited or even necessary a priori (García Amado, 2000, p. 311).

[4] Habermas and García Amado concentrate on the institutional mechanisms that can guarantee equal participation for all those affected in the political and law-making process. The empirical study carried out by Steiner et al compares how law-making deliberation takes place under different types of democratic governments: consensus vs competitive democracy, strong vs weak veto powers, presidential vs parliamentary stems, etc.

[5] I do not claim that these authors adhere to a maximalist approach to legislation, but that their work contributes (knowingly or unknowingly) to a maximalist, discourse based, research programme.

legislative and political discussions, none of them has attempted to tackle the central problem of a maximalist theory of legislation: How do we evaluate argumentation used to justify and criticise the ends pursued by a bill?[6] In order to answer this question, it is necessary to identify the type of arguments that are conventionally used to discuss the legitimacy of legislative goals.

The aim of this paper is to begin an inventory of goal-(de)legitimising argument schemes used in law-making practices. "Argument schemes" are argument forms that represent conventionalised inferential structures of arguments (Walton & Reed, 2002). They can be deductive or presumptive. Deductive schemes guarantee the acceptability of the conclusion whenever the premises are acceptable; presumptive schemes do not guarantee the acceptability of the conclusion, but make it plausible: they create a "presumption" in favour of the acceptability of the conclusion. Because this presumption is reversed when there is evidence to the contrary, presumptive schemes should be carefully and critically examined. This is the reasons why argumentation theorists, together with identifying and representing the internal structure of presumptive argument schemes, specify for each scheme a set of relevant critical questions. The argument schemes I have included in my preliminary inventory are presumptive in nature. They are weak forms of argument taken by themselves, but can be crucial, on a balance of considerations of which they are part. I shall concentrate on the representation of their internal structure and leave the formulation of the relevant critical questions for a future paper.

2. PRESUMPTIVE SCHEMES AND THE METHOD OF RATIONAL RECONSTRUCTION

In the literature on presumptive argument schemes (e.g., Perelman & Olbrechts-Tyteca, 1958/2000; Hastings, 1963; Walton, 1996; Walton, Reed, & Macagno, 2008), it is often left implicit the method by which the schemes are "identified" as conventionalised inference structures. To

---

[6] Even though these authors agree that instrumental argumentation is insufficient to justify a normative (legal or political) claim, and that the justification of the goals pursued by a course of action (legal or political) is necessary, they do not offer enough guidance as to how to establish the acceptability of those goals. Fairclough and Fairclough go a step further when they propose that goals should be justified by reference to values. However, such observation does not take us far enough.

demonstrate their conventionality, scholars normally proceed by first proposing a scheme and then giving an example, or a couple of examples, of arguments instantiating the scheme. The reverse order is also possible. Of course, the examples *per se* do not fully demonstrate the conventionality of the schemes, their generalised or contextually-bound acceptance. The proof of their conventionality is expected to be completed by the reader, who is supposed to recognise the instantiation of the scheme in many other cases. Such an approach to the identification of schemes is, I believe, akin to what Habermas has labelled the "method of rational reconstruction": a method which involves making explicit and theoretically systematising "the intuitive, pre-theoretical knowledge of competently speaking, acting and judging subjects" (Habermas, 1992b).

To begin the inventory of goal-(de)legitimising schemes I shall use the method of rational reconstruction. Examples provided to illustrate the instantiation of the schemes are taken from Chilean law-making debates over a bill introduced by the government of Sebastián Piñera in 2011 (HL Nº 20.634). The debates took place the 22nd of December of 2011, in the Education Committee of the Chamber of Deputies, and the 23rd of January of 2012, in the plenary session of the same chamber. The general aim of the bill was to reform a state-guaranteed student loan program for higher education known as "CAE" ("Crédito con Aval del Estado"). The specific objectives of the bill were, first, to reduce the interest rate for students under the CAE loan program from 6% to 2%; second, to broaden the range of financial institutions authorised to provide CAE credits to students (until then, only banks were authorised); and, third, to set the level of debt payments at 10% of the graduate income (up till then, debts were not contingent upon a graduate's income).[7] The bill was introduced a few months after an unprecedented wave of social protest, against the government and the state of the educational system, and in which students played a leading role.

3. GOAL-(DE)LEGITIMISING SCHEMES IN LEGISLATIVE DISCUSSIONS

To identify goal-(de)legitimising schemes we first need to define the type of standpoint these schemes are supposed to justify. I propose to formulate that standpoint as a normative statement with the form 'Goal G of the bill should (not) be brought about'. I will take "goal" to mean 'a

---

[7] The third measure was not originally in the bill; it was introduced in the Treasury Committee debate which preceded the debates analysed in this paper.

future state of affairs desired by an agent'. Future state of affairs include, on my account, circumstances that may or may not involve agency. This means that future state of affairs include actions to be realised in the future.

The goals *of a bill* are normally mentioned explicitly in legal memoranda, explanatory notes, ministerial speeches accompanying the introduction of a bill or amendments agreed upon later in the legislative process. Law-making discussions sometimes centre on what the (actual) goals of a bill are. These discussions, however, turn on a different type of standpoint – a factual standpoint – which I will not be considering here: 'Goal G is (not) the goal of the bill'.[8]

Goal-statements are sometimes formulated as evaluative statements with the form: 'Goal G is good (bad)' (see, e.g., Walton, Reed, & Macagno, 2008, p. 325). For a bill to be legitimate, however, its goals should not only be "good" or desirable (in the sense of 'worth being desired'). Legislative goals should also be a social priority and they should be feasible to realise. Both these requirements are presupposed by the normative formulation of the standpoint 'Goal G of the bill should (not) be brought about'.

Which schemes, then, can provide presumptive support for such type of standpoint? At least four schemes: the argument scheme from consequences, the value-based scheme, the model (and anti-model) scheme, and the social demand scheme.

3.1 Scheme from consequences

One way of justifying and refuting the disputed legitimacy of the goals of a bill is by referring to the practical consequences of bringing them about. This form of argumentation is based on the *argument scheme from consequences* (Walton, 1995).

The scheme has a positive and a negative variant. In law-making debates, the positive variant can be used to justify the legitimacy of a bill's goal by mentioning the desirable consequences of bringing about such goal. Consequences should be desirable to all those affected by the bill or desirable to some specific group of those affected, but not undesirable to the rest.

---

[8] Cf. Walton, Reed, & Macagno (2008, pp. 323-325), who represent the goal premise of several schemes –the practical inference, the value-based practical reasoning, and the argument scheme from goal– with the factual statement 'I/person P have/has goal G'. A similar formulation of the goal premise of practical argumentation can be found in Fairclough & Fairclough, (2012, p. 45).

> 1 Goal G of the bill should be brought about
> 1.1a Bringing about goal G leads to consequence C
> 1.1b Consequence C is desirable to all those affected by the bill or desirable to some specific group of those affected, but not undesirable to the rest
> 1.1a-b' In principle, one should bring about those legislative goals which lead to consequences desirable to all those affected by the bill, or desirable to some specific group of those affected, but not undesirable to the rest

Example 1 illustrates the application of the positive variant of the scheme. The argument in the example was presented by the government in the bill's legislative memorandum and quoted by congressman Carlos Montes in the plenary debate of the bill (HL Nº 20.634, p. 126):

> (1) 1 The CAE funding scheme should be continued
> 1.1a CAE has been a very important instrument for expanding higher education coverage in lower income segments
> 1.1b (Expanding higher education coverage in lower income segments is desirable)
> 1.1a-b' (In principle, one should bring about those legislative goals which lead to consequences desirable to all those affected by the bill, or desirable to some specific group of those affected, but not undesirable to the rest)

In the negative variant of the scheme from consequences, the legitimacy of a bill's goal is attacked by referring to the undesirable consequences of bringing about such goal:

> 1 Goal G of the bill should not be brought about
> 1.1a Bringing about goal G has consequence C
> 1.1b Consequence C is undesirable at least for some specific group within society
> 1.1a-b' In principle, one should not bring about legislative goals which have undesirable consequences at least for some specific group within society

In example, 2, congressman Gabriel Silber provides an argument based on the scheme from negative consequences (HL Nº 20.634, p.130):

(2) 1 Insurance companies should not be authorised to provide credit to students
    1.1a These actions may lead to greater deregulation
    1.1b (Greater deregulation is undesirable)
    1.1a-b' (In principle, we should not bring about legislative goals which have undesirable consequences at least for some specific group within society)

3.2 Value-based scheme

Another way of justifying or disputing the legitimacy of the goals of a bill is by reference to some set of values.[9] The argument scheme underlying this form of argumentation has been labelled "value-based" scheme (Walton, Reed, & Macagno, 2008). As in the previous case, the scheme has positive and negative variants.

In the positive variant, it is claimed that a legislative goal is legitimate on the basis that the goal instantiates a set of assumedly positive values. Values should be positive to all those affected by the bill or positive to some specific group of those affected, but not negative to the rest. The scheme can be represented as follows (adapted from Walton, Reed, & Macagno, 2008, p. 321):

1 Goal G of the bill should be brought about
    1.1a Bringing about goal G instantiates set of values V
    1.1b Set of values V is positive to all those affected by the bill or positive to some specific group of those affected, but not negative to the rest
    1.1a-b' In principle, if a legislative goal instantiates a set of positive values to all those affected by the bill or positive to some specific group of those affected, but not negative to the rest, then the goal should be brought about

In example 3, congresswoman María José Hoffman provides an argumentation based on the positive variant of the value-based scheme (HL Nº 20.634, p.146):

---

[9] A set of values may contain only one value.

(3) 1 We should enable students under the CAE scheme to pay the same interest rate as students who are under the Solidarity Fund Scheme (2%)
  1.1a This is only fair
  1.1b (Fairness is positive)
  1.1a-b' (In principle, if a legislative goal instantiates a positive value to all those affected by the bill, or positive to some specific group of those affected, but not negative to the rest, then the goal should be brought about)

According to Walton, Reed & Macagno (2008), in the negative variant of the value-based scheme, a speaker retracts her commitment towards a certain goal because it instantiates a value that the same agent evaluates negatively. A contextually adapted version of the negative variant of the scheme can be represented as follows:

1 Goal G of the bill should not be brought about
  1.1a Bringing about goal G instantiates set of values V
  1.1b Set of values V is negative, at least to some group of those affected by the bill
  1.1a-b' In principle, if a legislative goal instantiates a set of negative values at least to some group of those affected by the bill, then the goal should not be brought about

On the surface, the notion of a "negative value" might seem an oxymoron. However, the notion is not an oxymoron if we consider that: (1) values are not always universal; (2) the agent who puts forward the argument judges "negatively" some abstract concept which another agent regards as a value. Thus, an agent can hold, for example that "nationalism is a negative value", meaning 'it is a value for some, but not for me'.[10]

In the debate over higher education, participants did not present argumentation based on the negative variant as defined by Walton, Reed and Macagno (2008). However, they did present arguments which

---

[10] Nevertheless, as Perelman & Olbrechts-Tyteca (1958/2000, p. 81) observed, most of our differences of opinion are not about whether some value is positive or negative, but about value hierarchies in concrete circumstances. Values such as freedom, fairness, equality, for example, are generally regarded positive values; differences arise when we need to give priority to one value over another.

were based on some negative variant of the scheme. In this negative variant, an arguer claims that a goal should not be brought about because it is inconsistent with a set of presumably positive values, at least to some group of those affected by the bill under discussion:

> 1   Goal G of the bill should not be brought about
> 1.1a   Bringing about goal G is inconsistent with set of values V
> 1.1b   Set of values V is positive to at least some specific group of those affected by the bill
> 1.1a-b'   In principle, if a goal is inconsistent with a set of positive values to at least some specific group of those affected by the bill, then the goal should not be brought about

This second type of negative variant underlies the argument presented by congressman Sergio Aguiló, as can be seen in example 4 (HL Nº 20.634, p. 133):

> (4)   1   We should not reform the CAE funding scheme, but put an end to it
> 1.1a   The CAE funding scheme is unfair
> 1.1b   (Fairness is a positive value)
> 1.1a-b'   (In principle, if a goal is inconsistent with a set of positive values to at least some specific group of those affected by the bill, then the goal should not be brought about)

3.3 Model and anti-model schemes

Legislators can also defend and attack the legitimacy of the goals of a bill by reference to models. Perelman & Olbrechts-Tyteca (1958/2000, pp. 362-368) labelled the scheme underlying this form of argumentation the argument scheme from "model" and they identified, together with a positive variant, a negative variant which they referred to as the "anti-model" scheme. (See also Walton, Reed, & Macagno, 2008, p. 315).

    Perelman & Olbrechts-Tyteca define the "model" appealed to in the positive variant as "persons or groups whose prestige confers added value on their acts" (1958/2000, p. 363). In law-making debates, models usually refer to presumably successful nations or presumably prestigious international organisations. In developing countries such as Chile, for example, the models referred to in legislative debates are typically developed nations –Scandinavian countries, in particular– and international organisations, such as the OECD. In principle, legislators could also see a model in successful local government. The positive variant of the scheme from model can be represented as follows:

1. Goal G of the bill should be brought about
   1.1 Goal G has been brought about by an agent A, different from this legislature
   1.1' Agent A is a model for this legislature with respect to some variable V relevant to the case (e.g., economic, social, political or cultural development)

The anti-model scheme, as applied to law-making contexts, refers to some national, international or local government institutions which is considered unsuccessful or not prestigious. In Chile, right-wing politicians commonly refer to the governments of presidents Evo Morales in Bolivia and of Hugo Chávez and Nicolás Maduro in Venezuela as anti-models. This negative variant can be represented as follows:

1. Goal G of the bill should not be brought about
   1.1 Goal G has been brought about by an agent A, different from this legislature
   1.1' Agent A is an anti-model for this legislature with respect to some variable V relevant to the case (e.g., economic, social, political or cultural development)

In the debate over higher education that I have been analysing, lawmakers did not put forward anti-model argumentation. Congressman Gabriel Silver, however, presented an argument which referred to a presumed positive model to criticise the legitimacy of a goal of the bill. This second type of negative variant of the model scheme has the following form:

1. Goal G_1 of the bill should not be brought about
   1.1 Goal G_1 is inconsistent with goal G_2, brought about by an agent A, different from this legislature
   1.1' Agent A is a model for this legislature with respect to some variable V relevant to the case (e.g., economic, social or cultural development)

Hence, in this variant, it is claimed that a goal should not be brought about because the goal is inconsistent with the goal brought about by some model. Example 5 presents a reconstruction of congressman Silber's argument (HL Nº 20.634, p. 131):

(5)   1   Students should not pay 10% of their income when repaying their loan
       1.1   In most OECD countries, students repay around 5% of their income
       1.1'   (OECD countries are a model for this country in relation to the financing of higher education)

3.4 Social demand scheme

Finally, the goals of a bill can be justified and criticised by a vast array of schemes referring to the goals of the represented. The force of these schemes hinges on the democratic principle that legislators ought to represent the citizens' interests. One such scheme would justify or criticise the goal of the bill by reference to the majority's interests. Such scheme resembles the *argument from popular opinion* studied by Walton, Reed and Macagno (p. 311). I would like to focus, however, in another scheme within the array of schemes referring to the goals of the represented and which I shall label the "social demand scheme". The scheme differs from an argument from popular opinion in at least two respects: the scheme involves the performance of a specific speech act – a demand– and the speaker who performs the demand is a social group which is not necessarily a majority. In fact, the social group performing the demand could be a minority group. The group is, essentially, a political stakeholder: a group which has a particular interest or stake in the legislation under discussion.[11] As in the previous cases, the argument is weak unless it is complemented with other arguments (for example, the argument that the demand of the group is consistent with bringing about the goals of the majority).

The social demand scheme has two variants. In the positive variant, the validity of the bill's goal is justified by showing that the goal meets a social demand:

---

[11] In this sense, the scheme resembles the *position-to-know ad populum* argument studied by Walton, Reed and Macagno (2008, p. 311).

1. Goal G of the bill should be brought about
   1.1 Goal G meets social demand D of political stakeholder S[12]
   1.1' In principle, goals that meet social demands of political stakeholders should be brought about

In example 6, congresswoman María José Hoffmann uses the positive variant of the social demand scheme:[13]

(6) 1 We should make it easier for students to repay their loans under the CAE programme
    1.1 Making easier for students to repay their loan meets their demands
    1.1' (In principle, goals that meet social demands of political stakeholders should be brought about)

In the negative variant, it is argued that a goal of the bill should not be brought about because the goal is inconsistent with a social demand:

1. Goal G of the bill should not be brought about
   1.1 Goal G is inconsistent with social demand D of political stakeholder S
   1.1' In principle, goals which are inconsistent with the social demands of political stakeholders should not be brought about

In example 7, CONFECH (Confederation of Chilean Students) uses the negative variant of the social demand scheme:

(7) 1 We should not reform the CAE funding scheme, but put an end to it
    1.1 CAE programme is incompatible with students' demands

---

[12] Note that a political stakeholder may claim to represent not only a specific social group but also citizens at large. In that case, there could be two different types of arguments at stake: the argument from social demand and the argument from the majority's interest.

[13] In this discussion, the argument referring to the students' demand was on various occasions accompanied by the argument that it was not only a demand of the students but of Chilean citizens' at large. In this way, the argument from social demand and the argument from the majority's interest were combined, and strengthened each other.

1.1.1	Students have demanded free education for all
1.1.1'	(A loan scheme, such as CAE, is incompatible with a system that provides free education for all)
1.1'	(In principle, goals which are inconsistent with the social demands of political stakeholders should not be brought about)

4. CONCLUSIONS

My aim in this paper has been to begin an inventory of argument schemes that can be used to rationally justify or dispute the validity of the goals pursued by a bill. In these concluding remarks, I would like to mention three ways in which the inventory can be further developed. First, the inventory is clearly not exhaustive; other forms of argument – for example, those referring to constitutional principles– should be included in the list. Second, as I mentioned at the beginning of this paper, the schemes in the inventory are presumptive: they create only a *prima facie* reason to accept the conclusion. Thus, whether a real-world argument based on one of these schemes is reasonable, will depend on whether the arguer can provide appropriate answers to the critical questions relevant to each scheme. Hence, for the schemes to have an evaluative value, they ought to be complemented with a matching set of critical questions. Finally, to identify the schemes of my inventory, I have used qualitative methods. This method can be strengthened by quantitative research measuring the degree of conventionality of each of the schemes within a given legislature.

ACKNOWLEDGEMENTS: This research was supported by CONICYT FONDECYT/INICIACIÓN Nº 11160149.

REFERENCES

Atienza, M. (1997). *Contribución a una teoría de la legislación*. Madrid: Civitas.
Atienza, M. (2005). Reasoning and legislation. In L. J. Wintgens, *The theory and practice of legislation* (pp. 297-317). Farnham: Ashgate.
Bulygin, E. (1991). Teoría y técnica de la legislación. In C. Alchourrón, & E. Bulygin, *Análisis lógico y derecho*. Madrid: CEC.
Fairclough, I., & Fairclough, N. (2012). *Political discourse analysis*. London and New York: Routledge.

García Amado, J. A. (1988). *Teorías de la tópica jurídica.* Madrid: Civitas.
García Amado, J. A. (2000). Razón práctica y teoría de la legislación. *Derechos y libertades: Revista del Instituto Bartolomé de las Casas* (9), 299-318.
Habermas, J. (1992a/1996). *Between facts and norms: Contributions to a discourse theory of law and democracy.* Cambrdige, CA: MIT Press.
Habermas, J. (1992b). *Postmetaphysical thinking. .* Cambridge, MA: MIT Press.
Habermas, J. (2005). Concluding comments on empirical approaches to deliberative politics. *Acta Politica, International Journal of Political Science, 40*(3), 384–392.
Hastings, A. (1963). *A Reformulation of the modes of reasoning in argumentation* (unpublished doctoral dissertation). Evanston, IL: Northwestern University.
Karpen, U. (1986). Zum gegenwärtigen Stand der Gesetzgebungslehre in der Bundesrepublik Deutschland. *Zeitschrift für Gesetzgebung, 1*, 5-32.
Kelsen, H. (1945/2007). *General theory of law and state.* New York: Clark.
Marcilla Córdoba, G. (2013). Razón práctica, creación de normas y principio democrático: una reflexión sobre los ámbitos de la argumentación legislativa. *Anales de la Cátedra Francisco* Suárez, *47*, 43-83.
Oliver-Lalana, A. D. (2008). Los argumentos de eficacia en el discurso parlamentario. *Doxa, 31*, 533-566.
Oliver-Lalana, A. D. (2013). Rational lawmaking and legislative reasoning in parliamentary debates. In L. J. Wintgens & A. D. Oliver-Lalana, *The rationality and justification of legislation* (pp. 135-184). Dordrecht: Springer.
Perelman, C., & Olbrechts-Tyteca, L. (1958/2000). *The new rhetoric: A treatise on argumentation.* Notre Dame: University of Notre Dame Press.
Steiner, J., Bachtiger, A., Sporndli, M., & Steenbergen, M. R. (2004). *Deliberative politics in action: Analyzing parliamentary discourse.* Cambridge: Cambridge University Press.
Walton, D. (1995). *A pragmatic thoery of fallacy.* Tuscaloosa: University of Alabama Press.
Walton, D. (1996). *Argumentation schemes for presumptive reasoning.* Mahwah, NJ: Erlbaum. .
Walton, D., & Reed, C. (2002). Argumentation schemes and defeasible inferences. In *Working Notes of the ECAI 2002 Workshop on Computational Models of Natural Argument.* Lyon, France.
Walton, D., Reed, C., & Macagno, F. (2008). *Argumentation schemes.* Cambridge: Cambridge university Press.
Wróblewski, J. (1979). A model of rational law-making. *ARSP, LXV*(2), 187-201.

31

Ethos and Inference
Insights from a Multimodal Perspective

JÉRÔME JACQUIN
University of Lausanne
jerome.jacquin@unil.ch

While the inferential dimension of ethos has been studied extensively, its relationship with multimodality, i.e. the fact that linguistic devices used in verbal interaction combine with other semiotic resources such as gestures or shifts in gaze direction, remains largely unknown. Stepping from a language-oriented approach to argumentation, the paper describes a theoretical framework for the multimodal analysis of ethos in argumentative talk-in-interaction. An example taken from a video-recorded corpus of French public debates is provided.

KEYWORDS: ethos, gaze direction, gesture, index, inference, multimodality, Peirce, talk-in-interaction.

1. INTRODUCTION

While addressing the well-known relationship between ethos and inference, this short paper takes a fresh stance, that of multimodality, i.e. the fact that communication is most of the time multimodal, creating meaning by the combination of various semiotic resources. This paper thus examines ethos and the opposition between *showing* and *telling* from a linguistic, semiotic and multimodal perspective on argumentative talk-in-interaction (Doury, 1997; Jacquin, 2014; Plantin, 1996). How do verbal and non-verbal indexes combine in a way to be inferentially interpreted as one ethos? In order to answer this question, this paper uses extracts taken from French video-recorded public debates.

Section 2 gives background information about ethos as an inferential phenomenon and Section 3 integrates multimodality in this framework. Section 4 provides an example of where multimodal indexes converge in elaborating an ethos of knowledgeability. Section 5 discusses the results and further lines of research.

2. ETHOS AS INFERENCE

2.1 From a rhetorical perspective

Together with logos and pathos, ethos is part of what Aristotle calls the "proofs". Ethos consists of the use of the character or image of the orator in order to "inspire confidence":

> There are three things which inspire confidence in the orator's own character -- the three, namely, that induce us to believe a thing apart from any proof of it: good sense, good moral character, and goodwill. (Aristotle, 1954 II-1, 1378a)

Moreover, from an Aristotelian perspective, ethos and the aforementioned qualities "should be achieved by what the speaker says, not by what people think of his character before he begins to speak" (Aristotle, 1954 I-2, 1356a). In other words, ethos is a *verbal achievement* or performance and it should be distinguished from reputation – i.e. what other scholars call "prediscursive ethos" – and explicit self-attribution. *Manifesting* "good sense, good moral character, and goodwill" is a way for the orator to increase the persuasion factor of their discourse.

2.2 From a semiotic perspective

Being the more or less intentional "construction of an image of the self in discourse" (Amossy, 2014, p. 303), ethos is a complex inference that recipients derive from "indexes" (Peirce, 1932), "symptoms" (Berrendonner, 1981; Ducrot, 1984) or "contextualization cues" (Gumperz, 1992). For example, speakers must not say "I am competent in international finance", but should instead *display* such competence, by quoting statistics or using specific lexicon as indexes of their knowledge and abilities. As has been frequently noted, ethotic indexes operate at different levels of analysis, ranging from prosody and lexical choices to grammatical structures and speech acts (e.g. Bonnafous, 2002; Doury & Lefebure, 2006).

Recent studies state the importance of going beyond the classical, logocentric perspective on ethos: as an inference drawn from the rhetorical performance, ethos is not only verbally anchored but it is also *embodied*, i.e. indexed by body postures, gestures, clothing, ... (e.g. Constantin de Chanay & Kerbrat-Orecchioni, 2007; Poggi & Vincze, 2009; Streeck, 2008; Turbide, 2009). However, no theoretical nor analytical link between ethos and inference has been explicitly drawn from such a multimodal perspective. The next section, inspired by studies previously published in French by Jacquin & Micheli (2013) and Jacquin (2014, Chapter 9), addresses the challenge of tackling the semiotic diversity of the indexes participating in the construction of ethos.

3. MULTIMODAL ETHOS: A FRAMEWORK

Starting from the opposition between *show* and *tell*, one could intuitively think that body shows and speech tells. It is, however, a shortcut and linguists, even while disagreeing on how and where to draw the dividing line (see Jacquin & Micheli, 2013), have demonstrated that there is *speech that tells* and *speech that shows* (Berrendonner, 1981; Ducrot, 1984; Nølke, 2001; Recanati, 1979).

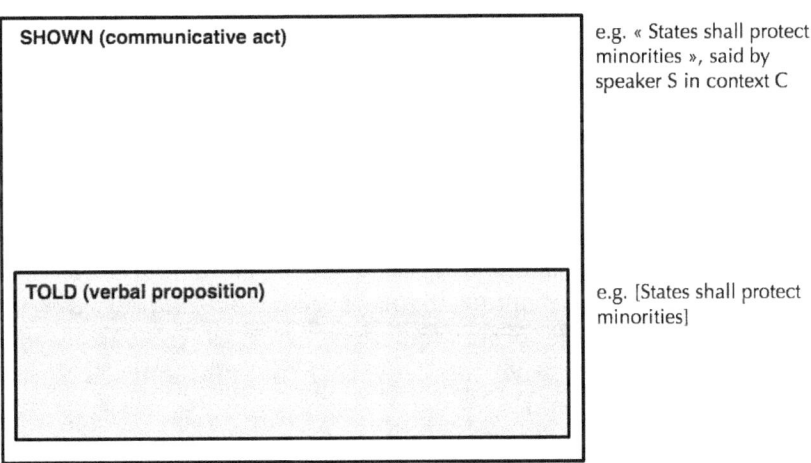

Figure 1 – What is told as part of what is shown

Based on Berrendonner's pragmatic insights (Berrendonner, 1981), Figure 1 suggests that what is *told* is part of what is *shown*. When considering the utterance "States shall protect minorities" said by speaker S in context C, what is *shown*, or what is immediately perceptible, is the communicative act. What is actually *told* in this communicative act is the verbal proposition, or verbal content that is uttered, i.e. that states shall protect minorities. That is consistent with the use of negation as a criterion for distinguishing between *show* and *tell* (Ducrot, 1984; Nølke, 2001): only the content uttered/told can be denied (i.e. [states shall not protect minorities]), not the communicative act consisting of uttering it.

As suggested in Figure 2, that implies that there is a first distinction to draw between *show* and *tell*, which is based on the semiotic support that is used.

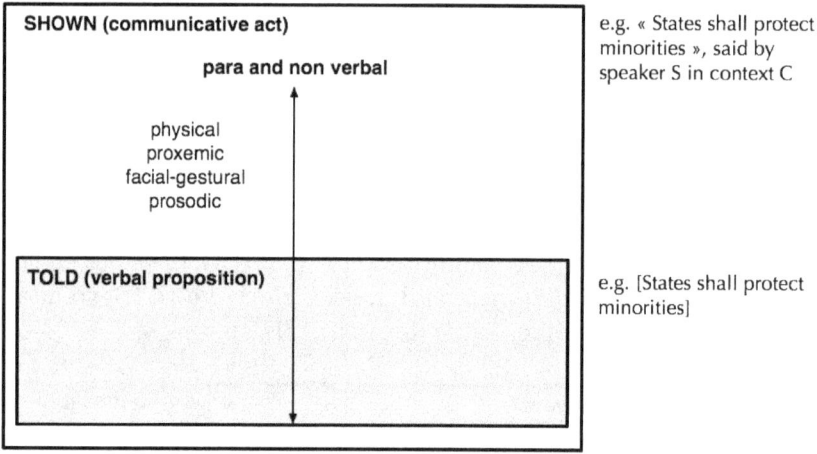

Figure 2 – Show vs. Tell based on semiotic support

In other words, and when considering verbal interaction, there is a distinction between the content being *shown* by embodied resources attached to the communicative act, such as gestures and facial expressions, and the content being specifically *told* by the verbal proposition. While the orator's clothes are part of what is shown through the communicative act, the verbal content [states shall protect minorities] is what is told.

But since what is told (i.e. the verbal proposition) is part of what is shown (i.e. the communicative act), what is *told* can also *show* something, i.e. it also works in an inferential way, as an index for something else, for example a property of the speaker. As outlined in Figure 3, there is thus a second level where the distinction between show and tell operates.

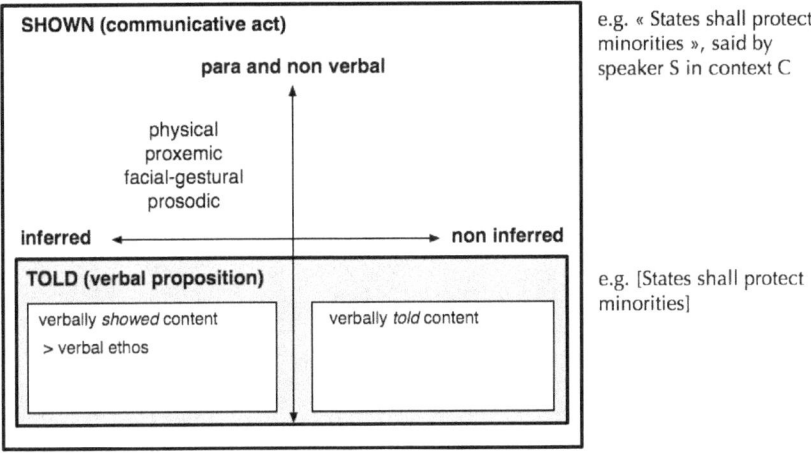

Figure 3 – Show vs. Tell based on the type of interpretation

Applied to the example, that means that what is told, i.e. [states shall protect minorities], can be inferentially interpreted as a way of constructing and showing an ethos of empathy. Multimodal ethos in verbal interaction thus consists of the combination of what is bodily and verbally shown through the communicative act, as illustrated by Figure 4.

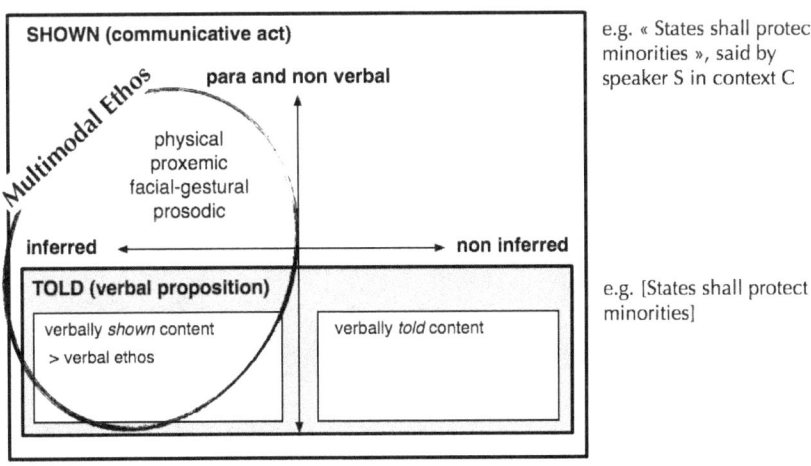

Figure 4 – Multimodally-shown ethos

4. MULTIMODAL ETHOS: AN EXAMPLE

The example analysed hereafter is taken from a video-recorded corpus of eight public debates held in the French-speaking part of Switzerland

and, more precisely, from a debate about the legal power of ecological associations.[1] In the extract, a member of the public (PUB5) takes the floor to give a general overview on environmental law.

Extract / REC-ECO / 00:36:21.000
```
1 PUB5  oui\ (.) y a peut-être juste une chose qu'il faut
        yes\ (.) there's maybe just one thing that I need to
2       préciser par rapport au droit de l'environn'ment#1 qui
        clarify regarding the environmental #1 law which is a
3       est un peu particulier#2 c'est que c'est un #3 un (.)
        little bit special  #2 it is  it is a #3 a  (.) law
4       droit qui fait appel à ce qu'on appelle #4 les concepts
        which is based on what one calls  #4 the undefined
5       juridiques indéterminés\ #5 (.) tu as parlé: #6
        legal concepts\  #5 (.) you talked  #6 about balance
6       d'équilibre (.) de la beauté du paysage des trucs comme
        (.) about the beauty of the landscape things like that\
7       ça\ #7 (..) et ça donne
        #7 (..) and it gives
```

---

[1] More information about this corpus can be found in Jacquin (2017).

| GUEST | PUB5 | PUB5 | GUEST |

| #Im1 (camera 1) | #Im1 (camera 2) |

#Im1
Just after having mentioned "environmental law" (line 2), PUB5 looks at the guest and raises his grasping gestures made with his left hand.

| #Im2 | #Im3 | #Im4 |

#Im2-4
Looking down, PUB5 repeatedly raises and lowers his left hand while classifying "environmental law" (lines 2-5).

| #Im5 | #Im6 | #Im7 |

Gazing back at the guest, PUB5 stabilizes the gesture at the end of the classification (image 5). He then rotates his hand (image 6) before returning to the same position (image 7).

At line 1, PUB5 starts by categorizing his turn as a "clarification" ("*préciser*") about environmental law, which is defined as relying on "undefined legal concepts" ("*concepts juridiques indéterminés*"). As shown by images 2-4, this "argument from verbal classification" (Walton, 2008, p. 129) is accompanied by a shift in gaze direction from the guest to the table and by a metaphoric grasping gesture that PUB5 repeatedly raises and lowers, as if the speaker has grasped the environmental law itself (see the 'bowl' configuration in Calbris, 2011).² Moving from the argument from verbal classification about the specificity of environmental law to the examples previously given by the interlocutor (see the reported speech *"tu as parlé de"*, "you talked about", at line 5), PUB5 repeats the grasping gesture from line 5 to line 7. The stabilization of the gesture at line 7 is intertwined with the end of a three-part list ("about balance", "about beauty of the landscape", "things like that"), which is a classical rhetorical device to project the discursive completion of an argumentation (e.g. Atkinson, 1984; Heritage & Greatbatch, 1986; Hutchby, 1997).

Multimodal indexes converge in constructing an ethos of knowledgeability: the explanation consisting of a definition and examples is combined with a grasping gesture that highlights the image of a speaker who knows what he is talking about.

5. DISCUSSION AND CONCLUSION

The aim of this short paper was to tackle the relationship between ethos, inference, and multimodality. Starting from an analytical framework initially published in French, (i) I discussed the fact that even if ethos is shown and not told, what is told can show something, including participating in the discursive and interactional construction of an ethos, and (ii) I identified the different kind of indexes involved in a multimodally-elaborated ethos.

A situation of converging indexes has been exemplified by an extract of public debate where a speaker combines different semiotic resources to display an ethos of knowledgeability. The framework and the analysis suggest that divergence between indexes is also theoretically possible, even if this situation is more complex, less intuitive and also less studied than convergence. There is clearly work to be done at this level.

---

² "The [facing downwards] spread-out fingers of one hand encircle [an] abstract entity" (Calbris, 2011, p. 313). Discussing metaphorical grasping gestures from a cognitive perspective, Gibbs (2008, p. 294) states "Thus, gesturing a grasping motion with one hand may both reflect some natural conceptualization of the idea of a concept, u t may also help a speaker ver ally articulate the idea of 'grasping a concept' as in 'I ust couldn't grasp that concept'."

ACKNOWLEDGEMENTS: An earlier version of the article benefited from comments and suggestions of participants at the 2nd European Conference on Argumentation held in Fribourg in June 2017. The author is most grateful to Jay Woodhams for revising his English text.

REFERENCES

Amossy, R. (2014). 2008. Argumentation et analyse du discours: Perspectives théoriques et découpages disciplinaires. Argumentation et analyse du discours [online], 1, selected paragraphs 1-18. Anonymous translator. In J. Angermuller, D. Maingueneau, & R. Wodak (Eds.), *The Discourse Studies Reader: Main currents in theory and analysis* (pp. 298–304). Amsterdam: John Benjamins.

Aristotle. (1954). *Rhetoric*. (W. R. Roberts, Ed.). New York: Modern Library.

Atkinson, J. M. (1984). Public speaking and audience responses: Some techniques for inviting applause. In J. M. Atkinson & J. Heritage (Eds.), *Structures of social action: Studies in conversation analysis* (pp. 370–409). Cambridge/Paris: Cambridge University Press/Maisons des Sciences de l'Homme.

Berrendonner, A. (1981). *Éléments de pragmatique linguistique*. Paris: Minuit.

Bonnafous, S. (2002). La question de l'ethos et du genre en communication politique. *Actes du premier colloque franco-mexicain en information et communication*. Retrieved from http://www.cerimes.fr/colloquefrancomexicain/

Calbris, G. (2011). *Elements of meaning in gesture*. Amsterdam: John Benjamins.

Constantin de Chanay, H., & Kerbrat-Orecchioni, C. (2007). 100 minutes pour convaincre: L'éthos en action de Nicolas Sarkozy. In M. Broth, M. Forsgren, C. Norén, & F. Sullet-Nylander (Eds.), *Le français parlé des médias* (pp. 309–329). Stockholm: Acta Universitatis Stokholmiensis.

Doury, M. (1997). *Le débat immobile. L'argumentation dans le débat médiatique sur les parasciences*. Paris: Kimé.

Doury, M., & Lefebure, P. (2006). "Intérêt général", " intérêts particuliers" : La construction de l'ethos dans un débat public / Version des auteurs. *Questions de Communication*, 9, 47-71.

Ducrot, O. (1984). *Le dire et le dit*. Paris: Minuit.

Gibbs, R. W. (2008). Metaphor and gesture. Some implications for psychology. In A. Cienki & C. Müller (Eds.), *Metaphor and gesture* (pp. 291–301). Amsterdam: John Benjamins.

Gumperz, J. J. (1992). Contextualization and understanding. In A. Duranti & C. Goodwin (Eds.), *Rethinking context: Language as an interactive phenomenon* (pp. 229–252). Cambridge: Cambridge University Press.

Heritage, J., & Greatbatch, D. (1986). Generating applause: A study of rhetoric and response at party political conferences. *American Journal of Sociology*, 92(1), 110–157.

Hutchby, I. (1997). Building alignments in public debate : a case study from British TV. *Text*, *17*(2), 161–179.
Jacquin, J. (2014). *Débattre. L'argumentation et l'identité au coeur d'une pratique verbale*. Bruxelles: De Boeck.
Jacquin, J. (2017). Embodied argumentation in public debates. The role of gestures in the segmentation of argumentative moves. In A. Tseronis & C. Forceville (Eds.), *Multimodal argumentation and rhetoric in media genres* (pp. 239–262). Amsterdam: John Benjamins.
Jacquin, J., & Micheli, R. (2013). Dire et montrer qui on est et ce que l'on ressent : Une étude des modes de sémiotisation de l'identité et de l'émotion. In H. Constantin de Chanay, M. Colas-Blaise, & O. Le Guern (Eds.), *Dire / montrer. Au coeur du sens* (pp. 67–92). Chambéry: Université de Savoie.
Nølke, H. (2001). *Le regard du locuteur 2 : Pour une linguistique des traces énonciatives*. Paris: Kimé.
Peirce, C. S. (1932). Elements of Logic. In C. Hartshorne & P. Weiss, Eds., *Collected Papers* (Vol. 2). Cambridge: Harvard University Press.
Plantin, C. (1996). Le trilogue argumentatif. Présentation de modèle, analyse de cas. *Langue Française*, 112, 9–30.
Poggi, I., & Vincze, L. (2009). Gesture, gaze and persuasive strategies in political discourse. In M. Kipp, J.-C. Martin, P. Paggio, & D. Heylen (Eds.), *Multimodal corpora. From models of natural interaction to systems and applications* (pp. 73–92). Berlin: Springer.
Recanati, F. (1979). *La transparence et l'énonciation*. Paris: Seuil.
Streeck, J. (2008). Gesture in political communication: A case study of the democratic presidential candidates during the 2004 primary campaign. *Research on language and social Interaction*, *41*(2), 154–186.
Turbide, O. (2009). *La performance médiatique des chefs politiques lors de la campagne électorale de 2003 au Québec* (doctoral dissertation). Université Laval, Québec.

APPENDIX

Transcript conventions[3]

/ \	Rising and falling intonations
:	Prolongation of a sound
-	Abrupt interruption in utterance
(.) (..) (...) (n)	Pauses (1/4, 1/2, 3/4 second; n = seconds)
MAIS	Emphasis
[YY YYYY]	Overlapping speech
&	Extension of the turn after an overlap
=	Latching
(it; eat)	Speech which is in doubt in the transcript
XX XXX	Speech which is unclear in the transcript
((laughs))	Annotation of non-verbal activity
#1 #im1	Picture 1

---

[3] Adapted from ICOR, v. 2013 (http://icar.univ-lyon2.fr/projets/corinte/bandeau_droit/convention_icor.htm; last accessed on July 2016), and Mondada (2007).

32

Ad Populum Arguments in a Political Context

HENRIKE JANSEN
Leiden University Centre for Linguistics
h.jansen@hum.leidenuniv.nl

An appeal to the opinion of a lot of people or even to the majority of people – also known as *ad populum* argumentation – is often regarded as argumentation that is inherently fallacious. Nevertheless, politicians today often refer to 'the will of the people' and present this will as a relevant factor for decision-making in a democratic society. This paper addresses some considerations regarding the rationality of this type of *ad populum* argumentation.

KEYWORDS: *argumentum ad populum*, bandwagon variant, descriptive vs. prescriptive standpoint, fallacy, linguistic devices, political context, popular opinion, pragma-dialectical approach

1. INTRODUCTION

The *argumentum ad populum* is a well-known type of reasoning in argumentation theory; nevertheless, it has not yet received a great amount of attention in the literature. Many textbooks do mention the argument, but their accounts are most often rather short, boiling down to the conclusion that the fact that a lot of people hold a certain standpoint cannot guarantee the correctness of that standpoint. As an illustration, Copi & Cohen (1990, p. 104) cite Bertrand Russell, who says that because of the 'silliness of the majority of mankind, a wide-spread belief is more likely to be foolish than sensible.' One could therefore think that there is not much to remark about this type of argument other than that it is simply a fallacy. However, some authors say that a distinction should be made between rational and irrational instantiations of *ad populum* argumentation, e.g. Douglas Walton in his 1999 monograph called *Appeal to popular opinion*. In this paper I will explore this presumption.

I will start with a discussion of the question of whether *ad populum* arguments could be rational under certain conditions. To this end I will provide an overview of the norms that are mentioned by different authors in the literature. In this overview I will explicitly take into account the type of standpoint that is supported with the *ad populum* argument. Where ideas or concepts can be clarified by a translation into pragma-dialectical terminology (van Eemeren & Grootendorst, 1992, 2004; van Eemeren, Grootendorst & Snoeck Henkemans, 2002), I will do this. My conclusion is that authors who hold the view that *ad populum* arguments are always fallacious adhere to an epistemological concept of this type of argument, which I believe does not occur in real discourse. Empirical research is needed in order to examine the drafting of naturally occurring *ad populum* arguments and the commitments that result from a particular formulation.

Before I proceed, it should be noted that the label '*ad populum* argument' is also in use for an appeal to the emotions of the public. This is the so-called 'mob appeal' (Walton, 1992, p. 82), referring to argumentation that makes use of expressive language and other devices calculated to excite an audience. The variant that is discussed in this paper concerns the argument that, because a large number of people hold a certain opinion, we should accept that opinion. This variant of *ad populum* argumentation is labelled the 'bandwagon' variant (Minot, 1981, p. 230; Freeman, 1995, p. 266; Govier, 2010, p. 162).[1]

2. SINGLE VERSUS COMPLEX *AD POPULUM* ARGUMENTATION

Many examples and descriptions of an *ad populum* argument in the literature concern the so-called 'factual' type. This involves a variant in which a descriptive type of standpoint is supported. In such a standpoint, the proposition is of a factual nature, which means that it indicates an actual state of affairs – it indicates how things are (i.e. according to the arguer), either in the present, the past or the future. In this variant standpoints are defended that either read 'X is the case', 'X was the case', or 'X will be the case'.

It can be deduced from the literature that it is not only descriptive standpoints that are connected to *ad populum* arguments. They can also support prescriptive standpoints, for example in contexts such as advertising and deliberation (Minot, 1980, p. 230; Nolt, 1984, p. 250; Johnson & Blair, 2006, p. 179; Govier, 2010, p. 161). After all, the

---

[1] Literally, this argument is called, according to Govier (*ibidem*): 'the fallacy of jumping on a bandwagon'.

aim of an *ad populum* argument used in these contexts is to ensure that the product is bought or that the course of action is followed. Thus, in these contexts we can expect a prescriptive type of standpoint, such as 'Buy X', or 'Action X should (or: should not) be carried out'.

A great deal of the literature on *ad populum* arguments conveys the opinion that these arguments are always fallacious, notwithstanding the type of standpoint they support. This view is often illustrated with the fact that history provides many examples of many people believing matters that afterwards appeared to be false. For example, for a long time people believed the earth was flat (Kahane, 1984, p. 56) and in a certain period Hitler's ideas were very popular (Toulmin, Rieke & Janik, 1984, 146). However, one might wonder whether *ad populum* arguments cannot sometimes provide some reason, some indication to accept a standpoint, albeit provisionally. As for the factual type – the type containing a descriptive standpoint – one could wonder whether it does not say something that the large majority of climate scientists claim that there is evidence for global warming. To just brush aside such a claim would be too simplistic. Similar cases are discussed by Walton (1999, pp. 201-205; 2006, pp. 91-93), who concludes that they consist of a combination of *ad populum* arguments and other types of argument, such as *position to know arguments* or *appeals to expert opinion*. It is the combination with these other types of argument that makes these instantiations of *ad populum* argumentation more reasonable.

In reply to Walton's analysis of these examples, however, Godden (2008) argued that these do not match what he calls the 'basic form' of *ad populum*. According to Godden, the cases discussed by Walton are not just plain *ad populum* arguments where '(...) the argument from popularity alone provides good reasons to accept its conclusion' (*ibidem*, p. 109). In pragma-dialectical terms, Godden distinguishes between complex *ad populum* argumentation, i.e. the cases discussed by Walton, and single *ad populum* argumentation, where the appeal to the view of a lot of people is the only element of the argumentation. According to Godden, Walton's examples draw their rationality from additional premises that are presupposed in the argumentation, namely that the group of people referred to are experts.[2]

---

[2] In this respect, Godden also mentions another example, i.e. one that is discussed in Walton, Woods & Irvine (2004, p. 37). In that example, the argument would actually consist of an instance of inference from the best explanation.

My (pragma-dialectical) analysis of the appeal to climate scientists makes that clear:[3]

1.
It is highly likely that global warming is a fact

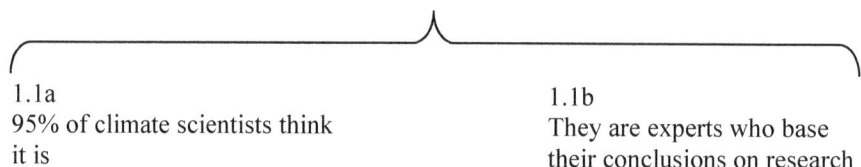

1.1a
95% of climate scientists think it is

1.1b
They are experts who base their conclusions on research

As Godden argues, and I agree with him, the cases discussed by Walton (and likewise my example about climate scientists) cannot negate the judgement that *single ad populum* arguments supporting a descriptive standpoint 'do not provide an adequate reason for accepting the conclusion' (*ibidem*, p. 108).

What does Godden's criticism imply for *ad populum* arguments supporting a *prescriptive* standpoint (cases that Godden did not consider)? Both Walton (1999, p. 34 ff.) and Kahane (1984, p. 57) write that it could be reasonable for an individual to decide that she wants to buy a bestseller or to see a blockbuster, just because she knows that her taste coincides with popular taste or with majority taste.[4] Or because she wants to be able to join in conversations about books or films that everyone is talking about. Indeed, these may be good reasons for an individual to take this action. But, again, this is not because of the *ad populum* itself, but because of the additional reasons provided. In these examples, too, it is not the appeal to popularity alone that supports the prescriptive standpoint 'I should buy the book' or 'I should see the movie'. Other considerations such as wanting to join in conversations are part of the argumentation as well, as is apparent from the following reconstruction:

---

[3] See for the pragma-dialectical method for reconstructing argumentation Van Eemeren, Grootendorst & Snoeck Henkemans (2002, Ch. 5).

[4] See also Walton (1999, Ch. 2) for more cases involving a prescriptive standpoint.

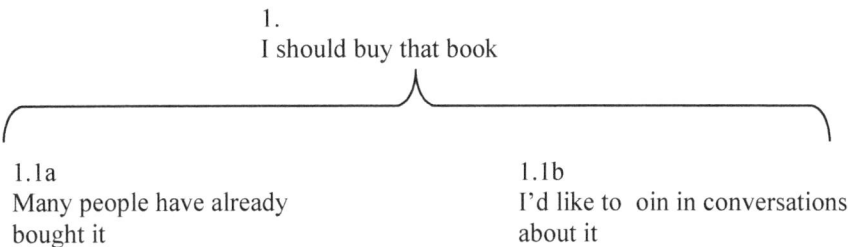

It may thus be concluded that there is a group of *ad populum* arguments that are not fallacious. This group involves complex argumentation, in which the reason that a lot of people endorse the standpoint supports the standpoint in combination with an additional reason, that 'bolsters' the argumentation (Walton, 2006, p. 92) and therefore removes the fallaciousness.

3. NORMS FOR AD POPULUM ARGUMENTATION IN A POLITICAL CONTEXT

One might conclude from the above that a single *ad populum* argument is always fallacious. That an opinion is held by many people can never guarantee its correctness. Nevertheless, one could wonder whether this is also true for 'political' *ad populum* arguments containing a prescriptive standpoint concerning a policy proposal. After all, should there not be some relation between how people in a society think about certain societal developments on the one hand and actual policy making on the other? Johnson & Blair (2006, p. 179) remark, for instance, that it would be strange if public opinion were largely to deviate from the laws that shape society. Minot (1980, p. 230) expresses it even more strongly: '(...) in a democratic society, the desire of the populace is the means for deciding an issue' [underlining is original]. These remarks imply that *ad populum* arguments can be legitimately used by arguers in a discussion on policy proposals.

    One of the authors paying particular attention to *ad populum* appeals used by politicians is Andone (2015). She discusses argumentation in which a European politician refers to the political barometer in order to sustain her standpoint that more should be done about equal rights for men and women:

> Gender equality is a long-lasting commitment for the EU. The results of a recent Eurobarometer show that <u>nearly all Europeans</u> agree that equality between women and men is a

fundamental right, and <u>a large majority of citizens</u> believe that tackling inequality between women and men should be a priority for the EU. [underlining is mine, HJ]⁵

Andone has some doubts with regard to the rationality of the *ad populum* appeal in this particular example, but it is her view that rational instantiations of popularity appeals in a political context are possible.⁶ She (2015, p. 1) strengthens this claim by citing some authors from political and social theory who have stated that an appeal to the majority is an important element in democratic decision-making (Lomax Cook, Barabas & Page, 2002; Holzinger, Reinhard & Biesenbender, 2014).

A similar view seems to be held by Oswald & Hart (2013, p. 10) when they discuss an example of a British politician referring to the will of the people in order to sustain the standpoint that immigration numbers should be limited:

- <u>The majority of British people</u> (...) are united on this issue.
- <u>Everyone</u> wants new people to settle as long as numbers are limited.
- Talk to <u>people</u> and whatever their background, religion or the colour of their skin – they ask the same thing: 'Why can't we get a grip on immigration?' These are the <u>people who are always ready to welcome genuine refugees to Britain, or families who want to work hard and make a positive contribution to our country</u>.
- But I've lost count of the times that <u>British people of ethnic backgrounds</u> have told me that firm but fair immigration controls are essential for good relations [underlining is mine, HJ].

---

⁵ Retrieved from http://www.europarl.europa.eu/RegData/seance_pleniere/compte_rendu/revise/2015/03-09/P8_CRE-REV(2015)03-09_XL.pdf, last accessed, 24.04. 2018.

⁶ If I understood Andone correctly, the strategic function of *ad populum* appeals used by politicians is that they are a means to evade the burden of proof. That is, when a politician appeals to majority opinion, the commitment to this opinion lies with this majority, and not with the politician who referred to it. Because the politician's audience is part of the majority appealed to, the likelihood of criticism being expressed is minimised. Andone argues that further research should make clear whether the politician's evasion of the burden of proof is a fallacious practice.

Although these authors provide a negative assessment of their example, they, too, seem committed to the idea that *ad populum* appeals could be rational under certain conditions. One of these conditions is that there should be evidence for the claim that a lot of people hold the alleged standpoint.

It is Walton's book on appeals to popularity where some considerations can be found concerning this 'democratic' popular appeal. These are the following:

- Questions can be raised with regard to the reliability of opinion polls (Walton, 1999, pp. 2-10, p. 125). This is an interesting elaboration on the criterion – mentioned by Oswald & Hart (see above) – that the number of people referred to should be sustained with evidence. According to Walton, a lot can go wrong in such a survey, i.e. whether a representative sample has been carried out, or whether the way the questions have been formulated is unambiguous and not misleading.
- In decision-making, it is important to exercise caution with regard to the tyranny of democracy (*ibidem*, p. 11; this concern was already mentioned by Toulmin, Rieke and Janik in 1984, p. 146).[7] After all, the idea that public opinion should be reflected in the laws of a society is not the only aspect of a democracy. Another characteristic is that minorities should not be discounted by majority opinion.
- Nothing changes so quickly as public opinion, which makes it a less reliable foundation on which to base legislation (Walton, *ibidem*, p. 11).[8]

To these considerations I would like to add a consideration taken from an essay by Jan-Werner Müller on populism (2016, p. 41). Müller says that it is often politicians themselves who are responsible for public opinion – for what the public thinks or feels. Müller gives the example of the following quote, uttered by Turkish President Erdogan in July 2016: 'What do my people want? They want the death penalty!' Erdogan started the discussion about the death penalty himself and has been a strong proponent of such a policy.

Looking at this list of considerations regarding the rationality of *ad populum* argumentation supporting a prescriptive standpoint, it

---

[7] Walton draws this from Tocqueville; Toulmin, Rieke and Janik from J.S. Mill.
[8] This concern is also drawn from Tocqueville.

strikes me that they can easily lead to a negative evaluation of political *ad populum* arguments. In principle it seems to be a good idea to take public opinion into account when discussing policy making. However, there are a lot of pitfalls, which do not speak in favour of the use of *ad populum* arguments in deliberations on particular issues. In my view, there is a lot to say for Walton's final judgement about these arguments, which is that they can never be the only reason for following a certain policy. They could provide *some* weight supporting a policy proposal, but independent, substantive reasons should always be considered and weighed (see also Anderson, 1979, pp. 720-721). The conclusion must then be that *ad populum* argumentation is often very weak, but not fallacious per se. It may function as an indication for accepting a prescriptive political standpoint, but this indication should, amongst other things, be considered in the light of different interests and the rule of law.

It should be noted that the situation I have been discussing here is whether an *ad populum* argument could be a good argument to use in a political *discussion*, e.g. when it is used by one politician against a fellow politician or in an attempt to convince an audience. My remarks about the inadequacy of an *ad populum* argument in such a discussion does not have any consequences for the legitimacy and sufficiency of a majority vote in institutionalised decision-making, for example by means of a referendum.

4. THE VIEW THAT *AD POPULUM* ARGUMENTATION IS ALWAYS FALLACIOUS

In the last section, single *ad populum* argumentation supporting a prescriptive standpoint in a discussion on policy making is regarded as a weak but not per se fallacious argument. This view competes with another approach, also supported in argumentation theory, that any *ad populum* argument, no matter what kind of standpoint it supports, is always fallacious. Johnson & Blair (2006) seem to represent this approach by making a distinction between *ad populum* arguments on the one hand and the application of democratic principles on the other. One such principle is that in certain situations it can be agreed by the participants that decisions – e.g. about how to proceed – should be based on a majority vote. Whereas this principle consists of a *procedural* rule with regard to majority opinion, an *ad populum* argument is based on a *criterial* rule, i.e. that majority opinion is a criterion of the 'truth or plausibility' of that opinion (*ibidem*, p. 178). Only the latter is fallacious.

A similar distinction is made between *ad populum* appeals and applying 'the principle of popular sovereignty' (Johnson & Blair, *ibidem*, p. 179). The latter refers to the idea that laws should be based on the will of the populace. It is applied when a majority vote in an election gets particular political consequences, e.g. law-making. According to Johnson & Blair, such an application is not the same as an *ad populum* appeal, because in the latter case, the standpoint is defended that the voters' opinion is the 'correct' view:

> But the vote does not prove that the majority was correct in its opinion. It would be quite consistent to acknowledge the authority of the vote while maintaining that the majority was wrong. (*ibidem*)

In conclusion: an application of the democratic principles described above does not imply a judgement about the soundness of the decision gained by a majority vote, whereas an *ad populum* argument does. An *ad populum* argument is supposed to imply that the fact that a lot of people prefer a certain decision makes this decision 'correct'. This interpretation of *ad populum* arguments is also expressed in Johnson & Blair's formalisation of these arguments as 'Everyone believes that P is true, therefore P is true' (*ibidem*, p. 176). (Although the authors do not say this explicitly, one should presumably read 'correct' instead of 'true' in situations where a prescriptive standpoint is at stake.)

In my opinion, this view is problematic. It entails that in a discussion about policies any appeal to the opinion of the majority constitutes a fallacious *ad populum* argument. The fallaciousness stems from the fact that in such an argument the majority is supposed to function as a criterion of the correctness of the standpoint. But why would we attribute this interpretation of an *ad populum* argument to an arguer? Why can't we interpret an *ad populum* in a discussion in a procedural way, as I did in the former section? When an arguer – e.g. a politician in parliament or just someone having a discussion with a friend – defends the standpoint that the government should adopt policy X because a lot of people think it is the best course of action, why can't she just mean that the view of a large group of people is a relevant consideration in political decision-making? I really don't see why this arguer would be committed to the view that the fact that the amount of people holding a preference for this policy makes this policy the 'correct' one in a criterial sense.

Moreover, Johnson & Blair's view becomes all the more problematic if we look at how *ad populum* arguments are formulated in real discourse. When do these arguments ever fit an abstract scheme like 'Everyone believes that P is true, therefore P is true'? Does this kind of *ad populum*, by which it is concluded that the standpoint is true or correct, exist at all? First of all, arguers often leave the standpoint of *ad populum* argumentation implicit, as – paradoxically – Johnson and Blair have observed themselves (*ibidem*, p. 177).[9] But even if the standpoint has been explicitly expressed, how can we know whether the arguer was committed to a criterial or a procedural interpretation of her appeal to a lot of people? In a corpus study that I carried out on the speeches of Geert Wilders, a Dutch anti-immigration politician, Wilders never concludes from his appeals to the people that they render his standpoint correct.[10] On the contrary, one way in which Wilders refers to the will of the people is by stating his own view and then adding that he is not the only one who has this opinion, as in 'I want fewer Moroccans, and that's what many other Dutch people think'. Or he labels himself expressly as the spokesman for a lot of people, as in 'On behalf of all those millions of Dutch people that you no longer represent, I say ...'. Another technique is to not even mention his own opinion any more, but simply to state that a lot of people find this or that, according to opinion polls, e.g. 'Two-thirds of the Dutch people are against immigration'. A final technique looks like the former but the difference is that the actual number referred to remains vague: 'A lot of people are against immigration'.

It is my impression, not only from this corpus study but also from examples that I gathered from other contexts, that arguers formulate *ad populum* appeals in such a way that they can never be pinned down on a line of reasoning saying that because a lot of people agree with them, this makes their standpoint true or correct. In order to examine this claim, we need to shift our attention from abstract schemes like 'Everyone believes that P is true, therefore P is true' to naturally occurring arguments and find out how *ad populum* arguments are formulated in real discourse. Presumably, many real-life instantiations of *ad populum* argumentation can be regarded as the result of strategic manoeuvring with presentational choice (cf. van Eemeren, 2010, Ch. 4). The strategy lies in the fact that, on the one hand,

---

[9] As an illustration, they give the example of a discussion where one of the discussants maintains that smoking marihuana should be forbidden because it is harmful. In this discussion, the other discussant could reply by saying: 'Oh, come off it! Nobody believes that nowadays!' In doing so, the standpoint that marijuana is not harmful is left to the interpretation of the hearer.

[10] The corpus consists of speeches by Geert Wilders published on his party's website pvv.nl.

the ways these arguments are formulated makes it hard to accuse the arguers of committing a fallacy. They can always respond to such an accusation by saying that they did not intend the number of people to imply that their standpoint is correct. They can always say that they meant that it does at least say something about a particular matter if so many people have a certain opinion, and that it would be wrong to ignore this opinion. On the other hand, these formulations are strategic because the reference to a lot of people evokes our natural inclination to join the majority. Oswald & Hart (2013, p. 6) explain that '*[a]d populum* operates on the back of conformity bias', which means that joining the group gave us evolutionary advantage and is therefore engrained in our cognitive capacities. As Jackson (1995) has already observed, this human predisposition would explain our tendency to be misled by *ad populum* arguments.

5. CONCLUSION

What can we conclude about the rationality of *ad populum* arguments? For an answer we have to make a distinction between single and complex *ad populum* argumentation. When additional reasons are added in order to justify the appeal to the majority because of specific circumstances, the argumentation, which has then become complex, does not have to be fallacious per se. With regard to single *ad populum* argumentation, we have to distinguish between the kinds of standpoints that it defends. *Ad populum* arguments supporting a *descriptive* standpoint are always fallacious. In my view, this judgement does not hold for a *prescriptive* standpoint. It is true that there are several drawbacks regarding the use of majority opinion in deliberation. However, *ad populum* arguments in a political context cannot be dismissed as fallacious per se. I concluded on the basis of a survey of the literature – mostly Walton (1999) – that *ad populum* arguments supporting a prescriptive standpoint in a political context can be regarded as insufficient but still rational. This means that the standpoint they support needs additional, substantive support.

     A discussion of Johnson & Blair's approach (2006), claiming that *ad populum* arguments are also fallacious if they defend a prescriptive standpoint, showed that these authors use a criterial interpretation of the appeal to a lot of people. It is not very realistic, however, to assume that *ad populum* arguments in this understanding can be encountered in real discourse. Even if arguers believe their inciting standpoints to be true or correct, they cannot be held to be committed to such a belief if they did not say this explicitly. In order to test this claim, an empirical investigation is needed of what arguers actually do say when uttering an *ad populum* argument. I believe that it is highly useful to shift our attention to how arguers draft *ad populum* argumentation in real discourse. An analysis of a small corpus of speeches by the Dutch politician Geert Wilders yields the impression that different categories can be distinguished demarcated by variation in linguistic means.

Follow-up research should be conducted to examine systematically whether the four categories that I distinguish in this paper are exhaustive. Furthermore, for each of these categories it should be ascertained to which line of reasoning an arguer can actually be held committed. To this end, we need to invoke the help of linguistics. A linguistic approach could open new ways for developing criteria to meet the urgent need for evaluation of *ad populum* arguments. In light of increasing populism in international politics, I believe it is necessary to get to grips with the value of appeals to 'the will of the people'.

REFERENCES

Anderson, C. (1979). The place of principles in policy analysis. *The American Political Science Review, 73*(3), 711-723.

Andone, C. (2015). Engagement et non-engagement dans les appels à la majorité par des hommes politiques. *Argumentation et analyse du discours*, 15, 13 pp. Retrieved from https://aad.revues.org/2021#tocto1n3.

Copi, I. M., & Cohen, C. (1990). *Introduction to logic* (8th ed.). New York/London: Macmillan.

Eemeren, F. H. van (2010). *Strategic manoeuvring in argumentative discourse: Extending the pragma-dialectical theory of argumentation*. Amsterdam, Philadelphia: John Benjamins.

Eemeren, F.H. van & Grootendorst, R. (1992). *Argumentation, communication and fallacies. A pragma-dialectical perspective*. Hillsdale, NJ etc.: Erlbaum Publishers.

Eemeren, F. H. van, & Grootendorst, R. (2004). *A systematic theory of argumentation: The pragma-dialectical approach*. Cambridge: Cambridge University Press.

Eemeren, F. H. van, Grootendorst, R., & Snoeck Henkemans, A. F. (2002). *Argumentation: Analysis, evaluation, presentation*. New York, London: Routledge.

Freeman, J. B. (1995). The appeal to popularity and presumption by common knowledge. In H. V. Hansen & R. C. Pinto (Eds.), *Fallacies: Classical and contemporary readings*. University Park: Pennsylvania State University Press.

Godden, D. (2008). On common ground and *ad populum*: Acceptance as grounds for acceptability. *Philosophy and Rhetoric, 41*(2), 101-129.
Govier, T. (2010). *A practical study of argument* (7th ed.). Belmont, CA: Wadsworth. Retrieved from http://www.kalabakas.dk/files/Books/A%20Practical%20Study%20of%20Argument.pdf
Holzinger, K., Reinhard, J., & Biesenbender, J. (2014). Do arguments matter? Argumentation and negotiation success at the 1997 Amsterdam Intergovernmental Conference. *European Political Science Review*, 6(2), 283-307.
Jackson, S. (1995). Fallacies and heuristics. In F. H. van Eemeren, R. Grootendorst, J. A. Blair & C. A. Willard (Eds.), *Analysis and evaluation. Proceedings of the Third ISSA Conference on Argumentation, volume II* (pp. 257– 269). Amsterdam: Sic Sat.
Johnson, R. H., & Blair, J. A. (2006). *Logical self-defense.* New York etc.: Idea Press.
Kahane, H. (1984), *Logic and contemporary rhetoric: The uses of logic in everyday life* (4th ed.) Belmont, Calif.: Wadswort.
Lomax Cook, F., Barabas, J., & Page, B.I. (2002). Invoking public opinion. Policy elites and social security. *Public Opinion Quarterly, 66*, 235-264.
Minot, W.S. (1981). A rhetorical view of fallacies: Ad hominem and *ad populum*. *Rhetoric Society Quarterly*, 11(4), 222-235.
Müller, J.W. (September 2, 2016). Trump, Erdogan, Farage: The attractions of populism for politicians, the dangers for democracy. *The Guardian*. Retrieved from https://www.theguardian.com/books/2016/sep/02/trump-erdogan-farage-the-attractions-of-populism-for-politicians-the-dangers-for-democracy.
Nolt, J.E. (1984). *Informal logic: Possible worlds and imagination.* New York: McGraw-Hill Book Company.
Oswald, S., & Hart, C. (2013). Trust based on bias: Cognitive constraints on source-related fallacies. OSSA Conference Archive. 125. Retrieved from https://scholar.uwindsor.ca/ossaarchive/OSSA10/papersandcommentaries/125.
Toulmin, S., Rieke, R. D., & Janik, A. (1984). *An introduction to reasoning.* Macmillan: University of California.
Walton, D. N. (1999). *Appeal to popular opinion.* University Park: Pennsylvania State University Press.
Walton, D. N. (2006). *Fundamentals of critical argumentation.* Cambridge: Cambridge University Press.
Woods, J., Irvine, A., & Walton, D. (2004). Argument: Critical thinking, logic and the fallacies (2nd ed.). Toronto: Pearson/Prentice Hall.

33

View on Inference from Argumentation Tiers Perspective

IRYNA KHOMENKO
Department of Logic, Taras Shevchenko National University of Kyiv, Ukraine
khomenkoi.ukr1@gmail.com

In this paper I present my reflections on inferential structure of argument from argumentation tiers perspective. I differentiate four tiers (logical, dialectical, rhetorical, and cognitive) and describe their key features. I pay special attention to inferential issues comparing logical and cognitive tiers of argumentation. Here I rely on the results of my experiment with *The Stag Hunt Task* from Rousseau's "A Discourse on Inequality".

KEYWORDS: inference, real argument, levels and tiers of argumentation, logical tier, dialectical tier, rhetorical tier, cognitive tier, The Suppression Task, The Stag Hunt Task

1. INTRODUCTION

Contemporary developments of argumentation theory can be presented as a plenty of theoretical and practical approaches. Scholars working in various fields (Philosophy, Logic, Psychology, Linguistics, Political Science, Law etc.) try to study topical issues of argumentation.

This situation, on the one hand, leads to numerous interesting results. However, on the other hand, the problem of inconsistency of terminology takes place among scholars. In this regard unfortunately the key terms studied in the argumentation field do not have strict definition and usage so far. This fact ensures that the same term in different contexts can be used with different meanings.

Thus, we fairly often can meet definition of argumentation such as a form of interaction, in which arguers resolve a conflict of opinions using real arguments.

The questions immediately arise: what are such arguments, what are their key features, what are distinctions between *real* and *non-real* arguments? However, in spite of numerous papers, books, and textbooks published over the last years, consensus as to what a real argument is has not been achieved. The only agreement is that argumentation theorists unit around the point that a real argument is a type of reasoning which is not a subject matter of formal logic. This point could be illustrated with Johnson's quote from early writings about of one of the vices of formal logic as:

> virtual disappearance from the mandate of logic of the focus on real argument (Johnson, 2000, p.105).

Govier also parts company with Johnson and claims that:

> What is strange is that in view of these substantial gaps between real arguments and the subject matter of formal logic, formal logic is still widely regarded as having something to offer to the nonspecialist (Govier, 1987, p. 2).

> To speak of informal logic is not to contradict oneself but to acknowledge what should be obvious: that the understanding of natural arguments requires substantive knowledge and insight not captures in the rules of axiomatized systems (Govier, 1987, p. 204).

In other words, real arguments are not the subject matter of formal logic, but rather what scholars call *informal logic, practical logic, philosophy of argument, theory of argument, applied epistemology, theory of reasoning*, and *theory of critical thinking*.

For my present purposes, it is important to define the term *real argument* more precisely than as a gap of formal logic. In order to do this, firstly, it should be pointed out that scholars have opposing views on this issue.

Thus, informal logicians conduct in-depth study of what they call *real* arguments. Besides they use various labels for this kind of reasoning. Among them are real, natural, everyday, actual, real-life, ordinary, mundane, marketplace argument. For example, Groarke in the quote below use two labels for such argument:

> In keeping with the emphasis on real argument, I will discuss musical argument in the context of examples of actual argument (Groarke, 2002, p. 419).

However it could be noted there is a point of view according to which such argument is not a theoretically significant and useful reasoning. For instance, Goddu assumes the follows:

> Perhaps appeal to "real" arguments has a legitimate pedagogical use. For example, appeal to the rough and ready distinctions between natural and contrived or everyday and specialized might ground the pedagogical goal of making the relevance and practical utility of the subject material obvious to students... I have merely been arguing that the notion has no useful theoretical role to play in a general theory of argument (Goddu, 2009, p. 12).

However we have witnessed plenty attempts to produce definition of real argument in argumentation community. For example, according to Blair and Johnson real argument is:

> actual natural language arguments used in public discourse, clothed in their native ambiguity, vagueness and incompleteness (Johnson and Blair, 1994, p.6).

Groarke thinks that real arguments are:

> the arguments found in discussion, debate and disagreement as they manifest themselves in daily life (Groarke, 2017).

According to Govier:

> an actual argument is simply an argument, a piece of discourse or writing in which someone tries to convince others (or himself) of the truth of a claim by citing reasons on its behalf. I speak of actual arguments because I do not wish to speak of the contrived arguments—series of statements constructed by logicians to illustrate their principles and techniques (Govier, 1987, p. 4).

Fisher asserts that:

> real arguments—not the 'made-up' kind with which logicians usually deal. They originate from various sources ranging from classic texts to newspapers (Fisher, 1988, p. 15).

Even though it seems that I have provided more than enough definitions for now, still, in my view, none of them is clear enough. That is why I will proceed with studying real arguments. For the sake of clarity and brevity, I would like to focus on their six key features which could help distinguish them from other types of reasoning. In my view, it can be described in the following way.

i. Unlike formal logic, which uses primarily artificial language, real argument is expressed by natural language. Scholars study arguments as: "they occur in natural language discourse rather than formal languages of the sort that characterize propositional logic, the predicate calculus, modal logic, etc. (languages which have rigorously specified syntax, semantics and grammar, and clearly defined proof procedures." (Groarke, 2017)

ii. Real argument is a dialogical argument. Here arguing requires at least two arguers. They express to each other divergent points of view on certain question and at the same time should keep in mind objections, which they may have.

iii. Real argument relates to everyday communication. In this regard the artificial reasoning from textbook on logic is not relevant to real arguments.

iv. Real argument mostly is a defeasible argument. We can see that some arguments, which we take to be good, are not sound by reflecting on examples of perfectly acceptable arguments whose premises are not all true, or whose inferential step is not deductively valid.

v. One of the key features of real argument is its incompleteness. Arguers often do not use all premises and conclusions in such arguments. Some of them do it on purpose of confusing the opponents, but sometimes this case occurs when arguers do not have sufficient skills to express their thoughts clearly.

vi. Real argument is dependent on the context of utterance.

vii. Real argument may be presented as a complex reasoning. It consists of two arguments: object argument and meta-argument.

2. LEVELS AND TIERS OF ARGUMENTATION

Moving on, let's turn now to the consideration of argumentation levels. In my view, based on the analogy from formal logic where object language differs from meta-language, real argument can be analysed on two levels: object level and meta-level. I belief they may be defined using the distinction between object argument and meta-argument. Let us look closer at this proposition.

Begin with explication term *object level*. Generally, it refers to reasoning about such objects as historical events, social events and politics, news in mass media and social networks, advertising, corporate and governmental communications, personal exchange and practical problems. Such reasoning can be called *object-argument*.

I see object argument as a set of statements that seeks to justify a conclusion by supporting it with premises; to defend it from objections; or both goals.

With regard to the components of object argument, I believe that we can use the traditional approach here: object argument can be considered as a system composed of premises and a conclusion. Conclusion is a statement that is based on other statements, called *premises*. Both notions are mutually interdependent and hang upon the context of argumentation. Thus, it can be stated that object argument is a claim-reason complex.

The next item on our agenda is to explain the term *meta-level*. Here I use it with the following meaning: meta-level of argumentation relies on meta-arguments. Meta-argument I see as reasoning about one or more object-arguments. Object argument in particular discussion is a subject matter of certain meta-argument.

I consider two types of meta-arguments. The first is interpretation of object argument. It can be seen as a description of construction or reconstruction of object argument details in order to ensure their understanding. While we talk about own argument, we concentrate on its construction. In case when we analyse arguments of others, we focus on its reconstruction. Another type of meta-argument is object argument's evaluation, namely the assessment of its merits. Method of critical questions can be used for the evaluation of such arguments.

Moving on let's consider now tiers of argumentation. First of all I would like to talk about various criteria for distinction of argumentation tiers. For example, I would like to specify such characteristics as perspectives and orientations. Let's bring our attention to them and begin with an approach that was proposed by Wenzel J. W. (1990).

He distinguishes three perspectives on argumentation: logical, rhetorical and dialectical. From his point of view logic, rhetoric, and dialectic are three different ways of argumentation understanding. The first related to the argumentation as a text product, the second – to the process as a result of which this product is created, the third – to the regulatory procedure of this process. In other words, we deal with product-based, process-based, and procedure-based approaches.

It should be noted that Wensel's standpoint has been considered by scholars as a standard position on this issue. However in the last few years a lot of them have criticized the approach about perspectives on argumentation. Among them are Blair (2012), Goddu (2011), Johnson (2009), Jorgensen (2014), Kock (2009) and others.

Based on these works, the ideas of other researchers and generalizing their results I propose to use such characteristic for distinguishing argumentation tiers as their orientations. From my point of view logical tier concerns the arguer text and reasoning contained in it. Rhetorical perspectives can be explicated considering argumentation with the focus on audience and its reception. Finally dialectical tier appears only when the opponent is in argumentation process. This point involves argumentation identification as a dispute between proponent and opponent according to certain rules.

To sum up, it may be noted that logical tier is a textual-oriented tier of argumentation, which includes meta-arguments about inferential structure of object arguments, isolated from arguers, audience and context of argumentation. Rhetorical tier is an audience-oriented tier of argumentation, which includes meta-arguments related to the audience reception of argumentation. Dialectical tier is an opponent-oriented tier of argumentation, which includes meta-arguments related to object argument's defence from possible criticism of other arguers and rules for the dispute procedure.

3. THE SHIFT TO THE COGNITIVE TIER OF ARGUMENTATION

I believe we can also analyse argumentation in a deeper way, adding one more of their tiers, namely cognitive. What is this tier? In order to answer the question let's pay attention again to the logical tier. Until the middle of the 20th century, the dominant approach to the study of argumentation was logical, or rather formal-logical one: "Logic was for millennia seen as the skeleton of argument" (Hample, 2006, p.2).

Logic is primarily concerned with reasoning and we can find in numerous text books on logic its definition as the science of reasoning. In order to summarize the traditional view of logic let's look at such quote.

> For over two millennia, since the days of Aristotle and Euclid, the notion of formal logic has figured centrally in conceptions of human reasoning, rationality, and adaptiveness. To be adaptive, the story goes, we must be rational about ends and means, truth and evidence. To be rational, we must reason about what means suit what ends, what evidence supports

what conclusions. And to so reason, we must respect the canons of logic (Perkins, 2002, p.187).

What are these canons of logic? They are known to all from text books on logic. With logical point of view, argumentation should be understood as a text with arguments produced by some person and available for logical reconstruction. You start by accepting certain premises; you then accept intermediate conclusions that follow from the premises or earlier intermediate conclusions in accordance with certain logical rules of inference. Let's not forget that in order to be available for logical analysis these arguments must be 'squeezed' in one or another logical form of reasoning, primarily deductive one. Finally you end by accepting new conclusions that you have inferred directly or indirectly from your original premises.

One problem with the traditional picture is that many people in daily life not always argue like that. Over the past half century in cognitive psychology, a lot of experimental data accumulated (Declerck & Reed (2001), Evans (1998), Fiddick, Cosmides & Tooby (2000), Ford (1995)). They testify to the discrepancy of the logical theory with the practice of reasoning, namely conditional, counterfactual and even syllogistic.

Moving on, let's turn now to the some situations confirm the fact that a person in daily life tends to be guided not only by the canons of logic, but by the context, cultural stereotypes and others effects. It could be pointed out all of these points relate to human's mental presentation of the world.

For my present purposes now I propose to examine the experimental results called *The Suppression Task*. The author of this experiment is Byrne R. and her results were published in the paper "Suppressing Valid Inferences with Conditionals" (Burne (1989)).

Everybody who has at least some knowledge of formal logic knows inference rules like modus ponens. It is often defined as an argument of the form: if A implies B and A holds, we can conclude that B holds.

In this regard such question arises: is this logical inference rule at the same time a mental inference rule for personal usage in reasoning with conditional sentence?

Byrne begins with reasoning having the form of modus ponens (Example 1).

> Premise 1: If she has an essay to write then she will study late in the library.

Premise 2: She has an essay to write.
Conclusion: She will study late in the library.

As a result, 96% of subjects conclude that she will study late in the library.

Then Byrne replicated these results by presenting a second conditional containing an alternative condition for the same consequent (Example 2).

Premise 1: If she has an essay to write then she will study late in the library.
Premise 2: She has an essay to write.
Premise 3: If she has a textbook to read she will study late in the library.
Conclusion: She will study late in the library.

She gets the same result. 96% of subjects chose the conclusion: *She will study late in the library*.

Further Byrne extends the experiment by introducing a second conditional with an additional requirement that must also hold (Example 3).

Premise 1: If she has an essay to write then she will study late in the library.
Premise 2: She has an essay to write.
Premise 3: If the library stays open she will study late in the library.
Conclusion: She will study late in the library.

Now only 38% of subjects made modus ponens inferences, compared to 96% when a simple conditional or two conditionals with alternative conditions were presented.

Byrne concludes that either there are no mental rules for the valid inferences, or that

...suppression by itself tells us nothing about the existence or non-existence of rules of inference in the mind. (Byrne, 1989, p.76)

She admits that the results can still be explained in terms of mental rules if one assumes that the joint representation of both sentences makes the application of inference rules impossible. Her proposal is that:

Formal theories, therefore, need to be supplemented with a detailed account of the process of interpretation, because premises of the same apparent logical form are represented in different ways depending on their meaning. (Byrne, 1989, p.77)

Further, she rejects mental rules in favour of the mental model theory. In this regard it should be noted that within consideration of cognitive tier of argumentation I prefer use the term *mental presentation* or *mental picture*.

Here, the idea is that people represent all possible states of affairs which are consistent with their world knowledge and the content of the sentences. A conclusion of real argument is only endorsed if there is standpoint which contradicts it.

4. INFERENCE AND THE STAG HUNTER TASK

At the final part of my talk I would like to show some application of my ideas based on the story called *The Stag Hunter*. What is it about? This story is briefly told by Rousseau, in "A Discourse on Inequality":

> If it was a matter of hunting a deer, everyone well realized that he must remain faithful to his post; but if a hare happened to pass within reach of one of them, we cannot doubt that he would have gone off in pursuit of it without scruple. (Rousseau, 1984)

The Stag Hunter became a game in modern logic and game theory. Scholars consider it as a strategic game, which has such characteristics.

> Each of a group of hunters has two options: she may remain attentive to the pursuit of a stag, or she may catch a hare. If all hunters pursue the stag, they catch it and share it equally; if any of the hunters devotes her energy to catching a hare, the stag escapes, and the hare belongs to the defecting hunter alone. Each hunter prefers a share of the stag to a hare. (Osborne, 2003, p.20)

Then researchers, using formal methods, build the payoff matrix. It shows that it would be better that all hunters prefer the stag to a hare.

You can learn more about this information for example in these books (Mann, Sandu, and Sevenster (2011), Osborne (2003), Skryms (2004)).

Now let's look at *The Stag Hunter* not from standpoint of mathematical and logical strategies, but from cognitive orientation, considering story of hunt as an argumentation.

In this regard it would be noted that Rousseau's story of the hunt leaves many questions open. Among them: What are the values of a hare and of an individual's share of the deer, given a successful hunt? What is the probability that the hunt will be successful if all participants remain faithful to the hunt? Might there be such situation that two deer hunters decide to chase the hare?

As we can see all of these issues and responds on them could be related to the cognitive presentation of the hunters. Their arguments are about deciding to hunt deer or hare depending on their understanding of the values of hare and deer, possibilities of a successful hunt etc.

In order to justify the importance of taking into account cognitive tier in argumentation analysis I try to conduct an experiment with my master students. I proposed them some arguments about the story *The Stag Hunter*. They begin with the argument having the form of Modus Ponens (Example 4).

> Premise 1: If the hunter wants to get more benefit, he will hunt the deer.
> Premise 2: He wants to get more benefit.
> Conclusion: He will hunt the deer.

As a result 94% of students chose the conclusion from Modus Ponens.

Then I added additional premises If he believes that others hunt the deer, he will hunt the deer (Example 5).

> Premise 1: If the hunter wants to get more benefit, he will hunt the deer.
> Premise 2: He wants to get more benefit.
> Premise 3: If he believes that others hunt the deer, he will hunt the deer.
> Conclusion: He will hunt the deer.

In this case 94% of subjects chose this conclusion. However among them 87% chose the conclusion *He will hunt deer*, and 7% of students include additional beliefs complimenting the content of conclusion. For example, *He will hunt the deer only if he knows that the other hunters also hunt the deer*.

The next case of adding further complimentary premises is (Example 6):

Premise 1: If the hunter wants to get more benefit, he will hunt the deer.
Premise 2: He wants to get more benefit.
Premise 3: If he does not belief that others hunt the deer, he will hunt the deer.
Conclusion: He will hunt the deer.

Here we obtain the following results: 94% of subjects chose *He will hunt the deer*. Among them 47% - *He will hunt the deer*, and 47% of students include additional beliefs in the conclusion.

When I mention a hare in the additional premises (Example 7):

Premise 1: If the hunter wants to get more benefit, he will hunt a deer.
Premise 2: He wants to get more benefit.
Premise 3: If a hare appears during the hunt, he will hunt a deer.
Conclusion: He will hunt a deer.

The next result appears: 94% of subjects chose the ordinary conclusion. Among them 67% - *He will hunt a deer* and 27% include additional beliefs.

Finally we can see one more argument example with hare (Example 8).

Premise 1: If the hunter want to get more benefit, he will hunt a deer.
Premise 2: He wants to get more benefit.
Premise 3: If he believes that he can catch a hare, he will hunt a deer.
Conclusion: He will hunt a deer.

Here 80% of subjects chose the ordinary conclusion. Among them 60% - *He will hunt a deer* and 20% use additional beliefs in the conclusion.

We can see summary of my experimental results in this table.

Case	Choice in conclusion *hunting a deer*	Choice of conclusion from MP	Choice of conclusion from MP + additional beliefs
Case 1	94	94	
Case 2	94	87	7
Case 3	94	47	47
Case 4	94	67	27
Case 5	80	60	20

Table 1. – The results of experiment

5. CONCLUSION

In this paper I have presented my reflections on inference from argumentation tiers perspective. In conclusion I would like to summarize the main points of my paper.

I consider argumentation as a form of dialogical interaction, where arguers aim is to resolve a conflict of opinions using real arguments.

In my view the key features of such argument can be described in the following way. It is expressed by natural language; it is a dialogical argument; it relates to everyday communication; it mostly is a defeasible argument; one of the key features of real argument is its incompleteness; it depends on the context of utterance; it is a complex reasoning that includes two sub-reasoning: object argument and meta-argument.

Real argument can be analysed on two levels: object level and meta-level. The first level relies on object arguments, meta-level – meta-arguments. Meta-arguments represent such tiers of argumentation as logical, disputing, rhetorical, and cognitive. The first is a neutral-oriented, the second is an arguer-oriented, and the third is an audience-oriented, and the forth orients to mental presentation or mental pictures of arguer.

At the final part of my talk I show some application of my ideas based on the story called *Stag Hunter*. I conducted an experiment with my master students, where I proposed them some arguments about it. The main objective of this experiment was to compare inferential issues from view of logical and cognitive tiers of argumentation.

To sum up let's highlight that experimental results witness:

Firstly, logical tier of argumentation provides the orientation on ideal rational arguer but cognitive tier supplies the considerations of real arguer.

Secondly, within of a logical tier, we deal with the logical inference. Within cognitive one we can tell about belief-management inference.

Thirdly, in the first case from the premises follows the context-independent conclusion, in the second case - context-dependent one. Such conclusion includes additional beliefs of arguers.

Finally I would like to note that, in my opinion, it would be better to classify such kind of cognitive distortions and to mention about it in the textbook of logic for students.

REFERENCES

Blair, J. A. (2012). Rhetoric, dialectic, and logic as related to argument. *Philosophy &Rhetoric*, 45(2), 148-164.
Byrne, R. (1989). Suppressing valid inferences with conditionals. *Cognition*, 31(1), 61-83.
Declerck, R., & Reed, S. (2001). *Conditionals: a comprehensive empirical analysis.* Berlin: Mouton de Gruyter.
Evans, J. (1998). Matching bias in conditional reasoning: Do we understand it after 25 years? *Thinking and Reasoning*, 4(1), 45–82.
Fiddick, L., Cosmides, L., & Tooby, J. (2000). No interpretation without representation: The role of domain-specific representations and inferences in the Wason selection task. *Cognition*, 77(1), 1–79.
Fisher, A. (1988). *The logic of real arguments.* Cambridge: Cambridge University Press.
Ford, M. (1995) Two modes of mental representation and problem solution in syllogistic reasoning, *Cognition*, 54(1), 1–71.
Govier, T. (1987). *Problems in argument analysis and evaluation.* Dordrecht-Holland: Foris Publications.
Goddu, G. (2011) Is argument subject to the product/process ambiguity? *Informal Logic*, 31(2), 75-88.
Goodu, G. C. (2009) What is a "real" argument? *Informal Logic*, 29(1), 1-14.
Groarke, L. (2002). Are musical arguments possible? In F. H. van Eemeren, et al. (Eds.), *Proceedings of the 5th International Conference on Argumentation.* Retrieved from http://rozenbergquarterly.com/issa-proceedings-2002-are-musical-arguments-possible/.
Groarke, L. (2017). Informal logic. In *Stanford Encyclopedia on Philosophy*, http://plato.stanford.edu/entries/logic-informal/ [accessed 2nd January 2017].

Hample, D. (2006). *Arguing: Exchanging reasons face to face.* New York: Routledge.
Johnson, R. H. (2000). *Manifest rationality.* Mahwah, NJ: Lawrence Erlbaum and Associates.
Johnson, R. H. (2009). Revisiting the logical/dialectical/rhetorical triumvirate. OSSA Conference Archive. 84. Retrieved from https://scholar.uwindsor.ca/ossaarchive/OSSA8/papersandcommentaries/84.
Johnson, R. H., & Blair, J. A. (1994). Informal logic: Past and present. In R. H. Johnson & J. A. Blair (eds.). *New essays in informal logic* (pp. 1-19). Winsdor, Ontario, Canada.
Jorgensen, C. (2014). Rhetoric, dialectic and logic: The wild-goose chase for an essential distinction. *Informal Logic, 34*(2), 152-166.
Kock, C. (2009) Choice is not true or false: The domain of rhetorical argumentation. *Argumentation, 23*(1), 61-80.
Mann, A. L., Sandu, G., & Sevenster, M. (2011). *Independence-friendly logic: A game-theoretic approach*, Cambridge, UK: Cambridge University Press.
Osborne, M. J. (2003). *An introduction to game theory.* Oxford: Oxford University Press.
Perkins, D. N. (2002) Standard logic as a model of reasoning: The empirical critique. In D. M. Gabbay, R. H. Johnson, H. J. Ohlbach & J. Woods (Eds.), *Handbook of the logic of argument and inference: The turn towards the practical* (pp. 187-224). Amsterdam: Elsevier.
Rousseau, J. (1984). *A discourse on inequality.* (M. Cranston, Trans.). New York: Penguin Books.
Skryms, B. (2004). *The stag hunt and evolution of social structure.* Cambridge: Cambridge University Press.
Jackson, S., & Jacobs, S. (1980). Structure of conversational argument: Pragmatic bases for the enthymeme. *Quarterly journal of speech, 66*(3), 251-265.
Tindale, C. W. (2015). *The philosophy of argument and audience reception.* Cambridge: Cambridge University Press.
Wenzel, J. W. (1990). Three perspectives on argument: Rhetoric, dialectic, logic. In J. Schuetz & R. Trapp (Eds.), *Perspectives on argumentation: Essays in honor of Wayne Brockriede* (pp. 9-26). Prospect Heights: Waveland.

34

Studying the Process of Interpretation of School Tasks: Crossing Perspectives

ALARIC KOHLER
University of Neuchâtel
Alaric.Kohler@hep-bejune.ch

TEUTA MEHMETI
University of Neuchâtel
teuta.mehmeti@unine.ch

In this paper we analyse situations of misunderstanding, by the mean of two analytical models: The pragma-dialectical and Grize's Natural Logic. Both analyses focus on a student's answer, first to an item of mathematics from PISA survey, and second to a paper-and-pencil exercise in mechanics at college. These examples provide candidate methodologies for investigating processes of interpretation about specific tasks in particular educational contexts, which may be approached as situated and socially negotiated inference processes.

KEYWORDS: education, interpretation, knowledge-oriented argumentation, natural logic, discourse analysis, perspectivism, points of view, pragma-dialectical model, situations of misunderstanding, socio-cognitive approach

1. INTRODUCTION

We are interested in the relation between the expected interpretation of educational tasks and the actual interpretation by students performing the task. In educational settings, it is indeed common for a task designer to set specific expectations in terms of task's interpretation and in terms of what students should produce as answers or solutions. However, students do not always succeed in inferring the designers' intentions and expectations. In this case, the responsibility of this failure is generally attributed to the students, and considered as a lack of knowledge or skill. Yet, before attributing students' failure in a task to their lack of knowledge or skill, one must verify wherever the task has been understood in the same way as intended. Otherwise, there is a risk to attribute a cognitive deficit to students who are actually answering a different question or problem. In this case, the failure of the task is due to a *situation of misunderstanding* rather than to a lack of cognitive ability.

In this paper we will analyse such situations of misunderstanding, by the mean of two analytical models that allow for detailed descriptions of the mismatch between the expected inferences and the actual inferences made by students. For each analytical approach, we will present one example.

The first example provides an analysis of a student's answer in an item of mathematics from the survey PISA, an international survey that aims to assess students' competencies at the end of compulsory school. The analysis is inspired by the pragma-dialectical model proposed by Van Eemeren and colleagues and serves to shed light on the diversity of students' arguments as opposed to the arguments expected by PISA designers. The second example provides an analysis of the interpretation of a paper-and-pencil exercise in mechanics by a college (or high-school) student. Grize's model of logico-discursive operations permits a micro-scale description of the inferences of the student leading to a situation of misunderstanding.

These examples display candidate methodologies for an investigation of the process of interpretation about specific tasks and in specific situations, responding to a theoretical concern raised long ago (e.g. Perret-Clermont, 1980), and many times since. It was observed, for instance, that students may provide the expected answers or solutions and still interpret the question or problem differently from the task's designer (i.e. teacher, survey designers). The meaning of language and other signs, such as graphs or mathematical symbols, cannot be taken for granted when several interlocutors are involved. Each one may have a different interpretation of the same signs, and probably will.

A psychological investigation of interpretation processes can only be carried in relation to specific tasks and specific situations as the meaning is not contained in the signs interlocutors are interpreting, contrary to the information processing metaphor assumes. The interpretation process itself may be approached as situated and socially negotiated inference process. In this sense, argumentation theories are useful, but must also be adapted to the specificity of a psychological investigation of (inter)subjectivity, e.g. articulating different perspectives on the same task.

2. FROM INFERENCE PROCESSES TO SITUATIONS OF MISUNDERSTANDING

The study of inference processes is generally addressed with formal logic, e.g. in cognitive science, or in reference to argumentation theories. In both domains, the analysis generally refers to norms and rules of the "good reasoning". Yet, as Vergnaud points out, "the argumentation, as an activity, is not only made of statements and explicit arguments, but also of implicit identifications which concern the meaning given to what the other says" (2015, p.390). Analysing interlocutors' interpretation is an additional challenge for researchers (Plantin, 2011). Taking up this challenge can contribute to the reconceptualization of argumentation Plantin (2011) calls for, far from a rationalist approach pinpointing fallacies. In this paper, we would like to start with a descriptive approach, interested in the subjective *points of view*, in the sense developed by Piaget (Mounoud, 2000). We thus aim at describing specific interpretations or inferences about the meaning of discourse or other signs, which are situated in specific moments of a process of communication. This focus raises a challenge, both methodological and theoretical: How can we describe subjective inferences about meaning without a mere assimilation to formal models?

A close observation and analysis of critical incidents such as misunderstandings makes the investigation possible, since it works as a revealer of the subjective points of view on what is supposed to be shared in the interlocutory situation, through their discrepancies from a *third person point of view* (Kohler, 2015). In order to do so, the misunderstanding is not approached as a merely linguistic or pragmatic phenomenon, but as situational. The situation is the most micro level of analysis at which cognitive processes such as inferences can be analysed while taking the subject's point of view in consideration. Taking childrens' points of view in consideration, Inhelder and Piaget (e.g. 1955) have shown that they use a qualitatively different rationality than the one described by formal models or used by experts among adults. This rationality is not necessarily flawed, and it is precisely one of the most interesting outcomes of *genetic epistemology* to show that what makes sense for the child, regarding a specific situation, is sometimes quite different from adult or formal logic, and nevertheless coherent. For instance, children do not always consider contradictions problematic in the sense it would require a solving in favour of one of the contradicting standpoints. These standpoints can co-exist in a child's reasoning, at a given point of his development, and only lead to further development under specific conditions. Among these conditions, social features of the situation play an important role, as firstly demonstrated experimentally with research on socio-cognitive conflicts (Perret-Clermont, 1979), and later developed as a *socio-genetic psychology* (Psaltis, Duveen & Perret-Clermont, 2009 ; Psaltis & Zapiti, 2014).

Inhelder and Piaget set a perspective which consisted in being interested in the actual reasoning of the child – his own "logic" in Piagetian's terms – independently from any adult-centred judgment about its rationality. The theoretical tools used by Piaget to describe this "logic of the child", i.e. Boole's logic and predicate logic, were not made for it and laid the ground to critique which misunderstood his scientific project (Apostel et al., 1963). If the theoretical tools were not fitted with the function Piaget made them play in his work, we have reason to think that the perspective itself is not only valuable – in particular for education – but remains a challenge for future research. Indeed, not only the child uses various rationalities but also any layperson: If formal logic is relevant for computer programming, choosing your menu at the restaurant may indeed require quite a different rationality. Hence, the relevance of inference depends not only on formal criteria, but also on the situation and on the actual question or problem reflected upon. We propose to contribute to this approach here, yet making use of more recent theoretical models in the field of argumentation for the description of the particular subjective inferences made by specific subjects in given situations. Our researcher third person point of view will be based on chosen models or theory of argumentation, taking the situation for scale of analysis, in order to include cultural, historical and social processes which fully contribute to relevance of inferences. Rigotti proposes the concept of *reasonableness* (2006, p. 519) for approaching the normative question of argumentation in relation to specific contexts of use.

This paper aims at describing actual reasoning and inferences of children on school tasks, in order to better understand what happens when the children's answers or performances to a task are not matching the expectations of the school system (e.g. Kohler, 2015; Greco, Mehmeti, Perret-Clermont, 2016). Therefore, we are interested on argumentation as a process, which underlines a dialogical approach (e.g. Kuhn & Udell, 2003; Kuhn & Udell, 2007; Plantin, 2005)

3. METHODOLOGY: A DESCRIPTIVE APPROACH

We have raised the issue of the relation between psychology, logic and argumentation. Grize (1982) comments this point as particularly problematic in the western tradition: The attempts of Pascal, Boole or others to study the "law of thought" raised the fear of "psychologizing", i.e. of a reductionism of the mind to psychological investigations, and was strongly opposed by scholars such as Frege. This issue is not one of argumentation theory nor of psychology, it is rather an issue of epistemology, and thus requires a response *in epistemology*. For the sake of this study, our response to this issue consists in making our epistemology explicit as perspectivist (Kohler, Lordelo et Carriere, 2017; Kohler et Donzé, 2017). In the perspectivist epistemology, scientific models and studies constitute specific perspectives on the object under investigation. A specific perspective may be more or less relevant for a specific research question, yet most often several perspectives are better than one. Moreover, perspective as such can be the object of the investigation: Can a specific model or theory contribute to shed light on what remained out of the scope of previous studies?

This perspectivist epistemology motivates a combined presentation of two case-studies using two different theoretical models for a similar research question. This confrontation of perspective on the same object of study not only provides descriptive results about the particular cases we have analysed, and allows for a methodological discussion about the specificities of each models when used for the description of situations of misunderstanding. This paper is only a first contribution to a discussion of the specificity of each perspective.

4. FIRST ANALYSIS

4.1 Procedure of analysis

This first analysis aims to describe separately two different points of view on a same task. In the task description provided by the designers, the task should deliberately promote one specific type of reasoning or argumentation. Yet, from the student's point of view the task can lead to multiple interpretations. The designer's point of view is reconstructed through an *a priori* analysis of the task (e.g. Artigue, 1988; Sensevy & Mercier, 2007), based on the Pisa item, the related comments from the designers, and the evaluation criteria. The students' points of view are reconstructed with an argumentative analysis inspired from van Eemeren and colleagues' analytical approach (e.g. van Eemeren & Grootendorst, 1992, 2004; van Eemeren, Grootendorst, Jackson, & Jacobs, 1993; van Eemeren, Grootendorst, & Snoeck Henkemans, 2002; van Eemeren, Houtlosser, & Snoeck Henkemans, 2007): It allows a careful description and reconstruction of students' reasoning underlying their answers. The comparison of the two points of view allows for the identification of misunderstanding.

A first step of the analysis consists in taking into account the designers' pre-established evaluation of what are considered good answers. It is followed by an *a priori* analysis of this evaluation scheme, and with the identification of students' argumentation. This argumentative reconstruction notably allows to integrate unexpressed premises (van Eemeren & Grootendorst, 1992; van Eemeren et al., 2002, Gerritsen, 2001), that are implicit elements in the argumentation. For van Eemeren and Grootendorst (1992, p. 60) "to establish precisely what someone who has advanced argumentation can held to if the argumentative is analysed as a critical discussion, an analysis must be carried out both at a pragmatic and at a logic level". They add that "if in the argumentation, parts of the arguments are implicit, then a logical analysis is indispensable" (ibid.). This is particularly relevant in cases where the interlocutors' statements are assessed only on their explicit answers (i.e. in written tests at school). Indeed, it has the potential to reveal the interlocutors' reasoning which, in turn, allows to identify whether it follows the expected and valued reasoning or not. In this sense, we insist particularly on the *analytical overview* rather than on the stages of a critical discussion. Other authors interested in knowledge-oriented argumentation and school situations have underlined the need to adapt the pragma-dialectical approach in the study of such situations (e.g. Baker, 2015; Greco, Miserez-Caperos, & Perret-Clermont, 2015).

4.2 Summary of the situation and presentation of the data

The data we present are part of an on-going research and consist of written answers from 159 students aged from 13 to 15 years old who answers to different items inspired by one specific item from the PISA survey. The official item "Robberies" (see figure 1) concerns a mathematical problem, in which students are invited to answer to a question around a fictive situation, presenting a journalist who makes a statement on a graph that contains data on the increase of robberies.[1]

---

[1] PISA item, retrieved from https://www.ge.ch/recherche-education/doc/pisa/codification-maths.pdf.

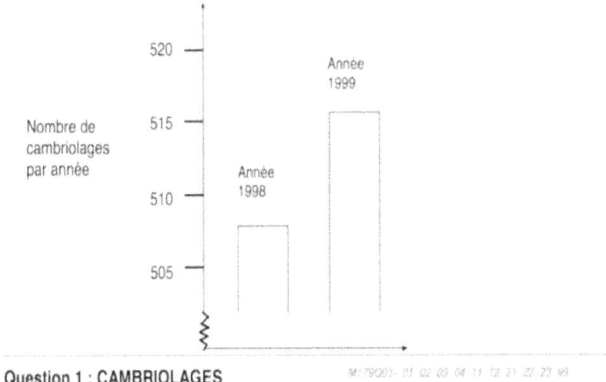

Figure 1 – Original P

Here is a translation (by the authors) of the item:

(1) ROBBERIES
During a TV show, a journalist shows this graph and says:
"This graph shows that there was a huge increase in the number of robberies, between 1998 and 1999."
Question 1: ROBBERIES
Do you consider that the journalist's statement is a correct interpretation of this graph? Justify your answer with an explanation..

This item is considered by PISA designers as a particularly difficult one, and there is an important rate of low achievement in it (OECD, 2009, p.300). Following the designers' analysis, such an item

> (...) involves the analysis of a graph and data interpretation (...) The competencies that are essential for solving this problem are understanding and decoding of a graphical representation in a critical way, making judgments and finding appropriate argumentation based on mathematical thinking and reasoning (although the graph seems to indicate quite a

big jump in the number of robberies, the absolute number of increase in robberies is far from dramatic; the reason for this paradox lies is the inappropriate cut in the y-axis) and proper communication of this reasoning process (OECD, 2004, p. 82).

Although such a description of the requirements of the task sheds light on multiple cognitive demands, we will see that the coding scheme to assess students' answers does not really allow for such a critical stance from the students' part.

We start by presenting the designers' coding scheme for students' answers (see figure 8 and 9 in Appendix) and present our *a priori* analysis of it. First, if we observe what PISA designers consider as an answer allowing to get full points, we may observe that they privilege students' answers that explicitly use arithmetical and geometrical models, on incomplete data. Students who do not use such models are considered as failing the item. On the other hand, a careful *a priori* analysis of the item itself allows observing that multiple interpretations are possible from the students' points of view. Indeed for instance they may interpret the task as requiring from them : a) to take the journalist' statement as a starting point, and seek for arguments to defend the statement "there is huge increase"; b) to counter the journalist' statement and/or interpretation of the graph; c) to propose a personal opinion d) to interpret the role of the journalist as a (un)legitimated authority in terms of interpreting/commenting the graph e) to decide on which criteria to assess the journalist' interpretation.

In argumentative terms, our hypothesis is that this may correspond to an uncertainty for the interlocutors, here the students, towards the issue to discuss. Indeed, it is not *a priori* clear for the student what is the issue at stake, and as we will notice in a student's answer, this can lead to students' cognitive efforts to decipher the issue at stake. In the following case, the student's efforts to do so are visible in the *analytical reconstruction* we made of her answer. The overall data show different examples of students' reasoning which is not aligned on the designers expectations for high achievement in this item. The reconstruction of their argumentation, and more particularly the identification of *unexpressed premises*, reveals that students may sometimes even adopt the designers' expected reasoning, but as the designers' assessment is only based on the explicit answers, such answers can nevertheless get low credits. The case we chose to present illustrates particularly well the student's efforts to provide arguments in regard to two different standpoints, probably induced by the multiple meaning inferences allowed by the task proposed.

4.3 Student's point of view

Trying to answer question 1 in the mathematical problem (see figure 1), the following student (see figure 2 for her complete answer) first approves the journalist's interpretation, answering thus to the question of whether there is or not an increase of robberies. She then goes on by answering by "no", giving arguments in regard to the qualification of the increase as "huge" and to more contextual aspects such as the growth of inhabitants. Then, she concludes that the journalist's interpretation is not correct as "it depends on which country or which city we are. Because there will be more robberies in a big city than in a small one or even less in a village".

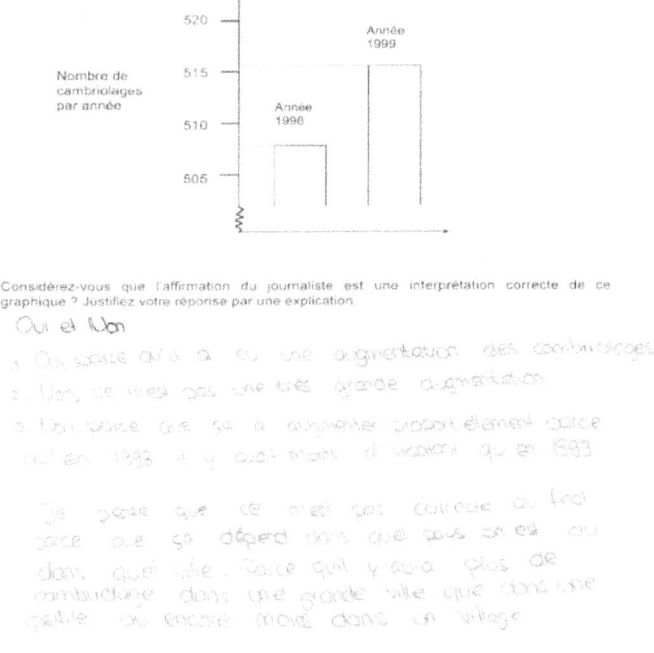

Figure 2 – Student's complete answer[2].

Here is a translation (by the authors):
    (2) Yes and No
    1.   Yes, because there has been an increase in the number of robberies.

---

[2] This excerpt and a first analysis of it have previously been presented for another purpose (see Mehmeti, 2016).

2. No, it is not a huge increase.
3. No, because it has increased proportionally because in 1998 there were less inhabitants than in 1999.

I think that finally it is not correct because it depends in which country or which city we are. Because there will be more robberies in a bid city than in a small one or even less in a village.

In terms of what counts as a good answer according to the designers, this student's answer would not allow her get the total points. However, we can observe that the student makes many efforts to argue on the qualification of the increase as a "huge" one, calling to unexpressed premises such as the necessity to know much more about other meaningful factors or contextual aspects (see figure 3).

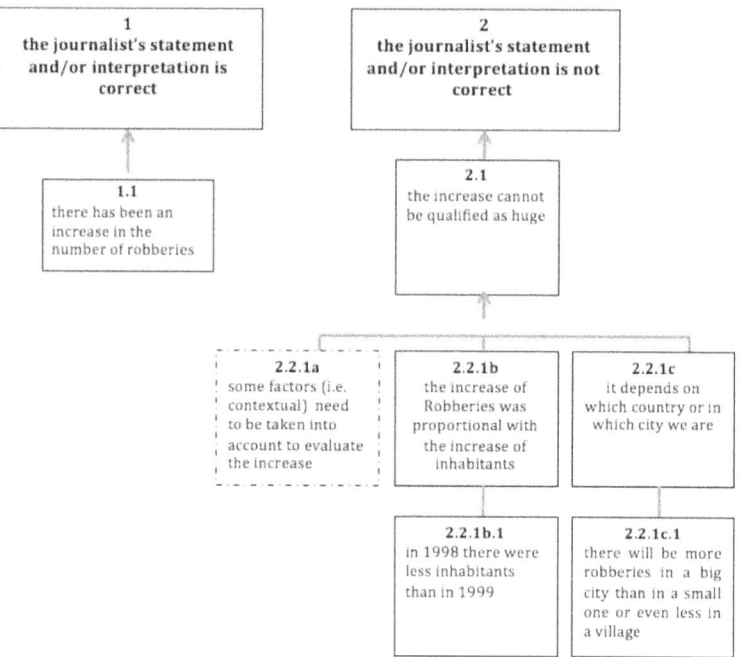

Figure 3 – Reconstruction of students' argumentation (in English) – unexpressed premises are indicated in unbold framed boxes

While such argumentation illustrates the student's efforts to decipher the *issue*, it allows to show that her answer probably does not follow the mathematical modelisation that is expected by the designers and valued as such.

5. SECOND ANALYSIS

5.1 Procedure of analysis

In this analysis, the interlocutors inferences will be described with *logico-discursive operations* (Grize, 1996). Since the actual thinking processes are to ever remain out of reach of direct observation, it is important to stress that the set of operations of Natural Logic constitutes a model. It provides the analyst with a detailed and functional descriptive language for setting hypotheses about the cognitive activity taking place within and through a discourse. In order to study misunderstanding, it is precisely the quality of its descriptiveness that is of particular interest. It allows a rather simple methodological approach of interpretation, which we resume to three steps here.

Firstly, the analyst identifies a *critical incident* (see for instance: Hughes, Williamson & Lloyd, 2007) in reference to the context, which could be the clue to a situation of misunderstanding. In the example below, the critical incident is an answer to an exercise that is considered wrong from the teacher's point of view, and that is not easily explained by a common mistake. Several reasons can lead to a wrong answer, yet a situation of misunderstanding is one of them.

Secondly, a detailed examination of the collected data (mostly written and oral discourse) supports the reconstruction of the "micro-histories" (Tartas, Perret-Clermont & Baucal, 2016) from several points of view. Yet, in this case, the several moments of the micro-history – the "learning phases" - are not controlled experimentally. They are chosen after data collection and according to the actual opportunities provided by the course of events and the natural setting. The result of this analysis can lead to either drop the hypothesis of a situation of misunderstanding, or to identify the specific element(s) from discourses and more generally from the situation which appear *divergent* from two (or more) points of view. In the example below, the reconstruction of the student's point of view leads to the hypothesis that she has considered two exercises related while the teacher did not.

Thirdly, a micro-scale analysis with Natural Logic provides a description of the misunderstanding, which is still related to the specific content of the various interlocutors' discourse, and to the situation of interlocution. This description allows to *state*, with the concepts and language of Natural Logic, the difference in meaning from the two (or more) points of view. This *statement* remains just a hypothesis, yet the overall examination of the context, situation, and discourse at a micro-scale, may greatly support its plausibility. When accepted, the hypothesis works both as a description of the logico-discursive activity of the interlocutors, and as a scientific argument for the researcher's interpretation of the situation under study as 'a situation of misunderstanding'. Moreover, depending on the context, such description can support comments concerning issues of concern. For instance, in the example below, the acceptance of the analysis supports comments on potential obstacles to learning related to the exercise layout, the various signs and discourse used to guide the student cognitive activity, and about what exactly are the expected learning processes in this particular case.

5.2 Summary of the situation and presentation of the data

During physics lessons, second grade college (US: high-school) students have paper-and-pencil exercises to complete individually. Since the overall instruction and its layout are relevant to the analysis, we reproduce below the entire exercise sheet[3] (see figure 4).

---

[3] The text originally in French has been replaced by the translation by the authors, without changing the layout, mathematical expressions and drawings.

1.2 The force of gravitation

The classical law of universal gravitation has been formulated by Newton in 1685 on the ground of experimental knowledge of his time in order to explain the movement of Planets around the Sun.

This law states that the modulus of the gravitational force F between two bodies of mass m and M, separated by a distance r is:

$$F = G \frac{mM}{r^2}$$

The distance r is the distance between the center of gravity of the two massive objects and G is called the constant of universal gravitation $G = 6.67 \cdot 10^{-11}\ Nm^2/kg^2$.

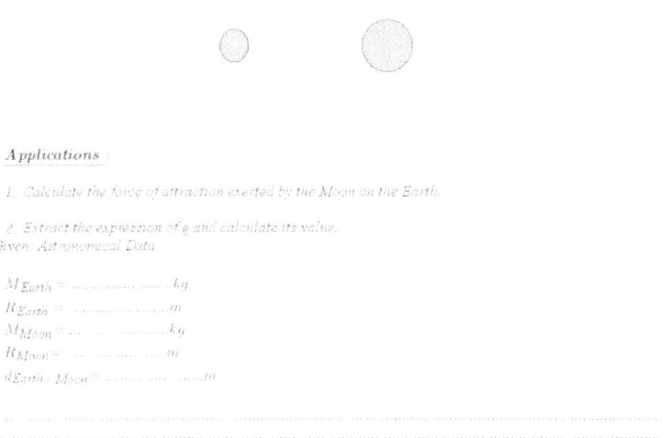

Applications

1. Calculate the force of attraction exerted by the Moon on the Earth.

2. Extract the expression of g and calculate its value.
Given: Astronomical Data

$M_{Earth} = \ldots\ kg$
$R_{Earth} = \ldots\ m$
$M_{Moon} = \ldots\ kg$
$R_{Moon} = \ldots\ m$
$d_{Earth-Moon} = \ldots\ m$

Figure 4 – The complete exercise sheet (translation by the authors).

The pedagogical objective of these exercises, from the teacher's point of view, is for students to use the knowledge presented earlier. This analysis is focusing on the answer of one student (17-y.o., female) to several questions on the same paper sheet. The data used for the analysis consists in the discourse of the teacher (recorded during an interview), and in the answer of the student written directly on the exercise sheet (see figure 5 and 6).

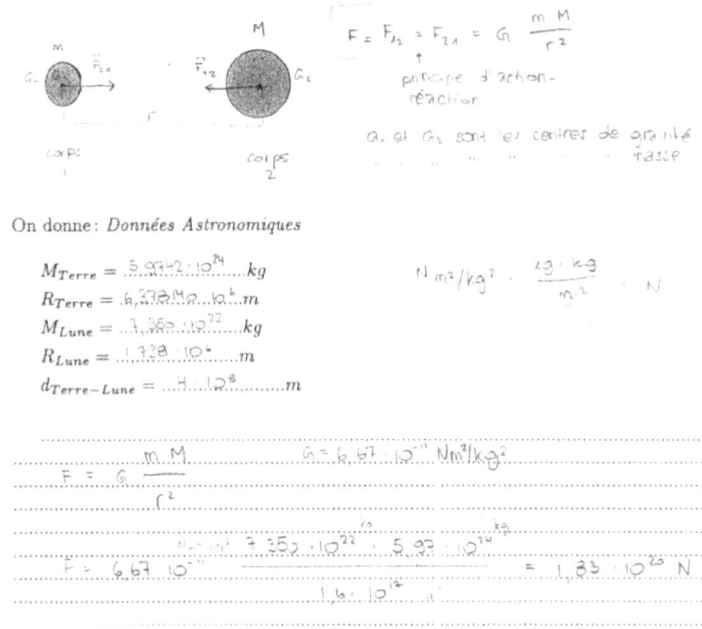

Figure 5 – Extract of the exercise sheet filled by the student (untranslated).

In her answer to the first question (see figure 5), the student uses the drawings of the two Planets to represent forces with arrows[4]. Alongside the sketch, the student poses the equality between the attraction force exerted on the Moon by the Earth and the attraction force exerted by the Earth on the Moon. She uses the mathematical expression of the universal gravitational force to calculate this force, introducing the astronomical data (copied from a reference book) into the equation. The answer is given in *Newton*, the unit of measure of forces. This first answer only interests the analyst for its influence on the way the student answered the second question. It is an element of the situation that is relevant for the analysis, yet it is not what the misunderstanding is about.

The aim of question 2 is for the teacher to make students find the value of *g*, using the mathematical expression of the universal gravitational law. The small letter *g* represents here the constant acceleration of falling objects (without friction) on the surface of the

---

[4] This type of sketch is part of the teaching of Physics, and students are encouraged to use it as a tool for solving problems.

Earth, and is known by student to be worth 9.81m/s^2. To answer the second question, the student starts by using the same mathematical expression of the universal gravitational law, replacing immediately *m* and *M* with, respectively, the mass of the Earth (M_T) and the mass of the Moon (M_L). She continues by stating the equality between the force of attraction between the Moon and Earth she just specified, with "M_L•g", i.e. the multiplication of the mass of the Moon with *g*, in reference of the second law of Newton (F=m•a). By stating this equality, the student actually chooses to calculate *g* as the acceleration of the fall of the Moon on Earth. Figure 6 presents the student's answer to question 2.

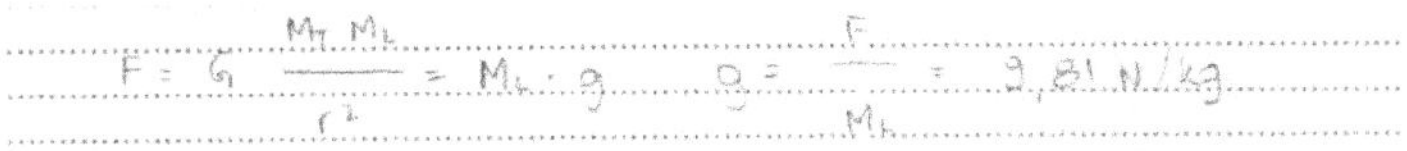

$$F = G\frac{M_T \cdot M_L}{r^2} = M_L \cdot g \qquad g = \frac{F}{M_L} = 9{,}81 \, N/kg$$

Figure 6 – The written answer of the student to question 2 (untranslated).

Let us briefly examine the answer expected by the teacher to question 2. The mathematical expression of the universal gravitational law can be used for any object falling on the surface of the Earth by keeping *m* undefined, replacing *M* by the mass of the Earth, and replacing *r* with the radius of the Earth, i.e. the distance between the centre of gravity of the mass *M* and the approximate location of the fall of the object of mass *m*. The mathematical expression is represented in figure 7.

$$F = G\frac{m \, M_{Earth}}{r_{Earth}} = 6{,}67 \, 10^{-11} \frac{m \, 5{,}97 \, 10^{24}}{(6{,}38 \, 10^6)^2} = 9{,}78m$$

Figure 7 – The mathematical expression of the answer expected by the teacher.

Using Newton's second law and reducing all forces applied on the falling object to the gravitational force, we can state the equality F=9.78•m=m•a, and hence a=9.78m/s^2. The consequence of the choice of r as the radius of the Earth is that, *g* is only (approximately) worth 9.81m/s^2 and constant *near the surface* of the Earth, and cannot be calculated from the fall of the Moon on Earth.

Now the question is where the student's interpretation has gone wrong. What is the specific inference this student may have done in her interpretation of question 2, which is inconsistent with the teacher's schematization?

5.3 Student's point of view

We are now making hypotheses about the student's inferences in the interpretation of the exercise. The first interesting fact to notice is that her numerical answer seems correct: She writes "$g=9{,}81 N/kg$". Yet, this result does not match the calculation she proposes with "F/M_L". If, according to her proposal, we replace m by the mass of the Moon, M by the mass of the Earth, r by the distance between the Moon and the Earth, the worth of g would be 0,00239 rather than 9,81. She probably simply skipped the calculation, writing the answer straight away.

The calculation is nevertheless not the main issue here. In order to check the plausibility of the hypothesis of a situation of misunderstanding, we need to reconstruct her reasoning. More specifically, the relevant question may be: Why has the student decided to find the value of g from the fall of the Moon onto the Earth? The hypothesis we will support here, is that the partly joined layout for question 1 and 2 has supported an inference from the student, in her interpretation of the task, that both questions were related.

Another element of the situation of interlocution may play a role in this interpretation. Since in Newtonian mechanics students are induced to treat object as *mathematical dots*, the decision they should make here in the modelling of the fall of any object on Earth as to take the radius of the Earth for r may appear contradictory to the habitual representation of objects as having neither volume nor surface. Yet, it is of course only a limited understanding of the concept of mathematical dots and its use for modelling in Mechanics that is at stake, since it is precisely by modelling the Earth as a dot, topographically corresponding to its gravity centre, that r can be conceived as the distance between the centre (the position of the mathematical dot) and the surface of the Earth (where the fall of the object takes place).

5.4 Description of a situation of misunderstanding

The third step of the analysis consists in a precise description, with Natural Logic, of the discrepancies between the teacher's point of view on the problem and the student's point of view. We have made the hypothesis that the difference consists in a logico-discursive operation relating exercise 1 and exercise 2. In order to make hypotheses about the interpretation of question 1 and 2, it is necessary to first describe the questions with logico-discursive operations. The written verbal content on the exercise sheet can be described as three logico-discursive operations of *determination*[5], and one operation of *configuration*, written below[6].

Question 1
- δ_1 ±to calculate, {attraction force exerted by the Moon on the Earth} ⇒ (– that the attraction force exerted by the Moon on the Earth to calculate)
  - d_1: (– that the attraction force exerted by the Moon on the Earth to calculate)

Question 2
- δ_2 ±to abstract, {expression of g} ⇒ (– that the expression of g to abstract)
  - d_2: (– that the expression of g to abstract)
- δ_3 ±to calculate, {value [of g]} ⇒ (– to calculate the value [of g])
  - d_3: (– to calculate the value [of g])
- τ_1 (d_2,d_3) ⇒ d_2 –and→ d_3
  - t_1 : d_2 –and→ d_3

We cannot take for granted the inference of a relation between question 1 and question 2. Indeed, it is not uncommon in school context

---

[5] In order to simplify the presentation, operations introducing object-class (α) and predicate (η) have not been written since the contribution of these operations to the schematization is visible through more complex operations (δ and τ).

[6] The operations are denoted with Greek letters (e.g. δ), and subscript numbers have been added for further reference to each particular operation (symbolized by '⇒'). Implicit parts of the discourse that needed to be mentioned are written in square brackets, e.g. [implicit discourse], and are limited to the strict minimum. The result of the operation (after the '⇒') are denoted with Latin letters (e.g. d) and corresponding subscript. This notation allows to distinguish between the process of transformation of the discourse (operations) and the result of it (products), which are specific part of the schematization (*objects of discourse, determinations, configurations*, etc.). Grize (1996) clearly distinguishes the 'détermination' from the concept of 'proposition' in any propositional logic, which is abstracted from any content. Since he conceived his logic as a logic of action, we can read the conventional notation (– that …) as 'a decision that …' or (– to …) as 'a decision to …', where the dash ('–') stands for the action of determining the value of the predicate (or leaving it open in a question), e.g. ±calculate, and of its association with the object-class, e.g. {attraction force exerted by the Moon on the Earth}.

to deal with a list of exercises on the same sheet, which are not related one to another. The hypothesis of an implicit operation of configuration [τ_2 (d_1, d_2, d_3) ⇒ d_1 −?→ (d_2 −and→ d_3)] consists in an inference in the interpretation of the task that may not be common to the student's point of view and to the teacher's point of view. Moreover, the quality of this configuration, written '?', is at least partially implicit and must be reconstructed from clues in the situation and context. In order to evaluate the plausibility of a divergence in the logico-discursive operations from the two points of view, we will attempt to formulate these operations from both points of view before comparing them.

From the teacher's point of view, there is a relation between question 1 and question 2. The clues for supporting this reconstruction are not only the presence of the two questions on the same sheet, which in itself is insufficient, but the common theoretical presentation of the universal law of gravitation, and the fact the astronomical data necessary for answering questions 1 are placed after question 2, on the page layout. In reference to Physics, it is expected for students to use some of the astronomical data for both questions, and that the universal gravitational law is used as a common reference for both questions. Moreover, the (partly) implicit operation τ from the teacher's point of view we are trying to hypothesize is (partly) verbalized with the title "applications". Rather than a direct implicit operation τ between d_1 and d_3, we propose to describe two implicit configurations which are analogical to each other. To express these implicit configurations, we have to state the determining operation written at the top of the page:

- δ_0 ±to be, {modulus of the gravitational force F between two bodies of mass m and M separated by a distance r between the centres of gravity}, {$F=GmM/r^2$} ⇒ (− that modulus of the gravitational force F between two bodies of mass m and M separated by a distance r between the centres of gravity be $F=GmM/r^2$)
  - d_0: (− that modulus of the gravitational force F between two bodies of mass m and M separated by a distance r between the centres of gravity be $F=GmM/r^2$)

The two partially implicit configurations can be described:
- τ_2 (d_1, d_2, d_3) ⇒ d_0 −application→ d_1
  - t_2 : d_0 −application→ d_1
- τ_3 (d_1, d_2, d_3) ⇒ d_0 −application→ (d_2 −and→ d_3)
  - t_3 : d_0 −application→ (d_2 −and→ d_3)

This stated, the relation between question 1 and question 2 from the teacher's point of view can be described as an analogical operation with d_0. It is only analogical and not identical, because the actual use of d_0 is different in both exercises. Yet, the logico-discursive operation – application→ is sufficiently vague to encompass both uses. It may be the that the student has not interpreted the concept the same way the teacher has, if she has at all read and interpreted the title "applications".

From the student's point of view, the relation between question 1 and 2 seems rather like a continuation. A clue of this hypothesis is that she uses her schematization to answer questions 1 as a starting point to answer question 2. Moreover, she uses d_0 identically in both answers, to express the force exerted by the Moon on the Earth (and vice-versa). The student's inference about the relation between question 1 and 2 remaining totally implicit, we only propose here a hypothesis based on the situational clues and the overall context. This hypothesis can be written:

$$[\tau_4 (d_1, d_2, d_3) \Rightarrow d_1 - [\text{can be used for, ...}] \rightarrow (d_2 -\text{and} \rightarrow d_3)]$$

This logico-discursive operation allows the student to skip the modelling phase for schematizing an answer to question 2. Instead of thinking of the Earth, the distance between the centre of gravity and the surface in order to set a new value to r, the student starts with the same model of the attraction between the Moon and the Earth. Hence, it may be that the misunderstanding of τ_2 and τ_3 as τ_4, is catalysed by the lack of understanding, skill or practice of modelling. Despite writing "centres of gravity" on her exercise sheet (see Figure 5), the student has not understood this part of the schematization communicated by the teacher.

5.5 Synthesis

According to this analysis, the divergence between the teacher's point of view and the student's point of view, is not in an operation τ and its absence, but rather in the qualitative specificity of the configuration relating questions 1 and 2. For the teacher, there are two partly implicit analogical configurations d_0 -application→... , while for the student there is an operation between d_1 and (d_2 -and→ d_3), which remains totally implicit, and probably quite vague. In Natural Logic, operations of configuration are usually qualified with the actual discourse, in order notably to keep the ambiguity as it is stated. For instance, an operation '-and→' can take several meaning which often are quite far from the formal logic "∩". Grize (1996) provides example where '-and→' takes rather the meaning of a contiguity, a sequential suite. The partially undefined meaning, or vagueness of such operation is of particular interest for the analysis of situations of misunderstanding.

6. DISCUSSION

In this paper, we aimed to understand how students points of view may fail to match the points of view of their teachers or evaluators, focusing on possible situations of misunderstanding. Both our analyses provide evidence for supporting the hypothesis of a situation of misunderstanding in the examples provided. The first analysis leads to question the context in which the argumentation takes place, and notably stresses the lack of clues for the student's inference. More particularly, the student's point of view can be shown to simultaneously respond to various issues. A similar result emerges from the second analysis: while the relation between exercise 1 and 2 can be precisely described with a logico-discursive operation from the teacher's point of view, the analysis shows that the corresponding inference from the student's point of view is not only different, but also less precisely defined. In both analyses, the situation of misunderstanding is elicited with discrepancies in the inferences made for interpreting the tasks. Yet, the focus is slightly different. In the first analysis, the reconstruction of the two points of view focuses on the interpretation of the issue, while in the second analysis the reconstruction of the student's point of view guides the analysis towards inferences on a partially implicit relation between the various parts of the symbolic content of the exercise sheet, and about its layout. On one hand, the analytical approach based on the pragma-dialectical model allows to stress the specificities of the educational context, where the issue is confused by an asymmetrical relationship where the students has to show his or her skills rather than develop an argumentation of his own. Moreover, we can make the hypothesis that in such asymmetrical situations, typical of tests, the different stages of a critical discussion may be easily disrupted. Grize's logico-discursive operations, on the other hand, lead the focus towards semiotics, the peculiar use by one student of signs and the sheet's layout for drawing inferences about the meaning of the school task.

Although we are still working on this tentative to use analytical models as a mean to study the interpretation of school situations and the potential difficulties for children to answer to the school's requirements, we can already underline some interesting results from these first observations. Indeed, the first analysis raises the various possible inferences about the issue of the argumentation. The diversity of issues to be discussed about the chosen Pisa item is important, despite a highly-constrained and framed interlocutory situation, in the context of a cross-country comparison for which state-of-the-art technical knowledge has been used in order to avoid any interpretative difficulties. The second analysis raises discrepancies in the inferences about the layout of the exercise sheet, which leads the student to engage in a cognitive activity far from the teacher's expectation.

Altogether, these observations contribute to show divergence in interpretations from different points of view on school tasks with a methodology of analysis based on explicit models which are providing a framework for making hypotheses about the particular interpretations of the agent involved in the situation and discussing it, while keeping the particular qualitative features of the discourse and situation under study.

ACKNOWLEDGEMENTS: Special thanks to Prof. Anne-Nelly Perret-Clermont for her support and encouragements to participate in various projects on argumentation, in particular "Argumentation Practices in Context – Argupolis" (ProDoc Doctoral Program, Swiss National Science Foundation, 2008-2012).

REFERENCES

Apostel, L., Grize, J.-B., Papert, S., and Piaget, J., (1963). *La filiation des structures.* Paris: Presses Universitaires de France.
Artigue, M. (1988). Ingénierie didactique. *Recherches en didactique des mathématiques*, 9(3), 281-308.
Baker, M. (2015). The integration of pragma-dialectics and collaborative learning research. Dialogue, externalisation and collective thinking. In. F. H. van Eemeren & B. Garssen (Eds.), *Scrutinizing argumentation in practice* (pp.175-199). Amsterdam: John Benjamins Publishing.
Eemeren, F. H. van, & Grootendorst, R. (1992). *Argumentation, communication, and fallacies: A pragma-dialectical perspective.* Hillsdale, NJ: Lawrence Erlbaum Associates.
Eemeren, F. H. van, & Grootendorst, R. (2004). *A systematic theory of argumentation : The pragma-dialectical approach.* Amsterdam: The press syndicate of the university of Cambridge.
Eemeren, F. H. van, Grootendorst, R., Jackson, S., & Jacobs, S. (1993). *Reconstructing argumentative discourse.* Tuscaloosa, Alabama: The University of Alabama Press.
Eemeren, F. H. van, Grootendorst, R., & Snoeck Henkemans, A. F. (2002). *Argumentation: Analysis, evaluation, presentation.* Mahwah (NJ)/London: Erlbaum.

Eemeren, F. H. van, Houtlosser, P., & Snoeck Henkemans, A.F. (2007). *Argumentative indicators in discourse. A pragma-dialectical study.* Dordrecht: Springer.

Gerritsen, S. (2001). Unexpressed premises. In F. H. van Eemeren (Ed.), *Crucial concepts in argumentation theory* (pp.50-80). Amsterdam: Sic Sat.

Greco Morasso, S., Mehmeti, T., & Perret-Clermont, A. N. (2016). Getting involved in an argumentation in class as a pragmatic move: Social conditions and affordances. In. D. Mohammed and M. Lewinski (Eds.), *Argumentation and reasoned action: Proceedings of the First European Conference on Argumentation*, (Vol. II, pp.463-478). London: College Publications.

Greco Morasso, S., Perret-Clermont, A.-N., & Miserez, C. (2015). L'argumentation à visée cognitive chez les enfants. In N. Muller Mirza and C. Buty (Eds.), *L'argumentation dans les contextes de l'éducation* (pp. 39-82). Bern: Peter Lang.

Grize, J.-B., (1982). *De la logique à l'argumentation.* Genève: Librairie Droz S.A.

Grize, J.-B., (1996). *Logique naturelle & communications.* Paris: PUF.

Hughes, H., Williamson, K., and Lloyd, A. (2007) Critical incident technique. In Lipu, S. (Ed.), *Exploring methods in information literacy research* (pp.49-66). Wagga Wagga, N.S.W: Centre for Information Studies, Charles Sturt University.

Inhelder, B., & Piaget, J., (1955). *De la logique de l'enfant à la logique de l'adolescent.* Paris, FR: Presses Universitaires de France.

Kohler, A. (2015). Elements of natural logic for the study of unnoticed misunderstanding in a communicative approach to learning. *Argumentum. Journal of the Seminar of Discursive Logic, Argumentation Theory and Rhetoric, 13*(2), 80-96.

Kohler, A., Lordelo, L. & Carriere, K. (2017). Researching research: Three perspectives for a hint of perspectivism. In G. Sullivan, (Ed.), *Proceedings of the 15th Biennal Conference of the International Society of Theory of Psychology* (pp.215-223). Coventry: Captus University Publication.

Kohler, A. & Donzé, T. (2017). De la pensée qu'il faut apprendre ou formater, à l'apprentissage de la pensée : quelques éléments d'une épistémologie perspectiviste pour un usage scolaire. In M. Lebrun, (Ed.), *Et si l'école apprenait à penser...* (pp. 59-84). Bienne: Éditions HEP-BEJUNE.

Kuhn, D., & Udell, W. (2003). The development of argument skills. *Child Development, 74*(5), 1245-1260.

Kuhn, D., & Udell, W. (2007). Coordinating own and other perspectives in argument. *Thinking & Reasoning, 13*(2), 90-104.

Mehmeti, T. (2016). *Students argumentation in a standardized test: Case study with a released item from PISA.* Presented at Students Perspectives to the Tasks and Demands of Schooling: Symposium Conducted at EARLI SIG 10, 21 and 25 Joint Conference, Reflective Minds and Communities, University of Tartu.

Mounoud, P. (2000) Le développement cognitif selon Piaget. Structures et points de vue. In O. Houdé & C. Meljac (Ed.), *L'esprit piagétien: Hommage international à Jean Piaget* (pp. 191-211). Paris: PUF.
OECD. (2004). *Learning for tomorrow's world - first results from PISA 2003.* OECD. Retrieved from http://www.oecd.org/education/school/programmeforinternationalstudentassessmentpisa/34002216.pdf
OECD. (2009). *Take the test sample questions from OECD's PISA assessments.* OECD. Retrieved from https://www.oecd.org/pisa/pisaproducts/Take%20the%20test%20e%20book.pdf.
Perret-Clermont, A.-N., (1979). *La construction de l'intelligence dans l'interaction sociale.* Berne: Peter Lang.
Perret-Clermont, A.-N. (1980). Recherche en psychologie sociale expérimentale et activité éducative. *Revue Française de Pédagogie*, *53*, 30-38.
Plantin, C. (2011) Pour une approche intégrée du champ de l'argumentation: Etat de la question et questions controversées. In V. Braun-Dahlet (Ed.), *Ciências da linguagem e didática das línguas sciences du langage et didactique des langues: 30 ans de coopération franco-brésilienne* (pp.181-207). São Paulo: FAPESP.
Plantin, C. (2005). *L'argumentation: Histoire, théories et perspectives.* Paris : PUF.
Psaltis, C., Duveen, G., and Perret-Clermont, A. N. (2009). The social and the psychological: Structure and context in intellectual development. *Human Development*, *52*(5), 291-312.
Psaltis, C., and Zapiti, A., (2014). *Interaction, communication and development: Psychological development as a social process.* London: Routledge.
Rigotti, E. (2006). Relevance of context-bound loci to topical potential in the argumentation stage. *Argumentation*, *20*(4), 519-540.
Sensevy, G., & Mercier, A. (Eds.) (2007). *Agir ensemble: L'action didactique conjointe du professeur et des élèves.* Rennes: Presses universitaires de Rennes.
Tartas, V., Perret-Clermont, A.-N., and Baucal, A. (2016). Experimental micro-histories, private speech and a study of children's learning and cognitive development / Microhistorias experimentales, habla privada y un estudio del aprendizaje y el desarrollo cognitivo en los niños. *Infancia y Aprendizaje*, *39*(4), 772-811.
Vergnaud, G. (2015). Argumentation et conceptualisation: Commentaires. In N. Muller Mirza& C. Buty, (Eds.), *L'argumentation dans les contextes de l'éducation*, Berne: Peter Lang.

APPENDIX

ROBBERIES SCORING 1

[Note: The use of NO in these codes includes all statements indicating that the interpretation of the graph is NOT reasonable. YES includes all statements indicating that the interpretation is reasonable. Please assess whether the student's response indicates that the interpretation of the graph is reasonable or not reasonable, and do not simply take the words "YES" or "NO" as criteria for codes.]

Full Credit

Code 21: No, not reasonable. Focuses on the fact that only a **small part** of the graph is shown.
- Not reasonable. The entire graph should be displayed.
- I don't think it is a reasonable interpretation of the graph because if they were to show the whole graph you would see that there is only a slight increase in robberies.
- No, because he has used the top bit of the graph and if you looked at the whole graph from 0 – 520, it wouldn't have risen so much.
- No, because the graph makes it look like there's been a big increase but you look at the numbers and there's not much of an increase.

Code 22: No, not reasonable. Contains correct arguments in terms of ratio or percentage increase.
- No, not reasonable. 10 is not a huge increase compared to a total of 500.
- No, not reasonable. According to the percentage, the increase is only about 2%.
- No. 8 more robberies is 1.5% increase. Not much in my opinion!
- No, only 8 or 9 more for this year. Compared to 507, it is not a large number.

Code 23: Trend data is required before a judgement can be made.
- We can't tell whether the increase is huge or not. If in 1997, the number of robberies is the same as in 1998, then we could say there is a huge increase in 1999.
- There is no way of knowing what "huge" is because you need at least two changes to think one huge and one small.

Partial Credit

Code 11: No, not reasonable, but explanation lacks detail.
- Focuses ONLY on an increase given by the exact number of robberies, but does not compare with the total.
- Not reasonable. It increased by about 10 robberies. The word "huge" does not explain the reality of the increased number of robberies. The increase was only about 10 and I wouldn't call that "huge".
- From 508 to 515 is not a large increase.
- No, because 8 or 9 is not a large amount.
- Sort of. From 507 to 515 is an increase, but not huge.

[Note that as the scale on the graph is not that clear, accept between 5 and 15 for the increase of the exact number of robberies.]

Code 12: No, not reasonable, with correct method but with minor computational errors.
- Correct method and conclusion but the percentage calculated is 0.03%.

Figure 8 – Scorring sheet for the item "Robberies", part I (PISA, OECD).

No Credit

Code 01: No, with no, insufficient or incorrect explanation.
- No, I don't agree.
- The reporter should not have used the word "huge".
- No, it's not reasonable. Reporters always like to exaggerate.

Code 02: Yes, focuses on the appearance of the graph and mentions that the number of robberies doubled.
- Yes, the graph doubles its height.
- Yes, the number of robberies has almost doubled.

Code 03: Yes, with no explanation, or explanations other than Code 02.

Code 04: Other responses.

Code 99: Missing.

Figure 9 – Scoring sheet for the item "Robberies", part II (PISA, OECD).

35

Modes of Inference in Aristotle's Concept of the Enthymeme

MANFRED KRAUS
University of Tübingen
manfred.kraus@uni-tuebingen.de

Based on a comparative analysis of the descriptions of the enthymeme in Aristotle's *Rhetoric* and *Prior Analytics* respectively, it will be demonstrated that the Aristotelian concept of the enthymeme incorporates a number of different modes of reasoning, including deductive, inductive and abductive, valid and defeasible modes, and that a unified theory of the Aristotelian enthymeme can be developed that covers the accounts of the *Topics* and *Rhetoric* just as well as that of the *Analytics*.

KEYWORDS: abduction, analytics, Aristotle, deduction, enthymeme, induction, Peirce, rhetoric, syllogism, topics

1. INTRODUCTION

As this conference has set itself the general theme of 'Argumentation and Inference', it may be appropriate to talk about a very old type of argument, which many argumentation scholars may meanwhile regard as boring and not presenting any interesting new aspects, but which in my view is highly pertinent because of the fact that it incorporates a fair number of different types of inference, namely Aristotle's concept of the enthymeme.

I will begin by making two preliminary remarks. My first remark, which nowadays should hardly be necessary to make, is: I will *not* talk about incomplete syllogisms and missing premises. For more than a hundred years, scholars have worked hard to refute the long-standing opinion that Aristotle's enthymeme was simply a syllogism with one premise suppressed. It should meanwhile be clear that this so-called theory of *syllogismus truncatus* does not represent Aristotle's original theory but is a later misinterpretation (Burnyeat, 1994, pp. 44-46; Green, 1995, pp. 27-32), which is not to say that it is not also one legitimate definition of enthymeme, which has been popular for a long time in history and is still being employed by many argument theorists. It is just not Aristotle's own theory.

The second remark may be more surprising. Contrary to our present-day perception, Aristotle's concept of the enthymeme was not at all a typical representation of an ancient mainstream theory. It was on the contrary a rather bold innovation that greatly diverged from what may be called the standard understanding of Aristotle's own times, as it emerges for instance from the definition given in the so-called *Rhetoric to Alexander* (10, 1430a23-37), a work roughly contemporary to Aristotle's *Rhetoric*. On this standard account, which persisted throughout antiquity and likewise emerges in classical Roman and late ancient rhetorical theory, the enthymeme was regarded not as a deductive argument, but as a pungent refutative argument that drew on inconsistencies within an opponent's words or actions.

In this paper, I will first review the different accounts of the enthymeme Aristotle gives in different works and point to apparent breaks and inconsistencies between them. Next, I will talk about earlier attempts to account for these inconsistencies and about the position of the *Rhetoric* between topical and syllogistic concepts of argumentation. I will then take a closer look at Aristotle's accounts in *Rhetoric* and *Prior Analytics* and especially at the examples he presents there for different types of enthymemes. This will finally lead to an attempt to harmonize both concepts by applying a theory of modes of inference developed by Charles S. Peirce, and by proposing a reading of the seemingly syllogistic account in *Rhetoric* book I on a non-syllogistic theoretical basis adopted from the *Topics*.

2. ARISTOTLE'S MULTIPLE ACCOUNTS OF THE ENTHYMEME: BREAKS AND INCONSISTENCIES?

Aristotle never gives any etymological or other explanation of the word 'enthymeme'. Nor do any of his predecessors or contemporaries. A general understanding of the word as such appears thus to be somehow presupposed (see Bons, 2002, p. 19).

Aristotle discusses enthymemes in several of his works, but particularly in two relevant contexts: on the one hand in his *Rhetoric*, and on the other near the end of the *Prior Analytics* (II 27), in which he expounds his advanced theory of inference. In the *Rhetoric*, moreover, elaborate treatments of the enthymeme are found in two different passages and from different points of view: While in chapter 2 of book I Aristotle presents a methodical and systematic discussion of the enthymeme within the framework of his general concept of the theory of means of persuasion, in chapters 22-25 of book II he gives extended lists of types of enthymemes and apparent enthymemes based on so-called *topoi* or argument schemes (he lists 28 topoi of enthymemes in chapter 23, and 9 topoi of 'apparent' enthymemes in chapter 24). Such topical enthymemes are, for instance, the inference from bigger to smaller ("if not even the gods do know this, human beings will know even less") or from smaller to bigger ("if a child can lift this weight, a grown-up man should be able to lift it even more easily") or the *argumentum e contrario* (if predicate p applies to subject A, the opposite predicate non-p will apply to the opposite subject non-A; e.g. "if war causes evils, peace will bring about good things").

A quite embarrassing difficulty, however, arises from the fact that there are on the one hand manifest affinities between the accounts in *Rhetoric* I 2 and in *Prior Analytics*, but on the other hand also obvious discrepancies between the respective treatments in *Rhetoric* I 2 and II 22-25. The latter account seems to accord rather with the *Topics*. This appears to disrupt the consistency of the *Rhetoric*.

3. EARLIER ATTEMPTS TO SOLVE THE PROBLEM

Aristotle's *Rhetoric* is a highly intricate work, in which not only theoretical elements from several other disciplines merge (Rapp, 2005), but which also has a very complex genetic history. Aristotle may have sketched a first draft of it in the fifties or forties of the fourth century, while he was a member of the Platonic Academy, but may have revised and supplemented it later on, especially so after he had returned to Athens in the 330s, which is only natural in view of the status of the text as a lecture manuscript (see Rapp, 2002, I, pp. 178-191; 314-319).

The classical approach to explain the discrepancies between a 'topical' interpretation of the enthymeme in book II and an 'analytical' reading in book I therefore has been the assumption of the coexistence of different genetic layers within the *Rhetoric*. This solution has been suggested most prominently by Friedrich Solmsen (1929) and Myles Burnyeat (1994). According to this theory, under the influence of the new discoveries made in the *Analytics*, Aristotle would have replaced an earlier 'topical' conception – similar to the account of inference given in the *Topics* – by a more 'syllogistic' analysis in book I, yet would have left traces of the topical conception in the final chapters of book II. However, it is precisely in chapters 22-25 of book II that one finds allusions to historical events of the 330s, which – unless they are later interpolations – would suggest a late date of those passages (Rapp, 2002, I, p. 180). On the other hand, the derivation of rhetoric from dialectics, as expounded at the beginning of book I in the famous statements on rhetoric being a counterpart (*antistrophos*) or side shoot (*paraphues*) of dialectics, would seem to be conceptually closer to Aristotle's juvenile study of dialectics in the *Topics*, which would rule out any relation or link to the much later *Analytics*. Cross-references to the *Analytics* that are found especially in passages dedicated to the enthymeme from signs in the first chapters of book I can more easily be explained as later interpolations (or intrusions of marginal notes) (Rapp, 2002, I, pp. 189-191). It may therefore not be necessary to declare, as Solmsen (1929, pp. 13-27) and Burnyeat (1994, pp. 31-38) have proposed, the entire passage on the enthymeme to be a later addendum or revision made under the influence of the syllogistic theory of the *Analytics* (see Rapp, 2002, I, p. 332; II, pp. 202-204).

Quite contrary to this, in his magisterial commentary of 2002, Christof Rapp insists that the passages on the enthymeme in the first chapters of book I can and should be understood entirely on the basis of topical principles, and that any influence of the syllogistic theory of the *Analytics* should be shunned (Rapp, 2002, II, pp. 241-248). His main arguments concern the uniformity of the definitions of *syllogismós* in *Topics*, *Rhetoric* and *Analytics*, none of which includes the particular requirements of a fully developed categorical syllogism, and the admissibility of one-premise arguments.

Here is a list of definitions of the enthymeme from various works by Aristotle:

i. *Topics* VIII, 164a3-6: Records of discussions should be made in a universal form, even though one has argued only some particular case: for this will enable one to turn a single rule into several. A like rule applies in rhetoric to enthymemes.

ii. *Rhetoric* I 1, 1355a6-8: the enthymeme is a kind of deduction (*syllogismós*).

iii. *Rhetoric* I 2, 1356b4-6: I call an enthymeme a rhetorical deduction (*syllogismós*), and an example a rhetorical induction.

iv. *Rhetoric* I 2, 1357a13-17: the enthymeme and the example are concerned with things which may, generally speaking, be other than they are, the example being a kind of induction and the enthymeme a kind of deduction (*syllogismós*), and deduced from few premises, often fewer than the regular deduction.

v. *Rhetoric* I 2, 1357a30-33: enthymemes are derived from probabilities and signs.

vi. *Rhetoric* II 26, 1403a17-18: an element (*stoikheion*) or a topic is a heading under which many enthymemes fall.

vii. *Prior Analytics* II 27, 70a10: an enthymeme is a deduction (*syllogismós*) from probabilities or signs.

From this list of definitions of the enthymeme in Aristotle it is evident that in book I of the *Rhetoric* (ii-iv) the concept of *syllogismós* or deduction is dominant, but also that there is some link between (vi) the account in book II and (i) the one single remark on enthymemes in the *Topics*, insofar as the subsumption of single enthymemes under general argument schemes is emphasized. We further see that the shared connection of enthymemes with probabilities and signs establishes another close-knit link between (v) *Rhetoric* I and (vii) *Prior Analytics*.

In the *Rhetoric*, Aristotle certainly defines the enthymeme as "a kind of *syllogismós*" (*syllogismós tis*, *Rhet*. I 1, 1355a8). But, according to Rapp, *syllogismós* does not necessarily mean the fully developed syllogism of the *Analytics*, with exactly three propositions and exactly three terms. For the definitely pre-syllogistic *Topics* already uses the same definition of *syllogismós* as a statement, "in which, from things posited, something other than these things necessarily follows through the things posited" (*Top*. I 1, 100 a 25–27). This definition is exactly reproduced in the *Analytics* (*Anal. Pr.* I 1, 24b18-20), and the *Rhetoric* only adds that the conclusion may also follow only "as a rule" (*Rhet*. I 2, 1356b16-18).

According to Rapp, the *syllogismós* in the *Topics* as well as in the *Rhetoric* is a simple deductive premise-conclusion argument, in which neither the number of premises nor the number of terms nor their linguistic expression is restricted in any way (Rapp, 2002, II, pp. 59-67). Even one-premise arguments are admissible (as is almost the rule in the *Topics*). The inferential force of this deductive argument is entirely based on a topical relation. On this account, the enthymeme of the *Rhetoric* proves to be a subspecies of the dialectical *syllogismós* of the *Topics*, differing only by its object (things which "may be other than they are", on which hence disagreement and debate is possible) and by its audience (listeners with no academic education and limited intellectual grasp). In a word, it is "a dialectical *syllogismós* in rhetorical application" (Rapp, 2002, II, p. 229). Yet if the number of premises is undetermined, formal deficiency by the missing of a premise cannot be its defining mark. Aristotle does occasionally recommend succinct formulation of enthymemes (see Sprute, 1982, pp. 130-132), yet only in consideration of the limited capacities of the audience. It is only in comparison with an apodictic *syllogismós* that an enthymeme is deficient in that it proceeds from generally accepted instead of evident or scientifically proven premises and can therefore only yield probable conclusions. But as an important fact we can assess that Aristotle bases the enthymeme firmly on a deductive pattern, which, as we have seen, is a bold innovation.

By basing the account of *Rhetoric* I likewise on topical ground, Rapp manages to save the unity of the *Rhetoric* and to render any discussions about dates and layers gratuitous, but at what price? His interpretation opens up another gap, since it fails to explain the evident parallelism of the examples presented for enthymemes from signs in *Rhetoric* I and in the *Analytics*.

4. ARISTOTLE'S EXAMPLES IN *RHETORIC* I AND *PRIOR ANALYTICS*

Let us therefore now turn to the accounts in *Rhetoric* I and *Prior Analytics*. As the sources of enthymemes, in both works Aristotle names probabilities (*eikóta*) and signs (*semeía*) (*Rhet.* I 2, 1357a30-33; *Anal. Pr.* II 27, 70a10). Both terms also occur in the *Rhetoric to Alexander*, yet as independent means of persuasion alongside the enthymeme. About enthymemes from probabilities, Aristotle does not bother much. In the *Rhetoric* he does not even give an example, and in *Prior Analytics* he just as much as mentions that propositions such as "envious people are malevolent" may be used as probable premises, because they are accepted by a majority of people. It is easy to supply, though, that what is meant by an enthymeme from probability is a deductive inference from a probable proposition to an equally probable conclusion (by way of a formally valid inference, as we may suppose).

This looks like a neat dichotomy. Yet since the second category is again subdivided into three different modes, what we actually get is a four-fold taxonomy.

Those three subtypes of enthymemes from signs, however, are classified in *Prior Analytics* according to the three Aristotelian syllogistic figures, but in the *Rhetoric* according to their character as valid or defeasible inferences and according to the criterion of inference from universal to particular or from particular to universal.

If we now look more closely at the examples by which Aristotle illustrates the three subtypes in the two treatises respectively, their similarity is absolutely striking.

> viii. *Rhetoric* I 2, 1357b10-21:
> A) From particular to universal:
> Type 1: Wise men are just, because Socrates was both wise and just.
> Type 2: (a) This man is ill, because he has a fever [and those who have a fever are ill]. – (b) This woman has had a child because she has milk [and women who have milk have had a child].
> B) From universal to particular:
> Type 3: This man has a fever, because he breathes hard [and those who have a fever do breathe hard].
>
> ix. *Prior Analytics* II 27, 70a11-24:
> 1st figure: Women who have milk are pregnant; this woman has milk; therefore she is pregnant.
> 3rd figure: Pittacus is wise; Pittacus is good; therefore wise men are good.
> 2nd figure: Pregnant women are pale; this woman is pale; therefore she is pregnant.

The examples in the *Rhetoric* are partly expressed in abbreviated form (bracketed parts supplemented). Nonetheless the parallels are obvious. One of the examples of type 2 is practically identical to the one for the 1st figure in the *Analytics*; in type 1 (3rd figure) it is only Socrates who is replaced by Pittacus, and the predicate 'just' by the fairly similar one 'good'. The examples of type 3 and 2nd figure respectively are clearly structurally analogous and both taken from a medical context. Can all these be mere coincidences? Should we imagine that Aristotle, when he first devised his *Rhetoric*, by mere chance happened to pick exactly those three structural types of examples of enthymemes from signs that would later in the *Analytics* turn out to fit so nicely with his three syllogistic figures? Rapp has no explanation for this. Against Rapp, I therefore hold that at the time he wrote *Rhetoric* I, Aristotle must at least have had some inkling, however blurred, of different modes of inference, although I agree with Rapp on the point that he may not yet have described them in terms of syllogistic figures.

5. ARISTOTLE'S EXAMPLES AND PEIRCE'S MODES OF INFERENCE

Let us look at the examples in detail: Type 2 is the only one that yields a valid and necessary conclusion, provided that the premise is true (which, I think, Aristotle invariably presupposes in all examples). An inference of that kind Aristotle calls a *tekmérion* (which may be translated as 'infallible sign' or 'proof'). This creates a problem, since what emerges is not a probable argument but a perfectly valid deduction. This has always been a standard argument of those who believed that only the *syllogismus-truncatus* model could account for any difference between this type of enthymeme and a syllogism proper. Yet the difference is easy to see: the enthymeme of this type employs singular terms ('this man', 'this woman'), which is prohibited in a syllogism proper on Aristotle's account. But as a rhetorical argument, an enthymeme necessarily must aim at a singular case (Sprute, 1982, p. 76). Singular terms, even proper names, are equally used in types 1 (Socrates, Pittacus) and 3 ('this man').

For an appropriate understanding of this threefold taxonomy, I would propose to apply a discovery made in the 1860s by the American pragmatist Charles Sanders Peirce, not least in the course of reflections on Aristotle's logic (e.g. Peirce 1878/1932). Peirce realized that when in a classical deductive categorical syllogism we switch propositions (which Peirce describes in terms of Rule, Case and Result), what we obtain are two other modes of inference, one of which aptly describes the procedure of an inductive argument (Case, Result; therefore Rule), while the third one (Result, Rule; therefore Case) gives a reasonable explanation for an observed fact, a mode Peirce initially called hypothesis, but later termed abduction.

Peirce's famous example was that of beans taken from a bag. But the same procedure also works for the standard example of Aristotelian syllogistics: From the general rule that human beings are mortal and the knowledge that Socrates is a human being, we infer deductively that Socrates must be mortal. Inversely, we can infer inductively from the observation that Socrates is mortal and that he is a human being, that human beings in general are mortal. Finally, if we observe that Socrates is mortal, and if we happen to know that human beings in general are mortal, we may tentatively infer (by way of abduction) that Socrates may probably be a human being. The latter inference, like the preceding inductive one, is of course defeasible, since the Socrates in question might as well be, say, a dog; we may always be mistaken.

I am using terms such as 'deductive' and 'inductive' in a purely structural sense here, and not, as they are often used, in a normative sense, whereby 'deductive' would describe a valid inference, and 'inductive' an invalid or defeasible one. In that sense, also deductive inferences can be invalid, as soon as negations and quantifiers get involved. It was Aristotle, who in the *Analytics* ascertained that only four out of 64 possible combinations of a first figure syllogism yield a valid conclusion (*Anal. pr.* I 4, 25b26-26b33).

But how about this application?

x. Rule: Women who have milk have had a child.
   Case: This woman has milk.
   Result: This woman has had a child.

xi. Result: Socrates is just.
   Case: Socrates is wise.
   Rule: Wise men are just.

xii. Result: This man breathes hard.
   Rule: People who have a fever breathe hard.
   Case: This man has a fever.

It is easy to see how Peirce's three modes square with Aristotle's three types of enthymemes from signs. If we look at (xi), the enthymeme of type 1, we observe that this in fact quite exactly describes the scheme of an inductive inference, and – because of the singular term involved – precisely in its rhetorical variant, namely the example. Socrates (or Pittacus) are just outstanding examples that are used to suggest a general relationship between two predicates. This type of inference, as Aristotle himself observes, is clearly refutable, even if the conclusion is true, since the syllogism is not universal: for although Pittacus is good, it is not therefore necessary that all other wise men must be good.

From this point of view, type 3 (xii) is an exemplification of an abductive inference, that is of an inference from a symptom observed to its subsumption under a general rule that would best explain it. This type of inference is always defeasible, as Aristotle again observes; for even if pregnant women are pale, and this woman is also pale, this does not necessarily mean she must be pregnant. But in spite of its general defeasibility, on the other hand, in practical life this type of inference is highly expedient and efficient, especially for physicians or judges, for determining an appropriate medical treatment for an illness, or for reconstructing the most plausible sequence of events.

If we look at them this way, types 1 and 3 would not be proper deductions at all, but actually inductions and abductions. Only type 2 (x) describes a real deduction (and a valid argument). Yet this fits very well with Aristotle's own remark that the other two types are actually non-deductive (*asyllogistoi*), since they are open to falsification by counter-examples. He may have intuitively perceived that their deductive character is disputable. As he states, even if all premises are true, these kinds of enthymemes from signs at best yield probable (but still highly plausible) conclusions. Yet when employed as enthymemes, inferences of type 1 and 3 *pretend* to be deductive, since type 1 draws universal conclusions from a single example (whereas it is demonstrated in the *Analytics* that 3rd-figure syllogisms may validly only yield particular conclusions; *Anal. pr.* I 6, 29a16-18); and type 3 draws positive conclusions (whereas in the 2nd figure only negative conclusions can be valid; *Anal. pr.* I 5, 28a7-9).

Such an interpretation of the description of the enthymeme in *Rhetoric* I, which as it were anticipates the later findings of the *Analytics*, very nicely fits with Aristotle's otherwise irritating statement in book II, chapter 25, that enthymemes arise from *four* sources, namely probabilities, infallible signs (*tekméria*), fallible signs (*semeía*) and examples. Enthymemes from probabilities are clearly those from *eikóta*, enthymemes from signs of type 2 represent infallible *tekméria*-enthymemes, type 3 those from fallible signs, and type 1 those from examples (see Figure 1 below). If this reading is sound, Aristotle actually manages to incorporate *all* kinds of rhetorical inferences into his overall concept of the enthymeme, deductive inferences as well as inductive or abductive.

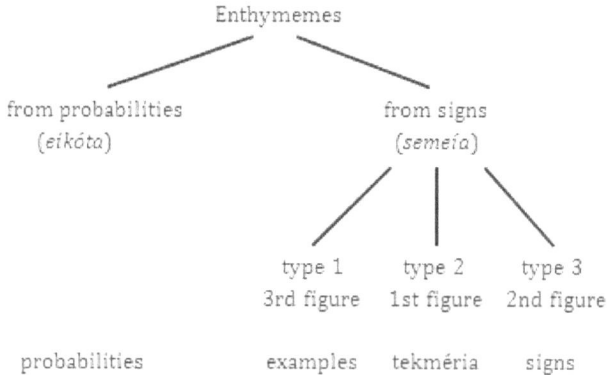

Figure 1 – Subtypes of enthymemes in *Rhetoric* I and *Prior Analytics*

6. A TOPICAL READING OF *RHETORIC* I 2

This said, to save Rapp's thesis, we still need to demonstrate how the three kinds of inferences from signs can be described without reference to syllogistic theory, on topical grounds alone, and as one-premise arguments. I think that such a description is indeed possible.

Scholars have often wondered how Aristotle's classification of enthymemes from signs according to inference from particular to universal and vice versa should be understood, especially why the infallible enthymemes of type 2 should be categorized as arguing from particular to universal. In a paper from 1988, Hermann Weidemann has therefore even proposed to swap types 2 and 3 (including introductory phrases) in Aristotle's text, so that only type 1 would be arguing from particular to universal, and both types 2 and 3 from universal to particular.

But this is probably the clue to the entire interpretation. In the *Topics*, which works predominantly on one-premise arguments, the inference warrant is usually the extensional relationship between two terms (genus to species, species to genus, definition to definiendum etc., see Primavesi, 1996, pp. 104-106; 150-151). Oftentimes the question is whether the replacement of one term in subject or predicate position by a related, but extensionally different one keeps the validity of a proposition sound or not. If we apply this principle to the examples in the *Rhetoric*, we see how it can work.

In the infallible sign enthymeme of type 2 'this man' is a species of the genus 'feverish', and 'feverish' is a species of the even wider genus 'ill'. From the proposition 'this man has a fever', by mere extension of the predicate, it is directly inferred that 'this man is ill'. This inference is sound, since, as we learn from the *Topics*, an extension of the predicate from species to genus will preserve the validity of the proposition.

In type 1, it is (correctly) assumed that Socrates is a species of the genus 'wise men', and also that 'wise men' is a species of the genus 'just men'. Now again, from the proposition that 'Socrates is wise' it is directly inferred, this time by extension of the *subject* from 'Socrates' to its genus 'wise', that wise men are just. Yet this inference is unsound, since extension of the subject does *not* necessarily preserve the validity of the proposition, because the assumed hierarchy of terms is an overly optimistic assumption. In reality, 'wise men' and 'just men' may overlap only to a certain extent, in which Socrates happens to be included. Hence at best it can be soundly inferred that *some* wise men will be just. Particular conclusions are sound within this type, yet universal ones are not. But type 1 also works by way of extension of a term (the subject in this case). This is probably why Aristotle classifies it as an inference from particular to universal. Hence not only is Weidemann's reordering of the text unnecessary, but it spoils the whole idea.

Now for type 3: It is assumed that 'this man' is a species of the genus 'breathing hard', and that 'being feverish' is also a species of the genus 'breathing hard' (alongside many other species, such as, e.g. having just done a long-distance run, being asthmatic or the like). From the proposition 'this man breathes hard' it is directly inferred, this time by restriction of the predicate, that 'this man has a fever'. But again, this inference is unsound; for restriction of predicate terms does affect the validity of a proposition. 'Being feverish' and 'this man' may well both be proper subsets of 'breathing hard', yet nonetheless still be mutually exclusive. What would be possible is to say that if this man did not breathe hard, it would be obvious that he did not have a fever (since feverishness is one of the species of 'breathing hard'). This type, in contrast to the other two, works on restricting a term; hence it works from universal to particular. Aristotle does not give any example for the restriction of a subject term; yet that would be a logically sound procedure as well.

If this reading is correct, it is perfectly possible to interpret Aristotle's description of enthymemes from signs in *Rhetoric* I on purely topical terms and as immediate one-premise arguments from one proposition to another. The *Topics* actually offers descriptions of such (sound) arguments from species to genus or from genus to species (e.g. in II 4, 111a14-20 or 111b17-23; Primavesi, 1996, pp. 151-152; 160-165). For parallels to the unsound inferences of type 1 and 3, however, one has to look into the *Sophistical Refutations*, which offer instructive analogies to both types (5, 166b37-167a20; 167b1-20). In modern fallacy theory, these are known as fallacies of "affirming the consequent" and "hasty generalization" respectively (Hamblin, 1970, pp. 35–37, 46-47). Hence, from the point of view of a dialectician, Aristotle classifies those two types as fallacies, whereas in the *Rhetoric* he happily embraces them as enthymemes because of their persuasiveness and practical usefulness in cases, which "may also be other than they are", and in which perfect validity of inference can therefore only be attained exceptionally.

7. CONCLUSION

I hope to have demonstrated that, in spite of the apparent breaks and inconsistencies, it may be possible to establish a unified theory of

Aristotle's concept of the enthymeme that manages to retain and explain the parallelisms of the *Rhetoric* both to the *Topics* and to the *Analytics*, and at the same time does justice to Rapp's arguments about a basically topical and non-syllogistic character of the *Rhetoric* by disclosing a way of describing the 'syllogistic' passages in *Rhetoric* I on the basis of topical argument structures alone.

REFERENCES

Bons, J. A. E. (2002). Reasonable argument before Aristotle: The roots of the enthymeme. In F. H. van Eemeren & P. Houtlosser (Eds.), *Dialectic and rhetoric: The warp and woof of argumentation analysis* (pp. 13-27). Dordrecht: Kluwer.

Burnyeat, M. F. (1994). Enthymeme: Aristotle on the logic of persuasion. In D. J. Furley & A. Nehamas (Eds.), *Aristotle's Rhetoric: Philosophical essays* (pp. 3-55). Princeton, NJ: Princeton University Press.

Green, L. D. (1995). Aristotle's enthymeme and the imperfect syllogism. In W. B. Horner & M. Leff (Eds.), *Rhetoric and pedagogy: Its history, philosophy, and practice. Essays in honor of James J. Murphy* (pp. 19-41). Mahwah, NJ: Erlbaum.

Hamblin, C. L. (1970). *Fallacies*. London: Methuen.

Peirce, C. S. (1878). Deduction, induction, and hypothesis. *Popular Science Monthly*, 13, 470–483. (1932). In C. Hartshorne & P. Weiss (Eds.), *Collected Papers of C. S. Peirce* (Vol. II, pp. 372–388, Ch. 5). Cambridge: Harvard University Press.

Primavesi, O. (1996). *Die Aristotelische Topik. Ein Interpretationsmodell und seine Erprobung am Beispiel von Topik B*. München: C.H. Beck.

Rapp, C. (2002). *Aristoteles, Rhetorik: Übersetzt und erläutert*. Berlin: Akademie-Verlag.

Rapp, C. (2005). Zur Konsistenz der Aristotelischen Rhetorik. In J. Knape & T. Schirren (Eds.), *Aristotelische Rhetoriktradition* (pp. 51-71). Stuttgart: Steiner.

Solmsen, F. (1929). *Die Entwicklung der aristotelischen Logik und Rhetorik*. Berlin: Weidmann.

Sprute, J. (1982). *Die Enthymemtheorie der aristotelischen Rhetorik*. Göttingen: Vandenhoeck & Ruprecht.

Weidemann, H. (1988). Aristoteles über Schlüsse aus Zeichen („Rhetorik' I 2, 1357b1-25). In R. Claussen & R. Daube-Schackat (Eds.), *Gedankenzeichen: Festschrift für Klaus Oehler zum 60. Geburtstag* (pp. 27-34). Tübingen: Stauffenburg.

36

"I Think Any Reasonable Person Will Agree...": A Corpus and Text Study of Keywords in Irish Political Argumentation

DAVIDE MAZZI
University of Modena and Reggio Emilia, Italy
davide.mazzi@unimore.it

This paper brings a corpus and discourse perspective to bear on the investigation of the broader argumentative implications of keywords in the context of 20th-century Irish politics. On the basis of two corpora including Michael Collins' papers and Eamon de Valera's speeches and statements, a keyword-in-context analysis was performed. Results provide evidence of the persuasive power of keywords as signposts leading to a better understanding of culturally shared rules of inference in political discourse.

KEYWORDS: political argumentation, keywords, phraseology, corpus, text, Ireland.

1. INTRODUCTION

In their capacity as means of access to a shared body of knowledge, keywords have been researched from a variety of oft-interrelated perspectives. First of all, they have been investigated as cultural keywords, by focusing on their role as tools to gain access to the inner workings of a culture (Williams, 1976; Liebert, 2003; Wierzbicka, 2006). Secondly, keywords have been investigated as indicators of propositional content, namely as those pointing to the conceptual structure of a text and its overall aboutness, as it were (Scott, 1998; Bondi, 2010). Thirdly, they were dealt with as items allowing analysts to identify lexico-grammatical patterns and schemas across a wide range of discursive and/or disciplinary areas (Hunston & Francis, 1998;

Stubbs, 1996 and 2001). Finally, keywords have been discussed in terms of their overall argumentative implications.

As far this last research strand is concerned, for instance, keywords were defined by Rigotti and Rocci (2005, p. 131) as words that act as *termini medii* in enthymematic arguments, operating at the same time "as pointers to an endoxon or constellation of endoxa that are used directly or indirectly to supply an unstated major premise". In an attempt to define the structure of reasoning that underlies the connection between a standpoint and its supporting arguments, keywords have also been incorporated into the Argumentum Model of Topics (AMT) (Rigotti, 2008; Rigotti & Greco Morasso, 2010). By delving into the relationship between implicit premises of a material nature – chiefly, endoxa – and the level of explicit lexical choices, this approach proved most effective in shedding light on keywords as those terms "that activate cognitive frames from which endoxa are then drawn to be used in the argumentation" (Bigi & Greco Morasso, 2012, p. 1142).

In the interest of bringing such analytical insights as close together as possible, this study is aimed at bringing a corpus and discourse perspective (Baker, 2006; Fetzer, 2014; Baker & McEnery, 2015) to bear on the study of keywords in argumentation. More specifically, the research is intended to combine quantitative keyword analysis (Bondi & Scott, 2010) with the more essentially qualitative perspective provided by AMT. In order to accomplish this purpose, a preliminary study of keywords in political argumentation was undertaken, with Irish political discourse in historical rather than contemporary terms as a case in point. In that regard, Michael Collins and Eamon de Valera's discourse was chosen as the object of the investigation, not merely due to their undisputed political stature in the Irish context, but also because in spite of an ever growing body of research on their historical significance (Costello, 1997; Hart, 2005; Ferriter, 2007), their profile as arguers is still waiting to be fully elucidated (Mazzi, 2016).

The rest of the paper is organised as follows. In Section 2, corpus design criteria are discussed, and the methodological tools are introduced: this will allow for a presentation of the dataset as well as a preliminary review of the procedure(s) through which the data were studied. Section 3 then presents the findings of the study, which are eventually discussed in the light of the relevant literature in Section 4.

2. MATERIALS AND METHODS

The study centred on the two key figures behind the establishment of the modern Irish State in the first half of the twentieth century (Ryle Dwyer, 2006). First of all, Michael Collins (1890-1922), the revolutionary reader and dedicated organiser, Director of Intelligence of the Irish Republican Army during the Anglo-Irish War of Independence (1919-1921), member of the Irish delegation that signed the historical Anglo-Irish Treaty that would lay the foundations of the Irish Free State from January 1922, Chairman of the Provisional Government Cabinet and first Minister for Finance of the new-born Irish State. Secondly, Eamon de Valera (1892-1975), formerly the Irish Volunteer and dogmatic Republican who opposed the terms of the Anglo-Irish Treaty signed by Collins and others on the grounds that it undermined Irish national aspirations, by partitioning the island of Ireland and severing the six counties of the North from the rest of the Irish State; and then, the statesman who would serve as both *Taoiseach* [Prime Minister] and *Uachtarán* [President of Ireland] until the end of a political career of unquestionable longevity.

In order to analyse the discourse of both leaders, two corpora were collected. The first one, the so-called Mick-Corpus, includes the eleven essays published by Collins in the volume *The Path to Freedom*. The second corpus, hereinafter referred to as the Dev-Corpus, includes 126 of de Valera's best known speeches and statements, as collected by his own personal secretary Maurice Moynihan (1980). The Mick-Corpus amounts to 32,335 words altogether, whereas the Dev-Corpus contains 288,254 tokens.

From a methodological perspective, the study inevitably had to grapple with the long-standing problem of how to identify keywords (Scott, 2010). For the sake of clarity, the research embraced and adapted O'Halloran's (2009, p. 25) notion of "corpus-comparative statistical keywords" as those items "being statistically more frequent in a text or set of texts than" in a corpus used for comparative purposes "known as the reference corpus". In particular, the two corpora presented above were used as each other's reference corpus, in order to extract Collins' and de Valera's distinctive keywords.

Once a keyword list was thus generated for each corpus through the linguistic software package *AntConc 3.2.1* (Anthony, 2006), three main steps were followed. First of all, any pattern of semantic proximity across each speaker's top-fifteen keywords was identified (Gramley & Pätzold, 2004). Secondly, the relevant keywords were extracted and concordanced, i.e. analysed in context (Sinclair, 2004; Römer & Wulff,

2010), in order to examine their preferred collocational and phraseological patterns. Finally, on the basis of a manual text-based analysis, the relationship between collocational patterns and the inferential configuration of key arguments was highlighted for both Collins and de Valera.

3. RESULTS

By comparing the Mick-Corpus and the Dev-Corpus, the top-fifteen keywords were identified for each of them. These are displayed in Table 1 below:

Corpus	Top-15 keywords
Mick-Corpus	Freedom, English, civilization, British, Gaelic, Irish, national, Ireland, had, Treaty, Nation, strength, free, spirit, succeeded.
Dev-Corpus	I, that, you, if, think, am, going, your, time, know, want, should, today, Constitution, Bill.

Table 1 – Top-15 keywords in Mick-Corpus and Dev-Corpus

First of all, it is interesting to note that a few keywords in the Mick-Corpus appear to denote abstract ideas and concepts, i.e. 'freedom', 'civilization', 'strength', 'spirit'. Secondly, some of the keywords in the Dev-Corpus may be described as indicating cognitive and/or volitional acts, namely 'think', 'know', '(am) going (to)', 'want'. By virtue of the prominent semantic proximity across those terms, they were extracted for the purpose of the exploratory keyword analysis attempted here for both Collins (Section 3.1) and de Valera (Section 3.2).

3.1 Collins' keywords in argumentation

As an illustrative example of the study of Collins' keywords, 'freedom' and 'civilization' can definitely be taken as a case in point. Even from the restricted sample of concordance lines displayed in Table 2, to begin with, it may be observed that 'freedom' tends to collocate with items qualifying it in terms of its overall extent, e.g. 'complete', 'full', 'crumb of':

Concordance lines of 'freedom'
...individual and national **freedom** - <u>of the fullest and broadest character</u>; freedom to think...
...<u>Complete</u> national **freedom** can now be ours, and...
...freedom we have won to achieve <u>full</u> **freedom**.
....<u>The complete fulfilment</u> of our full national **freedom**...
...that period Ireland would, it was hoped, [...] accept <u>the crumb of</u> **freedom** offered by...

Table 2 – Concordance lines and collocational patterns of 'freedom'

This trend applies to about 10% of the corpus occurrences of the keyword. Moreover, it underlies a striking opposition between the idea of full and accomplished freedom generated by the first concordance lines in Table 2, and the utterly negative notion of incomplete freedom embodied by 'the crumb of freedom' of the last line.

As the careful study of the context surrounding these occurrences reveals, on the one hand, Collins borrows the concept of freedom as a goal to be accomplished to the fullest and broadest extent from William Rooney (example 1 below). The journalist, poet and advanced-nationalist urged the Irish public opinion to embrace more radical views of Irish nationhood, by rejecting any form of "slavish loyalty to the British crown" and the "horror of the very name of revolution" (Rooney, 1909, p. 99) that had characterized the Irish parliamentary tradition since Daniel O'Connell:

> (1) He [William Rooney] interpreted the national ideal as 'an Irish State governed by Irishmen for the benefit of the Irish people'. He sought to impregnate the whole people with 'a Gaelic-speaking Nationality'. 'Only then could we win freedom and be worthy of it; <u>freedom - individual and national freedom - of the fullest and broadest character</u>; freedom to think and act as it best beseems; national freedom to stand equally with the rest of the world' (Collins, *Freedom within grasp*).

On the other hand, the 'crumb of freedom' is the image used by Collins to refer to what the British Parliament was ready to offer Ireland through the Government of Ireland Act 1920, which would go down in history as the Partition Act severing the mainly Protestant North from the overwhelmingly Nationalist South. This piece of legislation (cf. 2

below) was in fact a Home Rule act aimed at giving Ireland limited autonomy, while at the same forcing a partitioned island to continue to be part of the United Kingdom:

> (2) The Act was probably intended for propaganda purposes. It might do to allay world criticism to draw attention away from British violence for a month or two longer. At the end of that period Ireland would, it was hoped, [...] accept the crumb of freedom offered by the Act. Britain, with her idea of the principles of self-determination satisfied, would be able to present a bold front again before the world. (Collins, *Partition Act's failure*)

The contrast between full and partial, unaccomplished freedom is interestingly echoed by the use of 'civilization'. In a remarkable 60% of its occurrences instantiated in Table 3 below, the keyword was detected to collocate with items sharing a semantic preference of revival, as it were. This is demonstrated by the noun 'revival' along with a wide range of verbs including 'refresh oneself in', 're-awaken' and 'reconstruct'.

Concordance lines of 'civilization'
...our goal and the revival of our Gaelic **civilization**.
...and refresh ourselves in our own Irish **civilization**, to become again the Irish men and...
...of the last twenty years or more that it [Gaelic civilization] has re-awakened. ...
...now living hopes for a better **civilization**.
...dark world, to reconstruct our ancient **civilization** on modern lines, to avoid the errors...
...miseries, the dangers, into which other nations, with their **false civilizations**, have fallen.

Table 3 – Concordance lines and collocational patterns of 'civilization'

From Collins' perspective, what is to be reconstructed is invariably ancient Gaelic civilization, which was distinctive to Irish society before the English set foot on the island in the twelfth century. As can be seen in (3), this view of Irish civilization goes back to an ideal of socially cohesive society devoted to the cultivation of the mind and national pastimes (GAA, 1887; Murphy, 1948).

(3) The Irish social and economic system was democratic. It was simple and harmonious. The people had security in their rights, and just law. And, suited to them, their economic life progressed smoothly. Our people had leisure for the things in which they took delight. They had leisure for the cultivation of the mind, by the study of art, literature, and the traditions. They developed character and bodily strength by acquiring skill in military exercise and in the national games. The pertinacity of Irish civilization was due to the democratic basis of its economic system, and the aristocracy of its culture. Gaelic **civilization** was quite different. [...] Spiritually and socially they were one people. Each community was independent and complete within its own boundaries. The land belonged to the people. (Collins, *Distinctive culture, ancient Irish civilization*)

In passages such as this, Collins seems to draw on a notion of civilization as "an achieved condition of refinement and order" (Williams, 1976, p. 58), resolutely opposed to the 'false civilization' conjured up in the last concordance line of Table 3 and attested for 16.6% of the corpus entries of 'civilization'. Not surprisingly, this is the civilization that England was accused of imposing upon Ireland during more than five centuries of colonial rule. A civilization that later in the same essay Collins defines as a "misfit", a "garment" rendering Ireland "mean, clumsy, and ungraceful", while exposing its defects and giving it no scope to display its good qualities (Collins, 1922, p. 118). This is yet again a vivid picture Collins might have borrowed from the Irish nationalist imagery. After all, in his comprehensive account of the philosophy of the Gaelic League, the leading organisation designed to revitalize the ancient Irish language, Corkery (1948, p. 12) emphasizes the widely held view that "Ireland can of course continue to live its life in English – only, however, a *constricted* sort of life" [my emphasis].

Taken together, these findings point to the argumentation constructed by Collins around keywords. In more detail, Collins appears to dissociate (Van Rees, 2009, p. xi) as much between a notion of complete freedom and one of limited, unaccomplished freedom, as he does between true (Irish) civilization and false (English) civilization. Against this backdrop, the two keywords provide a link to the interlocutors' shared values, and as such they activate the appropriate frames (Chong & Druckman, 2006; Scheufele & Iyengar, 2014). In turn, from the relevant frames endoxa are drawn to be used in the argumentation. In this case, the endoxon at work may well be formulated as follows: complete freedom and true Irish civilization are the only worthwhile ends. This is less the result of speculation than what clearly emerges from the leading nationalist authors shaping up the Irish public opinion during Collins' lifetime (cf. Sheehy, 1980, Harkness, 1988 and McMahon, 2008). These include not only Rooney, but also Arthur Griffith and D.P. Moran. First of all, Griffith (1920, p. 1) called for an economically independent Ireland to stop England from having "the sole monopoly and absolute control of our trade, which to her is a great advantage". Secondly, Moran (1906, p. 80) advocated a full-fledged Irish-Ireland policy on the grounds that the Irish "are all in a state of general affectation playing up to a civilization that is not natural to them".

Resting on these foundations, the implicit component of much of Collins' argumentation broadly takes this form:

> MP: Worthwhile ends must be pursued.
> mp: Complete freedom and true civilization are worthwhile ends.
> C: They must be pursued.

Moving back from the implicit to the explicit part of Collins' reasoning, complete freedom and true civilization can be pursued and eventually achieved by accepting the terms of the Anglo-Irish Treaty signed by the Irish delegation in London in December 1921. This is justified by Collins through pragmatic argumentation, more specifically by means of its "Variant I" (Van Poppel 2012, p. 99) schematized below:

1 Action X should be performed
1.1a Action X leads to Y and Y^1
1.1b Y and Y^1 are desirable
1.1a-1.1b' (If X leads to Y and Y^1, and Y/Y^1 are desirable, then Action X should be performed)

In the above scheme, Action X is the ratification of the Anglo-Irish Treaty, whereas Y is the achievement of complete freedom, and Y^1 is the restoration of true Gaelic civilization. The effectiveness of pragmatic argumentation, urging the Irish people to embrace the Treaty in that it would lead to highly desirable effects, is arguably reinforced by the framing operations and endoxic elements activated by the use of keywords documented earlier on.

3.2 De Valera's keywords in argumentation

In order to discuss the argumentative implications of de Valera's use of keywords, the two key verbs 'know' and 'think' are worth looking at. By again starting from the patterns and contexts of use of key terms, there is remarkable continuity in the deployment of 'know' and 'think' on de Valera's part, as can be appreciated from Tables 4 and 5 below.

Concordance lines of 'know'
...task. You probably all **know** by now how much I believe in thoroughness of...
...of the school. We all **know** how much one teacher may mean to a whole locality.
... and deny her people imports to the extent that we all **know** of. We do...
...spare us to the end. As you all **know**, our history has been one of such active struggle...
...by infamous methods. We all **know** that his warning had the fullest foundation. We...
...undoubtedly lead to disaster. We all **know** that there is a body in this country with ...
...was sympathetic to communism. Everybody **knows** that is fundamentally untrue and false.

Table 4 – Concordance lines and collocational patterns of 'know'

Concordance lines of 'think'
...is a monarchy? <u>I do not</u> **think** <u>any constitutional lawyer of repute would attempt</u>...
...it is legitimate to borrow for. <u>I do not</u> **think** <u>anybody will seriously contest</u> that ...
...what was in mind. <u>I do not</u> **think** <u>anybody here will deny</u> that we were right...
...That narrow interpretation <u>I do not</u> **think** <u>had occurred to anybody</u> when the...
...A Chinn Comhairle, <u>I do not</u> **think** that <u>any good purpose would be served</u> by...
...for the country as a whole. <u>I do not</u> **think** that <u>any good purpose is to be served</u> by...

Table 5 – Concordance lines and collocational patterns of 'think'

First of all, the concordance sample in Table 4 shows that in 7.4% of its 352 entries, 'know' collocates with pronominal and/or indefinite entities ('we', 'you', 'everybody'). Secondly, 3.5% of the over 500 tokens of 'think' in the Dev-Corpus are represented by negative forms reporting the speaker's views ('I do not think'). Moving beyond the surface, however, the common ground between the two verbs appears to lie elsewhere. Behind the use of both verbs, in particular, lies de Valera's intention to portray himself as a prime example of man of common sense, he who embeds the mindset of the ordinary Irishman, a kind of 'reasonable man' (Wierzbicka, 2006, pp. 103-138) *par excellence*. Chunks such as 'As you all know' or 'We all know' in Table 4, let alone 'I do not think anybody will seriously contest' or 'I do not think anybody here will deny' in Table 5, are essentially instrumental in enabling de Valera to putatively argue from the position of the Irish people's spokesperson. In that regard, the passage reported in (4) below may serve as a typical example.

The extract is taken from one of de Valera's best known speeches, delivered before the Irish Parliament in the wake of Francisco Franco's victory in the Spanish civil war. As the Head of the Irish Government, de Valera had been put under increasing pressure to take steps to accord Franco formal diplomatic recognition. Failure to do so, the argument from opposition benches went, would constitute tangible proof that de Valera's government was sympathetic to communism.

(4) We do not always think alike, but I think Deputy MacDermot has replied sufficiently to the speech made from the opposite benches to warrant, if I care to use it, my allowing the amendment and all [to] go to the House without any word from me. <u>I think any reasonable person</u> who has listened to him or who will read what he has said <u>will agree that</u> the point of view which he has expressed is the right and proper point of view. [...] There has been, from the opposite benches, a continued effort since 1931 to try to mend the fortunes of their party and to build up a case for Fascist organisations on the ground that this Government was sympathetic to communism. <u>Everybody</u> **knows** that that is fundamentally untrue and false. <u>I do not</u> **think** <u>there is anybody in the country</u> who looks at the Government's action dispassionately <u>who will not be satisfied that</u> the Government have no more use for communist philosophy than has any member of the Opposition. The question is: how are we best to defend our philosophy and the philosophy of the vast majority of the Irish people in our attitude towards life and prevent organisations with a completely and fundamentally different policy from making inroads here? These grounds of prudence on which it is customary to act do not need to be explained. <u>I think they are self-evident to everybody</u>. It would be ridiculous, obviously, to give recognition to a government that was unable ultimately to maintain itself. *It certainly would not lead* – I am talking in general – *to cordial relations between the government of one country and another if the relations had to be with the restored government.* [...] I do not think there has been any attempt at all to show that it would, in fact, help General Franco that this recognition should be accorded. We are told that the cause of Christianity demands it. [...] I do not know that recognition on our part would involve any grievous consequences to us as a people, but *if we take it on the high grounds of protection and help for Christianity, then I think at least that we ought to hesitate when the head of Christianity has not deemed it wise or prudent to give the recognition that we are asked to give.* (de Valera, *Ireland and the civil war in Spain*, 1936)

Example (4) is a lengthy one, but it was worth including as much of it as we can see above by virtue of its overall explanatory potential regarding de Valera's argumentation. First of all, the first half of the passage includes both phraseology reflecting the collocational patterns identified in Tables 4 and 5 above – i.e., 'Everybody knows that' and 'I do not think there is anybody is this country…who will not be satisfied that' – and other items sharing similar semantic features, cf. 'I think any reasonable person…will agree that' and 'I think they [those grounds of prudence] are self-evident to everybody'. The reiterated use of such forms allows de Valera to basically frame his upcoming proposal as 'the reasonable one'. In turn, this secures the activation of the relevant endoxon on the interlocutors' mind, which may be phrased as follows: reasonable proposals should be accepted. As a result, the implicit component of de Valera's argumentation broadly takes this form:

> MP: Reasonable proposals should be accepted.
> Mp: Our policy is reasonable.
> C: It should be accepted.

In the second place, focusing more closely on the second part of the passage inherent in the explicit argument structure, it can be seen that de Valera's policy is the non-recognition of Franco's Government – cf., "It would be ridiculous, obviously, to give recognition to a government that was unable ultimately to maintain itself". More precisely, the view that Franco's Government should not be recognised represents the standpoint of multiple argumentation. Viewed pragma-dialectically, the standpoint is supported by two mutually independent argument schemes, namely pragmatic argumentation and causal argumentation, as in Figure 1 below:

S: Franco's Government should not be recognised.

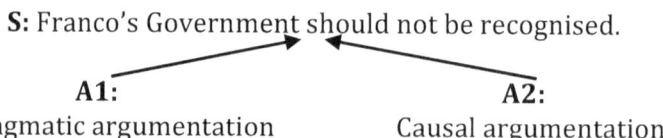

A1:
Pragmatic argumentation

A2:
Causal argumentation

Figure 1 – Argument structure of de Valera's speech
Ireland and the civil war in Spain

To begin with, de Valera's pragmatic argumentation can be accounted for as an instance of "Variant IV" (Van Poppel, 2012, p. 101) going back to the first italicised fragment in (4) – i.e., "It certainly would not lead…to cordial relations between the government of one country and another if the relations had to be with the restored government". In the schematisation below, Action X is the recognition of Franco's Government, while Y is the desirable consequence of establishing and maintaining cordial relations with the legitimate government. In other words, de Valera points out, should Spain's legitimate government ever be restored, recognising Franco would prevent us from attaining the worthy end of cultivating harmonious diplomatic relations with it:

1 Action X should not be performed
1.1a Action X does not lead to Y
1.1b Y is desirable
1.1a-1.1b' (If X does not lead to Y and Y is desirable, then Action X should not be performed)

As regards the second scheme supporting de Valera's standpoint, the presence of causal argumentation can be detected in the last italicised fragment in (4) – "if we take it on the high grounds of protection and help for Christianity, then I think at least that we ought to hesitate when the head of Christianity has not deemed it wise or prudent to give the recognition that we are asked to give". In this case, the fact that Ireland did not recognize Franco before the Holy See ever did is what causes the Irish Free State to be in a position to portray itself as a country championing Christianity's best interests. The scheme, where Y is the safeguard of Christianity, Z is the choice of not granting recognition to a State before the Holy See does and X is the Irish Free State, is reproduced below following Van Eemeren and Grootendorst (1992, p. 97):

> Y is true of X
> because Z is true of X
> and Z leads to Y.

Overall, the impact of the causal argument, and more generally of the whole pattern of multiple argumentation discerned above, seems to be maximised by the strategic use of the keywords 'know' and 'think' outlined at the outset. It is their distinctive use in context, as highlighted by their phraseology, that underlies the activation of the appropriate frame and the endoxon behind de Valera's argumentation. In turn, the power of the frame and the solidity of the resulting endoxon are secured in the light of both the strength of proposals presented as 'reasonable', i.e. by definition thoughtful and well-balanced, and the interlocutors' shared values.

As a matter of fact, it should not be forgotten that that was the Irish Free State in the 1930s, a country that a year after de Valera's speech in the aftermath of Spain's civil war would enact *Bunreacht na hÉireann*. This is a Constitution to which de Valera was to make a significant contribution, and which despite a number of amendments over the following decades, still retains a lasting Catholic imprint. After all, the Preamble to the Constitution (2012: 2) still has marks of the country's religious heritage, suffice it to think of forms such as "In the Name of the Most Holy Trinity..." or "We, the people of Éire, humbly acknowledge all our obligations to our Divine Lord, Jesus Christ...", which must have made de Valera's causal argument even more persuasive.

4. CONCLUSIONS

The keyword analysis presented in the prior section lent fresh insights at two main levels. As regards the findings themselves, first of all, corpus data have enriched our understanding of Collins and de Valera, by adding to the profile generated by historians of the two characters as the dominant figures of Irish nationalist politics in the age of Independence. On the one hand, Collins has often been perceived as the pragmatist, the man who made a substantial contribution to the Irish delegation's attempt to broker a sensible compromise through the Anglo-Irish Treaty. At the same time, his discourse has been shown by keywords to draw on an evocative nationalist imagery and as such, to be driven by noble ideals of freedom and civilization. On the other hand, de Valera has on many an occasion been described as the doctrinaire Republican, the hardliner who rejected the Anglo-Irish Treaty even after its ratification through democratic means, and therefore fostered the political climate eventually leading to the Irish civil war (1922-1923). At the same time, his discourse appears to reflect an increasingly statesman-like profile, whereby he aims at serving as the Irish people's true spokesperson and as a man acting mindfully and reasonably.

From a methodological point of view, secondly, the analysis indicates the major advantage of corpus-comparative statistical keywords as a genuinely data-driven method to engage in "the business of implicit premise recovery" (O'Halloran, 2009, p. 44) on the basis of terms whose keyness is more objectively established by software. More specifically, data tend to confirm the function of the selected keywords in argumentative text where, consistent with the literature, they were observed to "provide a link to the interlocutors' context and their shared values" (Bigi & Greco Morasso, 2012, p. 1142). Finally, the results in the whole of Section 3 seem to advocate a close integration of available approaches – notably, AMT and corpus linguistics – to achieve a number of desirable aims: to begin with, the combination of qualitative and quantitative evidence; secondly, the need to identify and focus on not merely key words, but also key phraseology in context as the actual starting point to accurately define the inferential configuration of arguments; finally, the opportunity to rely on a method to empirically check on culturally relevant corpora – as were the two datasets used here – whether the implicit premises attributed "to the arguer are indeed recoverable or at least partially justified in the cultural common ground" (Rocci & Wariss Monteiro, 2009, p. 95).

REFERENCES

Anthony, L. (2006). *AntConc 3.2.1.* <http://www.laurenceanthony.net>
Baker, P. (2006). *Using corpora in discourse analysis.* London: Continuum.
Baker, P., & McEnery, T. (2015). *Corpora and discourse studies. Integrating discourse and corpora.* London: Macmillan.
Bigi, S., & Greco Morasso, S. (2012). Keywords, frames and reconstruction of material starting points in argumentation. *Journal of Pragmatics, 44*(10), 1135-1149.
Bondi, M. (2010). Perspectives on keywords and keyness: An introduction. In M. Bondi & M. Scott (Eds.), *Keyness in texts* (pp. 1-18). Amsterdam: Benjamins.
Bondi, M., & Scott, M. (Eds.) (2010). *Keyness in texts.* Amsterdam: Benjamins.
Bunreacht na hÉireann/Constitution of Ireland (2012) [1937]. Retrieved from https://www.constitution.ie/Documents/Bhunreacht_na_hEireann_web.pdf.
Chong, D., & Druckman, J. N. (2006). A theory of framing and opinion formation in competitive elite environments. *Journal of Communication, 57*(1), 99-118.
Collins, M. (1922). *The path to freedom.* Dublin: Talbot Press.
Corkery, D. (1948). *The philosophy of the Gaelic League.* Dublin: Conradh na Gaeilge.
Costello, F. (1997). *Michael Collins in his own words.* Dublin: Gill and Macmillan.

Eemeren, F. H. van & Grootendorst, R. (1992). *Argumentation, communication, and fallacies.* Hillsdale: Lawrence Erlbaum Associates.
Ferriter, D. (2007). *Judging Dev: A reassessment of the life and legacy of Eamon de Valera.* Dublin: Royal Irish Academy.
Fetzer, A. (2014). *I think, I mean* and *I believe* in political discourse: Collocates, functions and distribution. *Functions of Language, 21*(1), 67-94.
GAA (1887). *The Gaelic Athletic Association for the preservation and cultivation of national pastimes.* Dublin: Cahill.

Gramley, S., & Pätzold, K.M. (2004). *A survey of modern English*. London: Routledge.
Griffith, A. (1920). *England's Irish philanthropy*. Washington: Friends of Irish Freedom National Bureau of Information.
Harkness, D. (1988). State and nation in independent Ireland. In The Princess Grace Irish Library (Ed.), *Irishness in a changing society* (pp. 123-131). Gerrards Cross: Colin Smythe.
Hart, P. (2005). *Mick. The real Michael Collins*. Oxford: Macmillan.
Hunston, S., & Francis, G. (1998). Verbs observed: A Corpus-driven pedagogic grammar. *Applied Linguistics, 19*(1), 45-72.
Liebert, W. A. (2003). Zu einem dynamischen Konzept von Schlüsselwörtern. *Zeitschrift für Angewandte Linguistik, 38*, 57-75.
Mazzi, D. (2016). *The theoretical background and practical implications of argumentation in Ireland*. Newcastle upon Tyne: Cambridge Scholars Publishing.
McMahon, T. (2008). *Grand opportunity: The Gaelic revival and Irish society*. New York: Syracuse University Press.
Moran, D. P. (1906). *The philosophy of Irish Ireland*. Dublin: Duffy & Co.
Moynihan, M. (1980). *Speeches and statements by Eamon de Valera*. Dublin: Gill & Macmillan.
Murphy, G. (1948). *Glimpses of Gaelic Ireland*. Dublin: Fallon.
O'Halloran, K. (2009). Implicit dialogical premises, explanation as argument: A corpus-based reconstruction. *Informal Logic, 29*(1), 15-53.
Rigotti, E., & Rocci, R. (2005). From argument analysis to cultural keywords (and back again). In F. H. van Eemeren & P. Houtlosser (Eds.), *Argumentation in practice* (pp. 125-142). Amsterdam: Benjamins.
Rigotti, E. (2008). Locus a causa finali. *L'analisi linguistica e letteraria, 16*(2), 559-576.
Rigotti, E., & Greco Morasso, S. (2010). Comparing the argumentum model of topics to other contemporary approaches to argument schemes: The procedural and material components. *Argumentation, 24*(4), 489-512.
Rocci, A., & Wariss Monteiro, M. (2009). Cultural keywords in arguments. The case of *interactivity*. *Cogency, 1*(2), 65-100.
Römer, U., & Wulff, S. (2010). Applying corpus methods to written academic texts: Explorations of MICUSP. *Journal of Writing Research, 2*(2), 99-127.
Rooney, W. (1909). *Prose writings*. Dublin: M.H. Gill.
Ryle Dwyer, T. (2006). *Big fellow, long fellow: A joint biography of Collins and De Valera*. Dublin: Gill & Macmillan.
Scheufele, D. A., & Iyengar, S. (2014). The state of framing research: A call for new directions. In K. Kenski & K. Hall Jamieson (Eds.), *The Oxford handbook of political communication theories*. New York: Oxford University Press. Retrieved from http://www.oxfordhandbooks.com/view/10.1093/oxfordhb/9780199793471.001.0001/oxfordhb-9780199793471-e-47.
Scott, M. (1998). *WordSmith tools 3.0*. Oxford: Oxford University Press.

Scott, M. (2010). Problems in investigating keyness, or clearing the undergrowth and marking out trails... . In M. Bondi & M. Scott (Eds.), *Keyness in texts* (pp. 43-57). Amsterdam: Benjamins.
Sheehy, J. (1980). *The rediscovery of Ireland's past: The Celtic Revival 1830-1930.* London: Thames and Hudson.
Sinclair, J. (2004). *Trust the text. Language, corpus and discourse.* London: Routledge.
Stubbs, M. (1996). *Text and corpus analysis. Computer-assisted studies of language and culture.* Oxford: Blackwell.
Stubbs, M. (2001). *Words and phrases. Corpus studies of lexical semantics.* Oxford: Blackwell.
Van Poppel, L. (2012). Pragmatic argumentation in health brochures. *Journal of Argumentation in Context*, *1*(1), 97-112.
van Rees, A. (2009). *Dissociation in argumentative discussions. A pragma-dialectical perspective.* Dordrecht: Springer.
Wierzbicka, A. (2006). *English: Meaning and culture.* Oxford: Oxford University Press.
Williams, R. (1976). *Keywords: A vocabulary of culture and society.* New York: Oxford University Press.

37

Arguing Inter-Issue in Public Political Arguments

DIMA MOHAMMED
ArgLab-IFILNOVA, FCSH, Universidade Nova de Lisboa, Portugal
d.mohammed@fsch.unl.pt

> This paper is about a particular aspect of the open-ended character of public political arguments. At any point in time in the public sphere, there are countless controversies roaming and issues being addressed. A vigilant political actor crafts her arguments carefully trying to keep under control the contributions these arguments make to the different issues present. In this paper, I explore the challenge of examining the strategic shape and rational quality of such arguments.
>
> KEYWORDS: multiple issues, commitments, public political arguments, reasonableness, strategic manoeuvring

1. INTRODUCTION

The main characteristics of political argumentation identified by David Zarefsky (2008) tell us a great deal about the complexity of this intriguing practice. Political argumentation (i) lacks time limits, (ii) lacks a clear terminus, (iii) features a heterogeneous audience, (iv) is characterised by open access and (v) is driven by an urgent need to act. The complexity of the practice is not just a challenge for arguers, who as Zarefsky tells us, have no option but to manoeuvre strategically as they argue. It is also a challenge for argumentation scholars who strive to develop the theoretical tools that can provide meaningful examination of political argumentation.

Important advances in the examination of political argumentation have been realised by means of integrating rhetorical insights (e.g. van Eemeren & Houtlosseer, 1999; Tindale, 2004), as well as institutional considerations (e.g. Goodnight, 2010; van Eemeren, 2010) and political considerations (Fairclough & Fairclough, 2012). Yet, crucial aspects of the complexity remain challenging. Tindale (2015) focuses on the role that the audience plays in the construction of the rhetorical argument. Lewinski and Aakhus (2013) put their effort in addressing the complexities that arise from the multiple parties typically involved in public argumentation. Mohammed (2016a) investigates ways for accounting for the multiple goals pursued in a public political argument. Each in their own way, these are a few of the scholars who are addressing the complexities that result from the open-ended, almost limitless, character of a public political argument. In this paper, I address a particular aspect of the open-endedness, namely the multiple issues that are usually simultaneously addressed when people argue in the public sphere.

At any point in time, there are countless controversies roaming and issues being addressed in the public sphere. It is not uncommon that when one makes an argument addressing one issue, the same argument may be a contribution to another issue too (Mohammed, 2016c). A vigilant political actor would craft her arguments carefully trying to keep under control the contributions these arguments make to the different issues present. In this paper, I will examine such cases where political agents manoeuvre strategically (van Eemeren, 2010) across multiple issues. My goal is two-fold. On the one hand, I would like to give an account of the inter-issue manoeuvring, and on the other hand, I aim at discussing the challenges facing scholars when examining it.

I will start with a case of what I am calling inter-issue manoeuvring. Following that, I discuss a proposal for examining this type of manoeuvring, that is to reconstruct the argument as a series of simultaneous discussions. The discussion will reveal an important challenge to the proposal, namely the difficulty of determining what issues are at stake in any public argument. In order to explore the challenge two more cases are discussed and further preliminary proposals are made. The proposals highlight more of what the analyst needs to capture but leaves us with further new challenges.

2. MANOEUVRING INTER-ISSUE: TRYING TO KEEP THE ARGUMENT UNDER CONTROL

A good example of a vigilant public arguer manoeuvring inter-issue is Barack Obama discussing terrorism and the measures to take in combating it. Aware of the many controversies intertwined in discussing the measures to take in fighting terrorism, Obama makes his word choices carefully and strategically trying to keep under control the contributions that his arguments make to the different issues present.

Let's look at what has become an iconic example of this, namely his repeated refusal to refer to ISIS, AlQaeda and other terrorist groups as "radical Islamic terrorists" when discussing the measure taken to combat these groups.

Obama came under increasing pressure for this choice. Eventually, he had to explain it, on multiple occasions. For example, in June 2016, in remarks circulated after "a counter-ISIL meeting", he argued that the use of the term "radical Islam" is not just unnecessary for the fight against terrorism, but it is also counterproductive as it would alienate Muslim allies in the Middle East (White House, 2016). On another occasion, during a CNN presidential town hall debate, he was asked to defend his choice by a Gold Star mother. In responding to her, he explained that he wouldn't use the term because that would lump in "murderers with billions of peaceful Muslims around the world" (CNN, 2016). The different explanations make explicit an effort aimed at manoeuvring between several issues. Such manoeuvres are usually left implicit, but in this case, as the choice became controversial, the manoeuvring had to become explicit, which is helpful for us, scholars examining it.

In explaining his choice, three main issues become clearly relevant. The first issue, the one which is usually the main issue being addressed, is about measures to combat terrorist groups (*issue i*). In it, president Obama had expressed and defended several positions. For example, in September 2014, in a speech particularly on "combating ISIS and terrorism" he argued that "Our objective is clear: we will degrade, and ultimately destroy, ISIL through a comprehensive and sustained counter-terrorism strategy" (White House, 2014). Later, in June 2016, in the remarks following the "counter-ISIL meeting", he started by saying "At the outset, I want to reiterate our objective in this fight. Our mission is to destroy ISIL" (White House, 2016). It is in discussing these issues that the President chose to avoid using the term "Islamic terrorism". A second important issue, which we now know was present when discussing measures to combat terrorism is the need to secure support from allies, such as for example from the Iraqi army whose role in the fight is crucial (*issue ii*). A third important issue also at stake here is the treatment of Muslims, particularly Muslim Americans (*issue iii*). Obama had been vocal in opposing discrimination against Muslim Americans. His refusal to use the expression "Islamic terrorism" became more persistent as proposals were being made for discriminatory measures, such as a pledge to make a Muslim register.

The choice that Obama makes not to use "Islamic terrorism" when referring to ISIS, AlQaeda and other terrorist groups is clearly significant for all these issues. The term "Islamic terrorism" has become associated with a view of Islam as inherently violent. Using it when justifying decisions on how to combat ISIS (*issue i*) can alienate potential allies and therefore it can have negative implications when attempting to convince potential allies to join the battle (*issue ii*). Using the term may also have negative consequences when considering the question of discrimination against Muslim Americans (*issue iii*). The view that Islam is inherently violent, typically associated with the phrase "Islamic terrorism", renders every Muslim a potential terrorist. Considering every Muslim a potential terrorist is usually used to justify adopting measures that are discriminatory against Muslims, such as the pledged Muslim Register. Using "Islamic terrorism" when defending decisions on how to combat ISIS (*issue i*) can be a ground for justifying discriminatory measures against Muslim Americans, which is something Obama would want to distance himself from (*issue iii*).

What we have here is a case of a position advanced in discussing one issue that is strategically formulated in such a way as to provide support for the arguer's cases in relation to other issues. Obama's refusal to use the term "Islamic terrorism" when discussing what measures to adopt in combatting ISIS (*issue i*), is made in order to distance oneself from the view of Islam as inherently violent. This distance is important for Obama when it comes to the issue of securing alliance for combatting terrorism (*issue ii*) as well as to the issue of how to treat Muslim Americans (*issue iii*).

3. MULTIPLE ISSUES DISCUSSED IN 'SIMULTANEOUS DISCUSSIONS'

In previous work, I have suggested to reconstruct public political arguments as a series of simultaneous discussions. The idea here is that in a public argument arguers are typically pursuing multiple goals (Mohammed, 2013, 2016a), which requires them to address several issues simultaneously (Mohammed, 2016a, 2016b). As the Obama case shows, arguers may respond to that by manoeuvring strategically between several issues. The strategic design of any particular move may involve choices that are strategic for more than just one issue. An adequate analysis needs to reflect that, and reconstructing the argument as a series of simultaneous discussions is meant to do it.

To reconstruct an argument as a series of simultaneous discussions is a suggestion that runs in the opposite direction of what seems like common wisdom among argumentation scholars. This common wisdom amounts basically to going as analytic and elementary as possible in reconstructing argumentative practice and examining it.

For example, van Eemeren et al. recommend the analyst to break multiple disputes into series of single ones (van Eemeren & Grootendorst, 1992, p. 20; van Eemeren, Grootendorst & Snoeck Henkemans, 2002, p. 8; van Eemeren, Houtlosser & Snoeck Henkemans, 2007, p. 22). Also, traditionally, argumentation theorists have advised in favour of treating a divergent argument, which is an "argument that plays a role in supporting more than one conclusion", as several separate arguments "having the same basic reason but leading to different conclusions" (Thomas, 1973, p. 36, Freeman, 1991, p. 234). There is no doubt that breaking a multiple dispute into elementary ones, or for that matter breaking a divergent argument into several separate ones, helps the analyst get a clear analytic view of the argumentative situation. Nevertheless, so much of the strategic design involved would go unnoticed if elementary disputes and single arguments were analysed in isolation (Mohammed, 2016b, 2016c).

Think of Obama's choice not to use the term "Islamic terrorism" when defending drone strikes as a measure to take out terrorist commanders. If we would only consider the issue of drone strikes, the choice does not make sense. It may be even counter-productive if we consider that it could alienate that segment of his domestic audience who would have liked him to use the term. The strategic design is revealed only when considering other issues, for example the discriminatory measures called for, which are supported by the view embedded in the term. Considering that, the analyst can see that the design is meant to distance Obama from the view that Islam is inherently violent in order to maintain his opposition to anti-Muslim discriminatory measures. In order to capture the strategic design of such argumentative choices, it is crucial to reconstruct the argumentative exchanges as several simultaneous dialectical discussions.

In order to avoid any misunderstanding, let me clarify that a discussion here is not to be understood as the actual encounter that occurs at a specific time and place. It is rather a dialectical analytic reconstruction of the real-life encounter, defined in terms of a standpoint about an issue and the argumentation advanced in support of it. The discussions are considered simultaneous because of those argumentative moves which are relevant to more than just one standpoint.

Two main features characterise discussions which may be reconstructed as simultaneous. First, there is at least one argumentative move that plays a role in the defence or attack of more than one standpoint. In other words, there are several claims with shared premises, or there are some choices made in defending or attacking one claim that are relevant to the defence or attack of another ... etc. Second, for two discussions to count as simultaneous, the claims discussed cannot be one subordinate to the other. This is necessary because otherwise, what we have are just simple cases of serial arguments (Freeman, 1991), also known as subordinative argumentation (Snoeck Henkemans, 1992).

I would also like to emphasise that reconstructing an argument as a series of simultaneous discussions involves both distinguishing the different elementary discussions, as well as considering the common elements between them. That is considering the multiple roles that some argumentative choices play. As the Obama case shows, this latter is crucial for appreciating the full scope of the strategic manoeuvring involved in crafting a public argument.

The Obama case is a good example, from the analyst's perspective, also because Obama was pressured into 'justifying the manoeuvre' and making explicit which issues he considered when making his word choice. But this is quite unusual. A manoeuvre is hardly a manoeuvre once explained let alone justified. Typically, the issues that an arguer has in mind when she manoeuvres strategically remain implicit. Consequently, the analyst has much less guidance in determining which issues are at stake. An important challenge facing the reconstruction of a public argument as a series of simultaneous discussions is in fact how to know what issues are at stake. As the next cases examined will show, this is no easy question.

4. WHAT IS AT STAKE IN A GIVEN PUBLIC POLITICAL ARGUMENT?

4.1 There is "no reason to be alarmed"

The second case is a series of tweets sent by the US president Donald Trump in the aftermath of the terrorist attacks that hit London in early June 2017 (The Independent, 2017b). The tweets are reproduced below:

> Example (1) The "no reason to be alarmed" tweets
>
> [i]   @realDonaldTrump, 4 June 2017, 12:17
> "We need to be smart, vigilant and tough. We need the courts to give us back our rights. We need the Travel Ban as an extra level of safety!"
>
> [ii] @realDonaldTrump, 4 June 2017, 12:19

"We must stop being politically correct and get down to the business of security for our people. If we don't get smart it will only get worse",

[iii]   @realDonaldTrump, 4 June 2017, 12:31
"At least seven dead and 48 wounded in terror attack and Mayor of London says there is 'no reason to be alarmed'!"

Tweet [i] include the tweeter's first reaction to the news of the attacks, namely taking advantage of the occasion to reiterate a position in favour of his controversial travel ban (*issue i*). As it is usual for tweets, the reconstruction of the complete argument requires going beyond the 140 characters of the tweet itself into the context. In defending the travel ban, the attacks in London can be understood as evidence that the terrorist threat is serious and that there is reason to be alarmed. Therefore, there is a need to act smart, vigilant and tough, and take pro-active measures such as the travel ban. The argument in tweet [i] may then be understood as in the stricture below

Tweet [i] || Issue (i): in favour of pro-active anti-terrorism measures, or more specifically, in favour of the travel ban

1   We should have the travel ban as an extra level of safety
    1.1a   We need to be smart, vigilant and tough
    (1.1b) (There is reason to be alarmed by the terrorist threat)
        (1.1b.1)   (There is another terrorist attack on London)

The understanding of the argument is confirmed in a statement made by a White House spokesperson later. In it, the argument becomes clearer when the link between the London attacks and the travel ban is made explicit: there are "constant attacks going on not just there but across the globe", therefore "there is a reason to be alarmed", and "to start putting national security and global security at an all-time top".

Tweet [ii] comes just two minutes after Tweet [i]. In it, the tweeter makes use of the situation in order to reiterate another of his previous positions, this time against political correctness. Here too, we can consider that the terrorist attacks in London are taken as evidence that the terrorist threat is serious, and reconstruct the argument against political correctness as in the structure below:

Tweet [ii] || Issue (ii) Against Political correctness

2 We must stop being politically correct
    2.1a   We must get down to the business of security for our people
        (2.1a.1)   (There is reason to be alarmed by the terrorist threat)
            (2.1a.1.1)   (There is another terrorist attack on London)
    (2.1b)   (Political correctness is obstructing the efforts of securing our own people)

But is that all? It's hard to be sure it is, but then when Tweet [iii] comes out, it becomes clear it isn't.

In Tweet [iii], which comes quite quickly, the US president attacks Sadiq Khan, the mayor of London. The attack is based on a reference to the statement the Mayor had just made. In it Khan had said: "Londoners will see an increased police presence today and over the course of the next few days. There's no reason to be alarmed". The attack on the mayor - "At least seven dead and 48 wounded in terror attack and Mayor of London says there is 'no reason to be alarmed'!"- is a clear case of a straw man fallacy. While Khan said there was "no reason to be alarmed by police presence", the tweeter made it look like the Mayor had said that there was "no reason to be alarmed by terrorism". Leaving the straw man aside, what is at stake in this tweet? what is the issue here? Is it still issue (i) in favour of the travel ban and pro-active anti-terrorism measures? Not really; why mention the London Mayor if the point is about pro-active anti-terrorism measures? It seems the issue here is about the London Mayor. Trump and Khan have had a history of attacks and counter attacks[1] that justifies this interpretation. The argument here may be reconstructed as in the structure below:

Tweet [iii] || Issue (iii) criticising the London mayor

(3) (Sadiq Khan is unfit to be London Mayor)
    3.1a Sadiq Khan says there is no reason to be alarmed
    3.1b There is reason to be alarmed
        3.1b.1 at least 7 dead and 48 injured by a terrorist act

---

[1] See the Atlantic (2017) for a "brief history of Trump's Feud with Sadiq Khan.

But is that all? Why would the president of the United States care to express an opinion about the London Mayor in the wake of terrorist attacks hitting the city? In one way, this may be understood as part of a larger discussion about issue (ii) political correctness. Khan and other typically left-leaning politicians have previously been accused of minimising the threat of terrorism, and of avoiding pointing out to the problem that Islam poses, motivated by political correctness. A notorious case, has been one that involved several people including Nigel Farage, the founder of the UK Independence Party and Trump's 'friend', as well as Donald Trump Jr., the president's son. Khan was repeatedly criticised for allegedly claiming that terror attacks are 'part and parcel' of living in a big city. The criticism was presented as an example of the political correctness that leads politicians to minimise the threat of terrorism and to avoid pointing out to the problem that Islam poses.[2]

From that perspective, the criticism in tweet [iii] can be understood as part of a more specific criticism of the London Mayor, something as in the structure below:

Tweet [iii] || Issue (iii) criticising the London mayor

(4) (Sadiq Khan is being inappropriately politically correct in order to avoid pointing out to the real problem, namely that there is inherent relationship between Islam and terrorism)
4.1a Sadiq Khan says there is no reason to be alarmed
4.1b There is reason to be alarmed
    4.1b.1 at least 7 dead and 48 injured by a terrorist act

Or even contributing to this discussion about political correctness as in the structure below:

---

[2] This was another case of a taking the Mayor's words out of context. In his statement, Khan had said that "it's part and parcel of living in a great global city you've got to be prepared for these things, you've got to be vigilant, you've got to support the police doing an incredibly hard job". While the Mayor's point was that being prepared for terrorism is necessary as "part and parcel" of living in a big city, the distortion alleges surrender and acceptance of terrorism in itself as "part and parcel" of living in a big city.

Tweet [iii] || Issue (ii) Against Political correctness

2 We must stop being politically correct
   2.1a We must get down to the business of security for our people
  (2.1b) (Political correctness is obstructing the efforts of securing our own people)
        2.1b.1a Motivated by political correctness, Sadiq Khan says there is no reason to be alarmed
        2.1b.1b There is reason to be alarmed
                4.1b.1b.1 at least 7 dead and 48 injured by a terrorist act

But are we as analysts really justified in attributing such positions to the tweeter and in reconstructing the tweets as involving simultaneous contributions to these issues? One thing is the positions that were attributed to the tweeter beyond the words expressed in the tweets are all positions that have been expressed either by the tweeter himself on other occasions or by what we may call his argumentative allies. Roughly speaking, argumentative allies of an arguer A are those arguers who have publically expressed positions which are similar to A's. Furthermore, the positions attributed are somehow necessary in order to make sense of the tweets. Even more, they seem like they do form a pattern of elements that belong together. For example, criticising Sadiq Khan, or for that matter any left-leaning or liberal politician, for minimising the threat of terrorism has been a recurrent move advanced in the context of attacking political correctness.

But are we justified in attributing a claim to an arguer on the basis of an argument that supports that claim? Can we say that, in the absence of evidence to the opposite, an arguer who advances an argument (x) is committed to support a standpoint (y) if in the public discourse x has been used to justify y? This may seem a bit too far, but we cannot also shy away from attributing standpoints to arguers simply because these standpoints are left implicit. In fact, we may get some guidance from observing how arguers deal with similar situations. Doing that, we realise that vigilant arguers who do not want such attribution make the effort to avoid it. Our next case is an example.

4.2 "It's time to stop saying ISIS has nothing to do with Islam"

Our third case is about an arguer who is aware of the different positions his words may be taken to support, and who makes an effort to prevent that. The arguer is Justin Welby, the Archbishop of Canterbury, and the argument came as part of a lecture he gave at the Catholic Institute of Paris, as he was awarded an honorary doctorate (Welby, 2016). In his lecture, the Archbishop discussed the challenges facing Europe today and called for including "a theological voice" as part of the response. He talked about mass disenchantment, the rise of anti-establishment politics and about religiously motivated violence. Predictably, the Archbishop argued that an adequate response to these challenges needs to involve a central role for religion.

In making the case for including a theological voice as part of the response to religiously motivated violence, the Archbishop said:

> This requires a move away from the argument that has become increasingly popular, which is to say that ISIS is 'nothing to do with Islam', or that Christian militia in the Central African Republic are nothing to do with Christianity, or Hindu nationalist persecution of Christians in South India is nothing to do with Hinduism. Until religious leaders stand up and take responsibility for the actions of those who do things in the name of their religion, we will see no resolution.

What is interesting here is the reference the Archbishop makes to religiously motivated violence outside of Europe. The Archbishop's point is that religion is part of the problem and must therefore be part of the solution. In Europe today, religiously motivated violence is mainly the one committed in the name of Islam. Strictly speaking, in order to make his point, it would suffice the Archbishop to say "This requires a move away from the argument that has become increasingly popular, which is to say that ISIS is 'nothing to do with Islam'". Mentioning the violence committed by Christian militia in the Central African Republic or Hindu nationalist in South India is irrelevant to the Archbishop's point. So, why does the Archbishop bring these in?

The answer to the question lies in understanding the potential argumentative uses of the assertion that *it's time to stop saying that ISIS is 'nothing to do with Islam'*. In other words, in seeing the different simultaneous discussions that the argument can contribute to. We first have the issue directly addressed by the Archbishop, namely the role of religion in general in our world today (*issue i*). Here, the argument goes along the line that it is "essential to recognise extremists' religious motivation in order to get to grips with the problem" (Welby, 2016). Considering the public discussions about terrorism and Islam, one realises that the assertion can easily be understood also as a contribution to a more specific discussion, namely one about the relationship between Islam and violence (*issue ii*). In this discussion, the argument works in favour of a link particular between Islam and violence, i.e. to a view of Islam as a particularly violent religion. The reference to the violence committed by Christian and Hindus is meant to curb the contribution of the Archbishop's argument to this discussion (*issue ii*).

In order for his assertion that *there is a need to stop saying that ISIS is nothing to do with Islam* not to be understood as support to the view that *Islam is a particularly violent religion*, the Archbishop mentions violence committed by Christian and Hindus. In later occasions, he is even more explicit in opposing this view. In June 2017, for example, he said that "London attack link to Islam as Christians killing Muslims is linked to Christianity" and that "religious scripture has been 'twisted and misused' by people to justify violence throughout history" (The Independent, 2017a). This is a yet another good example of an inter-issue manoeuvring: an arguer formulates his position in relation to one issue in such a way as to avoid commitments in relation to another issue.

Given the public presence of the issues, the support some arguments provide for certain standpoints is usually already established. As a result, the manoeuvring seems necessary in order to avoid being attributed an undesirable position. It seems that in cases where an argument has become publically associated with a certain standpoint, and in the absence of evidence to the opposite, the analyst may attribute the standpoint to an arguer who advances the argument. This would be in line with an intuitive interpretation as well as with the effort made by arguers when they would rather avoid an unintended undesirable commitment. A standpoint that is attributed to an arguer on the basis of an argument that has become publically associated with the standpoint may be referred to as a *standing standpoint*. If argument x has become publically associated with standpoint y, an arguer advancing x may be attributed commitment to y as a standing standpoint, unless there is evidence to the opposite.

5. DISCUSSION

Reconstructing a public argument as a series of simultaneous discussions and examining the inert-issue manoeuvring is important for an analysis that is fair to the argumentative practice. The examples examined in this paper show a few ways this happens.

The main challenge that faces such a reconstruction is central to the open-ended character of public political arguments which calls for it.

The main question that is still in need for investigation is how do we determine which issues are at stake in any given situation? The challenge is made more difficult as a result of the implicitness typical or argumentation.

The cases examined in this paper show that the boundaries of any discussion need to remain rather fluid. In order to interpret the argumentative meaning, it is necessary to go beyond any single encounter. But where do we stop remains unclear. Of course, this is by no means a novel challenge facing analysts. There has always been a need to consider context when examining argumentative practice. But with the new technologies of mass media and social networks, it is becoming harder to draw the limits of any single encounter.

The boundaries are fluid in many ways. They are temporally and spatially fluid. For example, when analysing a twitter argument, do we consider single tweets? The tweets of the day? The tweets about a certain subject or with a certain hashtag? May we include a later press conference where the tweets were discussed? There is no clear-cut definitive answer to these questions. The boundaries are hard to establish also in terms of the participants. How do we decide who is part of a certain argument, especially when the temporal and spatial boundaries are fluid? For example, can Trump be considered committed to a standpoint on the basis of what Nigel Farage had said? Are commitments transferrable? In principle, no, but if Trump is aware of Farage's arguments and claim, and he uses the same arguments that Farage had used without presenting an alternative standpoint, how else can we make sense of the relevance of these arguments? In dealing with the cases discussed in this paper, I suggested to identify "argumentative allies" and "standing standpoints". In the absence of evidence to the opposite, an arguer can be attributed a standing standpoint (y) when she advances an argument (x) that has become publically associated with standpoint (y). The attribution is even more justified when an argumentative ally, that is someone who has publically expressed similar positions, has advanced the argument x therefore y.

As a result of the difficulty of drawing the boundaries of argumentative encounters, the analyses end up being tentative, with several equally plausible interpretations and assessments depending on where the limits are drawn. The fluid boundaries are a challenge but they are not necessarily a problem. After all, having multiple plausible interpretations of an argument is empirically backed once the perspective of the recipients of the argument is taken into account: different people may and do understand the same argument differently. For us, argumentation scholars, the challenge is to find theoretically sound ways to deal with these plausible interpretations. An important task here is to develop a rationale for where to draw the boundaries of an argumentative encounter depending on the purpose of any particular analysis.

ACKNOWLEDGEMENTS: work has been supported by grants of the Portuguese Foundation for Science and Technology (FCT): SFRH/BPD/76149/2011 and PTDC/MHC-FIL/0521/2014.

REFERENCES

A brief history of Trump's feud with Sadiq Khan (June 5, 2017), *The Atlantic*. Retrieved on 30 September 2017 from https://www.theatlantic.com/international/archive/2017/06/khan-trump/529191/.

CNN. *Obama: Why I won't say 'Islamic terrorism'*[video file]. Retrieved on 30 September 2017 from: http://edition.cnn.com/videos/politics/2016/09/29/president-obama-town-hall-radical-islam-sot.cnn.

Eemeren, F. H. van (2010). *Strategic maneuvering in argumentative discourse: Extending the Pragma-Dialectical theory of argumentation*. Amsterdam: John Benjamins.

Eemeren, F. H. van, & Grootendorst, R. (1992). *Argumentation, communication, and fallacies: A Pragma-Dialectical perspective*. Hillsdale, NJ: Lawrence Erlbaum.

Eemeren, F. H. van, Grootendorst, R., & Snoeck Henkemans, A. F. (2002). *Argumentation: Analysis, Evaluation, Presentation*. Mahwah, NJ: Lawrence Erlbaum.

Eemeren, F. H. van, & Houtlosser, P. (1999). Strategic manoeuvring in argumentative discourse. *Discourse Studies*, *1*(4), 479–497.

Eemeren, F. H. van, Houtlosser, P., & Snoeck Henkemans, A. F. (2007). *Argumentative indicators in discourse: A Pragma-Dialectical study*. Dordrecht: Springer Netherlands.

Fairclough, I., & Fairclough, N. (2012). Political discourse analysis: A method for advanced students. London: Routledge.

Freeman, J. B. (1991). *Dialectics and the macrostructure of argument: A theory of structure*. Berlin: Foris; London: Routledge.

Goodnight, G. T. (2010). The metapolitics of the 2002 Iraq debate: Public policy and the network imaginary. *Rhetoric and Public Affairs*, *13*(1), 65-94.

Lewinski, M., & Aakhus, M. (2013). Argumentative polylogues in a dialectical framework: A methodological inquiry. Argumentation, *28*(2), 161–185. doi:10.1007/s10503-013-9307-x

Mohammed, D. (2013). *Pursuing multiple goals in European parliamentary debates: EU immigration policies as a case in point. Journal of Argumentation in Context, 2(1), 47-74.*

Mohammed, D. (2016a). Goals in argumentation: A proposal for the analysis and evaluation of public political arguments. *Argumentation, 30*(3), 221-245. doi: 10.1007/s10503-015-9370-6

Mohammed, D. (2016b). Not just rational, but also reasonable: Critical testing in the service of external purposes of public political arguments. In D. Mohammed & M. Lewiński (Eds.), *Argumentation and reasoned action: Proceedings of the 1st European Conference on Argumentation, Lisbon, 2015* (Vol. I, pp. 499-514). London: College Publications.

Mohammed, D. (2016c). "It is true that security and Schengen go hand in hand": Strategic manoeuvring in the multi-layered activity type of European parliamentary debates. In R. von Borg (Ed.), *Dialogues in Argumentation* (pp. 232-266). Windsor Studies in Argumentation.

Mortimer, C. (2017, June 5). Archbishop of Canterbury Justin Welby: London attack link to Islam as Christians killing Muslims is linked to Christianity. *The Independent.* Retrieved on 30 September 2017 from: http://www.independent.co.uk/news/uk/home-news/archbishop-canterbury-justin-welby-london-attack-islam-twisted-misused-muslim-faith-a7772916.html.

Snoeck Henkemans, A. F. (1992). *Analysing complex argumentation: The reconstruction of multiple and coordinatively compound argumentation in a critical discussion.* Amsterdam: Sic Sat.

Stone, J. (2017, June 4). Donald Trump hits out at Sadiq Khan and 'political correctness' after London Bridge terror attack. *The Independent.* Retrieved on 30 September 2017 from http://www.independent.co.uk/news/uk/politics/donald-trump-london-bridge-terror-attack-sadiq-khan-muslim-political-correspondent-islam-isis-a7771966.html

Thomas, S. N. (1973). *Practical reasoning in natural language.* Englewood Cliffs: Prentice Hall, Inc.

Tindale, C. W. (2004). *Rhetorical argumentation: Principles of theory and practice.* Thousand Oaks, Calif: Sage Publications.

Tindale, C. W. (2015). *The philosophy of argument and audience reception.* Cambridge: Cambridge University Press.

Welby, J. (2016). Archbishop Justin Welby on 'the common good and a shared vision for the next century'. Retrieved on 30 September 2017 from http://www.archbishopofcanterbury.org/articles.php/5809/archbishop-justin-welby-on-the-common-good-and-a-shared-vision-for-the-next-century

Wenzel, J. W. (1990). Three perspectives on argument: Rhetoric, dialectic, logic. In J. Schuetz & R. Trapp (Eds.), *Perspectives on argumentation: Essays in honor of Wayne Brockriede* (pp. 9-26). Prospect Heights: Waveland.

White House (2014). President Obama: "We Will Degrade and Ultimately Destroy ISIL". Retrieved on 30 September 2017 from https://obamawhitehouse.archives.gov/blog/2014/09/10/president-obama-we-will-degrade-and-ultimately-destroy-isil.

White House (2016). Remarks by the President after counter-ISIL meeting. Retrieved on 30 September 2017 from https://obamawhitehouse.archives.gov/the-press-office/2016/06/14/remarks-president-after-counter-isil-meeting.

Zarefsky, D. (2008). *Strategic* maneuvering in political argumentation. *Argumentation, 22*(3), 317-330.

38

Inference and Argument: Normative and Descriptive Dimensions

ANDREI MOLDOVAN
University of Salamanca, Spain
mandreius@usal.es

This paper addresses the complex issue of the nature of speech acts of arguing. It aims to discuss this question from a comparative perspective, by looking at the lessons that argumentation theory can draw from the philosophical literature on the nature of the speech act of assertion. By taking a Gricean account of assertion as a model, it considers the merits of a similar Gricean account of the speech act of arguing.

KEYWORDS: speech act of arguing, Gricean account, intentions, descriptive, justification, reasons, manifest.

1. INTRODUCTION

Given the intricate nature of the question for the correct analysis of the speech act of arguing, the present paper discusses only a few of the relevant issues. The conclusions reached are meant to be tentative, and subject to future investigation into the topic, which needs to address subtleties and problems that this paper omits, and offer a more definitive proposal.

In discussing the nature of *arguments*, many authors have pointed out that we need to distinguish different concepts of argument, among which: argument as an object, and, on the other hand, argument as a speech act. For instance, Johnson (2009, p. 3) writes:

> The distinction between product and process seems to me fairly secure. It has a longstanding history here and in other disciplines. In logic, for instance, the term 'inference' is

understood as ambiguous as between the process of drawing an inference and the inference that results from that process.

In Simard-Smith and Moldovan (2011) we have argued that the word 'argument', as used in ordinary talk, is not, strictly speaking, ambiguous between two literal senses, one referring to an act of arguing, and another referring to the abstract object that is the content of that act. Using various tests for ambiguity available in the literature, we have argued that, literally, the word 'argument' does not refer to a speech act, but instead to the *content* of such an act (apart from other unrelated meanings that it might have). This conclusion, however, does not imply a denial of the claim that *there are* acts of arguing, and that such acts deserve careful study. This paper aims to go in that direction.

The question 'what is the nature of the speech act of arguing?' presupposes that there are speech acts of arguing. But this is not obviously so, at least not if we take 'speech act' in the technical sense in which it is used in the philosophical literature. In this technical sense, the literature distinguishes, following Austin (1962, p. 101), between the *force* and the *content* of a speech act. The *force* of an utterance of a sentence characterizes the kind of move in a "language game" that a speaker typically makes when uttering this sentence (Green, 2014, §2.1). The act of asserting that p differs from the act of requesting that p, or promising that p and so on, in as much as they have a different force. Here p stands for the propositional content of the speech act, i.e. that which is asserted, required, promised etc.

Is there a speech act of arguing? While Austin (1962, p. 102) explicitly characterizes arguing as an illocutionary act, it might be thought, *pace* Austin, that it refers to an activity of a more complex nature, which involves making various speech acts, for instance, assertions. It might be thought that acts of arguing are units of discourse of a different level from speech acts such as asserting or promising, and belong to a classification of types of discourse to which *explaining* a fact, *telling* a story, *describing* an object, or *building a theory*, also belong. This is what various authors have suggested. For instance, David Hitchcock defines 'argument' as a speech act (a definition which he later rejects, but which might still be interesting as a definition of 'arguing') as follows: "an argument is a claim-reason complex consisting of an act of concluding (which may be of any of the five main types in Searle's taxonomy of speech acts) and one or more acts of premissing (each of which is an assertive)". (Hitchcock, 2007, p. 6). A similar definition of 'argument' we find in van Eemeren and Grootendorst (1984, pp. 19-35, 39-46). For them an argument is a "constellation of speech acts": "The constellation of statements S1, S2, (..., Sn) consists of assertives in which propositions are expressed... Advancing the constellation of statements S1, S2, (..., Sn) counts as an attempt by S to justify [or to refute] O to L's satisfaction." (1984, p. 43), where O is an opinion, S is the speaker, and L the listener. Each statement of the constellation is a speech act itself, and has a particular illocutionary force. However, according to van Eemeren and Grootendorst (1984), the constellation itself also has an illocutionary force. The authors distinguish between the illocutionary acts at the level of sentences (made by uttering individual sentences), and illocutionary acts at a higher level of the discourse. As a result, they write, "sentences uttered in argumentation in fact have two illocutionary forces simultaneously" (1984, p. 32) So, according to van Eemeren and Grootendorst (1984), speech acts of arguing are *complex* acts, constituted by other speech acts. Bermejo-Luque (2011, p. §3.3) also takes this approach and introduces the useful label "second order speech-acts" for acts of arguing. Bermejo-Luque (2011, p §3.3.2) suggests that "explain that p", "answer that p", "reply that p", etc. are also second order speech-acts. The complex act of arguing, she writes, contains various first order speech acts, which can be classified in two categories: acts of *adducing* premises, and acts of *concluding*. I tentatively assume this view in what follows, although I am confident that much of what I have to say makes sense even if one disagrees with this view.

2. THE NATURE OF ARGUING

The next question we need to address is whether a unitary analysis of all acts of arguing is possible. Does arguing have an essence, specific function, or constitutive aim? Many authors take arguing to have a variety of functions, including, but not limited to, the rational resolution of disagreement. As Kauffeld (1998, p. 439) points out, "the idea that argumentation ideally aims at a mutually satisfactory resolution of disagreement [...] runs at least back to Plato's Socrates". But Mohammed (2016, p. 223) notes that in answering the above question attention must be paid to the distinction between *collective* goals of the activity of arguing and *individual* goals that arguers set for themselves when engaging in this type of activity. Rational resolution of disagreement is, in this sense, a collective goal, and not an individual one. My purpose in this essay is to consider the plausibility of a Gricean account of the speech act of arguing, and for that reason, as it will become clear in what follows, the relevant question we need to ask concerns the existence of a universal individual goal of arguing.

Ralph Johnson identifies two main individual goals or functions of arguing: to persuade rationally (2000, p. 160), and "to arrive at the truth about some issue" (2000, p. 158). Frank Jackson (1987) has suggested two functions: that of convincing, and what he calls "the teasing-out function", i.e. the function of helping us notice the relevance of certain facts to our concern and the existence of certain inferential relations that we did not notice before. Austin (1962, p. 102-3) writes that *convincing* is the characteristic aim of arguing. And, as O'Keefe (2012, p. 20-21) shows, there is a long tradition of characterizations of arguing that takes conviction or persuasion to be its intrinsic feature.[1]

However, there is much widespread disagreement with the idea that persuasion is a universal goal of argumentation. Mohammed (2016, p. 223) shows that many argumentation theorists take persuasion to be a "secondary" or "extrinsic" goal, meaning that it is a derived one, parasitic on the "primary" or "intrinsic" goal of arguing. The list includes van Eemeren and Grootendorst (1984, p. 48), Hamblin (1970, p. 241), Johnson (2000, p. 160), Perelman and Olbrechts-Tyteca (1969, p. 4), among others. Indeed, the intention to persuade does not seem to be a necessary condition for arguing. To take a simple counterexample, consider the case of a philosopher who argues for *p*. He might not care much whether he changes the adherence of the audience to the thesis he proposes, but instead might be primarily concerned with showing to the philosophical community his valuable argumentative abilities.

Marianne Doury (2012) explicitly argues against persuasion as a universal goal of arguing. According to her, there is an open-ended list of possible aims that people might have when arguing, which include: to help elaborate a position; to convey the reasons why one holds a particular position; to silence the opponent; to have the last word; as a means for decision making, either in everyday ordinary situations, or in institutional settings; or simply a way of "occupying the floor". Doury (2012) notes that argumentation is sometimes part of the process by which one forms a certain position on a topic, and expresses it, which amounts to constructing one's identity: "A person defines oneself by one's beliefs, by the reasons one has for those beliefs, by the fact that one takes the responsibility of expressing those beliefs and by the way

---

[1] The traditional distinction between conviction and persuasion, O'Keefe (2012: 21) notes, is that between a discourse that affects the mental states of the audience and one that affects the behaviour or the will, respectively. See O'Keefe's paper for a detailed discussion of how these terms should be understood.

one chooses to verbalize them, among others. In this sense, argumentation fulfils also an *identifying function*." (2012, p. 109)

Doury concludes that, "If language users argue, they do so because this activity serves a purpose, but this purpose cannot be determined a priori and without reference to the context..." (2012, p. 108) Kauffeld (1998, p. 440) also suggests that the list of purposes is wide open. He writes: "But must we suppose that all modes of argumentation are subordinate to a single ideal end?" In a similar vein, Goodwin (2002) writes:

> Design theorists [...] do not *assume* that argumentative transactions are necessarily either functional or cooperative. For one thing, these transactions are so varied that it seems rash to assign them any one social function—unless perhaps the function of "giving some order to our interactions"; and even then, only if we included war and anarchy as instances of "order" (2002, p. 8).

Scott Jacobs (1989, p. 345) argues along the same lines that it is a mistake to take acts of arguing to be "a homogeneous class of speech act with a specifiable illocutionary force".[2]

I have no knock-down argument against this kind of scepticism. However, I find the claim that there is no universal purpose (or a closed class of purposes) for all acts of arguing counterintuitive. As competent speakers, we do have the ability to recognize acts of arguing, and to distinguish them from other linguistic acts, such as acts of explaining, with which they are usually contrasted. If this ability has a rational basis, then there must be a property or a cluster or properties of speech acts in virtue of which we identify acts of arguing. Given that such criteria must be part of our linguistic competence, the project of finding the correct analysis of arguing by offering a rational reconstruction of these acts should be possible. Moreover, it seems intuitive that we should look for this universal feature in relation to the purposes that

---

[2] A parallel could be drawn here with the vast literature on the speech act of assertion. Herman Cappelen writes that: "What philosophers have tried to capture by the term 'assertion' is largely a philosophers' invention. It fails to pick out an act-type that we engage in and it is not a category we need in order to explain any significant component of our linguistic practice" (Cappelen 2010). The view also reminds of Charles Travis (1997) and John Searle (1978), both of which argue that there is no literal meaning, understood as a constant semantic feature of linguistic expressions that is not context-dependent.

acts of arguing have, and not in relation to their inferential structure, which they often share with explanations.

It might also be the case that the present methodology is unsatisfactory, and that the way in which the nature of arguing should be discovered is not by directly considering possible candidates for a definition and imagining counterexamples to them. Such an approach is in line with the philosophical tradition of conceptual analysis, but recent developments in philosophy of language, such as experimental semantics and pragmatics, suggest that it might not be the best approach. Hamblin considers that a direct approach to questions of definition is not the best route to take: "there is little to be gained by making a frontal assault on the question of what an argument is. Instead, let us approach it indirectly by discussing how arguments are appraised and evaluated." (1970, p. 231) However, without carefully considering the merits of different definitions of a phenomenon, one might end up implicitly assuming without argument a particular definition. As O'Keefe (2011, p. 20) points out with respect to certain discussion in argumentation theory, they "might easily become confused because of hidden definitional disagreements". If this is so, then the present discussion should be going in the right direction.

Moreover, the situation does not seem so desperate for the defender of the project. As Dima Mohammed has shown, "Justification, or something closely related to it such as manifest rationality, is a goal of argumentation about which scholars seem to agree despite their different terminological choices" (2016, p. 224). She finds that this goal is acknowledged in Toulmin (2003, p. 12), van Eemeren and Grootendorst (1984, Ch. 2), Johnson (2000, p. 1), Bermejo-Luque (2011, p. 53; 2015, p. 2), and others. Even Goodwin and Innocenti (2016), although they reject functionalist approaches to analysing arguing, seem to accept this point:

> In order to affect an audience in any way (to persuade them, to induce them to alter their standpoint, etc.) a speaker first has to make a reason apparent. It is not possible to change A's relationship to C without making a reason apparent, although it is possible (as we have shown) for S to make a reason apparent without trying to change A's relationship to C. Making reasons apparent is thus a task pragmatically necessary for any audience effect. Therefore, if there is a

function of argumentation, making reasons apparent is more likely it. (2016, p. 10)[3]

The phrases "offering justification" and "giving reasons" are closely related. But "justifying" and "offering justification" seem preferable to "giving reasons" for the present purposes. First of all, explaining is also a form of giving reasons, as Marraud (2017, p. 5) notes (although this is another use of the word "reason", closer in meaning to causes, which are given in explanations). Second, one can give reasons for c, but finally decide to reject c for stronger reasons. In such cases we would say that one is giving reasons for c, but not that one is offering a justification of c, or that one is arguing for c. Arguing for c and justifying c seem closer, in this sense.

A further qualification is required in relation to the terminology: "justification" and "justifying" should be understood here not in the sense of *acquiring* justification for a claim, but in the sense of presenting it, making it available to others, offering it. It is a *social* sense of "justifying", one that involves an interaction with other people who are provided with the justification the speaker possesses (or at least thinks she possesses) for a claim. This social sense is made clear in van Eemeren and Grotendorst's (1984) formulation of the essential condition of the speech act of pro-argumentation, which gives "the quintessence of the illocutionary act" (1984, p. 21): it consists of advancing statements in "an attempt by S to justify O to L's satisfaction" (1984, p. 43). This indicates that they have in mind a social, interactive, dimension of justification.

3. APPROACHES TO THE SPEECH ACT OF ASSERTION

Turning now to the literature on the speech act of assertion, the first observation is that philosophical analyses of assertion can be divided into two classes: *descriptive* and *normative* accounts. Normative accounts define assertion by appeal to the rule that is essential for the evaluation of the speech act in question. They identify a norm that is constitutive to such acts, which individuates it and distinguishes it from any other speech act. Williamson (2000, p. 243) proposes the knowledge rule, which reads:

(KR) One must: assert *p* only if one knows *p*.

---

[3] However, they avoid talk of functions, and simply choose to say that making reasons apparent is what an argument *is* (2016: 11).

Descriptive accounts appeal to the psychological facts about the mind of the speaker or the hearer, and the details of the communicative situation in which the speech act occurs. Such a descriptive account can be derived from Grice's (1957, 1969) analysis of non-natural meaning. Grice's famous analysis reads as follows: "A meant$_{NN}$ something by x' is (roughly) equivalent to 'A uttered x with the intention of inducing a belief by means of the recognition of this intention'" (1969, p. 76).[4] This is an analysis of the concept of speaker meaning (or what a speaker means by x, where x is something that she does, for instance, an utterance of a sentence, or a gesture) in terms of the characteristic intention with which x is performed. The concept of speaker meaning bears some similarity to the notion of a locutionary act,[5] while the inducing of a belief is a perlocutionary concept. So, the Gricean analysis might be seen (roughly) as a perlocutionary analysis of a locutionary act. One could easily obtain from this a perlocutionary analysis of the *illocutionary* force of the act of asserting, all the more so because the characteristic perlocutionary effect of an assertion is forming a belief. On this basis, the following analysis is formulated (though not endorsed) in Pagin (2015, §3.1):

(GA) S asserts that p by the utterance u iff there is a hearer H such that

    i.   S intends u to produce in H the belief that p
    ii. S intends H to recognize that (i)
    iii. S intends H to believe that p at least partly for the reason that (i).

There is an impressive amount of literature dedicated to discussing the Gricean analysis of non-natural speaker meaning. Most of the objections to Grice's analysis of meaning apply to the above analysis of assertion as well. They can be classified in two groups: objections to the *sufficiency* of conditions (i) to (iii), and objections to the *necessity* of one or more of these conditions. Due to space limitations, I mention

---

[4] There are other, more sophisticated, formulations of the analysis in Grice's work, but the differences are not essential to our purposes here.

[5] Austin writes: "The act of 'saying something' in this full normal sense I call, i.e. dub, the performance of a locutionary act." (1962, p. 96) The expression 'what S means to say by x' picks out a subspecies of the kind that 'what S means by x' picks out. So, at least for cases in which x is a linguistic utterance, the act of speaker meaning bears some similarity to a locutionary act.

here only one particular objection which I take to be significant. It has been argued that condition (iii) is too restrictive: in many (and probably, most) cases of speaker meaning the speaker does not intend the hearer to form a particular belief *on the basis of* recognizing the speaker's intention to do so (Grice, 1969, p. 107). As Searle (1969, p. 47) points out, if I read a book I may have many reasons to believe what is said there but it is not one of my reasons that I recognize the author's intention to believe what she says. And, usually, it is also not the author's intention that her readers form a belief on such a basis. Cases of arguing are particularly interesting in this sense: when someone is arguing in favour of a thesis *p*, she usually intends to convince the audience that *p* on the basis of the strength of her argument, and not on the basis of the audience's recognition that she has this intention. The latter does not play the role of a *reason* in obtaining the desired effect. Therefore, condition (iii) is not fulfilled.

There are attempts to reject this objection to the analysis, but I am unsure as too whether they are successful. One such move makes recourse to Grice's distinction between justificatory reasons and explanatory reasons, and argues that in condition (iii) "reason" is used in the latter sense. Romero (forth.) develops this idea on the basis of Grice (2001) and Warner (2001, viii). An alternative move is simply to discard condition (iii), and this is what Neale (1992, pp. 547-9) proposes, although for different reasons. However, according to Grice (1957), (iii) is needed in order to distinguish the sense of 'meaning' that refers to a non-natural phenomenon from the sense of the word that refers to a natural (causal) phenomenon. I come back to this objection below, in the context of the discussion of a Gricean account of arguing, which I propose in the next section.

4. A GRICEAN ACCOUNT OF ARGUING

In the previous section I have mentioned the Gricean analysis of assertion GA, inspired in Grice's analysis of speaker meaning. Let us now intend to do the same for the act of arguing. The characteristic intention or goal of arguing is that of *offering justification* for a claim c to a hearer H, as the discussion in section 2 suggests. As a result, a tentative attempt to formulate a Gricean analysis of the speech act of arguing (call it GAr), which is a complex, second order, speech act, will take the following form:

(GAr) S argues that *c* by asserting $p_1, p_2, ... p_n$ and *c* iff there is a hearer H such that:

i. S intends to offer p_1, p_2, ... p_n as justification for c to H.
ii. S intends H to recognize that (i)
iii. S intends to offer justification for *c* partly on the basis of H's recognition that (i).

Due to space limitations, I can only make here some very quick remarks on the proposal. A first comment concerns condition (ii). Condition (ii) relates harmoniously with what I take to be a fundamental characteristic of the practice of argumentation, i.e. what Johnson (2000) calls its "manifest rationality". Johnson writes: "To say that the practice of argumentation is characterized by manifest rationality is to say that it is patently and openly rational" (2000, p. 163). The requirement of manifest rationality is that the participants in the practice of argumentation "agree not to do anything that would compromise the substance or appearance of rationality" (2000, p. 163). The example of the judge (given on the same page) is illuminating: she should do all that is possible not only to be as fair as possible but also to look as fair-minded as possible. Condition (ii) of GAr resonates with this idea: the arguer's intention to persuade must be manifest. So, GAr captures one fundamental dimension of manifest rationality, i.e. that arguing is a manifest attempt at reasoning.

The second comment is that the mentioning of "by asserting p_1, p_2, ... p_n" in the analysis is needed in order to distinguish acts from arguing from acts in which reasons are given for a claim without arguing, such as in a case of *showing evidence* for c. If you ask me whether it's raining and I am merely pointing towards the window so that you realize that indeed it is raining, I am offering you reasons for the belief that it is raining, and I intend you to recognize that I am doing so. However, we would not say that I have argued that it is raining. Adding the condition (ii) does not help with this problem: adding condition (ii) turns *giving reasons* into *manifestly giving reasons*, or "making reasons apparent", as in Goodwin and Innocenti (2016), or "manifest rationality", as in Johnson (2000). The present analysis avoids this difficulty by making reference to the *asserted* propositions that forms the structure of the reasons.[6]

A third and last comment relates to Searle's (1969, p. 47) objection to condition (iii) discussed above. According to Searle, an

---

[6] However, a further kind of cases might still be problematic. Bermejo-Luque (2015) writes: "by saying that my name is Alex in adequate circumstances, I can... give a reason to believe so by communicative means... Yet, if we don't want to say that mere assertions count as argumentation, it seems, again, that we will need something else..." (2015, p. 3)

objection both to the Gricean account of speaker meaning, and to that of assertion is that the speaker does not aim to produce a belief in the audience on the basis of the recognition of her intention to do so, but rather entirely on the basis of the reasons given. Does it also affect GAr? When I presented the paper at the ECA 2017 I thought so, and I opted for a version of the analysis that eliminated condition (iii) as formulated above.[7] However, the comments to my talk at ECA helped me realize that, in this case, this is not sufficient reason to reject condition (iii).[8] According to the analysis, the arguer does not intend that H's recognition of S's intention to provide justification for c function itself as justification for c. Propositions p_1, p_2, ... p_n are meant to be sufficient justification for c. Instead, S intends that H's recognition that S intends (i) (i.e. to offer justification for c) function as a reason for the *offering* of the justification, and not directly for c. S cannot succeed in making an offer (of whatever kind) unless S's intention is recognized. So, the recognition of the intention to offer a justification plays a role in the realization of the action. If the offering of reasons for c is not recognized as such then the audience is likely to miss the point, and might take the speaker to be doing something different (e.g. offer an explanation, make a proposal etc.).

An alternative analysis on which the intended effect is not to offer justification but to convince the audience H that c (or, alternatively, to obtain her commitment to c) is affected by this objection, because typically the arguer does not aim to convince H (or obtain her commitment) on the basis of H's recognition that this is what she is trying to do, but only on the basis of strength of the reasons offered. But, convincing is not, according to GAr, the characteristic aim of arguing.

ACKNOWLEDGEMENTS: The research that led to this paper has been supported by the Spanish Ministry of Economy and Competitiveness, EXCELENCIA program, project no. FFI2016-79317-P, "El acto de habla de argumentar y su lugar en la Teoría de la argumentación". Special thanks to Paul Simard-Smith; Jane Goodwin and Lilian Bermejo-Luque for their valuable comments on the topic, as well as to the rest of audience of the talk on which this paper is based, which was given at ECA 2017.

---

[7] This version of my proposal is discussed in Marraud (2017, p. 4).

[8] I especially want to thank Jean Goodwin for a discussion of this point, which helped me reconsider my position.

REFERENCES

Austin, J. L. (1962). *How to do things with words* (2nd ed.). Oxford: Oxford University Press.
Bermejo-Luque, L. (2011). *Giving reasons: A linguistic-pragmatic approach to argumentation theory*. Dordrecht: Springer.
Bermejo-Luque, L. (2015). Giving reasons does not always amount to arguing. *Topoi*:1-10.
Cappelen, H. (2011). Against assertion. In J. Brown & H. Cappelen (Eds.), *Assertion: New philosophical essays*. Oxford: Oxford University Press.
Doury, M. (2011). Preaching to the converted. Why argue when everyone agrees? *Argumentation*, 26(1), 99–114.
Eemeren, F. H. van & R. Grootendorst. (1984). *Speech acts in argumentative discussions*. Dordrecht-Holland: Foris Publications.
Goodwin, J. (2002). One question, two answers. OSSA Conference Archive. 40. Retrieved from
    https://scholar.uwindsor.ca/ossaarchive/OSSA4/papersandcommentaries/40
Goodwin, J. & Innocenti B. (2016). The pragmatic force of making reasons apparent. In D. Mohammed & M. Lewinski (Eds.), *Argumentation and Reasoned Action: Proceedings of the First European Conference on Argumentation* (Vol. 2, 449-462). London: College Publications.
Green, M. (2015). Speech Acts. *The Stanford Encyclopedia of Philosophy*. E. N. Zalta (Ed.), Retrieved from
    http://plato.stanford.edu/archives/sum2015/entries/speech-acts/
Grice, H.P. (1957). Meaning. *The Philosophical Review*, 66(3), 377-88.
Grice, H. P. (1969). Utterer's meaning and intention. *Philosophical Review*, 78 (2), 147-177.
Grice, H.P. (1989). *Studies in the way of words*. Cambridge, MA: Harvard University Press.
Hamblin, C. L. (1970). *Fallacies*. Newport News, VA: Vale Press.
Hitchcock, D. (2007). Informal logic and the concept of argument. In D. Jaquette (Ed.), *Philosophy of Logic* (pp. 101-129), Amsterdam: Elsevier.
Jackson, F. (1987). Petitio and the purpose of arguing, in conditionals (Ch. 6). Oxford: Basil Blackwell.
Jacobs, S. (1989). Speech acts and arguments. *Argumentation*, 3(4), 345-365.
Johnson, R. (2000). *Manifest rationality*. Mahwah, NJ: Lawrence Erlbaum Associates.
Johnson, R. (2009). Revisiting the Logical/Dialectical/Rhetorical Triumvarate. OSSA Conference Archive. 84. Retrieved from
    https://scholar.uwindsor.ca/ossaarchive/OSSA8/papersandcommentaries/84.
Kauffeld, F. J. (1998). The good case for practical propositions: Limits of the arguer's obligation to respond to objections. In F. H. v. Eemeren, R. Grootendorst, J. A. Blair & C. A. Willard (Eds.), *Fourth ISSA conference on argumentation* (pp. 439-444). University of Amsterdam: SICSAT.

Marraud, Hubert (2017) "¿En qué consiste argumentar? Argumentar, argumentación y argumento", talk given at IV Congreso Iberoamericano de Filosofía de la Ciencia y Tecnología, Salamanca, 3.07.2017.

Mohammed, D. (2016). Goals in argumentation: A proposal for the analysis and evaluation of public political arguments. *Argumentation, 30*(3), 221-245.

Neale, S. (1992). Paul Grice and the philosophy of language. *Linguistics and Philosophy*, 15(5), 509-59.

O'Keefe, D. J. (2012). Conviction, persuasion, and argumentation: Untangling the ends and means of influence. *Argumentation, 26*(1), 19-32.

Pagin, P. (2015). Assertion. *The Stanford Encyclopedia of Philosophy*, E. N. Zalta (Ed.), URL = http://plato.stanford.edu/archives/spr2015/entries/assertion/.

Perelman, C. & Olbrechts-Tyteca, L. (1958). *Traité de l'argumentation: La nouvelle rhétorique*. Paris, FR: Presses Universitaires de France.

Pinto, R. C. (2009). Argumentation and the Force of Reasons. *Informal Logic, 29*(3), 268-295.

Romero, E. (forth.). "Significado" de Grice. In D. Pérez Chico (Ed.), *Cuestiones de la filosofía del lenguaje*, PUZ.

Toulmin, S. E. (1958). *The uses of argument*. Cambridge: Cambridge University Press.

Travis, C. (1997). Pragmatics. In B. Hale & C. Wright (Eds.), *A companion to the philosophy of language* (pp. 87-107). Oxford: Blackwell.

Searle, John R. (1969). *Speech acts: An essay in the philosophy of language*. Cambridge, UK: Cambridge University Press.

Searle, J. R. (1978). Literal meaning. *Erkenntnis, 13*(1), 207–224.

Simard-Smith, P. & Moldovan, A. (2011). Arguments as Abstract Objects, *Informal Logic*, 31(2), 230-261.

Warner, R. (2001). Introduction: Grice on reasons and rationality. In R. Warner (Ed.), *Aspects of Reason* (pp. vii-xxxviii). Oxford: Clarendon Press.

Williamson, T. (2000). *Knowledge and its Limits*. Oxford: Oxford University Press.

39

Inferring Argumentative Patterns in Polylogues about Energy Issues

ELENA MUSI
Columbia University
em3202@columbia.edu

MARK AAKHUS
Rutgers University
aakhus@rutgers.edu

This paper proposes a scalable methodology for the study of argumentative patterns in online polylogues. Our linguistically informed corpus-based procedure is based on the annotation of the three dimensions of argument structure, argumentation schemes and lexical features. The devised methodology is applied to the analysis 6 discussion threads from the subreddit Changemyview pertaining to oil drilling and fracking issues. Merging the three analytic levels we discuss recurrent argumentative patterns in the energy context.

KEYWORDS: [argumentative patterns, polylogues, *Changemyview*, fracking]

1. FROM DISCOURSE FRAMING TO ARGUMENTATIVE PATTERNS

Studies about public controversies focused on fracking and oil drilling issues have addressed how energy practices and policies are perceived, promoted, and justified by different stakeholders from a variety of scholarly perspectives (Boudet et al., 2013; Metze, 2014; Jaspal & Nehrlich, 2014; Cotton et al., 2014).

One unifying approach across these media and policy studies is the identification of *discursive framing*, which is "how actors define, select and emphasize particular aspects of an issue according to an overarching shared narrative and set of assumptions" (Bomberg, 2010, p. 75). These studies emphasize the role of media framing with some success at analysing the "media packages" – that is, the way that framing and reasoning devices are bundled together in particular ways that guide interpretation of a situation and offer narratives or underlying rationales about action to be taken (Ihlen & Nitz, 2008; Rein & Schön, 1996). As Ihlen and Nitz (2008) point out, these studies are often good at identifying the surface, lexical features of framing that name situations but are often less successful in making the analytic moves that articulate the reasoning in framing in way that demonstrates what frames are privileged in particular social-political contexts.

The notion of argumentative patterns seems to conflate the thematic and the reasoning aspects underlying frame analysis. Argumentative patterns are, in fact, defined as constellations of "argumentative moves in which a particular kind of argumentation structure or a particular combination of argument schemes is exploited in defence of a particular type of standpoint" (van Eemeren & Garssen, 2013, p. 7). They, thus, constitute meaningful units of analysis to solve those conceptual issues that make the uncovering of frames a difficult task. Argumentative patterns have been studied in institutionalized contexts where particular issues are established and known and preferable lines of making and defending standpoints with particular acceptable lines of reasoning are developed. As shown by studies collected in the Special Issue of Argumentation "Argumentative patterns in discourse" (2016), they have been identified in the legislative, the legal, the medical and the scientific domains. For example, Snoeck-Henkemans (2016) found that in over-the-counter medicine advertisements standpoints aimed at guaranteeing the consumers' safety are of the prescriptive type backed up by a pragmatic argument. Starting from assumptions of domain specific institutional preconditions these analyses are hardly generalizable to public controversies continuously expanding in online fora.

At the state of the art, a systematic, reproducible and scalable methodology to identify argumentative patterns to more fully exploit the empirical and analytic potential of the argumentative pattern concept is lacking. Developing such methods would make an important contribution to argumentation research that would also contribute to improved understanding of framing in policy controversy. The main novelty of this study is to provide such a methodology combining insights from Linguistics, Informal Logic and Argumentation Theory. Since the main point here is about procedures for inferring argumentative patterns from text, the paper begins with a description of the data and methods. The results from applying the method are presented followed by a discussion of implications of the approach proposed here for argumentative patterns relative to developing theory in argumentation research about polylogues in argument fields, spheres, and communities and for application in related work about framing in environmental communication. [1]

2. IDENTIFYING ARGUMENTATIVE PATTERNS

The method proposed here identifies the distribution of argumentative patterns within a corpus of argumentative text. Our units of observations for reconstructing argumentative patterns include claims-premises for the structural elements of argumentative patterns, argument schemes for the inferential elements, and heteroglossic linguistic strategies for the presentational. The method enables discovery of the relative distribution of these elements and the identification of argument patterns within a corpus, regardless contextual peculiarities. Moreover, taking into account heteroglossic strategies, it boils down the identification of the stages of a critical discussion to the lexical level, being suitable to account for polylogical settings where argumentation stages may appear blurred. The proposed procedure is exemplified analysing 6 discussion threads from the subreddit Changemyview centred around oil drilling and fracking.

2.1 Multilevel annotation to identify argumentative patterns

As a blueprint to recover argumentative patterns we have devised a multilevel annotation taking into account claims-premises configurations, argumentative schemes and heteroglossic linguistic strategies. Such a choice aims to cover the structural, the inferential and the presentational dimensions.[2] These three analytic levels are necessary and complementary for the identification of argumentative patterns: different types of standpoints select different types (i.e. factual unassailable premise) and numbers of explicit premises as well as inferential relations that can be used to back them up. Besides the semantic types of propositions which make up the standpoints (e.g. predictions) and the speech acts they express (e.g. directives, assertions etc.), the topics of discussion further inform the constraints: claims

---

[1] Although the whole paper has been the result of a continuous process of interaction between the two authors, Elena Musi is the main responsible of sections 1 and 3, while Mark Aakhus of sections 2 and 4.

[2] With the term "presentational" we refer to linguistic strategies used by the speaker to position his stance with respect to that of other arguers.

asserting environmental consequences are, for example, likely to be presented as shared knowledge drawn from relations of the causal type.

In annotating the argument structure we have followed the guidelines proposed by Stede & Peldzus (2013, 2016) that distinguish claims – propositions that expresses an assertive speech act towards which the user is committed at the moment f utterance – major claims – major claims – propositions expressing the man stance in a text – premises – propositions which express reasons in support of claims – linked premises, propositions which bear supportive force only if considered conjoined (Freeman, 2000, 2011).

For the analysis of argument schemes connecting premises to claims/claims to major claims, we have, instead, adopted the annotation guidelines proposed by Musi, Ghosh and Muresan (2016), based on the Argumentum Model of Topics taxonomy of inferential relations (Rigotti & Greco Morasso, 2010). The guidelines have been empirically verified in terms of inter-annotator agreement. It has to be remarked that the AMT conception of argument schemes is particularly suitable to account for argumentative realities: argument schemes are not conceived as abstract rules reasoning but as inferential relations combing not only procedural, but also material premises. The latter are premises of contextual nature which allow the procedural component to become relevant to the specific situation: the datum coincides with the new information provided in the dialogue; the endoxon is common ground knowledge about the context.

Turning to the presentational level, we have observed the distributions of modals and evidentials which acquire, as underlined in Appraisal Theory, an rhetorical function since they "present it [a proposition] as but one proposition among a range of potential alternatives and thereby to open up dialogic space for any such alternatives" (Martin and White, 2005, p. 110).

We have, therefore, taken into account how speakers use modals and evidentials to express and negotiate their positions 'engaging' in multiple voice interactions following White's typology of heteroglossic strategies (2003):

- Concurrence: the speaker agrees with the hearer's points of views or shares knowledge with him presenting it as common ground
- Pronouncement: it expresses "intensifications, authorial emphases or explicit authorial interventions or interpolations" (White 2003, p. 269)
- Endorsement: external voices' opinions are construed by the speaker as correct or highly warrantable

- Concession: the speaker counters a point of view by first agreeing with some aspects of it
- Counter-expectancy: the speaker rejects reasonable viewpoints drawn from expectations

(1) While [there are [certainly]$_\text{evidential\_concession}$ some negative aspects to fracking]$_\text{claim}$, the contamination of ground water being the most concerning. [I think that fracking is a good option for energy production]$_\text{Major claim}$

argument scheme from causes

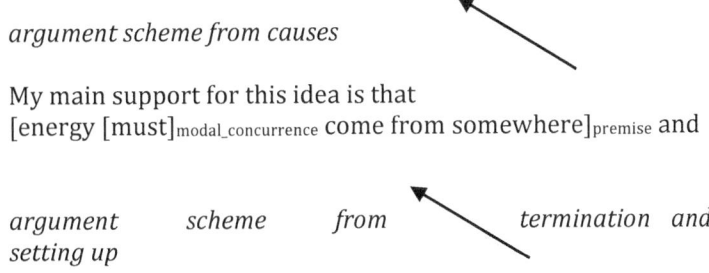

My main support for this idea is that
[energy [must]$_\text{modal\_concurrence}$ come from somewhere]$_\text{premise}$ and

argument scheme from setting up

from termination and

[the natural gas that is produced through fracturing is much cleaner and produces energy much more efficiently than coal or oil]$_\text{premise}$

The proposed modular methodology enables to not only highlight the key issues making up the disagreement space in energy frames, but to understand how they relate to each other in terms of more or less sound reasoning lines.

2.2 Argumentative context: CMV fracking corpus

We have collected our data from the subreddit ChangeMyView, a discussion forum "dedicated to the civil discourse of opinions, and is built around one simple idea: in order to resolve our differences, we must first understand them"[3]. Besides constituting a natural environment for the argumentative study of polylogues – multiple positions are pursued among multiple players appealing to multiple venues – ChangeMyView guarantees the presence of argumentation. It, in fact, resembles the design of a critical discussion as well as respects its core rules: advocating for his view to be changed, the original poster

---

[3] This descriptive definition of the subreddit as well as the submission and comment rules are available at the CMV wiki (https://www.reddit.com/r/changemyview/wiki/index).

opens up a confrontation stage from which the opening stage and the argumentation stage develop in the unfolding of arguments. Moreover, the "submission" and the "comment" rules resemble the main rules to conduct a critical discussion (Van Eemeren, Grootendorst & Snoeck Henkemans, 2002, p. 182-183).

As a sample we have selected 6 threads containing the terms "fracking"/ "horizontal drilling" / "hydraulic fracturing":

n.	thread title	comments	tokens	tokens per comment	users	Delta points
1	Bernie Sanders' energy policy is based in fantasy	59	4640	78.64	20	0
2	I believe that fracking is a good option for energy production and should be pursued. I consider myself an environmentalist	21	2478	118.00	7	0
3	I think the Keystone XL Pipeline should be built for environmental reasons, and protesters are fighting against their own interests	15	2230	148.67	11	0
4	The United States should not drill any oil	49	8600	175.51	15	0
5	I think that fracking is a practice that should be globally banned	6	956	159.33	3	0
6	If you benefit from resource extraction based economy then you cannot condemn fracking. If you want to protest fracking the true way to protest is to not consume...	5	626	125.20	5	0

Figure 1 – Sample of analysis selected from *Changemyview*

In our selection process we tried to include the widest possible variety of arguments about energy issues in the drilling/fracking context. We have, therefore, selected 2 threads pros oil drilling/fracking (threads 2, 3), 2 cons oil drilling/fracking (threads 4, 5) and 2 centred around energy policies to be put into action by policy makers (thread 1) or by the general audience (thread 6). Moreover, the length of the threads as well as the ratio of comments per user is quite varied.

As to the topics of discussion we have manually observed how they are lexically introduced and described. In line with traditional frame semantics (Fillmore, 1976) we assume that lexical units recall and activate cognitive and evaluative frames in the addressee's minds. Therefore, the frequency of specific words in discourses about a certain issue constitutes a key to access relevant frames. In corpus linguistics keywords are defined as "words which are significantly more frequent in a sample of text than would be expected, given their frequency in a large general reference corpus" (Stubbs, 1996, p. 24).

In order to check the significance of observed frequent words for the energy domain we have compared them to clusters of keywords in a control corpus of news about fracking collected through *Google Alerts* starting from February 2012 (ca 600.000 tokens).

3. RESULTS OF THE ANALYSIS

3.1 Distribution of argumentative components: schemes, epistentials, keywords-themes

From the annotation of the argumentation structure it has emerged that the number claims (222, 19 of which are *majorclaims*) outstands the number of textually expressed premises (136). Focusing on the premises-claims, claims/major claims inferential relations, the distribution of argument schemes is visualized in Figure 2

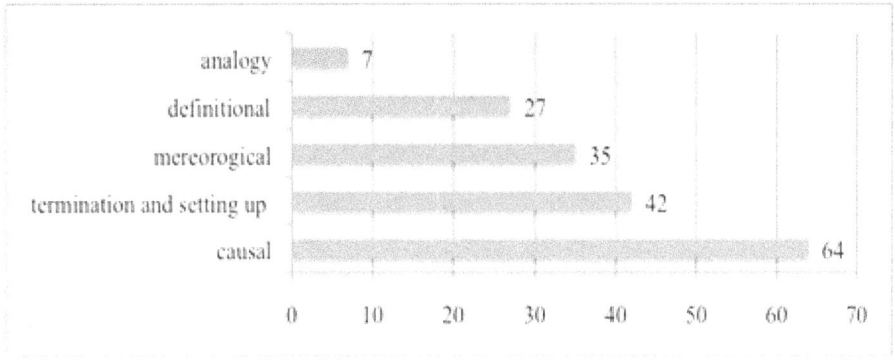

Figure 2 – Overall distribution of argumentation schemes

Turning to the analysis of rhetorical strategies, we have looked at epistentials to point out how argument schemes are performed and negotiated in interaction. The distribution of epistentials in the propositions functioning as premises and claims connected by different argumentation schemes in visualized in Figure 3:

epistentials	definitional	mereological	causal	termination and setting up	analogical	TOT
necessarily	1	0	0	0	0	1
obvious	0	0	1	0	0	1
show	1	0	0	0	0	1
clear	0	0	2	0	0	2
cannot	0	0	3	0	0	3
must	3	0	0	0	0	3
impossible	0	0	5	0	0	5
should	2	0	1	13	0	16
need	1	0	21	2	0	24
TOT	8	0	33	15	0	56

Figure 3 – Distribution of epistentials per argument schemes

From Fig. 3 it emerges that the epistentials occurring in our sample entail a high degree of speaker's commitment expressing necessity (e.g. "must", "necessarily"), impossibility (e.g. "impossible"), epistemic certitude (e.g. "obvious") or direct evidentiality (e.g. "show").

The manual observation of the topics discussed *qua* issues reveals four main cores: environmental impact, alternative forms of renewable energy, supply of energy demand and economic outcomes brought about by fracking and oil drilling. To assess the inter-genres generalizability of the attested narratives we have retrieved keywords from a control corpus to check whether the semantic domains of the observed frames are similar to the key terms in the control corpus about fracking. *Keyness* identifies terms that are unusually frequent in a control corpus. To calculate the *keyness* value the control corpus on fracking gathered through Google alerts was compared to a reference corpus created from 400 editorials randomly selected from the *New York Times* in the last two years. The 50 keywords showing higher keyness value in the fracking corpus are the following:

frequency	keyness	keyword
834	2879.74	fracking
1753	2623.466	gas
1265	1654.812	oil
741	1149.516	fracturing
702	1088.386	hydraulic
783	998.512	energy
827	949.49	water
556	873.029	shale
642	705.888	industry
510	688.8	drilling
433	609.63	wells
428	527.992	natural
410	516.395	environmental
350	464.502	regulations
402	407.91	production
365	385.791	county
549	308.62	well
173	271.644	chemicals
183	267.695	exploration
191	265.768	coal
595	251.59	state
160	240.293	uk
162	228.036	colorado
144	226.108	epa
165	220.057	sand
151	218.212	petroleum
192	216.215	environment
202	210.76	clinton
145	208.952	moratorium
199	206.593	prices
128	200.985	methane
286	198.444	local
331	196.187	use
124	194.704	extraction
258	190.391	process
176	185.696	resources
129	184.294	drill
297	183.733	information
123	182.72	fluid
123	182.72	waste
116	182.143	wastewater
114	179.002	operators
174	172.196	operations
129	171.516	maryland
108	169.581	groundwater
183	168.048	site
279	165.7	companies
105	164.871	contamination
169	162.076	development

Figure 4 – Keywords in the control corpus

As shown in Fig. 4., the keywords from the fracking corpus reveal a semantic domain that confirms the relevance of three out of four main observed topics in the CMV threads: environmental issues corresponds with keywords of "natural", "environmental", "environment", "contamination", "wastewater", "groundwater", and "methane"; supply of energy demand with keywords "energy" and "production"; and economic outcomes with keywords "prices", "companies", and "development". The fracking control corpus appears to have two additional themes not present in the CMV threads that refer to the activities of oil drilling and fracking (i.e. "oil", "gas", "shale", "petroleum", "extraction", "process") and around energy extraction policies in terms of regulations (i.e. "regulations" and "epa"). However, in the control corpus the relevance of alternative sources of energy does not emerge from the keywords' analysis, which may be due to the fact that natural gas can be conceived as an alternative source of energy itself.

3.2 Argumentative patterns

The analysis of the overlap among the structural, the inferential and the presentational levels of analysims has resulted in the identification of five argumentative patterns in the CMV threads about fracking and oil drilling. Throughout the analysis, argument schemes have turned out to be a key unit of observation. At a procedural level, presupposing the presence of premises/claims relations, they allow to identify what makes an argument. They, moreover, instruct about the type of standpoints at issue since they constrain the semantic type of proposition they are formed by. At a contextual level, reconstructing the material premises of argument schemes according to the *Argumentum Model of Topics* entails mapping what information is presented as new – datum – and what aspects are instead considered as part of the domain-specific common ground knowledge – endoxon – In this latter regard the analysis of keywords plays a major role suggesting the larger frames that are called upon by the arguers (Bigi & Greco Morasso, 2012).

3.2.1 Argumentative pattern type 1: Defending Predictive, Interpretive Standpoints

The most frequent argumentative pattern shows standpoints that are for the 80% of cases interpretative propositions of the predictive type. This type of standpoints is mainly attested where the topic is centred around the economic effects of policies regarding oil drilling/fracking are discussed (see 2). That of foreseeing future events represents, in fact, a central speech act in the economic domain (Walsh, 2004; Donohue, 2006).

Alternatively, predictions are used to ponder on feasibility of alternative energy sources (see example 3). In both cases the employed argument schemes are of the causal type.

(2) ["The oil industry is huge, the US gets tax dollars and other economic benefits from oil drilling"]Premise. ["Forcing the oil companies to shut down plants would cause widespread unemployment and make the markets associated with oil very unstable"]Claim

(3) ["it is expected to be only 0.5% of the countries (http://www.eia.gov/forecasts/steo/report/renew_co2.cfm) energy"]Premise. ["it cannot come close to meeting the energy demands of this country"]Claim

As exemplified in example (2), when efficient cause is at stake, the scheme proceeds from the cause to the effect. The other type of encountered causal argument scheme is based on inferential relations from means to goals:

(4) ["I work closely with a lot of scientists that are working on batteries and it would take a miracle for someone to develop the technology and manufacturing methods for cost-competitive batteries in that time"]Premise. ["Attaining 50% renewables in 20 years is **impossible**"]Claim

In example (4) the author is commenting on Bernie Sander's energy plan which reckons on the goal of attaining 50% renewables in 20 years. He qualifies this goal as *impossible* due to the lack of suitable means in the planned time, relying on an inferential rule such as "if the means required to attain a certain goal are lacking, the goal is not achievable". As shown in table 5., causal argument schemes, both of the efficient cause and the final goal type, are mainly associated to epistentials expressing necessity or impossibility (e.g. *cannot, impossible*), which convey a high degree of speaker's commitment over the embedded propositions.

These markers occur in the standpoints, which are, thus, presented as non-defeasible since drawn from premises epistemically highly reliable: both in example (3) and in example (4) the reliability of the predictions expressed in the premises is based on the *ethos* of the sources of information featuring a high level of expertise. As an effect, the rhetoric function played by epistentials is a particular type of *concurrence*: even though not necessarily shared by other users, the truth of the claim is presented as a belief shared with a community of experts and, thus, likely to achieve hearers' consensus or to change the mind of users having a different stance. Differently from what happens in cases of endorsement, the speaker does not only construes third party's opinions as warrantable, but he uses them as a starting point to draw his own inferences.

3.2.2 Argumentative pattern type 2: Defending Recommendations

The second most frequently encountered argumentative pattern features directive speech acts in the form of **recommendations** as standpoints. The reasonings in support of these standpoints are argumentation schemes from "**termination and setting up**": the premises express evaluations about positive or negative consequences of actions regarding oil drilling and fracking. The topics stem from the **economic** (see 5) to the **environmental** (see 6) aspects of fracking and oil drilling:

(5) ["To not invest in those things is pure folly that will leave us behind"]Premise. ["Building those plants and developing grid-scale energy storage is where we need to be putting our money."]Claim

(6) ["Water that is used (and highly contaminated) is kept out in open pits before transportation away from the site. This poses a **threat** to wildlife living nearby (should they drink it) and there's the added eventuality of that dirty water contaminating the water table (which it will, because not even our sewers can\"t stop this 100%)"]Premises. ["I think that fracking is a practice that **should** be globally banned"]Claim.

They are frequently introduced by the deontic modal *should* or by necessity epistentials such as the verb *need*. In a heteroglossic perspective, they can convey *Pronouncement* or *Concurrence* depending on constructional features of the subject. In example (5) the subject of the verb coincides with the agent of the embedded proposition and takes the form of the *inclusive* pronoun "we": the author presents the need of putting money on the development of oil drilling plants as a shared task that actively involves the American community – concurrence. In 6), instead, the agents of the advocated event ("banning fracking") is left implicit, while the recommendation is construed as a proposition in the scope of the propositional attitude indicator "I think" which marks the standpoint as an authorial intervention – pronouncement.

3.2.3 Argumentative patterns type 3: Defending evaluations

A third recurrent cluster of argumentative moves is centred around standpoints in the form of evaluative propositions expressing categorizations. The argument schemes at work are of the definitional type. As clarified by the lexical semantics of these predicates (i.e. *counterproductive, safe*) or by the context of utterance (i.e. *good for the economy*) the positive or negative value of oil drilling and/or fracking are decided on the basis of sustainability in terms of energy demand (see 7) as well as environmental impact (see 8), taking into account other possible alternative energy sources:

(7) ["The panels require rare earth metals. These **must** be mined and there is a limited quantity in the world"]Premise ["So solar is not a completely non consumptive energy source"]Claim

(8) ["It is **absolutely impossible** for fracking water to pollute the environment during the fracking process"]Claim. [The water shelf is only a couple hundred feet under the ground. Fracking occurs at over 10,000 ft. underground by cracking open shale and retrieving the natural gas out of the pores in the shale. There are many, many layers of rock between this and the water shelf. It is impossible for anything at that depth to miraculously travel upward through these layers about 2 miles and leech into the water supply [...]]Premises. ["The point of this long winded post is that fracking in and of itself is a very useful tool that is perfectly safe"]MajorClaim

As shown in examples (7) and (8), the premise/claim in support of the claim/major claim list definitional properties which allow to attribute to solar energy the predicate "not a non-consumptive energy sources", and to fracking the predicate "perfectly safe". Table 5. shows that the argumentative components linked by definitional inferential relations mainly contain deontic modals (cf. 2) or epistemicals (cf. 3) that entail a high degree of commitment. They are generally positioned in the premises, where they express **Concurrence**: in example 7) the deontic modal verb *must* frames the mining process as a *sine qua non* requirement for the construction of solar panels, presenting it as part of the common ground.

In example 8), the predicative construction *it is impossible*, modified by the intensifier *absolutely*, work as an epistential, assuming an evidential inferential function next to an epistemic modal one: the impossibility for fracking water to pollute the environment during the fracking process is drawn as an inference from a set of factual subsequent premises. The evidential value shifts the authorial emphasis lexically conveyed by the modal value – endorsement – towards and intersubjective positioning: the hearer, having access to the premises, can go through the same inferential path undergone by the speaker, concurring with his impossibility judgement.

In a rhetorical perspective, the use of markers of *Concurrence* in propositions functioning as premises has the function of presenting the evaluations as intersubjective and, thus, more easily acceptable.

3.2.4 Peripheral argumentative patterns

Other recurrent standpoints are interpretative propositions commenting on general facts. They are drawn from premises through inductive (mereorogical argument schemes) and analogical argument schemes. In particular, mereological argument schemes connect premises and claims that do not contain explicit markers of rhetorical relations.

This behaviour is due to the type of inductive reasoning at stake, which draws generalization regarding economic costs of oil drilling as well as environmental consequences frequently inferred from the authors' personal experience (see example 9)

> (9) ["all the pipe to be used for the project and all or most land designated for its construction is already sourced and set aside, waiting for government approval"]Claim. ["I live near one of the pipeyards that holds about one tenth of the total pipe needed for the pipeline, and that facility alone holds about 70 acres of stacked pipe"]Premise.

The mereological reasoning at stake is not examples of statistical, but of *rhetorical induction* (Aristotle, APr. 68b 15): in both cases the cited states of affairs constitute παράδειγμα, namely examples which are highly representative as well as relevant since stating unassailable facts.

The lack of epistentials working as rhetorical strategies is also characterizing the few encountered analogical argument schemes:

> (10) ["Stopping eating chocolate will not end the slavery - positive action will end slavery."]$_{Premise}$. ["if it happens to be horrible it is appropriate to oppose it without going \"off-grid\" or otherwise taking extreme efforts to avoid using products made with energy that directly or indirectly adds to frackers' profits"]$_{Claim}$

As exemplified by (10), the comparison between the state of affairs expressed in the premise and that expressed in the claim is recurrently used to show the (un)reasonableness of strategies envisaged in view of specific goals.

4. CONCLUSION AND FUTURE WORK

In this study we have proposed a systematic methodology to retrieve argumentative patterns in polylogues. We have devised a multilevel annotation scheme to account for the structural – distribution of claims and premises – , the inferential – distribution of argument schemes – and the presentational dimensions of arguments – distribution of heteroglossic strategies – that promises to be scalable to different contexts. The cross-analysis of these units of observation has highlighted the presence of recurrent patterns: the core ones appear to be i) predictions about effects brought about by energy policies presented as results of inter-subjectively acknowledgeable causal chains as well as ii) recommendations about stopping or setting up oil drilling and fracking on the basis of evaluations marked as belonging to a shared common ground. These two reasoning types are typical instances of the Aristotelian praxis-oriented argumentation called "practical reasoning" (Walton, 1990) at an institutional level: the participants of the discussions cannot themselves realize the advocated actions, which are bound to public consent as well as policy makers' choices.

It has to be noticed that the identified thematic narratives seem to largely converge with the frames identified by Bomberg (2015) in her study about anti and pro shale discourse in British broadsheet newspapers. Such a similarity suggests that the storylines elaborated in newsmaking draw upon issues perceived as inherently relevant by the audience and deserve further investigations in a cross-national perspective. We plan to apply our analysis on different text genres at scale to trace back the influence played by activity types in shaping argumentative patterns. While the automatic search for clusters of epistentials as well as quantitative thematic analysis are easily achievable, the accurate automatic identification of claims/premises and, especially, arguments schemes still constitute a challenge for Argument Mining. Our multilevel annotation can be used to create annotated corpora to feed computational models; moreover, the attested argument patterns can facilitate classification tasks in the energy domain.

On a different note, we aim to fine-tune our methodology to account for argumentation in polylogues as a process. Our results mirror, in fact, trends in the arguments produced by different users, but do not show what issues emerge as pragmatically relevant in the disagreement space: it could, in principle, be the case that, even if broadly used by users, argument patterns exploiting causal chains never become the target of disagreements being, thus, poorly controversial. This further step calls for devising a macroscope enabling to highlight what discourse moves become especially controversial in the polylogical argumentative process.

ACKNOWLEDGEMENTS: This paper is based on work supported partially by the Early Post Doc SNFS Grant for the project "From semantics to argumentation mining in context: the role of evidential strategies as indicators of argumentative discourse relations"

REFERENCES

Bigi, S., & Morasso, S. G. (2012). Keywords, frames and the reconstruction of material starting points in argumentation. *Journal of Pragmatics*, *44*(10), 1135-1149.

Bomberg, E. (2017). Shale we drill? Discourse dynamics in UK fracking debates. *Journal of Environmental Policy & Planning*, *19*(1), 72-88.

Boudet, H., Clarke, C., Bugden, D., Maibach, E., Roser-Renouf, C., & Leiserowitz, A. (2014). "Fracking" controversy and communication: Using national survey data to understand public perceptions of hydraulic fracturing. *Energy Policy*, *65*, 57-67.

Cotton, M., Rattle, I., & Van Alstine, J. (2014). Shale gas policy in the United Kingdom: An argumentative discourse analysis. *Energy Policy*, *73*, 427-438.

Donohue, J. P. (2006). How to support a one-handed economist: The role of modalisation in economic forecasting. *English for Specific Purposes*, *25*(2), 200-216.

Eemeren, F. H. van (2010). *Strategic maneuvering in argumentative discourse: Extending the pragma-dialectical theory of argumentation* (Vol. 2). John Benjamins Publishing.

Eemeren, F. H. van, & Garssen, B. (2013). Argumentative patterns in discourse. OSSA Conference Archive. 42. Retreived from https://scholar.uwindsor.ca/ossaarchive/OSSA10/papersandcommentaries/42.

Eemeren, F. H. van (Ed). (2016). Argumentative patterns in Discourse. Special Issue of *Argumentation*, *30*(1).

Freeman, J. B. (2000). What types of statements are there? *Argumentation*, *14*(2), 135-157.

Freeman, J. B. (2011). *Argument Structure: Representation and theory* (Vol. 18). Dordrecht: Springer.

Henkemans, A. F. S. (2016). Argumentative patterns in over-the-counter medicine advertisements. *Argumentation*, *30*(1), 81-95.

Ihlen, Ø., & Nitz, M. (2008). Framing contests in environmental disputes: Paying attention to media and cultural master frames. *International Journal of Strategic Communication*, *2*(1), 1-18.

Jaspal, R., & Nerlich, B. (2014). Fracking in the UK press: Threat dynamics in an unfolding debate. *Public Understanding of Science*, *23*(3), 348-363.

Lawrence, J., & Reed, C. (2015, June). Combining argument mining techniques. In *ArgMining@ HLT-NAACL* (pp. 127-136). Denver, Colorado.

Martin, J. R., & White, P. R. (2003). *The language of evaluation* (Vol. 2). Basingstoke: Palgrave Macmillan

Metze, T. (2017). Fracking the debate: Frame shifts and boundary work inDutch decision making on shale gas. *Journal of Environmental Policy & Planning*, *19*(1), 35-52.

Peldszus, A., & Stede, M. (2013). From argument diagrams to argumentation mining in texts: A survey. *International Journal of Cognitive Informatics and Natural Intelligence (IJCINI)*, *7*(1), 1-31.

Rigotti, E., & Morasso, S. G. (2010). Comparing the argumentum model of topics to other contemporary approaches to argument schemes: the procedural and material components. *Argumentation*, *24*(4), 489-512.

Schon, D. A., & Rein, M. (1995). *Frame reflection: Toward the resolution of intractable policy controversies*. Basic Books.

Tindale, C. W. (1999). *Acts of arguing: A rhetorical model of argument*. SUNY Press.

Walsh, P. (2004). Investigating prediction in financial and business news articles. In R. Facchinetti and F. R. Palmer (Eds.), *English modality in perspective: Genre analysis and contrastive studie*s (pp. 81-100). Frankfurt am Main: Peter Lang.

White, P. R. (2003). Beyond modality and hedging: A dialogic view of the language of intersubjective stance. *TEXT*, *23*(2), 259-284.

Apologia in a Networked Society: The Case of Volkswagen's Polylogical Challenge

CASSANDRA OLIVERAS-MORENO
Rutgers University
c.oliveras@rutgers.edu

MARK AAKHUS
Rutgers University
aakhus@rutgers.edu

MARCIN LEWIŃSKI
Universidade Nova de Lisboa
m.lewinski@fcsh.unl.pt

Prompted by an emissions crisis of epic proportion, this study focused on Volkswagen's early performances of apologia in strategic online venues including the company's web portals and social media. VW's responses across multiple venues for multiple audiences reveal strategies for structuring places for argument to happen or not. We develop an account of polylogical disagreement management by reformulating classic notions of apologia and stasis to explain VW's performance of crisis management under the conditions of networked society.

KEYWORDS: apologia, crisis communication, disagreement management, polylogue, social media, stasis theory

1. INTRODUCTION

Volkswagen had built a solid reputation around the world for its products and services as a socially responsible company leading on environmental sustainability with innovative developments such as clean diesel (e.g., VW was consistently ranked in the top 15 of the most reputable companies worldwide by the Reputation Institute from 2011-

2015). Indeed, in the lead essay of Volkswagen's 2014 Sustainability Report, Martin Winterkorn, CEO of Volkswagen AG, explains that

> we have learned that our business is no longer just about technical aspects like horsepower and torque. We have learned that sustainability, environmental protection and social responsibility can be powerful value drivers (Volkswagen, 2014, p. 12).

However, on September 18, 2015, the U.S. EPA (Environmental Protection Agency) and CARB (California Air Resources Board) shattered these claims with news that Volkswagen was being investigated for violations of the U.S. Clean Air Act. The carmaker was accused of installing a "defeat device" that would significantly lower the emissions level during official laboratory tests. The press release outlined how EPA and CARB would work "to ensure that the affected cars are brought into compliance, to dig more deeply into the extent and implications of Volkswagen's efforts to cheat on clean air rules, and to take appropriate further action" (CARB, 2015).[1] The defeat device impacted nearly 11 million vehicles worldwide and ultimately resulted in a record $15bn legal settlement in the U.S. The proportion of Volkswagen's emissions crisis raises fertile questions for the field of argumentation and the practice of crisis communication: How does one begin to account for such egregious violations, and how have today's argumentative practices evolved in the networked society?

Volkswagen responded to these accusations of guilt on September 20, 2015 with a public apology made by Winterkorn. He and the Board of Directors appealed directly to their customers and employees in denouncing the actions leading to the violations, vowing to do whatever was necessary to reverse the damage caused in order to re-establish trust step by step[2] (Volkswagen, 2015). Noteworthy in this case is Volkswagen's broader performance of corporate apologia – or line of defence – following this initial showing of regret. The company's crisis response engaged a variety of stakeholders across many venues – not only in the media and online platforms but also in legal proceedings and hearings with elected representatives. Volkswagen used multiple venues to manage the potential for disagreement with Volkswagen's initial broken promise about the cars it was delivering and its attempts to repair that. From early on, the company appeared to strategically

---

[1] For a detailed timeline of the unfolding of VW's emissions crisis, See Figure 1.

[2] Retrieved from www.media.vw.com

shield their primary online media sites from the crisis. The lack of comments on Winterkorn's official video apology[3] (Volkswagengroup), despite tens of thousands of views, was an early signal, as comments were disabled on all videos pertaining to the crisis posted on The Volkswagen Group's YouTube page. VW's website[4] ran TDI emissions related messaging across the top of its page in the months that ensued, redirecting to the microsite www.vwdieselinfo.com for information. No further mention of the crisis was made on the brand's home page. Much like The Volkswagen Group portal, the VW Diesel Info microsite provided an informational platform that anticipated but did not invite stakeholder questions. The central website, www.vw.com, was maintained foremost as a sales platform focused on VW's brand and product fleet with select TDI updates posted on the U.S. Media site. The Volkswagen Group, www.volkswagenag.com, representing the company's overarching organizational body, acted as the central repository for press announcements pertaining to the emissions crisis online. While the emissions crisis precipitated a large-scale polylogue among players both on- and off-line, Volkswagen's crisis communications deliberately sought to contain it.

It is argued here that Volkswagen's effort to win back the public's trust comprises a complex strategy of crisis management sensitive to the dynamics and opportunities of the new media ecology. In particular, VW reinvents the classic rhetorical theory concept of the *stasis* of jurisdiction to realize a contemporary version of *apologia* where the symbolic control of crisis is exercised in the strategic design of interaction to structure the expression of differences and pursuit of disagreements the aggrieved party has with the offender. Two notable engagements with stakeholders involved the use of their Facebook (FB) site in relation to the two microsites they developed to engage affected customers: The Goodwill Package[5] launched in November 2015 and the Settlement site[6] launched in October 2016. The FB site and, in particular, these two microsites digitally materialize the communicative act of apology within a broader performance of apologia by Volkswagen to manage the expression of differences and pursuit of disagreement with the company. VW thus exercised symbolic control of the crisis through the design of devices for argumentation that shape argumentative process.

---

[3] Retrieved from www.youtube.com/volkswagengroup

[4] www.vw.com

[5] vwgoodwillpackage.com

[6] vwcourtsettlement.com

2. APOLOGIA IN ARGUMENTATIVE POLYLOGUE

Volkswagen's use of social media and microsites in its crisis communication strategy offers an important case to better understand argumentation in contemporary controversies. The often professed rhetorical approach to crisis communication (Frandsen & Johansen, 2017; Hearit, 2006; Millar & Heath, 2004) can be given two different readings, captured in the short characterization of a crisis response as a "persuasive narrative". Hearit argues that "[c]entral to this 'counter description' of an apologetic response [to the accusation of wrongdoing] is the use of a persuasive narrative that attempts to alter the interpretation of the alleged act" (2006, p. 5). One reading is that of controlling a *narrative*: striving for *our* story to prevail over competing harmful accounts, for steering the story in *our* direction, or simply for cutting the narrative off altogether, so that it belongs to the past - there is simply nothing really to talk about anymore. Symbolic control of the crisis seeks to "delimit discussion" by "strategically naming" the organization's actions (Hearit, 2006, p. 81). Another reading is that of managing *persuasion* through argumentation: the chief goal of crisis communication is thus "to neutralize the argumentative force of the initial charges of the wrongdoing" (Hearit, 1994, p. 115). This is done by all kind of strategies defining *any* argumentation: producing more or less complex reasons for our case, rebutting opponents' charges and counter-arguments, addressing actual and possibly forthcoming critical questions, etc. (see Schuetz, 1990). So, "to delimit discussion" may mean tightly controlling the narrative or the content of arguments and counter-arguments.

The delimiting of discussion perspective from the rhetorical approach to crisis management requires updating to match the conditions of the new media ecology of the networked society. In regard to controlling the narrative, for instance, Chewning (2014) examines how BP's crisis response strategy, in the aftermath of the Deepwater Horizon oil spill, treated FB as a bulletin board not as a means for fostering stakeholder dialogue but to link to news coverage in selective ways to control the narrative. Thus, BP was generating "inter-media dialog" that co-opted narratives by "both inserting itself into the traditional media narrative, when favourable, and incorporating this narrative into its own representation of events" (Chewning, 2015, p. 78). In regard to managing persuasion, Hunter, Menestral, and Bettignies (2008), examine how Danone's crisis strategy misunderstood the rules of the new media ecology by attending to traditional media rather than the Internet and stakeholder media of the time. Danone refused to engage in direct dialogue with its stakeholders, especially the investor community, and its adversaries, especially anti-globalist advocacy groups. By avoiding stakeholder media in favour of traditional media, Danone missed the opportunity to engage through dialogue and conflict management strategies with stakeholder specific, community based watchdog media to address the argumentative force of multiple competing perspectives.

While each study addresses a different reading of the rhetorical approach to crisis response, both highlight the shifting realities of controversy in the new media ecology and the invention of new strategies to manage disagreement among many competing positions of many players in the controversy. When considering these cases from the perspective of argumentation theory following Wenzel (1979), it can be seen that each study emphasizes a different sense of argument. The BP case emphasizes the management of arguments as products while the Danone case emphasizes the management of arguing as a process. What is presumed but not articulated in both studies, and which is more broadly characteristic of such analyses, is the management of argumentation as procedures, which Wenzel (1979, p. 84) defines as "methodology for bringing the natural unreflected processes of arguing under some sort of deliberate control". These dialectical considerations are ordinarily construed as places where argumentation happens in a disciplined way such as in legal, scientific, and policy proceedings. While much argumentation theory focuses on modelling normatively preferable procedures, the communicative work of organizations to digitize their communicative actions in the new media ecology presents a variety of means for managing differences and pursuing disagreement in particular ways that discipline argumentation through the use or construction of digitized procedures where stakeholders are engaged (Aakhus, 2013; 2017).

Indeed, *place* is an important, although overlooked, strategic consideration for argumentation as recognized in stasis theory developed in ancient rhetoric. Stasis is a strategic line of defence an arguer can choose, typically in the context of a criminal trial. The first three classic staseis of Hermagoras (Braet, 1987; Hohmann, 1989; Kennedy, 1994, pp. 98-101) are: (1) "conjecturing" about the fact at issue (Did X kill Y?), (2) "defining" the crime (Was it murder or manslaughter?), and (3) "qualifying" the action (Are there any extenuating circumstances?). The fourth classic stasis is "transferring" jurisdiction to a different tribunal and is concerned with the contextual appropriateness of the tribunal/institution designated to adjudicate the case (Gross, 2004). According to Hohmann, a complete consideration of the fourth stasis would include "preliminary questions" such as "whether the issue at hand is a legal issue, whether there is any issue at all, what kind of argumentation should be used, what is the proper time and place for discussing the issue, etc." (1989, p. 176, fn. 17).

Within the rhetorical perspective on crisis communication, the stasis theory is a relevant and surprisingly up-to-date source of inspiration. Hearit (2006) distinguishes three basic strategies of crisis response. One is to deny charges altogether: the alleged wrongdoing simply didn't happen (cf. stasis 1). Second is to admit the fact but explain it away: it happened but it wasn't of our making and in general doesn't qualify as a culpable act we're accused of (cf. stasis 2). Finally, an organization can admit the fact and agree on its definition, thus take the blame and apologize, but subsequently strongly focus on the reparative actions which mitigate or make up for the past wrongs (cf. stasis 3). In a surprising, and as we argue entirely unjustified omission, the last stasis of transferring jurisdiction (*translatio*) is dropped here.

While indeed this last stasis can be dismissed as a distraction from the actual content of the case by recourse to "petty legalism" (Kennedy, 1963, p. 308), it is a powerful argumentative strategy in need of further acknowledgement (Braet, 1987; Gross, 2004). On the one hand, *translatio* is a "foreign matter" in legal argumentation, for "[i]t does not deal with the primary substantive issue at all, but rather shifts the ground [...] to the act of bringing the suit" (Hohmann, 1989, p. 176). On the other hand, *translatio* is, in an important sense, *the* primary issue: "once a procedural obstacle has been raised, the substantive issues need no longer be discussed. As a matter of economy, procedural issues are therefore explored first in legal proceedings" (Ibid.).

This, we argue, is particularly the case today. The networked society, built on ever evolving capacity of media, information, and communication technology, is noteworthy for the opportunities it presents for innovating the venues where and when argument happens. Aakhus and Lewiński (2017) develop a polylogical theory of argumentation that gives explicit attention to how *place* matters in the expansion and contraction of disagreement involving many positions pursued among many players in a controversy. Three ways of managing place were identified: *venue shopping*, where players actively seek the most favourable place to handle differences; *venue entrepreneurship*, where some players strategically alter key rules of engagement; and *venue creation*, where some players construct entirely new means to engage others in managing disagreement (Aakhus & Lewiński, 2017, pp. 195-199). The latter category of the communicative work of organizations regarding venue creation was highlighted as requiring more empirical attention. Therefore, adding argumentative place management to Hearit's three types of crisis responses is an important and, as our analysis shows, fruitful endeavour: it reveals the argumentation design involved in managing the very conditions for argumentation rather than just argumentation itself.

Here then a continued updating of the rhetorical perspective on crisis management can be developed by looking further than the management of positions and players in controversies to the places where differences are managed and disagreements pursued. Here it is examined how the practices of delimiting of discussion happen by controlling the venue of expression so that the way disagreement management itself is conducted thus constructs particular preferred content rather than other dispreferred content by the organization doing the communicative work of venue creation.

3. POLYLOGICAL DISAGREEMENT MANAGEMENT THROUGH VENUE CREATION

The case of Volkswagen's corporate apologia that follows the company's initial response to the infamous "dieselgate" is examined. VW's uses of differing media in following up their initial apology with their broad strategy of apologia illustrate the strategic uses of *places* for managing disagreement and the conduct of argumentation in the controversy. VW's maneuvering in the new media ecology during the initial unfolding of its emissions crisis including video, microsites and web pages, advertisements, press conferences, speeches, and social media were a testament to the importance of the stasis of place and the practices of venue shopping, venue entrepreneurship, and venue creation.

Two distinct venues are examined for their close relationship and starkly contrasting argumentative functions: Volkswagen's FB page and microsites designed by Volkswagen for customers affected by the crisis (http://vwgoodwillpackage.com/ and http://vwcourtsettlement.com). The FB page was analysed for posts coinciding with two flashpoints in the crisis that triggered significant argumentative activity (see Figure 1). Notably, these arrived on the only two occasions that VW posted about the TDI issue following their initial apology—approximately one month after news broke and one year later. Comment and reply threads were examined to assess how, and if, VW was directly engaging stakeholders.

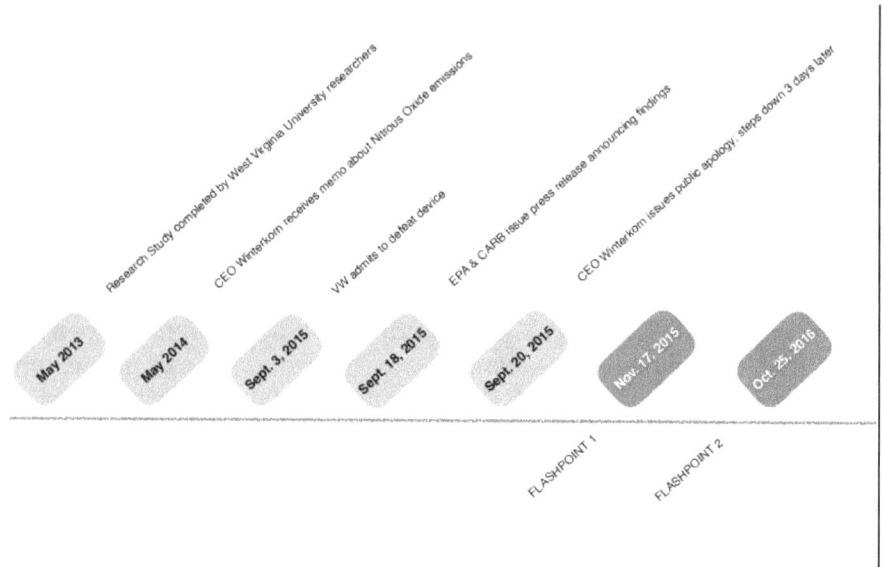

Figure 1 – Timeline of Volkswagen TDI Emissions Crisis with key Flashpoints

VW's apologia included published claims to "make things right" supported by action steps on their FB page directing consumers toward a form of compensation. The websites supporting these posts established a trusted, and intentionally limited, source for stakeholders to migrate toward. Here Volkswagen was in step with Benoit's (2015) image repair strategies, having worked to reduce the offensiveness of the event through bolstering and differentiation while issuing corrective action on both occasions. The following findings illustrate the rhetorical tension inherent to these venues during the crisis and the symbolic control that new media affords. While Volkswagen very publicly managed their emissions crisis, they did not fully legitimate social media as an appropriate place to deeply engage in apologia by inviting direct stakeholder argumentation. They maintained its design foremost as a "social" safe space for promoting the brand.

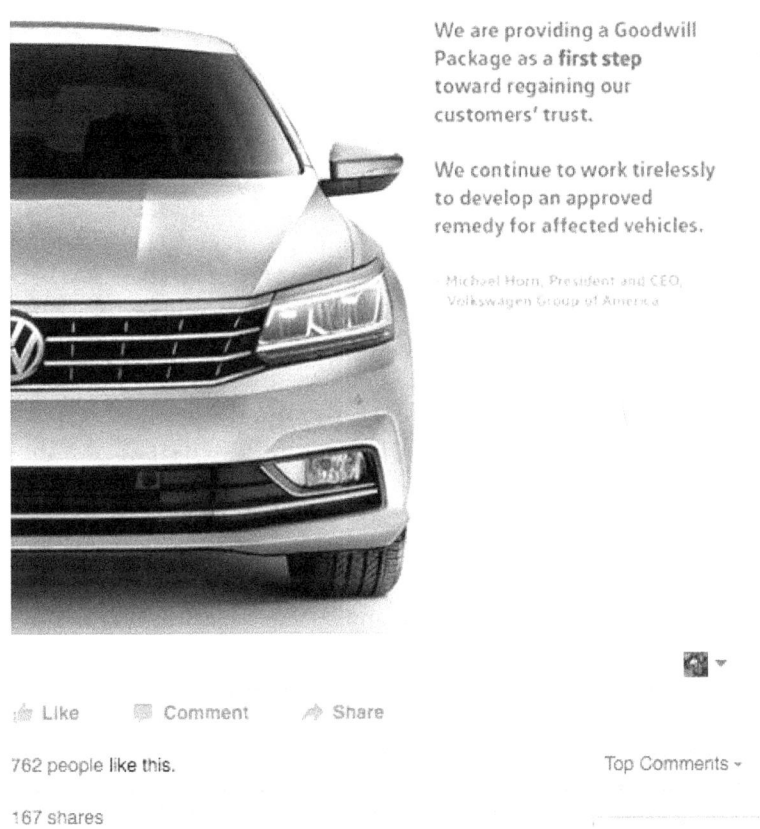

Figure 2 – Goodwill Package Facebook Post, November 17. 2015[7]

---

[7] Retrieved from *www.facebook.com/vw*

3.1 Flashpoint 1: Facebook and Goodwill Package Site

Early in its emissions crisis, Volkswagen's strategy centred around acknowledging the harm done to its customers and offering initial compensation – quite literally – in the name of Goodwill. VW ran a print campaign in over 30 U.S. publications announcing this first step toward repairing the situation (D'Orazio, 2015), reinforced by social media as seen in Figure 2.

The first flashpoint (Fig. 2) featured an abbreviated version of the print ad, guiding affected TDI consumers to a website for instructions on how to redeem their Goodwill Package. This post marked the first occurrence of Volkswagen offering compensation to those affected within social media that enabled stakeholder feedback.

The website, www.vwdieselinfo.com/goodwill_package, functioned as an official staging place for disseminating information regarding the claim. The layout of the page placed clear emphasis on eligibility and practicalities such as "Program Overview; Goodwill Package FAQ; Program Rules; and Cardholder Agreements". No chat or comment features were found in this portal at the time of analysis; a FAQ section anticipated questions from customers. Unlike the BP case, which employed inter-media dialog to intentionally guide users to outside media outlets in an effort to shape their narrative (Chewning, 2015), Volkswagen used social media as a bridge to their own corporate sites.

The Goodwill website delimited discussion by strategically naming Volkswagen's actionable first steps toward re-establishing trust. Consumers directly impacted by the crisis would receive straightforward and accessible restitution. At this seminal point in the crisis, this venue was created to move consumers away from the centre of the controversy toward a venue designed to maintain focus on managing the grievance of individuals harmed in a particular way, monetarily. Environmental grievances, for example, are not addressed in this space.

Flashpoint 1 introduced the dialectical, demonstrating how users took advantage of this "Goodwill" opportunity to question the promise of VW's originating post by expressing their frustrations. In response, the company further shaped this disagreement space by selectively responding to the comments posted. "Top" comments as determined by Facebook included the following topics: eligibility for package; uncertainty about right to participate in future litigation; poor post-sale customer service; and questions about turnaround time for goodwill compensation. VW replies were largely formulaic, consisting of further apology and acknowledgement of the question or complaint, followed by a direct link to the FAQ page. In select instances, users were directed offline to speak with a representative by phone. Several consumers reported turning directly to dealerships in search of answers. Having provided one answer to each "top" comment by redirecting stakeholder attention to what they considered a more appropriate venue, Volkswagen then retreated, engaging in no further conversation on that thread. Subsequently, these conversation threads would be taken up amongst other players including other affected customers, members of the general public, and VW loyalists.

Volkswagen's performance of apologia on FB was a key first step toward repair, echoing Hearit's third stasis and reinforcing the brand's appearance of transparency and responsiveness. While clearly speaking to 2.0 TDI owners, this message was not dyadic in structure. Here, "Goodwill" messaging was an integral part of a campaign designed to reach a variety of stakeholders (Regulatory Agencies, general public, other VW consumers). Volkswagen engaged in venue creation through the use of their Goodwill site as an official and centralized portal receiving traffic from multiple mediums such as direct mail, print ads, and its corresponding FB post. Both venues provided answers, but only FB allowed for public questioning of the Goodwill proposition. On a practical level, this flashpoint opened up a space for polylogical disagreement that would serve as a preview for how the company would manage litigation to come.

3.2 Flashpoint 2: Facebook and Settlement Site

Figure 3 illustrates the heralding of a settlement for 2.0 TDI vehicles; Volkswagen's only subsequent emissions related FB post, one year following the Goodwill package.

Moving consistently toward resolution, owners and lessees were encouraged to file claims via a dedicated microsite offering consumers user friendly access to the terms and conditions of the settlement.

The microsite, www.VWCourtSettlement.com, greeted affected customers with a succinct explanation of the settlement agreement, including actions such as paying for environmental remediation and promoting zero emissions technology. This was directly followed by diagrams of buy back options. Having had over one year to anticipate questions, VW designed what appeared to be a streamlined road map for guiding stakeholders through the process, step by step. Frequently Asked Questions, Video "how to" and a claims portal were accessible, as were links to a variety of court documents for more information. Much like their utilization of the "Goodwill" platform in Flashpoint 1, Volkswagen directed users toward a tightly controlled venue designed for self–service and no argumentative exchange. Those with additional questions about the settlements were instructed to read comprehensive court documents posted on the site, visit the FTC website, or lastly to phone Volkswagen directly via a toll-free number. While one could upload claims via the online portal, no chat, email, or comment features were found on this site at time of analysis.

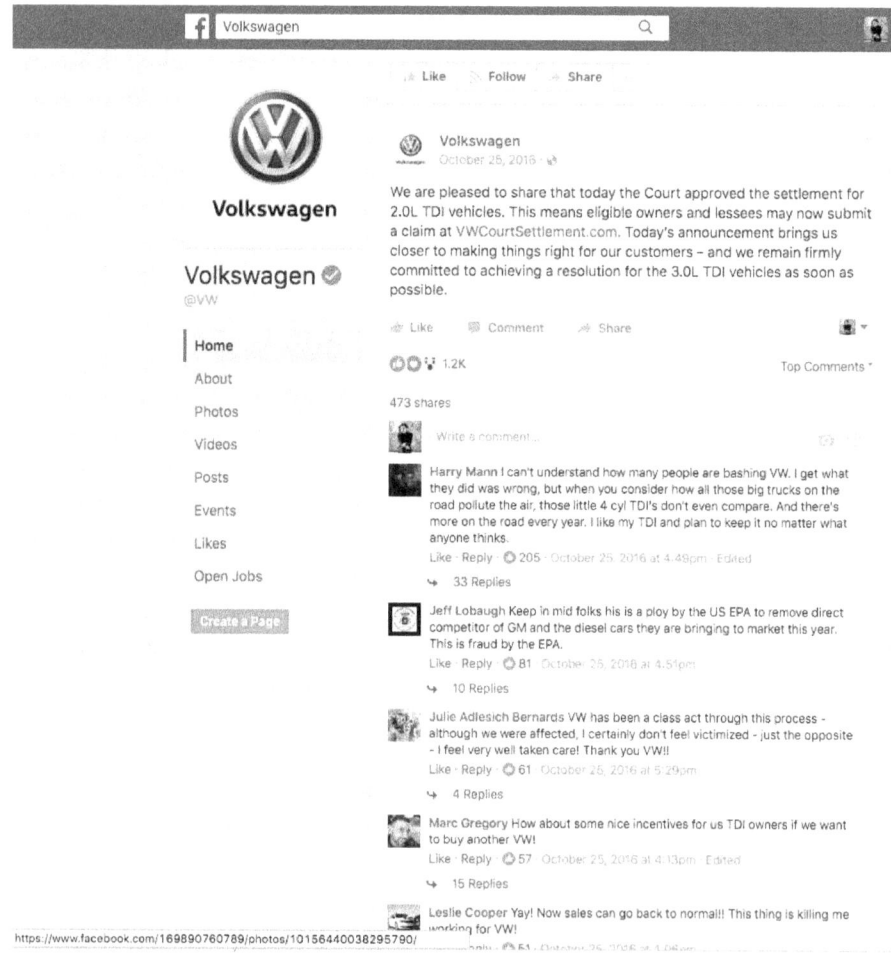

Figure 3 – Settlement Facebook Post, October 25, 2016[8]

---

[8] Retrieved from www.facebook.com/vw

Unlike its earlier Goodwill counterpart, the top five comments appearing below VW's settlement post were all positive. Notably, VW's decision to release the settlement announcement on Facebook shaped a space in which stakeholders were enabled to take up a line of argumentation in the company's defence, as well as voice disagreement. Active players in this polylogue included Volkswagen, TDI owners, the public, the courts, VW loyalists, VW haters, VW Dealership employees, VW investors, and conspiracy theorists (see Figure 4).

The "top" ranking comment with highest engagement yielded 205 likes and 33 replies including this one:

> I can't understand how many people are bashing VW. I get what they did was wrong, but when you consider how all those big trucks on the road pollute the air, those little 4cyl TDI's don't even compare. And there's more on the road every year. I like my TDI and plan to keep it no matter what anyone thinks.

This comment was followed by users attacking the regulatory agencies and deflecting responsibility from VW. Later in this thread, one participant posed: "Even if there is an agenda, was there any deception? Was there a lie? I think that is what is under scrutiny". The public disagreement during these flashpoints emerged as more a battle among stakeholders than with VW itself. Within the larger polylogue of the settlement post, each distinct comment thread became its own polylogue, with collapsed comments hidden in plain sight.

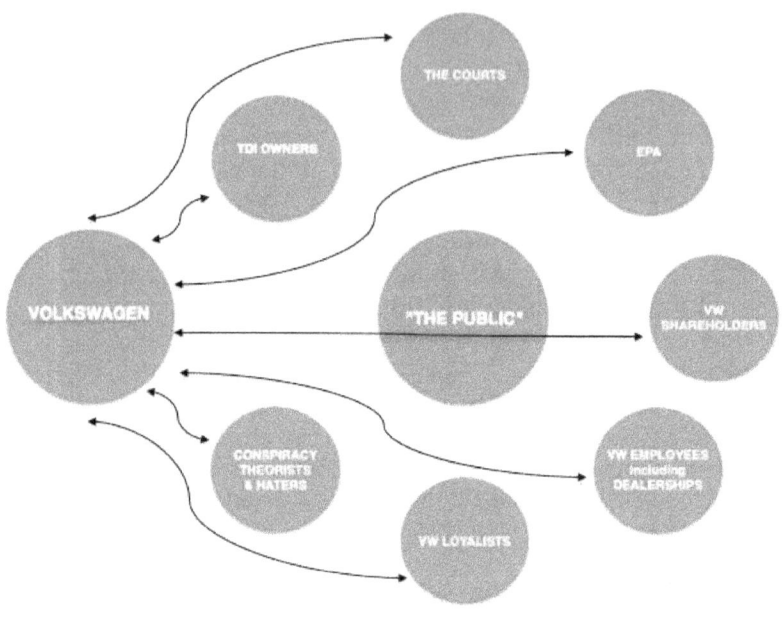

Figure 4 – Stakeholder Diagram illustrating Volkswagen's dyadic communications in the midst of their wider crisis polylogue

Hundreds of less visible negative consumer comments later followed, with users signalling that Volkswagen's settlement site may not have been as user friendly as it first appeared. The most frequent complaints addressed inefficiencies in the venues created by VW, such as the inability to make edits via the claims portal without being bumped to the back of the line; poor VW communication with its dealerships; processing delays that were diminishing buyback value of vehicles; and wait times of up to several hours for phone assistance. During Flashpoint 2, consumers again exhibited a pattern of spanning venues, returning to FB from the microsite to lodge critiques against Volkswagen's design for "making things right".

Finally, in stark contrast to Flashpoint 1, the company did not reply to any comment made on their settlement post. While silence holds the potential for being interpreted as passivity or relinquishing of control over defining and shaping the world (Brummett as cited by Hearit, 1994), here the opposite appeared to be in play. Rather than treat Facebook as a synchronous customer service venue, Volkswagen used it as a platform to push messaging directed to a microsite under the careful control of its brand. Given the ongoing litigation the company faced at this time, the communicative act of "pleading the fifth" on Facebook served a strategic function for driving consumers to the more private, dyadic claims portal and off-line to dealerships for answers specific to their situation. Volkswagen did not legitimate doubts or critiques raised by consumers with active responses during this flashpoint, as their defence at this stage had just concluded in the tribunal of an actual courtroom. VW had both an obligation and incentive for directing consumers to www.VWCourtSettlement.com as a way of facilitating the settlement process and closing the narrative.

4. DISCUSSION

Our analysis shows that the more traditional argumentative approach to crisis communication can take us a long way toward understanding VW's strategies of apologia. According to *The Guardian's* website (Neate, 2016), the company had initially denied the charges (thus implementing Hearit's first strategy of apologia), a strategy it had to amend given EPA's insistence and the evidence it brought to light. Only when the issue became public after EPA's announcement, did VW issue an apology within its broader apologia. Subsequently, it tried to control the damage by scapegoating some of its "rogue" engineers who violated the company's rules by installing the defeat device (Hearit's second strategy) and by stressing the reparative actions of "making things right" that are meant to mitigate or make up for the past wrongs (Hearit's third strategy). That is to say, the company did comprehensively engage in a strategic control over the content of its arguments. However, based on our analysis, we show how this approach is incomplete without also duly considering the strategic control over the places and conditions for argumentation. In the polylogical reality of networked communication, VW has done as much or possibly more to *design the procedures* for argumentation than to design the products and processes of argumentation, an exclusive focus of most existing research. For instance, Marsh (2006) constructs his three-part "syllogism of apologia" by specifically excluding the fourth stasis (p. 43).

A significant transference of control can be seen throughout this case. VW did not engage in a polylogical multi-party discussion with its stakeholders, an opportunity asynchronous online discussions clearly give (Lewiński, 2013). In particular, it did not engage in an open, collective criticism of its actions through complex argumentation covering the multiple threads opened up for discussion by Facebook users (see Lewiński, 2010). However, as VW worked instead to promote more private transactions on their dedicated websites for "making things right", many returned from those private spaces to the originating Facebook posts to publicly vent about a lack of clarity or sufficient customer support on them. At times, TDI customers managed the crisis in their own right, supporting one another and answering questions. In other instances, supporters of VW rallying on the company's behalf performed in ways that paid advertising could not. Both examples unfolded under Volkswagen's design. While VW and TDI owners were central voices, other active players in these flashpoints included the courts, VW loyalists, VW haters, VW Dealership employees, VW investors, conspiracy theorists, and the general public (see Figure 4). The fluidity surrounding argumentative forces between Volkswagen and said stakeholders illustrates the complexity of polylogical exchanges in a mediated world. These findings let us enter a broader discussion over the role of argumentation in various versions of *apologia*, the classic genre of public discourse aimed at defending one's actions and positions against accusations of wrongdoing (Hearit, 1994, 2006; Schuetz, 1990). VW's crisis communication strategies reveal innovations in the apologia genre that remain invisible, or even contradictory, to the classic accounts of apologia in crisis communication. The apparent underlying strategy by Volkswagen goes beyond the rhetorical-symbolic framing of their past and future actions typically advised in crisis communication literature.

Our polylogical analysis challenges (1) the emerging prescriptions that extol the normative values of dyadic dialogue in organizational communication between the (alleged) wrongdoer and *some* source of accountability (government agencies, "the public", "the media"); and (2) the emphasis on symbolic control in crisis management discussed by chief proponents of an argumentative approach to crisis communication (e.g., Hearit, 2006). These are based on the presumption of the simplest model of dialogue between two parties – the apologist and the wronged ones – that is commonly used to define, analyse and evaluate apologias. The alternative is to model apologia in terms of a *polylogue*, a dialogue between many parties, over various positions and across multiple places (Aakhus & Lewiński, 2017; Lewiński & Aakhus, 2014; Lewiński, 2014). Indeed, limiting discussion to but two sides is not a requirement of argumentative rationality; rather, as the case here suggests, it is a strategic choice. It is precisely the polylogical perspective that brings this strategy to light (Lewiński, 2016).

Furthermore, through its argumentative media strategy, VW conceptualized the wrongdoing and the adequate reparative action in terms of an individual harm rather than the collective harm. The polylogue was shifted to the private sphere involving a personal matter between VW and its consumer resolved via financial compensation. However, it addresses the concerns and values of only some stakeholders – VW car owners – but all human beings were victims with a stake. After all, the scandal was triggered by concerns for environmental protection and air quality, something clearly not limited to VW's customers. Problems with fraudulent advertising, frustrated hopes for a "clean diesel" car, VW's dealers' lost profits and legal problems, shareholders' failed investments – and more! – all make up an intrinsic part of this polylogue. However, rather than opening up these various lines of disagreement and engaging various dissenting positions and players (stakeholders) in a transparent polylogue, VW chose to argue through a web of private, bilateral relations / negotiations (See Figure 4). While we have specifically focused on VW's social media platform – thus omitting various forms of deal-making behind the closed doors – a form of strategic exclusion is sufficiently evident here. What is a crucial implication for argumentation theory, is that the polylogical perspective lets us see that what seems to be a perfectly reasonable dialogue, can be a strategically truncated, and thus highly spurious, polylogue.

REFERENCES

Aakhus, M. (2013). Deliberation digitized: Designing disagreement space through communication-information services. *Journal of Argumentation in Context*, *2*(1), 101–126.
Aakhus, M. (2017). The communicative work of organizations in shaping argumentative realities. *Philosophy & Technology*, *30*(2), 191–208.
Aakhus, M., & Lewiński, M. (2017). Advancing polylogical analysis of large-scale argumentation: Disagreement management in the fracking controversy. *Argumentation*, *31*(1), 179–207.
Benoit, W. L. (2015). *Accounts, excuses, and apologies: Image repair theory and research* (2nd ed.). Albany: State University of New York Press.
Braet, A. (1987). The classical doctrine of "status" and the rhetorical theory of argumentation. *Philosophy and Rhetoric*, *20*(2), 79–93.
Chewning. L. V. (2015). Multiple voices and multiple media: Co-constructing BP's crisis response. *Public Relations Review*, *41*(1), 72–79.
D'Orazio, D. (2015, November 15). Volkswagen apologizes for emissions scandal with full-page ad in dozens of papers. *The Verge*. Retrieved from http://www.theverge.com/transportation/2015/11/15/9739960/volkswagen-apologizes-with-full-page-ad-in-dozens-of-newspapers
EPA. (2015, September 18). EPA, California Notify Volkswagen of Clean Air Act

Violations / Carmaker allegedly used software that circumvents emissions testing for certain air pollutants. Retrieved from https://www.epa.gov/newsreleases/epa-california-notify-volkswagen-clean-air-act-violations-carmaker-allegedly-used

Frandsen F., & Johansen W. (2017). *Organizational crisis communication: A multivocal approach*. London: Sage.

Gross, A. G. (2004). Why Hermagoras still matters: The fourth stasis and interdisciplinarity, *Rhetoric Review*, *23*(2), 141-155.

Hearit K.M. (2006). *Crisis management by apology: Corporate response to allegations of wrongdoing*. Mahwah, NJ: Lawrence Erlbaum.

Hearit, K. M. (1994). Apologies and public relations crises at Chrysler, Toshiba, and Volvo. *Public Relations Review*, *20*(2), 113–125.

Hohmann, H. (1989). The dynamics of stasis: Classical rhetorical theory and modern legal argumentation. *American Journal of Jurisprudence*, *34*(1), 171-197.

Hunter, M. L., Le Menestrel, M., & De Bettignies, H. C. (2008). Beyond control: Crisis strategies and stakeholder media in the Danone boycott of 2001. *Corporate Reputation Review*, *11*(4), 335-350.

Kennedy, G. A. (1963). *The art of persuasion in Greece*. Princeton, NJ: Princeton University Press.

Kennedy G. A. (1994). *A new history of classical rhetoric*. Princeton, NJ: Princeton University Press.

Lewiński, M. (2010). Collective argumentative criticism in informal online discussion forums. *Argumentation and Advocacy*, *47*(2), 86–105.

Lewiński, M. (2013). Debating multiple positions in multi-party online deliberation: Sides, positions, and cases. *Journal of Argumentation in Context*, *2*(1), 151-177.

Lewiński, M. (2014). Argumentative polylogues: Beyond dialectical understanding of fallacies. *Studies in Logic, Grammar and Rhetoric*, *36*(1), 193-218.

Lewiński, M. (2016). Shale gas debate in Europe: Pro-and-con dialectics and argumentative polylogues. *Discourse & Communication*, *10*(6), 553-575.

Lewiński, M., & Aakhus, M. (2014). Argumentative polylogues in a dialectical framework: A methodological inquiry. *Argumentation*, *28*(2), 161-185.

Marsh, C. (2006). The syllogism of apologia: Rhetorical stasis theory and crisis communication. *Public Relations Review*, *32*(1), 41–46.

Millar D. P., & Heath, R. L. (Eds.) (2004). *Responding to crisis: A rhetorical approach to crisis communication*. Mahwah, NJ: Lawrence Erlbaum.

Neate, R. (2016, March 2). VW CEO was told about emissions crisis a year before admitting to cheat scandal. *The Guardian*. https://www.theguardian.com/business/2016/mar/02/vw-ceo-martin-winterkorn-told-about-emissions-scandal

Schuetz, J. (1990). Corporate advocacy as argumentation. In J. Schuetz & R. Trapp (Eds.), *Perspectives on argumentation: Essays in honor of Wayne Brockriede* (pp. 272–284). Prospect Heights: Waveland.

Volkswagen. (2015, September 20). Statement of Prof. Dr. Martin Winterkorn, CEO of Volkswagen AG. Retrieved from http://media.vw.com/release/1066/

Volkswagen. (2015, September 22). *Video statement Prof. Dr. Martin Winterkorn.* [Video File]. Retrieved from https://www.youtube.com/watch?v=wMPX98_H0ak

Volkwagen. (n.d.) [Court Settlement Website]. Retrieved March 24, 2017, from https://www.vwcourtsettlement.com

Volkwagen. (2014). Sustainability Report 2014. Retrieved from http://sustainabilityreport2014.volkswagenag.com/sites/default/files/pdf/en/Volkswagen_SustainabilityReport_2014.pdf (pp.12)

Volkswagen (n.d.). In *Facebook*. [Fan Page]. Retrieved December 3, 2015, from https://www.facebook.com/VW/

Volkswagen. (n.d.) [Company Website]. Retrieved December 3, 2015 from https://www.volkswagen.com

Volkswagen. (n.d.) [Goodwill Package Website]. Retrieved December 3, 2015 from https://www.vwgoodwillpackage.com

Wenzel, J. W. (1979). Jürgen Habermas and the dialectical perspective on argumentation. *Journal of the American Forensic Association, 16*(2), 83-94.

41

The Strategic Use of Examples in Supporting a Positive Evaluation of a Political Group

AHMED OMAR
Ain Shams University
ahmedomar198023@gmail.com

With the help of the extended pragma-dialectical theory, this paper aims to analyse and evaluate how Egyptian political columnists, arguing in favour of the feasibility of political change before the Arab Spring, maneuvered strategically by argumentation from examples in supporting a positive evaluation of the Egyptian people as a whole, in view of the institutional preconditions of political columns and the specific rhetorical predicament a columnist may face in this type of situations.

KEYWORDS: strategic maneuvering, argumentation from example, representativeness, sufficiency, resorting to hierarchically arranged examples

1. INTRODUCTION

In my doctoral dissertation (Omar, 2016), I was interested in analysing and evaluating how Al Aswany, a prominent Egyptian columnist, maneuvers strategically in supporting the feasibility of ousting President Mubarak by means of narrative and fictional techniques and forms before the revolutionary uprising of 2011.[1] Mainly using examples of pro-change individuals and groups, Al Aswany supports one

---

[1] The concept of strategic maneuvering is the outcome of extending the standard pragma-dialectical theory of argumentation. Arguers are assumed to maintain a delicate balance between committing to the dialectical norms of reasonableness and seeking for rhetorical effectiveness. Arguers selects topical choices and employs presentational devices in adaptation to the audience demand to maintain such an equilibrium (van Eemeren & Houtlosser, 2002, 2005, 2006; van Eemeren, 2010)

of the feasibility-related topics, i.e. that the Egyptian people are no more submissive, indifferent or inactive (henceforth: the "active people" topic).

I aim to analyse and evaluate how Al Aswany maneuvers strategically with argumentation from example to support the "active people" topic. In Section 2, I shall explain the rhetorical exigency challenging the columnist by demonstrating the "bad" image of the Egyptian people circulated in the pro-regime media. Section 3 will be dedicated to discussing different insights into argumentation from examples. In Section 4, I shall elaborate the specific argumentative predicament a political columnist faces when evaluating a political group as a whole positively. In Section 5 and 6, the illustrative cases of "Egyptian Awakened" and "The Coming Civil Disobedience on April6th." will be analysed respectively.[2]

2. THE IMAGE OF THE EGYPTIAN PEOPLE AND POLITCAL ACTIVISM IN MUBARAK'S ERA

As a developing country, Mubarak's Egypt depended so much on consent to control the people since the cost of using coercion only would be relatively high.[3] The autocratic regime of Mubarak did not only attempt at convincing the people that their leader was great or exceptionally qualified, but also that they would never be able to make a change because they are helpless and inactive, and more importantly, innately bad.

The pro-regime media and popular arts painstakingly blamed the people, and more specifically the youth, for their over-dependence on the government. For example, the unemployment problems, for example, were depicted as a result of the youth's insisting on working in the public sector, not the private one, in fields closely related to their education, not others. Film plots were praising the success of young people who travel abroad starting from point zero until becoming rich. The problem of price rise was not discussed in relation with economic deficits, but the people were blamed for over-consumption. Traders were blamed too for causing the problem by their greediness.

---

[2] These two columns were published in *Al Shorouk*, a Cairene elite newspaper. The first is translated by Jonathan Wright (Aswany, 2011). The second is translated by Ahmed Omar. Both texts can be found in this link: https://drive.google.com/file/d/0B12L9vn90kGQWmMwWlRhc3pwQzg/view

[3] For more on Gramsci's distinction between coercion and consent, see: Rucco (2012, 61)

Most importantly, creating a negative image of the people was extended to encompass political aspects. Concurrently with election times, the Egyptian people were rebuked for their abstention from political participation, instead of exposing the unfair conditions of voting.

Politically active Egyptians were stereotyped and viewed as a weird, isolated group of people. The stereotype of a political activist manifested itself in a male university student who is shown by the pro-regime media as highly individualistic, and quite intellectual, more often than not a socialist, leftist, or an Islamist. The Egyptian cinema repeatedly depicted political activists exclusively as university students who are quite affluent and thereby have time and financial capacity to show interest in political activism. For a pro-change advocate who aims to convince the audience of the acceptability of the "active people" topic before 2011, it would not be helpful to cite actions of this small, isolated group. The representativeness and sufficiency of these actions would be unlikely acceptable.

3. ARGUMENTATION FROM EXAMPLE: AN OVERVIEW

Aristotle classified examples as instances in support of a general principle, which are used to convince the audience of the acceptability of a thesis, or inducing it to imitate a role model. He regarded the use of example as a species of inductive proof that stands in the service of a rule which, although not absolute, shadows the principles of formal logic (Demon, 1997, pp. 129-130; Arthos, 2003, pp. 321-322).

According to Perelman and Olbrechts-Tyteca, argumentation by example is one of the relations that establish the structure of reality by resort to the particular case. The particular case can play a wide variety of roles: as an example, it makes generalization possible; as an illustration, it provides support for an already established regularity; as a model, it encourages imitation (Perlman & Olbrechts-Tyteca, 1969/1971, pp. 350-357).

Pragma-dialecticians distinguish, depending on the kind of the relationship that links the standpoint to the argument/s, three main types of argumentation, that each have their own argument scheme (van Eemeren & Grootendorst, 1992, pp. 96-102; van Eemeren *et al.*, 2007: 137). In symptomatic argumentation, a property, class membership, distinctive characteristic, etc. referred to in the argumentation also applies to the thing, person, or situation referred to in the standpoint (van Eemeren, *et al.*, 2007, p. 154). A basic subtype of symptomatic argumentation is argumentation from example in which "separate facts are represented as special cases of something general: on the basis of specific perceptions a generalization is made" (Garssen, 1997, p. 11).

In order to test the validity of a symptomatic argumentation, the antagonist can stir the following two questions: Are the separate cases indeed representativeness? And have sufficient separate cases been considered? (van Eemeren *et al.*, 2007, p. 155).[4]

4. THE ARGUMENTATIVE PREDICAMENT CONFRONTING AL ASWANY

In political communication, examples are frequently used to support a positive or negative evaluation of a political actor. Selections of what he or she did are advanced as a defence backing such an evaluation. When it comes to evaluating a political group as a whole, the actions of members are provided as examples which are generalized to convey how this group typically acts.

Using the pragma-dialectical terminology, this kind of argumentative practices can be described as follows: the protagonist supports his standpoint "Evaluation Y is true of political group X" by putting forward arguments from example "Evaluation Y is true of group members $X_{1,2,3...n}$". These examples together provide a coordinative support of the generalization included in the standpoint. It is coordinative because "each argument by itself is too weak to conclusively support the standpoint" (van Eemeren *et al.*, 2007, p. 65) because an evaluation of one group member cannot be commonsensically generalized to apply to a whole group, especially if it consists of a big number of people. In order to be considered by the antagonist as acceptable, these examples must be representative and sufficient.

When examples are given to evaluate a group of people as a whole, the question of representativeness becomes difficult to answer: to what extent can the antagonist consider the action(s) attributed to a group member representative of how a whole group act? Let alone the concomitance relation between the action and the evaluation.

In order to answer the representativeness question appropriately, the protagonist may evoke actions of the typical group members as examples. The protagonist may advance many examples in an attempt to answer the question of sufficiency appropriately. How

---

[4] Freely and Steinberg suggest a more detailed list of critical questions (Freeley & Steinberg, 2005: pp. 176,177). However, their suggested questions can be viewed as an articulation of the questions on representativeness and sufficiency.

many examples is sufficient is a debatable question. Perelman and Olbrechts-Tyteca pay attention to the role of variety in strengthening sufficiency: "when one wishes to clarify a rule with many different applications, it is good to provide examples that are as different as possible, as by doing so it can be shown that the differences are without importance on this occasion" (Perelman & Olbrechts-Tyteca, 1969, p. 353).

A political column comes into being as a simultaneous implementation of two genres: information-dissemination by providing data concerning a current event, and (indirect) deliberation by using this data in order to affect the political decisions made by readers, even on the long term (Omar, 2016, p. 67-71). In order to analyse and evaluate how a political columnist maneuvers strategically with argumentation from example, the following question should be answered:

> (A) What actions of which individuals can be cited as optimally representative and sufficient examples which yield the evaluation concerned maintaining the delicate balance between reasonableness and effectiveness?

In the specific case of Al Aswany addressing the "active people" topic, the question can be paraphrased as follows:

> (B) To what extent did Al Aswany, addressing the "active people" topic, succeed in selecting and framing certain aspects of current events that function as much representativeness and sufficient as possible?

In the next two sections, I shall focus on the selection of certain aspects and framing them respectively as topical choices and presentational devices adapted to the demand of the audience Al Aswany addresses.

5. THE "EGYPT AWAKENED" CASE

This column tackles the return of ElBaradei, a potential candidate for presidency and a prominent political change promoter, to Egypt and how he was warmly welcomed by thousands of Egyptians in Cairo airport. Al Aswany notices that, in spite of the threats of Mubarak's security agencies for those who intended to welcome ElBaradei, thousands of Egyptians gathered to welcome him. Al Aswany writes:

> The vast and impressive popular reception that Egyptians organized for Mohamed ElBaradei's return to Egypt conveys several important messages: First, from now on, no one can accuse Egyptians of being passive, submissive to injustice, disengaged from public affairs, or any other of those claims that no longer reflect the reality of Egypt.

The phrase "conveys several important messages" indicates the use of symptomatic argumentation. It makes a symptomatic link between the current event Al Aswany comments on, the impressed welcoming of ElBaradei, and a positive political evaluation of the Egyptian people implied in the phrase "no one can accuse Egyptians of being passive, submissive to injustice, disengaged from public affairs". At first sight, Al Aswany may be considered violating the freedom rule (van Eemeren & Grootendorst, 1992, p. 108; 2004, p. 190,191) by stating that "no one can accuse..." as he might seem preventing his reader from advancing a counter-standpoint. However, taking into consideration that a political column may be written in a quite emotionally 'hot' style, Al Aswany wants to convey that "it is not reasonable, in view of the information given in the opening stage, to accuse the Egyptian people of being...", but in an exaggerated manner.

The thousands of Egyptians who conquered their fear of suppression were presented by Al Aswany as a particular case that can be generalized to establish a rule concerning the Egyptian people as a whole. Al Aswany anticipates the critical questions of the antagonist regarding the representativeness and sufficiency of the cases he cites. Al Aswany utilizes his position as an eyewitness and a political activist to make a distinction among the participants between political activists and ordinary people to strengthen the representativeness of the cases cited:

> The thousands of Egyptians who conquered their fear [...] were not professional politicians, and most of them did not belong to political parties. They were very ordinary Egyptians. [...] they came from different provinces and different social classes. Some of them came in luxury cars and many came by public transport. They included university professors, professionals, students, farmers, writers, artists, and housewives, Muslims and Copts, women with and without veils and some wearing niqab.

Al Aswany positions most of those who conquered their fear as ordinary Egyptians, the most representative. He thus attempts to make it plausible that something in the "typical" Egyptian character have changed. The following three fragments/scenes are relevant to Al Aswany's response to the critical question of representativeness in this sense:

> ... an old woman came up to me and asked to speak to me in private. I took her aside and in a low voice she asked me, "Do you think the government will do anything to harm Dr. ElBaradei?" When I assured her that this was most unlikely, she sighed with relief and said, "May God protect him."
> ... I will not forget the man who came with his wife and their pretty little girl with two plaits, who sat on his shoulders carrying a picture of ElBaradei.
> ... I will not forget the dignified woman in the hijab, the good-hearted Egyptian mother who brought with her several packets of fine dates. She opened them one after the other and started to give them out to people standing around that she did not know.

These three scenes are aimed at illustrating the "typicality" of participants in a vivid and lively way: they are shown as mothers and fathers who act tenderly and caringly. Putting an emphasis on the ordinariness of the participants also helps in answering the critical question of sufficiency satisfactorily. Rooting his argumentation in the aforementioned distinction between the ordinary people and the "stereotyped" politically active people, Al Aswany makes use of what Perelman and Olbrechts-Tyteca called "resort to the hierarchically arranged examples": "Instead of merely cumulating a number of different examples, a speaker will sometimes strengthen the argumentation by example by restoring to the double hierarchy argument which makes *a fortiori* reasoning possible" (Perelman & Olbrechts-Tyteca, 1969, p. 354).

They quote from Aristotle's Rhetoric: "[Everyone honours the wise.] Thus, the Parians have honoured Archillochus, in spite of the bitter tongue; the Chians Homer, though he was not their countryman; the Mytilenaeans Sappho, though she was a woman; the Lacedaemonians Chiton ..., though they are the least literary of men" (11, 23, 1398b).

The behaviour of an old woman, a pair of parents, and a traditional mother is framed as tender, intimate and family-like to crystalize the ordinariness of the examples. Al Aswany implies that if those who were the most ordinary could conquer their fear, *a fortiori* those who occupied higher nodes (university students and unmarried young people for example) in the ordinary citizens Vs. politically active hierarchy would likely act in a similar way. Being constrained by the space of a political column, Al Aswany uses "compressed" small number of examples that can implicitly convey a bigger number. Sufficiency is also enhanced by illustrating diversity in the examples cited: the participants were diverse in gender, religious commitment, career, etc.

The same technique, i.e. "resort to the hierarchically arranged examples", is applied by presenting the current event cited: welcoming ElBaradei at the airport. By framing the welcoming as unsafe and surrounded by threats, Al Aswany implies that the Egyptian people would be perhaps even more politically active when it comes to safer activities (e.g. voting which Egyptians were historically blamed to be apathetic about). He writes:

> [The] Interior Ministry detained several young people simply for urging Egyptians to go out and welcome him. The security agencies also made it clear that they would not allow Egyptians to rally to greet ElBardei at the airport and announced they had mobilized eight thousand riot police to deal with anyone who gathered there.

The answers to question (B) (See: Sec. 4) can now be given. Al Aswany successfully employs the relatively wide space of maneuvering available in the macro-context of political columns. He benefits from the fact that a political column implements not only the genre of (indirect) deliberation, but also, and primarily, the genre of information-dissemination. He adds in the opening stage much information as material starting points that he will make use of later. Al Aswany strategically selects particular aspects of a current event (ElBaradei's arrival at Cairo airport) to strengthen his thesis that the Egyptian people have changed to be no longer politically inactive. This process of selection is similar to what a novelist or a short stories writer does when designing one of the narrative scenes or drawing one of the characters: a novelist concentrates, not on all the potential details of a scene or traits of character, but mainly on the details relevant to his plot. As an arguer, and by mentioning some personal details implying typicality and ordinariness, Al Aswany attempts to make his advanced generalization incontestable. The details mentioned points out that the cited persons/examples are typically inactive according to the historically established frame of reference of his audience, and, however, acted in an active way in an exceptionally unsafe situation. Al Aswany aims at implying that others who are less typical Egyptians would do even more in safer occasions, and hence answering the anticipated critical questions on sufficiency of the examples cited in a satisfactorily way.

6. "THE COMING CIVIL DISOBEDIENCE ON APRIL 6TH." CASE

"The Coming Civil Disobedience on April 6th." was written to promote for the coming civil disobedience (April 2009) and to convince the audience that its success is guaranteed. Al Aswany opens the column with an anecdote on one of the labour leaders which he heard from a female Swedish writer. The story shows that this worker could maintain his dignity and honour in spite of poverty as he refused to take money from the writer for hosting her in his house. She concludes the story:

> This Egyptian worker has taught me an unforgettable lesson. I have been thinking a lot of what he did, and came to a conclusion: *a man who suffers from this great extent of poverty, oppression, and injustice and nevertheless is still able to maintain his courage and dignity – this man will inevitably triumph* [My italics A.O.].

The italicized phrase repeats the conclusion of the column, summarizing how Al Aswany thinks of the future of Egypt:

> The civil disobedience of the coming April 6th. makes me optimistic about the future of Egypt, and as the Swedish writer said: "*A man who suffers from this great extent of poverty, oppression, and injustice and is still able to maintain his courage and dignity – this man will inevitably triumph*" [my italics A.O.]

As an anti-regime columnist and activist, Al Aswany cannot be optimistic about the future of Egypt unless he feels optimistic about the progress of anti-regime activism in the sense that the anti-regime activities can be expected to be fruitful. His words convey his belief that the supporters of political change will triumph in the sense of forcing the regime to initiate a democratization process or even ousting the regime itself. The Swedish writer's words relate to Al Aswany's optimism only if the meaning of her words are viewed as denoting the moral nature of all (or at least most) supporters of political change in Egypt. In view of the context concerned, her phrase can be rephrased as "These political activists who suffer from a great extent of poverty, oppression, and injustice and are still able to maintain their courage and dignity – these people will inevitably triumph".

Similarly, the column confers this moral value on the young activists of April 6th. Movement.[5] Al Aswany writes:

> I have met some of these young people [...] and found myself wondering: from where did they derive their legendary courage?! How could they take to the streets confronting the riot police (the Egyptian occupation army), be harshly beaten, dragged off, detained and tortured in state security premises, and then released more determined to change their country?!

Both the opening anecdote on El Mahallah's labour leader and the aforementioned comments on the youth of April 6th. Movement draw an image of the supporters of political change as free humans who are willing to determine their destiny in spite of the circumstances they are living in. This image goes against the presupposition that human behaviour is always a result of the conditions in which a man lives. The Swedish writer, and Al Aswany in turn, presupposes that the hospitality of the labour leader cannot be an automatic result of the poverty he lives in. Similarly, the toughness, brevity, and awareness of these young people cannot be automatically explained from the circumstances because of the cruel abuse they are subjected to.

In terms of the ideal model of a critical discussion, Al Aswany, playing the role of a protagonist, puts forward an implicit standpoint with a propositional content related to a feasibility issue. This issue is concerning the ability of the supporters of political change to push forward democratization. The propositional content conveyed is presented with a highly strong force "Supporters of political change will inevitably triumph" (1). The standpoint at issue is supported by an implicit argument concerning the exceptional moral character of these supporters as people who maintain their own choices in spite of hindering circumstances (1.1). This argument is in turn supported as a sub-standpoint by two examples: of a labour leader (told by the female Swedish writer) and of the youth of the April 6th. Movement (1.1.1a – 1.1.1b). The audience of Al Aswany, playing the role of an antagonist, views these two cases as only examples. The readers know well that the supporters of political change in Egypt include other groups and

---

[5] An Egyptian activist group established in Spring 2008 to support the workers of El Mahallah El Kobra, an industrial town, who were planning to strike on April 6th. 2006. The movement was famous for the use of social media (Facebook, Twitter, Flicker, etc.) and innovative use of nonviolent tactics. For more information, see:
https://en.wikipedia.org/wiki/April_6_Youth_Movement

individuals than this labour leader and the youth of April 6th. Movement. Al Aswany does not explicitly state whether each of these two cases can stand on its own to support the sub-standpoint at issue. However, it is more reasonable to assume that the columnist intends to advance these two cases as coordinatively supporting the sub-standpoint since only one case is hard to be expected to be acceptable conclusive evidence. Neither a labour leader alone nor a small group of young activists alone can be expected to be acceptable as a generalizable case.

The sub-standpoint 1.1 cannot be considered acceptable unless the audience views these two examples as generalizable cases. Anticipating the critical questions related to the representativeness and sufficiency of the examples cited, Al Aswany must attempt to answer these questions as much satisfactorily as possible. The research question of this chapter can be rephrased as follows: what results can be achieved by selecting these two examples? And how does Al Aswany frame them in his attempt to make the examples cited representative and sufficient?

With regard to the anecdote, Al Aswany phrases the anecdote aiming to strengthen the representativeness of the case highlighted in the anecdote by bestowing typicality and anonymity on its protagonist. Although the Swedish writer says she was hired to make a lengthy interview with one of the labour "leaders", the word "leader" is emptied of any potential connotations of empowerment. By referring to the protagonist of the anecdote as "this worker" twice, "the worker's [daughter]" once, and "this Egyptian worker" once, he is positioned as an ordinary worker not a labour leader. In addition, his name is not mentioned.

What is foregrounded in the anecdote as its core message is the contradiction between the circumstances and the resulting behaviour: poverty and showing hospitality to others; poverty and keeping an aspiration for freedom. The details narrated are an attempt to hyperbole both: in spite of the enormous poverty, hospitality is unexpectedly showed immediately to a foreign stranger; in spite of the urgent need for money, this worker insistently refused a relatively big amount of money (given that the euro is a hard currency in Egypt).

If a typical worker who is poorer, less empowered and only modestly educated can keep his dignity and aspiration for freedom, it is expected that the richer, more empowered and well-educated will do *a fortiori* the same, or even more. Similar to what happened in the "Egypt awakened" case, the technique of "resorting to hierarchically arranged examples" is applied. In an attempt to strengthen the sufficiency of the example cited, and therefore raising the acceptability of the standpoint put forward, a detailed example, carefully selected, is cited to imply that other examples supporting the same claim do really exist.

Knowing that the audience will evaluate the example of the labour leader as an argument that is by itself too weak to conclusively support sub-standpoint 1.1, Al Aswany provides another example: the youth of the April 6th. Movement. He writes:

> Whoever would have believed that young people in twenties would call for a civil disobedience through Facebook and Egypt would positively respond end to end?! [...] How could they take to the streets confronting the riot police (the Egyptian occupation army), be harshly beaten, dragged off, detained and tortured in state security premises, and then released more determined to change their country?! How and when do these sons and daughters feel the love of Egypt?! They were born in the era of Camp David, comprehensive corruption, decayed education, and superficial media; the era of contempt of national issues and ridiculing the notion of dignity.

In Sec. 2, I have explained that pro-regime media established the stereotype of the opposition political activist as "a male university student highly individualistic, and slightly intellectual, more often than not a communist, leftist, or an Islamist". The members of the April 6th. Movement may be considered as matching this stereotype. Yet, this is only partially true. Indeed, the members are "mostly young and educated Egyptians", but "most of them had never been involved with politics before joining the [Facbook] group"[6]. Therefore, it was hard to categorize the movement decisively (before 2011 and even afterwards); it was considered trans-ideological. "In its official statement, the April 6 movement takes pains to emphasize that it isn't a political party. But the movement has provided a structure for a new generation of Egyptians, who aren't part of the nation's small coterie of activists and opinion-makers". In addition, the movement is not predominantly masculine. It is remarkable that two of its most famous and prominent figures were female: Israa Abdul Fattah and Asmaa Mahfouz.[7]

---

[6] Samantha M. Shapiro wrote a lengthy commentary about the movement in the *New York Times* on 22 January 2009. For more information, see: http://www.nytimes.com/2009/01/25/magazine/25bloggers-t.html

[7] The first one was arrested by the Egyptian security after El Mahallah's April 6th. massive demonstration as one of its main promoters. For more information,

Al Aswany positions these supporters of political change as young family members making use of linguistic means, i.e. repeating the adjective "young people" twice, mentioning their age rate "in the twenties", and referring to them as "boys and girls" and "sons and daughters". The connotation is perhaps that they are still needy or potentially weak. He also puts an emphasis on the social conditions they were raised in by listing the factors that could have automatically led to indifference concerning politics and national dignity. In addition, they are subjected to high degree of abuse that could have automatically led to feeling afraid of participating in opposition political activities. It is an attempt to frame these actions as free and exceptional choices and thus stress the contrast between the circumstances and the resulting behaviour.

Al Aswany utilizes the image of youth established in the Middle Eastern countries in general. According to Swedenburg, youth are imagined as a threat because they are "vulnerable innocents". Middle Eastern regimes maintained and enhanced this imagination because it justifies different kinds of patriarchal guardianship over young people. Youth are not portrayed as inherently bad guys, but as so weak and naïve that they can be easily deceived and utilized by evil and dangerous forces. (Swedenburg, 2012, p. 287).

Viewed by readers as "vulnerable" innocents, young people are much less expected to resist the impact of external conditions on their behaviour. To sum up, Al Aswany, again, applies the technique of "resorting to hierarchically arranged examples" evoked in presenting the case of the labour leader's anecdote. He attempts to imply that if the younger, care-needy and highly vulnerable people can resist the circumstances they are living in, the less vulnerable are *a fortiori* expected to do so, or even more. Therefore, the sufficiency of the example cited will be strengthened by mentioning a small number of examples that imply the existence of much more.

7. CONCLUSION

For an arguer who aims to evaluate a political group positively in a space-limited macro-context (e.g. the political column), applying the

---

see:https://www.youtube.com/watch?v=Y5pKlnFLjcY. The second one was widely known for her help in sparking the 2011 revolutionary uprising through a video blog. For more information, see: https://en.wikipedia.org/wiki/Asmaa_Mahfouz

technique of "resorting to the hierarchically arranged examples" can be a successfully strategic maneuver. In the case of arguing in favour of the "active people" topic, framing the cases cited as most "typical" and "ordinary" implies that the generalization is incontestable as it surely applies to other unmentioned examples that are less typical and ordinary (i.e. more politically active).

REFERENCES

Arthos, J. (2003). Where there are no rules or systems to guide us: Argument from example in hermeneutic rhetoric. *Quarterly Journal of Speech*, *89*(4), pp. 320-344.
Aswany, A. (2011). *On the state of Egypt.* (J. Wright, Trans.). Cairo/NY: The American University in Cairo Press.
Demon, K. (1997). A paradigm for the analysis of paradigms: The rhetorical exemplum in ancient and imperial Greek theory. *Rhetorica*, *15*(2), pp. 125-158.
Eemeren, F. H. van (2010). *Strategic maneuvering in argumentative discourse: Extending the pragma-dialectical theory of argumentation.* Amsterdam: John Benjamins.
Eemeren, F. H. van, & Grootendorst, R. (1992). *Argumentation, communication, and fallacies*: *A pragma-dialectical perspective.* Hillsdale, New Jersey: Lawrence Erlbaum.
Eemeren, F. H. van, & Grootendorst, R. (2004). *A systematic theory of argumentation: The pragma-dialectical approach.* Cambridge: Cambridge University Press.
Eemeren, F. H. van, & Grootendorst, R. (2002). Strategic manoeuvring. Maintaining a delicate balance. In F. H. van Eemeren & Houtlosser, P. (Eds.), *Dialectic and rhetoric: The warp and woof of argumentation analysis.* (pp. 131-160). Dordrecht: Kluwer Academic.
Eemeren, F. H. van & Houtlosser, P. (2005). More about an arranged marriage. In C. A. Willard (Ed.), *Critical problems in argumentation*: Selected papers from the 13th Biennial Conference on Argumentation Sponsored by the American Forensic Association and National Communication Association August, 2003 (pp. 345-555). Washington, DC: National Communication Association.
Eemeren, F. H. van, & Houtlosser, P. (2006). Strategic maneuvering: A synthetic recapitulation. *Argumentation* 20(4), pp. 381-392.
Eemeren, F. H. van, Houtlosser, P., & Snoeck Henkemans, A. F. (2007). *Argumentative indicators in discourse: A pragma-dialectical study.* Dordrecht: Springer.
Freeley, A. J., & Steinberg, D. L. (2005). *Argumentation and debate: Critical thinking for reasoned decision making.* Belmont, CA: Thomson Wadsworth.

Garssen, B. (1997). *Argumentatieschema's in pragma-dialectisch perspectief: Een theoretisch en empirisch onderzoek* (doctoral dissertation). University of Amsterdam.

Omar, A. (2016). *Strategic maneuvering in supporting the feasibility of political change: A pragma-dialectical analysis of Egyptian anti-regime columns* (doctoral dissertation). University of Amsterdam.

Perelman, C. & Olbrechts-Tyteca, L. (1969). *The new rhetoric: A treatise on argumentation.* (J. Wilkinson & P. Weaver, Trans.). London: University of Notre Dame Press.

Roccu, R. (2012) *Gramsci in Cairo: Neoliberal authoritarianism, passive revolution and failed hegemony in Egypt under Mubarak, 1991-2010* (doctoral dissertation). London School of Economics.

Swedenburg, T. (2012). Imagined Youths. In R. Mahdi, & P. Marfleet, P. (Eds.), *Egypt: Moment of change.* (pp. 285-294). Cairo: The American University in Cairo Press.

42

It Ought To Be Therefore It Is: On Fallaciousness of So-Called Moralistic Fallacy

TOMÁŠ ONDRÁČEK
Department of Corporate Economy, Masaryk University, Czech Republic
ondracek.t@gmail.com

IVA SVAČINOVÁ
Department of Philosophy and Social Sciences, University of Hradec Králové, Czech Republic
iva.svacinova@gmail.com

> The problem of moralistic fallacy, crossing the gap from ought-propositions to is-propositions, is considered with regard to four questions: Should we consider all ought-propositions (or is-propositions) in the same manner? Is the ought-is move an inference or is it just a case of a practical assumption? Is this move fallacious in any discussion? To address these questions, we use the pragma-dialectical theory, where the ought-is relation is argumentatively considered as a relation between propositions in reason and standpoint.
>
> KEYWORDS: argument scheme, moralistic fallacy, ought-is inference, pragma-dialectics

1. INTRODUCTION

In 1978, Bernard David Davis published an article in Nature (Davis, 1978) entitled *Moralistic Fallacy*. In it he argued against forbidding knowledge based on fear and moral attitudes. He pointed out that arguments for this prohibition are founded on a wrong inference, on the derivation of an *is* from an *ought*. With regard to Hume and Moore, he called this the *moralistic fallacy* (MF).

After Davis, other theoreticians of science referred to this problem. Ridley (1998) called this the reverse naturalistic fallacy. He presented examples connected to political correctness (Ridley, 1998, p. 258), where a fact is upheld because it ought to be the case regarding our political view. Pinker (2003) associated this fallacy with the concepts of the Noble Savage and the Blank Slate. According to him, many believe that "[n]ature, including human nature, is stipulated to have only virtuous traits (no needless killings, no rapacity, no exploitation), or no traits at all, because the alternative is too horrible to accept" (Pinker, 2003, p. 162). Other examples can be found in scientific literature itself (e.g. Gould, 1996; d'Arms & Jacobson, 2000; Rushton & Jensen, 2005; Stroebe, Postmes & Spears, 2012; Gorelik & Shackelford, 2017).

Hence, even though MF can be seen as a fallacy which cannot occur because it is so silly that no one could commit it, there is a plenty of evidence that this fallacy or at least this label is used in today's scientific and academic papers. Thus, it is astounding that not much attention has been paid to it in argumentation studies. We would like to reduce this shortage and present at least some introduction to our research on this topic. We start with a simple question: What is MF in terms of argumentation theory?

Some preliminary notes must be made. We understand MF as a label invented and used by theoreticians of science and scientists. Thus, we will address the label itself and concrete examples on which the label was put by insiders. We also do not want to treat fallacies in a way described as "traditional", i.e. as an argument which "*seems to be valid* but *is not* so" (Hamblin, 1970, p. 12). We limit our study to pragma-dialectical theory. Thus, we ask the question: What is MF in pragma-dialectical theory? Even though we are not able to deal with the problem of MF in all its aspects, we believe that this is a convenient starting point for future studies.

2. IDEA OF MF

MF is a label invented, introduced and used by scientists and theoreticians of science to point out a flaw in argumentation (see e.g. Davis, 1978; Ridley, 1998; Pinker, 2003). It serves as an argument against the opponent who let himself or herself be misled by presuppositions which should not be taken as a sound foundation for scientific reasoning on facts.

For example, Gould and other proponents of the argument that there is no correlation between race and intelligence are accused of flawed reasoning by Rushton and Jensen (2005). They present several examples of misinterpreting evidence, of giving preference to a moral view over the obtained data. They even go so far as to say that such a theory is "a degenerating research paradigm" (Rushton & Jensen, 2005, p. 329). Gorelik and Shackelford (2017) point to a flaw in the inference that "suicide [is] unnatural or evolutionarily maladaptive due to one's belief that it is immoral" (Gorelik & Shackelford, 2017, p. 287). There are also other examples in other fields, even in scientific theory itself regarding the myth of self-correcting science (Stroebe, Postmes & Spears, 2012, p. 677). In general, MF is a case when someone "derive[s] an 'is' from an 'ought'" (Davis, 1978, p. 390) or simply when "'ought' implies 'is'" (Pinker, 2003, p. 162).

MF is seen as a flaw if it fulfils two conditions. The first one concerns an argumentative standard, which can be explicit or implicit. The argumentative standard defines what is a permissible and desirable mode of a particular (flawless) argumentation. In case of MF, the standard is given by the scientific context in which, roughly saying[1], the only way to argue for facts is by other facts. The second condition relates to the type of used assertions. A conclusion (standpoint) must be an assertion of a fact, and be considered as such. The argument/support/premise must be given by a statement expressing a moral or ethical judgement, and be considered as such.

2. ARGUMENTATION THEORY AND RESEARCH QUESTIONS

We choose pragma-dialectical theory as a convenient theory for the analysis of scientific discourse for several reasons. It is influenced by K. R. Popper's critical rationalism (cf. van Eemeren & Grootendorst, 2004, p. 16) which is a cornerstone of today's description of science[2]. It provides tools for description and rules for dialogue aiming to resolve a difference of opinion in a rational manner. Pragma-dialectics operate on an ideal model which might be hard to follow in real-life discussions, but which should be the style of scientific discussions.

There are two perspectives which we use in our analysis of MF: *emic* and *etic*. The emic perspective considers the point of view of the parties in a given discussion. It is participant-centred. Thus, it is the internal perspective of those who participate in the described argumentative situations. The emic perspective follows the

---

[1] In foundationalism, prevailing epistemological approach in science, there can be statements which are not founded on any other statements at all. But these kinds of problematic statements are not present in the given examples.

[2] However, it can be seen rather as an idealization and current approaches to science differ in many ways.

understanding of situations as it is shown by these actors regarding their communication. An example of such an understanding is labelling, in our case labelling by MF. On the other hand, the etic perspective is the external perspective of an analyst who should combine his or her understanding of the material and theoretical background to make a systematic description. It is a perspective which is theory-driven. An analyst in pragma-dialectics has an etic perspective, aiming for "identifying as adequately as possible every aspect of an argumentative discourse or text that is relevant to the resolution of a difference of opinion" (van Eemeren & Grootendorst, 2004, p. 74). The etic perspective cannot be without the emic perspective, because it needs its insights.

The MF label, as used by a party in real examples, is a case of such an insight from an emic perspective. Regarding its usage, it was already noted that a party uses this label to point out some negative phenomena, some flaw in the reasoning of the other party considering the context of the scientific discussion. We, as analysts, are trying to describe this act from the perspective of an ideal model of critical discussion. We are trying to find out what type of theoretical phenomena these concrete phenomena are and if there even is a correlating theoretical concept which can cover the usage of this emic label. This can give us new insights in many ways. If there is such a concept we can name it and further theoretically analyse it. If there is not, then we can try to find out whether it is our theory which is somehow wrong or whether it is an ambiguous or even mistaken usage on the side of the acting parties.

The ideal model of critical discussion is formulated and structured like an argumentation discourse if it was purely oriented towards resolving a difference of opinion. It is not an exact description of the actual argumentation. Hence, this model specifies the stages of a critical discussion which must be realised to resolve a difference of opinion, and verbal moves which are used in each of these stages. The basis of the ideal model is the idea that a difference of opinion can be resolved only if the involved parties agree upon an acceptance or rejection of a controversial claim. That means that one party has to be persuaded by argumentation of the other one to accept or dismiss the claim (cf. van Eemeren et al., 1993, p. 25; 2007, p. 9).

There are four analytical stages distinguished in an ideal model: the confrontation stage, opening stage, argumentation stage and closing stage. The relevant stage for using MF is the argumentation stage where argumentation is put forward and critically tested. We will be examining the types of statements used in MF and their connection regarding the argumentation scheme as presented in pragma-dialectics.

Pragma-dialectics defines ten rules[3] which must be kept in the course of a discussion to successfully achieve the resolution of a conflict of opinion. Breaking one or more of these rules is considered as an obstruction to achieving this goal, a fallacy. The ideal model serves not only as a heuristic tool, which enables a reconstruction of an argumentative discourse, but also as a critical tool for an evaluation of the discourse (van Eemeren et al., 2002, p. 27).

The use of the ideal model of critical discussion allows us to formulate the following research questions at two levels:

Level of (theoretical) description of MF
    A. What types of statements are used in MF?
    B. What types of argument schemes are realized in MF?
Level of evaluation of MF
    C. Are the considered cases of MF always cases of fallacies?
    D. Are the considered cases of MF of one type of a fallacious move?

3. METHODS

In order to answer the research questions, we need to find an adequate *explicate* of the emic concept of MF in the conceptual system of pragma-dialectics. According to Carnap, explication is a process of replacing an inexact concept (explicandum) with a more exact concept (explicate). The process of explication has two phases. First, the explicandum needs to be identified as clearly as possible. Subsequently, an explicate needs to be introduced. This requires an explicit specification of rules for the use of the explicate in terms of the target theory (cf. Carnap, 1950; Brun, 2016, p. 1215).

The emic concept of MF is defined by insiders, a community of scientists and theoreticians of science, as the inference of an is-proposition from an ought-proposition. They also apply this concept to concrete arguments and thus provide us, analysts, with examples of its use. Therefore, we have two types of explicandum: a general definition and examples for which we can formulate the explicates.

4. EXPLICATION OF MF BASED ON GENERAL DEFINITION

---

[3] We use the numbering and phrasing the rules from van Eemeren et al. (2002).

The concept of MF is defined in scientific discourse as deriving factual conclusions from values, moral positions, etc. (cf. Davis 1978, Ridley, 1998). The explicandum of MF based on this general definition can be formulated as follows:

> X is Y. [IS proposition]
> because X ought to be Y / Z. [OUGHT proposition]

There is a general problem to determine which particular type of proposition is used in a given text. Consider this example (0) (cf. Schurz, 1997, p. 10):

> (0)   The weather ought to be fine tomorrow.

Is this statement normative or descriptive (factual)? It is impossible to decide this without the knowledge of the context of its use. Sometimes even with this knowledge we still cannot determine it, because the clues given in the context might be misleading. Hence, in our analysis we determine the type of statement from an emic perspective. In this perspective, we simply consider the stipulation given by insiders.

In an etic perspective, in pragma-dialectics, there are generally three possible types of propositions distinguished: prescriptive, normative, and descriptive. The prescriptive proposition is an incitement to do something, the normative proposition attributes an ethical or aesthetic quality to something, and the descriptive proposition is the claim of the factual state of affairs (cf. van Eemeren 2010, p. 1-2). Jean Wagemans proposes in his elaboration of pragma-dialectics a general form of the types of propositions:

Type of standpoint/proposition	Form of standpoint/proposition
Prescriptive/proposition of policy	Act, policy (A) should be carried out.
Descriptive/proposition of fact	Person, event, thing, act, policy (E) has empirical property P.
Normative/proposition of value	Person, event, thing, act, policy (E) is/*should be* judged as J.

Table 1 – General form of the types of propositions (Wagemans, 2016, p. 7)

Having these options to find the counterparts of is- and ought-propositions in pragma-dialectics, the adequate candidate for the is-proposition seems to be the descriptive proposition. However, given the term "should" in both the prescriptive and the normative proposition, we can consider as a candidate for the ought-proposition both the normative and the prescriptive proposition. Considering these possibilities, it seems adequate to explicate the ought-is relation as a relation between the normative and the descriptive proposition, or the prescriptive and the descriptive proposition:

Ought-is explicated as a normative-descriptive relation	Ought-is explicated as a prescriptive-descriptive relation
1 Person, event, thing, act, policy E (X) has empirical property P (Y).	1 Act, policy A (X) has empirical property P (Y).
1.1 Person, event, thing, act, policy E (X) should be judged as J (Z).	1.1 Act, policy A (X) should be carried out (Z).

Table 2 – Explication of ought-is as pairs of propositions

If we are interested in what types of argumentation schemes can be used in the pairs thus captured, we need to identify the connecting principle between these pairs of propositions. Pragma-dialectics conceives of an argument as consisting of three components: standpoint, argument and connecting premise. The last component expresses the argument scheme which is "a more or less conventionalised way of representing the relation between what is stated in the argument and what is stated in the standpoint" (van Eemeren & Grootendorst, 1992, p. 96). It can be understood as a general representation of the specific justifying relation between argument and standpoint. The sustainability of the argument scheme can be tested by means of the so-called critical questions.

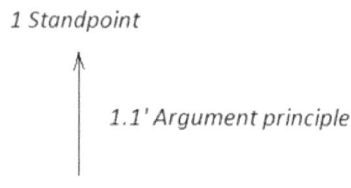

Figure 1 – General pragma-dialectical concept of the argument scheme

In determining what types of schemes can be implemented by given pairs of propositions, it is relevant to consider the distribution of the referents and predicates in the argument. In both the considered explicates of MF, they are pairs of propositions with the same referents but different predicates. This can be captured in the following general form (cf. Hitchcock, Wagemans, 2011, p. 186):

1 X is Y.
1.1 X is Z.

If we want to determine the argument principle, then we are looking for a proposition that grasps the possible relationship between two different predicates Y and Z. In pragma-dialectics, we can identify two schemes that capture such a relationship: symptomatic and causal. In general, we can capture them as follows:

1.1' Z is symptomatic of Y.
1.1' Z causes/is caused by Y

The MF in the conceptual framework of pragma-dialectics can be therefore identified as the following four possible types of arguments:

Ought-is explicated as normative-descriptive	Ought-is explicated as prescriptive-descriptive

(a)
1 Person, event, thing, act, policy E has empirical property P.
    1.1 Person, event, thing, act, policy E should be judged as J.
    1.1' Being judged as J is symptomatic for having empirical property P.
	(c)
	1 Act, policy A has empirical property P.
	    1.1 Act, policy A should be carried out.
	    1.1' Obligation to be carried out is symptomatic of having empirical property P.

(b)
1 Person, event, thing, act, policy E has empirical property P.
    1.1 Person, event, thing, act, policy E should be judged as J.
    1.1' Being judged as J causes/is caused by having empirical property P.
	(d)
	1 Act, policy A has empirical property P.
	    1.1 Act, policy A should be carried out.
	    1.1' Obligation to be carried out causes/is caused by having empirical property P.

Table 3 – Explication of argument principle implemented by pair of ought-is

As we showed, MF concerns primarily the argumentation stage, i.e. the situation in which the protagonist puts forward an argument in support of the standpoint. From this point of view, we are interested in whether by submitting the possible explicates of the moralistic fallacy there is a violation of rules that are directly relevant to the course of the argumentation stage. The rule that could be considered as directly relevant for capturing the moralistic fallacy seems to be rule 7 (cf. van Eemeren & Grootendorst 1992, p. 159, van Eemeren et al., 2002, p. 130):

> 7. A party may not regard a standpoint as conclusively defended if the defence does not take place by means of an appropriate argumentation scheme that is correctly applied.

Specifically, pragma-dialecticians consider the so-called argument ad consequentiam as the fallacy in which the facts are inferred from moral judgements. They conceive of it as an inappropriately used causal scheme (van Eemeren et al., 2002, p. 131):

> Another well-recognized unsound way of arguing is to appeal inappropriately to a causal relation. The mistake of *confusing facts with value judgments* is a fallacy that is traditionally known as the *argumentum ad consequentiam*. In support of a standpoint with a factual proposition, an argument is advanced that is normative because it points out undesirable effects of the standpoint: "It isn't true, because I don't want it to be true" or "It's true, because I want it to be true." An example of *ad consequentiam* is: It can't be raining, because that would mean we'd have to cancel our picnic.

We can reconstruct the example of argument ad consequentiam as follows:

1 It will not rain. [Descriptive]
(1.1a) (Rain leads to cancellation of picnic.)
1.1b The cancellation of a picnic is considered undesirable. [Normative]
(1.1ab') (Being judged as undesirable causes not raining.)

An argument thus reconstructed can be seen as the case (b) of explicates of MF. Rule 7 is violated because the protagonist will probably not be able to adequately justify the sustainability of the 1.1ab' and answer critical question 1 connected with the causal scheme: Does the assessment of the cancellation of a picnic as undesirable in fact lead to not raining? Namely, the protagonist cannot satisfactorily explain how the assessment of a situation as undesirable can cause a factual state of affairs like the state of weather.

Let us summarize. Based on the explication of a general definition of MF in the pragma-dialectical framework, we see MF as four possible types of argument, at least one of which is explicitly considered as a fallacy by pragma-dialectics, an argument ad consequentiam.

5. EXPLICATION OF MF BASED ON EXAMPLES

The second source of evidence for our explication can be taken from the examples of the use of MF, i.e. arguments labelled as MF by insiders in an emic perspective. We approach them through a pragma-dialectical reconstruction and evaluation due to the standards of critical discussion. We present analyses of three examples which are used as typical cases or basic illustrations of MF.

5.1 The violence case

The Violence Case is a typical example of MF. It also reflects a frequent pattern occurring in other cases of MF (cf. Gorelik & Shackelford, 2017; d'Arms & Jacobson, 2000). Pinker's (2003, pp. 307-308) formulation of this case is as follows:

> (1a) Recall Ashley Montagu's UNESCO resolution that biology supports an ethic of "universal brotherhood" and the anthropologists who believed that "nonviolence and peace were likely the norm throughout most of human prehistory." In the 1980s, many social science organizations endorsed the Seville Violence Statement, which declared that it is "scientifically incorrect" to say that humans have a "violent brain" or have undergone selection for violence.

Rushton and Jensen (2008, p. 638) offer a similar formulation:

> (1b) An example of the moralistic fallacy: Claiming that, because warfare is wrong, it cannot be part of human nature.

In an emic analysis of MF as an ought-is relation, we could briefly paraphrase the example as follows:

> Violence is not a part of human nature [IS proposition]
> because violence ought to be wrong [OUGHT proposition].

A pragma-dialectical reconstruction reveals that it is the relationship between the normative proposition in the argument and the descriptive proposition in the standpoint. The example uses the symptomatic argument scheme:

> 1 Violence is not a part of human nature. [Descriptive]
> 1.1 Violence is judged as morally wrong. [Normative]
> (1.1') (Characteristics judged as morally wrong are characteristically not a part of human nature.)

Regarding the rules of critical discussion, the argument seems to violate rule 7, inappropriate use of the argument scheme. For the protagonist, it is probably not possible to prove that there is a relevant symptomatic relationship between the moral judgement and the empirical property in this case. he would probably not be able to respond successfully to critical question 1: Is judging something as morally wrong indeed typical of not being part of human nature?

5.2 The equality case

The second example of MF is chosen from the page collecting fallacies and promoting good reasoning (Moralistic Fallacy, 2016):

> (2) Men and women ought to be equal. Therefore, women are just as strong as men and men are just as empathetic as women.

The emic analysis could be formulated as follows:

> Women and men are equal in strength and empathy [IS proposition]
> because they ought to be equal. [OUGHT proposition].

With pragma-dialectical analytical tools, we can reveal two arguments in this example. Both are pairs of the normative and descriptive proposition connected by a symptomatic scheme:

> 1 Women and men are equally strong. [Descriptive]
> 1.1 Woman and men are judged as equal (in rights). [Normative]
> (1.1') (To be judged as equal (in rights) indicates equality in strength.)
>
> 2 Men and women are equally empathetic. [Descriptive]
> 2.1 Woman and men are judged as equal (in rights). [Normative]
> (2.1') (To be judged as equal (in rights) indicates equality in empathy.)

If we focus on examining the arguments in terms of standards of critical discussion, it seems that both arguments violate two rules. First, rule 10 is violated. According to this rule, parties must not use any formulations that are insufficiently clear or confusingly ambiguous (cf. van Eemeren et al., 2002, p. 136). In phrasing the argument "Men and women ought to be equal", the use of the word "equal" without specification gives the impression that it is the same type of equality which is referred to in the standpoint (strength/empathy).

According to van Eemeren, fallacies of unclarity or ambiguity occur "not only by themselves, but also—even often—in combination with violations of other discussion rules. Lack of clarity sometimes accompanies a fallacy and enhances its effect" (van Eemeren et al., 2002, p. 136). In this case, the fallacy of ambiguity is combined with a violation of rule 7, an inappropriate use of the argument scheme. The protagonist probably will not be able to give a plausible explanation of the relationship between the rights and empathy or strength of men/women.

5.3 The one-way street case

The third case presented as an illustration of MF is taken from another page promoting critical thinking:

> (3) Have you ever crossed a one-way street without looking in both directions? If you have, reasoning that people shouldn't be driving the wrong way up a one way street so there's no risk of being run over from that direction, then you've committed the moralistic fallacy. Sometimes things aren't as they ought to be. Sometimes people drive in directions that they shouldn't. The rules of the road don't necessarily describe actual driving practices. (Moralistic Fallacy, 2009)

The emic naïve analysis that leads insiders to capture this illustration as a case of MF could be paraphrased as follows:

> People in a one-way street drive only in one direction [IS proposition]
> because they ought to drive in one direction [OUGHT proposition].

The pragma-dialectical reconstruction reveals that it is a pair of two descriptive propositions using the symptomatic argument scheme:

> 1 People in a one-way street drive only in one direction. [Descriptive]
> 1.1 People in a one-way street are supposed to drive in one direction based on road traffic rules. [Descriptive]

(1.1') (It is characteristic for road traffic rules that they are (generally) followed.)

A reconstruction of the argument in pragma-dialectics reveals no violation of the rules of critical discussion. The commentary to the example, however, brings to light the reasons why the authors consider this as a fallacious move: "Sometimes things aren't as they ought to be. Sometimes people drive in directions that they shouldn't. The rules of the road don't necessarily describe actual driving practices" (Moralistic Fallacy, 2009). We can say that the authors doubt the sustainability of the propositional content of proposition 1.1. According to them, the statement that people in a one-way street are supposed to drive in one direction based on the traffic rules is not necessarily true in all cases. From the point of view of the ideal model of critical dialogue, they are in the role of the antagonist in the argumentation stage critically testing the argumentation. However, it does not mean that the argument is a fallacy. Continuing a critical discussion would show (depending on the additional argumentation put forward by the protagonist) whether premise 1.1 is sustainable or not.

6. CONCLUSION

What is a moralistic fallacy? To answer this, we deployed pragma-dialectical theory and tried to explicate and analyse an emic definition and selected examples of MF. Regarding the used theory, we specified the original question into four sub-questions at two levels, the first two (A. and B.) at the level of theoretical description and the other two (C. and D.) at the level of evaluation.

(Question A.) What types of statements are used in MF? Based on an explication of an emic definition of MF, we considered the descriptive proposition as an explicate of the is-proposition, and the normative or prescriptive proposition as two possible explicates of the ought-proposition. A further investigation of the individual examples of MF displays the pair of descriptive and normative propositions as the prominent one.

(Question B.) What types of argument schemes are realized in the MF? We identified four possible types of arguments that use a causal or symptomatic argument scheme. A comparison with examples leads to the conclusion that insiders consider cases of MF prominently as arguments utilizing a symptomatic scheme connecting the normative and the descriptive proposition. Most cases of MF follow option (a) from four possible explicates:

1 Person, event, thing, act, policy E has empirical property P.
1.1 Person, event, thing, act, policy E should be judged as J.
1.1' Being judged as J is symptomatic of having empirical property P.

In this paper, we presented one exception, The One-Way Street example, where a symptomatic scheme with two descriptive propositions was identified. However, this is quite a rare case, a deviation from ordinary use, hence it is reasonable to exclude this particular case in the process of explication of MF.

(Question C.) Are the considered cases of MF always cases of fallacies? It was shown that the most probable candidate in the conceptual system of pragma-dialectics is ad consequentiam. It is a violation of rule 7 of the critical discussion, the fallacy of an inappropriately used causal argument scheme. But this type of fallacy does not overlap with empirical cases of MF considered by insiders. The investigation of illustrations displays a violation of rule 7, however, it is a case of *inappropriately used symptomatic argumentation*.

(Question D.) Are the considered cases of MF of one type of fallacious move? In cases which turned to be fallacious it seems so. Rule no 7 is violated.

We can conclude in pragma-dialectical terms that the insiders are willing to consider multiple phenomena as cases of MF. However, it seems appropriate to limit the label of MF to the unique phenomenon covering the most frequent cases. The comparison of explicates of the emic definition and specific instances of MF leads us to conclude that MF can be described from the point of view of pragma-dialectics as a fallacious use of a symptomatic argument scheme connecting a normative argument with a descriptive standpoint.

This paper was meant to be an initial study of the problem of MF with the use of analytic theory of argumentation. Much more work needs to be done and many more questions are still left to be addressed.

REFERENCES

d'Arms, J., & Jacobson, D. (2000). The moralistic fallacy: On the 'appropriateness' of emotions. *Philosophical and Phenomenological Research*, *61*(1), 65-90.

Brun, G. (2016). *Explication as a method of conceptual re-engineering.* Erkenntnis, *81*(6), 1211-1241.
Carnap, R. (1962). *Logical foundations of probability.* Chicago: University of Chicago Press; London: Routledge and Kegan Paul.
Davis, B. B. (1978). The moralistic fallacy. *Nature, 272*(5652), 390-390.
Eemeren, F. H. van (2010). *Strategic maneuvering in argumentative discourse: Extending the pragma-dialectical theory of argumentation.* Amsterdam/Philadelphia: John Benjamins Publishing Company.
Eemeren, F. H. van, & Grootendorst, R. (2004). *A systematic theory of argumentation: The pragma-dialectical approach.* Cambridge: Cambridge UP.
Eemeren, F. H. van, & Grootendorst, R. (1992). *Argumentation, communication, and fallacies: A pragma-dialectical perspective.* Hillsdale, New Jersey: Lawrence Erlbaum Associates.
Eemeren, F. H. van, Grootendorst, R., Jackson, S., & Scott Jacobs. (1993). *Reconstructing argumentative discourse.* Alabama: The University of Alabama Press.
Eemeren, F. H. van, Grootendorst, R., & Snoeck Henkemans, A. F. (2002). *Argumentation: analysis, evaluation, presentation.* Mahwah, New Jersey, London: Lawrence Erlbaum Associates.
Eemeren, F. H. van, Houtlosser, P., & Snoeck Henkemans, A. F. (2007). *Argumentative indicators: A pragma-dialectical study.* Dordrecht: Springer.
Gorelik, G., & Shackelford, T. K. (2017). Suicide and the moralistic fallacy: Comment on Joiner, Hom, Hagan, and Silva (2016). *Evolutionary psychological science, 3*(3), 287-289.
Gould, S. J. (1996). *The mismeasure of man.* New York / London: WW Norton & Company.
Hamblin, C. L. (1970). *Fallacies.* Stuffolk: Methuen & Co Ltd.
Hitchcock, D., & Wagemans J. H. M. (2011). The pragma-dialectical account of argument schemes. In E. T. Feteris, B. Garssen & A. F. Snoeck Henkemans (Eds.), *Keeping in touch with pragma-dialectics: In honor of Frans H. van Eemeren* (pp. 187-206). Amsterdam/Philadelphia: John Benjamins Publishing Company.
Moralistic Fallacy. (2009). *Logical Fallacies.* Retrieved 2017-06-19, from http://www.logicalfallacies.info/relevance/moralistic/.
Moralistic Fallacy. (2016). *Logically Fallacious.* Retrieved 2017-06-19, from https://www.logicallyfallacious.com/tools/lp/Bo/LogicalFallacies/128/Moralistic-Fallacy.
Pinker, S. (2003). *The blank slate: The modern denial of human nature.* New York: Penguin Books.
Ridley, M. (1998). *The origins of virtue: Human instincts and the evolution of cooperation.* New York: Penguin Books.
Rushton, J. P., & Jensen, A. R. (2008). James Watson's most inconvenient truth: Race realism and the moralistic fallacy. *Medical Hypotheses, 71*(5), 629-640.

Rushton, J. P., & Jensen, A. R. (2005). Wanted: More race realism, less moralistic fallacy. *Psychology, Public Policy, and Law, 11*(2), 328-336.

Schurz, G. (1997). *The is-ought problem: An investigation in philosophical logic* (Vol. 1). Springer Science & Business Media.

Stroebe, W., Postmes, T., & Spears, R. (2012). Scientific misconduct and the myth of self-correction in science. *Perspectives on Psychological Science, 7*(6), 670-688.

Wagemans, J. H. M. (2016). Constructing a periodic table of arguments. In OSSA conference archive. Retrieved from http://scholar.uwindsor.ca/ossaarchive/OSSA11/papersandcommentaries/106/.

43

Pragmatic Inference and Argumentative Inference

STEVE OSWALD
University of Fribourg, Switzerland
steve.oswald@unifr.ch

I offer a theoretical discussion of the relationship between pragmatic inference (inference about meaning) and argumentative inference (inference about the acceptability of a premise/conclusion relationship). The discussion (i) compares an argumentative view on meaning construction and an interpretative view on argument evaluation, (ii) argues that pragmatic inference can constrain argumentative inference, and (iii) assesses the complexity of an account of argumentative exchanges seen through the lens of the inferential tasks they involve.

KEYWORDS: pragmatics, inference, argumentation

1. INTRODUCTION

If, following Pinto, we define inference as "the mental act or event in which a person draws a conclusion from premises" (Pinto, 2001, p. 32), we can identify at least two different types of inferences that are at play in argumentative processes. These are pragmatic inference (henceforth PI) and argumentative inference (henceforth AI); the former is concerned with processes of (naïve) comprehension while the latter is here taken to denote mechanisms of argumentative processing, and evaluation in particular.

In recent years, Macagno and Walton (henceforth MW) have proposed to combine both types of inference in a model meant to explain how, in their view, an account of pragmatic phenomena (such as implicature or presupposition) can benefit from the input of argumentation theory. In doing so, they attempt to explain the rise of implicit meaning in communicative practices as the result of (typically abductive) argumentative processes. The general orientation of this piece of research is thus to address typically pragmatic phenomena through the lens of argumentation theory. I wish here to question the rationale of this endeavour and propose, instead, that, as long as we are interested in accounting for these processes from a psychologically plausible perspective, the combination of insights should go in the opposite direction. I.e., there are better grounds to use pragmatic theory to account for argumentative phenomena than the other way around.

In section 2, I discuss the general features that need to be taken into account in order to define the notion of inference. In sections 3 and 4, I draw on relevant literature both in pragmatics and argumentation theory to characterise each type of inference according to the features identified in section 2. In section 5 I discuss the two directions in which PI and AI can interact both at the theoretical level and at the methodological level. I conclude by assessing the merits of a cognitive pragmatic approach to argumentation.

2. INFERENCE

It proves difficult to find a clear and concise definition of inference for two reasons: (i) few articles or thematic glossary entries are devoted to the term, and (ii) the term itself is polysemic.

One of the first acceptations of the term 'inference' qualifies it as a piece of information. Gerrig & Zimbardo's (2001) glossary of psychological terms featured on the American Psychological Association's website, for instance, defines it as **"[m]issing information** filled on the basis of a sample of evidence or on the basis of prior beliefs and theories" (*APA*, my bold). The first meaning listed by the *Oxford English Dictionary* is similar in that it defines it as "[a] **conclusion** reached on the basis of evidence and reasoning" (*OED*, my bold). What these two definitions have in common is the idea that the term inference is used to denote the *result* of a reasoning process. In other words, an inference is some sort of propositional content which can be derived from the consideration (and combination) of other propositional contents (evidence), through some form of reasoning. This conceptual acceptation is not the one that will be retained here.

Next to this first meaning, the entry of *inference* in the *OED* lists a second meaning: "[t]he process of inferring something". Under this understanding, inference ceases to denote a result and instead denotes the process by which pieces of information are combined in order to derive (other) information. In this sense, inference becomes a cognitive process and is thought of in procedural terms: the term thus denotes a particular cognitive procedure of information management. This is the sense of *inference* that this paper addresses.

If we turn to philosophical research, in one of the few philosophical papers entirely devoted to the notion of inference, Brown notes the following, as he focuses on the use of the term:

> (...) to say 'I infer' or 'He infers' is to expound one's views, together with **an indication of why one holds them**, or to ascribe views to someone else, together with an indication of why he holds them. (...) [I]nferring is a matter of holding views and having reasons for them." (Brown, 1955, p.354, my bold)

From this definition it appears that (i) we use the term *inference* to refer to the views that we hold (and interestingly to why we hold them, see also Hanna's definition in what follows), (ii) inference is concerned with justification, and (iii) in principle inference can be about different things, as the contents of the inference are not specified. Echoing this construal, Hanna adopts the following working definition of inference:

> (...) a cognitive process leading from the mental representation of the premises of a deductive, inductive, or abductive (abduction = inference-to-the-best-explanation) argument to the mental representation of the conclusion of that argument, where the cognitive transition from the representation of the premises to the representation of the conclusion is governed by some rule-based standards of cogency, such that if all the premises are believed by a cognizer or cognizers and if the cognitive transition from representing the premises to representing the conclusion is also believed by that cognizer or those cognizers to be cogent, then, other things being equal, the conclusion will also be believed by that cognizer or those cognizers. (Hanna, 2014, p.89-90)

Interestingly, both philosophical definitions stress an *argumentative* nature of inference, as they both include considerations about its purpose, namely, under their view, justification: Brown puts at the forefront of his claim the idea that through inference people offer reasons for holding the views they hold and Hanna goes as far as calling inference a process which leads a cogniser to believe the conclusion of an argument based on its premises and some standard of cogency.

While this construal makes perfect sense in an argumentative perspective focused on the practice of exchanging reasons or resolving differences of opinion, it fails to do justice to other kinds of inference such as PI, whose purpose is crucially not justification. As a consequence, Brown's and Hanna's definitions are closer to AI than to

PI. In order to be able to compare PI and AI, I therefore propose to keep the procedural dimension of inference but to loosen our working definition to define it as the cognitive process by which one piece (or set) of information is combined to another piece (or set) of information in order to derive a third piece (or set) of information.

I suggest that this working definition is loose enough but also adequate enough to allow us to characterise different types of inference with the following three features:

i. the nature of the combination or relationship involved
ii. the goal (or purpose) and scope of the inference
iii. the input and the output of the inference

In what follows I characterise PI in terms of these three features (section 3) and AI as well (section 4).

3. PRAGMATIC INFERENCE (PI)

The notion of pragmatic inference and its theorising owes a great deal to Grice's pioneering pragmatic work on meaning (collected in his posthumous 1989 book),[1] as his model of communication as a cooperatively rational conversational undertaking provided the building blocks of contemporary pragmatics.

The starting point of his theory is the idea of semantic underdeterminacy, which follows from his (1957) distinction between natural and non-natural meaning. The idea of semantic underdeterminacy rests on the clear-cut distinction between what is said and what is meant. Grice provides a principled account (consisting of a cooperative principle and 4 conversational maxims which further specify what conversational cooperation should amount to) of how conversational participants are able to figure out that what speakers mean is many times different and more specific than what they say. Grice names this type of implicit meaning *implicatures*.

Crucially, in his model, Grice considers that reaching an (naïve) interpretation of someone's utterance involves *inferring* speaker meaning, which in turn boils down to recognising (and thereby

---

[1] For the sake of clarity and given the readership of this collective book, let me state that here pragmatic inference is by no means equivalent to practical or pragmatic argumentation. While the latter refers to a type of argumentation in which consequences are considered to support a claim, the former has nothing to do with justification, as it refers the pragmatic notion of inference, concerned with meaning and naïve interpretation.

fulfilling) a communicative intention. Implicatures are thus defined as contents which are implicitly speaker-meant and inferentially hearer-derived. Here we see that the construal of inference offered by Grice is not of an argumentative nature, as PI simply denotes the process by which implicit contents may be reached.

That is to say that PI is an inference about meaning. If we turn to the three features that need to be specified in order to characterise the type of inference we are dealing with, here is what we can say about PI:

i. the relationship involved in PI is one of non-demonstrative deduction (following Sperber & Wilson, 1995, pp. 65-71).
ii. the goal (or purpose) and scope of the inference is interpretative and seeks to secure an interpretation of speaker meaning.
iii. the input of PI consists in verbal and contextual material and its output is a representation of speaker meaning.

Let us take an example to illustrate this:

(1) Laszlo: "Hey, wanna go to the movies tonight?"
Nina: "(sigh) I have an exam tomorrow morning."

In order to understand Nina's utterance as a refusal, Laszlo should combine the content 'Nina has an exam in the morning' with background assumptions such as 'people with exams the next day usually spend the previous evening studying'. Notice that the inference can be defeated, for instance when the major premise does not apply, for example in case Nina might be a party animal with little regard for academic performance. This is why cognitive pragmaticians in the footsteps of Sperber & Wilson (1995) call this general structure non-demonstratively deductive.

Although PI is many times characterised as inference to the best explanation (see e.g. Allott 2010 and Geurts 2010), this does still not include argumentative concerns: when speakers naturally convey implicatures in ordinary conversation, they are not engaged in an argumentative discussion about what they exactly mean. However, the inference addressees perform to derive the implicature may very well be defined as an inference to the best explanation:

> The speaker has said something that on the face of it is irrelevant (or false, or over/under-informative, long-winded etc.). What is the best explanation for this? In many cases the best explanation will be that the speaker intended to convey something more, an implicature. (Allott, 2010, p.94)

Under this view, speaker meaning derivation is akin to some form of explanation in which the *explanans* is the identification of communicative intentions, and in this sense, figuring out what someone means is finding the best explanation as to why they uttered what they uttered in context. But notice that this is not yet sufficient to call this inference an argumentative inference, to which I now turn.

4. ARGUMENTATIVE INFERENCE (AI)

In order to characterise AI, I will draw on the recent argumentative theory of reasoning (Mercier & Sperber 2009, 2011, 2017), which offers a novel take on the emergence of reasoning in the human species and incorporates a full account of AI. According to this model, the set of argumentative tasks we humans can be faced with is cognitively dealt with by an argumentative module (a set of processes dedicated to the management of argumentative data for argumentative purposes). The module is said to be responsible for the production and the evaluation of arguments.

Crucially, its function is to perform the AIs that are required in the production of arguments, typically when we are defending a standpoint, but also the AIs that are required to evaluate the arguments that others offer us. The inferential nature of the operations generated by the argumentative module is evident when we consider Mercier & Sperber's description of its workings: "what the argumentative module does then is to take as input a claim and, possibly, information relevant to its evaluation and to produce as output reasons to accept or reject that claim" (Mercier & Sperber, 2009, p.154).

Moreover, the argumentative module delivers, through inference, "a representation of a relationship between a conclusion and reasons to accept it" (*ibid.* p.155). This is to say that AI is not only responsible for the generation of arguments, but also that it functions with some sort of normative standard against which the acceptability of the link between premises and conclusions is measured. This is why it can be said that AI is about the acceptability of a justificatory link.

I can now characterise AI in terms of the three features specified above:

i. the type of relationship at play can vary along normative standards and types of argument schemes.
ii. the goal (or purpose) of AI is evaluative when it comes to reception and seeks to assess the quality of argumentation.

iii. the input of AI is the representation of speaker meaning (i.e., the output of PI), and its output is an evaluative representation of the relationship between a conclusion and reasons to accept it.

AI, unlike PI, is thus about assessing the quality of the relationship between premises and conclusions. This is why AI can be said to fulfil an evaluative role. PI, on the other hand, fulfils an interpretative role by delivering a representation of speaker meaning. Notice, however, that AI and PI are closely related: from the above characterisation it is clear that AI takes as input the output of PI. This will be of capital importance when we consider the relationships between both types of inference.

5. RELATIONSHIPS BETWEEN PI AND AI

Argumentation theory and pragmatics have a long-standing relationship which unfolds both at the level of their sometimes converging respective objects of study and at the scholarly level, since the two disciplines often interact in existing models of argumentation.

Some representative examples of this mutual cross-disciplinary are the following:

- Pragma-dialectics (van Eemeren & Grootendorst 2004), for instance, follows principles from both speech act theory, in its definition of argumentation as a complex speech act with associated felicity conditions (which the 10 rules of the critical discussion can be taken to embody) and Grice's account of rational communication through the postulation of a "communicative principle", which is adapted from Grice's (1989) cooperative principle and associated conversational maxims.
- Walton's pragmatic account of fallacies (1995) adopts a pragmatic approach as well by conceiving of fallacious argumentation as the situation in which argumentative moves force dialectical shifts which are normatively problematic; moreover, argument schemes, the building blocks of argumentation in Walton's model, are characterised as "pragmatic structures that display the form of an argument" (Walton, 1995, p.xii).
- Some rhetorical approaches regularly take on board and discuss insights from pragmatic theories because these are concerned with meaning reception: Tindale's construal of audience (2015, 1992), for example, is thoroughly based on a discussion of Sperber & Wilson's

notions of cognitive environment and mutual manifestness, as well as of Grice's model of meaning.
- Oswald's work (2007, 2011, 2014, 2016a, 2016b) systematically navigates the pragmatics/argumentation theory interface on a methodological and theoretical front, by discussing cognitive pragmatics tools meant to assist the analyst in the reconstruction of argumentative discourse and by developing a model of rhetorical effectiveness grounded on the cognitive mechanisms taken to regulate information processing.

While argumentation theory usually deals with questions related to argument evaluation and argument quality and pragmatics is traditionally concerned with issues of meaning, both can be taken to share one concern, namely curiosity for the mechanisms by which representations end up entertained by individuals through communication: (cognitive) pragmatics seeks to describe and explain why and how people understand each other's utterances and argumentation theory seeks to describe and explain why and how people end up accepting or rejecting the claims that others support with arguments. In both processes, quite minimally, the main issue is about explaining why and how a given representation becomes part of the individual's cognitive environment. The purpose of the next two sections is therefore to try to assess whether, from a cognitive perspective, it makes sense to consider that argumentative processes feed interpretative processes, and the other way around. In other words, I will now consider whether we may use AI to derive PI (MW's position) or whether we may use PI in deriving AI (the position defended here).

5.1 Influence of AI on PI (?)

In recent years, MW have published a series of papers and a book (e.g., 2013, 2017) in which they consider what an account of pragmatic notions such as implicature and presupposition stands to gain from the input of argumentation theory. Specifically, they postulate that implicatures should generally be viewed as inferences to the best explanation meant to resolve conflicts of presumptions. In turn, these resolutions are said to be potentially shaped by different argument schemes depending on the context.

While the construal of implicature in terms of inference to the best explanation is far from new (it was already present in Grice's seminal account of implicature (1967), see Hobbs (2008) for a discussion, and Allott (2010) and Geurts (2010) for illustrations), the kind of directionality between AI and PI that MW have in mind may turn out to be problematic on several counts.

In their own terms, they are out to propose an account of implicature – that is, an account of PI – that rests on the possibilities and affordances of AI:

> This **account of implicature** shows a crucial relationship between interpretation and dialogue theory in two key respects. First, implicatures need to be explained in terms of dialectical relevance. And second, **they need to be analyzed as implicit arguments**, involving a pattern of reasoning leading from a specific premise to a conclusion. Such pragmatic and linguistic phenomena can be therefore integrated and developed within dialectical argumentation theory, and can be starting points for **developing argumentation theory into a theory of textual interpretation**. (Macagno & Walton, 2013, p.211, my bold)

This leaves little doubt as to the directionality the authors postulate between AI and PI: AI is to be used as an explanatory instrument for an account of PI. This is what can legitimately be inferred from their view of argumentation theory as an account that becomes the tool to develop "a theory of textual interpretation" (*ibid.*).

To support this reading, we can also observe that since pragmatics has always sought to provide theories of textual interpretation, MW's contribution can only be understood as a contribution that argumentation theory has to offer to the field of pragmatics. In order to define how interpretation should be conceived from an argumentative perspective, they state, to that effect, that "interpretation is an argumentative activity that is carried out based on presumptions and breaches, or rather clashes, of presumptions." (Macagno & Walton, 2013, p.208).[2] And since implicatures have routinely been considered to be inferences to the best explanation for 50 years now, MW's contribution to the study of implicature has to be

---

[2] Notice also here the parallel between what M&W offer and traditional accounts of indirectness like Grice's (1989) and Searle's (1969), which both postulate that implicit meaning has to be worked out through the recognition of a breach of some conversational standard (for Grice, this involves maxim flouting and for Searle the exploitation of speech act felicity conditions). Just like its illustrious predecessors', M&W's account therefore seems to be out to explain how meaning is derived.

appreciated in terms of the typological value of the argument schemes that may be used to realise and verbalise abductive inferences.

Now, there is a fundamental question that in my view needs to be addressed: under this account, does what MW offer still qualify as AI? In other words, are we here dealing with genuine argumentation (as defined in argumentation theory)? I think three arguments can support a negative answer to this question.

First, inference to the best explanation is more explanatory than argumentative, and this is something M&W acknowledge:

> [o]n our perspective, implicatures are indirect speech acts of a kind (Bach 1994, 13), whose presumptive meaning differs from the intended one. Such a discrepancy, caused by a conflict of presumptions, need to be resolved through a process of **explanation**." (Macagno & Walton, 2013, p.208, my bold)

This means that the kind of inference performed while figuring the best explanation for a speaker's utterance, which should lead the addressee to identify speaker meaning, is not meant to convince anyone of anything, unlike AI, but to supply a possible (and ideally the best) explanation to the speaker's communicative behaviour.

Second, the goal of AI and that of PI are by definition at odds, in the sense that accepting an intended standpoint as a result of evaluation is not the same process as reaching an intended interpretation: to ground the distinction, let us just observe that one can obviously understand a speaker's claims and arguments without being convinced by them. Successful argumentation requires acceptance but successful interpretation does not. To give but an example, speakers do not want to convince hearers that the interpretation they reach is the right one in the same way that speakers want to convince hearers that they are the best candidates in an election. Given that the goals of AI and PI differ, the question of elucidating how exactly an account of AI can help accounting for PI therefore remains open.

Third, MW's proposal is framed as an account of interpretation, but it remains yet unclear how the features of AI can transfer to fulfil the goal of PI. An account of interpretation primarily needs to postulate mechanisms of comprehension, not mechanisms of conviction – even if these may otherwise interact at many levels. Yet, whether interpretation necessarily includes an argumentative component still needs to be justified. Macagno seems to be aware of this caveat when he notes that

> [t]his model can be considered argumentative *lato sensu* (cf. Van Eemeren & Grootendorst, 2004): it describes a dialogical process of reasoning in which the speaker invites the interlocutor to draw a specific inference in order to reconstruct the purpose of a move." (Macagno, 2012, p.262)

Here the definition of interpretation is phrased in terms of an addressee's reconstruction of the purpose of a move; in this sense, the inference to the best explanation is indeed the mechanism by which interpretation unfolds. And yet, I fail to see how we can still call this AI: these inferential (or reasoning) patterns (i.e. argument schemes) seem to be stripped of any defining argumentative feature: they are neither borne out of disagreement, nor do they appear to be generated to solve any genuine and interpersonal dialectical disagreement. Furthermore, they do not seem to be accompanied by any persuasive intention – at least no dialogical persuasive intention. Even if one could loosely consider that what goes on in an addressee's mind when he tries to figure out speaker meaning is some sort of resolution of difference of opinion between competing interpretations, and that the process by which he decides which one was intended ends up convincing him of the rightness of one of the interpretations, this is far from what argumentation theorists, and cognitive psychologists, for that matter, routinely take AI to be.

5.2 Influence of PI on AI

In order to characterise the relationship between AI and PI, under the view considered in what follows, I venture that we need to take into account the psychological plausibility of the inferential mechanisms involved, in particular the way they might be thought to depend on each other.

The first argument to support the idea that PI comes prior to AI is found in an observation from Sperber *et al.* (2010, p.367), who state that "comprehension of the content communicated by an utterance is a precondition for its acceptance". This means that before you can assess whether some claim follows from the premises that are provided in its support, you need to be able to represent and understand the content of both the premises and the claim, as, save logical (formal) validity, nothing can be gained from an argumentative articulation of meaningless statements. This first observation seems to plead for a picture in which PI is necessary for AI, but not the other way around. And in fact, this follows from our earlier characterisation of PI and AI: AI takes as input the output of PI, which seems to suggest that AI operates at a point where PI has finished its job.

The second line of argument to support the directionality from PI to AI comes from rhetorical scholarship. Rhetoricians have long observed that the way you frame your argument can have significant impact on its chances of success: as way of illustration, take the following contrasting argumentations:

(2) Abortion should be illegal because it is the act of murdering babies.
(3) Abortion should be illegal because it is the act of murdering embryos.

Argumentations like (2) have typically been observed to speak to conservative audiences, which are usually more likely to consider that abortion is a crime. Focusing on (2), it moreover seems that talking about the act of murdering babies is lexically coherent with the construal of murder as the act of intentionally killing a person (babies are persons). (3), on the other hand, fails to establish this level of lexical coherence, because an embryo is, precisely, not yet a person. The difference between both arguments is thus a lexical one, and this difference can have consequences on the output of the addressees' AI, depending on their cognitive environment. To take another example (see Oswald 2011), equating the project of developing a state-funded national insurance system to replace an expensive and over-competitive private market with a "health tax", as in (4) below, might go a long way in terms of persuasion, since it allows the speaker to take on board the negative connotation of 'tax' in order to negatively frame the state-funded project:

(4) In fact, the introduction of a premium calculated according to income is equivalent to the introduction of a health tax.

Tindale (1992) develops a similar point as he considers how arguments might fare depending on the addressee's cognitive environments. Typically divisive issues which draw on ideological representations are likely to undergo such processes, as in (5) below (reproduced from Tindale, 1992, p.183):

(5) The Roman Catholic Church does not endorse the use of contraceptives, and therefore University students' health plans should not subsidize birth control pills.

Tindale notes that (5) is likely to have contextual effects in the cognitive environment of Catholic university students, while none at all in non-Catholics. This, in my view, clearly illustrates that a representation an addressee is able to draw from his interpretation of the speaker's utterance (i.e., the output of PI) will have a subsequent effect on AI in terms of persuasiveness.[3]

It therefore seems that what one understands from a message may in fact influence how one will evaluate the argumentation contained in the message. This is to say that PI influences AI, and this is a straightforward consequence of the fact that AI partly operates on the result of PI as an input.

6. CONCLUSION

The links between Argumentation Theory and Pragmatics are evident but quite complex. In particular, I have tried to show that two different types of inference, distinguishable in terms of their input/output and their goal and scope, are at stake when we combine insights from the two disciplines: PI, which is an inference about meaning, and AI, which is an inference about a justificatory relationship.

While I have advocated that PI influences AI, and not the other way around, two nuances must be added. First, there is obviously a sense in which AI can still be considered to feed an account of PI, and that has to do with the typological benefits of AI in terms of identification of argument schemes and types of inference. However, should we adopt this position, then the contribution of argumentation theory would need to be understood as being restricted to offering a reservoir of reasoning or inference patterns (*à la* argument schemes). And if this is the case, then there is little to be gained from argumentation theory as a whole in the study of cognitive pragmatics, given that its only import would lie in its argument scheme list – which is arguably not a defining disciplinary contribution.

Secondly, the alternative option, namely to consider that AI feeds on PI, should be understood as targeting the reality of cognitive phenomena. In this respect, using an account of PI to study some

---

[3] We could also note, in passing, that this is also what lies at the core of the extended pragma-dialectical model of argumentation (see van Eemeren 2010) under the different constraints that strategic manoeuvring can exploit, and which stipulate different ways in which argumentative messages can be phrased to increase their chances of rhetorical success.

features of AI can be beneficial because it is concerned with a psychological question, that of rhetorical effectiveness, and puts on a par the specificities of AI and PI as cognitive inferences to try to understand how they are related at the level of cognitive processing.

REFERENCES

Allott, N. (2010). *Key Terms in Pragmatics*. London: Continuum.
Brown, D. G. (1955). The Nature of Inference. *The Philosophical Review*, 64(3), 351–369.
Eemeren, F. H. van (2010). *Strategic Maneuvering in Argumentative Discourse: Extending the pragma-dialectical theory of argumentation*. Amsterdam: John Benjamins Publishing Company.
Eemeren, F. H. van, & Grootendorst, R. (2004). *A Systematic Theory of Argumentation: The Pragma-dialectical Approach*. Cambridge: Cambridge University Press.
Gerrig, R., & Zimbardo, P. (2001). *Psychology and Life* (16 edition). Boston: Allyn & Bacon. Glossary available on (30.10.2017) http://www.apa.org/research/action/glossary.aspx?tab=9..
Geurts, B. (2010). *Quantity Implicatures*. Cambridge: Cambridge University Press.
Grice, H. P. (1957). Meaning. Philosophical Review, 66(3), 377–388.
Grice, H. P. (1989). *Studies in the Way of Words*. Cambridge, Mass.: Harvard University Press.
Hanna, R. (2014). What is the nature of inference? In V. Hösle (Ed.), *Forms of Truth and the Unity of Knowledge* (pp. 89–101). Notre Dame, Indiana: University of Notre Dame Press.
Hobbs, J. R. (2006). Abduction in Natural Language Understanding. In L. R. Horn & G. Ward (Eds.), *The Handbook of Pragmatics* (pp. 724–741). Oxford: Blackwell Publishing.
Horn, L., & Ward, G. (2006). *The Handbook of Pragmatics*. Oxford: Blackwell Publishing.
Hösle, V. (2014). *Forms of Truth and the Unity of Knowledge*. Notre Dame, Indiana: University of Notre Dame Press.
Macagno, F., & Walton, D. (2013). Implicatures as Forms of Argument. In A. Capone, F. L. Piparo, & M. Carapezza (Eds.), *Perspectives on Pragmatics and Philosophy* (pp. 203–225). Dordrecht: Springer.
Macagno, F., & Walton, D. (2017). *Interpreting Straw Man Argumentation: The Pragmatics of Quotation and Reporting*. Dordrecht: Springer.
Mercier, H., & Sperber, D. (2009). Intuitive and reflective inferences. In J. Evans & K. Frankish (Eds.), *In Two Minds: Dual Processes and Beyond*. Oxford: Oxford University Press.
Mercier, H., & Sperber, D. (2011). Why do humans reason? Arguments for an argumentative theory. *Behavioral and Brain Sciences*, 34(2), 57–74.
Mercier, H., & Sperber, D. (2017). *The Enigma of Reason*. Cambridge, Mass.: Harvard University Press.

Oswald, S. (2007). Towards an interface between Pragma-Dialectics and Relevance Theory. *Pragmatics & Cognition*, 15(1), 179–201.

Oswald, S. (2011). From interpretation to consent: Arguments, beliefs and meaning. *Discourse Studies*, 13(6), 806–814.

Oswald, S. (2014). It is Easy to Miss Something you are not Looking for: A Pragmatic Account of Covert Communicative Influence for (Critical) Discourse Analysis. In C. Hart & P. Cap (Eds.), *Contemporary Studies in Critical Discourse Analysis* (pp. 97–119). London: Bloomsbury Publishing.

Oswald, S. (2016a). Rhetoric and cognition: Pragmatic constraints on argument processing. In M. Padilla Cruz (Ed.), *Relevance Theory: Recent developments, current challenges and future directions* (pp. 261–285). Amsterdam: John Benjamins Publishing Company.

Oswald, S. (2016b). Commitment Attribution and the Reconstruction of Arguments. In F. Paglieri, L. Bonelli, & S. Felletti (Eds.), *The Psychology of Argument. Cognitive Approaches to Argumentation and Persuasion* (pp. 17–32). London: College Publications.

Pinto, R. C. (2001). *Argument, Inference and Dialectic*. Dordrecht: Springer Netherlands.

Searle, J. R. (1975). Indirect speech acts. *Syntax and Semantics*, 3, 59–82.

Sperber, D., & Wilson, D. (1995). *Relevance: Communication and Cognition*. Oxford: Blackwell.

Tindale, C. W. (1992). Audiences, relevance, and cognitive environments. *Argumentation*, 6(2), 177–188.

Tindale, C. W. (2015). *The Philosophy of Argument and Audience Reception*. New York: Cambridge University Press.

Walton, D. N. (1995). *A pragmatic theory of fallacy*. Tuscaloosa: University of Alabama Press.

Inference | Definition of inference in English by Oxford Dictionaries. (n.d.). Retrieved October 30, 2017, from
https://en.oxforddictionaries.com/definition/inference

44

Challenging Judicial Impartiality: When Accusations of Derailments of Strategic Manoeuvring Derail

H. JOSÉ PLUG
University of Amsterdam
h.j.plug@uva.nl

Impartiality is one of the core values underlying the administration of justice. A complaint about a lack of impartiality of a judge may be filed on the grounds of the judge's behaviour or his verbal behaviour. In this paper I will analyse complaints that concern the judge's use of rhetorical questions during court hearings. I will explore what role these complaints may play in the strategic manoeuvring of a party who seeks the judge's disqualification.

KEYWORDS: Derailment of strategic manoeuvring, Impartiality, Legal argumentation, Strategic manoeuvring, Rhetorical question

1. INTRODUCTION

To ensure the proper application of the law in an impartial, just and fair manner is essential to the public's trust in the office and the authority of the judge. In much the same way as in other European countries, the legal system in the Netherlands has several statutory provisions that provide a party to a proceeding with the right to seek a judge's disqualification if a party has doubts about the impartiality of the judge. A complaint about the judge's lack of impartiality may be filed on the grounds of the judge's behaviour or his verbal behaviour. This paper discusses complaints about the lack of judicial impartiality that are based on the judge's verbal behaviour, in particular on his use of rhetorical questions.

Recent research shows that the number of cases in which Dutch judges are challenged is growing; a tendency that may also be observed in court practices in other European countries. This increase may be related to the changing (communicative) role of the judge. Traditionally, the role of a judge is to apply the law and his performance is confined to being 'the mouthpiece of the law'. A reorientation on the judge's role, however, demands for the judge to be more active and more communicative during court hearings (Plug, 2016). Although this reorientation may have a positive effect on perceived procedural justice by the parties, it may increase the judge's risk of being criticised for his verbal behaviour. If this critique does indeed result in a disqualification request, the question that needs to be answered by the court that decides on matters of presumed partiality is whether or not a lack of impartiality may be inferred from the judge's verbal behaviour in a legal procedure.

In this paper, criticism on the verbal behaviour of the judge will be discussed by making use of the extended pragma-dialectical argumentation theory (Van Eemeren, 2010). From this theoretical perspective the argumentative moves by both the judge and the parties in a legal procedure are considered instances of strategic manoeuvring. I will examine what role criticism on the verbal behaviour of the judge can play in the strategic manoeuvring of a party that seeks the judge's disqualification. I will illustrate this role by means of the analysis of an example of a complaint about the judge's verbal behaviour that concerns his use of a (alleged) rhetorical question in a case where his impartiality is challenged.

In the following section, section 2, I will give an argumentative reconstruction of the disqualification procedure as a sub-discussion that precedes the decision on the judge's impartiality. In section 3.1, I will analyse the way in which the use of a rhetorical question may have an argumentative function in legal activity types. In section 3.2, I will determine how the judge's use of a rhetorical question may result in a complaint about a (presumed) lack of impartiality and, as such, in an accusation of a derailment of strategic manoeuvring on the part of the judge. On the basis of this analysis I will discuss theoretically motivated criteria for the analysis and evaluation of these accusations.

2. A DISQUALIFICATION PROCEDURE: SUB-DISCUSSION AND SPECIFIC ACTIVITY TYPE

The right to have access to an impartial judge is enshrined in international, European and national codes and legislation.[1] If the judge

---

[1] See for example the Bangalore Principles of Judicial Conduct (value 2: 2.5) and the European Convention of Human Rights (art. 6).

himself foresees there may be doubts about his impartiality, he has the options of exemption or withdrawal. A reason for exemption and withdrawal may, for example, be a personal relationship with (one of) the parties or lawyers. Exemption is considered to be a means to prevent situations that might give rise to doubts about judge's impartiality. The judge can decide to withdraw (recuse) before a session, but also during a session.

If the judge sees no reason for exemption or withdrawal, but a party to the proceedings has doubts about his impartiality of the judge to whom his case is submitted, this party may start a disqualification procedure.[2] In the Netherlands, the disqualification panel that decides on the request for recusal consists of judges from the judiciary in which the initial trial takes place. The panel decides on whether or not the request is admissible before it decides on the (im)partiality of the judge.

From a pragma-dialectical perspective, an attack on the judge for not being impartial or biased might appear, at first glance, as an *argumentum ad hominem*. Since from this theoretical perspective both parties and the judge can be conceived as participants to a rational resolution-oriented discussion they may be expected to manoeuvre strategically within the boundaries of reasonableness. This means that they balance their dialectical and rhetorical objectives and respect the rules for critical discussion (van Eemeren, 2010). According to these discussion rules, an attempt to exclude a participant, in this case the judge, from the discussion by means of a personal attack could be regarded as a violation of the Freedom Rule: "Discussants may not prevent each other from advancing standpoints or from calling standpoint into question." On this ground, the contribution to the discussion by the party to the process who claims that the judge has no right to speak would then be considered a fallacious discussion move or a derailment of strategic manoeuvring.

However, when taking into account the institutional conditions of the activity type of legal proceedings that allow for the possibility to discuss the impartiality of the judge and to exclude the judge from the discussion, an attack on the judge for not being impartial should not necessarily be considered a fallacious discussion move or a derailment of strategic manoeuvring. Yet, the prerequisite to legitimately bring forward the accusation that the judge is not impartial, is that the accusation becomes subject of a separate procedure, a disqualification procedure. Subsequently, the disqualification panel in its turn, will have to evaluate the argumentation underlying the accusation.

---

[2] In this paper, 'disqualification' and 'challenge' are used interchangeably.

This disqualification procedure can be reconstructed as a sub-discussion that is situated in the opening stage of the main discussion on the initial case at hand, in which the protagonist and the (institutional) antagonist of a standpoint should reach an agreement about the distribution of the (institutional) discussion roles. After the difference of opinion that is subject of the sub-discussion has been resolved, the main discussion can be continued. Since the sub-discussion has its own institutional goal and its own institutional characteristics, it can be regarded as a specific activity type of legal proceeding. In this sub-discussion, the proposition 'p1' that is subject of the dispute concerning 'Judge X should be disqualified in case Y'. The division of the discussion roles in the sub-discussion, too, is institutionally determined. The party to the initial legal process who doubts the judge's impartiality takes the position of a protagonist (+/p1) in the sub-discussion. The discussion role of the judge who is accused of not being impartial, may be reconstructed as that of an (institutional) protagonist who takes the opposite position regarding proposition 'p1': (-/p1). The discussion role of the disqualification panel is that of an (institutional) antagonist, whose position regarding the propositions (+ p1) and (- p1) is: ?/ +p1 and ?/ –p1.

Subsequently, the institutional characteristics of the activity type of a disqualification procedure also define part of the argumentation that the party which challenges the judge has to bring forward in the sub-discussion. The general explicit or implicit argumentation underlying the disqualification request is determined in prevailing legislation as well as jurisprudence (case-law).[3] The criteria that determine the argumentation underlying complaints on a lack of impartiality were first developed in the so-called Hauschildt case, that was brought before the European Court of Human Rights. In the decision, the Court, when it discussed the concept of impartiality, distinguished between a subjective test and an objective test. The judge has to fail only one of the two tests to be disqualified from the case at hand. The objective test focuses on the circumstances of the case that may give rise to the fear of partiality. The subjective test concerns the state of mind and the attitude of the judge on the case at hand and the parties involved in the case. A judge's attitude or prejudiced state of mind may be deduced from his (verbal) behaviour during court hearings.

---

[3] In jurisprudence by the European Court of Human Rights (EHRC) and the Dutch High Court criteria are formulated as to how impartiality should be determined.

2.1 Arguments in a sub-discussion on the verbal behaviour of the judge

In the Netherlands, complaints about the (verbal) behaviour of the judge during court hearings are one of the most frequent reasons to request for the judge to be disqualified.[4] These complaints may refer to remarks suggesting that the judge is convinced of the defendant's guilt or to an informal choice of words that suggests the judge's sympathy for the standpoint of one of the parties.[5] Whether or not the judge's verbal behaviour in a particular case fulfils the condition of impartiality is part of the subjective test. The guiding principle for the evaluation of the argumentation for these claims is that the so-called personal impartiality of a judge must be presumed until there is proof to the contrary. The proof should thus be provided by argumentation that justifies a lack of personal impartiality on the part of the judge.

In case the judge's verbal behaviour constitutes the basis to doubt his impartiality, it may be expected that the argumentation consists of a quotation of the judge's choice of words during court hearings for instance or of a reference to the judge's use of rhetorical figures such as a simile, a hyperbole or a metaphor. These examples of verbal behaviour may be considered specific characteristics of a lack of personal impartiality.

The argumentation structure of the sub-discussion on the disqualification of the judge based on complaints about the judge's verbal behaviour may be represented as follows:

---

[4] See for example Knapen (2012) and Hammerstein (2014).

[5] In Dutch case-law (2007 – 2010) complaints about the judge's non-verbal behavior concern, for example, eye-rolling, reading while (the legal representative of) the accused party is speaking, a cynical or irritated tone of voice.

1.   Request R to disqualify Judge X should be granted

1.1   Judge X is not impartial in case C

1.1.1   Judge's X impartiality doesn't successfully pass the subjective test

1.1.1.1 Judge's X verbal behaviour during court hearing in case C proves a lack of personal impartiality (Y)

1.1.1.1.1a   1.1.1.1.1b
Judge X said Z   Z indicates a lack of impartiality (Y)

Figure 1 – The argumentation structure of a sub-discussion on impartiality based on the judge's verbal behaviour

In figure 1 the request for the disqualification of the judge is solely based on complaints about his verbal behaviour. However, in legal practice this ground for doubt on judicial impartiality may be combined with other complaints about the (verbal) performance of the judge. In these cases, the argumentation underlying a disqualification request consists of complex argumentation that, from a maximally argumentative analysis, should be reconstructed as multiple argumentation (Plug, 2002).

3. COMPLAINTS ABOUT THE USE OF RHETORICAL QUESTIONS

One of the recurring complaints about the judge's verbal behaviour in Dutch court hearings concerns the use of rhetorical questions. In the following I will analyse how these complaints are part of the argumentation underlying these requests. After that I will demonstrate in what way complaints about the use of rhetorical questions may be considered acquisitions of derailments of strategic manoeuvring, and how, on the other hand, these complaints themselves run the risk of being considered (or qualified as) a derailments of strategic manoeuvring.

3.1 The argumentative function of rhetorical questions in legal activity types

A rhetorical question (Erotesis) is considered a question that does not require an answer, because the answer is implied or obvious in a given context. The question form is not used for information gathering but may be used to make an assertion.

Argumentation theoretical studies on rhetorical questions focus on their role in argumentative contexts. In these studies by, for example, Slot (1993), Ilie (1994, 1995), van Eemeren, Houtlosser & Snoeck Henkemans (2007) and Snoeck Henkemans (2006), three main functions of rhetorical questions are distinguished. The first function is that of putting forward a standpoint. The second function is that of a rhetorical question as a means of putting forward an argument. The third function is that of using a rhetorical question of proposing something as a common starting point in the opening stage of a discussion.

These three functions can be recognized in legal literature in which the use of rhetorical questions in argumentative courtroom discourse is discussed.[6] Much of the literature on courtroom practices in common law systems focuses on the use of rhetorical questions by attorneys and prosecutors but not so much on that of judges. Practices that are studied are, for example, the closing argument or the summation (Danet, 1980) that attorneys and prosecutors present before the jury and the judge (Pascual, 2006, Tiersma, 2000, Ruskin, 2014). Here, the focus lies on rhetorical questions that are used in specific activity types of legal procedure that consists of a monologue or 'implicit' discussion, as van Eemeren and Grootendorst (1992, p.43) call it. In these 'implicit' discussions, the defendant or the prosecutor, the protagonist(s), make an attempt to counter (possible) doubt or criticism of the jury or the judge, but the role of the antagonist(s), the jury or the judge, remains implicit.

An example presented by Tiersma (2000, p. 185-186) may illustrate how a rhetorical question can be a means to create common ground and may thus function as a common starting point. The example concerns the 1995 trial of O.J. Simpson, where the prosecutor Clark uses a series of rhetorical questions in her closing argument that she immediately answers herself.[7] In her closing argument, Clark evaluates the role of Mark Fuhrman, the detective who testified regarding his discovery of evidence in the Simpson case, as follows.

---

[6] The use of rhetorical questions in legal writing is studied by, for example, Sala (2010).

[7] In rhetoric this strategy of raising a question followed by an immediate answer is known as the figure of speech hypophora (also referred to as anthypophora or antipophora), see, for example, Lanham (1991).

(1) Prosecutor: Did he lie when he testified here in this courtroom saying that he did not use racial epithets in the last ten years?
Yes.
Is he a racist?
Yes.
(...)
But the fact that Mark Fuhrman is a racist and lied about it on the witness stand does not mean that we haven't proven the defendant guilty beyond reasonable doubt.

In this example the prosecutor uses rhetorical questions as a means to present both the propositions 'F did lie' and 'F is a racist' as common starting points for the implicit discussion with the members of the jury. Therefore, these propositions can, in principle, be used as arguments. However, even though the propositional content of the (potential) arguments is deemed acceptable, the prosecutor considers their justificatory force to be insufficient to outweigh the justificatory force of the argument ('it is proven') in favour of the conclusion that the defendant is guilty beyond reasonable doubt. Acknowledgement of the common starting points as expressed by the rhetorical questions should thus not lead to the conclusion that the defendant is innocent.

An example of an explicit courtroom discussion by Tracy (2016), demonstrates how the judge may use a rhetorical question to put forward a standpoint as well as an argument. Although in common-law systems it is usually the attorney who asks the questions, in, for example, small claims court and in sentencing hearings it is the judge's task to do so. The example concerns the 'In re Marriage Cases' (2008) on the issue of revoking the right of same sex-marriage in California. When discussing what should happen to same-sex couples who had already married, the judge and the attorney brought forward the following.

(2) Judge (C):   [But-but Mr. Star, is that really fair to the people that depended on what this court said was the law? Upended their lives. Changed their property responsibilities with their spouses. Um is it really fair to throw that out?
      Attorney (S): Your-your honour, first of all I would quarrel with, quote, throwing it out, our position is more, if I may say so, specific with respect to…
(Excerpt 1.29, CA8 Justice Chin, Attorney Starr, from Tracy, 2016)

Here, the judge uses the first rhetorical question to implicitly convey the standpoint that '[the proposal of the attorney to restore the rights of couples in same-sex marriages] is unfair for the people who depended on what the court said was the law.' He justifies his standpoint by the arguments that '[these couples] upended their lives' and 'changed their property responsibilities with their spouses.' By means of the second rhetorical question, the judge complements these two arguments with the implicit argument that 'to throw that out is not fair.'

The attorney responds to the second rhetorical question and criticises the implicit argument for being a straw-man fallacy: the judge unjustly attributes to him the standpoint that he 'throws out' the negative consequences of his claim as presented by the judge. In pragma-dialects this discussion move is considered a violation of the discussion rule (3) that stipulates that 'Attacks on standpoints may not bear on a standpoint that has not actually been put forward by the other party' (Standpoint Rule). This violation species on what grounds the strategic manoeuvring may derail when a rhetorical question is used to present an indirect argument.

3.2 Justifying complaints about rhetorical questions in disqualification requests

In example (2) the attorney criticises the judge's verbal behaviour immediately after the judge has presented his argumentation by means of rhetorical questions. An immediate response enables the participants to discuss the (supposedly) derailed discussion move, to readjust or 're-rail' the move and then continue the initial discussion.

In a disqualification procedure, criticism regarding the judge's use of a rhetorical question is put forward only after the court hearing in which the rhetorical question was presented. The criticism on the rhetorical question constitutes the basis to doubt the judge's impartiality and is subject to evaluation by the disqualification panel. If the criticism concerns an accusation of a derailment of strategic manoeuvring on the part of the judge, the disqualification panel needs to decide on the (supposedly) derailment and its consequences. The argumentation for the outcome of the panel's evaluation is published in a judicial decision.

The following pragma-dialectical analysis of a case that was brought before a Dutch disqualification panel, demonstrates the complications that may occur regarding the justification and evaluation of complaints on rhetorical questions may occur. In this case, support for the applicant's disqualification request included the following.

> (3) At the hearing, the judge has proven not to be impartial (...) In reaction to the applicant representatives' appeal to the ten-day term, as referred to in Article 8:58 of the General Administrative Law (AwB), the Judge asked: What is it that you mean to achieve by that? According to the applicant, this is a rhetorical question.
> (Court Den Haag, April 3, ECLI:NL:RBDHA, 2015, 4564)

Starting from the general argumentation structure that was presented in section 2.1, the applicant's argumentation that criticises the judge's use of a rhetorical question in example (3) may be reconstructed as follows.

1. Request R to disqualify Judge X should be granted
   |
   (1.1-1.1.1)
   |
   1.1.1.1 Judge's X verbal behaviour during court hearing in case C proves a lack of personal impartiality (Y)

   1.1.1.1.1   (1.1.1.1.1')
   Judge X used a   using a rhetorical
   rhetorical question (Z)   question (Z) indicates
                                     (personal) partiality (Y)

   1.1.1.1.1.1
   "What is it that you mean to achieve by that?"

Figure 2 – The argumentation structure of a sub-discussion on impartiality based on criticism regarding a rhetorical question

The disqualification panel that decided on this case, allowed the complaint about the judge's use of a rhetorical question. The panel argued in much the same way as the applicant who filed the disqualification request.

(4) (...) it is up to the judge to decide, on the grounds of the principle of due process, whether the documents will actually be allowed, upon which the judge may ask to motivate the request. Since it has not been contradicted by the judge, the disqualification panel is of the opinion that it has been sufficiently established that the judge did this by way of a rhetorical question. By using this stylistic device the judge may have created the impression that he was not wholly unbiased as to this case.
(Court Den Haag, April 3, ECLI:NL:RBDHA, 2015, 4564)

In both the examples (3) and (4), the justifying relationship between the argument and the (sub) standpoint is based on symptomatic argumentation. In this type of argumentation, the argument can be regarded as an indication or a sign that something is the case. In other words, the judge's expression of a rhetorical question is regarded as a sign or an indication that the judge is not impartial. The general scheme for symptomatic argumentation (Van Eemeren et al. 2007, 154) may be applied to the argumentation in the example as follows.

Standpoint   Y is true of X (a lack of personal impartiality (Y) is true for judge X)
Because: Z is true of X (the use of RQ1 (Z) is true for judge X)
And: (the use of RQ1 (Z) indicates (personal) partiality (Y))

This scheme that may serve as the departing point for the evaluation of the argumentation, reveals the two premises the evaluation should focus on. The evaluation regarding the first premise would be whether an utterance of the judge could indeed be interpreted as a rhetorical question. The justification that is given in example (3) is confined to the quotation of the question that was asked by the judge during court hearing. This reference to the exact wording of the question is not unimportant because – as was originally the case here – the judge could have been misquoted. However, the quotation, which may also be considered as symptomatic argumentation in its own right, does not suffice as an argument in support of the conclusion that the judge's question may be interpreted as a rhetorical question.

In the first place, the argument wrongfully implies that the identification of a question as being a rhetorical question is unproblematic, although many studies on rhetorical questions (Frank, 1990; Ilie, 1994; Schmid-Radefeldt, 1977; Slot, 1993) in their search for clues as to the identification of rhetorical questions have demonstrated the opposite. Without further justification, an argument that consists of a quotation of the disputed question may be considered a fallacious discussion move. By creating the impression that the judge's question is deliberately misinterpreted, the strategic manoeuvring derails due to a violation of the Language use rule (Rule 10) that determines that 'discussants may not use any formulations that are insufficiently clear or confusingly ambiguous and they may not deliberately misinterpret the other party's formulations.'

In the second place, the argument as expressed in the first premise leaves implicit what indirect speech act the rhetorical question amounts to. The characteristics of the activity type of a disqualification procedure in which a rhetorical question is criticised, make it relevant to determine if the rhetorical question may be considered as an assertion that, in principle, could function as a standpoint. Given that in a legal procedure the discussion role of the judge may be regarded as that of an institutional antagonist, in a disqualification procedure it should be justified that the judge is committed to an assertion that might function as a standpoint.

The analysis of the complications concerning the first premise in the argumentation scheme indicates that the argumentation brought forward to justify that an utterance of the judge could indeed be interpreted as a rhetorical question, should be extended. The approach to the analysis of rhetorical questions by Snoeck Henkemans (2006) which is based on speech act theory, may serve as a starting point for this extension. In this approach, a rhetorical question can be seen as indirect speech act because if taken literally, it violates two of the rules for communication. In the first place, the addresser already knows the answer, so the question is superfluous. Secondly, the question is insincere, since the addresser does not expect to get an answer from the addressee. Based on these criteria, a proposal for the extension (in italics) of the argumentation concerning the first proposition in the scheme can be modelled as follows.

>1.1.1.1.1 Judge (X) used rhetorical question RQ1
>>1.1.1.1.1a Judge (X) expressed question (Q1)
>>1.1.1.1.1b Q1 should be interpreted as assertion (A)
>>>*1.1.1.1.1b.1a Q1 is superfluous when taken literally*
>>>*1.1.1.1.1b.1b Q1 is insincere when taken literally*
>>>*1.1.1.1.1b.1c Interpreting Q1 as assertion (A) rectifies the superfluity and the insincerity of Q1 when taken literally*

The evaluation of the second premise pertains to the justificatory force of the argument that is expressed in (1.1.1.1.1'): The judge's rhetorical question indicates a lack of personal impartiality. Here, the critical questions associated with the argumentation scheme, as developed in pragma-dialectics (Van Eemeren, 2010), focus on the symptomatic relationship on which the argumentation is based. The question that is relevant, is whether bringing forward a rhetorical question is indeed indicative for not being impartial. When taking into account the institutional characteristics of a legal procedure, bringing forward a rhetorical question does not automatically indicate a lack of personal impartiality. However, if the rhetorical question could be regarded as a (negative) standpoint concerning the case at hand, this rhetorical question could indeed indicate a lack of personal impartiality on the part of the judge. Therefore, in anticipation of the critical question concerning the symptomatic relation between the argument and the (sub) standpoint, the argumentation could be extended as follows.

(1.1.1.1.1') RQ1 (Z) indicates a lack of personal impartiality (Y)
 (1.1.1.1.1'.1a) RQ1 (Z) functions as an assertion that conveys a (negative) standpoint of the judge (X)
 (1.1.1.1.1'.1b) The judge's standpoint concerns the case the judge is already acting on.
 (1.1.1.1'.1a/b') An assertion by the judge that conveys a (negative) standpoint concerning the case he is acting on proves a lack of (personal) impartiality (Y)

This argumentative overview provides a model for the analysis of the argumentation in a disqualification request concerning complaints on the judge's use of rhetorical questions. The model aims to provide more insight into the scope of the arguments that may be relevant for the interpretative decisions concerning these complaints. These insights may therefore be of help for the presentation as well as for the evaluation of the argumentation underlying this type of complaints pertaining to the judge's verbal behaviour.

4. CONCLUSION

In this contribution I set out to demonstrate that a disqualification procedure may be regarded as a sub-discussion if viewed from a pragma-dialectical perspective, which considers a legal procedure as an argumentative discussion. Due to institutional constraints that govern the justification of decisions in a disqualification procedure, such a procedure may be regarded as a specific argumentative activity type. Based on institutional sources (i.e. legislation and case law) I reconstructed the general argumentation structure of the justification in this activity type in cases in which the litigant's criticism is directed at the judge's use of rhetorical questions. In pragma-dialectical terms, this criticism may be considered as an accusation of a derailment of strategic manoeuvring on the part of the judge: by making use of a rhetorical question that indicates partiality, the judge may be accused of excluding the (argumentation of) the complaining litigant from the discussion.

 However, if the accusation of a derailment of strategic manoeuvring is based on no more than a quote of the judge's rhetorical question, this accusation, in itself, could be considered as a derailment of strategic manoeuvring on the part of the litigant who challenges the judge or of the disqualification panel. If the interpretation of a judge's question as a rhetorical question that indicates partiality is not supported by arguments that justify this interpretation, the arguer himself is under the suspicion of abusing the ambiguity of the question form and, thus, evading the burden of proof.

Based on speech act theory, I presented an extended model that anticipates critical questions regarding complaints about the use of (alleged) rhetorical questions as part of symptomatic argumentation underlying disqualification requests. This model may serve as a starting point for the analysis and evaluation of the justification of (decisions on) these requests.

REFERENCES

Danet, B. (1980). Language in the legal process. *Law & Society Review*, *14*(3), 445-564.
Eemeren, F. H. van (2010). *Strategic maneuvering in argumentative discourse. Extending the pragma-dialectical theory of argumentation*. Amsterdam/Philadelphia: John Benjamins.
Eemeren, F. H., & R. Grootendorst (1992). *Argumentation, communication and fallacies: A pragma dialectical perspective*. Hillsdale, NJ: Lawrence Erlbaum Associates.
Eemeren, F. H. van, P. Houtlosser & A. F. Snoeck Henkemans (2007). *Argumentative indicators in discourse. A pragma-dialectical study*. Dordrecht: Springer.
Frank, J. (1990). You call that a rhetorical question? Forms and functions of rhetorical questions in conversations. *Journal of Pragmatics*, *14*(5), 723-738.
Hammerstein, A. (2014). 'Onpartijdigheid in het geding', *Trema*, *26*(5), 148-154.
Ilie, C. (1994). *What else can I tell you? A pragmatic study of English rhetorical questions as discursive and argumentative acts* (doctoral dissertation). University of Stockholm.
Ilie, C. (1995). The validity of rhetorical questions as arguments in the courtroom. In F. H. van Eemeren, R. Grootendorst, J. A. Blair. & C. A. Willard, (Eds.), *Proceedings of the Third ISSA Conference on Argumentation: Special fields and cases* (Vol. 5, pp. 73-88). Amsterdam: Sic Sat.
Knapen, M. (2012). 'Advocaten ontdekken wraking', *Advocatenblad*, 1, 18-23.
Lanham, R. A. (1991). *A handlist of rhetorical terms*. Berkeley: University of California Press.
Pascual, E. (2006). Questions in legal monologues: Fictive interaction as argumentative strategy in a murder trial. *Text & Talk*, *26*(3), p.10-21.
Plug, H. J. (2002). Maximally argumentative analysis of judicial argumentation. In: F. H. van Eemeren (Ed.) *Advances in pragma-dialectics*. Amsterdam: Sic Sat / Newport News, Virginia: Vale Press, p. 261-270.

Plug, H. J. (2016). Administrative judicial decisions as a hybrid argumentative activity type. *Informal logic*, *36*(3), 333-348.

Plug, H. J. (accepted). The analysis of argumentation underlying complaints about a lack of judicial impartiality.

Sala, M. (2010). Interrogative forms as professional identity markers in legal research articles. In G. Garzone & J. Archibald, (Eds.), *Discourse, identities and roles in specialized communication* (pp. 301- 320).Bern: Peter Lang

Schmid-Radefeldt, J. (1977). On so-called 'rhetorical questions'. *Journal of Pragmatics*, *1*(4), 375–392.

Slot, P. (1993). *How can you say that? Rhetorical questions in argumentative texts*. Amsterdam: Ifott.

Snoeck Henkemans, A. F. (2006). Manoeuvring strategically with rhetorical questions. In F. H. van Eemeren, J. A. Blair, C. A. Willard and B. Garssen (Eds.), *Proceedings of the Sixth Conference of the International Society for the Study of Argumentation* (ISSA)(pp. 1-11). Amsterdam: Rozenberg.

Tiersma, P. M. (2000). *Legal language*. Chicago: University of Chicago Press.

Tracy, K. (2016). *Discourse, identity and social change in the marriage equality debates*. New York: Oxford university press.

45

The "Neoliberal Agenda": How Portuguese Parties Use The "Neoliberalism" Concept to Argue Against Austerity

VERA RAMALHETE
IPRI-NOVA
vramalhete@fcsh.unl.pt

MARCO LISI
FCSH/IPRI-NOVA
mlisi@fcsh.unl.pt

To fully understand the recent economic crisis, it is necessary to observe how political actors have framed the narratives and the political discourse. This paper focuses on the Portuguese case by analysing the use of "neoliberalism" in argumentation, namely in the discourse against austerity. Drawing on parliamentary debates between 2009 and 2017, this study aims to examine the frame in which "neoliberalism" is used, particularly by left-wing parties, unveiling the strategic component behind this rhetoric tool.

KEYWORDS: crisis, neoliberalism, political parties, rhetoric

1. INTRODUCTION

In 2011, Portugal signed a memorandum agreement with three international institutions – the so-called "troika" – i.e. European Commission (EC), European Central Bank (ECB) and International Monetary Fund (IMF) – committing to implement fiscal consolidation policies and structural reforms, in exchange for a bailout loan. The memorandum agreement followed several packages of austerity measures implemented by the Socialist Party (PS) government, as a response to the financial crisis. The 2011 legislative elections marked the defeat of the socialist government and the victory of the neoliberal right (Freire & Pereira, 2012). The coalition government between PSD (Social Democratic Party) and CDS-PP (Social Democratic Centre-

Popular Party) argued that Portugal had been living "above its possibilities" and started an "adjustment period".

Austerity measures were needed due to a "state of exception", caused by "the irresponsibility of the past", the right-wing parties argued (Fonseca & Ferreira, 2015). The "discourse surrounding reforms was one of 'necessity', the 'lack of alternative' to gain 'credibility'" (Moury & Standring, 2017, p. 27). Not only was the right thing to do, it was the only possible way to "save the country from a disaster" – "there is no alternative" (TINA), alerted PSD and CDS-PP leaders (Fonseca & Ferreira, 2016, 2015).

Started by PS centre-left government (2009-2011), that approved several packages of austerity measures with the support of right-wing PSD, which deepened these reforms and went "beyond troika's programme" (Fonseca & Ferreira, 2016, 2015; Moury & Standring, 2017). In this context, how did the opposition respond? What strategies did left-wing parties use to refute the "lack of alternative" speech?

Portugal, one of the European countries most severely affected by the crisis, is an important case study. The international context, along with internal factors, originated a severe economic, financial, political and social crisis, initiating almost a decade of austerity measures. With the proliferation of the TINA discourse in the public debate and the media, with painful austerity measures presented as inevitable in the European Union (EU) (Queiroz, 2016), it is important to observe the use of the term "neoliberal" in the political debate.

Drawing from different contributions from argumentation theory, rhetoric studies and political discourse analysis, we analyse examples of speeches in the national parliament using the "neoliberalism" concept, looking at the arguments made by MPs. We apply content analysis to parliamentary debates during plenary sessions, between October 2009 and March 2017, to observe the use of this concept by Portuguese parties. To complete this approach, we analyse some speech examples in more detail, in order to better evaluate the strategic use of the term "neoliberalism" by political parties.

We argue that "neoliberalism" was used by Portuguese parties – mostly the left-wing ones – to argue against austerity during the crisis. Left-wing Portuguese parties used the term "neoliberal" to highlight the ideological choice of the policies being pursued by the government and supported by right-wing parties and the EU institutions.

Left-wing Portuguese parties used practical reasoning and arguments from values and from negative consequences (Fairclough &

Fairclough, 2013; Hansen & Walton, 2013; Walton & Macagno, 2015) to argue against austerity policies and to advocate for policy changes. Furthermore, the imprecise use of the concept allowed the parties to adapt it to different circumstances. Parties try to redirect and intensify attitudes of the audience against austerity policies through the use of persuasive definitions (Stevenson, 1938), emotive meaning and by appealing against the values the word "neoliberal" implies.

2. CRISIS, NEOLIBERALISM AND POLITICAL DISCOURSE

2.1 Crisis

The story of the recent crisis is known and gathers a large consensus (Bermeo & Pontusson, 2012; Morlino & Quaranta, 2016). The bankruptcy of the American bank Lehman Brothers, connected to the subprime crisis, marked the beginning of the period that has become known as the Great Recession. Starting in the USA, soon the crisis spread to Europe. The economic turmoil begun as a bank crisis, but it triggered "a crisis of public debt, then a crisis of investment; a related social crisis then affected the European periphery" (della Porta et al., 2016, p. 8).

The extent of the consequences of the Great Recession turned the crisis into a particular relevant subject, motivating studies from different angles, and grounded in different theories and methodologies. Political studies have been trying to observe the impact of the crisis in political representation, electoral outcomes and democratic consolidation, as well as reconstructing the context of political and policy decisions during the periods of external intervention (Cordero & Simón, 2016; Freire & Lisi, 2016; Freire, Lisi, Andreadis, & Viegas, 2014; Hernández & Kriesi, 2016; Kriesi, 2014; Moury & Freire, 2013; Moury & Giorgi, 2015; Moury & Standring, 2017; Veebel & Kulu, 2014).

The response to the crisis varied across countries but started in an initial stage by favouring countercyclical policies. With growing concerns around Greek debt, a new "consolidation stage" emerged after 2009, marked by cuts in spending and austerity policies – and a "neoliberal approach". According to a number of scholars, the Great Recession can be interpreted as the stage of late neoliberalism, that is, a crisis of the turn towards the free market within a Polanyi-like double movement in capitalist development (della Porta et al., 2016, p. 8).

The economic crisis also generated a crisis of legitimacy and responsibility, greatly affecting political support, as more and more groups in society felt themselves non-represented within institutions that were increasingly considered as captured by big business. (...) The effect has been a dramatic acceleration of trends towards declining party membership, loyalty and identification as well as decreases in conventional forms of participation and (especially) institutional trust (della Porta et al., 2016, p. 8).

In this context, parties and governments needed to respond, namely by trying to blame external actors for the crisis (Cordero & Simón, 2016; Vasilopoulou, Halikiopoulou, & Exadaktylos, 2014). Citizens dissatisfaction with the European Union – although support for the euro inside the euro area appeared to remain high (Hobolt & Wratil, 2015) – further stimulated the strategy of attacking EU institutions and EU governs.

Governments need to justify their choices vis-à-vis voters, while incumbents have been challenged by non-mainstream parties for popular support. The Great Recession may have not only eroded the support of incumbent parties, but of all the mainstream parties (Hernández & Kriesi, 2016, p. 204). For the left-wing parties, this context accentuated the need to attack "neoliberals". Left mainstream parties had to differentiate themselves from the incumbent and the "neoliberal agendas", which encouraged attacks by radical left parties towards mainstream parties, by giving them an opportunity to gain more support.

2.2 Neoliberalism

The crises "have revived interest in studying the emergence, consolidation, crisis, and resilience of neoliberalism" (Ban, 2016, p. 8). The definition of "neoliberalism" however is problematic (Ban, 2016; Peck, 2013; Venugopal, 2015), encompassing different phenomena throughout the times and spaces (Saad-Filho & Johnston, 2005).

Neoliberal theories depart from classical liberal and neoclassical economic theories originally based on the Austrian school, Chicago school and German ordoliberalism theories. Later, since the 1970s, the concept of neoliberalism changes and builds more on monetarism, supply-side, rational expectations and public choice theories, which advocate a greater role for financial markets, trade and financial openness. Governments must pursue fiscal discipline; and public finances and economic policies in general should be benchmarked by market credibility.

The rise of neoliberalism is also marked by the transformation of "labor market accords, industrial relations systems, redistributive tax structures and social welfare programs" (Campbell & Pedersen, 2001). It is also usually characterised by "reduced state intervention into economic affairs" (Campbell & Pedersen, 2001), but this definition "sits uneasily with the extensive interventions of governments to correct and nudge the markets" (Ban, 2016).

The relation between neoliberalism and globalisation is also complex. Some authors conceptualize globalisation as an extension and consequence of neoliberalism (Litonjua, 2008; Saad-Filho & Johnston, 2005); other associate the effects of globalisation to the spread of neoliberal politics. Hanspeter Kriesi (Kriesi et al., 2006, 2008, Kriesi et al., 2012) argued that globalisation (or denationalisation) transformed the national political space in Western European countries by creating a new "integration-demarcation cleavage" opposing the winners and losers of globalisation.

From this viewpoint, the implementation of neoliberal policies has been associated to the politicisation of the cultural dimension – that gain importance with the globalisation phenomenon – dividing a cosmopolitan and nationalist position (Kriesi et al., 2012). Combined with the traditional economic dimension, this new cleavage creates four major coalition groups: the "neoliberal-cosmopolitan" (right-wing parties and business interests) and "neoliberal-nationalist" (right-wing populists) on the neoliberal side, opposing to the left-wing parties and unions on the "interventionist-cosmopolitan" and the "interventionist-nationalist" coalitions (Kriesi et al., 2012, pp. 20–22). Although this new division has reshaped the political space in many West European countries, it is worth noting that Portugal has remained immune to the emergence of new populist actors, while the party system has displayed a remarkable stability.

The concept of neoliberalism is thus frequently "incoherent" (Venugopal, 2015, p. 166), "ill-defined" (Mudge, 2008, p. 703), and "confusing" (Turner, 2008, p. 2). Most important, is an "insult word" (Ban, 2016, p. 8), and is never used to identifying oneself.

> Instead, neoliberalism is defined, conceptualized and deployed exclusively by those who stand in evident opposition to it, such that the act of using the word has the twofold effect of identifying oneself as non-neoliberal, and of passing negative moral judgment over it. Consequently, neoliberalism often features, even in sober academic tracts, in the rhetorical toolkit of caricature and dismissal, rather than of analysis and deliberation (Venugopal, 2015, p. 179).

Neoliberalism definition therefore carries "an emotive judgement", frames the reality and contains an implicit argument from values – "the process of reasoning leading the interlocutor to consider a certain state of affairs as desirable and consequently as a reason of action" (Macagno & Walton, 2008, p. 539). The ambiguity of the word can be used strategically, to adapt to the situation, without the need of discuss the new implicit redefinitions (Macagno & Damele, 2013, 2015).

In this sense, politicians – "who learn to be nimble in the use of emotive language in order to gain advantages over their opponents" (Macagno & Walton, 2014, p. 4) – can use the word "neoliberal" "to modify the emotive meaning denotation of a persuasive term in a way that contains an implicit argument from values" (Macagno & Walton, 2008, p. 525). To redirect the audience perception, politicians can give a new conceptual meaning to the word without changing its emotive meaning ("persuasive definition") or change the emotive meaning without altering its descriptive meaning ("quasi-definition") (Stevenson, 1938, 1944).

2.3 Discourse

To fully understand what happened, we also need to observe what has been said during the crisis – how politicians framed the events and the policies chosen, which arguments were used. The analysis of political discourse has been a useful instrument contributing to a better comprehension of the crisis (Fairclough & Fairclough, 2013; Fonseca & Ferreira, 2015; Moury & Standring, 2017; Tekin, 2014). As Fairclough (2005, p. 23) put it, "in dealing with neo-liberalism we are dealing centrally with questions of discourse".

Political discourse is fundamentally argumentative and with practical and strategical objectives. In Fairclough and Fairclough words,

> politics is most fundamentally about making choices about how to act in response to circumstances and goals, it is about choosing policies, and such choices and the actions which follow from them are based upon practical argumentation (Fairclough & Fairclough, 2013, p. 1).

In the next section, we analyse the political discourse about neoliberalism in the Portuguese Parliament during the crisis.

3. ANALYSIS

To analyse how Portuguese parties use the word "neoliberalism", what do they mean and what is their political intention when employing the term, we adopt a content analysis approach. First, we start by conducting a word count with the terms "neoliberal", "neo-liberal", "neoliberals", "neo-liberals", "neoliberalism" and "neo-liberalism" in the transcripts of the plenary sessions from parliamentary debates in the democratic regime. The term "ultraliberal" and its derivatives were later added to the research, since we observed that it occurred in a related matter and similar meaning, with some significant expression[1].

We analyse these occurrences in parliamentary debates between October 2009 (the beginning of XI legislature) and March 2017 (the latest transcripts available at the date when we performed the empirical analysis)[2]. For each occurrence, we observe the expression and concept of neoliberalism used, as well as the author (party and MP or member of the government), the target, the theme and context. This analysis allows us to choose pertinent examples to observe the argument in more detail using argumentation theory lenses.

This time frame allows us to compare the period before, during and after troika's intervention in Portugal, and to observe differences in parties' attacks with different incumbent parties. This period covers three different governments: a minority centre-left (PS) government from 2009-2011, a right wing (PSD/CDS-PP coalition) government (2011-2015) and a left government (PS government with parliamentary support from radical left wing parties BE, PCP and PEV, from 2015 onwards).

The term "neoliberalism" (and its derivatives, such as neoliberal or neoliberals) was used 1527 times by Portuguese parties at parliamentary debates in the democratic regime. Since 1975, the word was used in 774 plenary sessions at the Portuguese parliament. Figure 1 shows the growing yearly use of the word neoliberalism, which peaks in 2009. Privatisations, changes in the labour legislation, and cuts in pensions and wages marked the discussion that year, with the radical left (BE, PCP, PEV) accusing PS of being ambivalent about neoliberalism – proclaiming anti-neoliberal sentences but conducting neoliberal policies.

---

[1] 42 times in a total of 328. Other neoliberalism related concepts, such as "ordoliberalism", were not used in these debates.

[2] The transcripts are available at Portuguese parliament website: http://debates.parlamento.pt/ .

Figure 1 – Word count for "neoliberalism" in parliamentary debates, by year
Source: Portuguese Parliament

During the period covered by this study, however, the use of the word registers a descendent tendency, when we look at the results per year. The substantive neoliberalism and the adjective neoliberal are referred 328 times between 2009 and 2017, in 190 parliamentary debates, which represents 24% of the plenary sessions that have taken place during this time.

Table 1 shows the use of the word "neoliberalism" by legislature (word count with total number of occurrences, absolute and relative number of debates in which the word is referred and average number of debates with the presence of the word per legislative session). As expected, the period corresponding to the right-wing coalition government, between 2011 and 2015 – the only complete legislature in the case in analysis –, registers higher frequency of "neoliberalism". 23% of the debates had a discussion about neoliberalism. Neoliberalism was a topic of discussion, on average, in 27 debates per legislative session. A number slightly higher than the average per session in the legislature before, that preceded the bailout; and significantly higher than the average during the next legislature (14.5). The legislature before the "memorandum" was also rich in the use of the term "neoliberalism", with 33% of the debates during this two-year government with this word. On average, we can find mentions of neoliberalism in 26 debates per session.

Legislature	Word count	Debates	% Debates	Average of debates per legislative session
XI (2009-2011)	112	52	33%	26
XII (2011-2015)	175	109	23%	27.3
XII (2015-2017)	41	29	19%	14.5
Total (2009-2017)	328	190	24%	23.8

Table 1 – Use of the term "neoliberalism" concept, by legislature. Word count: number of times the word "neoliberalism" and related expressions are used in the parliamentary debates; Debates: number of debates with the expression "neoliberalism"; % Debates: percentage of debates that have the expression "neoliberalism" regarding the total number of debates in the period under study; Average of debates per legislative session: average number of debates with "neoliberalism" by legislative session. Source: Portuguese Parliament

The legislature after troika's intervention and supported by all left-wing parties in parliament – only partially analysed here in this work given the dates –, still has a relative high use (19% of the debates; 14.5 debates on average by legislative session), as shown in table 1. This is due to the fact that left-wing parties (including incumbent PS) continue to use this term to attack the previous coalition government's actions and ideas in order to defend their own programme. "Neoliberalism" is still used by left-wing parties to argue against austerity, but this time parties use this term to justify the policies that are being implemented and the need to reverse the prior "neoliberal agenda".

Furthermore, the EU is a constant target, during all the period analysed here. The "neoliberal path/course" that EU has undertaken, or the "neoliberal Europe" are common expressions used by left-wing MPs. In fact, the EU, PS, PSD and CDS-PP, alone or combined, are the ones accused of being "neoliberals" or following the "neoliberal model" – the second most common expression, along with the "neoliberal agenda", and after the expression "neoliberal politics". Although residual, other "neoliberal" actors attacked are regulatory authorities, especially Portuguese Central Bank, financial markets and speculators.

PSD is the most attacked, followed by its right-wing coalition partner CDS-PP. PS is both a target and the author of these attacks. PCP and BE, radical left parties, are never accused of being neoliberals.

Indeed, neoliberalism is a term used almost exclusively by the left. PSD and CDS-PP only use these terms to defend themselves from accusations of being neoliberals or conducting neoliberal policies[3].

The main argument used in the Portuguese parliament by the left-wing parties is that Portugal needs an alternative, because the government strategy is wrong as it is based in the wrong goals and values – "to apply a neoliberal agenda". Despite the government's claim that "there is no alternative", opposition argues that there is in fact an alternative way, but the government (and/or the right-wing parties) does not (want to) see it, because "it does not fit their ideological prejudices" (Lopes, 2011) and due to its "neoliberal blindness" (Ferreira, 2012, 2014, 2015).

It is this blindness that prevents the right-wing parties to admit that their strategy is causing damages. In this sense, and to reinforce this claim, the opposition systematically presents the negative consequences that neoliberal policies are causing. This is particularly efficient, since "criticism that appeals to negative consequences that have already arisen (...) can conclusively falsify a claim" (Fairclough & Fairclough, 2013, p. 160)

PCP is the party that uses this concept more often – the party accounts for 40% of the use of the word "neoliberalism" (table 2). CDU, the electoral coalition composed by PCP and PEV, accounts for half (51%) of the occurrences, mainly in the sense described above. Using metaphors and opposing "the neoliberals" to "the people" – in a way characteristic of a populist discourse (Mudde, 2004; Rooduijn & Pauwels, 2011) – PCP argues for a political alternative, as can be seen in example 1. Example 2, an excerpt from a discourse about the crisis given by a PCP's MP during 25th of April celebrations, illustrates the use of metaphors by the communists.

(1) Portugal needs a political alternative that breaks the decline cycle (...) that the bankers and the so called

---

[3] There is an exception in 2011, few weeks before the government withdrawal, where PSD MP Pedro Duarte accuses the socialist government of having "a hidden neoliberal agenda that aims at the dismissal of many teachers in Portugal" (Duarte, 2011). This illustrates how PS was being perceived and characterised at the time, when even the right-wing parties were accusing the socialists of being neoliberals.

financial markets want to impose to our country and to our people. (Novo, 2010a)

(2) The crisis appears as the perfect opportunity for neoliberal butchers of all kinds to shave off some fat, but a lot of tenderloin gets cut as well. (Lopes, 2012b)

After PCP, the socialists come second in terms of the use of the concept, surpassing BE, as shown in table 2. Although this is slightly influenced by the use of the term to refute opposition's accusations of being a neoliberal government in the XI legislature (2009-2011), in the following periods PS uses the same argumentation strategy as the radical left. In the XIII legislature (from 2015 onwards), when it is the incumbent party, it is even tide with PCP (plus PEV).

	BE	CDS-PP	PCP	PEV	PS	PSD	Total
2009-2011	21%	1%	45%	8%	20%	5%	100%
2011-2015	10%	3%	38%	14%	31%	5%	100%
2015-2017	10%	3%	35%	8%	40%	5%	100%
Total	14%	2%	40%	11%	28%	5%	100%

Table 2 – Use of the concept divided by parties and legislatures
Source: Portuguese Parliament

In the case of PS, besides being used to call for a political alternative, this attack is also used to justify a political strategy of ceasing to support the government. In a speech delivered by Carlos Zorrinho, in November 2012, PS's parliamentary leader says that the government has neoliberalism in their nature – it "runs in their blood" – and that it does not care about the "tragic results from its policies" (Zorrinho, 2012). This way, PS tries to justify its intention to stop the support that it had previously been giving to the government, allowing for the application of the memorandum, while trying to avoid accusations of being irresponsible.

In the following legislature, António Costa improves this strategy. The Prime-Minister and PS's leader accuses the right-wing parties of having "exhausted their Christian democratic and social Cristian roots and converted to a very clear neoliberalism" (Costa,

2016). It is this "radical neoliberalism", this "radical derivative, from an ideological point of view, from the Portuguese right", clearly identified with PSD leader, Pedro Passos Coelho, that is "preventing large consensus in Portugal". That is why PS needs to govern with the support of the left-wing parties and not from PSD, because the party is not in the centre anymore – that is the implicit argument made by the Prime-Minister.

The term "neoliberalism" is used in a similar way by PS, PCP and BE. For the Portuguese left, neoliberalism is an ideology within capitalism, the opposite from socialism and associated to a conservative view. It favours less state and privileges the financial markets and the capital, against the people and citizen's interests – especially the weakest segments of society. In concrete, it means an "obsession with privatisations", a threat to both the public health and education systems and deregulation of labour market. It is the "finance dictatorship", obsessed with the debt and deficit.

However, for both BE and PCP this term has an even broader and more critical sense: it is contrary not only to socialism, but also to democracy and modernity. Moreover, PCP clearly highlights its association with globalisation, imperialism and loss of sovereignty and national interest.

Neoliberalism	PCP	BE	PS
Against the state (and welfare state); privatisations	x	x	x
Labour market deregulation	x	x	x
Against the workers, the citizens and the people	x	x	
Favours the capital and the markets	x	x	
"Finance dictatorship"; debt and deficit supremacy	x	x	
Capitalism	x	x	
Globalisation	x		
Imperialism	x		
Anti-democratic	x	x	
Anti-socialist	x	x	x
Conservative	x	x	x
Austerity	x	x	x

Table 3 – Use of the concept divided by parties and legislatures
Source: Portuguese Parliament

Even more importantly, the concept is associated with austerity. Left-wing parties argued that right-wing parties used the crisis as an excuse to impose its "neoliberal plan". From this viewpoint, austerity is a consequence of the neoliberal agenda and a component of neoliberalism.

PCP is the party that uses the concept of "neoliberalism" with a broader meaning – as shown in table 3 –, but PS is the party that uses the concept in a more flexible way, depending on the context and its political position.

This diffuse use of the concept by PS to suit its goals is highlighted by the opposition in the 2009-2011 legislature, where the government, while applying "neoliberal" policies, accuses the EU of being neoliberal, by trying to force Portugal to pursue inappropriate solutions for the crisis. "Does it make sense to criticise the European Commission of being ultraliberal and then practice policies of the same nature domestically?", asks a BE MP (Fazenda, 2010). Another MP, this time from PCP, points to the fact that the Prime Minister "does not think that is neoliberalism" to privatise state companies (Novo, 2010b), since he will go forward with the privatisation despite having attacked EU's "neoliberal perspectives" just the week before.

Right-wing parties also highlight the use of the concept of neoliberalism as a persuasive definition. PSD and CDS-PP say that the authors never explain what do they mean or even accuse them of not knowing what neoliberalism is – for instance in this sentence where right-wing CDS-PP's MP Diogo Feio says that "there are even accusations of the arrival of the "neoliberals" – but is never explained what that means" (Feio, 2009).

The word "neoliberalism" is often used by parties without defining it, but classifying a fragment of reality and implying a system of values, that indicates that they are right and the opponents are wrong. It is an effective and common argumentation manoeuvre of using emotive language (Macagno & Walton, 2014, p. 2).

4. CONCLUSION

This paper contributes to the study of the arguments and discourse used during the crisis, by drawing on the Portuguese case to examine how the term "neoliberalism" is used by political parties and leaders to justify and orientate their strategy and programmatic platforms. Our research strategy was twofold. First, we discuss and elaborate a framework for the analysis of the "neoliberalism" concept. Second, our empirical analysis focuses on parliamentary debates in Portugal from 2009 to 2017.

Left-wing Portuguese parties used the concept of neoliberalism to argue against austerity and respond to the non-alternative discourse dominant during the crisis. Left-wing parties use this concept to argue that the right-wing parties and the EU are pursuing the wrong goals, based on "wrong" (neoliberal) values. They say that the radical neoliberalism is preventing the government to see the consequences of austerity policies, namely the social gains achieved after democratisation and European integration. Austerity is a consequence of the neoliberal agenda and a component of neoliberalism. Therefore, one most apply an alternative policy, the opposition says, using practical reasoning.

The attacks against "neoliberalism" exist throughout all the period analysed (2009-2017). Left-wing Portuguese parties used these attacks to argue against cuts in the public spending, against privatisations, and austerity in general, before and during troika's memorandum period. After the end of troika's programme and even during the PS government supported by radical left parties that started in 2015, MPs continued using the concept of neoliberalism to attack past decisions and consequently arguing for the need of reversing that measures, as well as to attack European institutions and policies.

The word "neoliberalism" is often used in a vague, imprecise way that allows the parties to strategically use it to attack a set of different situations. This is particular useful for PS, that makes a more flexible and strategic use for the term, adapting the meaning of the word to its argumentation and political goals. This persuasive definition carries "an emotive judgement" and contains an implicit argument from values.

Unsurprisingly, PCP is the party that uses more this concept of neoliberalism, to attack all the mainstream parties and the EU. The party applies additional rhetorical strategies, such as the use of metaphors to illustrate the effects of the "neoliberalism". Also interesting is the use of a "populist rhetoric" opposing the neoliberals (the banks, the capitalists, the European Union, Germany) to "the people" and "the country". However, this is a topic that deserves to be explored in future research.

The findings in this exploratory study can be further developed in several ways. Parliamentary discourse has a particular meaning for democratic representation, but it would be interesting to compare the use of the term neoliberalism in other contexts, for instance, electoral campaigning discourses. It is plausible to expect a more intense use of this concept during electoral contests, as it is a rhetoric tool that can be easily employed to improve parties' performance. This can be one explanation for the high level of the use of "neoliberalism" in 2009, a year with three elections. Moreover, it would be also interesting to examine how this term is used in different documents, such as party manifestos.

ACKNOWLEDGEMENTS: This work was supported by Portuguese National Funds through the Fundação para a Ciência e a Tecnologia (FCT) in the framework of the project "PTDC/IVC-CPO/3098/2014". The authors would like to thank Giovanni Damele for his comments and suggestions.

REFERENCES

Ban, C. (2016). *Ruling ideas: How global neoliberalism goes local*. Oxford: Oxford University Press.
Bermeo, N., & Pontusson, J. (Eds.). (2012). *Coping with crisis*. New York: Russell Sage Foundation.
Campbell, J. L., & Pedersen, O. K. (2001). *The rise of neoliberalism and institutional analysis*. Princeton: Princeton University Press.
Cordero, G., & Simón, P. (2016). Economic crisis and support for democracy in Europe. *West European Politics*, *39*(2), 305–325.
Costa, A. (2016). *Reunião Plenária de 13 de Maio de 2016*. Lisboa: Diário da Assembleia da República.
Duarte, P. (2011). *Reunião Plenária de 3 de Março de 2011*. Lisboa: Diário da Assembleia da República.
Fairclough, I., & Fairclough, N. (2013). *Political discourse analysis: A method for advanced students*. New York: Routledge.
Fairclough, N. (2005). Neo-liberalism - A discourse analytical perspective. *Proceedings of Conference on British and American Studies* (pp. 1–18). Brno: Masarykova univerzita.
Fazenda, L. (2010). *Reunião Plenária de 15 de Julho de 2010*. Lisboa: Diário da Assembleia da República.
Feio, D. (2009). *Reunião Plenária de 5 de Março de 2009*. Lisboa: Diário da Assembleia da República.
Ferreira, J. L. (2012). *Reunião Plenária de 19 de Outubro de 2012*. Lisboa: Diário da Assembleia da República.
Ferreira, J. L. (2014). *Reunião Plenária de 22 de Outubro de 2014*. Lisboa: Diário da Assembleia da República.
Ferreira, J. L. (2015). *Reunião Plenária de 22 de Maio de 2015*. Lisboa: Diário da Assembleia da República.

Fonseca, P., & Ferreira, M. (2016). Paulo Portas e a legitimação discursiva das políticas de austeridade em Portugal. *Paulo Portas and the Discursive Legitimation of Austerity Policies in Portugal*, *51*(221), 886–921.

Fonseca, P., & Ferreira, M. J. (2015). Through 'seas never before sailed': Portuguese government discursive legitimation strategies in a context of financial crisis. *Discourse & Society*, *26*(6), 682–711.

Freire, A., & Lisi, M. (2016). Introduction. Political parties, citizens and the economic crisis: The Evolution of southern European democracies. *The Portuguese Journal of Social Science*, *15*(2), 173-193.

Freire, A., Lisi, M., Andreadis, I., & Viegas, J. M. L. (2014). Political representation in bailed-out southern Europe: Greece and Portugal Compared. *South European Society and Politics*, *19*(4), 413–433.

Freire, A., & Pereira, J. S. (2012). Portugal, 2011: The victory of the neoliberal right, the defeat of the left. *The Portuguese Journal of Social Science*, *11*(2), 179-187.

Hansen, H. V., & Walton, D. (2013). *Argument kinds and argument roles in the Ontario provincial election*. Rochester, NY: Social Science Research Network.

Hernández, E., & Kriesi, H. (2016). The electoral consequences of the financial and economic crisis in Europe. *European Journal of Political Research*, *55*(2), 203-224.

Hobolt, S. B., & Wratil, C. (2015). Public opinion and the crisis: the dynamics of support for the euro. *Journal of European Public Policy*, *22*(2), 238-256.

Kriesi, H. (2014). The Populist Challenge. *West European Politics*, *37*(2), 361-378.

Kriesi, H., Grande, E., Dolezal, M., Helbling, D. M., Höglinger, P. D., Hutter, P. S., & Wüest, P. B. (2012). *Political conflict in Western Europe*. Cambridge; New York: Cambridge University Press.

Kriesi, H., Grande, E., Lachat, R., Dolezal, M., Bornschier, S., & Frey, T. (2006). Globalization and the transformation of the national political space: Six European countries compared. *European Journal of Political Research*, *45*(6), 921-956.

Kriesi, H., Grande, E., Lachat, R., Dolezal, M., Bornschier, S., & Frey, T. (2008). *West European politics in the age of globalization*. Cambridge, UK; New York: Cambridge University Press.

Litonjua, M. D. (2008). The socio-political construction of globalization. *International Review of Modern Sociology*, *34*(2), 253-278.

Lopes, A. (2012). *Reunião Plenária de 22 de Dezembro de 2011*. Lisboa: Diário da Assembleia da República.

Macagno, F., & Damele, G. (2013). The Dialogical Force of Implicit Premises. Presumptions in Enthymemes. *Informal Logic*, *33*(3), 361–389.

Macagno, F., & Damele, G. (2015). The hidden acts of definition. Definition in Law: Statutory Definitions. Definition and Burden of Persuasion. In *Logic in the theory and practice of lawmaking* (pp. 225–251). Cham, CH: Springer.

Macagno, F., & Walton, D. (2008). The argumentative structure of persuasive definitions. *Ethical Theory and Moral Practice*, *11*(5), 525-549.

Macagno, F., & Walton, D. (2014). *Emotive language in argumentation*. Cambridge, UK: Cambridge University Press.

Morlino, L., & Quaranta, M. (2016). What is the impact of the economic crisis on democracy? Evidence from Europe. *International Political Science Review*, *37*(5), 618-633.

Moury, C., & Freire, A. (2013). Austerity policies and politics: The case of Portugal. *Pôle Sud*, *39*(2), 35-56.

Moury, C., & Giorgi, E. D. (2015). Introduction: Conflict and consensus in parliament during the Economic Crisis. *The Journal of Legislative Studies*, *21*(1), 1-13.

Moury, C., & Standring, A. (2017). 'Going beyond the Troika': Power and discourse in Portuguese austerity politics. *European Journal of Political Research*, *56*(3), 660-679.

Mudde, C. (2004). The populist Zeitgeist. *Government and Opposition*, *39*(4), 541-563.

Mudge, S. L. (2008). What is neo-liberalism? *Socio-Economic Review*, *6*(4), 703-731.

Novo, H. (2010a). *Reunião Plenária de 2 de Novembro de 2010*. Lisboa: Diário da Assembleia da República.

Novo, H. (2010b). *Reunião Plenária de 9 de Julho de 2010*. Lisboa: Diário da Assembleia da República.

Peck, J. (2013). *Constructions of neoliberal reason* (Reprint edition). Oxford: Oxford University Press.

della Porta, D., Andretta, M., Fernandes, T., O'Connor, F., Romanos, E., & Vogiatzoglou, M. (2016). *Late neoliberalism and its discontents in the economic crisis: Comparing social movements in the European periphery*. Cham, CH: Springer.

Queiroz, R. (2016). Neoliberal TINA: an ideological and political subversion of liberalism. *Critical Policy Studies*, *0*(0), 1–20.

Rooduijn, M., & Pauwels, T. (2011). Measuring populism: Comparing two methods of content analysis. *West European Politics*, *34*(6), 1272-1283.

Saad-Filho, A., & Johnston, D. (Eds.). (2005). *Neoliberalism: A critical reader*. London; Ann Arbor, MI: Pluto Press.

Stevenson, C. (1944). *Ethics and language*. New York: American Mathematical Society Press.

Stevenson, C. L. (1938). Persuasive definitions. *Mind*, *47*(187), 331-350.

Tekin, B. Ç. (2014). Rethinking the post-nNational EU in times of austerity and crisis. *Mediterranean Politics*, *19*(1), 21-39.

Turner, R. S. (2008). *Neo-liberal ideology: History, concepts and policies*. Edinburgh: Edinburgh University Press.

Vasilopoulou, S., Halikiopoulou, D., & Exadaktylos, T. (2014). Greece in crisis: Austerity, populism and the politics of blame. *JCMS: Journal of Common Market Studies*, *52*(2), 388-402. https://doi.org/10.1111/jcms.12093

Veebel, V., & Kulu, L. (2014). Against the political expectations and theoretical models: How to implement austerity and not to lose political power. *Baltic Journal of Economics, 14*(1-2), 2-16.

Venugopal, R. (2015). Neoliberalism as concept. *Economy and Society, 44*(2), 165-187.

Walton, D., & Macagno, F. (2015). A classification system for argumentation schemes. *Argument & Computation, 6*(3), 219-245.

Zorrinho, C. (2012). *Reunião Plenária de 13 de Setembro de 2012*. Lisboa: Diário da Assembleia da República.

46

Effect of Explicitness of Teachers' Arguments on Quality of Adolescent Students' Inferences in Science and History

CHRYSI RAPANTA
Universidade Nova de Lisboa
crapanta@fcsh.unl.pt

The study has a double focus: a) the dialogical quality of classroom natural discourse; and b) the argumentative structure of inferences hidden in such dialogues. The analysis proposed combines Walton's dialogue types, to distinguish between different pedagogical goals, and Toulmin's Argument Pattern, for the identification of argument elements. The study concludes with considerations regarding how teachers' reasoning preferences for certain dialogue moves influence the manifestation of data, backings, warrants, and rebuttals in young adolescents' inferences.

KEYWORDS: teacher-student interaction, adolescents, Walton's dialogue types, Toulmin Argument Pattern

1. INTRODUCTION

For many scholars (Kuhn, 1991; Kuhn, Katz, & Dean, 2004; Mercier & Sperber, 2011; Hornikx & Hahn, 2012) argumentation is the core of reasoning, and more precisely, of what has been called "informal" reasoning, in order to distinguish it from the formal deductive logic applied in inferential structures where only one conclusion is possible. Following informal reasoning standards, in most everyday and education-related issues, the legitimacy of an inference is based on the degree to which it can be justified (Voss & Means, 1991). Most of the time, argumentation is an interpersonal activity during which people (believe that) are being opposed, and they use evidence, weighing and integration in order to defend their points of view. Whether the focus is placed on the product of the argumentative reasoning, i.e. argument, or on the process of putting forward arguments in order to reach a goal, i.e. argumentation, it is expected that every produced argument takes into account the dialogical aspects of good reasoning (Kuhn, Shaw, & Felton, 1997; Kuhn et al., 2004), and every argumentation process leads to the production of legitimate arguments (Voss & Means, 1991).

In everyday life, whether people will apply argumentative reasoning or not depends on a series of factors, which can be related to the person, the context, or the issue itself, among others. Similarly, not all inferences that are produced, either in our minds or in a dialogue, are products of argumentative reasoning. First of all, there is the idea that argumentation is more than an inference, as it includes the representation of various components, e.g. claim, conclusion, premises, backing, etc., plus a mental activity at a meta-representational or meta-cognitive level, which is simply put the awareness of why and what is being argued about (Mercier & Sperber, 2011). Secondly, focusing only on the relation between a claim and a conclusion, there are at least two ways for this to happen: the intuitive way, in which a claim unconsciously leads to a conclusion, and the reflective way, in which the reasoner reflects on the conclusion through an examination of the reasons (s)he based it on. Skilled argumentation only represents an example of the latter (Mercier & Sperber, 2011), whereas recent theories of fallacies claim that fallacious arguments arise from heuristic inferences based on intuitive thinking without addressing a set of critical questions (Walton, 2010). This does not mean that intuitive inferences are per se irrational; if this were so, almost all human actions, initially based on intuition, would be by definition erroneous. However, as the unconscious nature of intuitive belief formation creates opportunities for logical flaws, an additional reflective process is necessary when we want to talk about welthought out, reasoned actions and decisions.

The present study focuses on the quality and structure of adolescents' arguments during classroom interactions and how teachers reasoning preferences, either for more intuitive or for more reflective inferences, influence the manifestation of data, backings, warrants, and rebuttals in students questions, answers, and discussions. More precisely, we were interested to know: (a) the frequency of reflective inferences, i.e. explicit arguments, from part of the teachers and the students, (b) any correlation between the degree of explicitness in teachers' and students' arguments, and (c) whether there are any differences in such degrees as a result of the curriculum contents taught.

2. THEORETICAL FRAMEWORK

This work has two main points of departure as establishment of its theoretical framework. The first one derives from the literature on explicit and implicit reasoning. My claim is that explicit arguments tend to be more reflective arguments, and I mainly base it on the following argument by Mercier and Sperber (2009):

> There is a major difference between accepting some representation as a fact, and accepting some claim because of explicit reasons. In the second case only, do we experience engaging in a mental act that results in a conscious decision to accept (p. 159).

According to these authors, the relationship between conscious and unconscious is the baseline of the relationship between explicit and implicit reasoning. In other words, when parts of reasoning are made explicit, it is because they form part of the consciously selected part to be presented of an inference or an argument. A great part of human reasoning is implicit, and much of it is unconscious or automatic. When this part is consciously omitted from a speaker or writer, it might mean at least two things: (a) it was consciously considered as obvious to the audience by the reader or writer, and thus it was omitted; or (b) it was subconsciously omitted by the reader or writer, rendering this part of reasoning an underlying presupposition, which is not always shared by the audience. Whether an analyst decides that it is the first or the second case largely depends on the type of reasoning in its overall context. According to Walton (1990), there are four main distinctive categories of types of reasoning: (a) monolectical versus dialectical, (b) alethic vs epistemic; (c) static vs dynamic; and (d) practical vs theoretical. When reasoning is dialectical, dynamic, epistemic and theoretical, the goal is to seek and co-construct evidence that justifies the truth of a proposition in relation to a knowledge base. In that case, establishing what part of this knowledge participants in the discussion already share is an essential, normative aspect. Classroom discourse when viewed as co-construction of meanings and concepts between the teacher and the students can be considered as a type of the reasoning described above. In this sense, making explicit the parts that relate to knowledge, including the existing knowledge on which new assumptions are drawn and elaborated, is considered a positive aspect of this type of dialectical interaction. Thus, identifying the level of explicitness in classroom discourse is an important aspect of its analysis from a constructive argumentative reasoning perspective.

However, not all reasoning needs to be made explicit, because not all reasoning forms part of the arguments put forward in a discussion (Walton, 1990). Reasoning is by its nature inferring, i.e. warrant-using (Walton, 1990), whereas argument can be either warrant-using or warrant-creating, or even both at the same time (Toulmin, 2003). This is why the way how warrant is used inside an inference can make us think of whether or not it is an argument. Generally speaking, when it is difficult to distinguish between warrant and data, because the warrant is implicit in the data, then most probably we have to do with an inference. When both data and warrant can be independently identified, either as part of the argument or by additional means (e.g. critical questions), then most probably we are dealing with an argument. The following is an example of explicit reasoning in a history classroom setting. Case 1 represents an inference, whereas Case 2 an argument.

Case 1	Case 2
The first World War brought Europe to a state of decline. Europe was destroyed and the USA was flourishing with a great production of cars, etc.	The first World War brought Europe to a state of decline, as its economy was destroyed. On the one hand, there was a huge inflation of the European currencies and on the other hand there was a technological boost evident in the USA.

Table 1 – Distinction between inference and argument.

My decision to characterize Case 1 as an inference and Case 2 as an argument was not only based on the structural elements distinction discussed above; it also took into account the specific conversational context in which this specific reasoning took place. Although the conversational setting is the same in both cases, i.e. a history teacher talking about World War I, the communicative function of the two pieces of discourse is different. In Case 1, the function is mainly expository with the second sentence clarifying or explaining the content of the first sentence, linked by a "Why is it so?" relationship (Kuhn, 2001); whereas in Case 2, the function is mainly persuasive, with the second sentence justifying the first sentence, linked by a "How do you know" relationship (Kuhn, 2001). As Walton (2005) puts it,

The distinction between reasoning and argument is explicitly based on the purpose that reasoning is used for in a dialogue exchange between two parties (pp. xii-xiii).

2.1 Importance of argument in education

From Dewey to Piaget and from Freire to Vygotsky, many philosophers and educators have asserted that learning works best when it is an active, creative process, designed through a participatory pedagogy and social learning perspective. It is for this reason that argumentation has been proposed as a method to promote critical thinking, dialogue and learning.

Traditionally the term 'argument' has been used to refer to a valid product of argumentative reasoning consisting of at least one claim and one premise, while the term 'argumentation' has been used to refer to the process by which arguments are dialogically and dialectically constructed (Schwarz & Shahar, 2017). As a socio-cognitive activity, argumentation requires a two-fold capacity, one more related to the social aspects of learning collaboratively through arguing together with others about an issue, and another more related to the cognitive gains as a result of the argumentative activity, as for example the subject-related knowledge gained in a classrooom as result of argumentative activities designed for specific learning purposes. This two-fold capacity has also been described in literature as "learn to argue" and "argue to learn" (Muller-Mirza & Perret-Clermont, 2009).

In any of the two cases, argumentation dialogue in educational contexts is important as it helps to make knowledge explicit, promotes conceptual change (or broadening or deepening), and allows for the co-elaboration of new knowledge to take place (Baker, 2003; Felton et al., 2015). In addition, it has been shown that exposure to dialogic argumentation significantly increases the quality and quantity of students' argumentative discourse, mainly arguments, counter-arguments, and rebuttals (e.g. Zohar & Nemet, 2002; Kuhn & Udell, 2003; Erduran, Simon & Osborne, 2004).

2.2 Why to care about teachers' arguments?

Not only do teachers influence students as being part of the dialectical activity represented by dialogical argumentation; their particular ways of arguing have an impact on students' arguments. Two types of evidence, in the field of argumentation and education, exist in this regard.

The first type of evidence refers to the finding of many studies confirming that the conscious use of argument structures by teachers leads to a major use of argument structures by their students. Erduran et al. (2004) describe a two-year intervention study in which twelve junior high school science teachers were taught how to promote students' argumentation in their classrooms. The post-test showed significant improvements in students' quality and quantity of arguments produced. Although the researchers have not linked the effects of teachers' talk on concrete student outcomes, they make a strong claim about the importance of teachers explicitly focusing on epistemic goals of argumentation. Examples of these goals are:

> talking and listening to others, conveying the meaning of argument through modelling and exemplification, positioning oneself within an argument and justifying that position using evidence, constructing and evaluating arguments, exercising counter-argument and debate, and reflecting upon the nature of argumentation (Simon, Erduran & Osborne, 2006; p. 265).

Venville and Dawson (2010) report an experimental study with middle-grade students in which the argumentation level (both quantity and complexity) of students in the experimental group significantly increased as a result of a short (three lessons) teacher intervention. These findings confirm previous studies (i.e. Zohar and Nemet, 2002; Christodoulou & Osborne, 2014) confirming that argument-based teaching has a positive impact on students' argumentation skills.

The second type of evidence relates to the fact that the argument goal made explicit by the teacher leads to different types of arguments produced by the students (e.g. Felton, Garcia-Mila & Gilabert, 2009; Nussbaum & Kardash, 2005).

3. THE PRESENT STUDY

The previous studies reporting some findings in relation to the effect of teachers' explicitness of argumentative discourse on students' own arguments were intervention studies, meaning that either the researchers implemented an argument-based intervention, or the teachers were instructed in order to be able to apply some type(s) of argumentation in their classrooms. In contrast, the interest of the present study was in naturalistic classroom discourse. More precisely, the main goal was to see how teachers' spontaneous argument discourse had an impact on students' arguments, with regard to a dependent variable called "explicitness". Explicitness was defined as the degree to which:

(a) Argumentation sequences are made explicit within the same topic-based dialogue.
(b) Arguments-as-products are made explicit as part of an argumentation dialogue sequence.
(c) Different speakers, including teacher and students, contribute in the dialectical construction of arguments.
(d) Different speakers, including teacher and students, contribute in the shift from one argumentation dialogue sequence to another.

A secondary goal of this research paper is to identify whether the disciplinary field played a role in the quality of classroom-based argumentation in the middle grades. Previous work by the author showed that disciplinary field plays a role in the quality of arguments constructed by university students (Rapanta & Walton, 2016). In the present study context, the issue addressed was whether there was a difference or not in the relationship between level of explicitness of teachers' and students' argumentation in natural sciences and social sciences.

3.1 Participants and design

The participants were six middle-grade Portuguese teachers; three of them from the disciplinary field of history and the other three from science. Each teacher participated with one classroom, giving a total of 171 students aged between 12 and 15 years old. Data were collected between October and December 2016 in a non-participant classroom observation mode. In total, 36 classhours were audio-recorded and fully transcribed resulting in a classroom discourse dataset of 9507 lines.

3.2 Data analysis

The dependent variable was the level of explicitness of both teachers' and students' arguments as manifested in: (a) the argumentation dialogue goal; and (b) the argument structure. The independent variable was the disciplinary field being history or science.
    The segmentation of the dataset was based on the identification of dialogue moves, defined as "text sequences that express the same dialogical intention (...) or that do not express a different dialogical intention" (Macagno & Bigi, 2017; p. 156). For the analysis, a two-level

coding was implemented: (a) an adaptation of Walton's types of dialogue (Walton, 2008) for argumentation-as-process; and (b) Toulmin's Argument Pattern (TAP) (Toulmin, 2003) for arguments-as-products. On one hand, TAP provides us with a simple reasoning structure very much used in educational research due to the connection of evidence (data, backing) with claims and the focus on warrant as a main vehicle for justifying which evidence-based theory is more sound than another. On the other hand, the identification of argumentation-based pedagogical dialogues applying the normative criteria proposed by Walton is a promising methodological tool in educational research.

3.2.1 Instruments

Table 2 shows an adaptation of Walton's types of dialogue for pedagogical contexts. From the seven types of dialogue proposed thus far in (Walton, 2008) and (Walton, 2011), the present study focused only on four of them as being the most relevant to the classroom discourse context under analysis. These were: information-seeking, inquiry, discovery, and persuasion dialogues.

TYPE	INITIAL SITUATION	MAIN GOAL	PARTICIPANTS' AIMS	SIDE BENEFITS
Information-seeking (IS)	Need of shared knowledge	Make background knowledge explicit	Build common ground Check knowledge Share information	Examine previous understanding Investigate difficulties/lack of knowledge
Inquiry (IN)	Need of examining evidence or interpretations of evidence	Find the strongest evidence or interpretation of evidence	Assess evidence Interpret evidence Compare evidence Coordinate evidence with claims	Better understanding of evidence Critical standing towards sources of evidence Acquiring technical terminology
Discovery (DS)	Need of possible explanations of a problem	Find the best hypotheses for testing or analysis	Define problems Choose criteria for testing Search for evidence	Stimulate creativity Establish an environment for problem solving Stimulate curiosity
Persuasion (PE)	Alternative explanations/ solutions	Find the best explanation or solution to a problem	Persuade others Support explanations with the strongest evidence available	Develop and reveal positions Build up confidence

Table 2 – Adaptation of Walton's dialogue types for pedagogical contexts

An example of a classroom discourse excerpt from the dataset coded using the four dialogue types is shown in the Appendix.

As far as the use of TAP is concerned, below is an example of coding from a science and from a history classroom. Both examples represent explicit argument structures in teachers' discourse. It is interesting to note that in sciences, the argument pattern follows a claim-data-warrant-rebuttal structure, whereas in history the sequence is warrant-data-claim.

Science example of argument	History example of argument
Is it still lithium (the element with the easiest ionic transformation)? No it is potassium (**claim**). Why? Because the valence electron is what? More distanced, do you see? (**data**). It is more distant, therefore less attracted, therefore it will give away its electron easier (**backing**). If an electron is more distanced from the nucleus, there is less attraction force, isn't it? (**warrant**). Potassium's valence electron has a stronger attraction to the nucleus, so it will need more energy to give its electron away (**rebuttal**).	So let's think.. During the war we need weapons, and to make weapons we need industries (**warrant**). So now? Will those industries still be necessary? No (**data**). So what do we need to do out of these industries? Modify the production (**claim**). An industry that used to produce weapons during the war, it now needs to produce other products in order to continue in function.

Table 4 – TAP elements coding in science and history classroom discourse [the transcripts were originally in Portuguese and translated by the author].

3.3 Findings

Our analysis consists of some descriptive statistics regarding to the type of dialogues and TAP elements emerged in the natural classroom discourse settings, and of some tests of correlation to identify any significant patterns in the inter-relations between teachers and students argumentative discourse.

Regarding the types of argumentation dialogue emerged, there were in total 103 sequences of information-seeking dialogues, 73 sequences of inquiry dialogues, 13 sequences of discovery dialogues and only 5 sequences of persuasion dialogues. Regarding the types of TAP elements emerged, there were in total: 53 claims, 70 data, 19 warrants, 14 backings, and 26 rebuttals. Table 5 shows the frequencies, means and standard deviations of both argumentation dialogues and TAP elements.

	Arg. Dialogue		TAP Element	
Frequencies	IS	103	Claim	53
	IN	73	Data	70
	DS	13	Warrant	19
	PE	5	Backing	14
			Rebuttal	26
	Mean	48.5	Mean	36.4
	Median	43	Median	26
	St. Dev.	47.34	St. Dev.	24.07

Table 5 – Frequencies, means, medians, and standard deviations of the argumentation dialogue sequences and TAP elements.

I was further interested in looking at whether each type of dialogue is mostly initiated by the teachers or by the students and if there was any significant difference in this regard according to the type of dialogue. The first hypothesis was confirmed with a chi-square test for goodness of fit, giving a significant result for teachers being responsible for all types of argumentation dialogue emerged (χ^2 = 76.722, p< 0.001). The second hypothesis was not confirmed, meaning that the significant role of teacher as dialogue initiator was not differentiated between the different types of dialogue. However, as seen below, the percentage of students initiating a type of dialogue as compared to teachers is higher for discovery and persuasion dialogues. Table 6 shows the distribution of frequencies of argumentation dialogues initiated by the teachers or by the students per each type of dialogue.

Initiated by	IS	IN	DS	PE	Total
Teachers	82	64	9	3	158
Students	21	9	4	2	36
Total	103	73	13	5	**194**
% stud vs. teachers	20.38	12.33	30.77	40	

Table 6 – Frequencies of argumentation dialogues initiated by the teachers or by the students.

Regarding the relation between argumentation dialogue types and manifested TAP elements, both by teachers and by students, we did not find any significant result. This means that with the given data, we cannot assume any strong interrelation between type of dialogue sequence and type of argument elements enacted in each type. Table 7 shows the frequencies of TAP elements distributed per argumentation dialogue type.

	Claim	Data	Warrant	Backing	Rebuttal
IS	14	17	7	1	5
IN	27	38	10	9	12
DS	4	4	0	4	6
PE	8	11	2	0	3

Table 7 – Frequencies of TAP elements per argumentation dialogue type.

I was further interested in identifying whether teachers' explicitness of argument elements had any significant impact on the manifestation of argument elements from part of the students. The hypothesis was that teachers were more explicit as compared to the students in the production of argument elements in certain types of dialogue. The Mann-Whitney U Test conducted for number of argument elements in total enacted in different types of dialogues was significant for the argument elements produced by teachers in inquiry dialogues (p value .03005, sig at $p<0.5$).

Finally, the disciplinary area, whether history or science, did not play a role in the number and types of argumentation dialogues emerged and the level of explicitness of teachers' or students' arguments.

4. CONCLUSION

The findings show that in naturalistic classroom settings argumentation dialogue of any type is mainly initiated by the teacher. It is further implied that teachers' explicitness of argument elements has a significant impact in inquiry argumentation dialogue, maybe in terms of transforming them into other types of dialogues, such as discovery or persuasion.

The disciplinary area did not play a significant role in how argumentation is made explicit in classroom discourse. However, it was observed that in the science classrooms some type of counter-productive use of explicitness takes place when teachers "give away" the right answer (or best explanation). This is mainly due to the difference between inductive and abductive argumentative reasoning, being the first being more encouraged in history whereas the second mainly forms part of scientific reasoning.

In future research we will: (a)Include a micro-level of explicitness regarding types of dialogue moves belonging to each sequence; (b) Identify types of arguments by using argumentation schemes that are more possible to appear in a type of dialogue sequence rather than in another; and (c) Do a pre-post teacher training comparison of the explicitness level of argument structures.

ACKNOWLEDGEMENTS: The author received funding by the Fundação para a Ciência e a Tecnologia (FCT) for her post-doctoral research project "Learning in communities of practice: an argumentative approach to educational praxis" (SFRH/BPD/109331/2015).

REFERENCES

Baker, M. (2003). Computer-mediated argumentative interactions for the co-elaboration of scientific notions. In J. Andriessen, M. Baker & D. Suthers (Eds.), *Arguing to learn. Confronting cognitions in computer-supported collaborative learning environments* (pp. 47–78). Amsterdam: Springer.

Christodoulou, A., & Osborne, J. (2014). The science classroom as a site of epistemic talk: A case study of a teacher's attempts to teach science based on argument. *Journal of Research in Science Teaching*, 51(10), 1275-1300.

Erduran, S., Simon, S., & Osborne, J. (2004). TAPping into argumentation: developments in the application of Toulmin's argument pattern for studying science discourse. *Science Education*, 88(6), 915–933.

Felton, M., Garcia-Mila, M., & Gilabert, S. (2009). Deliberation versus dispute: The impact of argumentative discourse goals on learning and reasoning in the science classroom. *Informal Logic*, 29(4), 417-446.

Felton, M., Garcia-Mila, M., Villarroel, C., & Gilabert, S. (2015). Arguing collaboratively: Argumentative discourse types and their potential for knowledge building. *British Journal of Educational Psychology*, 85(3), 372–386.

Hornikx, J., & Hahn, U. (2012). Reasoning and argumentation: Towards an integrated psychology of argumentation. *Thinking & Reasoning*, 18(3), 225-243.

Kuhn, D. (1991). *The skills of argument*. Cambridge, NY: Cambridge University Press.

Kuhn, D. (2001). How do people know? *Psychological science*, 12(1), 1-8.

Kuhn, D., & Udell, W. (2003). The development of argument skills. Child Development, *74*(5), 1245-1260.
Kuhn, D., Katz, J. B., & Dean, Jr, D. (2004). Developing reason. *Thinking & Reasoning*, *10*(2), 197-219.
Macagno, F. & Bigi, S. (2017). Analyzing the pragmatic structure of dialogues. *Discourse studies*, 19(2), 148-168.
Mercier, H., & Sperber, D. (2009). Intuitive and reflective inferences. In Evans, J. St. B. T. & Frankish, K. (Eds.), *Two minds: Dual processes and beyond* (pp.149-170). Oxford: Oxford University Press.
Mercier, H., & Sperber, D. (2011). Why do humans reason? Arguments for an argumentative theory. *Behavioral and brain sciences*, *34*(02), 57-74.
Muller-Mirza, N., & Perret-Clermont, A.-N. (Eds.). (2009). *Argumentation and education: Theoretical foundations and practices*. New York: Springer.
Nussbaum, E. M., & Kardash, C. M. (2005). The effects of goal instructions and text on the generation of counterarguments during writing. *Journal of Educational Psychology*, *97*(2), 157-169.
Rapanta, C. & Walton, D. (2016). Identifying paralogisms in two ethnically different contexts at university level. *Infancia y Aprendizaje, 39*(1), 119-149.
Schwarz, B. B., & Shahar, N. (2017). Combining the dialogic and the dialectic: Putting argumentation into practice in classroom talk. *Learning, Culture and Social Interaction*, *12*, 113-132.
Simon, S., Erduran, S., & Osborne, J. (2006). Learning to teach argumentation: Research and development in the science classroom. *International Journal of Science Education, 28*(2-3), 235-260.
Toulmin, S. E. (2003). *The uses of argument*. Cambridge, UK: Cambridge University Press.
Venville, G. J., & Dawson, V. M. (2010). The impact of a classroom intervention on grade 10 students' argumentation skills, informal reasoning, and conceptual understanding of science. *Journal of Research in Science Teaching*, *47*(8), 952-977.
Walton, D. N. (1990). What is reasoning? What is an argument? *The Journal of Philosophy*, *87*(8), 399-419.
Walton, D. (2005). *Fundamentals of critical argumentation*. New York: Cambridge University Press.
Walton, D. N. (2008). *Informal logic: A pragmatic approach* (2nd ed.). Cambridge: Cambridge University Press.
Walton, D. N. (2010). Why fallacies appear to be better arguments than they are. *Informal Logic*, *30*(2), 159-184.
Zohar, A., & Nemet, F. (2002). Fostering students' knowledge and argumentation skills through dilemmas in human genetics. *Journal of Research in Science Teaching*, *39*(1), 35–62.

APPENDIX

Excerpt of a discussion in a Portuguese 9th grade history class (the transcript was translated from Portuguese by the author).

Line	Speaker	Speech	
1	Teacher	I am kind of asking you, based on what you saw last year, anyone can explain to me what was the socialist proposal of Karl Marx about in the the 19th century?	IN
2	Teacher	You saw this last year, anyone remembers what was one of the big goals of that political philosophy? Teresa?	
3	Teresa	Not having age classes.	
4	Teacher	Not having?	
5	Teresa	Not having classes ... social divisions...	
6	Teacher	Ah social divisions, yes my dear, age classes was not possible.	
7	Teacher	Not having social divisions; and why were those social divisions existing?	
8	Teacher	In the 19th century the social difference were existing because of ... classes.	
9	Teacher	So the big objective was to abolish a society in which there were ...	
10	Vasco	classes	
11	Teacher	Abolish a society with existing social classes (writes on board), abolish the social classes in a way that they adapt to an ultimate state that would be the communism.	
12	Teacher	Anyone can tell me any ... or the principal characteristic of communism? A communist society ... what really defines a communist society? Yes?	
13	Filipa	A society where there is equality.	
14	Teacher	Where exists equality, where there are no social classes	
15	Teacher	Then if there are no social classes, there will not be necessary to have what? Oh you are reasoning! Raise your hand.	
16	Manel	A superior condition?	
17	Teacher	Alright. If there are no social classes, there are no layers, then no one is superior to no one, and therefore it is not necessary to have what? (*no reply*)	

18	Teacher	So, folks, I will make the question in another way, what is that grounds the existence of social classes? What is the basis for us to speak about social classes? Teresa.	
19	Teresa	The money that they have?	
20	Teacher	The money, which means the wealth that determines ... the diferences between the wealth of the different groups, it is one aspect.	
21	Teresa	The difference of salary.	
22	Teacher	The differences in salary, the differences in earnings, therefore the social class is based on the differentiation of wealth, wealth that results from any earning source, it can be the salaries, it can be the means... the assets that they have, earnings coming from their properties, etc. It's good. Yes, Maria?	
23	Maria	Does this mean that any work they do they earn the same?	DS
24	Teacher	In theory we could say this. Yes, the wealth.. Therefore this means what?	
25	Teacher	That there is not any wealth difference. If there is not any wealth difference given that they are all equal eh... Yes Maria?	
26	Maria	It doesn't make much sense, imagine a person who works more, would she receive the same as another person who works less? It doesn't make much sense.	
27	Teacher	We are not going to enter in many debates today, sorry but today I have to finish this..	
28	Teacher	Ah Maria says that it doesn't make any sense because... Can you please tell me why you think that it doesn't make any sense?	
29	Teacher	Who manages to measure that the work is bigger for one rather than for another or that is [] of that work for the society?	
30	Maria	The same work, both work in the field, one person works 8 hours and the other works 4 [] How do they do this, I don't understand how they will manage to define the same wealth if there are people who work more...	
31	Teacher	Because it is like this, it starts from the principle that in that ultimate state of the society, given that people no longer choose their profession in relation to ...	
32	Teresa	... they do what they like	
33	Teacher	Exactly. Go on, Teresa, you were saying	

		correctly.	
34	Teresa	People sometimes choose a job because they earn more, and this is where everything is earned in the same way, if they like to do it, they will do it	
35	Teacher	If they like to do it, they will do it .. On top of that, I will also reply you with an article that I read last year in Expresso.	
36	Teacher	Did you hear about the start-up congresso that is coming next week? Have you heard about that?	
37	Everyone	Yes	
38	Teacher	There are older colleagues of yours who are going there.	
39	Teacher	Anyway, there was that start-up, I don't remember now the name of the businessman but it is not relevant, and also I don't know very well in what area he was, but two or three years ago he founded this business, one of those business that use latest technologies, and he decided, he was a young mana round 30 years old, he decided to give the same salary to everyone working in bis business, from the top of administrators to the last doorman, everyone would receive the same.	
40	Teacher	Later on, but this must have been some time ago, I came across again in Expresso that the businessman ... that the business works with the major productivitiy and that no issue was caused by the fact that the doorman was receiving the same salary as the administrator. It was around 5000 $ per month.	
41	Teacher	Therefore, folks, this is for you to understand that things are changing, that here the goal was exactly this, that people were being fulfilled in their work, then naturally they were working for the common good, but now I am, it is obvious that ... and as they were working for the common good, there were also existing between people forming that society what? Conflicts?	
42	Everyone	No	
43	Teacher	It would be naturally harmonic. Therefore there was no need for what?	
44	Everyone	Wars	

45	Teacher	For wars, for policing to get a or b, then there was no need for a state, in the way we conceive it today.
46	Maria	Yes, but this is something...
47	Teacher	Utopic, I know. Yes but this was the definition of a communist society.
48	Teacher	The state already of happiness where there was not need of a state, of a state that had the power to regulate and to make .. to regulate security and to protect the private property like it happens today; is it clear?

47

On the Epistemic Basing Relation

JUHO RITOLA
University of Turku, Finland
juho.ritola@utu.fi

This paper will discuss the epistemic basing relation. I will briefly present two positions that have traditionally attracted the most attention in the literature. Next, I will discuss new argumentation by Boghossian (2014) that tries to advance the debate, and its criticism by Broome (2014) and Wright (2014). I will conclude in the negative: though the intuition of active rationality in inference is strong, we seem to have no credible account of it.

KEYWORDS: basing relation, inference, rule-following

1. INTRODUCTION

This paper discusses the nature of epistemic basing relation. This relation holds between a reason and a belief if and only if the reason is a reason for which an agent holds the belief in question. Typically, writers on this topic concentrate on indirect beliefs, i.e. beliefs held based on other beliefs, but this should not obscure the fact that the relation can also hold between sensory experience and a belief. Some writers, like Ram Neta (n.d.) understand it to hold between any rationally determinable condition (intention, action, judgment, emotion etc.) and the reason for which it is held. The basing relation is of primary importance to anyone who wants to better understand the nature of justifying inference. You may have good reasons for your belief, but if those reasons do not properly connect with your belief, your belief is not justified (by those reasons). Hence, proper basing is something that can turn a belief into a justified belief.

In this paper, I will first briefly present the two major answers to the question of what constitutes a proper connection between a reason and a belief: the causal and the doxastic accounts. I hold that a causal condition is necessary for basing. Yet, it also seems that a pure causal account cannot account for the intuition that inferring is something we ourselves actively do. Inferring is something for which we are responsible; it does not seem, intuitively, to be something that merely *happens* to us. By inferring, we exercise our rational capabilities. *Prima facie*, a fully rational inferential belief would seem to be something we hold, because we *take something* to be a good reason for that belief, not only because the reason caused us to believe it.

Traditionally, the debate between the doxastic and the causal account of basing was taken to be part of the debate between the epistemological internalists and the externalists. In recent years the blooming research on the concept of reason has brought new authors and suggestions to the topic of basing relation. In the third section, I will present Paul Boghossian's (2014) suggestion that the basing relation consists essentially of the reasoner taking the premises to be a reason for the conclusion, where this taking is to be understood as following a rule, and the criticisms John Broome (2014) and Crispin Wright (2014) have presented against this suggestion. Based on the discussion, it appears that we still do not have an account of reasoning beyond a certain causal relation between the premise-belief and the conclusion-belief.

2. THE CAUSAL AND THE DOXASTIC ACCOUNTS

As mentioned, the two main accounts of the basing relation are the causal account and the doxastic account. Though the discussion on these accounts has reached a considerable level of sophistication, for the purposes of this paper we need to be brief.[1]

2.1 Causal accounts of basing

A causal account of the basing relation holds that a belief's being based on another belief consists in the conclusion-belief[2] being caused and sustained by the premise-belief. This causality has been interpreted as a disposition (Moser, 1989; Evans, 2013), with the help of the concept of intervention (McCain, 2012), and counterfactuals (Swain, 1979, 1981, and 1985; Bondy, 2016).

---

[1] For a longer introduction and references, see Stanford Encyclopaedia of Philosophy (https://plato.stanford.edu/), s.v. "The Epistemic Basing Relation".

[2] I will concentrate on cases of inferential belief, that is, a belief being based on another belief.

The major problem of causal accounts is what is called the problem of deviancy:

(Deviant Causal Chains) Suppose I'm giving a talk. Suddenly, Donald walks into the room, which causes me very to be very nervous, which causes me to spill my coffee on my shirt, which causes me to believe that there is a stain on my shirt.

It does not seem appropriate to say that my belief that there is a stain on my shirt is based on my belief that Donald walked into the room. A reasonable response to this problem might be a dispositional account.

(Dispositional Theory): S's belief that p is based on m iff S is disposed to revise her belief that p when she loses m. (Evans, 2013, p. 2952).

I would not be disposed to lose my belief that there is a stain on my shirt if I lost the belief that Donald walked into the room. But if the existence of a disposition were agreed to be a necessary condition of basing, we would arguably accept causality as necessary condition of basing anyway.[3,4]

One could argue that the chain of events in the example is too long anyway. The premise-belief can only be something more immediate to the conclusion-belief. But this objection misses the persistency of deviancy. It can crop up in a single step; the length of the chain is irrelevant to the problem.[5] Boghossian (2014, p. 4) notes that a depressed person might infer regularly from "I am having so much fun" to "Yet, there is so much suffering in the world". We do not want to say that the second belief is based on the first. No matter how tight we make the connection between the premise-belief and the conclusion-belief, the problem is bound to recur. I do not think it should be the decisive question in the debate between a causal account and a doxastic one.

---

[3] I myself am persuaded by J.L. Mackie's arguments in *Truth, Probability, and Paradox* (1973) to that effect.

[4] There is a very advanced explication of basing through counterfactuals by Bondy (2016). That account might have merits in respect to the other causal accounts, but from the point of view this paper it is another causal account. I think causality is an important part of basing, but the focus of this paper is elsewhere.

[5] John Turri (2010) argues convincingly that in the case of *proper* basing causal relations are not sufficient, but I leave the issue of proper basing aside here.

More important point is that from a non-sceptical perspective, we want to be causally connected with the world. Although the issue of proper basing is not in our focus here, we want that cases of proper basing are cases where we connect with the actual world.[6] So, there is reason to commit to the idea that the basing relation is causal. In my view, the two accounts have different ambitions and accepting a causal account does not imply that one could not also think doxastic account has something to offer.[7]

2.2 Doxastic accounts of basing

A doxastic account of basing holds that a belief's being based on another belief consists in the agent being, to some degree, aware of the reason being a reason for the conclusion-belief.

> (Doxastic Criterion) The reasoner believes the truth of a reason is evidence for the truth of conclusion.

This criterion purports to describe the intuition that we are not just 'victims' or locations of causal processes. We *actively* draw inferences by seeing connections between a reason and a conclusion.

A common objection to a doxastic criterion is that it entails a metabelief to the effect that my belief that r is a reason to believe p. But

---

[6] This does not imply that proper basing, where the actual world is such that one *could not* connect with it (say, because it controlled by an evil demon), is not more valuable than improper basing, where the actual world is such that one could not connect with it.

[7] There is one more objection we ought to dispel, before moving on discussing the doxastic account. A much-discussed argument against the causal account is the case of the Superstitious Lawyer (Lehrer, 1971). The case purports to prove that one could base a belief on some line of reasoning without that line of reasoning being the causal reason for the belief in question. In the case, there is an alternative causal line of reasoning sustaining the belief in the innocence of the Lawyer's client, namely the Lawyer's superstition. Many philosophers remain completely unimpressed by this case (e.g. (Goldman (1979) finds it "unconvincing".) I agree with Harvey Siegel that the case begs the relevant question. In any case, I am interested in the cases of basing, and the very description of the case says that the line of reasoning could not make a difference: the emotional factors in Lehrer's case are such that the only thing that could motivate the Lawyer to believe her client is innocent is the Lawyer's unshakeable belief in the cards. According to a dispositional reading, the Lawyer's belief is based on the superstition, not on the line of reasoning, and that seems intuitively correct.

this is a mistake: it is the content of the belief that is the central factor, not 'the belief that p.' A doxastic account does not imply that the relevant belief is a metabelief *about* the basis-belief. Instead, the relevant belief is a belief about the content of the premise-belief being a reason to believe the conclusion-belief.[8] Suppose I believe that the trees are swaying, and on that basis come to believe that it is windy out there. My reason for believing the conclusion belief is not 'I believe that the trees are swaying.' Instead, my reason is that I believe of the trees that they are swaying; I am, in some sense that we are trying to specify, looking at the content and seeing that that content has consequences.

As clear as that distinction seems, the suspicion of a meta-belief has become so entrenched that it is worthwhile to translate the point to distinctions drawn in recent discussion on reasons. Neta (n.d.) makes the point succinctly. We have *normative* reasons, i.e. reasons that favour something else, and *explanatory* reasons, i.e. reasons that explain why something believes something. But further, explanatory reason need not be the reason *for which* you believe something. The fact that explains why Othello killed Desdemona is that he was jealous, but it is not the reason for which he killed her. What motivated him was the content of his (false) belief that Desdemona was being unfaithful. Similarly, the fact that I believe the trees are swaying figures in the explanation of *why* I believe it is windy out there, but it is not the reason *for which* I believe it is windy out there

Three common objections to a doxastic account are i) it requires too much of epistemic agent; ii) a doxastic account cannot be sufficient; and iii) cases of unconscious basing.

The first objection involves cases of pre-verbal children and animals. Surely we cannot require that they form the kinds of evidential beliefs that we are interested in, or that they are able to use the kinds of evidential concepts that support-relations involve. I think this objection is mistaken. As Sylvan (2016, p. 286) points out, treating something as a normative reason does not require that one has the concept of normative reason. Analogously, a cat might treat something as prey without having the concept of prey.

The second objection is that a doxastic account is not sufficient. Suppose a leader of a cult tells Rene, a member of the cult, that his belief in God is a good reason to believe everything else he believes. Rene

---

[8] I do not have the space to deal with Lewis Carroll's (1895) paradox. I will only note that we should we should not try to make the basing that inference (putatively) involves *merely* a premise, before we have examined what that basing might be. That would beg an important question too.

slavishly comes to believe this. Are Rene's beliefs now based on his belief in God? It would seem not. We need not take issue with this claim here. There is good independent reason to believe that basing involves causation and I do not want to advance the claim that a simple doxastic account is sufficient. The same goes for the cases of sub- or unconscious reasoning. It seems impossible to deny that there are cases where we infer something and do not, or even could not, have awareness of the relation. I do not deny that there are such cases as I accept causality as a necessary criterion of basing. But I do want to remark that the phenomenon of basing and our ability to bring this basing to consciousness are two different things.

To sum up: it seems feasible to claim that the basing relation is some type of causal relation. The interesting question is whether we should add some conditions that do justice to the kind of active and skilful reasoning in which we engage when trying to figure out what to believe.

3. REASONING AND THE TAKING-RELATION

I will now present and make some comments on a recent dialectic between three writers (Paul Boghossian, John Broome, and Crispin Wright (2014)) that seem to advance the debate on the nature of inference.

3.1 Boghossian on the taking-condition

Boghossian (2014) purports to give an account of inference as person-level, conscious, voluntary mental action.[9] He finds inspiration in Frege who wrote that "[t]o make a judgment because we are cognisant of other truths providing a justification for it is known as *inferring*" (1979, p. 3). Since we are interested in the general phenomenon of inferring, we should modify this slightly. We do not want to concentrate on reasoning from truths, and we do not want the success grammar built into the definition. So, we get:

> (Inferring) S's inferring from p to q is for S to judge q *because S takes* the (presumed) truth of p to provide support for q. (Boghossian, 2014, p. 4).

---

[9] Famously, Daniel Kahneman has made the distinction between System 1 and System 2 reasoning. Boghossian explains that here he is "…interested in reasoning that is System 1.5 and up" (2014, 2).

The essential part of this definition is the *taking*: S somehow processes what the presumed truth of p would mean and takes it to provide support for q. This leads into:

> (Taking condition) Inferring necessarily involves the thinker *taking* his premises to support his conclusion and drawing his conclusion *because* of that fact. (Boghossian, 2014, p. 5)

So, for example, I wake up remembering that:
It rained last night.

I then combine this with my knowledge that:

If it rained last night, then the streets are wet.

This could lead me to conclude:

The streets are wet.

The question before us is now how to understand the 'taking' involved in the (Rain)-example. Boghossian (2014, p. 6) notes the difficulties of taking as a 'full-fledged normative doxastic construal.' Consider:

> (FFNC) My judging (1) and (2) supports my judging (3).

This invites the objection of requiring too much sophistication from the inferring agents. Earlier, I contended that one need not have the normative concept of support for taking something to support something else. But if we construe the taking as a full-fledged belief that uses the concept of support, this response is no longer available; one now needs the concept to formulate the belief.[10] Boghossian (2014, p. 6) then considers a first-order rain belief as an explanation:

> (FORB) If it rained last night and if it rained last night, then the streets are wet, then, the streets are wet.

As he (ibid.) notes, this invites Carroll's famous regress: if we now ask how does this belief get us to the conclusion, another, more

---

[10] Though we should not read too much into this concept. I think typical adults, who have never been introduced to epistemology or logic, can formulate meaningful beliefs about consequences of their beliefs, when they direct their thoughts to these matters.

complex conditional is needed (having as antecedent this conditional and as consequent the conclusion). And so on. No reasoning can emerge from this regress.

The problem is that we are trying to understand just what exactly is basing, that is, how do we move from one state of belief to another state of belief, or add another state of belief to our existing beliefs. It seems wrong-headed to go about this by adding another belief.[11] Typically, there is such a linking-belief, and I am partial to thinking that the arguer has to be justified in this linking-belief, in order to be justified in the conclusion-belief. But just adding more belief just does not seem to *explain* the move itself.[12]

Boghossian considers an alternative. Perhaps a belief like (FORB) is always involved, but it acts as a *background* condition rather than a premise. But then we must explain what we mean by this background condition. The difference is the fact that the premise-belief is the one on which the conclusion is based, whereas the linking-belief is involved but not as the basis. But we are trying to understand basing itself, and if this is our explanation, "we seem to have helped ourselves to the very notion at issue" (Boghossian, 2014, p. 8).

Based on such considerations[13], Boghossian argues that we should construe the thought transitions that inference involves as *guided* by inference rules. The picture he propounds includes an occurrent intentional state (the Taking-condition), and gives an account of inference as following a rule. Consider:

> (Email Rule) Answer any email that calls for an answer immediately upon receipt!

Prima facie, following a rule seems interestingly similar to an inference. When we follow a rule, we take the fulfilment of the antecedent of a rule

---

[11] I tried to argue for this in my 2007 ISSA-paper "Justified Belief in the Link of an Argument", but there the discussion involved a rather curious argument by Cling (2003) and was directed towards understanding what epistemically good basing is.

[12] Metaphors have limited use in analytic philosophy, but I'll try one anyway. Imagine someone has an arrow and wonders how to get this arrow on target. Adding more belief seems like helping the person with an arrow by giving her another arrow.

[13] He discusses more than what I note here, for example an intuitional construal of (FORB). But for the purposes of this paper, we can go on to the suggestion he ends up recommending.

as a reason to do something, just as in reasoning we take some belief as a reason to believe something else. Furthermore, in order for us to be able to say that someone acts because she is following a rule, rather than just conforming to it, there must be some intentional state that encodes the rule. Following the (Email Rule) rule can be given the following description:

> Well, I have grasped the rule, and so am aware of its requirements. It calls on me to answer any email that I receive immediately. I am aware of having received an email and so recognize that the antecedent of the rule has been satisfied. I know that the rule requires me to answer any email immediately and so conclude that I shall answer this one immediately. (Boghossian, 2014, p. 13)

In other words, I grasp the rule, form a view that its trigger conditions are satisfied, and draw the conclusion that I must respond. The problem with this is that the intentional rule-following requires an inference, and inference requires rule-following. We should not expect a non-circular analysis of following a rule of inference. (Boghossian, 2014, p. 13-17)

Wright (2014, p. 30) argues forcefully that such an account cannot get off the ground. If in order for me to follow a rule R, I must make an inference$_1$, that is, to move from a belief that carries the content of R to another belief, I must do another inference, inference$_2$, to follow that rule, which again needs a further inference... As Wright notes, this seems to make following any rule into a supertask that we could not possibly do in every inference.

Boghossian considers and rejects three different alternative explanations. First, we could argue that the guidance by rules is to be explicated as a mere disposition to conform to R under appropriate circumstances. We have already noted that there is some reason to believe dispositions reduce to causal relations. So, I think Boghossian (2014, p. 15) is right to argue that just seems like regular causation.

Second, an intermediate picture could be to argue that there is a rule being encoded by another intentional state, but it is not consciously accessible or one that "he consults in figuring out what to believe" (2014, p. 15). When I consider (1) and (2), one sub-personal system recognizes the logical form of the of the premise, which in turn activates another sub-personal system that encodes the MP rule, which in turn issue in my believing (3). All that is needed that there is some representational system ensuring via a causal mechanism that a belief in (3) results from considering (1) and (2) (Boghossian, 2014, p. 15).

Perhaps this is so for System 1 reasoning but doubt remains whether that could be the case for the kind of System (2) thinking that we are after. This picture of active performance still seems to linger. Boghossian (2014, p. 16) argues that nevertheless, if you had no awareness of how that conclusion came to you, full rationality would still require you to ask yourself whether you endorse the conclusion that had come to you. And this would require that you lay bare you reasoning process by which the premises are supposed to have led to the conclusion.

I think this is an important point, but it does not fully meet the concerns that might be raised. If I have no clue how the conclusion came to me, I would have no reasoning process to lay bare anyway. If I can naturally point to some premises[14], then I can evaluate the epistemic value of those premises as propositions. But this evaluation can be importantly different than the actual basing that we are trying to elucidate. Moser held that

> S occurrently satisfies an association relation between E and P=$_{df}$ (i) S has a *de re* awareness of E's supporting P, and (ii) as a nondeviant result of this awareness, S is in a dispositional state whereby if he were to focus his attention only on his evidence for P (while all else remained the same), he would focus his attention on E. (1989, pp. 141-142)

The disposition has a condition for being activated the other way around too: it is not only that I was caused to believe P, when directing the evidential attitude to (3), I am also caused to think of (1) and (2). In the epistemic evaluation phase that Boghossian refers to, we are considering the premises as two different propositions P and (P then E). Note that here the focus is on the two different propositions, in *de dicto* form. We are no longer discussing the nature of *actual* basing that we started from. So, I think Boghossian fails to make an effective argument against the dispositional position. Broome elaborates the dispositional position and we will turn to that next.

3.2 Broome on reasoning

Broome (2014) is, like Boghossian, interested in identifying the kind of active reasoning that we seem to be able to do. To do this, we should be

---

[14] They need not just 'pop up' to me head. The can partially depend on me remembering the moment that I realized that (3) when considering (1) and (2) just like I might remember very vividly that some, say roses were pink, not red.

able to distinguish it from mere causation. Broome also thinks that reasoning is rule-following, but he accepts, unlike Boghossian, that rule-following can be blind. Reasoning is following a rule blindly, but the key difference, to separate this from mere causation, is to hold that the disposition in question is two-fold: it is the disposition to act in particular way (to follow a rule) and that way of acting to *seem right* to you. The rule that you blindly follow need not be the correct rule to follow; we want to accept that even when you are reasoning badly, you are still reasoning. But the dual nature of this disposition, its seeming right to you allows for the possibility of correction. You may check your reasoning, and this gives the reasoning the kind of active personal sense that we are after. Since the disposition involves the aspect of seeming right, you may later withdraw the conclusion, if you were to check the reasoning and it no longer seemed right to you. (Broome, 2014, p. 21-22)

Just how is this disposition put into effect then? Broome holds that subpersonal processes determine what seems right to you. In fact, subpersonal process determine everything we do: "[wh]en you sign a letter, subpersonal processes causal processes determine that your fingers move in just the way that forms your signature" (2014, p. 23) But Broome does add that in the case theoretical reasoning, you are still not just following a rule idly, you are giving the rule a strong endorsement:

> You would not believe the conclusion if you did not take the premise to imply the conclusion, or at least support it. You may not consciously believe that the premises support the conclusion. Even so, we may treat your disposition to believe the conclusion when you believe the premises, and for this to seem right, as itself implicitly taking the premises to support the conclusion. Since you believe the conclusion because of your disposition, you believe it because you take the premises to support the conclusion. So you satisfy Boghossian's taking condition (Broome, 2014, pp. 23-24).

So, Broome holds there is a linking-belief between the premise and the conclusion. But this is not what constitutes the reasoning. In his *Rationality through Reasoning (RtR)* (2013, ch. 13) Broome characterizes his account further:

> [...] something must distinguish a linking belief from a premise-belief. We now have a good way to make the distinction. Reasoning is an operation on the contents of the

premise-beliefs, whereas the content of the linking belief is not operated on. The distinction does not have to be made in terms of consciousness, so the linking belief may be either conscious or unconscious. (2013, p. 234)

But as noted, Broome also wants to hold to the picture of us actively drawing the conclusion (2013, pp. 234-235; see also fn. 5 ch. 13): reasoning is a rule-governed operation on the contents of your premise-belief. It is something you do, and it is only reasoning if the rule guides you (RtR, 2013, p. 237). But how do we know that we are being governed by the rule? In (RtR), Broome no longer emphasizes that the rule-following must be blind. Instead, he emphasizes that the process, executed by subpersonal processes, must seem right to you. But now a reader might raise a critical question. Why should the fact that something seems right to me be a sign of it being under my control? One might further ask what else can the 'seeming right' consists of than it seeming to the agent that in taking the premise to imply the conclusion, one did the right thing? I myself am attracted to the idea that Broome is advocating: that reasoning is a belief operating on another belief. But it seems to me that this operation, however it is carried out, is more likely to be the taking itself (whether conscious or not), than the dual disposition of a disposition to act in a particular way (perhaps blindly) and the disposition for this to seem right to you.

3.3 Wright and the simple proposal

Wright (2014) raises problems for any account that sees inference as rule-following. He argues that if the inference is going to be regarded as rule-following, whether person-level or not, there will have to be a state that *registers* the obtaining[15] of a support-relation between the premise(s) and the conclusion. Further, this registration state will have to control the movement; otherwise this would be just regular causation.[16] Arguably, we now face a dilemma. Either the registration state is general or it is specific to the inferential transition in question.

Suppose first it is general, i.e. that any transition of the appropriate kind is licensed when the system is in a state of acceptance

---

[15] Or represents there being such a relation. Remember that we do not want to make factual assumptions.

[16] Wright directs his remarks against Boghossian's taking-condition, but it seems to me that it applies also to Broome: Broome thinks his account implies the taking condition and that the rule guides the transition.

of the relevant kind of premise. Wright argues that the control exerted by the registration state would have to consist of the appreciation that a specific move now comes to the ambit of the registration state: transitions of such and such kind are mandated, this is a transition of such and such kind; so this transition is mandated. But then, we have represented the original inference as involving another inference, and a regress ensues. (Wright, 2014, p. 31)

Suppose now it is particular, that is, each registration state is specific to the inferential transition it controls. But now we have to note that even though our inferential capabilities are finite, they are open-ended in the way our linguistic competence is: we are able to handle new sentences as they come. Since this is the case, it would seem that we would have be able to access the specific inferential transition from some general informational states that encode our general inferential program. And now again an inference is presupposed by the account of inference. But then, if we are forced to accept that at some point the inference will need to made without any further processing, which seems to involve inference, what reason do we have to insist that inference as such is not such an operation? (Wright 2014, p. 32)

Based on such considerations, Wright holds that we have to accept what he calls the Simple Proposal:

> (SP) A thinker infers q from $p_1...p_n$ when he accepts each of $p_1...p_n$, moves to accept q, and does so for the reason that he accepts $p_1...p_n$. (Wright 2014, p. 33)

In Wright's view, then, an inference is *au fond* a basic mental action. If the dilemma Wright presents is unavoidable, we are, in effect, back to simple causation. But now consider the following scenario. I believe the sides of my office are all two meters long. I want to know what the surface area is. In case 1, I multiply two sides and start believing that the area is four square meters. In case 2, I add the two sides and start believing that the surface is four square meters. According to (SP), there are no such differences: we *cannot talk* of these two as different cases, because any inference is just believing *because of the premises*. Based on the case of Achilles and the Tortoise, one cannot complain that the two cases have different premises[17], because the inference is the operation I apply on the premises, not a premise. But it is possible to talk of the two

---

[17] The argument can be constructed as: p_1: The sides of my office are two meters long, and p_2: If the sides of my office are two meters long, its surface are is four square meters.

different cases. So, there must be some kind of specific mental operation I apply on the premise beliefs, some specific kind of case where a belief operates on other beliefs in some way. One suggestion could be that that operation is just the taking, but I have no space to defend that suggestion further here.

4. CONCLUSION

In this paper, I presented the traditional debate about the epistemic basing relation and some new attempts to do justice to the rational, person-level inferring of which we seem capable. Though both Boghossian and Broome make appealing suggestions, it seems we do not yet have a credible account of how one belief, a linking-belief (or awareness or however that is to be formulated) operates on another belief, the premise-belief.

REFERENCES

Boghossian, P. (2014). What is inference? *Philosophical Studies*, *169*(1), 1-18.
Bondy, P. (2016). Counterfactuals and epistemic basing relations. *Pacific Philosophical* Quaterly, 97, 542-569.
Broome, J. (2013). *Rationality through reasoning*. Chichester, UK: Wiley-Blackwell.
Broome, J. (2014). Comments on Boghossian. *Philosophical Studies*, *169*(1), 19-25.
Carroll, L. (1895). What the tortoise aaid to Achilles. *Mind*, *14*, 278-280.
Cling, A. D. (2003). Self-supporting arguments. *Philosophy and Phenomenological Research*, *66*(2), 279-303.
Evans, I. 2013. The problem of the basing relation. *Synthese*, *190*(14), 2943–2957.
Frege, G. (1979). *Logic*. In *Posthumous Writings* (pp. 1-8). Chicago: The Chicaco University Press.
Goldman, A. (1979). What Is justified belief", in G. Pappas *Justification and knowledge*. Dordrecht: D. Reidel.
Lehrer, K (1971). How reasons give us knowledge, or the case of the gypsy Lawyer". *The Journal of Philosophy*, *68*, p. 311–313.
Mackie, J. L. (1973). *truth, probability, and paradox: Studies in philosophical logic*. Oxford: Oxford University Press.
McCain, K. (2012). The interventionist account of causation and the basing relation. *Philosophical Studies*, *159*(3), 357–382.
Moser, P. (1989). *Knowledge and evidence*, Cambridge, UK: Cambridge University Press.

Neta, R. (n.d.). Basing is conjuring. Retrieved from http://mindsonline.philosophyofbrains.com/2015/session3/basing-is-conjuring/.
Ritola, J. (2007). On justified belief in the link of an argument. *Proceedings of the Sixth Conference of the International Society for the Study of Argumentation*, (pp. 1181-1184). Amsterdam: SitSac
Swain, M. (1979). Justification and the basis of belief", in G. Pappas (Ed.), *Justification and Knowledge* (25-49). Dordrecht: D. Reidel.
Swain, M. (1981). *Reasons and knowledge*. Ithaca, NY: Cornell University Press.
Swain, M. (1985). Justification, reasons and reliability. *Synthese*, *64*(1), 69–92.
Sylvan, K. (2016). Epistemic reasons II: Basing. *Philosophy Compass*, *11*(7), 377-389.
Turri, J. (2010). On the relationship between propositional and doxastic knowledge. *Philosophy and Phenomenological Research*, *80*(2), 312-326.
Wright, C. (2014). Comment on Paul Boghossian, "What is inference". *Philosophical Studies*, *169*(1), 27-37.

48

Dialogical Argumentation in Financial Conference Calls: the Request of Confirmation of Inference (ROCOI)

ANDREA ROCCI
Università della Svizzera italiana
andrea.rocci@usi.ch

CARLO RAIMONDO
Università della Svizzera italiana
carlo.raimondo@usi.ch

In this paper we explore the role of dialogical argumentation in the context of a specific genre: the financial communication in the earnings conference calls. Specifically, we focus on a dialectically peculiar argumentation move, the request of confirmation of inference, and on its role inside the dialogue between financial analysts and corporate executives.

KEYWORDS: Argumentation; Dialogue; Indirect Questioning; Earnings Conference Calls; ROCOI

1. REQUESTS OF CONFIRMATION OF INFERENCE IN EARNINGS CONFERENCE CALLS

Earnings conference calls (ECCs) are voluntary disclosures increasingly used by listed companies to communicate with the investors, the financial analysts, and the financial markets at large. They consist in teleconferences held by corporate managers with financial analysts, immediately following the publication of the press release containing the quarterly earnings announcement. The ECC is the proper occasion for the corporation managers to explain what they did and what they are willing to do and to provide the proper rationales for the quarterly/annual results that have been disclosed just before the call.

As observed by Lev (2012, p. 4), ECC are one of the few routine business operations that are never delegated; in fact the top management (CEO, CFO, CIO, COO) personally conducts all ECCs, together with the investor relations specialists. At the same time, Lev (2012, p. 47) observes that in the financial literature ECCs have been shown to be informative for the financial markets and therefore they should be able to convey important, new information which is in turn capable of influencing the stock prices and the volumes of stock trading: "Why else would investors and analysts continue to attend conference calls?" – he quips.

The informative content of the earnings conference calls is therefore relevant for retail and professional investors who want to understand in depth the economic results of a specific listed firm and how the results have been generated. Such a result reinforce the relevance and the importance of the earning conference calls' content from both the point of view of the information producers (the company and, more specifically, the company's top executives) and the point of view of the financial markets players who scrutinize the ECCs in search of a better understanding of their investment activity.

It is more difficult to understand *what* exactly in the content of the calls is useful for the markets, given that companies try to avoid disclosing new information items during the call, because substantial new disclosures would lead to a new written communiqué, additional legal hurdles and would leave the markets with the impression that their original quarterly results report was incomplete and not diligently prepared.

Which parts of the ECCs are relevant and which information is transmitted in ECCs is still debated (cf. Raimondo and Rocci, 2018, p. 302-303). The ECCs are normally composed of a monological presentation by the top management and a subsequent Q&A session between financial analysts and the company executives (Crawford Camiciottoli, 2010, 2013). This article will focus on the second part, examining in depth a certain type of questioning widely used by the financial analysts and how firm's executive are going to handle it.

During the Q&A phase of ECC, it is common for analysts to ask indirect questions which require the respondent to confirm or disconfirm a proposed inference. We named this move the "Request of Confirmation of Inference" (in short ROCOI). Interviewed analysts do mention this kind of move: "We ask for qualitative thoughts and insights into industry trends or specific business lines, just so that *we're also double-checking our own thought processes and that our models are solid*" (Brown et al., 2015, p. 19 our italic.)

To clarify what we are talking about, in examples (1) and (2) reported hereafter we report a first example of ROCOI.

(1) Curtis Rogers Woodworth - Nomura Securities Co. Ltd., Research Division
Mark, I was wondering if you could comment on some of the divergence we're seeing in growth rates among the portfolio. I mean, it seems like the bar markets are showing pretty meaningful year-on-year declines, whereas beam is flat and the sheet market's showing modest growth. Do you feel like that's indicative of just the fact that the non-res cycle is sort of stabilizing and maybe we're seeing, at least on the bar side, more weakness than on the industrial equipment side?

(2) Mark D. Millett - Co-Founder, Chief Executive Officer, President and Executive Director
Well, I'm not so sure that we're seeing our bar side depreciate like that. I think we're somewhat steady. I think the – generally, our order rate at the structurals is sort of steady to up incrementally.
Steel Dynamics (STLD) Second Quarter 2013 Earnings Conference Call, July 18, 2013 10:00 AM ET

The analysts Curtis Rogers Woodworth is asking for the confirmation of his argument about the dynamics of sub-sectors businesses in the steel industry, in order to be able to hopefully draw some conclusions for future trends. We reconstruct the argumentation structure supporting his tentative conclusion (cf. the modal *maybe*) in Figure 1, below.

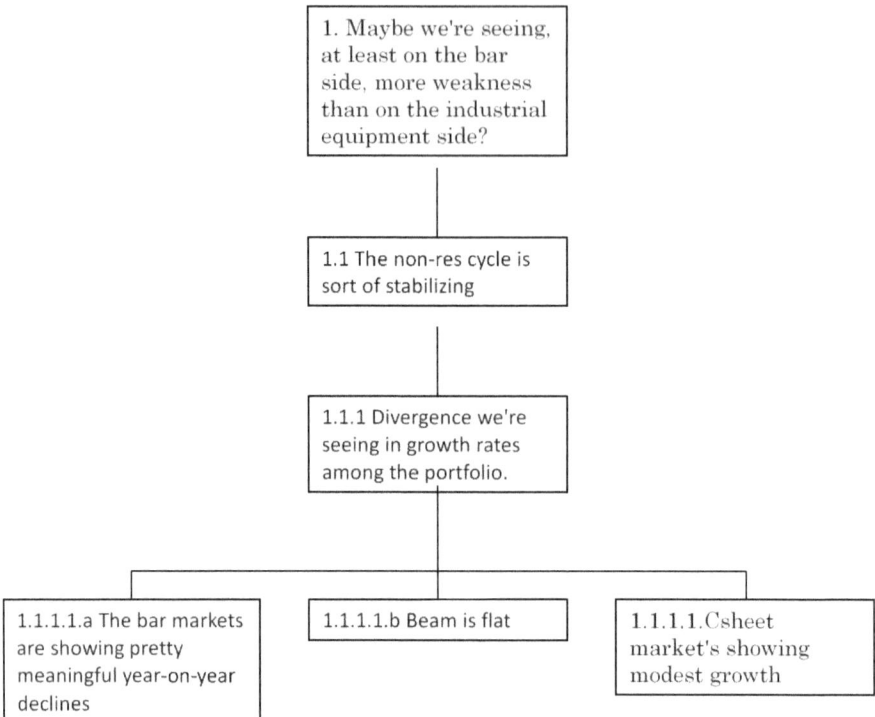

Figure 1 – Argumentation structure of the request of confirmation of inference in Example (1)

2. HYPOTHESES AND RESEARCH QUESTIONS

Following Palmieri (2018, p. 51) we maintain that in the process of argumentation "a piece of information becomes a relevant premise for decision-making or for an epistemic statement on which decisions are based" and, as a consequence, "understanding the inferential process by which a standpoint is justified sheds light on how financial information can concretely contribute to sound investment decisions".

The main hypothesis of this paper is therefore that dialogical argumentation is one of the factors that make ECCs informationally relevant and is a crucial determinant of the informational differences observed in prior studies. Argumentation does not necessarily involve the disclosure of new information. It always involves the use of old and/or new information as premises starting from which the standpoint is rationally justified. Argumentative inferences lead financial analysts to get new information about the firm, the firm's executives, and the executives' ability to understand their business and to account for their actions before their shareholders and the financial community.

From the viewpoint of argumentation theory, our research on ROCOI and other question-answer patterns contributes to the "argumentation in context" research agenda and, in particular, to the understanding of *argumentative patterns* (van Eemeren 2015). Here we understand argumentative patterns as recurrent sequences of argumentatively relevant moves that appear to be constrained by the main goal and by the set of institutional *rules, commitments* and *incentives* characterizing the activity type.

If we look at ECC as an activity type, it is not difficult to see how it can involve a critical discussion and what contextual constraints affect it. Companies, aiming at persuading investors to buy or keep their shares, can use argumentation to communicate information strategically by attempting to favour a particular inferential processing of the disclosed information. Executives thus have an incentive to take the role protagonists. Westbrook (2014) observes that preparing an ECC for the executives of a listed company is "like preparing a rhetorical case before a court of law in which the logic of an argument is presented in a way that shows a company has implemented sound business decisions in the face of the economic and market environment with which it has been dealing" (pp. 174-175).

Financial analysts, on the other hand, are expected to act on behalf of investors to enable them to obtain information and reach reasonable investment decisions. In principle, this should lead them to take the role of antagonists to critically test managers' standpoints. Given their obligations to shareholders, investors and regulators both parties are *de iure* committed to rationality in the discussion.

For a financial analysts, questioning firm's executives during conference calls is a sort of a play of delicate balances between opposing incentives: on a hand the analysts want to know the more as possible about the company, on the other hand they want to maintain their relationship with firm's executives in order to be able to get relevant information in the future. Additionally, they have to keep in mind that both their questions and executives' answers will be publicly available. In view of these considerations, certain empirical research questions concerning the behaviour of analysts arise:

> RQ1: To what extent analysts assume the role of antagonists in an argumentative discussion?
> RQ2: Do they challenge the protagonist to provide arguments when needed?

Here we argue that ROCOI is a prototypical *dialogical argumentative pattern* in ECC, aimed at creating a certain kind of argumentative confrontation and allowing analysts to play the role of critical antagonists to the managerial standpoints in a way that suits their strategic goals and the institutional preconditions of the activity type. In this connection, we raise a theoretical research questions and two empirical questions:

> RQ3: How does an argumentation pattern like ROCOI fit in a critical discussion?
> RQ4: What are the recurrent *argumentative indicators* associated with the ROCOI pattern?
> RQ5: What are the pragmatic, rhetorical, interactional and informational functions of the ROCOI argumentation pattern?

In the present, exploratory, study we examine qualitative and quantitative corpus evidence that suggests some preliminary answers to these questions, encouraging us to broaden and deepen the inquiry.

3. THE CORPUS

In order to answer these questions we rely on the annotation of a corpus of 8 ECCs (6 from 2013, and 2 from 2016), composed together by almost 98k tokens. On this corpus we apply a manual and a semi-automatic annotation of ECCs transcripts using the software UAMCorpusTool. UAMCT is a multi-layer hands-off corpus annotation tool by Mick O'Donnell[1].

To annotate the ECCs we employ an adapted version of the annotation scheme developed by Palmieri, Rocci and Kudrautsava (2015), which was used also in Budzynska, Rocci and Yaskorska (2014). The scheme categorizes both analyst questions and managerial replies.

For the questions the scheme is built according to their probable intended effect on the reply by corporate representatives, as shown in Figure 2, below.

---

[1] freely available at http://www.wagsoft.com/CorpusTool/index.html.

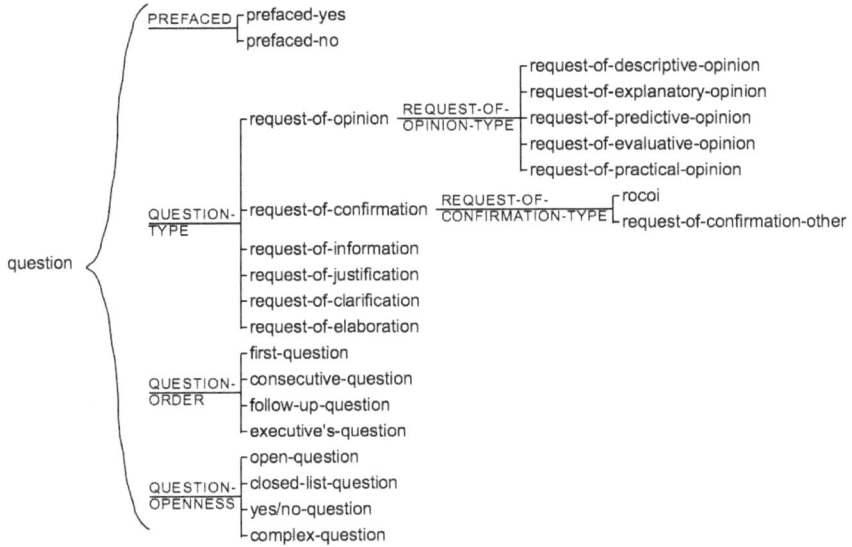

Figure 2 – UAM-CT annotation scheme for analysts' questions.

In other words, we have specified what kind of speech act the analyst requests (e.g. an opinion, a clarification, an elaboration, etc.). Therefore we have questions that are requests of elaboration, requests of justification, requests of clarification, requests of opinion, and requests of confirmation. Among the latter we distinguish between requests of confirmation of inference (the ROCOI) and simple requests of confirmation, asking to have confirmed something other than a reasoning. The opinions requested are annotated according to a semantic taxonomy distinguishing between *descriptive*, *predictive*, *explanatory*, *evaluative* and *practical* viewpoints. Coherently, the answers are annotated with a matching scheme. First of all, it is important to differentiate between those advancing justified standpoints and those containing mere, unsupported, opinions. We then use the same semantic taxonomy for the type of opinion or standpoint. For a full explanation of the annotation scheme, refer to Palmieri, Rocci and Kudrautsava (2015).

4. DESIGN OF ANALYSTS' QUESTIONS IN ECCS

It has been observed (Crawford Camiciottoli, 2009, p. 669), that, in contrast with what happens with journalists' questions in press conferences, indirect question frames are extremely common in ECCs, making up about two-thirds of the questions asked by analysts.

Typically, they include second person speech act verbs (*go back to, give more detail on, elaborate*), attitude verbs (*think, view, feel, expect*) and various adverbial expressions (*right, exactly*) to precisely shape the kind of communicative action that is requested from the corporate representatives, as shown in the following examples:

>Okay, but you're thinking of that as helping the core number grow, right?

>Okay, good. And then can you just go back to the avionics delays that you mentioned are driving that?

>And maybe give us a little more detail on exactly what's going on there?

Similarly to the ROCOI, the request for more details about a single fact or a reasoning, showcased by the last two examples, seems to be strategic in the effort to extract more information, in a way that is at the same time more polite and more demanding, leaving the burden of explaining the denial of an answer to the executive.

The abundant presence of such linguistic features made the functional annotation of questions described in the previous section more robust and reinforced the hypothesis that certain argumentatively relevant question designs give rise to conventionalized patterns associated to certain linguistic indicators.

5. RESULTS OF THE CORPUS ANNOTATION

From the annotation of the corpus made following the scheme shown in Section 3, we obtain some general descriptive characteristics of the whole corpus of ECCs. The corpus is formed by a total number of 266 questions the analysts asked to the corporate executives. More than half of the questions (58%) are preceded by an assertive *preface* (cf. Clayman and Heritage, 2002, p. 104) that lays the foundations for the intended question. Among all the questions almost half of them (47%) are requests of opinion, most of the time requests of predictive opinion (21%) and requests of evaluative opinion (18%).

This is coherent with arguing in the *interaction field* of finance (Rocci, 2014, p. 204), in which being able to precisely predict future development (linked to the predictive opinions) and being able to correctly assess the meaning of something that happened is very valuable and precious as it could lead each financial analyst to the gathering of a better understanding of the opportunity to invest in the specific company.

We then identified a number of requests of elaboration (18%), justification (9%), and clarification (3%). Finally, we identified 47 requests of confirmation, among which more than half (31 occurrences) are specifically ROCOI.

The direct challenge to defend a standpoint (i.e. the Request of Justification) seems to be less used than the ROCOI. The financial analysts tend to prefer a more indirect challenge that is performed using the ROCOI.

ALL CORPUS

Feature	n	%
prefaced-yes	155	58%
prefaced-no	111	42%
request-of-opinion	124	47%
request-of-explanatory-opinion	16	6%
request-of-predictive-opinion	56	21%
request-of-evaluative-opinion	48	18%
request-of-practical-opinion	15	6%
request-of-confirmation	51	19%
rocoi	31	12%
request-of-confirmation-other	20	8%
request-of-justification	23	9%
request-of-clarification	7	3%
request-of-elaboration	47	18%
TOTAL	266	100%

Table 1 – Analysts' questions in the corpus

We then zoomed on the ROCOI in order to understand for which kind of inference the confirmation is demanded in each single case.

ROCOI	N	%	N ecc	YES	NO	NA	other
ROCOI Inferred Description	5	16%	3	4	-	1	-
ROCOI Inferred Evaluation	3	10%	2	1	1		1
ROCOI Inferred Explanation	5	16%	3	2	1		2
ROCOI Inferred Practical Judgement	1	3%	1			1	
ROCOI Inferred Prediction	17	55%	8	10	6	1	
TOTAL	31	100%	8	17	8	3	3

Table 2 – A closer look at ROCOI in the ECC corpus

Table 2 shows the distribution of the kind of inference that is embedded in the ROCOI. Also inside the ROCOI, the role of the predictions seems to be pivotal, being 17 out of 31 ROCOI (55%) referred to predictive inference, i.e. inferences about the future of the company or about the future dynamics of processes related to the company business. We then found a 16% of ROCOI about inferred description and another 16% of ROCOI about inferred explanations. The ROCOI about inferred evaluation and inferred practical judgments are less frequent, respectively 10% and 3% in this corpus.

	N	%	ECCs
Mere descriptive opinions	4	10.26%	3
Mere explanatory opinions	4	10.26%	2
Mere predictive opinions	7	17.95%	5
Mere opinions (total)	15	38.46%	5
Justified descriptive standpoints	1	2.56%	1
Justified evaluative standpoints	4	10.26%	3
Justified explanatory standpoints	5	12.82%	3
Justified practical judgment standpoints	1	2.56%	1
Justified predictive standpoints	10	25.64%	6
Justified refusal to answer	3	7.69%	2
Justified answer (total)	24	61.54%	16
Grand Total	39	100.00%	21

Table 3 – Reactions to ROCOI

In Table 3 we report the analysis of the reactions to the ROCOI, namely if and how executives react to financial analysts when they ask a question in the form of a ROCOI. Almost two thirds of replies (24 cases) by corporate executive present supporting arguments (justified standpoints), while in more than one third of the cases (15 cases) only a mere opinion is provided, without any explicit supporting arguments sustaining the opinion. In a few cases (3 cases), managers refuse to answer the ROCOI providing arguments in support of their decision not to answer. These arguments refer to corporate disclosure policy or to the convenience for the firm to not disclose a specific piece of information. Consistently with the questions pattern, the predictive answers are the most common in the corpus. In order to explore in depth the interplay between the questions typology and the provided answers, we analyse the coherence between questions and answers.

	Total	Coherent Answers	Non Coherent Answers	Justified Standpoint	Mere opinion
ROCOI Inferred Prediction	17	16	1	12	5
ROCOI Inferred Description	5	4	1	2	3
ROCOI Inferred Evaluation	3	2	1	3	0
ROCOI Inferred Explanation	5	3	2	2	3
ROCOI Inferred Practical Judgements	1	1	0	1	0
TOTAL	31	26	5	20	11
Percentage		84%	16%	65%	35%

Table 4 – Coherence between ROCOI and reactions

Almost all replies to ROCOI (26 out of 31, the 84%) are coherent with respect to the type of opinion being asked. This seems to be consistent with a view in which the respondent is coherent with the dialogical commitment. Therefore the incoherent answers, namely the answers in which executives provide the analysts and the audience with a different kind of inference are a few, just 16%. Hence, the executives seem to tend to respond positively to the ROCOI the most times, replying negatively to ROCOI only one time in four (26%)

6. A DETAILED CASE STUDY

We selected one case for which we provide a full reconstruction of the argumentation structure both of the ROCOI and of the manager's answer.

> Bank of America Earning Conference Call Q2 2013, July 17, 2013, 8:30 am ET
>
> (3) Matt O'Connor, Deutsche Bank
> Within fixed income trading businesses, obviously June proved to be a tough month, I think, for a lot of folks, and it seemed like the trends were maybe a little bit weaker than were seeing elsewhere when we factor in some of the charges you had in the first quarter.
>
> (4) Bruce Thompson, Bank of America Corp, CFO
> A couple of things on the fixed income business. Let's look at it over year, because this business probably has the most seasonality with respect to the first

quarter. If you look at year over year, and if you looked within the businesses, within both the rates and currencies area as well as the different credit trading areas, which we look at investment grade, high yield, as well as our loan sales and trading, the performance year over year was actually pretty good.

Where we had weakness in the second quarter of this year is in three areas. The first is that we continue to run off the structured credit trading book and you had a pretty significant decline during the second quarter of í13 relative to the prior year from the continued runoff of that book. From a P&L perspective, it's largely runoff at this point, so we're not going to have to discuss that much going forward.

The other two areas on a relative basis that were weaker, we have a very significant business that's got number one market shares in the municipal finance space, and if you look at the prices in the spread widening, it was very dramatic during the month of June in the muni space. That negatively affected us. And then in the mortgage space, obviously the market widened out significantly there, and we had some lumpy items in the second quarter of 2012 as well.

So I think as you look at the quarter, and you look at the fixed income business, once again we run it as a holistic business between new issue and sales and trading. The new issue business had a great quarter. Those areas where the markets were good, rates and currencies, credit trading across the board, actually performed pretty well. In the three areas that I mentioned, one, because it's running off and two, given the market dynamics, didn't perform as well as we would have expected."

The first one is a case of a ROCOI in the form of a Request of confirmation of an inferred evaluation followed by a disconfirmation supported by a justified evaluative standpoint. The managerial argument elicited by the ROCOI is interesting: it first involves an *undercutter* (cf. Pollock, 1987) of the ROCOI's relevance, but then the manager moves on with a *rebutter* of the standpoint of the ROCOI. In Figure 3, we can see the analyst's argument paired with the undercutter: the CFO of Bank of America argues that the analyst is using the wrong benchmark comparing quarter by quarter instead than year by year.

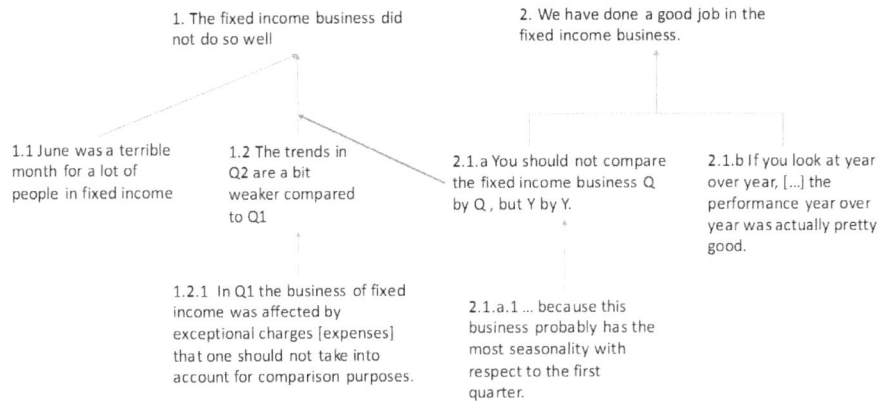

Figure 3 – Argumentation structure of examples (3) and (4) for the undercutter part.

The manager does not seem to entirely trust the effectiveness of the undercutter and goes with showing that also seen from a quarter by quarter angle the performance of their fixed income business is good for some items and that the negative performance of other items can be explained by external factors outside the managers' control (cf. Figure 4, below).

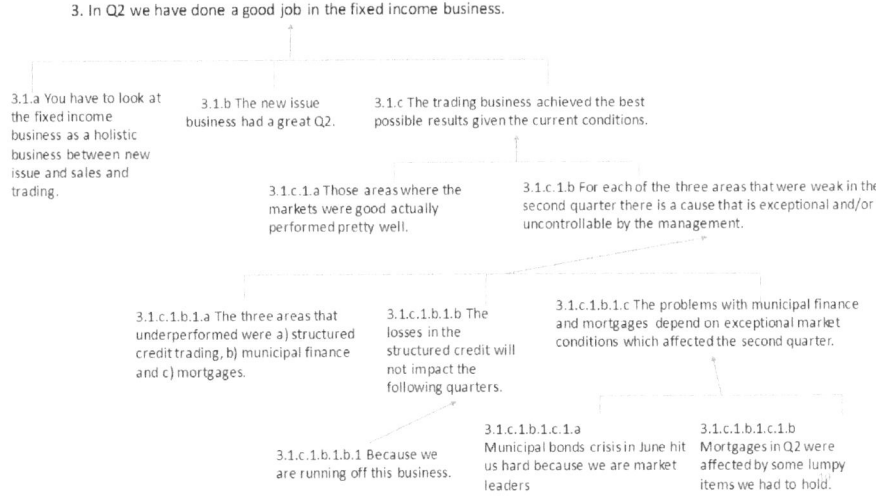

Figure 4 – Argumentation structure of example (4) for the rebutter part.

It is noteworthy here that the standpoint being defended by the manager does not concern the financial performance of the fixed income business – for that the disclosed numbers speak by themselves – but rather the *managerial* performance. It is also worth noticing how the manager's argument proceeds by disaggregating items in the fixed income business to let those that performed well shine and to explain away those that did not perform well.

7. LINGUISTIC INDICATORS OF ROCOI IN THE ECC CORPUS

A certain range of linguistic indicators is found to be repeatedly associated with the presence of the ROCOI. At the present stage of investigation, we have yet to establish how *statistically significant* these indicators are, i.e. how significantly more frequent they are in ROCOI compared to other question designs. We can however already establish that they are qualitatively *meaningful*, because (a) they appear repeatedly in ROCOIs and (b) their semantic and pragmatic function is coherent with the functional design of the ROCOI. We distinguish four categories of indicators of ROCOI:

i. Modal and evidential expressions: Obviously, I think, It seems like, Maybe, Probably, Normally, Perhaps, It sounds, It looks like, It's fair to say, Should, Certainly. This first group refers to a number of modal and evidential expression, which arguably have the double function (a) of signalling that the proposition to be confirmed is the more or less tentative result of the analyst's inferences (cf. Musi and Rocci 2017), and (b) they are likely used to hedge the meaning of the sentence in the effort of maintain a proper level pf politeness in the dialogue, also when the proposed ROCOI seem harsh or not desirable for the executive.

ii. Report of inferential operation: Putting [...] together, I am trying to see, am I just doing the math right?, I understand, I guess, We shouldn't anticipate, I'm wondering if I'm doing the math wrong. The expressions reported in this group referred to the inferential operation performed by the analysts.

iii. Explicit request of (dis-)confirmation: correct my thinking, what part of my logic would you like perhaps to correct?, just to make sure I've got [the data] about correct here, does that sound about right?, I just wanted to check if that's the case?, you don't think [...]?, is that the right line of thinking?, I'm wondering if I'm correct in my thinking there. This third group is specifically linked to the request of confirmation or disconfirmation of the inference.

iv. Hearsay and report: I think I hear, some lawyers say, Your disclosures say, as you just indicated, in line with where you would've thought, I know you highlighted, I know that [] is included in these numbers. These are also part of the evidential space. The analyst in a way shift the burden of his affirmation to the fact the same issue has already been presented previously.

8. DISCUSSION AND PRELIMINARY RESULTS

The preliminary inspection of ROCOI in the previous pages already enables us to answer the theoretical question RQ3, concerning their fit between their pragmatic functioning and the framework of the *critical discussion*. In terms of the Pragma-Dialectic theory of argumentation (van Eemeren and Grootendorst, 2004), ROCOI appear to create an argumentative confrontation. They do so in an indirect manner. By advancing a standpoint, albeit marked by a weak epistemic stance, the analyst is cast in the role of the *protagonist* and the manager in that of the *antagonist*.

More precisely, ROCOI have the capacity to create a *mixed confrontation*, where the antagonist becomes in turn protagonist of a negative standpoint on the same proposition. In fact, we can observe that managers, once requested to confirm or disconfirm the inference presented by the analyst cannot afford a purely antagonistic role. They are supposed not only to criticize the analyst's argument but also to propose an alternative view and back it up with arguments. This is particularly evident in the double line of argumentation of the manager in our little case study. The results of the annotation study also suggest that empirically ROCOI have a certain success in eliciting argumentation from managers.

More generally, the corpus data support also a partial answer to RQ1 and RQ2, painting the analysts as cautious and indirect antagonists. This does not necessarily mean that they are ineffective antagonists.

Our data, including the linguistic indicators (RQ4) we have reviewed, is consistent with the idea that ROCOI also play a politeness function (RQ5). By asking correction from manager's analysts cast themselves in a modest role, preserving the face of both participants and providing (nominal) escape routes.

It is quite clear, on the whole, that ROCOI also have an information probe function, performed by advancing hypotheses which they ask to confirm analysts seek to extend the limits of the information disclosed by the firm and the extent of managerial commitments. Finally ROCOI, may also have a bait function. In critical situations analysts can provide «lifeboat» benevolent interpretations (which they may not believe), while at the same time obliging the managers to support their credibility with arguments. We will explore further these functions in a series of in-depth case studies.

9. AGENDA FOR FUTURE WORK

Possible directions for future research and development include a number of possible streams of research. First, to analyse in detail the dimension of counter-argumentation of managers, distinguishing between rebutters and undercutters. Secondly, to develop a collection of analyses, showcasing the different functions of the ROCOIs. Third, building a comparative analysis of the usage of ROCOI, following different/opposite situations, i.e. crisis or sustained growth. Fourth, an argumentation mining perspective: a genre specific argumentation mining will be feasible, given the presence of particular argumentative indicators. Finally, it would be relevant to go deeper in the relationship between the financial communication and the financial fundamentals, in the effort to contribute to solve the dilemma about information and communication role in finance.

REFERENCES

Brown, L. D., Call, A. C., Clement, M. B., & Sharp, N. Y. (2015). Inside the 'black box' of sell-side financial analysts. *Journal of Accounting Research*, 53(1), 1–47. doi:10.1111/1475-679X.12067.

Budzynska, K., Rocci, A., & Yaskorska, O. (2014). Financial dialogue games: A protocol for earnings conference calls. In S. Parson et al. (Eds.), *Computational Models of Argument* (pp. 19-30). IOS Press.

Clayman, S., & Heritage, J. (2002). *The news interview: Journalists and public figures on the air*. Cambridge: Cambridge University Press.

Crawford Camiciottoli, B. (2009). 'Just wondering if you could comment on that': Indirect requests for information in corporate earnings calls. *Text and Talk*, 29(6), 661-81.

Crawford Camiciottoli, B. (2010). Earnings calls: Exploring an emerging financial reporting genre. *Discourse & Communication*, 4(4), 343–59.
Crawford Camiciottoli, B. (2013). *Rhetoric in financial discourse: A linguistic analysis of ICT-mediated disclosure genres*. Amsterdam/ New York: Rodopi.
de Oliveira, M. do C. L., & Rodrigues Pereira, S. M. (2017). Formulations in delicate actions: A study of analyst questions in earnings conference calls. *International Journal of Business Communication*.
Eemeren, F. H. van (2015). Identifying argumentative patterns: A vital step in the development of pragma-dialectics. *Argumentation, 29*(3), 1-23. doi:10.1007/s10503-015-9377-z.
Eemeren, F. H. van, & Grootendorst, R. (2004). *A systematic theory of argumentation: The pragma-dialectical approach*. Cambridge: Cambridge University Press.
Lev, B. (2012). *Winning investors over: Surprising truths about honesty, earnings guidance, and other ways to boost your stock price*. Boston, MA: Harvard Business Press.
Musi, E., & Rocci, A. (2017). Evidently epistential adverbs are argumentative indicators: A'corpus-based study. *Argument and Computation*, 8(2), 175-192. doi:10.3233/AAC-170023.
Palmieri, R., Rocci, A., & Kudrautsava, N. (2015). Argumentation in earnings conference calls: Corporate standpoints and analysts' challenges. *Studies in Communication Sciences*, 15(1), 120-32.
Pollock, J. L. (1987). Defeasible reasoning. *Cognitive Science*, 11(4) 481–518.
Rocci, A. (2014). The discourse system of financial communication. *Cahiers de l'ILSL*, 34, 201-21.
Rocci, A., & Raimondo, C. (2018). Conference calls: A communication perspective. In A. V. Laskin (Ed.). *The handbook of financial communication and investor relations* (pp. 293-308). Hoboken, NJ: JohnWiley & Sons.
Westbrook, I. (2014). *Strategic financial and investor communication: The stock price story*. Abingdon, Oxon:Routledge.

How to Create Rhetorical Exercises?

BENOÎT SANS
Université libre de Bruxelles (ULB) - GRAL
Benoit.Sans@ulb.ac.be

It is well-known that the Ancients taught rhetoric and argumentation thanks to various exercises (progymnasmata cycle, declamations, etc.) and it is tempting to use them to train contemporary students or pupils. However, Ancient exercises took place in a very different reality, not always relevant to nowadays teenagers. In this paper, I will show how, respecting the Ancient principles, we can adapt the Ancient pedagogical material and create new problems and exercises to train specific techniques.

KEYWORDS: argumentation, *controversia*, exercise, *ethos*, *logos*, *pathos*, practice, *progymnasmata*, rhetoric, *staseis*,

1. INTRODUCTION

Since 2013, I have been working on a research project that aims at reintroducing, in High Schools, rhetorical exercises inspired by those practised in Antiquity (Dainville & Sans, 2016; Ferry & Sans, 2014; Sans, 2017). A partnership has been made with a Belgian High School in order to test the Ancient exercises with contemporary pupils and to observe the effects of this training on their capacities and performances. The present paper intends to show how the Ancient pedagogical material can be adapted in order to create new exercises and train specific techniques.

Thanks to various sources, the Ancient rhetorical training is still well known, especially around the first centuries AD (Pernot, 2000, pp. 194-207). Somewhere during what we call Secundary or High School, the children were initiated to rhetoric by a *rhetor* through a set of exercises called *progymnasmata*, "preparatory exercises" (Gibson, 2008; Kennedy, 2003; Patillon, 2002; 2008; Webb, 2001). These exercises were organised so that the complexity would gradually increase, from basic writing exercises (e. g. a fable, a short narrative) to specific developments (e. g. description, praise) and complete argumentations (arguing for or against a law proposal). Thanks to these exercises they were ready to go to the second step, the declamations (Bonner, 1949; Russel, 1983; Sans 2015; Winterbottom, 1980), which were complete fictive speeches, supposed to be closer to real cases. There were two kinds of declamations: the *suasoria*, in which an action is either recommended or misadvised, and the *controversia*, which was a kind of trial simulation. Such exercises were in use long after Antiquity. It was then tempting to use them again in order to train our future citizens to rhetoric and argumentation.

Because it represented a typical and stimulating situation in which argumentation and rhetorical abilities are needed and which permitted twofold argumentation (Danblon, 2013, pp. 127-148; Ferry, 2013), I particularly focused on the *controversia*. The course was built around two kind of sessions: first, free sessions, devoted to the *controversia*, where the pupils were asked to argue, as best as they could, for both sides and to deliver their speeches in random pairs; the other pupils played the role of an audience and were asked to evaluate both performances according to rhetorical various criteria (relevance, consistency, reliability, emotions, etc.); and secondly, more theoretical sessions where I drew pupils attention on various rhetorical aspects or techniques through specific exercises. Yet, even before the beginning of the experiment, problems arose. Here is a typical example of a *controversia*:

> **Rule**: *let a tyrannicide opt for whatever he will.*
> An orator in a city ruled by a tyrant persuaded the tyrant to abdicate his tyranny. He asks for a reward as tyrannicide. (Sopatros, *Division of Questions*, 95, 21-98, 11)

Such exercises have often been ill-conceived: they often seem violent, complex and artificial (also in the way they were treated) and therefore useless. Moreover, even if one understands the usefulness of the *controversia* on a technical point of view, one faces a huge chronological and cultural gap: these exercises took place in an obsolete reality and a good classical background is needed to approach them. Unless the goal is to better understand the Ancient culture (Heath, 2007), acquiring this background could be a loss of time in the framework of an argumentation training. In order to take benefit from the advantages without the disadvantages, the Ancient pedagogical material had to be adapted. And that meant that we needed to know the reasons why rhetors trained their students with *controversiae* and how they built them.

2. IN THE RHETOR'S WORKSHOP: HOW TO CREATE A CONTROVERSIA?

To answer to the first question, we must go back to the Ancient treatises: *controversiae* illustrated the "theory of issues" (gr. *staseis*; lat. *status*), which is known thanks to Hermogenes (Heath, 1995; Patillon, 2009) Sopatros (Weissenberger, 2010) and many others (*Rhetorica ad Herennium*, Cicero, Quintilian,...). This theoretical system allowed the learners to identify a type of issue (Berry & Heath, 1997; Russell, 1983; Heath, 1997), in a list of thirteen (see table 1 in the appendix). The example quoted earlier, for instance, falls under the "definition" issue, because the discussion is not about the deed but about the way to name it and whether it is an instance of the alleged crime or not. This system also allowed to know the main arguments or argumentative "moves" available for each issue, which were called "headings" (gr. *kephalaia*; lat. *capitula*) (see the division of the definition issue in table 2). It has been experimented (Heath, 2007) and applied to current trials or debates (Kock, 2012). But it is also possible to build new examples from it.

The problem is that our sources gave us the theory, a lot of examples and achieved models, but no information about the creation process, probably because it was too obvious. Then, the researcher must rebuild these practical procedures from some disseminated clues and transform them into operational hypotheses, following a process similar to experimental archaeology or reverse engineering.

As teachers and professors who often give exercises to check whether students are able to apply the notions they learned, we may at least assume that with the *controversiae*, the rhetors wanted to see whether their students mastered the theory of issues and used it properly. An argument that may support this hypothesis is that the rhetors and manuals paid much attention to avoid several flaws that made the cases impossible to argue (gr. *asystatos*) or less acceptable (see table 3): in the *monomeres* issue, for instance, only one side can be argued; in the *kata to antistrephon* issue, both sides can use the same arguments. Through these examples, one understands that the aim of the rhetors was always to preserve space to balanced and fair argumentations on both sides. There relies the principle that guided the creation of new problems: to illustrate or to train to a specific issue or argumentation, one had at least to ensure that the corresponding arguments were possible and available for case (Sans, 2015). According to this principle, one is able to create new *controversiae* in a reality closer to the one of the pupils I worked with, to their everyday life, especially at school, like in the following example, called "The pupil with scissors":

> **Rule:** *it is strictly forbidden to bring weapons at school.*
> Before the beginning of a class, two pupils are violently arguing. One of them makes a rush at the other. Alarmed by the noise, a teacher enters in the classroom. The pupil who has made a rush turns back and goes back to his place. In this haste, scissors fall out his pocket. He is accused of attempting to use a weapon. He defends himself.

Or this other one called "Bulls' eye!"

> "**Rule**: *any aggression against another pupil or staff member may warrant expulsion*
> It was about 10:40 am when the young history teacher, who was hired this year, came to the schoolyard for surveillance. The pupils were playing basketball and the game seemed very tight. When the teacher turned his back, he was suddenly hit at the head by the ball, and lightly wounded. He easily identified the shooter: a gifted, but unruly pupil that he had punished many times for misbehaviour during his class. This time, the teacher accused him of aggression and demanded his expulsion."

These problems are built on the same issue than the first example. Of course, I may not pretend that such problems are equivalent to those used in Antiquity, especially on a psychological and cultural point of view; but what matters on a technical point of view, if the goal is to make produce arguments and argumentative strategies, is that the structure is preserved in order to reveal a similar discussion and consequently, the same argumentation possibilities. If one wants to create a definition issue, the deed attributed to one of the characters need to not exactly fit the law, but to be similar by some aspect. One can say for instance, in the scissors' case, that "scissors are a part of the school equipment and should not be considered as a weapon" (*horos*) or "that pupils are allowed to have scissors at school" (*antilepsis*), but also reply that "a weapon is simply something used to harm" (*anthorismos*) and "in that case, the scissors were intended to harm" (*metalepsis*). To sum up, by unveiling the structure, by adding or taking away oriented or ambiguous elements, one can open or close argumentation's possibilities or strategies (that means, other issues) and then, channel the rhetorical efforts.

3. TARGETING THE INTRINSIC PROOFS (LOGOS, PATHOS, ETHOS)

As the most advanced exercise of the *curriculum*, *controversia* was also made to give space to and to practice the previous parts of the training, including notions like the Aristotelian intrinsic proofs (*ethos, pathos, logos*). Consequently, it can be used as a frame within which these categories can be isolated and studied. But by using the same principle of reading a structure backwards, one can also design specific exercises to target specific techniques or proofs.

In the field of *logos* and argumentation, in order to stimulate the production of various arguments, the question to be asked is not how an argument is built, but what is needed to build an argument.

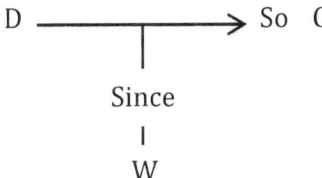

Looking back at Toulmin's (1958, p. 90) model, in the case of the above mentioned *controversiae*, the conclusion (C) is already known or chosen; and many of the warrants (W) one can imagine, like "if something has bad consequences, it should be avoided" or "if someone had the will and capacity to do something, she or he probably did it", are commonly available in the mind, especially in a familiar context or when various schemes have already been taught; most of the time, they are simply built through the connection between a data and a conclusion. Then, the only thing needed to build inferences and various arguments, is the data. Having a closer look at "Bull's eye", one can now see that specific data points have been inserted in order to practice various schemes or *topoi*: if one wants to practice the argument from consequences, one needs to put something that has potentially good or bad consequences (extreme punishment[1]); if one wants to create the possibility for the argument from the intention or causes, one needs something from which one can infer motivation (potential revenge); if one wants to test the *ad hominem* argument and make it available, one needs personal information about the characters ("The teacher is still

---

[1] In Ancient declamations, the penalty is often death or disinheritment. In adapted examples, I noticed that extreme punishment stimulate motivation and argumentation.

young and inexperienced. He probably took that decision too quickly to assert his authority"; "this pupil is still gifted at school and has a good potential, we should first think to his future"). The other rhetorical proofs can also be practiced in a similar way[2].

On the level of *pathos*, research in cognitive sciences (Ortony, Clore & Collins, 1988) have shown that emotions depend on the evaluation of the situation by the person who feels, and that this evaluation is linked to various parameters (like: Is it positive or negative? Does this concern me or another person? How do I feel about this person?). But many of these parameters can already be found in the definitions of the second book of Aristotle's *Rhetoric*. Pity, for instance, is defined as "a certain pain at an apparently destructive or painful evil happening to one who does not deserve it and which a person might expect himself or one of his own to suffer, and this when it seems close at hand" (*Rhet.* 2, 8, 1385b); and pity is linked to indignation, which is the emotion that one feels towards someone who commits injustice. As Micheli (2010) showed by analysing the past parliamentary debates in France on death penalty, these parameters can be adapted to speech or discourse. But the same parameters can also be learned and then used to try to provoke specific emotions by representing the elements of a situation in certain way. This is what I experimented with my pupils through the following exercise ("Marc and Veronica" or "The car accident"):

> Around 8 am, Marc, 45, salesperson in an appliance store, hit Veronica, a 35-year-old promising CEO. Veronica was not crossing on the crosswalk. She was having a phone conversation with a colleague at the moment of impact; she was not looking and did not see the car. She died before rescuers arrived. Marc was eager to take his children to school; he was driving at a speed of 47 km/h; the traffic light near the crossroad had just turned yellow.

This is not a *controversia*, but rather an accumulation of elements around one tragic event. Here, independently from any trial context, the pupils were asked to write two short reports, one that aimed at provoking pity or sympathy for the victim and one that aimed at provoking the same emotion for the driver. Here again, knowing the definitions and the aim they had to pursue, the pupils quickly understood how to create pity or indignation by using the available

---

[2] Analogy and example cannot be implemented in this way, but are often stimulated by a familiar context.

data. But in the second composition, some of the pupils tried to induce sympathy for Mark by simply blaming the victim, and received a bad appreciation from their classmates. It is not because the options are available, that there are equally interesting or appropriate (Ferry & Sans, 2015). On this occasion, they also learned to avoid inappropriate or aggressive strategies.

Finally, for what concerns the last intrinsic proof, *ethos*, I noticed that the pupils often expressed themselves severely and seemed unsympathetic when they assumed the role of a teacher. To make their students sensitive to this category, the Ancients recommended an exercise called *"ethopoiia"* or *"prosopopoiia"* (Amato & Schamp 2005; Hagen, 1966). This exercise, which was a part of the *progymnasmata*, consisted in imitating the *ethos*, sometimes associated to a *pathos*, of a given character in a given context through a first-person speech. Only a few information was given in a typical formula like "What words Hecuba might say when Troy was destroyed?"(Aphthonios, *Prog.* 2, 4, Patillon). Knowing that Hecuba was the wife of Priam, king of Troy, and lost many of her children in the war against the Greeks, like one can read in many famous plays, one could expect that in such circumstances, this character would have certain attitude and feelings according to her status, that were supposed to appear through her speech. As Aristotle (*Rhet.* I, 2, 1356a), Barthes (1970, p. 212) or Ducrot (1984, pp. 200-201) argued, the principle of *ethos* is that the orator by pronouncing his speech, by using certain arguments, expression and tone, says "I am this, I am not this". Following that principle, my pupils had to appear credible in their role and to make their classmates draw conclusions (by some kind of abductive reasoning) on the personality of their character (is he sympathetic? honest? etc.); here, the speech itself become the starting point of inferences. For this exercise, I used, among other scenarios, a classic movie and drama scene: the "good" and the "bad" cop interrogating a suspect.

All these exercises could not only be used to practice various arguments and rhetorical proofs. As arguing is a natural capacity, I also experienced that by giving the information needed to build a specific argument or proof, these exercises can also trigger inferences so that the learners will spontaneously produce the expected arguments (Sans, 2015); they are then a powerful pedagogical mean to discover argumentation and rhetoric and to develop a critical reflection from the learners' own compositions and performances.

4. CONCLUSION

To conclude, coming back to the objectives of the research project I mentioned in the beginning of this paper, the results were very encouraging (Dainville & Sans, 2016; Sans, 2017): in only a few sessions, pupils were better at identifying the points of discussion, spontaneously used the techniques they learned to make richer argumentations and to create emotions; they also became more conscious of their own image as they spoke, showing more amiability, determination and self-confidence. They maintained and reinforced these abilities in the next sessions. In the end, they were able to choose between several strategies and at the same time had developed their critical minds. However, after having learned how to find arguments and proofs (*inventio*), pupils were still facing difficulties with structure and consistence (*dispositio* and *memoria*). The next step of my research would be to work on these categories in similar way (why do we need an introduction or a narrative in a speech? In which cases?). But whatever the following, this experiment already helped to better understand the functioning of the Ancient exercises. It reminds us of this simple truth: good dishes are often made from good recipes.

REFERENCES

Amato, E., & Schamp, J. (2005), ἨΘΟΠΟΙΙΑ: *La représentation de caractères entre fiction scolaire et réalité vivante à l'époque impériale et tardive*. Salerno: Hélios.
Calboli Montefusco, L. (1986). *La dottrina degli "status" nelle retorica greca e romana*. Hildesheim: Olms-Weidmann.
Barthes, R. (1970). L'ancienne rhétorique. Aide-mémoire. *Communcations, 16*, 172-223.
Berry, D. H., & Heath, M. (1997). Oratory and declamation. In S. E. Porter (Ed), *Handbook of classical rhetoric in the Hellenistic period (330 B.-C.- A.D 400)* (pp. 393-420). Leyden: Brill.
Bonner, S. F. (1949). *Roman declamation in the late republic and early empire*. Berkeley, LA: University of California Press.
Dainville, J., & Sans, B. (2016). Teaching rhetoric today: Ancient exercises for contemporary citizens. *Educational Research and Reviews, 11*(20), 1925-1930.
Danblon, E. (2013). *L'homme rhétorique: Culture, raison, action*. Paris: Éditions du Cerf.
Ducrot, O. (1984). *Le dire et le dit*. Paris: Éditions de Minuit.
Ferry, V. (2013). The virtues of dissoi logoi. OSSA Conference Archive. 46. Retrived from

https://scholar.uwindsor.ca/ossaarchive/OSSA10/papersandcommentaries/46.

Ferry, V., & Sans, B. (2015). L'intelligence émotionnelle: Un art rhétorique. *Le langage et l'homme, 50*(2), 147-161.

Ferry, V., & Sans, B. (2014), Educating Rhetorical Consciousness in Argumentation. In *Proceedings of the 6th Annual Conference of the Canadian Association for the Study of Discours and Writing*, (pp. 96-112). St. Catharines, Ontario: Brock University.

Gibson, C. A. (2008). *Libanius' progymnasmata. Model exercises in Greek prose composition and rhetoric* (C. A. Gibson, Trans.). Atlanta: Society of Biblical Literature.

Hagen, H. M. (1966). *Ἠθοποιία. Zur Geschichte eines rhetorichen Begriffs*. Erlangen: Universität Erlangen-Nürnberg.

Heath, M. (2007). Teaching rhetorical argument today. In J. G. F. Powell (Ed.). *Logos: Rational argument in classical rhetoric* (pp. 105-122). London: University of London.

Heath, M. (1997). Invention. In S. E. Porter (Ed.), *Handbook of classical rhetoric in the Hellenistic period (330 B.-C.- A.D 400)* (pp. 90-119). Leyden: Brill.

Heath, M. (1995). *Hermogenes. On issues: Strategies of argument in later Greek rhetoric*. Oxford: Oxford University Press.

Kennedy, G. A. (2003). *Progymnasmata: Greek textbooks of prose composition and rhetoric*. Atlanta: Society of Biblical Literature.

Kock, Chr. (2012). A tool for rhetorical citizenship: Generalizing the status system. In Chr. Kock & L. Villadsen (Eds.). *Rhetorical citizenship and public deliberation* (pp. 279-295). Pennsylvania: The Pennsylvania State University Press.

Micheli, R. (2010). *L'émotion argumentée. L'abolition de la peine de mort dans le débat parlementaire français*. Paris: Éditions du Cerf.

Patillon, M. (2009). *Corpus Rhetoricum. T 2. Hermogène. Les États de la cause*. Texte établi et traduit par M. P. Paris: CUF.

Patillon, M. (2008). *Corpus Rhetoricum. T. 1. Préambule à la rhétorique. Aphtonios, progymnasmata. Pseudo-hermogène, progymnasmata* (M. Patillon Trans.). Paris: Collections des Universités de France.

Patillon, M. (2002). *Aelius Théon: Progymnasmata* (M. Patillon, Trans.). Paris: Collections des Universités de France.

Pernot, L. (2000). *La rhétorique dans l'Antiquité*. Paris: Librairie générale française.

Ortony, A., Clore, G., & Collins, A. (1988). *The cognitive structure of emotions*. Cambridge: Cambridge University Press.

Russell, D. A. (1983). *Greek declamation*. Cambridge: Cambridge University Press.

Sans, B. (2017). Des exercices anciens pour les citoyens de demain. Bilan d'un an d'enseignement de la rhétorique. *Enjeux, 91*, 114-135.

Sans, B. (2015). Exercer l'invention ou (ré)inventer la controverse. In V. Ferry, & B. Sans (Eds.). *Rhétorique et citoyenneté. Exercices de rhétorique*, 5. Retrieved from: https://rhetorique.revues.org/399.

Toulmin, S. E. (1958). *The use of argument*. Cambridge: Cambridge University Press.
Webb, R. (2001). The progymnasmata as practice. In Y. L. Too (Ed.), *Education in Greek and Roman antiquity* (pp. 289-315). Leyden: Brill.
Weissenberger, M. (2010). *Sopatri quaestionum divisio / Sopatros: Streitfälle. Gliederung und Ausarbeitung kontroverser Reden* (M. Weissenbeger, Ed & Trans.). Würzburg: Königshausen & Neumann.
Winterbottom, M. (1980). *Roman declamation*(M. Winterbottom, Ed.). Bristol: Bristol Classical Press.

APPENDIX

I	"Logical/rational" issues (gr. *staseis logikai* ; lat. *status/constitutiones rationales*)	
1	Conjecture (gr. *stochasmos* ; lat. *coniectura*)	The defence denies the alleged acts
2	Definition (gr. *horos* ; lat. *finitio*)	The defence concedes that it committed the acts alleged but denies that it is an instance of the crime of which it is accused
	Quality (gr. *poiotês* ; lat. *qualitas*)	
3	A. Counterplea (gr. *antilepsis* ; lat. *qualitas iuridicialis absoluta*)	The defence maintains that it had a right to act in the way it did
	B. Counterposition (gr. *antithesis, antithetikê* ; lat. *qualitas iuridicialis adsumptiva*)	The defence concedes that there was a *prima facie* wrong, but maintains that there are other factors which may counterbalance or negate that wrong. This can be done in four ways:
4	a. Counterstatement (gr. *antistasis* ; lat. *comparatio*)	The action was justified by some beneficial consequence
5	b. Counteraccusation (gr. *antegklêma* ; lat. *relatio criminis*)	The blame really relies with the victim
6	c. Transference (gr. *metastasis* ; lat. *remotio criminis*)	The blame really relies with some third party
7	d. Mitigation (gr. *syggnômê* ; lat. *concessio*)	The action can be explained in a way that reduces the defendant's culpability
8	C. Practical (gr. *praktikê* ; lat.	The issue is concerned

		qualitas negotialis)	with the evaluation of a proposed future action
9		Objection (gr. *metalepsis* ; lat. *translatio*)	The defence challenges the validity of the proceedings on the basis of some explicit legal provision
II		"Legal" issues (gr. *staseis nomikai* ; lat. *status/ constitutiones legales*)	
10		Letter and Intent (gr. *dianoia kai rhêton* ; lat. *scriptum et voluntas*)	One party tries to restrict the application of a law by appealing from its explicit content to its implicit intent.
11		Conflict of Law (gr. *antinomia* ; lat. *leges contrariae*)	The two parties dispute about which of two conflicting legal provisions should take precedence in the present case
12		Assimilation (gr. *syllogismos* ; lat. *ratiocinatio*)	One party tries to extend the application of a law beyond its explicit content by arguing that by implication it applies to similar cases
13		Ambiguity (gr. *amphibolia* ; lat. *ambiguitas*)	The two parties dispute about the interpretation of a law that is ambiguously expressed

Table 1 – Theory of issues (Heath, 2007; Russell, 1983)

Presentation (gr. *probolè*)	The prosecution uses the circumstances of the act in question to enhance its significance.
Definition (gr. *horos*)	The defence uses a strict definition to show that the act falls outside the category alleged by the prosecution
Counterdefinition (gr. *anthorismos*)	The prosecution proposes a looser definition which brings the act within the category...
Assimilation (gr. *syllogismos*)	and argues that the looser definition captures all significant features of the category.
Legislator's Intent (gr. *gnomê nomothetou*)	Each party argues that its definition reflects the underlying intention of the law.
Importance (gr. *pêlikotês*) & Relative importance (gr. *pros ti*)	Amplification is used to argue that what was done is inherently important, and more important that what was not done – or, on the other side, that what was done is inherently important and more important than what was.
Counterposition (gr. *antithesis*)	The act is defended by pointing to its beneficial consequences (counterstatement), shifting blame (counteraccusation or transference), or invoking mitigating circumstances (mitigation)
Objection (gr. *antilepsis*)	The prosecution tries to find some circumstance of the act which negates the counterposition...
Counterplea (gr. *metalepsis*)	...to which the defence responds by arguing that the act is legitimate in itself
Quality (gr. *poiotês*) & Intention (gr. *gnomê*)	An examination of the accused's life and character in general, and his (alleged) intention in the particular act in question
Conclusion (gr. *epilogos*)	Conclusion

Table 2 – Division of Definition issue (Heath, 2007; 1995)

I	Invalid causes (gr. *asystata*)	
1	Uncircumstantial (gr. *aperistaton*)	Purely arbitrary act (no bases for arguments)
2	Wholly equivalent (gr. *isazon*)	Identical arguments on both sides
3	One-sided (gr. *monomeres*) (sometimes assimilated to or distinguished from the issue « without colour », gr. *achrômon*)	The case can be made / argued only on one side
4	Insoluble (gr. *aporon*)	No resolution in principle (deadlock, *aporia*)
5	Reversible (gr. *kata to antistrephon*)	Arguments used by each side can be used as arguments against the same side
6	Implausible (gr. *apithanon*)	Incompatibility between character and acts (ex.: Socrates as a brothel-keeper)
7	Impossible (gr. *adunaton*)	Acts are impossible (ex.: Apollo lies).
8	Disreputable (gr. *adoxon* ou *aprepes*)	Indecent (ex.: a husband who prostitutes his wife)
II	Intermediary / not so good	Between invalid and valid causes
1	Ill-balanced (gr. *eteroppepes*)	One side has a far stronger case than the other
2	Flawed in invention (gr. *kakoplaston*)	Anachronism or factual error
3	Prejudiced (gr. *proeilêmmenon têi krisei*)	The jury's predisposition make the verdict a foregone conclusion

Table 3 – Invalid and intermediary causes (Heath, 1995; Russell, 1983)

50

Filling in the Gaps:
The Role of Audience Inference in Exigence and Ethos

BLAKE SCOTT
KU Leuven
blake.scott@student.kuleuven.be

I argue that the audience's active role in argumentation can be understood in terms of inferential contributions to the argumentative situation. I discuss two aspects of the argumentative situation that I claim must be inferentially established by the audience: exigence and ethos. I also examine how certain features of argument context, such as the medium of argumentation, constrain the audience's inferential contributions in ways that can either help or hinder constructive argumentation.

KEYWORDS: audience, rhetoric, inference, exigence, ethos

1. INTRODUCTION

In light of recent work on audience reception, (Tindale, 2015; Kjeldsen, 2016) in this paper I explore some of the problems that audiences face in receiving arguments. In so doing I hope to draw attention to a topic that has received less attention from argumentation theorists than perhaps it deserves, namely, the ways in which new media alter or modify the reception of argumentation. For the purposes of this paper, I will use the term "new media" to refer broadly to those forms of digital communication that have been made possible by the Internet.

My primary interest in exploring the effects of new media on argumentation is to challenge the following commonplace assumption—an assumption that is often taken as self-evident rather than an indication of a subject that requires further study. This assumption is that new media generally tend to promote or incentivize unproductive argumentation practices while preventing or inhibiting more constructive ones. To avoid confusion, by "constructive argumentation practices" I simply mean those practices which give the participants in an argument the ability to understand where their interlocutors are coming from, so to speak, and not whether any kind of consensus, agreement, or mutual understanding is actually achieved as a result. In particular, what I want to explore here is why it is that new media appear to encourage unproductive argumentation practices by looking at the unique constraints that these media provide to argumentative situations. I will here discuss some of the ways in which I believe that new media make it difficult for audiences to make accurate and appropriate inferences about certain constitutive features of argumentative situations. In this paper I will limit my discussion to only two such features of argumentative situations. These features are (1) the *exigence* motivating argumentation, and (2) the *ethos* of the arguer.

2. AUDIENCE RECEPTION AND NEW MEDIA

Rejecting the popular pessimism toward the spread of new technologies into our everyday lives, in *Smarter than you Think: How Technology is Changing our Minds for the Better* Clive Thompson reminds us that before the Internet came along, "most people rarely wrote anything at all for pleasure or intellectual satisfaction after graduating from high school or college" (Thompson, p. 48). Although Thompson's focus is not specifically directed toward argumentation, I think it is safe to conclude from the fact that these new technologies have led to an explosion of online writing, that much of this writing is argumentative in nature—as a cursory glance through one's Facebook feed will quickly reveal. If Thompson is right, then, contrary to the commonplace assumption mentioned above, it is not that new media have helped to deliver the final blow to a once thriving community of rational arguers. On the contrary, these new media have helped give birth to, and are now struggling to raise, a much larger and more engaged community of arguers than ever before seen; a period in which the writing and receiving of arguments has become a strikingly everyday phenomenon for more and more people.

From social media platforms and private forums to the dreaded comment section of online videos and news articles, the places where argumentation can now occur have expanded far beyond anything that was possible previously. Now well within the digital age, the introduction of the Internet into our everyday lives has enabled us not only to communicate in different ways but has also enabled us to argue with, and be in audience to, people which had been out of our immediate communicative reach just decades ago.

While many argumentation theorists have indeed acknowledged this extension of argumentation into the digital world, the worry that has motivated this paper is that our relatively young field of study continues to conceive of its object from a philosophical perspective that privileges a notion of argumentation best suited to traditional print media (e.g. academic articles, books, newspapers, etc.). Whether or not this worry is justified, it nonetheless invites the question as to whether our underlying philosophical understanding of argumentation is directly translatable to the kinds of argumentation that new media have made possible, or whether a closer look at new media themselves is needed.

Let us now look briefly at what makes new media distinct. One of the defining features of these digital media is their interactivity.[1] Interactivity, in brief, can be understood as a characteristic of particular media whereby the mediations between users and the medium itself are more quantitatively and qualitatively complex than those found in traditional ones. What is of particular relevance to the study of argumentation in the phenomenon of interactivity then are the ways that these interactive media involve a dramatic increase in audience activity. To confine my discussion to a single activity, I will here focus only on the inferential contributions of the audience.

Among many things that audiences do in argumentative situations, making inferences is undoubtedly one the most important. How are inferences to be understood here? In their recent book *The Enigma of Reason*, Hugo Mercier and Dan Sperber argue against a narrow conception of inference—which is usually restricted to a kind of explicit, higher-order reasoning process—for a more expanded one that includes a broader range of cognitive phenomena. Mercier and Sperber thus define inference generally as "the extraction of new information from information already available, whatever the process" (Mercier & Sperber, p. 53). According to this broader model, the notion of inference admits of a spectrum of conscious awareness that can be distinguished from other cognitive processes by the way in which the process of reasoning from premise(s) to conclusion is metacognized (Ibid., p. 66).

Following this understanding of inference, what I want to pursue here are some of the implications of this account with respect to the increased activity of audiences given the interactivity of new media. Or, more precisely, the ways in which audiences in digital spaces are uniquely engaged in active inferential processes that function to furnish

---

[1] See Warnick & Heineman (2012; pp. 51-61).

particular argumentative situations with relevant contextual information.

To open up this line of inquiry, I begin with the following question. What must audiences have to infer in argumentative situations? In any argumentative situation, i.e. irrespective of the medium through which it takes place, there are several structures that I would suggest are always present. First, there is always an event-situation relation. This is to say that particular argumentative situations always take place with reference to some event(s), or state of affairs that has given rise to the need for an argument in the first place. Second, there is always an arguer-audience relation. At the rhetorical level, this means that arguers and audiences will always have some view of their interlocutor in mind during the argumentation process.

Although these structures of argumentative situations may be more or less explicit to the parties involved, I would suggest that any situation we would want to call argumentative will necessarily involve these two relations. In other words, if we were to imagine ourselves asking some participants in an argument (either during or after) the following questions, they would be able to provide some account, however accurate, of the contents of these two structures: (1) what event(s) or state of affairs was it that brought about the need for you to have an argument in the first place? (2) Who is it that you take yourself to be producing arguments for (on the arguer side), or who is that you take yourself to be in audience to (on the audience side)?

In virtue of being able to answer these questions in any way at all, I am suggesting that our imaginary arguers have been engaged in the kinds of inferential processes as described by Mercier and Sperber that have helped them to establish for themselves the contents of these two necessary structures of argumentative situations. It is almost unimaginable that our imaginary arguers here would be unable to provide some kind of answer to these questions. To keep the structures I have mentioned clear, I will refer to them simply as (1) exigence and (2) ethos.

Now, while these structures are common to all argumentative situations, what is of more central importance is that the nature of the medium through which argumentation takes place will have some constraining or amplifying effect on the ease with which the arguers are able to answer these questions, i.e. are able to make these kinds of inferences concerning exigence and ethos. In many online spaces, for example, audiences must infer, from the available information given by the medium, relevant information concerning exigence and ethos in a way that is qualitatively different than in more traditional media. Yet, as with any situation, this information is critical for interpreting and evaluating the argumentation to which individuals find themselves in audience—particularly if we are committed to some version of the principle of charity.

In what follows I will discuss exigence and ethos briefly not in order to draw any strong conclusions about the merits of any particular medium of argumentation, which would demand empirical analyses of particular media, but rather to try and bring some clarity to this line of inquiry at a more general level. For as Perelman and Olbrechts-Tyteca point out, it is the task of the philosophy of argument to "isolate argument structures, the analysis of which must precede any experimentation" (Perelman & Olbrechts-Tyteca, p. 9).

Regarding exigence, I here follow the general thrust of Lloyd Bitzer's account of rhetorical situations. On this view, argumentation does not simply emerge ex nihilo but rather follows from a real or perceived exigence that gives rise to the need for some kind of discursive intervention. According to Bitzer, any exigence can be understood as an "imperfection marked by urgency...a defect, an obstacle, something wanting to be done, a thing which is other than it should be" (Bitzer, p. 6). Moreover, an exigence is specifically rhetorical "when it is capable of positive modification and when positive modification requires discourse or can be assisted by discourse" (Ibid., pp. 6-7). Most importantly, Bitzer claimed that: "In any rhetorical situation there will be at least one controlling exigence which functions as the organizing principle" (Ibid., p. 7).

Depending on the particular medium, the real exigence of an argument can be more or less difficult to determine. To take one example, consider the way in which platforms such a Facebook enable users to inferentially establish the exigence motivating their argumentation. Within Facebook arguments often take place among users in response to a shared piece of journalism. In other cases, users may engage in argumentation in response to a short post made by someone with whom they are closely or remotely connected via their network of friends. In such cases it is not so much that particular individuals are targeted by arguers as *the* specific audience, but rather that users may find themselves in audience by virtue of their own recognition that they have something to contribute to an ongoing discussion.

In either case, what is it that the audience is inferentially contributing? The first inferential step that the audience must engage in is to try and figure out where the arguer is "coming from", so to speak. This is to say that the audience must infer from the arguer's post or shared link where the arguer stands with respect to the exigence that gave rise to the argument in the first place. This first step, we might say, involves an inference regarding the social distance between oneself and the arguer on the basis of the standpoint taken by the arguer.[2]

In contrast to more static kinds of media such as direct speech or printed text, it seems to me that it is often more difficult to establish where arguers are "coming from" within new media. Some consequences of this, in many cases, are that much of the preliminary argumentation involves an effort among participants to clear up confusions and gaps between their respective points of departure, which can be made all the more difficult given the sheer amount of participants that are able to participate within these media: the logistics of who is being addressing can often be a major problem unto itself. Since new media platforms undoubtedly provide quick and easy access to a large and diverse audience, the complexity of mediations between an actual material event, say the announcement of a new government policy, or a reaction to an argument made by a Facebook friend in response to a piece of journalism, makes it difficult for arguers to establish a common ground upon which constructive argumentation can take place—especially when we take into consideration the different sources of information that users have at their disposal given their different political commitments, and personal affiliations, etc.

Thus, the phenomenon of "talking past one another", I would submit, is a consequence of a failure—in part due to the nature of the medium and in part due to the users—to adequately establish, at the very least, a tentative agreement on the exigence of the situation. In more rhetorical terms, the failure to adequately capture the exigence beneath the argumentation leaves the *stasis* of the argumentation ambiguous, and thus prevents arguer's from understanding one another's standpoints—the audience often fails to accurately infer the arguer's perceived understanding of the exigence. From the point of view of *The New Rhetoric*, such a state of affairs might be described as a rhetorical failure on the part of the arguer to choose adequate points of departure such that the possibility of persuasion might be realized. This point of view, however, given Perelman and Olbrechts-Tyteca's explicit intention to study print media, overemphasizes the responsibility of the arguer to the neglect of the medium itself. For when it comes to digital environments, where in most cases it would be practically speaking impossible to choose points of departure suitable to the entirety of one's

---

[2] My use of the term "social distance" draws on the work of Michel Meyer. (See Meyer, 2010; Turnbull, 2014).

audience, it seems to fall more on the side of the audience to read back from what is given by the arguer, the way in which the arguer understands the exigence of the situation.

In a similar way, the same problem emerges for ethos as well. Here, the inferential move on the part of the audience is one of having to interpret information about the arguer from two sources: the explicit discursive material and any available user information that the medium makes available. In all but the most tight-knight and well-developed argument fora, important information concerning the actual ethos of the arguer is difficult to determine.

Depending on the particular medium then, the degree of difficulty will vary in determining important features concerning the character of an arguer, such as their credibility, expertise, reliability, prior commitments, etc. Assuming that it is important to know the ethos of the person we are in audience to, it can be seen how media that make it difficult to make accurate inferences about an arguer's character will likely have some negative effect on the quality of the arguer-audience relation, i.e. the accuracy of the ethotic inferences between arguer and audience. Thus if I am correct about audiences having to make these inferences regardless of the quantity or quality of the information available to them, it follows that media which make this process more difficult will likely have a detrimental effect on the quality of the overall process of the argumentation, since the audience will have to make up this information deficit with (1) assumptions about the way in which the arguer understands the exigence motivating their argumentation, and, (2) assumptions about the ethos of the arguer. It seems to me that if constructive argumentation (as I have defined it) is our goal in arguing with one another, then inferences concerning these two important structures of argumentative situations ought never be left only to guesswork.

3. CONCLUSION

To conclude, I want to draw attention to four theses that, whether ultimately accepted or rejected, ought to orient future research on the role of new media in argumentation.

  i. *All argumentation is mediated.*

As a social practice, arguing never happens without mediation. It seems to me that the paradigm medium through which we have studied argumentation has been that of the print medium. As argumentation begins to take place more and more in digital spaces, it is worth asking whether or not some of the fundamental philosophical assumptions about the nature of arguments that we have inherited will need to be revised.

    ii. *Not all argumentative media are equal.*

While apparently trivial, the purpose of this claim is to emphasize the fact that different argumentative media amplify or intensify different aspects of argumentation. From a purely descriptive point of view, it is clear that an analysis of, say, a presidential debate would yield different kinds of emphasis depending on whether the debate was heard on the radio, watched on the television, or streamed online.[3]

    iii. *Some argumentative media enable audiences to make more accurate inferences about an arguer's point of departure.*

Some media make it more or less difficult for audiences to inferentially establish such features common to all argumentative situations as exigence and ethos. My hypothesis here, following Perelman and Olbrechts-Tyteca, is that constructive argumentation involves a certain appreciation of the point of departure of one's interlocutor. When an audience fails to see precisely where someone is "coming from" (i.e. how they understand the exigence, who they are, etc.), the process of argumentation confronts an obstacle: further argumentation is needed in order to correct for the initial misunderstanding, to fill in the initial arguer-audience gap. Given the temporal limits of argumentation, this can often detour the attention of those involved from addressing the original exigence.

    iv. *The fidelity of an argumentative medium can be assessed in terms of the degree to which it enables audiences to make accurate inferences about argumentative situations.*

When it comes to characterizing the quality of an argumentative medium, I propose that we can analyse a given medium in terms of its

---

[3] Consider Marshall McLuhan's famous analysis of the differences in the reception of the 1960 Kennedy-Nixon presidential debate depending on whether the audience had heard it over the radio or watched it on television. (See McLuhan, 2003)

ability to enable audiences to make accurate inferences about situational features of argumentation such as exigence and ethos, among others. I would call such media, "high fidelity". Consider the example of Facebook again. In some ways Facebook makes it easier for audiences to understand the exigence of an argument, as it enables users to make arguments in direct response to other arguments with an intuitive post and reply interface. When a user posts a news article, a video, or some other content, other users may reply directly to that content, thus making it easy (in some ways) to keep in mind the matter at hand. With respect to ethos, on the other hand, in some cases it can become rather difficult to determine exactly whom it is that one is arguing with. When arguing in the comments of public news articles, for example, it can be difficult to practice the appropriate rhetorical sensitivity to one's audience given the diversity of backgrounds, knowledge bases, and value hierarchies of one's interlocutors.

While these remarks have been admittedly cursory, I think that many of the issues I have touched upon have far-reaching theoretical implications concerning the way that we understand arguments as theoretical objects; how we understand argumentation as a social practice; and how we can approach the evaluation of arguments in ways that are appropriate to their nature and social function. Above all, I hope that any clarification that I have been able to provide in the direction of this line of inquiry will serve as a starting point for further research into the nature of new media argumentation.

REFERENCES

Bitzer, Lloyd. (1968). The rhetorical situation. *Philosophy & Rhetoric*, *1*(1), 1-14.
Kjeldsen, Jens E. (2016). Studying Rhetorical Audiences: A Call for Qualitative Reception Studies in Argumentation and Rhetoric. *Informal Logic*, *36*(2), 136-158.
McLuhan, Marshall. (2003). *Understanding media. The extensions of man: Critical Edition* (W. Terrance Gordon). Berkeley, CA: Gingko Press.
Mercier, Hugo and Sperber, Dan. (2017). *The Enigma of reason*. Cambridge, MA: Cambridge University Press.
Meyer, Michel. (2010). The Brussels school of rhetoric: From the new rhetoric to problematology. *Philosophy & Rhetoric*, *43*(4), pp. 403-429.
Perelman, Chaïm and Olbrechts-Tyteca, L. (1957). The new rhetoric. *Philosophy Today*, *1*(1), pp. 4-10.
Thompson, Clive. (2013). *Smarter than you think: How technologies are changing our minds for the better*. New York, NY: Penguin Books.

Tindale, C. W. (2015). *The philosophy of argument and audience reception.* Cambridge, NY: Cambridge University Press.

Turnbull, Nick. (2015). *Michel Meyer's problematology: Questioning and Society.* New York, NY: Bloomsbury.

51

Multimodal Argumentation in Factual Television

ANDREA SABINE SEDLACZEK
University of Vienna
a.sedlaczek@tele2.at

This paper contributes to the growing discussions about visual and multimodal argumentation in argumentation theory with a perspective of critical discourse analysis as well as approaches to semiotics and multimodality. Looking at the context of factual television programmes, I will explore the relationship between the semiotic manifestation of argumentation at the macro and micro level of complex multimodal texts and the cognitive inferential processes entailed in the interpretation of the argumentation by the recipients.

KEYWORDS: critical discourse analysis, inference, multimodal argumentation, multimodality, Peirce, semiotics, television

1. INTRODUCTION

The study of visual and multimodal argumentation has gained traction in the last 20 years, receiving increasing attention from a variety of approaches, investigating a wide range of communicative forms and media genres, and dealing with a number of pertinent issues that generally move between the material and cognitive aspects of multimodal argumentation (Groarke, Palczewski, & Godden, 2016; Kjeldsen, 2015). These three aspects – the theoretical and methodological approach, the text genre under investigation, and the theoretical issue at stake – are important starting points for the present paper. In this paper, I want to contribute to the study of multimodal argumentation from the perspective of multimodal critical discourse analysis, which connects argumentation theory with insights from critical discourse analysis (Reisigl & Wodak, 2016), approaches to multimodality (Jewitt, 2014), and the semiotic theory of Charles Sanders Peirce (Peirce, 1931–1958). Adopting this approach, I will investigate multimodal argumentation in audio-visual filmic texts in the context of factual television. In accordance with the conference theme – argumentation and inference – the question of inferences in multimodal argumentation will be at the core of this paper. More specifically, I will explore the relationship between the communicative manifestation of argumentation at the macro and micro level of complex audio-visual texts and the cognitive inferential processes entailed in the interpretation of the argumentation by the recipients.

This paper expands on a paper presented at the first ECA conference, in which I provided a first insight into my approach of multimodal critical discourse analysis for the investigation of multimodal argumentation in factual television (Sedlaczek, 2016a). The background of the research presented in this paper is a study on media discourses about climate change in the context of climate change mitigation initiatives on Austrian television (cf. also Sedlaczek, 2014, 2016b; Sedlaczek, 2017).

2. MULTIMODAL CRITICAL DISCOURSE ANALYSIS

The approach of multimodal critical discourse analysis propagated in this paper combines a discursive and a semiotic perspective on multimodal argumentation. The discursive perspective follows the Discourse-Historical Approach (Reisigl, 2014; Reisigl & Wodak, 2016), which sees discourse and argumentation as intricately linked. The DHA defines discourse as a set of context-dependent semiotic practices that are situated within various fields of social action and that are characterised by argumentativity and pluriperspectivity. Discourses involve the challenging and justification of validity claims of truth or normative rightness between various social actors who hold different points of view (Reisigl & Wodak, 2016, p. 27). Argumentation is thus situated in a context of dispute. Its basic purpose is to convince or persuade. The discursive perspective thus stresses the strategic aspect of argumentation. Argumentation involves strategic choices by the social actors in an argumentative exchange. These do not only involve thematic choices – *what* arguments are brought forward – but also choices of presentation and representation – *how* these arguments are presented and what semiotic means and realizations are used to construct the overall texts.

Looking at argumentation from a semiotic perspective, it can be defined as both a communicative and cognitive phenomenon. More specifically, argumentation is an abstract cognitive pattern of problem-solving (Reisigl, 2014, p. 70) that entails a functional relation between three basic elements: a claim or conclusion, one or several premises or arguments in support of this conclusion and a warrant or conclusion rule that links the premises with the conclusion. On the communicative surface level of texts, this abstract pattern becomes manifest in a functionally connected network of communicative acts (Reisigl, 2014, p. 70). In line with my multimodal approach, the label communicative acts encompasses both speech acts and other multimodal acts, such as the presentation of visual or auditory textual elements. The concept of multimodality, as it is increasingly discussed in discourse and communication research, grasps the way texts construct meaning with a variety of different semiotic means or resources (Jewitt, 2014). Refraining from a detailed reflection about the concept of mode in this paper, I mainly distinguish between verbal (spoken and written language), visual (static and moving images) and auditory (music and sounds) modes. In addition, other semiotic resources, such as gestures, prosodic features of the human voice as well as film-specific techniques (e.g. cinematography, editing) are understood as being important to consider in multimodal texts.

Following from the communicative manifestation of argumentation in texts, the recipient has to cognitively reconstruct the intended pattern in order to understand the argumentation. Owing to the fact that the concrete textual manifestation and the communicative intention behind it often do not correspond, inferences play a role in the process of interpretation in two basic ways. The first type of inferences are inferences about the meaning of the communicative acts themselves. These inferences, such as presuppositions or implicatures, can be called "pragmatic inferences" (Oswald, 2018). They can be distinguished from "argumentative inferences", which are inferences about the argumentative link between communicative acts, i.e. the link from premises to a conclusion via a conclusion rule. These inferences are especially relevant in enthymematic argumentation, in which some of the functional elements of argumentation are left implicit and have to be inferred cognitively (Reisigl, 2014, p. 72). Both pragmatic and argumentative inferences rely on contextual information as well as previous knowledge in addition to the explicitly textualised elements.

3. MULTIMODAL ARGUMENTATION FROM A SEMIOTIC PERSPECTIVE

I will now turn to a closer discussion of the link between the communicative and cognitive aspect of multimodal argumentation. Specifically, I will use the concepts from the semiotic theory of Charles Sanders Peirce to conceptualise the relationship between the communicative manifestation of argumentation on the surface level of multimodal texts and the cognitive inferential processes entailed in the interpretation of the argumentation by the recipients.

Previous contributions in the literature have already theorised certain aspects of visual and multimodal argumentation in reference to some of Peirce's semiotic concepts (e.g. Behr, 2005; Gross, 2011; van den Hoven, 2011, 2015). These contributions have mostly referred to his most famous conceptual triad of iconic, indexical and symbolic signs, and sometimes to his argumentatively relevant distinction between abduction, induction and deduction as types of inference. In this paper, I want to demonstrate that a more comprehensive consideration of Peirce's theory can provide further important insights concerning multimodal argumentation.

Peirce advances a triadic sign concept that involves a relationship between three central elements of a sign: (1) the material manifestation of the *sign*, (2) the reference of the sign to a material reality or *object*, and (3) the interpretation of the sign and the construction of meaning in the *interpretant* (CP 2.228, 2.274).[1] On each of these three poles, Peirce makes further triadic distinctions. These distinctions are founded in his phenomenological categories of Firstness as the category of possibility, Secondness as the category of (re)action and Thirdness as the category of mediation and law (Deledalle-Rhodes, 2007, p. 237). In the following discussion, I want to use Peirce's categories on the three sign poles to examine presentational, referential and cognitive aspects of multimodal representation and argumentation respectively. An overview of Peirce's categories and their meanings is given in Table 1. For succinct outlines of Peirce's sign model see Lefebvre (2007); Reisigl (2017).

---

[1] In line with accepted Peirce scholarship, references to the Collected Papers of Charles Sanders Peirce (Peirce, 1931–1958) are made with the abbreviation CP, followed by the numbers of the volume and paragraph.

Phenomenological categories: Sign poles:	**Firstness** *possibility*	**Secondness** *(re)action*	**Thirdness** *mediation/law*
Sign *presentational aspects*	qualisign/tone *potentiality*	sinsign/token *actuality*	legisign/type *habituality*
Object (immediate, dynamic) *referential aspects*	icon *similarity*	index *contiguity*	symbol *convention*
Interpretant (immediate, dynamic, final) *cognitive aspects*	rheme *term*	dicent *proposition*	argument *inference* (abduction, induction, deduction)

Table 1 – Categories of Peirce's sign model

The first pole of the sign can be used to investigate the presentational aspects of a multimodal text. Here, Peirce distinguishes between qualisigns, sinsigns and legisigns (Reisigl, 2017, p. 24). As has been mentioned before, audio-visual texts are composed of a variety of semiotic choices with a meaning potential. This connects to Peirce's concept of qualisigns – qualities of material signs that can become signs. All the qualities that verbal, visual or audio-visual texts consist of, such as verbal or visual elements, colour, layout, prosodic features, sounds etc., have the potential to function as signs and take on a meaning. Some of these qualities have a conventional or rule-based interpretation, which would make them legisigns or types in Peirce's terminology. As such, they are conventional resources that are particularly useful as strategic choices for the construction of arguments. Both qualisigns and legisigns are actualised on the communicative surface level of texts as sinsigns or tokens.

The second pole of the object refers to referential aspects of the relationship between sign and object. In reference to the object, Peirce distinguishes between the dynamic object as the external material reality and the immediate object as it is already represented in a sign (CP 4.536). Concerning the relationship between the sign and its dynamic object, Peirce distinguishes between iconic, indexical and symbolic signs. If the relationship of the sign to its object is based on similarity, it is iconic; if it is based on a material or deictic connection, it is indexical; and if it is based on a convention, it is symbolic (Lefebvre, 2007, p. 223). It is worth emphasising that this sign-object-relationship is functional and always depends on the specific usage of a sign in a communicative context. Accordingly, the same sign can be used in different communicative contexts to refer to a variety of objects, constituting different sign-object-relations (CP 2.230). A photographic image of a polar bear, for example, could be used as an iconic representation of an animal that can be identified as a polar bear. In a different communicative act, it could be employed as an indexical sign indicating the fact that there was a polar bear in front of the camera lens at the time the photo was taken. In other contexts, the same image could be used as a symbol for the object "climate change".

The third pole of the interpretant refers to cognitive aspects of sign interpretation. The interpretant is significant, because signs in Peirce's view only become signs, if they are interpreted (CP 2.308). The "meaning" of a sign thus only emerges as a context-dependent, actualised interpretation. Peirce distinguishes three kinds of interpretants: immediate, dynamic and final (CP 4.536). The immediate interpretant refers to the possibility of a sign being interpreted in a certain way. The dynamic interpretant is the actual effect a sign has on the interpreter. This effect could be an emotional response, a cognitive understanding or an action (for example as the effect of a persuasive text – for example in advertising – that manipulates the interpreter into doing something). The dynamic interpretant corresponds to what speech act theory calls the perlocutionary effect of a communicative act (Reisigl, 2017, p. 24). The final interpretant is described by Peirce as the interpretant that would be formed, if the sign would be given enough consideration. It is an ideal interpretation of how the sign represents and interprets its object. This final interpretant can be compared to what we as argumentation theorists or discourse analysts do when analysing and interpreting texts or providing argument reconstructions.

In relation to this final interpretant, Peirce distinguishes three types of sign-interpretant relations: rheme, dicent and argument (Lefebvre, 2007, p. 223ff.). A rheme is an interpretation of a sign as a sign of possible meaning, for example a single term taken out of context that has no truth-value in it. A dicent is a more complex sign that consists of at least two rhematic signs, one of which has to be an index. The dicent is a sign of fact that is linked to the world with an index and thus has a truth-value, for example a photograph or a proposition. An argument in Peirce's view is an even more complex sign that is a pattern of reasoning or inference based on a law.

Peirce distinguishes three types of inference: deduction, induction and abduction (CP 5.161). These can be linked to the sign-object relations: Deduction relies on symbols to draw logical conclusions based on a known rule. Induction is dependent on indices, i.e. multiple instances of evidence, from which a general rule can be inferred. Finally, abduction forms hypotheses based on icons (Merrell & Quiroz, 2010). Abduction does not only play a role in argumentative inferences, but in pragmatic inferences as well, which involve the construction of meaning from multimodal texts that involve iconic signs.

Peirce's concepts regarding the interpretant are especially relevant for multimodal argumentation, as they can inform previous discussions in the literature on how images or other non-verbal signs can function as premises or conclusions in an argument and whether visual or multimodal argumentation has to be propositional (Roque, 2015). Considering Peirce's theory, one can maintain a propositional view of argumentation, by taking a wide understanding of propositions (or "dicents" in Peirce's terminology). The central component of a dicent that ascribes it a truth-value is an index indicating the referent in the world. An index, however, does not have to be an explicitly expressed (verbal or visual) element. Contextual information or background knowledge can also provide an index (Lefebvre, 2007, p. 226). In addition to the index, a dicent includes a sign asserting some information about the indexed object. This sign is usually symbolic, but may also be iconic. In the latter case, it forms the basis of an abductive inference, in which the interpreter has a more active role in reconstructing a proposition out of the iconic perception as a plausible hypothesis, based on the available information and contextual knowledge (Reisigl, 2017, p. 25).

4. MULTIMODAL ARGUMENTATION IN A TV SPOT

In order to illustrate these theoretic discussions with concrete examples, I will analyse an instance of multimodal argumentation in an audio-visual text. The example is a TV spot from an initiative on Austrian public service television called "*Mutter Erde*" ("Mother Earth"), which launched a campaign intending to promote climate change mitigation. The campaign was titled "2 degrees are more than you think". This title refers to the goal of the Paris climate agreement of the United Nations Framework Convention on Climate Change to limit the increase in the global average temperature to below 2 °C above pre-industrial levels.[2] The TV spot presents a very carefully constructed multimodal argument that mainly relies on visual and auditory means.[3]

The spot is 1 minute and 45 seconds long and consists of five shots that present a clear argumentative chain: The first shot constructs the main claim, the next three shots present three connected arguments in support of this claim, and the last shot adds a further claim and a normative conclusion following from the first argumentation. The spot features a green background throughout that is reminiscent of a chalkboard. On this background, a hand is seen in close-up, writing and drawing with a white chalk and a red felt pen. Accompanying sound effects reinforce the impression of a hand writing on a chalkboard. The transition between shots is provided in the form of diagonal or horizontal swipes, with which a fresh chalkboard surface is displayed.

All these different choices of cinematography and montage are potential qualisigns in Peirce's terminology. When interpreting the meaning or argumentation of this spot, however, not all semiotic choices will be interpreted as equally relevant. While some elements contribute directly to the multimodal argumentation, others are only presentational choices that add to the overall effect of the spot. They will influence how the spot is perceived and how persuasive the argument is for the recipients. In Peirce's terms, these choices may lead to different dynamic interpretant effects.

The first claim of the spot is co-constructed with verbal and visual means. A visual metaphoric representation of the Earth as a melting scoop of ice cream in an ice cream cone is slid into view on the right-hand side of the frame, while on the left, the hand is writing the verbal statement "2 degrees are more than you think". The spatial co-occurrence between these two elements is a strong indexical link that establishes an interpretative connection between them. An additional highly relevant visual element is the red colour of the text "2 degrees". Colour is a potential qualisign that can assume different meanings. In the context of weather and climate, the red colour is a legisign, as it has a conventional meaning, denoting warm temperatures. The choice of the red colour adds to the meaning of the rhematic symbol "2 degrees" by identifying it as a warming of 2 degrees. Finally, the iconic visual metaphor of the Earth as a melting scoop of ice cream specifies both the

---

[2] See the Paris Agreement, Article 2, paragraph 1, online under: http://unfccc.int/resource/docs/2015/cop21/eng/l09r01.pdf.

[3] The spot can be watched online under the following link: https://www.muttererde.at/motherearth/uploads/2017/05/MutterErde_The menschwerpunkt-Klimawandel2.mp4

scope and significance of this warming, i.e. a warming of the global temperature that is having negative impacts on the Earth. The final propositional interpretant of this first claim, as it is formed from these relevant elements and background knowledge, could be thus reconstructed as follows: "A rise in the global average temperature of 2°C would have more negative impacts on the Earth than you think". The phrase "more than you think" establishes the argumentative context. The claim is positioned as a counter-claim to a belief that is assumed as common among the viewers of this spot, namely that a warming of 2°C cannot make much difference.

After establishing this claim in the first shot, the TV spot then presents three causal argument schemes in support of the claim. These arguments show three prospective impacts of a rise in the global average temperature of 2°C. Apart from the verbal symbol "+2°", the construction of these arguments relies solely on iconic and indexical visuals and sounds. The temperature change is displayed on a thermometer, which starts with a medium temperature in each of the three arguments. Using the white chalk, the hand alternatively draws houses above a wave line, crops and a person skiing on a white mountain (the last drawing is additionally accompanied by jovial alpine folk music). After adding a red bar to the thermometer and inserting "+2°" next to it, the hand uses the red felt pen to change the respective drawing, while sound effects help to clarify the intended meaning: In the first drawing, additional wave lines are added over the houses, accompanied by the sound of flowing water (see Figure 1). In the second drawing, clouds with lightning and snowflakes are drawn above the crops – accompanied by the sounds of thunder, rain and wind – and the crops are partly erased with an eraser. In the third drawing, the white mountain is covered in blades of grass and flowers, while birds are chirping, and the person skiing is changed to a person standing on the mountain and holding his skis in an upright position, with a question mark above his head.

Figure 1 – First causal argument scheme in the TV spot (own drawing)

Abductive (pragmatic) inferences are crucial to infer the intended meanings of these iconic representations (e.g. interpreting the wave lines as water or the white colour of the mountain as snow). Taking contextual information and previous knowledge into account, the final interpretants of the presented causal argument schemes could thus be reconstructed as follows: (1) "If the temperature increased by 2°C, houses would be flooded through rising water levels". (2) "If the temperature increased by 2°C, crops would be destroyed by severe weather". (3) "If the temperature increased by 2°C, the fun of skiing (in Austria) would end due to lack of snow".

Together, these hypothetical causal argument schemes constitute an argument by example that supports the claim from the beginning that a warming of 2°C would have more negative impacts than you think. (In order for this argumentative link to make sense, there has to be an intermediary inferential step, inferring that the examples given are negative impacts.)

From this argumentation, a further verbalized claim is added in the last shot: "It is in our hands!". To connect this enthymematic claim to the previous argumentation, further inferential processes are involved. On the one hand, the claim "it is in our hands" can be connected to the hand that was seen as effecting the warming and the subsequent impacts. This can lead to an inference that we humans are the cause of this warming. On the other hand, the claim can also lead to an inference that we humans can avert the hypothetical warming and consequences, which connects to the final normative conclusion of the spot (in reference to the initiative): "Protect the climate now with *Mother Earth*".

The above examination of a TV spot wanted to demonstrate the usefulness of Peirce's categories in an analysis of multimodal argumentation in complex multimodal texts, as they help to grasp the interplay between presentational choices, intended meaning and inferential processes in the construction of multimodal argumentation.

5. MULTIMODAL ARGUMENTATION IN FACTUAL TELEVISION

The analysis in the last section – while certainly not exhaustive – provided a detailed look at the microstructure of a short TV spot that featured a carefully constructed multimodal argumentation. Moving from the genre of television spots to factual television programmes (such as news reports or various documentary film genres), several crucial differences can be observed. Factual television programmes are usually longer than a TV spot and they display more complex structures on their macro as well as micro level. Thus, multimodal argumentation in factual television programmes will crucially depend on the way the filmic text is edited and organised on various levels.

To investigate the interaction between the macrostructure and microstructure of a filmic text, insights from film analysis and text linguistics are helpful. On its basic level, a film consists of single shots that are combined to form scenes and sequences. Sequences are combined to form stages, which realise larger structures of the filmic text as a whole (Iedema, 2001). On these different levels, the filmic text can display different structuring principles. Text linguistics distinguishes five main patterns of text formation: narration, argumentation, description, explication and instruction (Reisigl, 2014). These patterns combine on different levels of the micro and macrostructure of a (multimodal) text. The dominant pattern of a text usually depends on its specific genre. Following Iedema (2001) as well as Bordwell and Thompson (2008), factual or documentary film genres can be attributed a narrative or expository form. The expository form is further subdivided into a rhetorical form (with an argumentative pattern) and a categorial form (with a descriptive or explicatory pattern). While the specific form of the film determines the dominant pattern on its macrostructure, these different patterns of text formation also manifest themselves in the scenes, sequences and stages of the filmic text on different levels. Thus, they can also intersect, i.e. a narrative scene may be part of a larger argumentative structure, by functioning as the premise for a conclusion.

The principles of drawing inferences from the communicative manifestation of a multimodal text that I demonstrated with the example of the TV spot in the last section equally apply here, but on a broader level. While the analysis on the microstructure investigated the relationship between visual, verbal and auditory elements within a shot and regarded them as signs that can take on different meanings, an analysis of the macrostructure of a filmic text will examine the complex relationship between the visual, verbal and sound tracks across scenes, sequences and stages, which constitute more complex signs that can equally construct meaning.

I will briefly discuss the way multimodal argumentation becomes manifest on the macrostructure of a filmic text with an example of a documentary feature that was broadcast in a culture and arts programme on Austrian television. The feature discusses the growing use of the passive house technology as a measure for climate change mitigation. It combines descriptive scenes composed of alternating close, medium and long shots of the exterior and interior of passive houses with a continuing voice-over, accompanying music, and talking-head shots of architects and inhabitants of the passive houses shown. While the visual scenes and the verbal voice-over themselves are mostly descriptive and explanatory, they are embedded in an overall argumentative structure. In three stages, the feature first presents a single-family passive house and constructs it as a 'paradise' (accompanied by soft music), then discusses the ecological and aesthetic merits of the passive house technology (accompanied by a playful xylophone music), and finally introduces a passive house residential building, with which a vibrant urban lifestyle is emphasised (reinforced by the up-tempo song "The Time is now" by the English-Irish pop duo Moloko). The succession of these stages and their multimodal construction suggest a contrast between the two examples of passive houses given – a contrast that is made explicit and gains argumentative relevance in the verbal conclusion by the voice-over at the end of the feature:

> Rethinking is necessary, climate experts warn, because one can protect resources considerably better in the multi-storey residential building than in the single-family house in the green countryside. But is this even imaginable in Austria, the land of the home builders? (Eller & Huemer, 2012, own translation).

While a more detailed discussion of this documentary feature is beyond the scope of this paper (cf. also Sedlaczek, 2016b), it becomes clear that the argumentative value of the feature can be inferred from the interplay between the descriptive scenes at the microstructure and the argumentative stages at the macrostructure, involving visuals, music, interviews and the verbal voice-over.

6. CONCLUSION

My objective in this paper was to show the potential of a discursive and semiotic perspective for the analysis of multimodal argumentation in complex multimodal texts in the context of factual television. The discursive view focuses on the strategic aspects of communicative acts and the variety of semiotic choices that are combined in complex structures on the micro and macro level of audio-visual texts. The semiotic view employs the semiotic concepts of Peirce to reflect on the interplay of presentational, referential and cognitive aspects in the production and interpretation of multimodal texts. Accordingly, the discursive and semiotic approach of multimodal critical discourse analysis was used in this paper to examine the relationship between the communicative manifestation of argumentation in complex audio-visual texts and the cognitive inferential processes that are involved in the interpretation of the argumentation by the recipients.

REFERENCES

Behr, M. (2005). Argumentation durch Bilder: ein Aspekt politischer Ikonographie. In K. Sachs-Hombach (Ed.), *Bildwissenschaft zwischen Reflexion und Anwendung* (pp. 212–229). Köln: Von Halem.

Bordwell, D., & Thompson, K. (2008). *Film art. An introduction* (8th ed.). Boston, MA: McGraw-Hill.

Deledalle-Rhodes, J. (2007). The relevance of C. S. Peirce for sociosemiotics. *Sign Systems Studies, 35*(1–2), 231–248.

Eller, N., & Huemer, K. (Writers) & S. Pichl (Director). (2012). Unser Klima 2012: Energiewende / Klimahäuser, *Kulturmontag*. Wien: Österreichischer Rundfunk, ORF.

Groarke, L., Palczewski, C. H., & Godden, D. (2016). Navigating the visual turn in argument. *Argumentation & Advocacy, 52*(4), 217–235.

Gross, A. G. (2011). A model for the division of semiotic labor in scientific argument: The interaction of words and images. *Science in Context, 24*(4), 517–544.

Iedema, R. (2001). Analysing film and television: a social semiotic account of Hospital: An Unhealthy Business. In T. van Leeuwen & C. Jewitt (Eds.), *Handbook of visual analysis* (pp. 183–204). London: Sage.

Jewitt, C. (Ed.) (2014). *The Routledge handbook of multimodal analysis* (2nd ed.). London: Routledge.

Kjeldsen, J. E. (2015). The study of visual and multimodal argumentation. *Argumentation, 29*(2), 115–132.

Lefebvre, M. (2007). The art of pointing. On Peirce, indexicality, and photographic images. In J. Elkins (Ed.), *Photography Theory* (pp. 220–244). New York: Routledge.

Merrell, F., & Queiroz, J. (2010). Icons and abduction. *Signs, 4*, 162–178.

Oswald, S. (2018). Pragmatic inference and argumentative inference. In S. Oswald & D. Maillat (Eds.), *Argumentation and Inference: Proceedings of the 2nd European Conference on Argumentation, Fribourg 2017* (Vol. II, pp. 615–629). London: College Publications..

Peirce, C. S. (1931–1958). *Collected papers of Charles Sanders Peirce* (C. Hartshorne, P. Weiss, & A. W. Burks, Eds.). Cambridge: Harvard University Press.

Reisigl, M. (2014). Argumentation analysis and the discourse-historical approach. A methodological framework. In C. Hart & P. Cap (Eds.), *Contemporary critical discourse studies* (pp. 67–96). London: Bloomsbury.

Reisigl, M. (2017). Diskurssemiotik nach Peirce. In E. W. B. Hess-Lüttich, H. Kämper, M. Reisigl & I. H. Warnke (Eds.), *Diskurs – semiotisch. Aspekte multiformaler Diskurskodierung* (pp. 3–30). Berlin: de Gruyter.

Reisigl, M., & Wodak, R. (2016). The discourse-historical approach (DHA). In R. Wodak & M. Meyer (Eds.), *Methods of critical discourse studies* (3rd ed., pp. 23–61). London: Sage.

Roque, G. (2015). Should visual arguments be propositional in order to be arguments? *Argumentation, 29*(2), 177–195.

Sedlaczek, A. S. (2014). Multimodale Repräsentation von Klimawandel und Klimaschutz. Eine theoretische und methodologische Annäherung am Beispiel des Österreichischen Rundfunks. *Wiener Linguistische Gazette, 78A*, 14–33.

Sedlaczek, A. S. (2016a). Multimodal argumentation in a climate protection initiative on Austrian television. In D. Mohammed & M. Lewiński (Eds.), *Argumentation and reasoned action: Proceedings of the 1st European Conference on Argumentation, Lisbon 2015* (Vol. II, pp. 933–946). London: College Publications.

Sedlaczek, A. S. (2016b). Representations of climate change in documentary television: Integrating an ecolinguistic and ecosemiotic perspective into a multimodal critical discourse analysis. *Language & Ecology*. Retrieved from http://ecolinguistics-association.org/journal

Sedlaczek, A. S. (2017). The field-specific representation of climate change in factual television: A multimodal critical discourse analysis. *Critical Discourse Studies, 14*(5), 480–496.

van den Hoven, P. (2011). Iconicity in visual and verbal argumentation. In F. H. van Eemeren, B. Garssen, D. Godden, & G. Mitchell (Eds.), *Proceedings of the 7th Conference of the International Society for the Study of Argumentation: ISSA 2010* (pp. 831–842). Amsterdam: Sic Sat.

van den Hoven, P. (2015). Cognitive semiotics in argumentation. *Argumentation, 29*(2), 157–176.

52

Bakhtin at the White House:
The Argumentative Dimension of the Direct Address in the TV Series House of Cards

CARMEN SPANÒ
University of Auckland, New Zealand
carmenspano78@gmail.com

ANTONIO BOVA
Franklin University Switzerland, Switzerland
abova@fus.edu

CARLO GALIMBERTI
Università Cattolica del Sacro Cuore di Milano, Italy
carlo.galimberti@unicatt.it

DANIELA TACCHI
Università Cattolica del Sacro Cuore di Milano, Italy
daniela.tacchi@libero.it

ILARIA VERGINE
Università Cattolica del Sacro Cuore di Milano, Italy
ilaria.vergine1@gmail.com

In the field of the media studies, the articulation between diegetic and extra-diegetic has become a privileged place for the exploration of textual phenomena that characterize media objects. The direct address can be considered an ideal 'border place' for the articulation of the diegetic and extra-diegetic dimensions. This paper will analyze the use of the direct address in the TV series House of Cards (first season) from textual, argumentative and interlocutory points of view.

KEYWORDS: Audience Engagement, Direct Address, Illocutionary Analysis, Inference, Strategic Maneuvering

1. INTRODUCTION

The American TV series House of Cards (2013 – in production) is a Netflix Original conceived and produced by Beau Willimon and David Fincher. It is an adaptation of the homonymous British TV show broadcast by BBC in 1990, which was in turn based on the novel by Michael Dobbs. The plot of the series centres around the actions of Frank and Claire Underwood, a powerful couple who lives in Washington D.C. and whose main goal in life is to climb the ladder of power up to the top, regardless of what this process might entail. In the construction of its storylines, House of Cards is characterized by a significant use of the direct address technique. This element of the representation stands as the peculiar trait of the show, to the point that its use comes out to as significant in affecting the development of the plot, the depiction of characters' profiles and the audience mode of reception. The main protagonist Frank Underwood is the one who is responsible for repeatedly breaking the fourth wall - in the first four seasons of House of Cards, the privilege to address the audience directly is only his. Through Frank's asides, the series constructs a reality that is not limited to the diegetic boundaries of its representation; in this sense, the direct address can be described as an ideal 'border place' for the articulation of both the diegetic and the extra-diegetic dimensions of the text. Indeed, the distribution of information between characters and viewers is strictly dependent on the privileged relationship between the main protagonist and the public. As a discursive strategy employed to disseminate information as well as convey traits of the characters' personalities to an external entity, the direct address contributes to create a frame of expression that exists only for Frank and the spectators. Within this suspended time and space, the character distances himself from the world depicted on the screen and moves towards the world on the other side; this peculiar condition exemplifies the constant blurring of diegetic and extra-diegetic. Furthermore, the direct address is employed to convey the idea of politics as a form of a spectacle that finds in the argumentation its driving force – in this sense, it stands as a reminder of Frank's ability to manipulate. The direct address helps to re-create the basic conditions in which Frank exerts his fascination through the persuasive use of words, both within the universe of *House of Cards* by interacting with the other characters and outside of it when he talks to the audience. This dynamic calls attention to the use of the direct address as instrumental to the constant alternation and overlapping of diegetic and extra-diegetic.

On the basis of these features of the direct addresses in *House of Cards*, we will consider the inferences solicited and allowed by Frank Underwood's asides as a suitable object for the exploration of the argumentative processes that derive from their dialogic-conversational nature. In particular, we will highlight the interweave of the cognitive, psycho-social and cultural dimensions that characterize Underwood's interpellations. The practice of the direct address will be analysed with the final intent to: a) understand how reasoning, but more generally inference, influences and shapes the three aspects - topical selection, adaptation to audience demand, and presentational choices - of Frank Underwood's strategic maneuvering; b) clarify how Frank Underwood's mental/discursive representation of the audience plays a role in the processing of his argumentation with the final goal to simplify their understanding and their acceptance by the audience itself; c) investigate the illocutionary, psychosocial and cognitive dimensions of the inferences triggered by the argumentative elements of the direct address. These objectives will be achieved through the analysis of two scenes of direct address selected from the first season of *House of Cards*. In this paper we will give an account of the argumentative dimension (van Eemeren, 2010; van Eemeren, Grootendorst, 2004; van Eemeren, Houtlosser, 2002) of the two asides, even if a preparatory work for this analysis has been previously done on their textual (Danesi, 2002; Eugeni, 2008), discursive (Schreier, 2012) and interlocutory (Searle, Vanderveken, 1985; Trognon & Batt, 2010; Trognon et. al., 2011) levels, to better understand how all these layers of meaning production concur to articulate both the diegetic and extra-diegetic dimensions. Some traces of these analyses could be only found in the Conclusion, given the fact that they will be the object of future papers.

2. ANALYSIS

2.1. Discourse analysis

The direct address makes easier for the audience to understand both the diegetic events and the protagonist's point of view. In fact, the analysis of the verbal layer highlights how the main character interprets, understands and evaluates diegetic elements. In particular, three main aspects emerge from the discourse analysis of the entire *corpus* of direct addresses that belong to the first season. Firstly, Frank Underwood shows to the audience his thoughts regard to three themes: elements of the frame, elements about other characters and elements about himself. Secondly, the extra-diegetic audience perceives Frank

Underwood as a tourist guide that explains how a politic has to behave in order to achieve his goals in the politic world of Washington. Furthermore, the main character uses a linguistic style rich of metaphors in order to facilitate the comprehension of his point of view. The content of metaphors is typically linked to violent images, food and animals (i.e. *"she's as tough as a two-dollar steak"*). This depicts the protagonist's idea of being a strong predator, so the audience realizes Frank Underwood is a character that behaves actively in the diegetic world. Thirdly, through direct addresses, the protagonist gives an example of the phenomenon of social influence (Galimberti, & Lecci, 2017). Co-occurrences in the discourse illustrate the main respects that Frank Underwood associates to himself: authority, power, and control. In his opinion, manipulating others and having power depend on being able to manage or control the communication in relational situations. Relational aspects of power are fundamental to achieve goals.

2.2. Argumentative analysis

Frank Underwood, the main character of the show, wants to convince two different audiences that he is the right person to lead the country: the diegetic audience, i.e., the American people inside the show, and the extra-diegetic audience, i.e., the people watching the show. In this study, we will specifically focus on the extra-diegetic audience and on the argumentative strategies used by Frank to convince this particular audience. To do so, we selected and analysed two scenes from the first season of House of Cards in which Frank talks directly to the extra-diegetic audience.

> Example 1: Welcome to Washington (S01, E01)
>
> President-elect Garrett Walker. Do I like him? No. Do I believe in him? That's beside the point. Any politician that gets 70 million votes has tapped into something larger than himself, larger than even me, as much as I hate to admit it. Look at that winning smile, those trusting eyes, I latched onto him early on and made myself vital. After 22 years in Congress, I can smell which way the wind is blowing. Jim Matthews, his right Honourable Vice President. Former Governor of Pennsylvania. He did his duty in delivering the Keystone State, bless his heart. Now they're about to put him out to pasture. But he looks happy enough, doesn't he? For some, it's simply the size of the chair. Linda Vasquez, Walker's chief of staff. I got her hired. She's a woman, check. And a Latina, check. But more

important than that, she's as tough as a two-dollar steak. Check, check, check. When it comes to the White House, you not only need the keys in your back-pocket, you need the gatekeeper. As for me, I'm just the lowly House Majority Whip. I keep things moving in a Congress chocked by pettiness and lassitude. My job is to clear the pipes and keep the sludge moving. But I won't have to be a plumber much longer. I've done my time. I've backed the right man. Give and take. Welcome to Washington.

In this sequence, Frank wants to convince the extra-diegetic audience, i.e., the people watching the show, that he deserves a more prominent political role than the one he has now. We can describe Frank's standpoint by using his words: *"I won't have to be a plumber much longer"*. What are the arguments used by Frank to convince this audience? In this first excerpt, we have identified three different types of argument:

(1) "I've done my time".

The first argument used by Frank is based on the importance of being an experienced politician. He starts his direct address by introducing the President-elected Garrett Walker and informing the audience that he supported him during the presidential elections. Frank says that he spent 22 years in Congress, and after all this time he acquired a crucial political skill: *"I can smell which way the wind is blowing"*. It does not matter if he believes in President-elected Walker's qualities or not. According to Frank, what is important is that he supported the right man.

(2) "For some, it's simply the size of the chair".

The second argument used by Frank is based on the importance of being ambitious, and he immediately wants the audience to know that he is very much ambitious. After the President-elected Garrett Walker, he introduces the Vice President Jim Matthews. Franks describes him as a politician with no power: *"He did his duty in delivering the Keystone State, bless his heart. Now they're about to put him out to pasture. But he looks happy enough, doesn't he?"*. According to Frank, Vice President Matthews looks happy even if he doesn't have any power, since *"for some, it's simply the size of the chair"*. Frank tells the audience that the vice-presidency is not enough to give him the power he wants. He does not look for a nice, big chair. He wants the presidency, the real power.

(3) "When it comes to the White House, you not only need the keys in your back-pocket, you need the gatekeeper".

The third argument used by Frank is based on the importance, in a place like the White House and more in general in a city like Washington, of having control over people and information. After having introduced the President-elected Garrett Walker and the Vice President Jim Matthews, Frank adds a third character - Linda Vasquez, the White House Chief of Staff. Because of her role, Linda is described by Frank as the person that controls who goes through the office of the President. Linda is the gatekeeper of the White House, and Frank tells the audience that he has control over the gatekeeper, because he "*got her hired*".

Example 2: Speech in the church (S01, E03)

Good Morning. Thank you, Reverend. And thanks to the choir for that beautiful hymn. I want to read this morning from... No. You know what no one wants to talk about? Hate. I know all about hate. It starts in your gut. Deep down here. Where it stirs and churns. And then it rises. Hate rises, fast and volcanic. It erupts, hot on the breath. Your eyes go wide with fire. You clench your teeth so hard you think they'll shatter. I hate you, God. I hate you! Don't tell me you haven't said those words before. I know you have. We all have. If you've ever felt soul-crushing loss. There are two parents with us today who know that pain. The most terrible hurt of all, losing a child before her time. If Dean and Leanne were to stand up right now and scream those awful words of hate, could we blame them? I couldn't. At least their hatred I can understand. I can grasp it. But God's wantonness, his cruelty, I can't even begin to... My father dropped dead of a heart attack at the age of 43, 43 years old. And when he died, I looked up to God and I said those words. Because my father was so young, so full of life, so full of dreams. Why would God take him from us? Truth be told, I never really knew him, or what his dreams were. He was quiet, timid, almost invisible. My mother didn't think much of him. My mother's mother hated him. The man never scratched the surface of life. Maybe it's best he died so young, he wasn't doing much but taking space. But that doesn't make for a very powerful eulogy, now does it? I wept, I screamed, "Why God? How can I not hate you when you steal from me the person I most love and admire in this world? I don't understand and I hate you for it". The Bible says in Proverbs, "Trust in the Lord with all your heart and lean not on your understanding. Lean not on your own understanding". God is telling us to trust him. To love him despite our own ignorance. After all, what is faith if it doesn't endure when we are tested the most? We will never understand why God took Jessica, or my father or anyone. And while God may not give us any answers, he has given us the capacity for love. Our job is to love him without questioning his plan. So I pray to you, dear Lord. I pray to you to help us strengthen our love for you, and to embrace Dean and Leanne with the warmth of your love in return. And I pray that you will help us fend off hatred, so that we may all truly trust in you with all our hearts, and lean not on our own understanding. Amen.

Also in this second sequence, it is possible to identify two audiences for Underwood's speech. One is represented by the people within the diegesis (the story), that is the people who are in the church. The other is represented by the viewers of the TV series, that is the people who occupy the extra-diegetic space and are also the recipients of Frank's direct addresses (written with bold and italics style), constantly on that border that separates the diegetic from the extra-diegetic. Frank Underwood's figure stands on the pulpit and seems to be confronting the people in the church. The public remains silent for all the time during the politician's speech, but the people appear to be moved by his words. The idea of confrontation that the scene presents further reinforces the concept of separation between Frank and the other characters. He seems to be always looking for an audience, in the attempt to use his manipulative skills so as to prove, to himself and us, that he can convince people and influence them.

From an argumentative perspective, in this sequence, Frank wants to convince the extra-diegetic audience, i.e., the people watching the show, that he is a valuable person, not like his father. We can describe Frank's standpoint by using his own words: "*I am not like my father. I am his opposite*". We have identified three different types of arguments advanced by Frank to support this standpoint:

(4) "He was quiet, timid, almost invisible".

The first argument used by Frank is based on his father's personality. He describes his father as quiet and timid. He was so quiet and so timid that, according to Frank, he was almost invisible. Frank, on the contrary, is not invisible, since is neither quite nor timid. He stands on the pulpit in the church, capable of moving the feelings of the people who are listening to him. He is also quite theatrical in the course of this speech since his primary goal is to shock his audience(s) not only by displaying anger but also indignation, and by revealing the truth right at the climax – *Why would God take him from us?!* - of the flow of his utterances.

> (5) "My mother didn't think much of him. My mother's mother hated him".

In his second argument, Frank mentions what his mother - his father's wife - and his mother's mother - thought about his father. They were two persons close to his father, and therefore we can infer that they knew him very well. Accordingly, what they thought is relevant in assessing his father's value. Frank tells us that his mother didn't think much of her husband, while his mothers' mother even hated him. The reasoning behind this argument is that if even the persons that spend an entire life with you doesn't consider you as valuable, then it means that you are not.

> (6) "He wasn't doing much but taking space".

In his third argument, Frank tells the extra-diegetic audience that his father was living a not successful life. He was not doing much but just taking space. We can infer in this case that Frank's father was taking space from people that, on the contrary, aim to live a successful life, doing much. People like Frank, much active in life and very ambitious.

2.2.1 Frank Underwood's strategic maneuvering

In the analysis of Frank's argumentative strategy, we will now examine all the three fundamental aspects of an argumentative strategic maneuvering as described by van Eemeren and Houtlosser (2002, p. 140):

> *Topical potential.* The first aspect of maneuvering strategically to be considered is the selection of the most expedient moves to make one's position stronger. Frank's choice is that of making clear to the audience what his goal is, i.e., *I won't have to be a plumber much longer*, and who his rivals are, i.e., President-elected and Vice President. Similarly, he decides to immediately describe to the audience some of the qualities he will use to achieve his goal: experience, ambition, and control over people and information.

Audience demand. The second aspect of strategic maneuvering puts the emphasis on the role of the audience. The argumentative choices should comply with audience demand, namely, the addressee's preferences. The arguments used by Frank aim to reveal to the audience the deepest, darkest features of his personality and the real reasons behind his behaviours. For example, he immediately shares with the audience his secret goal, i.e., I won't have to be a plumber much longer, and that he doesn't matter if he believes or not in a person that he supported. The reasoning behind this choice is that if the audience is the only one to know who he is and what he wants, it will actually understand his behaviours, and accordingly it will like him even more. By doing so, Frank adapts his argumentative strategy to what he believes the audience's preferences are.

Presentational devices. The third aspect in maneuvering strategically pertains to the selection of appropriate presentational devices at the communicative and stylistic level. Indeed, strategic maneuvering is realized in delivering the discourse through appropriate communicative means. Frank uses metaphors that allow to easily understand the sense of his intentions and the motivations behind his plans. Moreover, the use of irony and sarcasm characterizes Frank's speech directed to the extra-diegetic audience. By doing so, he tends to distance himself from all the other characters of the show, building an exclusive relationship with the audience, while also presenting himself as an experienced, ambitious and powerful individual.

Altogether, the argumentative analysis of the two scenes from the first season of House of Cards in which Frank talks directly to the extra-diegetic audience brings to light how he wants to convince the audience that he is the right person to lead the country because of his personality, and not because of his political ideas that can contribute to making his country a better place. It is all about his personality, his own skills, and strengths, while there is no mention of what he wants to realize for his own country. *"It's all about the politicians, not about politics".*

2.3 Interlocutory analysis

Now we would like to deepen the level of our analysis, focusing on the illocutionary layer by means of an elementary application of the Interlocutory analysis (Searle, Vanderveken, 1985; Trognon & Batt, 2010). The principal aim of this analysis is the description of some illocutionary mechanisms on which argumentation is based, to, therefore, be able to enlighten how they aliment the cognitive and psychosocial processes that constitute the phenomenology of interaction. To perform this task, we will take into account the first example, *"Welcome to Washington"* (S01, E01) analysed in the preceding paragraph.

In Fig. 1 we present the structure of Frank's standpoint as already described, specifying the illocutionary nature of the statements presupposed by the three arguments that compose it.

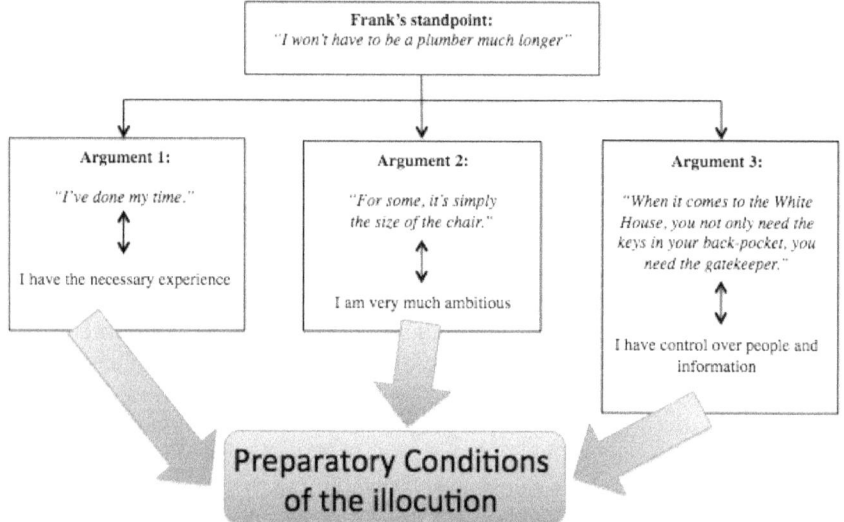

Figure 1 – Illocutionary nature of the three arguments

According to Searle and Vanderveken, the three statements constitute the 'preparatory conditions' of what is explicitly affirmed by the three arguments (1985, pp. 16-18). A preparatory condition is a state of affairs that must be presupposed by the speaker in employing a particular illocutionary force (in our cases asserting) and is a necessary condition for the non-defective employment of that force. In a real conversation, "in the performance of a speech act the speaker presupposes the satisfaction of all the preparatory conditions" (Searle, Vanderveken, 1985, p.17). In this case, the speaker, acting in a 'fictional conversation', needs to explicit the presuppositions useful to build the conversational context. Therefore, he triggers a process of co-construction of a 'cognitive environment' to be shared with the audience. In other words, by means of the modulation of the illocutionary values of his speech, Frank produce some effects on the cognitive layer and therefore on the psychosocial one. In fact, as the argumentative analysis has pointed out, here the cognitive effects 'stem' from the illocutionary choices and contribute to determine Frank Underwood's strategies in a twofold manner. First, by sharing his 'secret goal' with the audience, Frank puts himself on the border that divides the diegetic from the extra-diegetic dimension, accomplishing a double task: a) at the diegetic level, he contributes to build his own subjectivity; b) at the extra-diegetic level, he proposes to the audience the role of addressee of his asides. Second, by making use of irony and sarcasm, he exploits Grice's Maxim of Quality and therefore engages the audience in cooperative actions, establishing illocutionary relationships with strong relational fallouts on both the cognitive and the social layers (Vanderveken, 1990, pp.72-75). A further step in this direction can be performed by focusing on what we could call the 'pragma-semiotic root' of this sequence, that is the last remark "Welcome to Washington".

Let's take into consideration the illocutionary meaning of 'Welcome' as a verb. According to Searle and Vanderveken "to welcome somebody is to receive him with hospitality, and thus welcoming might be defined as an expression of pleasure or good feeling about the presence or arrival of someone. Welcoming... is essentially hearer-directed" (1985, p. 216). In the light of this consideration, we could say that audience became a reality composed by 'required guests'. In fact, *by saying* (illocutionary layer) "Welcome", Frank manifests his *intention* (illocutionary plus cognitive layers) to attribute a *double role* (illocutionary plus cognitive plus psychosocial layers) to the audience: a) a *dialogical role*, because since then the audience becomes an essential partner for the validation of his future direct addresses, leading them on a path that goes from extra- to intra-diegetic level; b) a *relational role*, because the audience becomes a real partner in the narrative process by virtue of a complementary path going from intra- to extra-diegetic dimension. In other words, by means of the arguments expressing his standpoint, Frank Underwood gives rise to an effective intertwinement among the actions accomplished by his talk (illocutionary layer), the discursive world (cognitive layer) and the roles he proposes to the audience (psycho-social layer).

3. CONCLUSION

On the basis of the preceding paragraphs, we can state that our analysis is intended to answer to questions raised in the media study field – i.e. the role of the direct address in TV series – by means of three kinds of technical tools: discourse, argumentative, and interlocutory analysis. As seen, the first one is specifically intended to prepare the way for the second and the third ones. In this paper we have particularly developed the argumentative analysis, focusing on how reasoning, inferences and strategic maneuvering give shape to a coherent discursive behaviour in direct addresses. The interlocutory analysis, for its part, has been used to describe some illocutionary mechanisms on which argumentation is based, enlightening how they aliment cognitive and psychosocial processes. Being conscious that this is only a first attempt to put in place a multi-technique approach that still needs to be developed and adapted to media studies field, nevertheless, we think that at least two (relevant) issues need to be presented at the end of this paper. First – as far as the definition of the research object is concerned – we think that our work shows that *argumentation* represents a relevant topic for today's media studies in audience reception (Galimberti & Spanò, 2017; Tindale, 2015). Allowing us to explore how the direct address technique gives birth to what we could call a 'mediated conversation', argumentation has shown to be a specific place where diegetic and extra-diegetic dimensions in TV series are articulated. And this is a clear gain for researchers who are interested in studying media also from an 'outer' perspective, not bounding themselves to the 'media text', but opening themselves to the 'media context', that is the social scene on which nowadays – via the social media – an important part of the fortune of TV series is determined. Second – as far as the methodological level is concerned – we believe that a non-reductionist stance (Bova, Arcidiacono & Clement, 2017; Galimberti & Trognon, 1996; Rigotti, Greco & Morasso, 2009; Trognon & Batt, 2010) is a consistent choice to approach psychosocial interaction via the study of inference in argumentative processes.

We also feel that the integrated approach we adopted in this study seems to have all that is needed for avoiding the reductionist traps and on the way to analyse the direct address as a mediatic, discursive, argumentative and illocutionary object worth to be explored to comprehend the role that it plays in the dynamics of audience reception processes.

REFERENCES

Bova, A., Arcidiacono, F., & Clément F. (2017). The transmission of what is taken for granted in children's socialization: The role of argumentation in family interactions. In C. Ilie & G. Garzone (Eds.), *Argumentation across communities of practice: Multi-disciplinary perspectives* (pp. 259-288). Amsterdam: Benjamins

Danesi, M. (2002). *Understanding media semiotics.* London: Arnold.

Eugeni, R., (2008) Grave danger: Il design dell'esperienza. In M. Pozzato, G. Grignaffini (Eds.), *Mondi seriali: Percorsi semiotici nella fiction* (pp. 51-69). Milano: Link Ricerca.

Eemeren, F. H. van (2010). *Strategic maneuvering in argumentative discourse.* Amsterdam/Philadelphia: John Benjamins

Eemeren, F. H. van, & Grootendorst, R. (2004). *A systematic theory of argumentation: The pragma-dialectical approach.* Cambridge: Cambridge University Press.

Eemeren, F. H. van, & Houtlosser, P. (2002). Strategic manoeuvring: Maintaining a delicate balance. In F. H. van Eemeren & P. Houtlosser (Eds.), *Dialectic and rhetoric: The warp and woof of argumentation analysis* (pp. 131-159). Dordrecht: Kluwer.

Galimberti, C. & Lecci, M. (2017), Autorità e influenza. Il punto di vista della psicologia sociale e alcuni possibili vantaggi per la ricerca storica. In M. P. Alberzoni & R. Lambertini (Eds.), *Autorità e consenso. Regnum e monarchia nell'europa medievale* (pp. 19-42). Milano: Vita e Pensiero.

Galimberti C., & Spanò C. (2017). *Intersubjectivity in media consumption as a result of the relation between texts and contexts: The cases of Game of Thrones*, Essais. École doctorale Montaigne-Humanités, Pessac.

Rigotti, E., & Greco Morasso, S. (2009). Argumentation as an object of interest and as a social and cultural resource. In N. MullerMirza, & A. -N. Perret-Clermont (Eds.), *Argumentation and education: Theoretical foundations and practices* (pp. 9–66). New York, NY: Springer.

Schreier, M. (2012). *Qualitative content analysis in practice.* London: Sage Publications.

Searle J., & Vanderveken D. (1985). *Foundations of illocutionary logic.* Cambridge: Cambridge University Press.

Tindale, C. W. (2015). *The philosophy of argument and audience reception.* Cambridge: Cambridge University Press.

Trognon, A., Galimberti, C. (1996). La virtù della discussione libera nelle decisioni di gruppo. In C. Regaglia & G. Scaratti (Eds.), *Conoscenza e azione nel lavoro sociale* (pp. 97-111). Roma: Armando.

Trognon, A., & Batt. M. (2010). Interlocutory logic: A unified framework for studying conversational interaction. In J. Streeck (Ed.), *New adventures in language and interaction* (pp. 9-46). Amsterdam: John Benjamins.

Trognon, A., Batt. M., Sorsana, C., & Saint-Dizier, V. (2011). Argumentation and dialogue. In A. Trognon, M. Batt, J. Caelen & D. Vernant (Eds.), *Logical properties of dialogue* (pp. 147-186). Nancy: Presses Universitaires de Nancy.

Vanderveken, D. (1990). *Meaning and speech acts: Principles of Language Use* (Vol. 1). Cambridge, UK: Cambridge University Press.

53

Arguments from Other Cases

KATHARINA STEVENS
University of Lethbridge
Katharina.stevens@uleth.ca

Arguers sometimes cite a decision made in an earlier situation as a reason for making the equivalent decision in a latter situation. I argue for two kinds of "arguments from other cases": those from precedent and those from parallel argument. They differ in their structures and conditions of cogency, even though they often look the same in presentation. Their similar appearance poses a risk of miss-evaluation and fallacious use, making a clearly theorized distinction important.

KEYWORDS: analogy, argument schemes, argument types, precedent

1. TWO KINDS OF ARGUMENTS FROM OTHER CASES

The fact that the decision to φ was made in one situation is sometimes used as a reason for making a decision to φ in another, similar situation. This is called case-to-case reasoning. In argumentation, an arguer may cite the decision to φ from an earlier situation as a premise in order to convince an interlocutor to make the decision to φ now. For ease of expression, I will call this an "argument from other cases" throughout the paper. I argue that there are two types of arguments from other cases. On the one hand, an arguer might use a normative argument by case-based analogy. On the other hand, she might use an argument by precedent.

A normative argument by case-based-analogy is used to show that making the advocated decision to φ in the current situation is as justified as making the decision to φ in the earlier situation. It supports the claim that the decision should be made by showing that it is an independently correct decision. By contrast, an argument by precedent is used to show that the advocated decision to φ should be made now just because the equivalent decision to φ was in fact made in the past. It supports the claim that the decision should be made by presenting the mere existence of the past decision to φ as a reason to decide to φ now. The difference between the argument types does not often appear as a difference in argument presentation. Both arguments use a past decision to φ to show that a present decision to φ should be made. The premises actually expressed are usually the fact that the decision was made, and the claim that the situation in which it was made is similar to the present situation. Conclusions are stated simply as the claim that some specific decision should be made now. This makes arguments from other cases susceptible for miss-uses and miss-evaluations. Unwary arguers might make - and unwary interlocutors may be persuaded by - an argument that relies neither on a legitimate claim for the relevance of the past decision, nor on a justified past decision.

2. TREATING LIKE CASES ALIKE – THE WEAK AND THE STRONG VERSION OF A PRINCIPLE

Case-to-case reasoning and arguments from other cases have a close relation to the principle that "like cases should be treated alike", which is widely regarded as a valid principle (e.g. Perelman and Olbrechts-Tyteca, 1969, p. 218; van Eemeren and Garrsen, 2014, p. 49, Garrsen, 2009). This principle is also often presented as the basis of common-law reasoning, and is used to justify the use of precedent in the law.[1]

---

[1] See, e.g. Jeremy Waldron (2012). He wants to defend the doctrine of precedent on *Rule of Law* grounds. He lists the principle to treat like cases alike among a number of reasons supporting the adoption of *stare decisis* that are regularly discussed in addition to rule-of-law-reasons. According to the doctrine of precedent, the fact that certain decisions were made in earlier cases is a reason to make them again in similar later cases, independently of the correctness of the earlier decision. However this applies to so-called *binding* precedent. Binding precedent is a precedent in a common-law system which was decided in the same jurisdiction and within the same doctrine as the present case by a court higher or as high as the present court. In common-law jurisdictions, a judge may only decide not to follow binding precedent if she can either distinguish (by claiming that there is a relevant difference between the cases); or if her court has the power to overrule and she finds the conditions for overruling have been met. However, these conditions are very strict: the precedent case's decision has to be wrong in such a way that following the precedent would result in extremely unjust outcomes. There is also *persuasive* precedent, precedent for which one of the conditions I listed is not fulfilled. In persuasive precedent, the merit of the decision does play a role because if a

However, in spite of its wide popularity, there is an ambiguity in what we mean when we cite it. In the next section I will characterize this ambiguity through what I see to be a weak and a strong reading of the principle. While the principle is universally valid on the weak reading, it is not on the strong reading.

2.1 The weak reading of the principle that like cases should be treated alike

The weak reading of the principle states that if φ-ing is the right choice for one case, then it is also the right choice for a case (relevantly) similar to it.[2] On this reading, the principle is simply a reformulation of the idea that reasons are general. If the existence of some element ε is a reason for φ-ing in one situation, then it is a reason for φ-ing *in general*. Accordingly, if a second situation shares all the elements based on which φ-ing was justified in a first situation (and there are no relevant differences), then a justification for φ-ing should also exist in the second.[3] In this weak formulation, it is a universally valid principle. But it does not provide a warrant to go from the mere fact that one situation *has* been treated in one way to the conclusion that the other *should* also be treated in that way. The decision in the first situation might have been wrong. Then the similarity of the two situations can at most be a reason to think that treating the second situation in a similar fashion would also be wrong (compare Adler, 2007; Guarini 2004; Bermejo-Luque, 2012).[4]

---

persuasive precedent is followed, the law is not applied, but rather extended (On analogy and extending the law see e.g. Canale and Tuzet, 2014). Unless otherwise stated, I will be talking about binding precedent where-ever I talk about legal precedent throughout the paper.

[2] That the two readings of the principles are different is not always clearly recognized by authors who discuss the principle to treat like cases alike. (See, e.g. Govier, 2014, p.320 f.)

[3] The weak version of the principle is what Levvis calls the "principle of relevant similarity". (Levvis 1991)

[4] See for example Marcello Guarini's choice of a "core scheme" for normative or classificatory arguments by analogy (Guarini, 2004, p161). This scheme features the conclusion that two objects, people or situations should be classified or treated in the same way rather than the conclusion that the target object, person or situation should be treated in some specific way:

    1. a has features f1, f2, ... fn
    2. b has features f1, f2, ... fn

The weak version of the principle therefore cannot support the legitimacy of arguments from other cases by itself. It can only do so together with the assumption that the decision to φ in the first case was the right or justified decision. Therefore it can only support normative arguments by case-based analogy. These arguments, in addition to the premise that the two cases are similar, also have a premise – or a background assumption[5] - stating that the decision made in the source-case was the correct one.

2.2 The strong version of the principle that like cases should be treated alike

Sometimes the principle is used in a stronger form, in order to say that the mere fact that some case *has* been treated by φ-ing is in itself a reason to φ again in similar cases.[6] This is the version of the principle that would be needed to show that arguments by precedent are generally legitimate. This stronger version of the principle that like cases should be treated alike is presumably the one referred to in the law when it is used to justify the doctrine of precedent. However, in this form, the principle seems to be justifying a move from *is* to *ought* indiscriminately, which gives reason to doubt its validity: Accepting the strong version of the principle would mean to accept that the mere *performance* of an all-things-considered wrong action φ would make φ-ing *less* wrong. But the *mere* performance of an action should not be able to alter its normative status.

However, an action can be performed in circumstances that give rise to additional normative reasons. For example, the action could be performed in such a context that people form an expectation on its basis. Imagine for example a teacher making a decision about one

---

      3. a and b should be treated or classified in the same way with respect to f n+1

Compare this scheme to the schemes for normative arguments by analogy that have been offered by other authors. These schemes include the premise that the source-case was treated correctly and move to the conclusion that the target-case should be treated in the same way. (See, e.g. Walton, Reed and Macagno, 2008, p. 316, or Waller, 2001)

[5] I avoid the possibly exhausting discussion of whether this - and the premises (assumptions) stating a second-order reason that I introduce in the next section - play the role of missing premises or background assumptions.

[6] Perelman seems to be aware that his and Olbrecht-Tyteca's principle of justice does not hold *always.* See, e.g. (Perelman 1980, p. 89/90).

student while others are watching. The way the students understand the role of the teacher might lead them to reasonably expect that the same decisions will be made about them in similar cases. Circumstances like these might give rise to additional reasons why the mere fact that the decision to φ was made in a similar situation is a reason for making the decision to φ now.

Joseph Raz introduced the widely-used distinction between first- and second-order reasons into the philosophy of practical reasoning and the philosophy of law. He pointed out that in addition to first-order reasons weighing directly on the decision whether to φ, there are also second-order reasons for or against treating certain facts as first-order reasons for or against φ-ing.[7] (Raz, 1975, p. 39 f.) For ease of expression, I will adopt his vocabulary here. Reasons like that people have formed expectations that cases similar to a past case will be treated by φ-ing can function as second-order reasons. They do because they are reasons for treating the combination of the *mere* fact that a past situation was treated by φ-ing and the fact that it is similar to the present situation as a reason for φ-ing now.

The strong version of the principle to treat like cases alike does not hold absent such additional second-order reasons. If it did, that would entail that the mere performance of an act changes its normative status. It follows that arguments by precedent are only legitimate if they have an additional acceptable premise (or background assumption) to the effect that a second-order reason for treating the combination of the other two premises as relevant for the current decision exists.

3 THE TWO TYPES OF ARGUMENTS FROM OTHER CASES AND RELEVANT SIMILARITIES

So far I have shown that both kinds of arguments from other cases share a similar structure. Both arguments rely on a premise stating that the

---

[7] To Raz, second-order reasons are reasons (not) to act on first order reasons for action. Daniel Whiting has recently presented arguments that suggest that such reasons might not exist. (2015) Instead, second-order reasons might really be first order reasons for certain mental acts: A "second order reason" might be a reason to perform the mental act of in- or excluding certain facts as first order reasons when deliberating whether to perform a certain act. Whiting does not want to call such reasons second-order reasons. By contrast, I think that they would still make for an interesting and important class of reasons, and that calling them second-order reasons would make their effects clear. In any case: whether second-order reasons have the nature Raz ascribes to them or the nature Whiting ascribes to them makes little difference for this paper.

decision to φ was made in a source-case, a similarity premise, and an additional (hidden or missing) premise or background assumption. The main difference I have identified between the two kinds of arguments is the content of the additional premise or background assumption. Is this the only difference? I argue that it is not. There is a further difference and it concerns the relevance of the similarities between the argument's source- and target-case.

In arguments by analogy, arguers use the similarity of the two analogues to support the conclusion that there is a further respect in which the analogues are similar – either a further shared property, or the way in which it should be treated or classified. It is therefore of great importance to know which similarities exactly are supposed to be the ones supporting this conclusion, which similarities are the relevant ones.[8]

In his often-cited paper "Argument by Analogy", Andre Juthe proposes an elegant solution to this problem (Juthe, 2005): Say that a similarity between a source-analogue and a target-analogue in an analogy consists of an element ε of the source analogue and an equivalent element ε^* of the target analogue. An argument by analogy supports the claim that there is a further similarity between the analogues. This further similarity is between a known element s in the source analogue and an equivalent element s^* that we may now claim for the target analogue (Juthe calls this the "Assigned-Predicate" (Juthe, 2005, p.4)). A similarity based on ε and ε^* is *relevant* if the possession of element ε in the source analogue is part of the determination of the element s in the source analogue. Juthe points out that "part of the determination" is meant to be read widely: the determining relation between ε- ε_n and s "can be any type of relation (including probable, causal, epistemic, normative, evaluative, resultant or supervinient)" (Juthe, 2005, p. 10).

3.1 Relevant similarities in normative arguments from case-based analogies

Normative arguments from case-based analogy are legitimated by the weak version of the principle that like cases should be treated alike

---

[8] I think that this is the basis for the debate regarding whether arguments by analogy include a missing a premise in the form of a conditional that lists the properties possessed by both analogues. For contributions to this discussion see: (Govier, 1989, Waller, 2001; Guarini, 2004; Shecaria, 2013 and Bermejo Luque, 2014)

according to which those cases should be treated alike to which the same reasons apply. In addition, they rely on the assumption that the decision made in the source-case was the correct decision.

In these arguments, the element s and s^* are the two decisions to φ in the source- and target-case respectively. Because the argument relies on a relationship of justification, the elements ε-ε_n that determined the element s in the source-case are those elements on the basis of which the decision to φ was justified. Therefore the relevant similarities in these arguments will be those that have, as one of the two corresponding elements, a property of the source-case on the basis of which the decision to φ was justified. The two cases are relevantly similar for the purpose of a normative argument from case-based analogy if each element of the source-case that is necessary to justify the decision to φ has a corresponding element in the target-case.

3.2 Relevant Similarities in Arguments by Precedent

For arguments by precedent referring to the *justification* for the decision to φ is not an option because these arguments use past decisions to φ in order to argue for a present decision to φ *independently* of whether the past decision was justified. Nonetheless, arguments by precedent *do* rely on the acceptability of the premise that the two cases are alike, so the relevance of the similarities between the two cases has to be evaluated somehow.

Now, it would be possible to try to amend the method of referring to justification when determining whether a similarity is relevant or not. We might say that those similarities are relevant which have, as one corresponding element, those elements ε-ε_n that are the basis for the *best available* justification for the decision to φ in the source-case. Alternatively, we might say that those similarities are relevant which have, as one corresponding element, one of those elements ε-ε_n that the decision-maker in the source-case *thought* justified the decision at the time.

I do not think that we should construct an account of what a relevant similarity is in the context of an argument by precedent by amending the account we have for normative arguments by case-based analogy. Instead, which similarities are going to be relevant within the context of an argument by precedent depends on the content of the second-order reason on which the argument relies. Compare the following two examples of arguments by precedent:

First, consider the most iconic kind of argument by precedent, the argument from legal, binding precedent in common-law jurisdictions.⁹ According to the (authoritative) doctrine of precedent, the decision made in a past case *authoritatively binds* a later court to make the equivalent decision in a later case if the cases are *legally the same*. Legal sameness is identified with the idea that the two cases have to be *similar* in legally relevant respects and have no legally relevant differences that would justify distinguishing the cases. In this context, the reason for the legitimacy of arguments by precedent is the existence of an authoritative rule. This rule determines that judges are bound by the decision of earlier judges, as they are laid down in precedent opinions.[10] Therefore, similarities between two cases will be relevant that have, as one corresponding element, those elements ε-ε_n that are *legally* relevant in the precedent case. The precedent judge has the authority to determine which elements of the precedent case are legally relevant. She does so by citing them in the precedent opinion as the factors on the basis of which the precedent decision was made. These are also the factors that belong to the case's *ratio decidendi* or *holding* (Schauer, 2012). It is important to see that these elements *might* be, but do not *have* to be those elements that together make the best case for the decision or those elements on the basis of which the precedent judge *really believed* that the decision was justified. Rather, they are the factors that the precedent judge decided to *authoritatively determine* as the factors on the basis of which the decision was made in the opinion.

Contrast this with another example. Say that the second-order reason for treating the existence of a past decision to φ in a similar case as a reason for φ-ing again is that the decision to φ caused expectations. Imagine you are in a teaching position of some kind, and have recently allowed a student to hand in a paper two weeks late. Through the channels of gossip, you had known that the student was dealing with some significant personal problems. So, when he broke out in tears,

---

[9] Importantly, I am referring to binding, not persuasive precedent. See footnote 2.

[10] Of course, there are further background reasons needed to justify the existence of this rule. But insofar as judges are bound by the doctrine of precedent, they need not refer to these background reasons when they evaluate arguments by precedent. They merely need to refer to the existence of the rule, and the fact that the rule is valid. (See, e.g. Schauer, 1991, p. 145 ff.) This is vital, among other reasons because there is no consensus between authors as to *why exactly* the doctrine of *stare decisis* in the law is justified. Rule of Law reasons, reasons of efficiency, and reasons of predictability are regularly cited, but the lists of reasons are not always the same, and there is disagreement about the relative importance of the reasons (See, e.g. Schauer, 1987; Duxbury, 2008 Chapter 5; Sherwin, 1999; Waldron, 2012)

your heart of stone was softened. This was the reason you granted the extension. But because you are not even really supposed to know about the personal problems, you have to keep these reasons to yourself in your official capacity. Now another student comes to you and asks for an extension. Like many students every year, she explains that she is simply overwhelmed with work and starts to cry. And she cites your earlier decision.

Here, if the argument by precedent is legitimate at all, it has to be on the basis of the reason that your earlier decision created expectations in your other students. How do you determine relevant similarity now? According to Juthe's account of relevant similarity in arguments by analogy, you should look for those elements $\varepsilon\text{-}\varepsilon_n$ in the precedent case that determined your decision. What is the relation here? Your students surely did not form expectations that take into account features of the precedent case which they do not know about, like that the earlier student had significant personal problems. So it cannot be justification. And you did not produce anything remotely similar to a judicial opinion. Your students formed expectations about cases that are similar to the way *they* perceived the precedent case. Now, you might want to limit the applicability of this second-order reason to expectations formed on the basis of a reasonable perception of the precedent case: The student has to have a reasonable, but not necessarily an accurate, understanding of which elements of the precedent case where relevant for your decision. It might be – and in practice is – quite hard to work out what these features are. Important, however, is that you cannot expect of the student to form her expectations to get an extension only if she, too, has significant personal problems. This is so because she could not have known that the student in the precedent case had these kinds of problems. Arguments by precedent may be legitimated through the existence of expectations caused by the earlier decision. Then the relevance of similarities has to be determined by identifying those elements of the precedent case that the expectation-former would (reasonably) have perceived as relevant for the precedent decision.

I think this shows that for arguments by precedent, it is implausible to think that there could be one unified method for determining which similarities between the cases are the relevant similarities.

4. THE EVALUATION OF ARGUMENTS FROM OTHER CASES

One of the most widely used methods for representing argument types and guiding their evaluation is the formulation of argument schemes, which represent patterns of reasoning accompanied by associated critical questions.[11] However, the attempt to formulate an argument scheme for arguments from other cases immediately runs into a problem: As I mentioned in the first section, the differences between the two kinds of arguments from other cases often do not make for a difference in argument presentation.

That the presentation of arguments by analogy is unrevealing is common to arguments by analogy in general. In a recent paper, Douglas Walton has defended the idea that we need *two* argument schemes for the basic argument by analogy instead of one. (Walton, 2014, p. 38). The first one is used to *recognize* arguments by analogy when they are used "in the wild", so to speak:

> Similarity Premise: Generally, case *C1* is similar to case *C2*.
> Base Premise: *A* is true (false) in case *C1*.
> Conclusion: *A* is true (false) in case *C2*.[12]

---

[11] This is the method proposed in Walton, Reed and Macagno (2008). To illustrate, here is an argument scheme for argument by expert opinion:
  Major Premise: Source E is an expert in subject domain S containing proposition A.
  Minor Premise: E asserts that proposition A is true (false).
  Conclusion: A is true (false).
  1: Expertise Question: How credible is E as an expert source?
  2: Field Question: Is E an expert in the field that A is in?
  3: Opinion question: What did E assert that implies A?
  4: Trustworthiness Question: Is E personally reliable as a source?
  5: Consistency Question: Is A consistent with what other experts assent?
  6. Backup Evidence Question: Is E's assertion based on evidence?
  (Walton, Reed Macagno,2008, p.310)

[12] The second scheme is supposed to guide a deeper analysis and evaluation of the argument, and Walton proposes to use the scheme proposed by P.J. Hurley (but also cites a version from I.M Copi and C. Cohen as well as the scheme Guarini proposed):
Entity *A* has attributes *a*, *b*, *c* and *z*.
Entity *B* has attributes *a*, *b*, *c*.
Therefore, entity *B* probably has *z* also.

I think that Walton's first scheme has to be altered only marginally to represent the form in which we usually encounter arguments from other cases:

> Similarity Premise: Generally, case *C1* is similar to case *C2*.
> Base Premise: The decision to φ was made in *C1*.
> Conclusion: The decision to φ should be made in *C2*.

However, in spite of the two arguments often *looking* the same when they are presented, they are quite different. Above, we dedicated two sections to the differences between the two kinds of arguments by other cases. I will now argue that these differences have important implications for the way in which arguments from other cases have to be evaluated. Therefore, it is important to represent these differences in the argument schemes for arguments from other cases.

4.1 Differences in evaluation

Four different kinds of considerations for the evaluation of arguments by analogy are repeatedly suggested in textbooks and scholarly papers.[13] I will discuss each consideration in turn: [14]

4.1.1. Is the conclusion true in the source-case?

I am here interested in arguments from other cases, and these arguments are practical. Therefore I will translate the question whether the conclusion is true of the source-case into the question whether the decision to φ was justified in the source-case. This question is clearly important to ask when it comes to normative arguments by case-based analogy. But is it also important to ask this question when dealing with arguments by precedent? Obviously not. After all, arguments by precedent rely on second-order reasons for treating the decision to φ in the source-case as a reason for deciding to φ again in a similar target-

---

[13] Here I use (Groarke, Little and Tindale, 2004), see also (Vaughnn and McDonald, 2013; Walton, 2014)

[14] I do not mean to claim that these are all the possibly helpful critical question for the evaluation of arguments by analogy. I do think, however, that these are the most important questions when it comes to the distinction between the two argument types. For further critical questions that might be important for the evaluation of arguments by analogy in general, see e.g. Andre Juthe's dissertation on the topic of argument by analogy (Juthe, 2016, unpublished manuscript).

case. It does not matter whether that decision should or should not have been made in the source-case. So instead of asking whether the decision to φ in the source-case was correct, the evaluator should ask whether there is a second-order reason that legitimizes the argument by precedent. This means that there is at least one completely different critical question depending on which kind of argument from other cases is being evaluated.

4.1.2 Are the two cases similar in sufficient relevant respects?

The question of whether there are the right kinds of similarities, and enough of them, is of course important for any argument that employs an analogy, including both kinds of arguments from other cases. But – under what circumstances is the answer to these questions affirmative? As I argued in section 3, which similarities between two cases are relevant might differ depending on whether the argument is a normative argument from case-based analogy or whether it is an argument from precedent. It is important to keep this in mind when the question is asked. For normative arguments from case-based analogy, the two cases will likely have to be similar with respect to the elements ε_1-ε_n that were necessary for the justification of the decision in the source-case. For arguments by precedent, on the other hand, which similarities are relevant, and whether there are sufficient relevant similarities, will depend on the kind of second-order reason that legitimizes the argument.

4.1.3 Are there relevant dissimilarities/differences between the two cases?

The question whether there are any relevant differences between the two analogues is, just like the question about similarities, one that will be important for any argument that relies on an analogy. Relevant differences can be divided into two groups. One kind of relevant difference is the *absence* of a relevant similarity. If an element ε_1 was a necessary part of the determination of the decision to φ in the source-case, then that there is no corresponding element ε_1^* in the target-case constitutes a relevant difference. For these differences, the same considerations as for relevant similarities can be applied when comparing the two kinds of arguments by other cases. However, a difference of this kind would have likely already influenced the evaluator's assessment of the argument at the stage of asking whether the two cases have sufficient relevant similarities.

More likely, the kind of difference an evaluator will be after is the second kind of difference: that the target-case has an additional element ε_x for which there is no equivalent in the source-case, and which is the basis for a reason *against* making the decision to φ. I am not sure whether there is any important implication of the differences between the two kinds of arguments by other cases to look out for here. It might be worthwhile to consider that where the second-order reason legitimating an argument by precedent refers to an expectation, the kinds of differences based on which the argument can be rejected might be limited. It might be dubious to reject the argument based on a difference the importance of which is too hard to understand for the person who formed the expectation.

4.1.4 Are there alternative source analogues that can serve in arguments with a different, incompatible conclusion?

There is also the question whether there is an equally good or better source analogue available from which an analogy can be constructed that will serve for an argument with a different, incompatible conclusion.[15] This question will be helpful in the evaluation of either kind of argument from other cases. Nonetheless, I should point out that it might be an especially interesting question where an argument by precedent cites a source-case in which the decision to φ was clearly wrong. Even if such an argument by precedent relies on a valid second-order reason and provides reason to φ in the target-case, the same source-analogue may sometimes also be used in a counter-argument by case-based analogy. Where the decision to φ was clearly wrong in the source-case and the two cases are relevantly similar, an argument may be construed that the decision to φ would also be wrong in the target-case. Which of the two case-based arguments is stronger would depend on whether the second-order reason legitimizing the argument by precedent is stronger than the reason based on the wrongness of the past decision.[16]

---

[15] This consideration gets mentioned for example by Walton (2014 p. 25) and Govier (2014, p. 338)

[16] However, it is very important to keep in mind that this will not work with all arguments by precedent. Legal arguments by binding precedent, for example, are legitimated by an authoritative rule that prohibits judges from taking into account the wrongness of the precedent decisions. As Grant Lamond put it: judges have to treat all precedent decisions *as if they had been correctly decided* (Lamond, 2005).

This section has shown that the difference between the two types of arguments from other cases should lead to one completely different critical question for each argument. In addition, the conditions under which the other three questions can be answered satisfactorily are different with respect to the two types of argument. This is especially true for the question about relevant similarities. These differences in evaluation, together with the differences between the structures of the two kinds of arguments, speak for establishing two different argument-schemes with two different sets of critical questions. Each of these argument-schemes is completed with their additional premise (or background assumption).

Normative Argument from Case-Based Analogy
(P1) Generally, case C1 is similar to case C2.
(P2) The decision to φ was made in C1.
(P3) The decision to φ was justified in C1
(C) The decision to φ should be made in C2.
(Q1) How was the decision to φ justified in C1?
(Q2) Are the two cases sufficiently relevantly similar with respect to the elements that justified the decision in C1?
(Q3) What might be relevant differences between C1 and C2?
(Q4) Are there alternative source analogues that can serve in arguments with a different, incompatible conclusion?

Argument by Precedent
(P1) Generally, case C1 is similar to case C2.
(P2) The decision to φ was made in C1.
(P3) Treating the existence of the decision to φ in C1 as a reason for deciding to φ in similar cases is justified.
(C) The decision to φ should be made in C2.

(Q1) What is the justification for treating the existence of the decision to φ in C1 as a reason for deciding to φ in similar cases?
(Q2) Are the two cases sufficiently relevantly similar with respect to the elements in C1 picked out by the justification?
(Q3) What might be relevant differences between C1 and C2?
(Q4) Are there alternative source analogues that can serve in arguments with a different, incompatible conclusion?

5. CONCLUSION: MISS-USES, MISS-EVALUATIONS, AND THE IMPORTANCE OF KEEPING THE DISTINCTION IN MIND

The fact that the two kinds of argument from other cases are rather different, but that their differences are rarely reflected upon and that they often *look* the same in presentation poses the risk of leading to miss-uses and miss-evaluations (maybe even fallacious uses, were malice is involved). The special problem here is that each kind of argument relies on a background assumption - but does not need the background assumption the other relies on. Arguments by precedent do not need a justified past decision, and normative arguments from case-based analogies work fine without additional second-order reasons. Of course, where there is neither a justified past decision *nor* an applicable second-order reason, an argument that cites a past decision to φ in a similar case cannot justify φ-ing in a present case. But legitimate arguments from other cases are so common, and the principle to treat like case alike has such a good reputation, that it is easy to forget this. Where it is forgotten, arguers and interlocutors might be seduced into committing an *is/ought* mistake (or fallacy), or a mistake (or fallacy) of *two wrongs make a right* (Govier 2014, p. 341).

In addition, the similarity of the two kinds of arguments might lead to a miss-judgement of the *strength* that an argument from other cases has. For example, a normative argument by case-based analogy might only show that a decision is merely as *permissible* as an earlier decision was. An additional second-order reason would be needed to argue further by precedent that the fact that the *permissible* decision was actually made now constitutes a reason to think that the decision *has* to be made. Confusing the two arguments can lead interlocutors into taking on duties they would not have to take on if the distinction was clear to them.[17]

Finally, the fact that the two arguments look so similar and that people are not often inclined to look closely at the exact kind of argument that is being advanced can be abused. Imagine a setting where interlocutors have made a bad past decision and are under scrutiny by an audience – a setting that appears often in politics: An arguer who knows that their interlocutor will find the advocated decision wrong, and who cannot argue for it on its own merits might instead cite another decision that the interlocutor made in the past. Even if no strong enough

---

[17] My father fell into this trap when he made the mistake of agreeing to pick my sister up late at night at a friend's house. He suddenly found himself frequently waiting for his daughters to be ready to come home until deep into the night.

second-order reason is available, this strategy might still work. This is so because in order to reject the argument from other cases altogether, the interlocutor would have to reject its possible form as a normative argument from case-based analogy. If she does not do this, the audience will take it that the interlocutor must believe the advocated decision in the present case to be correct –like her own past decision. Making the argument from other cases in these circumstances serves the purpose of bringing up a past decision and putting the interlocutor before a choice: Either she makes the decision advocated by the arguer now or she admits a wrong decision in the past. Where admitting to wrong decisions is associated with a loss of face, making such an argument can amount to choosing coercion.[18] This coercion works because the threat of loss of face does supply the interlocutor with a motivation to treat the existence of the past decision and the similarity of the cases as a reason for repeating that decision now. It is a bad, coercive motivation that takes the place of a valid second-order reason in a twisted version of an argument by precedent. The audience is convinced that a normative argument by case-based analogy was given, when in fact a twisted kind of argument by precedent was operating.

REFERENCES

Adler, J. E. (2007). Asymmetrical analogical arguments. *Argumentation*, *21*(1), 83-92.
Bermejo-Luque, L. (2012). A unitary schema for arguments by analogy. *Informal Logic*, *32*(1), 1-24.
Bermejo-Luque, L. (2014). Deduction without dogmas: The case of moral analogical argumentation. *Informal Logic*, *34*(3), 311-336.
Canale, D. and Tuzet, G. (2014). Analogy and interpretation in legal argumentation. In H. Jales Ribeiro (Ed.), *Systematic Approaches to Argument by Analogy* (pp. 227-242). Dordrecht: Springer.
Duxbury, Neil (2008). *The nature and authority of precedent*, Cambridge: Cambridge University Press.
Eemeren, F. H. van, & Garrsen, B. (2014). Argumentation by analogy in stereotypical argumentative patterns. In H. Jales Ribeiro (Ed.), *Systematic approaches to argument by analogy* (pp. 41-56). Dortrecht: Springer.
Garrsen, B. (2009). Comparing the incomparable. Figurative analogies in a dialectical testing procedure. In F. H. van Eemeren & B. Garssen (Eds.), *Pondering on problems of argumentation: Twenty essays on theoretical issues* (pp.133-140). Springer: Dordrecht.

---

[18] See Gilbert (1997) on the importance of face in argumentation.

Gilbert, M. (1997). *Coalescent argumentation*. Mahwah, NJ: Lawrence Erlbaum Associates.
Govier, T. (1989). Analogies and missing premises", *Informal Logic*, *11*(3), 141-152.
Govier, Trudy (2014). *A practical study of argument*, Wadsworth: Boston.
Guarini, M. (2004). A defense of non-deductive reconstructions of analogical arguments. *Informal Logic*, *24*(2), 153-168.
Groarke, L. and Tindale, C. (2004). *Good reasoning matters!: A constructive approach to critical thinking*. Don Mills, ON; New York: Oxford University Press.
Juthe, A. (2005). Argument by analogy. *Argumentation*, *19*(1), 1-27.
Lamond, G. (2005). Do precedents create rules? *Legal Theory*, *11*(1), 1-26.
Levvis, G. W. (1991). The principle of relevant similarity. *Journal of Value Inquiry*, *25*(1), 81-87.
Perelman, C. and Olbrechts-Tyteca, L. (1969). *The new rhetoric*. Notre Dame, IN: University of Notre Dame Press.
Perelman, C. (1980). *Justice, law and argument: Essays on moral and legal reasoning*. Dordrecht: D. Reidel Publishing Company.
Raz, J. (1990). *Practical reason and norms*. New York: Oxford University Press.
Schauer, F. (1987). Precedent. *Stanford Law Review*, *39*(3), 571-605.
Schauer, F. (1991). *Playing by the rules*. Oxford, England: Claredon Press; New York: Oxford University Press.
Schauer, F. (2012). *Thinking like a lawyer*. Cambridge, MA; London: Harvard University Press.
Sherwin, E. (1999). A defense of analogical reasoning in the law. *The University of Chicago Law. Review*, *66*(4), 1179 -1197.
Vaughn, L. and McDonald, C. (2013). *The power of critical thinking* (3rd Canadian ed.). Don Mills, Ontario: Oxford University Press.
Waldron, J. (2012). Stare decisis and the rule of law: A layered approach. *Michigan Law Review*, *1*(111), 1-32.
Waller, B. N. (2001). Classifying and analyzing analogies. *Informal Logic*, *21*(3), 199-218.
Walton, D., Reed, C., & Macagno, F. (2008). *Argumentation schemes*. New York: Cambridge University Press.
Walton, D. (2014). Argumentation schemes for argument from analogy. In H. Jales Ribeiro (Ed.), *Systematic approaches to argument by analogy* (pp. 23-40). Dordrecht: Springer.
Whiting, D. (2015). Against second order reasons. *Nous*, *1*(23), 1-23.

54

The Attraction of the Ideal Has No Traction on the Real: On Adversariality and Roles in Argument

KATHARINA STEVENS
University of Lethbridge
katharina.stevens@uleth.ca

DANIEL H. COHEN
Colby College
dhcohen@colby.edu

If arguers were exclusively concerned with cognitive improvement, arguments would be cooperative. However, we have other goals and there are other arguers, so the default is adversarial argumentation. We naturally inhabit the heuristically helpful but cooperation-inhibiting roles of proponents and opponents. We can, however, opt for more cooperative roles. The resources of virtue argumentation theory are used to explain when proactive cooperation is permissible, advisable, even mandatory – and also when it is not.

KEYWORDS: adversariality, cooperation, virtue argumentation, argument roles, argument choices.

1. INTRODUCTION

The dangers of too much adversariality are well known: it turns cooperative deliberations in search of reasoned outcomes into verbal competitions spiralling out of control into no-holds-barred quarrels.[1]

---

[1] Compare, for example, the way in which Walton and Krabbe illustrate the way downward from deliberation over negotiation to quarrel (Walton and Krabbe, 1995, p. 107) or Gilbert, who warns the readers of his textbook *Arguing with People* that they should be aware that more adversariality on their side will *always* be matched from the other side. (Gilbert, 2014, p. 66)

The question here is whether adversariality in an argument is *ever* warranted. We know there can be too much, but can there be too little?

Bailin and Battersby (2017) have recently given this question a definite "no". They argue that we need a new account of non-adversarial argumentation to spotlight the role of cooperation in argumentation. They criticise Cohen (2014), who does de-emphasise adversarial aspects, for not going far enough in the direction of cooperation.[2] For Cohen, some degree of adversariality can be justified, even for a virtuous arguer, by the context of an argument and her role in it. Bailin and Battersby reject the idea that choosing adversarial rather than cooperative moves can be an expression of argumentative virtue, and they object to casting arguers in the roles of proponents and opponents. Instead of roles, they prefer to think in terms of argument-tasks. Each arguer shares the responsibility for fulfilling every task, so a virtuous arguer always chooses to be cooperative. They base the claim on the *telos* of argumentation as epistemological betterment. If argumentation is for epistemological betterment, they argue, then it is a fundamentally cooperative enterprise. It follows that virtuous arguers, committed to this *telos*, will choose the cooperative move at each stage.

Their conclusion goes too far. Even granting that epistemic betterment is the *telos* of all argumentation, a virtuous arguer need only regard cooperation as a defeasible default rather than an obligatory choice. Bailin and Battersby fail to appreciate that argumentation is as much a social endeavour as an epistemological one.[3] We argue in various contexts, for various reasons, and, most important, with *other people*. Virtuous arguers respect the epistemic *telos* of argumentation and act correspondingly. However, they also respect the particular social situations in which they argue. Their moves, whether towards or away from cooperation, are informed by those situations. The cooperative move is often but not always the best one for the epistemic *telos* of argumentation. There is no mechanical algorithm for good arguing. Instead, argument roles provide handy rules of thumb for how

---

[2] Cohen in both 2013 and 2014 critiques the language of proponents, opponents, antagonists, and combatants to denote the participants in arguments (although in the end, he acquiesces to common usages). He goes so far as to describe the judges, juries, kibitzers, and other participants as "the extras" and "the supporting cast". Stevens 2016 embraces argumentative roles as a tool to understand virtue argumentation.

[3] In fact, they make the conscious choice to favour the epistemological aspect over the social aspect of argumentation (Bailin and Battersby, 2017, p. 7)

to argue but ultimately, it is an arguer's character that is most important, which is why virtue theories serve so well.

2. EVERYDAY ARGUERS BETWEEN COOPERATION AND ADVERSARIALITY.

In "adversarial argument", as we will use the term, arguers argue to win, i.e., to make their side in the argument the one that is adopted, chosen, or believed. We will use "cooperative argument" to describe dialogues in which the participants freely share information, inferences, and even strategies to advance the argumentation without regard to the effects on winning and losing. Arguers in the latter are partners in a shared endeavour instead of adversaries.[4]

Bailin and Battersby privilege the cooperative mode by reference to the epistemic betterment of the arguers as the overall *telos* of argumentation.[5] We agree.[6] Arguers commit themselves to some amount of cooperation merely by engaging in argument. After all, whether we argue in order to move from disagreement to agreement, from a problem to a solution, or from a question to an answer, we are not after *any* agreement, *any* answer, or *any* solution. We seek *reasoned* outcomes: *justified* agreement, *correct* answers – by rational means. Argumentation enhances the chance for success by taking into account diverse reasons and considerations. Cooperative argumentation, with its free exchange of ideas unaffected by partisan strategies, makes the connection to the epistemic *telos* of argument manifest. When we argue with one another, we *ipso facto* become partners in the shared enterprise whose goal is epistemic betterment.

---

[4] Or they are confederates, if they happen to find themselves on the same side. Adversarial argumentation differs from what Aikin calls "belligerent argumentation" in that belligerent argumentation is an extreme form. It is no-holds-barred argumentation in which arguers are willing to do anything necessary to win, even resorting to fallacies, intimidation and emotional violence. (Aikin, 2011, p. 250) Adversarial argumentation also differs from what Gilbert calls "eristic argumentation" which is connected to the "desire to achieve one's strategic ends at all, or at least some, moral costs if necessary" (1997, p. 43). Adversarial argumentation can be eristic, but does not have to be. One could argue adversarially by making moves that are strategically necessary but do not cross any moral boundaries.

[5] Bailin and Battersby 2017, p. 8. Cf. Aberdein, 2010 p. 173.

[6] Privileging the epistemic dimension of argumentation, especially as it applies to all the participants in an argument, is consonant with the view expressed in Cohen 2013 and Stevens 2016.

Nonetheless, adversariality remains central to the actual practice of argumentation because arguments typically arise from differences in our standpoints.[7] The demands of *completely* cooperative argumentation are too much for ordinary arguers. For example, full cooperation means pointing out stronger, and even winning, strategies that our interlocutors miss, even when it leads to losing the argument.

There is, then, a tension between the cooperative and adversarial aspects of argumentation with every-day arguers caught in the middle: Cooperative argument might be essential to the epistemic *telos* of *argumentation*, but that *telos* might not be part of the motivation for the *arguer*.[8] Arguers have their own interests and goals. When you are tired, in bed, and arguing with your partner about who should take the dog out, any cognitive benefits that stem from being rationally persuaded that it is your turn would be fortuitous. The epistemic pleasure of recognizing what fairness demands hardly outweighs the physical discomfort of having to get out of bed to walk the dog in the rain.[9] When arguers' motivations do not coincide with the *telos* of argumentation, the default is to argue *against* each other.

---

[7] Scott Aikin 2011 argues that adversariality is a necessary part of argumentation, and needs to be managed rather than removed. We agree with Hundleby 2013 that adversariality is not absolutely necessary. She points out that we may argue in absolutely cooperative ways (p. 254) But we recognize that people choose to argue (e.g., instead of fighting) in situations that are set up as adversarial, perforce making the argument also adversarial.

[8] Walton and Krabbe 1995, p. 66 distinguish the purpose of a dialogue type, and the goals of a dialogue's participants.

[9] That argument is better described as a persuasion dialogue or a negotiation than an inquiry into how to divvy up household chores fairly. Of course, arguers do not always or only argue to further personal interests. When we argue to figure out the best solution to a problem or to gain clarity on an issue, our goals in the argument align with the telos of argumentation. Those cases are closer to the paradigm that Bailin and Battersby have in mind: pure inquiry or its practical sister, pure deliberation. (Bailin & Battersby, 2017) However, empirically, this is more the exception than the rule for argumentative encounters. Gilbert 1997, p. 74/116 thinks that pure inquiry is so rare that building an argumentation theory around it would be idle.

3. VIRTUOUS ARGUERS AND THE IMPERATIVE TO COOPERATE

As self-interested arguers, we want to succeed; as epistemic agents, we are compelled to *get it right*.[10] Descriptively, we often decide how to argue according to what is more important to us in the moment. Normatively, we can ask what we *should* do. How would a virtuous arguer resolve this tension?

There is an easy, apparently straightforward analytic answer that grows directly out of what has already been said: A virtue is a disposition to act in accordance with a *telos*.[11] Perfectly virtuous arguers will argue in such a way as to further that *telos*. If we identify the *telos* with the cognitive betterment of the arguers, it follows that virtuous arguers will want to get it right. Therefore, they will be disposed to enhance the reasoning of their interlocutors, both as means to their own epistemic betterment and for the benefit of the other arguers. The *telos* of argumentation is directly connected to cooperative argumentation. Argumentative virtue apparently demands that we remind our partners of their earlier dog-walking efforts.

This is the answer that Bailin and Battersby favour. The epistemic *telos* of argumentation grounds their theory of argumentation. The goals that motivate us to argue and the social situations in which we argue are of secondary concern. Consonant with this, Bailin and Battersby criticise Cohen (2015) for acknowledging the role for roles in argument: reference to proponents and opponents, critics, judges, and audiences tacitly endorses the old adversarial model.[12] When argumentation is perfectly cooperative, roles are not necessary. Therefore, they would rather talk of *aspects* or *tasks* of argumentation. It is incumbent on every arguer to help with each task,

---

[10] See also Gilbert on heuristic and eristic argumentation 1997, chapter 3 and Williams 2004, ch.1.

[11] Admittedly, the neutral phrase "in accordance with" finesses several important debates. For helpful and relevant general accounts of virtue, see, e.g., Zagzebski 1996, Battaly 2008, Annas 2011; for the responsibilism-reliabilism debate, see, e.g., Code 1984, Greco 1999, Axtel and Carter 2008.

[12] Cohen 2014b, commenting on an earlier version of Stevens 2014, does indeed still refer to proponents and opponents, despite his own sustained critiques of the DAM account of argumentation beginning in 1995 and continuing in an unbroken chain that includes, notably, 2013 and 2014a, both of which problematize those roles. Cohen 2003 and Cohen 2017 offer justifications for the continued use of the conceptual apparatus of the adversarial metaphor despite its objectionable elements.

whether it is finding reasons, raising objections, or revising assumptions.

In the abstract, the analytical answer is a good one. In the abstract, we can live with the idea that virtuous argumentation requires that we deliberately lose those arguments that we think we *should* lose, even when we *could* win.[13] In the abstract, the virtuous path is clear: be cooperative; take responsibility for *all* parts of an argument; pursue all lines of reasoning, for and against. Virtuous arguers are partners, not proponents or opponents. However, while we can live, in the abstract, with the rigid demands of cooperative argumentation, the abstract is not where we live. The arguers with whom we argue are flawed human beings; the contexts in which we argue are full of complex contingencies; and, let's admit, we ourselves inevitably fall short of ideal virtue.

4. LIVING IN THE MATERIAL WORLD

A normative theory of argumentation that provides guidance only for ideal people in ideal circumstances is hardly ideal. Everyday arguments, arguers, and contexts are far from ideal and too complex for us to be able to rely on simple rules for virtuous arguing. Instead, we need to cultivate the virtues that will help us navigate difficult choices in complex situations successfully.

Three factors complicate the choice between cooperation and adversariality.[14] First, human reasoning is flawed; people suffer from

---

[13] Of course, in the abstract, there is no such thing as losing an argument. Every argument would be an inquiry or a pure deliberation, and all the arguers would be interested in nothing but the truth or the best solution.

[14] The discussion here is not exhaustive. Additional factors play a role, e.g., the *face-goals* that Gilbert 1997 discusses. Gilbert notes that arguers have goals related to their relationships to the other arguers and the way they will be perceived by them. These goals exist in addition to the strategic goals that motivate their arguing and their (possible) commitment to the epistemic *telos* of argumentation. They can interfere with both the *telos* of argumentation and the participants' strategic argumentative goals, even if the arguers are committed to the *telos*. It can be important to preserve face when we argue, especially if the arguers have ongoing interactions outside the argument. (See Gilbert on familiars, e.g. 1997, 2016) But preserving face is also necessary simply to stay in the argument as a full participant that others take seriously. If the participants of an argument cannot take each other seriously, arguing will be futile and no one will gain epistemically, no matter what their goals are. Therefore, the virtuous arguer should pay attention to the ways in which the

biases. Second, the contexts for argument introduce external obligations and restrictions. Third, arguers, even virtuous ones, have to interact with other arguers who might not be so virtuous.

4.1 Bias

The first factor complicating the choice between adversariality and cooperativeness are the biases that warp our reasoning. According to Mercier and Sperber 2011, reasoning evolved as a capacity to persuade opposed others, not for isolated inferences. We are more motivated to seek arguments to convince others than reasoning to reinforce accepted conclusions. The heuristics that serve well in the former case may turn out to be harmful biases in the latter. Our biases are less influential when we evaluate others' arguments than when we evaluate our own, so we do better when we are set against other reasoners. To some extent, competition elicits our best arguments and adversariality can further, rather than hinder, the *telos* of argumentation.[15] Nonetheless, elements of cooperation are necessary. Mercier and Sperber admit that there is a condition for the positive effects of competition. There has to be what they call a "felicitous context" for argumentation so that people are willing to admit when others are right, i.e. we need "arguments among people who disagree *but have a common interest in the truth*" (emphasis added) (2011, p. 65).

---

context of an argument affects the face-goals of the participants as well as the conditions under which these face-goals can be fulfilled. This could necessitate some adversarial behavior.

[15] Some draw the conclusion that adversariality, really, is what makes argumentation work. When Mercier and Sperber's 2011 paper appeared, David Zarefsky quickly pointed out that these findings imply that argumentation might be most useful when set in an overall adversarial context. (Zarefsky, 2012) Since adversarial contexts invite cheating, Zarefsky suggested that this incentivizes rules forbidding unfair means. If the epistemic *telos* of argumentation is best served through contests in which the strongest arguments survive, it would go a long way to justifying the adversarial model of argumentation. Stevens 2016 contrasts this model with the cooperative model. Mercier and Sperber's model does indeed predict that if people perform better in adversarial contexts than in solitary reasoning, then some dissensus in groups would lead to better outcomes. And that is so: groups outperform single persons when members disagree. However, when they agree – or when subgroups agree and animosity exists - Sunstein's polarization effects set in and people simply find more reasons for the views they already favor. (See, e.g. Sunstein, 2000)

The normative upshot of this is that arguers have built-in biases, but those biases can be mitigated, and even helpful, in group deliberations with disagreeing voices. Thus, idealized cooperative situations are not the optimal ones *for humans*, even when we are committed to the epistemic *telos*. We need to maximize cooperation without completely losing adversariality.

4.2 Context

Arguments are not abstract entities. They have specific arguers with personal histories in complex situations, and those contexts matter. First, the context of the argument might include obligations on the arguers that require a more adversarial posture. Courtrooms, negotiations, and formal debates all embody an adversarial structure as a means to the *telos* of argumentation. This structure has been adopted to mitigate the natural animosity that arises in highly charged situations. The other participants are counting on some degree of adversarial behaviour, so it is not detrimental. One arguer, deviating from the expected norm and staying purely cooperative in an attempt to be virtuous would destroy the carefully crafted equilibrium.[16]

Contexts also entail obligations that can override the imperatives to cooperate. Criminal defendants, for example, have an existential need (and therefore a legal right) for a committed defender. The attorney, by taking the case, incurs a *personal* obligation to her client, in addition to the obligations that come with the assigned role of defender. Similarly, it might be necessary to adopt some degree of adversariality when the stakes of the argument are very high and failing to convince the interlocutor might entail extreme costs (think of talking a drunk friend out of calling an ex-lover). Such external obligations can overrule epistemic imperatives. In general, the context in which an argument takes place may give virtuous arguers good reasons to adopt some measure of adversariality.

---

[16] Criminal trials are the most dramatic context of this kind. The procedural structure of trials, with their well-defined roles for attorneys and judges, serves its own version of the argumentative *telos* with a built-in presumption for innocence, in order to achieve justice. The parties involved are typically so invested – financially, emotionally, existentially – that asking them to argue cooperatively is asking for the impossible. To deal with this problem, the argumentative tasks have been divided up and assigned to judges, juries, and attorneys. These roles come with obligations that put limits on cooperative argumentation.

4.3 Other arguers

The most important contextual factor is the other arguers. Even if a virtuous arguer is not constrained by outside obligations, her interlocutors might be. Even if she has low stakes in the argument, the stakes might be high for others. And even if she is able to check her cognitive biases, her interlocutors might not. Even more problematic, she may confront vicious arguers.

For people to argue cooperatively, they have to trust each other.[17] Gascon argues that distrust can derail argumentation because it leads to incessant questioning both about the truth of what others assert and the cogency of the arguments they advance. (Gascon, 2016a) For cooperative argumentation to succeed, arguers need to be able trust that their fellow arguers portray their goals honestly and argue fairly. Vicious arguers pose a special problem for virtuous arguers because if one arguer becomes partisan while the other remains steadfastly cooperative and neutral, the balance is upset. Arguments on one side will get a disproportionate amount of the air-time to develop. Vicious arguers can take advantage of the trust that cooperative interlocutors give them. A virtuous arguer, committed to the epistemic *telos*, may have to engage an adversarial interlocutor with her own proportional measure of adversariality.[18]

5. ROLES IN ARGUMENT

The last section showed that even virtuous arguers may find themselves caught in the tension between cooperativeness and adversariality. Their commitment to the *telos* of argumentation shields them from selfish impulses to win and provides a presumption for cooperativeness. Nonetheless, virtue can lead to some degree of adversarial behaviour. At each turn, arguers face the question: How much cooperativeness or adversariality is called for right now? Reaching an answer is difficult. It involves ongoing analyses of changeable situations, of other arguers, and of the relative importance of the conflicting goals to win and to get it right.

Is constantly doing all this, plus thinking about the argument itself, too much? Perhaps, but being good ain't easy! However, we think it *can* be made easier, e.g., by taking advantage of the social nature of

---

[17] See, eg, Govier 1993 and Walton 1999.

[18] This is the argumentative analog to the simple, but surprisingly successful, "Tit-for-tat" strategy in Prisoner's Dilemma tournaments.

argumentation. Argumentation is typically a multi-party endeavour, so the *telos* of epistemic betterment is also achieved *together*, by the whole group of arguers. A virtuous arguer does not have to bring it about for everyone by herself. She just has to make positive contributions to the argument she shares with others.

Ideally, arguers could take equal responsibility for all argumentative tasks. The group will do better than a single reasoner because they can pool their information, combine their creative resources, and coordinate their critical resources, e.g., by using many eyes to spot mistakes in reasoning. As Bailin and Battersby note, "[i]t is important for successful argumentation that the various tasks be performed, but the division of labor is [...] incidental" (Bailin and Battersby, 2017, p. 5) But when stakes are high, others are untrustworthy, or contexts give rise to special obligations, the imperative to take full responsibility for the entire argument runs up against individuals' reasons to argue adversarially. The *telos* of argumentation still requires that all tasks be filled, but arguers can be unburdened by a division of labor. Argumentation requires a structure that can organize this division. Arguers, under time and resource-constraints, need tools to establish such structures fast and efficiently. Argumentative roles serve as those tools.

5.1 Roles and the division of labour

A role consists of a cluster of interconnected behaviours, expectations, goals, tasks and norms (e.g. Turner, 2001, p. 233 ff.; Biddle, 1979, p. 55 ff.; Turner, 2002, p. 171). We learn roles, including argumentative roles, by socialization. We all recognize the roles of proponent and opponent in arguments, and we know that proponents find arguments for a position, while opponents raise objections.[19] Many of us will be able to identify such additional argumentative roles as *speaker* and *audience*, *critic*, *adjudicator* and *devil's advocate*. Importantly, socialization also teaches us which sets of roles hang together and how to recognize when others have taken on a specific role. (Turner, 2001, p. 247, Biddle, 1979, pp. 64/76) When we realize that another has assumed a role in an argument, e.g., a proponent, we also recognize that if we want to participate in the argument, the natural way is to take on an associated

---

[19] Cohen 2014 draws on this shared knowledge when he points out that *none* of the argumentative roles we typically recognize seems to include pointing out missed opportunities as one of the key-associated tasks.

role, e.g., an opponent, thereby setting up a familiar structure.[20] Roles divide the argumentative labour by mediating between individual argumentative behaviour and conventional social structures.

In a common enterprise, tasks can be divided up into functional roles designed to help achieve the common goal. Such roles typically come in sets that form structures. (Turner, 2001, p. 235; Biddle, 1979, p. 70 ff.) Ralph Turner cites three bases for the functional differentiation of roles, two of which we think can be applied to argumentation[21].

First, roles are differentiated into tasks to reflect the pre-existing dispositions of the participants. This ensures that everyone concentrates on what they do best, enhancing the efficiency of the entire group. What makes this relevant for argumentation is Mercier and Sperber's research, according to which people perform better when they are at least weakly committed to a position that they have to defend against others (Mercier & Sperber, 2011). This speaks for dividing tasks into basic proponent and opponent roles.[22]

Where biases are strong because pre-commitments run deep, or where contextual obligations or trust-issues enter the picture, Turner's second principles applies: roles are further differentiated to avoid

---

[20] Research into dialogue types has revealed different ways that argumentative engagements can be structured. (See, e.g. Walton, 1998, or Walton and Krabbe, 1995) Bailin and Battersby's preferred structure is the inquiry, in which all arguers play the same role covering all argumentative tasks. (Bailin & Battersby, 2017) Another well-known structure is that of the critical discussion as the pragma-dialectical approach describes it, with well differentiated proponent- and opponent roles. (See, e.g. van Eemeren & Grotendorst, 2004 p. 51)

[21] Turner 2001, p. 236. The third basis is associated with a division of tasks because special skills and knowledge are required to perform each of them that take a lot of time and commitment to learn. Turner cites the roles of physician and attorney as examples: these roles are differentiated because the special skills needed to carry them out effectively are too great for a single person to acquire them all. This principle seems not to apply to ordinary argumentation. At least the basic skills associated with argumentation are so vital that everyone should acquire them.

[22] Several argumentation theorists (e.g., Aikin, 2008 and Zarefsky, 2012) make the related point that arguers do well to adopt the roles of proponents and opponents because competition is such a strong motivator. Lewinski 2017 defends the idea that (adversarial) models of dialogues in which participants act as proponents and opponents lead to rational decisions in practical reasoning – assuming proponents and opponents respect certain rules. He also cites Mercier and Sperber to emphasize the advantages of such division.

situations in which people have to deal with conflicting goals (Turner, 2001, p. 237). This allows individuals to focus on single tasks and minimizes the risk that someone may covertly favour one goal over another. The norms associated with these roles help keep arguers in check. It is harder to get away with fallacious moves when everyone is aware of your commitment and is watching.

In the best case, arguers use their knowledge about argumentative roles (and self-knowledge!) to adopt those roles for which they are best suited belonging to a structure appropriate for the specific argument and context. By exhibiting the behaviour associated with the role, they invite the other arguers to assume complementary roles. If all goes well, the established structure provides everyone with information about what to expect from others, which tasks to concentrate on and which norms their performance will be judged by. Roles allow individual arguers to concentrate on the issue and their individual tasks, trusting that the group as a whole is set up to realize the goal of epistemic betterment.

5.2 Problems with roles and the virtuous use of roles

The last section paints a pretty picture of how arguers can use roles to divide the argumentative labour, but there are problems with how we are socialized into pre-existing argumentative roles. It is too easy to fall into the conventional roles without questioning their appropriateness. We automatically and uncritically become proponents and opponents whenever we sense a disagreement. Thus, the very thing that makes roles effective for providing structure are sources of potential harm: roles are "contagious" because they come in sets. When an arguer adopts a role, it pressures other interlocutors to fall in line and play complementary parts. This can be a harmful in several ways. Two examples will suffice.

First, roles involve more than tasks, goals and norms. Argumentative roles, like all roles, are typified by clusters of associated behaviours. Unfortunately, typical adversarial-argument behaviour can be detrimental to the *telos* of argumentation, as many feminist critiques of adversariality have made abundantly clear: when we associate aggressive behaviour with the proponent and opponent roles, this can lead to argumentation that mostly serves to hurt feelings, deepen existing structures of power, and reaffirm prejudices.[23]

---

[23] See, e.g. Rooney (2010) and Hundleby (2013). Aikin points out that "Sessions at the American Philosophical association are regularly described as 'blood

Second, there is the problem of role-conflict (Turner, 2001, p. 245). People play several roles at a time; attempts to fulfil the expectations of one role can conflict with attempts to fulfil those of another. This can be especially true for women in adversarial argumentation-contexts, finding themselves between a rock and a hard place: The roles of proponents and opponents alike seem to ask for assertive or even aggressive behaviour, while their roles as women come with expectations of submissiveness, politeness, and quiet (Rooney, 2003).

We acknowledge that the way people are currently socialized into argumentative roles comes with important problematic baggage, but we disagree with Bailin and Battersby when they reject the use of task-differentiating roles altogether. The problems and complicating factors discussed in section 3 make completely cooperative argumentation unrealistic. When they do, the ways in which the argumentative labour is divided should *not* be incidental. Instead, it should be guided by the two bases for functional differentiation discussed above. When virtuous arguers encounter people who are biased or have legitimate reasons for being partisan, it would be unfair to expect of them the behaviour necessary for cooperative argumentation. Where a virtuous arguer finds that she herself is so committed, it would be disingenuous to pretend that she is not.

Could we mitigate the dangers associated with pre-formed roles by distributing the argumentative tasks when argumentative engagement is initiated?[24] Theoretically, such a solution is possible, but as before, we do not live in theory: real-time arguments come with time- and resource-constraints. Treating roles as heuristic tools turns them into valuable resources.

Like any social construct, roles are neither rigid nor unchangeable and seldom fully specified.[25] Rather, they provide rules of

---

sport', and many paper panellists have referred to the commentator as the paper's 'assigned assassin'." (Aikin, 2011, p. 255/56) The role of "opponent at an APA-session" is *manifestly* associated with behaviours detrimental to the *telos* of argumentation. These roles can be dangerous!

[24] For example, at the point of the pragma-dialectical "opening-stage" (See van Eemeren & Grotendorst, 2004).

[25] The fluidity of roles should not be overstated as Bailin and Battersby 2017 do. It is not the case that roles are fluid to the point of non-existence. For example, the role of opponent, in its most primitive form, simply contains all argumentative behaviour directed against the acceptance of a certain claim or thesis. The arguer does not leave the role simply because an objection has to be supported by an argument. Claiming that in this moment, the opponent

thumb for behaviour, norms, and expectations, which may, of course, need to be tailored to fit specific contexts. Accordingly, individuals not only role-take and role-play, but also *role-make*: we make roles fit the unique features of the context, including our own dispositions and talents (R. Turner, 2001, p. 235; J. Turner 2002, p. 174). The role of an opponent can be played in many ways, from aggressive to careful and even deferential. If existing roles are objectionable, even as vague as they are, they need not be taken as givens. The norms and behaviours associated with argumentative roles are indeed proper subjects for critical scrutiny by argumentation theorists.[26]

Virtuous arguers should neither ignore the existence of argumentative roles nor rigidly follow a rule of constant cooperativeness; rather they should adopt flexible, reflective relationships to them. The difference between virtuous arguers and others is that instead of unquestioningly falling into the conventional argumentative roles, virtuous arguers will use those roles, while keeping the *telos* of argumentation in mind, to the benefit of the argument.

A virtuous arguer is aware that if she adopts a certain argumentative role, others will respond accordingly. She will choose roles that enhance the chances for the epistemic betterment of all, *given the specific participants and the actual circumstances*. She knows that certain argumentative tasks require certain argumentative virtues more than others. So, she strives for self-knowledge and takes those argumentative roles she feels she will be able to fulfill well, enlisting the help of others to play the complementary roles. Virtuous arguers also acknowledge that because arguments are dynamic events, the roles are fluid. Where commitments to pre-established positions soften, an opponent can morph into an interlocutor, or perhaps even a co-inquirer. Arguments that start out as adversarial need not stay that way.

---

becomes a proponent is exploiting an ambiguity: In some theoretical contributions to argumentation theory, the opponent role is differentiated more strictly to include only the raising of objections (see, e.g. Wohlrapp, 2014, p. 86).

[26] To some degree, this is what happen in legal adjudication systems, i.e., the structures we have built to deal with high-stake, low-trust disputes with the greatest potential for adversariality run amok. Legal theorists continually and vigorously re-visit the argumentative roles for defense attorneys, prosecutors, civil lawyers and judges. See, e.g. Lon L. Fuller's discussion of the roles of adjudicators in "The Forms and Limits of Adjudication", where he carefully examines and evaluates the norms associated with judges (Fuller, 1978).

6. CONCLUSION

While we agree that too much adversariality can be harmful, we hope to have shown that it is not adversariality *per se* that is the problem, but its vicious forms. What makes adversariality vicious is not its connection to picking sides and argumentative roles. Choosing argumentative moves so that they will advance one position over others – arguing to win – is an action that can be the result of virtue as much as it can be the result of vice. To be sure, adopting the argumentative role of proponent or opponent shows a choice for some degree of adversariality. What counts, though, is what motivates the choice. When arguers viciously let selfish, partisan thinking dictate their choices about whether to be cooperative or adversarial, vicious adversariality results. Virtuous arguers, i.e., arguers whose character makes them more disposed to be motivated by the *telos* of argumentation, can also assume oppositional roles in arguments. They make choices based on their understanding of arguments as organic wholes – including the other participants, the epistemic parameters, and the social contexts within which they arise. It is this last characteristic of a virtuous arguer that is especially noteworthy; the awareness that she is part of a larger whole, much of which is not within her control, and the ability to let that awareness inform her behaviour in such a way as to enhance the entire argument.

REFERENCES

Aberdein, A. (2010). Virtue in argument. *Argumentation*, *24*(2), 165-179.

Aikin, S. (2011). A Defense of war and sports metaphors in argument. *Philosophy and Rhetoric*, *44*(3), 250-272.

Aikin, S. (2008). Holding one's own. *Argumentation*, *22*(4), 571-584.

Annas, J. (2011). *Intelligent virtue*. New York: Oxford University Press.

Axtel, G. & Carter, J. A. (2008). Just the right thickness: A defense of second wave virtue epistemology. *Philosophical Papers*, *37*(3), 413-434.

Bailin, S. and M. Battersby. (2017). DAMed if you do; DAMed if you don't: Cohen's "missed opportunities". OSSA Conference Archive. 90. Retrieved from https://scholar.uwindsor.ca/ossaarchive/OSSA11/papersandcommentaries/90.

Battaly, H. (2008). Virtue epistemology. *Philosophy Compass: Epistemology*, *3*(4), 639-663.

Biddle, B. J. (1979). *Role theory: Expectations, identities and behaviours*. Burlington: Elsevier Science.

Code, L. (1984). Towards a responsibilist epistemology. *Philosophy andPhenomenological Research*, *45*, 29-50.

Cohen, D. H. (2003). Just and unjust arguments. OSSA Conference Archive. 16. Retrieved from https://scholar.uwindsor.ca/ossaarchive/OSSA5/papersandcommentaries/16.

Cohen, D. H. (2013a). Virtue, in context. *Informal Logic*, *33*(4), 471-485.

Cohen D. H. (2013b). Skepticism and argumentative virtues: Sextus, Nagarjuna, and Zhuangzi. *Cogency*, *5*(1), 9-31.

Cohen, D. H. (2014). Commentary on von Radziewsky. In D. Mohammed and M. Lewinsky, (Eds.) . OSSA Conference Archive. 145. Retrived from https://scholar.uwindsor.ca/ossaarchive/OSSA10/papersandcommentaries/145.

Cohen, D. H. (2015). Missed opportunities in argument evaluation. In F. H. van Eemeren & B. Garssen (Eds.) *Reflections on theoretical issues inargumentation theory*. Switzerland: Springer International Publishing.

Cohen, D. H. (2017). Commentary on Michael Yong-Set's "ludological approach to argumentation". OSSA Conference Archive. 52. Retrieved from https://scholar.uwindsor.ca/ossaarchive/OSSA11/papersandcommentaries/52.

Fuller, L. L. (1978). The forms and limits of adjudication. *Harvard Law Review*, *92*(2), 353-409.

Gascon, J. A. (2016) Virtue and arguers. *Topoi*, *35*(2), 441-450.

Gilbert, M. (2016). Ethos, familiars and micro-cultures. In: F. Paglieri, L. Bonelli, & S. Felletti (Eds.), *The psychology of argument: Cognitive approaches to argumentation and persuasion* (pp. 275-285). London: College Publications.

Gilbert, M. (2014). *Arguing with people*. Peterborough, ON: Broadview Press

Gilbert, M. (1997). *Coalescent argumentation*. Mahwah, NJ: Lawrence Erlbaum Associates.

Govier, T. (1993). When logic meets politics: testimony, distrust and rhetorical disadvantage. *Informal Logic*, *15*(2), 93-104.

Greco, J. (1999). Agent reliabilism. In: J. Tomberlin (Ed.), *Philosophical perspectives 13: Epistemology*. Atascadero, CA: Ridgeview Press.

Hundleby, C. (2013). Aggression, politeness, and abstract adversaries. *Informal Logic*, *33*(2), 238-262.

Lewinski, M. (2017). Practical argumentation as reasoned advocacy. *Informal Logic*, *37*(2), 85-113.

Mercier, H. and D. Sperber. (2011). Why do humans reason? Arguments for an argumentative theory. *Behavioral and Brain Sciences*, *34*(2), 57-111.

Rooney, P. (2003). Feminism and argumentation: A response to Govier. OSSA Conference Archive. 77. Retrieved from https://scholar.uwindsor.ca/ossaarchive/OSSA5/papersandcommentaries/77

Rooney, P. (2010). Philosophy, adversarial argumentation, and embattled reason. *Informal Logic*, *30*(3), 203-234.

Sunstein, C. (2000). Deliberative trouble? Why groups go to extremes. *Jale Law Journal, 110*, 71-119.
Stevens, K. (2016). The virtuous arguer: One person, four roles. *Topoi, 35*(2):375-383.
Turner, R. H. (2001). Role theory. In: Jonathan H. Turner (Ed.) *Handbook of sociological theory*. New York: Springer: 233-254
Turner, J. H. (2002). *Face to face – Toward a sociological theory of interpersonal behaviour.* Stanford: Stanford University Press.
Walton, D. N. (1999). Ethotic arguments and fallacies: The credibility function in multi-agent dialogue systems. *Pragmatics and Cognition* (1): pp. 177-203.
Walton, D. N. (1998). *The new dialectic.* Toronto: Toronto UniversityPress.
Walton, D. N. and Krabbe, E. C. W. (1995). *Commitment in dialogue.* Albany, NY: State Univ. of New York Press
Williams, B. (2004). *Truth and truthfulness: An essay in genealogy.* Princeton, NJ: Princeton University Press.
Wohlrapp, H. (2014). *The concept of argument.* Dordrecht: Springer.
Van Eemeren, F. H. and Grootendorst, R. (2004) *A systematic theory of argumentation: The pragma-dialectical approach.* New York: Cambridge University Press.
Zagzebski, L. (1996). *Virtues of the mind.* New York: Cambridge University Press.
Zarefsky, D. (2012). A Challenge and an Opportunity for Argumentation Studies. *Argumentation and Advocacy, 48*(3), 175-178.

Cultural Disagreements and Legal Argumentation: An Educational Program in Middle Schools

SERENA TOMASI
University of Trento - Cermeg
serena.tomasi@hotmail.it

This paper reports a project on training young students in legal argumentation. The model of teaching is focused on the normative model of judicial debate called C.A.L.S., according to which argumentation is "a work of many hands". The logical checks used by the courts in cross-cultural disputes can contribute to the settlement of the cultural conflicts in schools, or help students to move from conflict or indifference to consent.

KEYWORDS: legal argumentation, education, cultural disagreements, rhetoric.

1. INTRODUCTION

This essay is built around the analysis and evaluation of a one-year research program on legal argumentation and education, locally developed in Trento (Italy) and co-funded by local institutions (University of Trento, Local government and Ca.ri.tro).

The research project is aimed at providing a set of instruments for the management of cultural diversity in schools and promoting a new approach for civil learning based on the rhetorical procedure traditionally performed in trials.

At the core of the project is the predominant conviction of the pervasiveness of argumentation and its capacity to directly influence all aspects of everyday life, even for young students faced with cultural diversity. It presumes a background in the essentials of argumentation theory, which is inherently interesting for providing insights about the uses and abuses of the distinctive types of arguments in everyday cultural disagreements.

The chief target of the research is to carry out a conflict resolution test in a middle school in the most multi-ethnic school district in Trento (Italy), applying argumentative techniques and, particularly, the 'critical test' used by the courts in cross-cultural disputes. The analysis of some particularly powerful Italian case studies based on cultural disagreements demonstrates that the Courts move strategically within the categories of religion and culture. The judge is unable to limit the study of the case to the legislation, taking into account other factors regarding cultural dynamics, applying critical questions to give relevance to the culture of one party in the trial, and checking the plausibility and consistency of the arguments. The critical test is an argumentative procedure by which judges finally find a reasonable and acceptable solution.

The practical interest of the project is immediately apparent: presenting a reasonable tool for handling everyday conflicts, by bringing to light the role of argumentation in the case of cultural disagreements. Cultural disagreements are the genuine effect of multiculturalism, which is the ground for the occurrence of divergences.

Multiculturalism is a broad category including the (ethical and social) conflicts generated by the strong presence of a multicultural dimension in society. Multiculturalism is also a very popular topic in political thought in response to cultural and religious diversity: generally speaking, a multicultural society is one where different cultures co-exist with racial differences, values, religions, and patterns of thinking; a multicultural society is one where people from different cultures, nationalities, and ethnic and religious groups live in the same area and where mutual differences are innate and may become the basis for discrimination. The debate on this topic is broad and it is not in our interest to try to tackle it.

A clarification is necessary. There is a distinction between:

a) A multicultural approach (see Walzer, 1997, 1998), which presumes immeasurable diversity across cultures. Each culture, including its own values, thoughts, and attitudes is unique and, therefore, different from others. In this view, it is not possible to imagine a policy of integration because it is not possible to assess a culture by external categories but only from inside.

b) An intercultural approach (see Taylor, 1993), which assumes that there is no exclusive diversity in multicultural society; therefore a cross-cultural dialogue is possible, promoting intercultural or inter-ethnic experience.

In this project, we assume the second sense of social and ethnical pluralism, presenting a model for fruitful mutual understanding. An intercultural approach suggests community mediation in very broad terms, promoting a mediation program functional to setting up an argumentative discussion, by rebuilding a common ground and overcoming mutual misinterpretations.

Within this framework, this paper will follow the four stages of development of the research program.

Firstly, a theoretical stage will define the model of legal argumentation for handling the conflict, comparing contemporary schemes of reasoning and presenting a normative model for judicial reasoning, elaborated by Cermeg (Research Centre on Legal Methodology established in Trento), and the so called C.A.L.S. (Cooperative Argumentative Legal Syllogism), which describes and explains argumentative speech in trials as a cooperative decision-making process.

Secondly, an empirical stage will more specifically map the context where the model has been tested; in this section we will present the classroom climate and the school environment of the sample.

Thirdly, in the practical stage, the point of view will be practically oriented to the application of the conflict resolution principles in the classes. This insight on the lesson plan is a peculiar trait of the study, depending on the day-to-day training in real-life schooling practice. We had the chance to experience in classrooms the extent to which a reasonable exchange of opinions taking place in an argumentative context can contribute to the settlement of the cultural conflict, or help move from conflict or indifference to consent.

Finally, the last stage will be devoted to the presentation and discussion of the results of the research program, making explicit some of the limits of this account and suggesting further research perspectives.

2. THEORETICAL PERSPECTIVES ON THE ROLE OF ARGUMENTATION

The thesis of this account is that judicial argumentation can have an important place in an educational program for handling controversial conversations, which need to be treated with caution because they involve cultural disagreements. The problem is that cultural disagreements have a much greater impact in social and ethical relations because they represent genuine and apparently unsolvable disagreements: they can be considered the result of misunderstandings of cultural differences or the outcome of divergent but legitimate interpretations. These divergences depend on more basic differences among ethical and political conceptions that stay in the background and characterize the interpersonal dimension and define the cultural context. To this end, argumentation theory has a cognitive value in 1) coping with challenges that arise from the cultural context; 2) enhancing the trickiness of arguments which rely on the prejudices in an audience; 3) explaining how cultural disagreement can be reasonably discussed.

A series of studies in argumentation have been considered in elaborating the educational program and teaching argument to students. The following theoretical contributions provide a necessary complement to the educational model: they are necessary in gaining a more systematic understanding of the role of argumentation in educational programs.

2.1 The pragma-dialectical approach

The pragma-dialectical model of critical discussion can be considered the first and fundamental investigation involved in setting the lesson plan for students .

As for the focusing interest, van Eemeren remarks that the fundamental aim of any argumentative discussion can be defined as reasonably resolving a difference of opinions (van Eemeren & Grottendorst, 1992, 2004). In particular, in pragma-dialectical terms, it is crucial for the correct development of the critical discussion that the parties recognize the issue and the mutual difference of opinions; then, the parties must adopt a peculiar status in relation to their role as participants in the argumentative discussion. If they are willing to engage in the critical discussion, they will be responsible for making the final decision by respecting a code of conduct. The great advantage of the code of conduct is that it is readily available argumentative knowledge, consisting of a list of ten rules. The normative side of the theory is developed in ten commandments which guide the disputants in setting up the discussion and acting in a reasonable way, without letting the process become a mere matters of wills or personal recriminations. The Ten Commandments are powerful tools and are therefore useful to know, both for guarding against wrong arguments and for using good arguments. It is not only a matter of logic but, fundamentally, of ethos: each party has the commitment to defend its own side of a controversy but, according to the common code, each party also has the commitment to enhance and not compromise the whole discussion. Therefore each party would try to eliminate bias in a critical discussion and honestly cooperate towards an acceptable solution.

The idea of adopting a normative grid can have a great impact in education in different ways: it presents a normative analysis of the condition under which conflicts can be correctly solved; it is also a tool for action and evaluation of the speech acts; moreover it recognizes the leading and fundamental role of the parties which have, at least, a common interest: the resolution of the disagreement.

Based on the valuable insights of the ruling, students in the training shared a code of conduct based on five simple rules. They assume a common task: to sketch their breakpoint and discuss it. Whoever did not respect the rule was admonished: the violation was clearly marked in a chart displayed in the classroom. Addressing the pragma-dialectical principles, the adopted rules for young students are:

- Rule of freedom, to encourage students to participate in the discussion, make their own comments, provide arguments, and reject or accept other points of view and questioning. The idea is that interaction can take form only if the participants can dialogue in a natural way.
- Rule of responsibility: young students are often 'hot-headed' and impulsive in expressing their thoughts. The principle of responsibility involves a reflection on their practice and opens the way for beneficial effects on the management of the conflict. This requirement implies accuracy in that each party must give arguments supporting a claim.
- Rule of relevance, which is centred on the quality of argumentation in order to be relevant. This critical guidance suggests avoiding certain types of emotional appeals, which have a much greater impact on the audience but may not contribute to the goal of the discussion (for instance, arguments ad hominem).
- Rule of language, which regards the way of speaking. It implies avoiding ambiguity and being willing to define the boundaries of the terms and make explicit what is unclear or unexpressed. The agreement would not otherwise be solid since the counterparty does not really understand what is meant. A clarification on this point is needed: the students involved in the educational program belong to the second generation of migrants. They are all Italian-

speaking and they do not have serious comprehension problems. Therefore in the training part, linguistic-cultural mediation by an interpreter was never necessary.
- Rule of conduct, which concerns the behaviour of students in the classroom. This rule throws light on the concept of interaction, which is inherently dialectical, in that it involves the other party and, therefore, a sequence of moves. A dialogue would be possible only by assuming a cooperative attitude, without provocation, incitement, or noise. This requirement also suggests that there is a pragmatic component of dialogue, which involves the need to rethink arguments in relation to the situation and the dynamic exchange.

Therefore, the present model takes a pragma-dialectical view as for the dynamic concept of discussion and its regulation. The normative standards proposed by the pragma-dialectical theory of argumentation are relevant in the educational model for the function assigned to arguing: they represent the felicity conditions for an ideal conflict resolution process. And so, during the interaction, the parties may change their attitude: at the beginning, the conflict is still in a significant phase but, by the end, the parties should show mutual understanding and respect.

2.2 The model of topics

Another point of view of argumentation is the one proposed by Eddo Rigotti and Andrea Rocci (Rigotti & Rocci, 2006) in the context model. They specify context as a constitutive aspect of communication, composed by two dimensions: the institutional and the interpersonal one.

On the one hand, there are institutional relations between discussants involved in communication: these relations are implemented by rules. But on the other hand, Rigotti and Rocci define the interaction as "that piece of social reality where the communication takes place". This connects with the other topic we want to deal with: the interpersonal level of communication. Each communication is affected by the individual's stories, their representations, and their frames of reference. The so-called cultural context influences the possible proceedings of the conflict: different ideas, different views and conceptions characterize the framework of discussion. Especially when working with ethnic conflicts, the cultural dimension has to be analysed.

The analysis of the state of affairs is relevant for the topical potential (the discovery of loci argomentorum) and for the presentational techniques aimed at the effectiveness of argumentation. This approach, by invoking the role of the subjects, investigates how the audience can be affected by emotions. The generation of emotions depends on the speakers involved: strategies triggering emotional effectiveness suggest that arguments can be taken only in strict connection with the concrete situation (Rigotti & Greco 2006, 2009; Greco, 2011).

Their contribution, which is focused on some innovative aspects of the doctrine of Aristotle, Cicero, Boethius, and the medieval scholars, is functional in that it points out the relational dimension of arguments which is inherent in every human interaction.

2.3 The use of rhetoric

How should we react to disagreement and what consequences may follow with other people? Like every reasonable person, we may always disagree with others but we must do it in a reasonable way. We must examine the issue and offer reasons for our positions and beliefs.

Practically oriented research papers by Jeanne Fahnestock and Marie Secor (Fahnestock & Secor, 1982; Fahnestock, 2011) represent an approach to teaching arguments without logic. The absence of classifications of syllogisms or patterns or rules of validity is deliberate. They suggest students come up with arguable topics from their own experience, readings, and activities. Moving from the assumption that argument pervades our lives, they recognize that the construction of an argument begins with determining the issue and proceeds with identifying the proposition to discuss. In order to provide arguments supporting the claim, they provide a set of questions to direct students in arguing: the division of steps is clear and lucid. They do not give general advice about inference, fallacies to avoid or analysis of issues, but provide questions, which represent categories of answers: What is it? How did it get that way? Is it good or bad? What should we do about it? Students are quick to go over the simplicity of this four-part division and make use of topics.

Following these suggestions, students should then be able to settle the matter, present their reasoning, succeed in dialectic moves, avoid snares, and be effective. Answering the first question helps students introduce a definition of the issue or basic elements about reality; answering the second question explains the causal relation and focuses on invention; answering the third question suggests proposals of arguments; answering the fourth question implies creative and ethical assumptions.

We assume that this way of thinking about arguments and teaching argument interesting and useful for young students who are not sufficiently prepared in logic. It provides a guide for building arguments, from simple structures to more complex discussions.

2.4 C.A.L.S.: a normative model in forensic rhetoric

The innovative aspect of the research program is the application of legal categories in the educational field. How do judges in trials mentally solve cultural disagreements in legal controversies? How do they compose the evident disagreement?

To help young thinkers in arguing, we present the argumentative techniques used by judges to manage cultural and religious conflicts. Through legal reasoning, people may recognize each other's way of life and accept these differences with respect and appreciation.

The legal context is a peculiar and fruitful context of study of argumentation for multiple reasons: firstly, the very nature of law, which is inherently controversial, has the potential to show the significance that argumentation may have in resolving social conflicts and developing social skills; secondly, the legal context is a more cogent context due to its pervasiveness in individuals' lives: everybody has experienced a legal situation and believes in 'justice' as a way to put an end to controversy.

Therefore the final proposal to the students was to imitate what judges and lawyers do. Judges do not have automatic answers but they build solid premises upon plausible arguments.

In a recent work, Italian constitutional scholar Ilenia Ruggiu (Ruggiu, 2012) has reconstructed Italian policymaking and case law on cultural issues (different topics are examined: the use of the crucifix, the cultural legitimacy of circumcision, etc.). Case by case, she identifies a few recurring topics and qualifies the judicial moves as mainly rhetorical, showing that judges do not limit their reasoning to the norms (which are often lacking in cultural controversies) but apply critical questioning in order to equally weigh the opinions.

Most of the recent research developed in Italy by the Cermeg (Research Centre on Legal Methodology, established in Trento; see Manzin, 2012) has been focused on the form of legal reasoning and the use of rhetoric in the decision process. Maurizio Manzin and Serena Tomasi (Manzin & Tomasi, 2014) presented a normative model of judicial debate called C.A.L.S., Cooperated Argumentative Legal Syllogism. They highlighted that in the logical construction of the decision, the judge is the ultimate but not the only builder, claiming the idea of legal argumentation as "a work of many hands", which implies a specific commitment to agreement for each party in the trial. Before an impartial judge, parties must set their positions, demonstrate their soundness and consistency, resist to objections, and persuade each other. The final decision does not depend only on logic or on a rational weighing of arguments, but must also take into account the parties' emotional and interpersonal involvement in the situation. Parties cooperate towards a common goal.

The ideal of the "automatic judge" (e.g. Montesquieu) is a dream: his work is not a lonely job because the context of communication is not monological (Manzin, 2013, 2014). We are still fond of the idea that judges come to the solution by a formal demonstration according to the scheme p, then q (iff p is true, q is true). This is not realistic at all: in a dialogic situation q always anticipates the justification, therefore q is true, iff p is true. Each party provides arguments, both on statutory interpretation and on evidence, then the judge collects the arguments and check them for plausibility, consistency, and contradictions (Manzin, 2013, 2016). The term for describing the management is "cooperated syllogism": on the one hand it supposes common work by the parties and the judge in building the legal syllogism, on the other it admits the use of conflicting strategies by the parties to persuade the judge.

Of course, these changes in the way of understanding the judicial discussion are relevant for education: thinking about trials, the first ideas that come to mind are 'strategy', 'struggle', 'tricks', which could lead to a misleading view of the judicial context. If we come to the idea of cooperation in debates in trials, we recognize the plausible situation that happens in real ordinary life and we suggest students, simulating a real pleading, adopt this attitude with positive consequences for the commitments and the final result.

As mentioned before, this is the notion of debate we favour in the present work for acquiring critical skills and promoting a critical solution of cultural conflicts in schools (Manzin, 2008, Tomasi, 2016).

3. THE EMPIRICAL COMPONENT

This section corresponds to the second stage of the research program, which has been mainly devoted to mapping the degree of pluralism in schools in Trentino.

Focusing on the situation in the local area, we collected data processed by local institutions, carrying out a deep statistical analysis on local middle schools. Then we conducted a specific and close examination on the selected school, where there are multiple ethnics groups. As far as the preferred school is concerned, we mapped the social composition of classes and the students' backgrounds and analysed dynamics of social integrations in the classroom.

Immigration in Trentino is a relatively recent phenomenon. The foreign student population has grown over the last ten years, and in the last two years it has been stable. The selected pilot school is situated in Gardolo, a suburban district of Trento. Gardolo middle school is the most cosmopolitan and multi-ethnic school in the district. In Gardolo, 60% of the students enrolled in school are immigrants. There are classrooms in which the majority of students are not Italian citizens.

Two classes in the final grade in the same school were selected to participate in the practical training for two months. The researcher, with the cooperation of the tutor (a teacher well-known to the students), was introduced to the class, promoted the surfacing of hidden cultural disagreements and undertook the practical program in legal arguing.

4. EXERCISE FOR THE RESOLUTION OF CULTURAL DISAGREEMENTS

The experimental part in classrooms was organized into three sections:

- a) a preparatory education course for teachers;
- b) a weekly cross-intercultural teaching program in classrooms
- c) an evaluation test (a cross-check including non-participating groups of students).

The contents of the lesson were focused on legal argumentation, to engage students and manage their cultural obstacles by means of argumentation as a regulated practice.

Students, especially young students, need explicit instructions in argumentation: we provide concrete examples for a more critical engagement, for outstanding and failing performances.

In the classroom, by adopting a set of argumentative rules, teachers and students shared responsibility for creating and stimulating a learning environment and critical attitude.

The activities with the students were prepared in advance, discussing with the teacher involved the theoretical insights of the research. We taught a 20hr course for teachers on argumentation (especially about analysis, recognizing premise and conclusion).

Then we moved into two classes for promoting discussion on cultural issues, proposing several exercises:

- Basic notions of law (substantive law, procedural law; international law and domestic law, sources of law, primary normative acts, especially constitutional acts)
- Exercise focused on critical readings (i.e. to encourage thinking about the ways in which migration is presented in the mainstream media: what stereotypes do they hide or do they reproduce? Do the critical considerations that emerge from the readings confirm or challenge these notions? Do they promote or compromise diversity and multiculturalism?)
- Exercise on presenting a standpoint and bringing to light the issue, following the Fahnestock method (Fahnestock & Secor, 1982), by encouraging students in constructing arguments answering 4 critical questions: What is it? How did it get that way (causal relation)? Is it good or bad? What should we do about it?
- Exercises focused on an ongoing scholarly conversation about an arguable issue, in which the students intend to take a well-defined stand (displaying the crucifix; carrying weapons for religious reasons, representation of Nativity scene during Christmas).
- Exercises on the uses of rhetoric by structuring arguments and expressing them by working out matters of style, voice and clarity.
- Group exercise to improve self-motivation, training in reasoning, peer collaboration. I reject the idea of performing antagonistic confrontation. Neither disputants in a trial are really antagonistic: they have to cooperate towards an acceptable and rational solution, even though in a strategic manner.
- Individual exercise for practicing oral argument, counter-argument, rebuttal, seeing argumentation as dialogic
- Exercise on evidence supporting the claim: provide a list of key critical questions to ask about evidence. These exercises concern discourse analysis and single-word operators and connectors that can show the argumentative features of a text.

Let's now consider the result of the teaching-test.

5. CONCLUSIVE REMARKS

We shall now discuss the prominent features that emerged when analysing the results of the research in applying the educational program.

After the training, students generally improve their knowledge about cultural diversity and acquire skills for discussing. The validation has been proved by the decision by the school management to take the training again.

This research may represent an advance in the study of argumentation and education for the sample and for the role of law in education for conflict resolution. Students learnt argumentative competence and knowledge about the sources of law, human rights, and the boundaries of protection of fundamental rights.

We highlight three rewarding aspects.

A crucial point has been the role of the researcher (and, therefore, of the teachers involved as tutors). They guide students towards the exploration of the origin of the conflicts. At the beginning students were not able to identify the issue: the major problem was getting people to take a stand. The researcher started by asking the students what the problem was, and what the difference of opinions was. In this way they specified their positions. Then the teacher/researcher encouraged the search for the source of information and asked people to return to present their standpoint in order to constantly check the consistency.

The aim has never been to increase the conflicting background, but to define a common ground, bringing the parties to a true confrontation. The teacher/researcher played a fundamental role in that they shifted the discussion to the options for conflict resolution. In this way, they act as a mediator opening a series of subdiscussions about the possible correspondence. The parties were never juxtaposed but connected to the same pragmatic goal. The teachers were trained in exercises focused on their coaching role and mediating role: they should facilitate interaction between, maintaining an impartial and neutral role.

The empirical analysis brought to light the impact of the topical potential: the selection of arguments and the topical choice of students reveal the weight of interpersonal contexts. The presence of different cultures in classes facilitates the construction of arguments about culture by referring to the personal experience of the classmates. The facts that have originated the conflict are, at the same time, the elements upon which they reach a consensus. It might be said that they found a solution on the basis of arguments they identify in their personal experience. The story of the personal relationships shaping the context of the dispute becomes a reason to solve the conflict or the premises for the resolution.

The questionnaire also proved to be a reliable instrument for evaluations. The assessment questionnaire was complex and composed of eight types of exercises, related to rhetorical skills, in particular to the use of ethos, pathos, and logos. The questionnaire was solved by students from four classes, different from the sample, equal in age, in the same school. The classes of the sample were successful; the other four classes were not entirely successful. Mostly they did not solve exercises on logic and the ones focused on explaining social issues such as integration, marginalization, and segregation. We can explain the result in that argumentative skills are not spontaneous but require specific program teaching.

For this reason the study of legal argumentation can be seen as a valid basis for establishing more reasonable and better-found communicative interaction in society, going beyond cultural differences.

In a comparative way, the research program could be improved by applying the test in a different school with a smaller number of foreign students. It would be useful to look at other groups: a different school environment may change the cultural dimension.

REFERENCES

Belvisi, F. (2000). *Società multiculturale, diritti, costituzione: Una prospettiva realista*. Bologna: CLUEB.
Canale, D. (2013). Il ragionamento giuridico. In G. Pino, A. Schiavello & V. Villa (Eds.), *Filosofia del diritto: Introduzione critica al pensiero giuridico e al diritto positivo* (pp. 316-351). Torino: Giappichelli.
Eemeren van, F. H. & Grootendrost, R. (1992). *Argumentation, communication, and fallacies: A pragma-dialectical perspective*. Hillsdale, NJ: Lawrence Erlbaum Associates.
Eemeren van, F. H. & Grootendorst, R. (2004). *A systematic theory of argumentation*. Cambridge: Cambridge University Press.
Eemeren van, F. H. (Ed.). (2009). *Examining argumentation in context: Fifteen studies on strategic manoeuvring*. Amsterdam: John Benjamins.
Eemeren van, F. H. 2010. *Strategic maneuvering in argumentative discourse: Extending the pragma-dialectical theory of argumentation*. Amsterdam/Philadelphia: John Benjamins.
Fahnestock, J., & M., Secor. (1982). *A rhetoric of argument*. New York: Random House.
Fahnestock, J. (2011). *Rhetorical style: The uses of language in persuasion*. New York: Oxford University Press.

Feteris, E. T. (1987). The dialectical role of the judge in a legal process. In J. W. Wenzer (Ed.), *Argument and critical practices: Proceedings of the fifth SCA/AFA Conference on Argumentation* (pp. 335-339). Annandale: Speech Communication Association.

Feteris, E. T. (1999). *Fundamentals of legal argumentation: A survey of theories on the justification of judicial decisions.* Dordrecht: Kluwer Academic Publisher.

Feteris, E. T. (2005) (Ed.). Schemes and structures of legal argumentation. *Argumentation, 19*(4). Special Issue.

Greco, S. (2011). *Argumentation in dispute mediation.* Amsterdam/Philadelphia: John Benjamins.

Manzin, M. (2008). Il cinismo giudiziario e le virtù del metodo. In P. Moro (Ed.), *Scrittura forense: Manuale di redazione del parere motivato e dell'atto giudiziale* (pp. 1-3). Torino: Utet.

Manzin, M. (2012). A rhetorical approach to legal reasoning: The Italian Experience of CERMEG. In F. H. van Eemeren & B. Garssen (Eds.), *Exploring Argumentative Contexts* (pp. 137-148). Amsterdam: John Benjamins.

Manzin, M. (2013). Taking judges seriously: Argumentation and rhetoric in legal decisions. In G. Kišiček, I.Ž. Žagar (Eds.), *What do we know about the world? Rhetorical and argumentative perspectives* (pp. 251-271). Studies in Argumentation 1. Windsor (Can.): Digital Library Dissertationes 25.

Manzin, M. (2014). *Argomentazione giuridica e retorica forense.* Torino: Giappichelli.

Manzin, M. (2016). Is the distinction between "cooperative" and "strategic" crucial for jurisprudence and argumentative theory? In D. Mohammed & M. Lewinski (Eds.), *Argumentation and Reasoned Action. Proceedings of the 1st European Conference on Argumentation, Lisbon 2015*, (Vol. I, pp. 129-133). London: College Publications.

Manzin, M., & Tomasi, S. (2015). Ethos and pathos in legal argumentation: The case of proceedings relating to children. In B. J. Garssen, D. Godden, G. Mitchell & A. F. Snoeck Henkemans (Eds.), *Proceedings of the 8th Conference on Argumentation of the International Society for the Study of Argumentation,* Amsterdam 2014 (pp. 930-941). Amsterdam: Rozenberg/Sic Sat.

Murphy, M. (2012). *Multiculturalism: A critical introduction.* New York: Routledge.

Pastore, B., & L. Lanza. (2008). *Multiculturalismo e giurisdizione penale.* Torino: Giappichelli.

Raz, J. (1998). Multiculturalism. *Ratio Juris, 11*(3), 193-205.

Rigotti, E., & Rocci, A. (2006). Towards a definition of communication context: Foundations of an interdisciplinary approach to communication. In M. Colombetti (Ed.), *The communication sciences as a multidsciplinary enterprise* (pp.155-180). Studies in communication sciences 6 (2). Anniversary Issue.

Rigotti, E., & Greco, S. (2009). Argumentation as object of interest and as social and cultural resource. In N. Muller-Mirza & A. N. Perret-Clermont (Eds.) *Argumentation and education: Theoretical foundations and practices* (pp. 9-66). New York: Springer.
Ruggiu, I. (2012). *Il giudice antropologo: costituzione e tecniche di composizione dei conflitti multiculturali*. Milano: Angeli.
Taylor, Ch. (1993). *Multiculturalismo: La politica del riconoscimento*. Milano: Anabasi.
Taylor, Ch. (1993b). *Reconciling the solitudes: Essays on Canadian federalism and nationalism* (G. Laforest, Ed.). Montreal: McGill-Queen's University Press.
Tomasi, S. (2016). Fairness and legal reasoning. Commentary on Konczol's Fairness, definition and the legislator intent: Arguments from epieikeia in Aristotle's rhetoric. In D. Mohammed & M. Levinski (Eds.), *Argumentation and reasoned action: Proceedings of the First European Conference on Argumentation, Lisbon 2015* (pp. 337-341). London: College Publications.
Walzer, M. (1997). The politics of difference: Statehood and toleration in a multicultural world. *Ratio Juris, 2*, 165-76.
Walzer, M. 1998. Sulla tolleranza. Roma-Bari: Laterza.

56

The Explicit/Implicit Distinction in Multimodal Argumentation: Comparing the Argumentative Use of Nano-Images in Scientific Journals and Science Magazines

ASSIMAKIS TSERONIS
University of Amsterdam
a.tseronis@uva.nl

The distinction between explicatures and implicatures as well as their varying degrees of strength acknowledged within Relevance Theory can help to capture the complex meaning-making processes underlying the interpretation of multimodal texts as instances of argumentation. These pragmatic insights will be used to compare the ways in which arguments about the revolutionary character and societal impact of nanotechnology are constructed by computer-generated images of the nanoscale on the covers of scientific journals and science magazines.

KEYWORDS: front covers, explicatures, implicatures, multimodal argumentation, nano-images, popular science magazines, science communication, scientific journals; visual flag

1. INTRODUCTION

As Bateman (2014) explains, the apparent effortlessness of perceiving visuals simply by having our eyes open and looking at them is revealed to be an illusion when we are confronted with visual representations with which we are not familiar: "Then just seeing is replaced by a more conscious attempt to work out what is going on" (p. 8). Computer-generated images that present structures or processes at the scale of a billionth of a meter, which are invisible without the use of special microscopes and visualization software, are one such type of image. These so-called nano-images, produced by electron microscopes or created by artists, are used on the covers of scientific journals and science magazines to construct arguments about the revolutionary character and societal impact of nanotechnology. The study of the front cover as a multimodal genre that employs images and text to promote the sales or readership of the magazine requires the combination of insights from multimodal discourse analysis and argumentation theory (Jewitt, 2014; Norris & Maier, 2016; Tseronis & Forceville, 2017). When studying multimodal argumentation, the analyst needs to pay attention not only to the content and form (verbal or visual) but also to the context, in order to reconstruct the standpoint and the arguments brought forward in support of it.

For a better understanding of what it is about image-text combinations that guides the receivers of multimodal texts along particular lines of interpretation than others, the analyst also needs to have recourse to the inference process that receivers undergo. Relevance Theory's (Sperber & Wilson, 1995) focus on how both the explicit and the implicit side of communication give rise to inference can allow for a comprehensive account also of the ways in which the information derived from the visual mode contributes to the meaning-making process, and thereby to the construction of a multimodal argument. Choices regarding the visual content and style can thus be shown to guide the viewer to infer implicatures or explicatures constrained by the viewer's encyclopaedic knowledge and the broader situational context. Nanoscientists and the general public, confronted with nano-images on the covers of scientific journals and science magazines, respectively, will interpret these differently given their respective background knowledge and expectations.

The goal of the paper is twofold: a) to explore the potential of concepts proposed by Relevance Theory for the argumentative analysis of multimodal discourse; b) to apply these insights to the argumentative analysis of a concrete multimodal genre, namely the front covers of science magazines and scientific journals, where nano-images are used. With reference to degrees of explicitness and implicitness it will be shown how the arguments reconstructed from the covers with nano-images in scientific journals and science magazines differ.

2. RELEVANCE THEORETICAL INSIGHTS FOR THE ANALYSIS OF MULTIMODAL ARGUMENTATION

2.1 Multimodal argumentation

In the last twenty years, argumentation scholars have shown interest in the role of images and other semiotic modes in argumentative communication (spoken or written) (Birdsell & Groarke, 1996; Blair, 2004; Groarke, 2015; Groarke et al., 2016; Kjeldsen, 2015; Tseronis, 2017; Tseronis & Forceville, 2017; van den Hoven, 2012). This visual turn in the field of argumentation studies has been partly facilitated by the acknowledgment that argumentation as a communicative activity is multimodal, in the sense that, more often than not, standpoints and arguments in support of them are communicated by a combination of semiotic modes, as is the case with communication in general (Kress, 2010). Multimodal argumentation can thus be broadly described as a communicative activity, in which more than one mode, other than the verbal (be it spoken or written), play a role in the procedure of testing the acceptability of a standpoint. Semiotic modes combined in multimodal discourses that seek to convince an audience include: written language, spoken language, static images, moving images, music, non-verbal sound, gestures, gaze, and posture.

The acknowledgement of the inherent multimodality of communicative practices has opened a cross-disciplinary dialogue between argumentation scholars and researchers in areas such as visual communication, semiotics, multimodal analysis, and media studies, among others, for a comprehensive study of situated and mediated argumentative practices (Tseronis & Forceville, 2017). The work carried out in the field of multimodal analysis in particular (Bateman, 2014; Jewitt, 2014; Jewitt et al., 2016; Sanz, 2015) has been of great relevance to this direction. In this area of discourse analysis, concepts have been developed for the study of the affordances of various semiotic resources, and, more importantly, of the ways in which meaning is conveyed by the interaction of the verbal and the visual (or other non-verbal) modes. These concepts can be fruitfully combined with analytical categories from argumentation and rhetoric in order to account for the argumentative interpretation of multimodal texts.

In line with the multimodal analysis of communication, argumentation theorists need to pay attention to both the content and the form of each mode. In addition, one needs to study not only each mode separately but also the way modes combine, while paying equal attention to both the (verbal / visual) content and the (verbal / visual) form. Finally, attention also needs to be paid to the context (argumentative situation, audience, genre) in which a multimodal text is produced and interpreted. Birdsell and Groarke (1996, p. 5) have stressed early on the important role of context when analysing and evaluating visual arguments by identifying three layers that need to be considered: the immediate visual context, the immediate verbal context, and the visual culture. In order to better account for the ways in which context affects the interpretation process, I have recourse in this paper to Relevance Theory, which is a cognitive pragmatic theory of communication that takes a dynamic approach to the way context guides and constrains the inference process on both the explicit and implicit side of communication (see also Tseronis, 2018).

2.2 The explicit-implicit distinction

There is a tendency in the literature on visual and multimodal argumentation to equate rather hastily the distinction between what is explicit and what implicit, respectively with the verbal mode and the visual mode. This view follows a more or less standard approach in pragmatics which equates explicit and implicit meaning with encoded and inferred meaning, respectively. In this view, an illocutionary act is performed explicitly when a performative is used that names the act performed, while it is performed implicitly when no such performative is used. Compare "*I promise* I'll be there" with "I'll be there". Within a speech act theory perspective, there is a further distinction proposed between direct and indirect speech acts, the latter having a grammatical form that is more closely associated with one type of act than the one for which it is used in a given context. Consider, for instance, the textbook example of the interrogative sentence "Can you pass the salt?" used to *request* the addressee to pass the salt rather than to *enquire* about his/her ability to do so.

Even visual semioticians consider that images do not constitute an explicit form of communication. Nöth (2011, p. 310), for one, writes:

> A picture cannot express explicitly whether it is used to ask a question, to give an order, to threaten, to make a promise, or to congratulate. Hence, a theory of pictorial acts can only be a theory of indirect pictorial acts. If a picture addresses its readers with the purpose of asking a question or warning against a danger, the pictorial question or warning can only be an indirect one.

With this comment, however, Nöth overlooks the fact that an image – like a verbal text – is never communicated in void, and that there are visual and other formal cues as well as knowledge of the situation and genre that allow one to distinguish when it is used to warn one from when it is used simply to inform or to create a humorous effect.

Contrary to this view, Novitz (1977, p. 92) stresses the fact that to be able to specify the proposition an image may convey, one needs to know how that image is used:

> My argument will be that, for the most part, it is in virtue of certain contextual features that we can reasonably say of a picture that it is used to indicate a flower and to attribute a certain structure to it. The context, one might say, provides certain clues as to how the picture is being employed: what illocutionary act it is being used to perform and what proposition it is made to express.

The equation of explicitness with the verbal mode and of implicitness with the visual mode is problematic in several respects:

- It neglects the fact that there is nevertheless something that is depicted in an image and thereby that some (visual) content is provided.
- It backgrounds the fact that there is inferencing going on also when an addressee seeks to understand a message conveyed entirely in the verbal mode.
- It reduces the explanatory and analytic potential of the distinction between implicit and explicit aspects of communication since almost all naturally occurring communication (verbal or non-verbal) ends up being analysed as indirect and implicit.

The cognitive and pragmatic approach to communication that Relevance Theory assumes (Sperber & Wilson, 1995; Wilson & Sperber, 2012; Wilson, 2017) acknowledges the role that context and inference processes play for the recovery of both the explicit and the implicit content of messages. As Wilson (2017, p. 92) puts it:

> For relevance theorists, 'explicit' is both a classificatory and a comparative concept: any communicated proposition with a linguistically encoded conceptual constituent is explicit to some degree, and the greater the proportion of decoding to inference, the more explicit it will be. On this approach, any

utterance can be made more explicit, and there is no such thing as 'full explicitness'.

According to Relevance Theory, utterance comprehension is an inferential process based on the assumptions derived from the explicit content of a message and further contextual information, guided by the principle of relevance. As Wilson (2017, p. 90) explains:

> interpreting an utterance is like solving a complex simultaneous equation, and the interpretation process is crucially seen as carried out in parallel rather than in sequence. It is not a matter of first identifying the explicit content, then supplying contextual assumptions and then deriving contextual implications (and other cognitive effects), but of mutually adjusting tentative hypotheses about explicit content, context, and cognitive effects, with each other and with the presumption of relevance, and stopping at the first overall interpretation that makes the utterance relevant in the expected way.

While originally the focus of relevance theorists has been almost exclusively on verbal communication, the theory's broad definition of ostensive communication makes it possible to apply relevance theoretical distinctions to other modes and media of human communication (see Yus, 1998, 2011; Ifantidou & Tzanne, 2006; Wharton, 2009; Desilla, 2013; Forceville, 2014; Forceville & Clark, 2014). The relevance theoretic comprehension heuristic described in the previous quote can thus also prove useful for arriving at the premises and the conclusion of an argument that is wholly or partially communicated by means of visual elements. The meaning of images is not simply encoded in what the image depicts or in the visual properties it has. It is the constant interaction with the viewer's cognitive environment at a given time and place that determines what a visual (verbal or multimodal) stimulus means. The information available to the viewer may vary depending on his/her own encyclopaedic knowledge but also on the time and place. The way the image-maker presents the message gives clues to the audience and guides them in order to reinforce or adjust their assumptions.

One can thus distinguish what an image depicts (the proposition or propositions communicated, as it were) from what one means to communicate by that image (its interpretation). In the first case, the viewer arrives at a conclusion about what the image depicts, based on inferences drawn from what is directly visually available (in terms of form and style, shapes, lines, colour etc.), and with reference to encyclopaedic and other background knowledge. In the second case, the viewer arrives at a conclusion about what the producer of the image means to communicate, based on inferences that make use of the conclusion about what the image depicts and broader encyclopaedic knowledge (paying attention to the situational context, who made it, addressing whom, in which medium/genre, at what given moment, etc.). Following the distinction between explicatures and implicatures proposed by Relevance Theory, I would then suggest that the former communicated assumptions be considered as 'explicatures' and the latter as 'implicatures'. The explicatures of an image would thus be the assumptions recovered from the visual content of an image (what is depicted) enriched from background knowledge and from any verbal or other visual co-text or context present. The implicatures of an image would be the assumptions recovered entirely on the basis of inferential processes making use of background knowledge and any co-textual or contextual clues (verbal, visual or other).

At this point, I need to acknowledge that the above proposal does not strictly follow from Relevance Theory's original identification of explicitness with the linguistically encoded meaning (or logical form) of an utterance (Sperber & Wilson, 1995, p. 182). In the case of visual and multimodal communication, it is in fact debated whether drawings, photos and paintings, for example, can be said to consist of minimal units of meaning encoded in such a way that guarantees their interpretation by all those who share that code. Forceville and Clark (2014) have taken up the challenge to explore the possibility that explicatures or 'explicature-like' assumptions are communicated by visuals, acknowledging that there are, after all, varying degrees of codedness in visual and non-verbal signs. While the theoretical and terminological issue is far from settled, suffice it to say that for the purposes of this paper the distinction between explicit and implicit content along the lines proposed by Relevance Theory (see Carston, 2009) is a pertinent one for the analysis of multimodal argumentation, in particular, for one more reason next to those listed above. By recovering the assumptions based both on the explicit and on the implicit side of what is communicated it also becomes possible to determine the propositions to which the image-maker can be committed, whether recovered from what is explicitly communicated or from what can be reasonably implicated, and thereby to assign varying degrees of strength to these commitments (see Tseronis, 2018). In this way, a systematic and theoretically-driven answer can be provided to the questions raised by critics of visual and multimodal argumentation, regarding what is communicated by images and multimodal texts, and how can the analyst attribute commitments to their makers.

3. NANO-IMAGES ON THE FRONT COVER

3.1 Front cover as a multimodal argumentative genre

The front cover is generally considered as the show window of a magazine that attracts the readers' attention and informs them about the stories featured in the inside pages. Held (2005, p. 173) defines front covers as a multimodal media genre "which announces, indicates and appraises subsequent texts inside the magazine". From an argumentation studies perspective, front covers can be described as an argumentative activity type that belongs to the domain of commercial communication whose institutional point is to promote products and services (on argumentative activity types, see van Eemeren, 2010).

In previous studies (Tseronis, 2015, 2017), I have distinguished two interrelated levels of argumentation put forward by a front cover: the primary level concerning the argumentation *of* the cover, where the standpoint advanced is the one inciting the public to buy the specific issue of the magazine; and the secondary level concerning the argumentation *in* the cover, which pertains to the standpoint that the editors of the magazine assume over the specific cover story. These two levels are presented together in the following structure:

1. Buy the specific issue of the magazine
1.1a The main story that the magazine covers is on issue X
1.1a' Issue X is a newsworthy / important / relevant topic for the reader
1.1b The magazine takes position P on the main story
      1.1b.1 …. [The argumentation *in* the front cover]
1.1b' The position of the magazine reflects its profile (investigative journalism, etc.) / resonates with the reader's position on current issues

While the above structure can be considered as typical of the front cover as an argumentative genre, in general, more specific argumentation structures can be described when considering the slightly different argumentative goals served by the popular science magazine cover compared to the scientific journal cover. In the former, the choices made regarding the content of the image and its visual style can be related to the communicative goal of increasing the sales of the particular magazine. In the latter, however, these choices can be primarily related to the claim regarding the ground-breaking and innovative research reported in the particular journal.

Compared to the front covers of news magazines where image and text can be reconstructed as conveying a certain editorial viewpoint on the main story (the argumentation *in* the cover), the front covers of scientific journals and popular science magazines do not appear to have this secondary level of argumentation. Image and text in these cases combine in order to attract the potential reader's attention to the specific issue and to the main story or article covered. The covers of scientific journals and popular science magazines can thus be understood as functioning as 'visual flags'.[1]

Considering the fact that popular science magazines and scientific journals address two different kinds of audience with the distinct (but related) purposes to inform the general public about potentially interesting applications of science and technology, on the one hand, and to inform the scientific community about current research, on the other, two argumentation structures can be distinguished:

Argumentation of the science magazine cover:
1. Buy this issue of the magazine
1.1 The main story that the magazine covers is on issue X (e.g. nanotechnology and nanoscience)
1.1' Issue X is a **newsworthy / important / relevant** topic for the reader

Argumentation of the scientific journal cover:
1. Read this issue of the journal[2]
1.1a The scientific research published in this journal (e.g. concerning a specific area within the field of nanotechnology and nanoscience) is **ground-breaking**
1.1b The results of the research (e.g. concerning a specific area within nanotechnology and nanoscience) published in this journal have **direct/positive applications** in society
1.1a-b' The research featured on the cover is a good example of both / either.

---

[1] The term was coined by Groarke (2002; see also Groarke & Tindale, 2013, pp. 143-158) to describe the function of visuals (potentially combined with text) to attract the viewer's attention to the argument.

[2] Assuming that scientific journals address scientists in academic and research institutions that already have online access to these journals, the inciting standpoint in this case is not so much about 'buying' the specific issue as it is about 'reading' the issue or a specific article within it.

In both types of publication, the choice of the image, layout and wording on the cover is of primary importance. It does not only reflect the ethos, as it were, of the specific publishing house, it also seeks to raise the reader's interest in the specific article or study (see Grampp, 2015). The image and the text together with information derived from the context help the reader to recover assumptions in order to assess the relevance (in the technical sense) of the premises regarding newsworthiness / importance or the ground-breaking character and the plausibility of the research (that is to help him/her arrive at the premise 'the story is newsworthy / important' or the premise 'the research is ground-breaking' or 'the research has societal applications', respectively, in the two argumentation structures presented above).

As explained in section 2.2, the recovery of the intended assumptions will depend not only on the extent to which the producer of the message has made these (more) manifest but also on the extent to which the reader is in a position to retrieve these from the cognitive environment. The strength of the explicatures will thus depend on how strong the manifestness of the intention of the image-maker is to communicate something (the more this is encoded, the more explicit) (see Clark, 2013, p. 212). The strength of the implicatures will depend on how much pragmatic inference is required in their recovery (the more infernceing, the weaker the implicatures will be). How familiar one is with related images, the scientific area, and what associations one can make will all play a role in recovering the explicatures and implicatures, and eventually in arriving at the interpretation about newsworthiness or the ground-breaking character of the research. In the following section, I focus on images of the nanoscale appearing on the covers of specialized scientific journals and popular science magazines.[3]

3.2 Nano-images

One area of research where visualization and computer-generated images have played an essential role early on, both for the scientific community and the general public, is nanotechnology. In this field of research, chemistry, physics, biology and engineering converge in order

---

[3] I am not interested here in explaining how these images are interpreted and processed in general, but only in their argumentative interpretation when they are used in a specific genre. That is why I am focusing on their use in the front cover, assuming that front covers have a certain argumentative goal as I have explained in this section.

to study matter at the scale of a billionth of a meter (Roco & Bainbridge, 2002). Strictly speaking, images of nanostructures (see Figure 1) cannot be said to represent or depict, because they visualize something which, despite its physical substance, remains invisible without the use of digital imaging tools and related software (Ruivenkamp & Rip, 2014).[4]

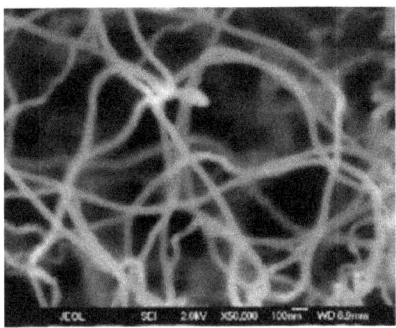

Figure 1 – Z.F. Ren, carbon nanotubes shown under electron microscope, 1998. (© Nano-Lab, Inc.)[5]

Despite their virtuality, or precisely because of it, these nano-images are used by various stakeholders (scientists, (non-)government agencies, science journalists, and science activists) in multimodal texts which communicate various views about nanotechnology. Next to their imaging function of visualizing matter at the nanolevel, nano-images such as the so-called Nanolouse (see Figure 2) exercise great evocative power by inviting the viewer not only to imagine how nanostructures actually look like but also to picture the future nanoworld promised by nanoscientists. It thus comes as no surprise that colour-enhanced, 3D computer generated images of the nanolevel feature on the front covers of popular science magazines and online image-galleries. Interestingly, similar images also appear on the covers of scientific journals specializing on nanotechnology (see Ottino, 2003).

---

[4] The scanning tunnelling and electron microscopes operate in a way that is akin more to 'touching' than actually 'seeing', and the data generated must be converted by computer software into a visual representation (Wickson, 2010).

[5] Available at: http://www.nano-lab.com/nanotube-image.html (last accessed 6 September 2017).

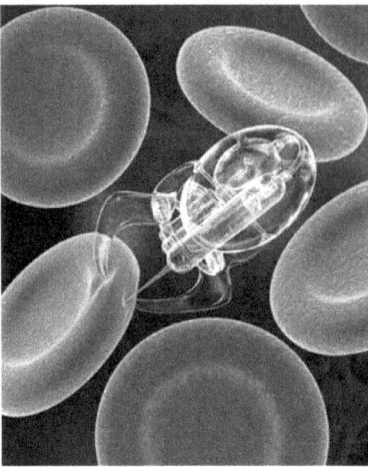

Figure 2 – The so-called Nanolouse; digital illustrator Coneyl Jay's impression of a nanobot in the blood taking test samples. Winner of the 2002 Visions of Science Award.[6]

The images on the cover may be of the grey-scale type representing the nanolevel visualized by a scanning tunnelling microscope (such as in Figure 1) or they may be an artist's rendition of that level (such as in Figure 2). For a typology of nano-images found in image galleries, see de Ridder-Vignone and Lynch (2012). For a typology of images appearing on scientific journals, see Grampp (2015). Choices in colouring, perspective, and overall composition can sometimes be related to a specific coding scheme that specialists are in a position to identify (for example, the use of 3D billiard-ball-like shape for atoms or the colouring of carbon atoms in light-grey; but see Ruivenkamp & Rip (2010) for more discussion about the fact that such coding is not really fixed in the case of nanotechnology). More often than not, the colouring and 3D perspective used on the front covers tend to create associations with more familiar scenes borrowed from the macro-world (landscapes, elements of the natural world, machines, etc.) or which appeal to both the general and the specialized audience's cultural knowledge. Besides the meaning conveyed by the visual content and the visual style of the particular nano-image, an extra layer of meaning is provided by the short text about the cover that accompanies

---

[6] Visit http://www.coneyljay.com/nanotechnology (last accessed 6 September 2017).

it, as well as by the way the image is contextualised (how it relates to the featured article of the specific issue and to other images in it).

4. POPULAR SCIENCE MAGAZINES VS SCIENTIFIC JOURNALS

As suggested in section 3.1, the front cover functions as a 'visual flag' that attracts the reader's attention to the story or research study featured inside. As such it serves the broader argumentative goal of the magazine or the journal, which is to increase the sales and citations, respectively. Furthermore, in section 3.2 it was observed that computer-generated, aesthetically pleasing and artistically rendered images of the nanoscale appear not only on the front covers of popular science magazines, addressing a general audience, but also on the covers of specialised scientific journals, addressing scientists working in the field of nanoscience or related fields. In this section, I compare the nano-images appearing on the front covers of science magazines with those appearing in scientific journals, and discuss the way these images are contextualized. The aim is to illustrate how the inference process described by Relevance Theory can help to account for the ways in which the readers of the front cover arrive at the premise regarding the value / contribution of the story/research article (see structure above), constrained by their own cognitive environment and the way the images on the cover are designed.

A quick look at the online archives of popular science magazines such as *Scientific American* and *Popular Science*, on the one hand, and scientific journals specializing on Nanotechnology and Nanoscience (such as *Nano Letters* and *Nanoscale*, or *Nature Nanotechnology*), on the other, makes it clear that in both types of publications computer-generated, and aesthetically pleasing images are used which, however, are not immediately familiar and self-explainable to a general or even specialized audience. Even the text that appears on the cover in the form of a caption or title does not always help to identify what it is one is looking at.

The biggest difference between the nano-images appearing on the covers of popular science magazines and those on the covers of scientific journals is that the latter are almost always produced by graphic designers and illustrators who collaborate closely with the authors of the specific research article that appears inside the journal. Most of the time, a text about the cover and its relation to the research reported inside the specific issue appears on the contents page to help identify the main topic, and even explain how the image was produced. This is the case, for example, with all the covers of *Nano Letters* (one of the first scientific journals to specialize on nanoscience and nanotechnology since 2001), and the covers of *Nature Nanotechnology* from 2008 onwards (the caption provided in the issues from 2006 until end of 2007 was barely informative). Other scientific journals, however, such as *Nanoscale*, provide no information about the image of the cover, except for referring to the article from which the artistic image was inspired.

Nano-images appearing on popular science magazines featuring a topic on nanoscience and nanotechnology tend to be simple, depicting a rather concrete item (the tip of the scanning tunnelling microscope, a cancer cell attacked, or nanoparticles). On the other hand nano-images on the covers of scientific journals are more complex, depicting a certain process in more detail. Grampp (2015) suggests that cover images in popular magazines are more allegoric and symbolic, while those of scientific journals are more indexical and iconic. Nevertheless, as he notes:

> the choice of indexical or iconic presentation forms has a certain price. It remains uncertain what exactly is being displayed or what is actually supposed to be described. Even though we are looking at concrete items of the external reality we do not exactly know what these concrete things are or what is being described by them (pp. 167-168).

When comparing the way the nano-image on the front cover is contextualised, that is the way it relates to the images and the text of the article appearing inside the magazine or journal, one also notes a number of differences. In the case of popular science magazines, the image is simply repeated in smaller or bigger size, cropped differently, while in the case of scientific journals one finds images and schematics inside the article which are even more concrete and complex, and have only a vague resemblance to the image of the cover.

To illustrate the above observations and to show how the readers' cognitive environment and the images' visual style could constrain the recovery of explicit and implicit content, I compare two covers of popular science magazines with two covers of scientific journals devoted on related topics. First, I compare the cover of the science magazine *Scientific American* and the cover of the scientific journal *Nature Nanotechnology* (see Figure 3 and Figure 4, below), which both depict the tip of the atomic force microscope, the instrument that together with the accompanying visualization software makes it possible to 'see' structures at the nanolevel (see also note 4, above).

Figure 3 – Cover of September 2001. Photograph by Felice Frankel, with technical help from J. Christopher Love.

Figure 4 – Cover of November 2012. Image: Eva Bieler, Marija Plodinec, Roderick Lim. Artwork: Martin Oeggerli/Micronaut. Cover design: Alex Wing.

In the *Scientific American* cover the magnified tip of the microscope is the only thing depicted, covering the half top part of the image. On the *Nature Nanotechnology* cover, the tip appears touching the surface of a sphere that stands out from an indeterminate mass covered with what could be described as grated cheese. While there is no text on the cover of the *Scientific American* that clearly identifies what is depicted, on the *Nature Nanotechnology* cover one reads "AFM in cancer diagnosis".[7]

---

[7] On the contents page of the *Scientific American* a similar image appears with the caption "Magnified tip of an atomic force microscope". The words 'Nanoprobes' and 'Atom-Moving Tools' that appear on the cover could be

Readers of *Scientific American* who may be knowledgeable about the fact that special microscopes are required to 'see' as it were the nanolevel by 'touching' the surface of atoms, could identify the image on the cover as the magnified tip of such a microscope. The rest, however, would have to make associations based on the shape, colour and framing of that image to arrive at assumptions about what it depicts. At best one could think of a highly stylized depiction of the swirl of a hurricane, but the difficulty of making any connection between hurricanes and the topic of the special issue being nanotechnology would leave one wonder about the relevance of the image. This said, the image together with the titles announcing the topics covered in this special issue would definitely raise the reader's curiosity. Nanotechnology is not only announced by the word 'nanotech' appearing in big white capital letters in the middle of the image, but also visually enhanced by the tip of the microscope directly pointing at it, which literally functions as a visual flag directing the reader's attention to the topic of this special issue.

The readers of *Nature Nanotechnology* would identify the image as depicting the tip of an atomic force microscope even without the verbal caption on the cover. However, it is unclear whether all readers would identify the orange sphere protruding from the red mass covered in white grated cheese as a cancer cell protruding from a malignant cluster, without the heading "AFM in cancer diagnosis".[8] What is important here is that, dissimilar to the image on the cover of *Scientific American*, a process is depicted with great detail that does not merely seek to raise the reader's interest but to illustrate the method carried out in the research article featured on the cover. While the sole image of the magnified tip of the atomic force microscope was enough to attract the general audience of a magazine such as *Scientific American*, the image on the cover of *Nature Nanotechnology* illustrates the actual use of the microscope in diagnosing and characterizing the progression of cancerous cells, something which is of importance for the specialized audience of this journal.

---

identified as referring to the image only by those familiar with this topic or after having read the articles inside.

[8] A long explanatory text on the inside page of the journal summarizing the main point of the article from which the image was inspired provides the necessary background for better understanding it. See http://www.nature.com/nnano/journal/v7/n11/covers/index.html (last accessed 7 September 2017).

As a second example, I compare the covers of the science magazine *Popular Science* and of the scientific journal *Science* depicting the precision-guided cure of cancerous cells (see Figure 5 and Figure 6, below).

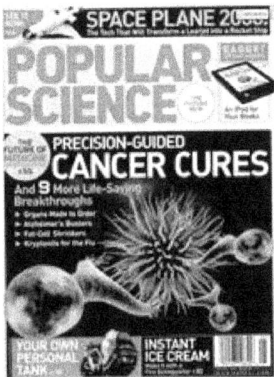

Figure 5 – Cover of August 2006. Cover illustration: Nick Kaloterakis.

Figure 6 – Cover of December 2013. Image: Valerie Altounian/Science.

The readers of *Popular Science* could effectively combine the information provided by the title "Precision-guided cancer cures" and the image to arrive at the conclusion that what is depicted on the cover is a cancer cell. However, it would take extra knowledge and reading of the short description about the cover on the contents page to identify the spherical objects surrounding the cancer cell as nanoparticles that can attack / cure cancer with precision. The image on the cover of

Science is more detailed when compared to that of *Popular Science* but not necessarily more self-explanatory. Even members of the specialised audience that *Science* addresses may need to read the short text about the cover in order to identify the pink peanut-like objects at the left side as antibodies, and the big spherical object at the right as a T cell, which once in contact with the antibodies will be activated to attack a tumour cell.[9]

From the above brief discussion of the two examples, some preliminary observations can be drawn about the argumentative use of nano-images in the front covers of scientific journals and science magazines with reference to Relevance Theory's distinction between explicit and implicit content. The image on the front cover of both types of publication functions first and foremost as a 'visual flag' that attracts the reader's attention, and thereby conveys a presumption of optimal relevance to the reader/viewer, namely that: a) The ostensive stimulus is relevant enough for it to be worth the addressee's effort to process it; and b) The ostensive stimulus is the most relevant one compatible with the communicator's abilities and preferences (Clark, 2013, p. 32-33). This means that the interpretation of the use of a particular image on the front cover depends both on the cognitive effort and processing abilities of the receiver, and on the degree to which the image-maker has chosen to make manifest his/her communicative intentions (ideally seeking to minimize the processing costs for the receiver). The first aspect can be related to the different backgrounds and interests of the readers of the respective publications: general audience with an interest in science and technology in the case of science magazines, compared to a specialized audience, knowledgeable about research in nanoscience and nanotechnology in the case of the scientific journals. The second aspect can be related to the detail and artwork that the images used in the two types of publication exhibit.

As discussed above, in the case of science magazines, the images are more generic and lack the details that illustrate specific processes and structures at the nanolevel. However, they still require extra processing effort from the viewers, given that they depict something that is not visible without the use of special instruments and visualization software, not to mention that nanotechnology is a topic with which even the audience of science amateurs is not really familiar

---

[9] It is worth noting here that *Science* does not address exclusively a specialised audience of nanoscientists but of scientists in general. The text about this cover can be found online at http://science.sciencemag.org/content/342/6165 (last accessed 7 September 2017).

with, because it is least discussed, compared to astrophysics and genetic biology, for example. The readers of science magazines would therefore most probably draw on associations with familiar patterns and structures from the macro world, such as landscapes and natural objects based on the colours and shapes they see in these images. As a result, more pragmatic inferences will be required on their part to arrive at an interpretation. Drawn by the aesthetic properties of these images, the readers of science magazines will thus recover implicatures concerning the wonders of new technologies and evaluative appreciations of technological progress and its applications.

The images on the front covers of scientific journals, on the other hand, are in principle the product of specially commissioned cover designers and illustrators, and they exhibit more detail. The expert viewers of these front covers are in a better position to recover explicitly communicated propositions based on what is depicted on the image, given their familiarity with the specific research area as well as with the use of colouring and other visualization schemes that are applied. This said, because of the artistic nature of some of the images appearing also on the covers of scientific journals, the cost for recovering the explicatures may at times be relatively high (for science experts who are not necessarily familiar with the specific area of nano-research, for example). In the end, the readers of scientific journals will be able to recover implicatures that go beyond the admiration of the wonders of technology, and concern the innovativeness and applicability of the research featured on the cover, basing their inferences on the explicatures recovered concerning what is being depicted on the cover.

5. CONCLUSION

In this paper, I have proposed applying Relevance Theory's distinction between explicitness and implicitness to instances of argumentative communication where image and text combine to convince the addressee of a certain claim. I have focused on the multimodal genre of the front cover whose argumentative goal can be described as seeking to promote the specific issue of a magazine. The front covers of science magazines and scientific journals, in particular, function primarily as 'visual flags' seeking to invite the reader to buy or read the specific issue, without expressing any further standpoint regarding the cover story/research. The focus on the use of nano-images, namely aesthetically enhanced images of nanostructures, raises interesting questions as to what is explicit and what implicit, and for whom, when comparing the different audiences that science magazines and scientific journals address.

From the observations drawn after a brief review of various covers of these two types of publications, and from the preliminary comparative study of four covers featuring related topics, it is suggested that the readers of science magazines would mainly recover implicatures concerning the wonders of nanoscience, while those of scientific journals would recover implicatures concerning the innovativeness of the nano-research reported, based on explicatures recovered from the content of what is depicted on the cover. These qualitative observations could form the basis for an empirical study concerning the interpretation and reception of nano-images in the front covers of these two types of publications. Such a study would not only compare the actual inferencing processes of different groups of readers, but would also help to answer questions about what claims and arguments actual audiences of these publications recover. Another direction for further research concerns the study of the dynamic ways in which the reader of a magazine may construct an argument while browsing through its pages, thereby making connections between the image on the cover, the explanatory text accompanying it, and the presentation of the featured article inside the magazine.

REFERENCES

Bateman, J. (2014). *Text and image. A critical introduction to the visual/verbal divide*. London: Routledge.
Birdsell, D. S., & Groarke, L. (1996). Toward a theory of visual argument. *Argumentation and Advocacy, 33*(1), 1-10.
Blair, A. J. (2004). The rhetoric of visual arguments. In C. A. Hill & M. Helmers (Eds.), *Defining visual rhetorics* (pp. 41-62). Mahwah, NJ: Lawrence Erlbaum Associates.
Carston, R. (2009). The explicit/implicit distinction in pragmatics and the limits of explicit communication. *International Review of Pragmatics, 1*(1), 35-62.
Clark, B. (2013). *Relevance Theory*. Cambridge: Cambridge University Press.
Desilla, L. (2012). Implicatures in film: Construal and functions in Bridget Jones romantic comedies. *Journal of Pragmatics, 44*(1), 30-53.
Forceville, C. (2014). Relevance Theory as model for analysing multimodal communication. In D. Machin (Ed.), *Visual communication* (pp. 51-70). Berlin: Mouton de Gruyter.
Forceville, C., & Clark, B. (2014). Can pictures have explicatures? *Linguagem em (Dis)curso, 14*(3), 451-472.

Grampp, S. (2015). The art of science. Making things popular with scientific journal covers. *Online Journal of Communication and Media Technologies, 5*(2), 157-180.

Groarke, L. (2002). Toward a pragma-dialectics of visual argument. In F. H. van Eemeren (Ed.), *Advances in Pragma-dialectics* (pp. 137-151). Amsterdam: Sic Sat/ Virginia: Vale Press, Newport News.

Groarke, L. (2015). Going multimodal: What is a mode of arguing and why does it matter? *Argumentation, 29*(2), 133-155.

Groarke, L., Palczewski, C. H., & Godden, D. (2016). Navigating the visual turn in argument. *Argumentation and Advocacy, 52*, 217-235.

Groarke, L., & Tindale, C. (2013). *Good reasoning matters* (5th ed.). Toronto: Oxford University Press.

Held, G. (2005). Magazine covers–a multimodal pretext-genre. *Folia Linguistica, 39*(1-2), 173-196.

Ifantidou, E., & Tzanne, A. (2006). Multimodality and relevance in Athens 2004 Olympic Games televised promotion. *Revista Alicantina de Estudios Ingleses, 19*, 191-210.

Jewitt, C. (Ed.). (2014). *The Routledge handbook of multimodal analysis* (2nd ed.). London: Routledge.

Jewitt, C., Bezemer, J., & O'Halloran, K. (2016). *Introducing multimodality*. London: Routledge.

Kjeldsen, J. E. (2015). The study of visual and multimodal argumentation. *Argumentation, 29*, 115-132.

Kress, G. (2010). *Multimodality: A social semiotic approach to contemporary communication*. London: Routledge.

Norris, S., & Maier, C. D. (Eds.) (2015). *Interactions, images and text. A reader in multimodality*. Berlin: Mouton de Gruyter.

Nöth, W. (2011). Visual semiotics: Key features and an application to picture ads. In E. Margolis & L. Pauwels (Eds.), *The SAGE handbook of visual research methods* (pp. 298-315). London: Sage Publications.

Novitz, D. (1977). *Pictures and their use in communication*. The Hague: Martinus Nijhoff.

Ottino, J. M. (2003). Is a picture worth 1000 words? Exciting new illustration technologies should be used with care. *Nature, 421*, 474-476.

de Ridder-Vignone, K., & Lynch, M. (2012). Images and imaginations: An exploration of nanotechnology image galleries. *Leonardo, 45*(5), 447-454.

Roco, M. C., & Bainbridge, W. S. (2002). Converging technologies for improving human performance: Integrating from the nanoscale. *Journal of Nanoparticle Research, 4*(4), 281-295.

Ruivenkamp, M., & Rip, A. (2010). Visualizing the invisible nanoscale study: Visualization practices in nanotechnology community of practice. *Science Studies, 23*(1), 3-36.

Ruivenkamp, M., & Rip, A. (2014). Nanoimages as hybrid monsters. In C. Coopmans, J. Vertesi, M. E. Lynch, & S. Woolgar (Eds.), *Representation*

in scientific practice revisited (pp. 177-200). Cambridge Massachusetts: MIT Press.
Sanz, M. J. P. (Ed.). (2015). *Multimodality and cognitive linguistics*. Amsterdam: John Benjamins.
Sperber, D., & Wilson, D. (1995). *Relevance: Communication and cognition*. Oxford: Blackwell.
Tseronis, A. (2015). Multimodal argumentation in news magazine covers: A case study of front covers putting Greece on the spot of the European economic crisis. *Discourse, Context & Media, 7*, 18-27.
Tseronis, A. (2017). Analysing multimodal argumentation within the pragma-dialectical framework: Strategic manoeuvring in the front covers of *The Economist*. In F. H. van Eemeren, & Wu Peng (Eds.), *Contextualising Pragma-Dialectics*. Amsterdam: John Benjamins.
Tseronis, A. (2018). Determining the commitments of image-makers in arguments with multimodal allusions in the front covers of *The Economist*: Insights from Relevance Theory. *International Review of Pragmatics, 10*, 243-269.
Tseronis, A., & Forceville, C. (Eds.) (2017). *Multimodal argumentation and rhetoric in media genres*. Amsterdam: John Benjamins.
Van den Hoven, P. (2012). Getting your ad banned to bring the message home? A rhetorical analysis of an ad on the US national debt. *Informal Logic, 32*(4), 381-402.
Van Eemeren, F. H. (2010). *Strategic maneuvering in argumentative discourse. Extending the pragma-dialectical theory of argumentation*. Amsterdam: John Benjamins.
Wharton, T. (2009). *Pragmatics and nonverbal communication*. Cambridge: Cambridge University Press.
Wickson, F. (2010). Images. In D. H. Guston (Ed.), *Encyclopedia of nanoscience and society* (Vol I, pp. 328-329). Thousand Oaks: Sage Publications.
Wilson, D. (2017). Relevance theory. In Y. Huang (Ed.), *Oxford handbook of pragmatics* (pp. 79-100). Oxford: Oxford University Press.
Wilson, D., & Sperber, S. (2012). *Meaning and relevance*. Cambridge: Cambridge University Press.
Yus Ramos, F. (1998). Relevance theory and media discourse: A verbal-visual model of communication. *Poetics, 25*(5), 293-309.
Yus Ramos, F. (2011). *Cyberpragmatics. Internet-mediated communication in context*. Amsterdam: John Benjamins.

Social Costs of Epistemic Vigilance and Premises in Arguments

CHRISTOPH UNGER
*Norwegian University of Science and Technology (NTNU),
Trondheim, Norway*
christoph.unger@ntnu.no

Implicit premises that are mutually manifest often escape the audience's epistemic vigilance. It has been suggested that such premises are discursive presuppositions functioning as backgrounds for the evaluation of relevance and that this is the reason that they escape veracity checking. I argue instead that this due to the fact that the rejection of mutually manifest premises incurs social costs, because this would narrow, rather than enhance, the mutual cognitive environment between communicator and audience.

KEYWORDS: epistemic vigilance, premises, mutual manifestness, relevance, discursive presuppositions

1. INTRODUCTION

Consider the following arguments:

(1) Parking fees for electric cars must rise, because so many people drive to work in electric cars nowadays.

This argument is in fact sound and rests on the following premises:

(2) a. Electric cars need parking space just like other cars.
b. If more people use cars to come to work, the available parking spaces are filled up with commuters' cars.
c. If the available parking spaces are filled up with commuters' cars, shoppers can't park any more near the shops they want to visit.

d. If shoppers can't park any more near the shops the want to visit, then they won't shop there anymore and the shops get in trouble.
e. If the shops get in trouble, the city must introduce countermeasures.
f. If parking fees rise, commuters won't commute by car to work.
g. Therefore, raising parking fees is a solution to the parking space problem that commuters' use of electric cars created.

Yet most naïve persons (yours truly included) will raise their eyebrows when confronted with (1), thinking that this is a strange argument. The reason is that when we hear (1) out of the blue, we tend to process this argument against premises such as the following:

(3) a. Electric cars do not have harmful emissions.
b. Driving cars that do not eject harmful emissions is always beneficial.
c. Driving cars that do not eject harmful emissions should always be encouraged.

Seen against premises such as (3), the argument in (1) is indeed dubious. However, the premises (3b) and (3c) are certainly not warranted. We realise this as soon as we think consciously about it. But when intuitively evaluating utterances such as (1), we tend to use these dubious premises as a self-evident fact.

Where do we get (3) from? We get this from much of public discourse in our society, so that this piece of information is part of the common ground we share. Information deeply entrenched in our common ground is often taken for granted rather than checked for its truth. De Saussure (2013) calls this the *presupposition fallacy*.

In this paper I want to discuss the question of what explains the presupposition fallacy. De Saussure (2013) argues that this fallacy is the result of a heuristic procedure that is applied to information which is considered as given, and which causes such information to be only superficially processed. I argue that the cases of presupposition fallacy discussed by De Saussure (2013) are somewhat different than cases such as (1), and that there is another explanation for the latter: the rejection of information considered as common ground involves *social costs*, because communication, and social relations in general, are enhanced when common ground is increased. Hence, avoiding to evaluate the factuality of premises treated as common ground is a natural result of rational communicative behaviour; it is neither a fallacy nor the result of limitations inherent in a heuristic procedure.

2. THE ISSUE: MUTUALLY MANIFEST PREMISES

2.1 German modal particle ja and manipulation

Presupposition fallacies do not only arise with texts that are overtly presented as arguments, but also with presumably non-argumentative utterances involving linguistic indicators of common ground. One case in point is the German particle *ja*. This particle in its modal use— i.e. between the finite and infinite part of the verb, typically unstressed— is widely recognised to indicate information regarded as common ground. Example (4) illustrates this use.

(4) Diese Jungs aus Turin rocken. Waste Pipes sind sympathisch, spielen gut und werden *ja* schon mal als die heimlichen Nachfolger von Led Zeppelin bezeichnet.
'These boys from Turin rock. Waste Pipes are congenial, play well and are *(MP)* sometimes already called the secret successors of Led Zeppelin' (A09/JAN.00035 St. Galler Tagblatt, 03.01.2009, S. 34; Hin und Weg) (Unger, 2016, p. 41).

This example comes from a concert review. The addressees of such a review are people interested in the kind of music and concerts under discussion. Such an audience must be assumed to either know already that the band *Waste Pipes* are rumoured to be successors of a famous precursor, or able to inferentially anticipate this information. In either case, the clause marked with *ja* conveys information that is manifest to the audience, and it is manifest to the concert reviewer that this information is manifest to her audience; in other words, it is mutually manifest information in the sense of Sperber and Wilson (1995).

Why should a communicator linguistically indicate overtly that she regards a certain piece of information as mutually manifest?

Communicators may want to indicate that they regard a certain piece of information as mutually manifest when they need to make sure that the audience will use exactly the intended piece of information from among the mutually manifest assumptions. However, the linguistic indication of mutual manifestness may be exploited for argumentative or manipulative purposes. Blass (2000) gives the following example from the weekly magazine *Der Spiegel*, an interview with the politician Oskar Lafontaine:

(5) S: Der Ansturm billiger Arbeitskräfte aus
         The storm cheap labour from
     Spanien, Portugal und Griechenland beginnt
     Spain Portugal and Greece starts
     gerade erst. Wie wollen Sie die
     just only how want you the
     Billigkonkurenz stoppen?
     cheap-competition stop
     'The storm of cheap labour from Spain, Portugal and Greece has just begun. How do you want to stop the cheap competitors?'

   L: Jedenfalls nicht durch eine Senkung der
     in-any-case not by a reduction of-the
     deutschen Löhne auf das portugiesische Niveau.
     German wages to the Portugese niveau
     Die Befürworter einer solchen Strategie fordern
     The supporters of-a such strategy demand
     Lohnsenkungen ja nie für sich,
     wage-deductions of course never for themselves,
     sondern immer nur für andere.
     but always only for others.
     'Under no circumstances by lowering the German wages to the Portuguese level. Those who are in agreement with that demand this lowering never for themselves, of course' (Example 17 from Blass (2000, p. 50)).

Many people would not agree with the claim that all who propose wage cuts propose this only for others, and L knows this, still he uses this claim as an uncontroversial assumption supporting his argument. *Ja* indicates that the audience should entertain the proposition expressed as if it were mutually manifest.

2.2 Pragmatic fallacies

On the face of it, the mutual manifestness fallacy in (5) is similar to what may be called pragmatic fallacies: the *Loaded Question fallacy* and the *Moses Illusion*.

(6)   Do you accept the proposition of law "The construction of minarets is banned?" (de Saussure, 2013).

People who respond positively to the question in (6) have to accept the following premises, each of which can easily be falsified. Yet many people unquestioningly accept these premises, as evidenced by the result of a popular vote (see de Saussure (2013) for more detailed comments):

(7)   a. There is a relevant number of minarets actually in place or projected.
       b. Minarets could modify Swiss landscapes
       c. Minarets are a threat of some type. (de Saussure, 2013)

(8)   It has been calculated that New Year's day 2000 will be a Friday the 13th. Do you think that this is a bad omen? (de Saussure, 2013, pp. 185-186)

People confronted with the question in (8) tend to answer the question without protesting that New Years Day can never be on the 13th day of any month.

(9)   Where should the survivors of a plane crash be buried if the crash site is exactly on the border between two countries?

(10)   How many animals did Moses put into the Ark?

De Saussure (2013) argues that these examples may be explained by the notion of 'shallow processing' as proposed by Allott (2005). Considering example (9), this explanation runs roughly as follows: relevance can be achieved easily by widening the concept SURVIVOR to SURVIVOR* referring to 'people involved in a plane crash'. The inference process behind this interpretation could be described as follows: an easily accessible assumption is that there are often no survivors in plane crashes. Moreover, it is easy to see that the question where to bury the dead can be a tricky one (hence relevant to wonder about) when the crash site is exactly between two countries. By assuming that the word *survivor* refers more generally to people involved in a plane crash, this relevance expectations can be satisfied. Consequently, the proposition 'The survivors of an air crash are buried in X' is not entertained at all.

But at closer inspection, there is a difference between cases such as (6), (1) and (5), on the one hand, and (8), (9) and (10) on the other. The latter group of examples may plausibly be explained by the shallow processing of concepts as suggested by Allott (2005). This is not plausible in the former group of examples, because the contextual assumptions or discursive presuppositions involved cannot be traced back to individual concepts. Rather, it seems that in this group of examples the audience does not scrutinise the questionable contextual assumptions claimed as being mutually manifest for their truth value.

3. EXPLAINING EPISTEMIC VIGILANCE ESCAPE

3.1 Mutual manifestness and communication

As is well known, the linguistically encoded meaning in an utterance under-determines the speaker's meaning. Rather, it serves as partial evidence that the audience takes into account to *infer* the speaker's meaning together with suitable contextual assumptions. Sperber and Wilson (1995) argue that this process is constrained our mind employs a simple heuristic in this process:

(11) a. Take the first hypothesis about explicit meaning, implicit import and intended context that is easily accessible;
b. Check whether the utterance, on this interpretation, provides improvements in knowledge (in terms of giving rise to new implications, corrections of mistaken beliefs, strengthening of previous assumptions, or improvements in memory organisation).[1] If these improvements in knowledge are at least sufficient to make it worth the audience's attention, accept the interpretation as the one intended by the speaker.

This heuristic is sensitive to mental processing effort. The more accessible potential contextual assumptions are, the more likely is that they are accessed. Communicators who want to be understood must

---

[1] On the inclusion of 'improvements in memory organisation' in the list of cognitive improvements in knowledge see Wilson and Sperber (2012, 62), Wilson and Sperber (2012, 271) and Wilson and Sperber (2004, 628) footnote 2.

therefore aim at making it easy enough for audiences to access the contexts they intend them to use. One way to achieve this goal is to select the intended context from the *mutual cognitive environment* of communicator and audience. The mutual cognitive environment of communicator and audience consists of those pieces of information that are *mutually manifest* to them. Pieces of information are *manifest* to an individual to the extent that the individual is capable in principle of representing mentally and accepting them as true or probably true. Pieces of information are mutually manifest to two individuals to the extent that they are manifest to both and that this fact is manifest to both individuals.

Pieces of information that are mutually manifest have the following properties: first, the communicator can be confident that the audience is able to access this information, and second, the audience can easily access this information. The larger the pool of information that is mutually manifest to communicator and audience, the better they are able to communicate. Sperber and Wilson describe this situation as follows:

> However, we want to argue that there is another major reason for engaging in ostensive communication, apart from helping to fulfil an informative intention. Mere informing alters the cognitive environment of the audience. Communication alters the mutual cognitive environment of the audience and communicator. Mutual manifestness may be of little cognitive importance, but it is of crucial social importance. A change in the mutual cognitive environment of two people is a change in their possibilities of interaction (and, in particular, in their possibilities of further communication). (Sperber & Wilson, 1995, 61-62)

This suggests that the larger the mutual cognitive environment between communicator and audience becomes, the better their prospects for further communication will be. Moreover, the more assumptions are *mutually* manifest to them, the more can they exploit this mutual stock of assumptions and become conscious of their social closeness. In case a communicator argues a claim C and overtly bases her argument around a premise taken from the mutual cognitive environment, then she can be confident that the audience will accept the claim. Doing otherwise would mean that the audience is rejecting a part of the mutual cognitive environment. Such an act amounts to a move that increases social distance between communicator and audience. All else being equal, this is presumably an undesirable move. Therefore, rejecting an argument explicitly based on mutually manifest premises involves a social cost for the audience, over and above the cognitive cost involved in assessing the inferential strength of the inferences used in the argument.

To enhance the mutual cognitive environment is a desirable goal from the standpoint of enhancing social relations and improving possibilities for communication. As a baseline, the mutual cognitive environment between two individuals who are physically co-present at a certain location consists of all those pieces of information that can be perceived by both of them. This is because it is manifest to us that others having the same biological capacities for perception are capable of representing these pieces of information and accepting them as true. Moreover, once it is manifest to both individuals that they share some social group membership, a certain set of cultural knowledge is mutually manifest. But how could the mutual cognitive environment be extended beyond that baseline?

It is tempting to suggest that verbal communication is an obvious means to this end. But communicators frequently do communicate ideas that are not manifest to them: communicators may tell lies, or simply pass on speculations. Moreover, they may have been mistaken in their beliefs, or have been misled by others, passing on false information. For these reasons, audiences cannot assume that information transmitted in verbal communication was intended to become mutually manifest.

For an audience to accept communicated information as (possibly) true, they have to believe that the communicator is competent (i.e. knows the information to be true, not making mistaken assumption, not mislead by others herself, etc.) and benevolent (i.e. has no intention to mislead the audience), and that the communicated ideas are not inconsistent with any strongly evidenced background knowledge they possess. When these conditions are met, verbal communication is maximally beneficial to us. Sperber et al. (2010) argue that the mind in endowed with a variety of dedicated mental mechanisms that audiences employ to evaluate whether these conditions apply. These mechanisms are heuristic devices that check for various properties such as speaker benevolence, speaker competence, speaker track record in reliability, and so on. Arguably the most powerful of these devices is the argumentation module that looks at claims that the audience is not prepared to accept at face value and checks their logical properties and inferential validity (Mercier & Sperber, 2009, 2017).

Given that extending the mutual cognitive environment is desirable from a social point of view, one should expect that only information that has been thoroughly evaluated by appropriate *epistemic vigilance* mechanisms be added to the mutual cognitive environment. However, this is not necessarily the case, as the examples above show. Obviously, audiences apply some sort of heuristic process that pre-empts full epistemic assessment to at least some communicated information that is accepted as mutually manifest.

I suggest that the following heuristic procedure is applied to check whether communicated information is mutually manifest:

(12) Accept as mutually manifest communicated information **I** with the following properties:
a) the communicator does not propose **I** as a claim of her own;
b) it is not manifest to the audience that other communicators have disputed/argued over **I**;
c) **I** is not inconsistent in the context selected by the comprehension procedure.

The rational for this heuristic is the following: most typically, when communicators convey mutually manifest information in their utterances, it is by way of a reminder, to ease access to information that the audience may infer or recall from memory themselves. Such information has been communicated many times before. Given that communicators can argue over its veracity if someone feels the need for it, the absence of such argumentation may indicate that it has passed the epistemic vigilance of audiences before, hence one may skip this process in order to free up resources for other cognitive processes. If furthermore there are no early indications of logical inconsistency in the currently relevant context, then one may proceed assuming that **I** is mutually manifest.

In a sense, this heuristic does exercise some epistemic vigilance: it takes the absence of awareness of dispute over the veracity of some information as motivating accepting it as probably true. But it is still a question why this heuristic should apply pre-empting more powerful epistemic vigilance mechanisms when so much is at stake.

I propose that the identification of mutual manifestness be counted as a positive cognitive effect. This means that the heuristic for the identification of mutually manifest information in (12) would be called upon by the comprehension heuristic, in parallel with heuristics identifying the other types of positive cognitive effects (contextual implications, contextual eliminations, straightening, re-organisation of memory). But since the effect of this heuristic includes some type of epistemic vigilance, it is likely that other, more cost intensive epistemic vigilance heuristics are not accessed.

4. PREDICTIONS

4.1 Uncritical mutual manifestness acceptance only in communicative settings

The hypothesis developed above predicts first of all that people may uncritically accept information presented as mutually manifest only in 'real' communicative settings in the social world. When people are given the task to analyse a given argument, the effect will disappear.

This prediction is in accord with intuitive, informal observations:

i. We are able to spot the phenomenon when we analyse verbal arguments;

ii. Linguists easily spot the phenomenon when studying the use of modal particles or other linguistic indicators of mutual manifestness;

It should be more or less straight forward to test this prediction experimentally. An interesting group of subjects to study are experts: argumentation theorists. Those people can easily spot uncritical acceptance of information presented as mutually manifest. But they probably use the device in their own communicative argumentation practice. And linguists won't stop using modal particles such as German *ja* in a 'manipulative' way in their spontaneous conversation just because they have studied the phenomenon. Indeed, the social dimension of the phenomenon should be more pronounced with experts than with lay people.

Since the mutual manifestness detection heuristic involves checking whether the audience is aware of the information having been disputed by others, one can expect that the more socio-culturally homogeneous communicator and audience are, the less likely it is that information that is verbally communicated but not presented as a novel claim will be flagged as disputable. This means that the more homogeneous communicator and audience are, the less likely it is that the audience spots improper use of information presented as mutually manifest.

4.2 Interaction with confirmation bias

Communicators are subject to the confirmation bias: they are good at providing arguments supporting their claims, but bad at providing counterarguments against their claims. When people work together in groups in such a way that they need to discuss the task and convince each other, group performance leads to much better reasoning results because the group members cross-check their respective confirmation biases. Heterogeneous groups do better than homogeneous ones (Mercier & Sperber, 2011, 2017).

Given that weaknesses in arguments may include illicit use of mutually manifest information, and that these problems are less easily checked by homogeneous communicator-audience groups (by the hypothesis developed in this paper), one should expect that homogeneous groups may be less effective at spotting problems with mutual manifestness in arguments than heterogeneous groups. Notice that the prediction is not merely that heterogeneous groups will perform better than homogeneous groups in improving arguments (this more general prediction can be expected on independent grounds: the more heterogeneous a group is, the more likely it is that the individual members come with different viewpoints and are therefore more likely to uncover and correct confirmation bias effects). Rather, the prediction is that homogeneous groups will do worse than heterogeneous groups in correcting a certain type of argumentational infelicity: the improper use of information presented as mutually manifest. Properly testing this prediction requires looking at the type of improvements in argumentation that groups apply, not merely at whether how many groups of each type (homogeneous-heterogeneous) have improved faulty arguments to completely logically correct ones.

4.3 No interaction with Moses Illusion and Loaded Question Fallacy

Since on the hypothesis developed in this paper the Moses Illusion and Loaded Question Fallacy do not involve effects of mutual manifestness detection, there should be no effects of the degree of group homogeneity on these effects. In other words, homogeneous and heterogeneous communicator-audience groups should be similarly efficient in spotting and correcting these fallacies.

5. CONCLUSION

I have made the observation that mutually manifest information or information presented as mutually manifest often appears to be accessed uncritically—it appears to resist epistemic vigilance. I have argued that although this phenomenon may look like a bug in the system, it is a feature of communication: rejecting information presented as mutually manifest would lead to the narrowing down of the mutual cognitive environment between communicator and audience, thereby reducing rather than advancing communicative possibilities. To explain this, I proposed that the mind's comprehension mechanism includes a heuristic for identifying mutually manifest information, which is activated by the search for positive cognitive effects that is raised in every act of ostensive communication. Positive cognitive effects include the identification of mutual manifestness. With this addition to relevance theory, I proposed to explain that the comprehension module's mutual manifestness identification heuristics tends to take precedence over more powerful mechanisms of epistemic vigilance.

ACKNOWLEDGEMENTS: I warmly thank Kaja Borthen, Didier Maillat, Steve Oswald, Dan Sperber, and the audience at ECA 2017 for helpful comments. This research was supported by the project "The Meaning and Function of Norwegian Tags (NOT)", financed by the Norwegian Research Council, project number 230782.

REFERENCES

Allott, N. (2005). The role of misused concepts in manufacturing consent: A cognitive account. In L. de Saussure, & P. Schulz (Eds.), *Manipulation and ideologies in the Twentieth Century: Discourse, language, mind* (pp. 147–168). Amsterdam: Benjamins Publishing.
Blass, R. (2000). Particles, propositional attitude and mutual manifestness. In Andersen, G., & Fretheim, T. (Eds.), *Pragmatic markers and propositional attitude* (pp. 39–52). Amsterdam: Benjamins.
de Saussure, L. (2013). Background relevance. *Journal of Pragmatics*, *59*(Part B), 178-189.
Mercier, H., & Sperber, D. (2009). Intuitive and reflective inferences. In Evans, J. S. B. T., & Frankish, K. (Eds.), *In two minds: Dual processes and beyond*. Oxford: Oxford University Press.
Mercier, H., & Sperber, D. (2011). Why do humans reason? Arguments for an argumentative theory. *Behavioral and brain sciences*, *34*(02), 57–74.

Mercier, H., & Sperber, D. (2017). *The enigma of reason*. Harvard University Press, Cambridge, Massachusetts.
Sperber, D., Clément, F., Heintz, C., Mascaro, O., Mercier, H., Origgi, G., & Wilson, D. (2010). Epistemic vigilance. *Mind & Language*, *25*(4), 359–393.
Sperber, D., & Wilson, D. (1995). *Relevance* (2nd ed.). Oxford: Blackwell. First edition 1986.
Unger, C. (2016). Degrees of procedure activation and the German modal particles ja and doch. *Studia Linguistica Universitatis Iagellonicae Cracoviensis*, *133*(1), 31–74.
Wilson, D., & Sperber, D. (2004). Relevance Theory. In Horn, L. R., & Ward, G. (Eds.), *The handbook of pragmatics* (pp. 607–632). Oxford: Blackwell.
Wilson, D., & Sperber, D. (2012). *Meaning and relevance*. Cambridge University Press.

58

Sets of Situations, Topics, and Question Relevance

MARIUSZ URBAŃSKI
Adam Mickiewicz University, Poznań, Poland
Mariusz.Urbanski@amu.edu.pl

NATALIA ŻYLUK
Adam Mickiewicz University, Poznań, Poland
Natalia.Zyluk@amu.edu.pl

Our research provides formal tools for analyses of inferential question processing involved in solutions to a specific class of abductive problems. We model this processing in terms of relations of sifting and funnelling. Definitions of these relations employ logic of questions, situational semantics, and topic relevance. As we show on the basis of *Mind Maze* gameplays, these relations account well for empirical data and allow for a comparative analysis of styles of such problem solving.

KEYWORDS: abduction, logic of questions, topic relevance, situational semantics

1. INTRODUCTION

In this paper we provide formal tools for analyses of question processing involved in solutions to a specific class of abductive problems. We model this processing in terms of relations of sifting and funnelling. Definitions of these relations employ logic of questions, situational semantics, and topic relevance. In the section 2 we briefly outline the source of our data. In the section 3 we characterize formal tools used in analysis of the data. On this basis, in the sections 4 and 5 the concepts of topic and question relevance, and of an admissible answer are introduced, respectively. In the sections 6 and 7 we give analyses of exemplary *20 questions* and *Mind Maze* gameplays in these terms. We conclude the paper by pointing at some directions for further research.

2. THE SOURCE EMPIRICAL DATA

We devised materials for this research on the basis of *Mind Maze* tasks. This is a game by Igrology in which, according to the manual, a gamemaster "describes a strange story and the players must determine why and how the story happened". Solution of each of the tasks is dependent on discovering key pieces of information (which are known to the gamemaster only) by asking auxiliary questions. Thus the task of the player is to process a sequence of questions, posed on the basis of a story's content and subsequent answers of the gamemaster. The players may collaborate in order to reach the solution. We modified original rules of the game, in order to allow for more cooperative behaviour of a gamemaster as well as to smoothen the process of data gathering. In particular, as in the original version, we allow for only yes-no questions to be asked, but with addition of two admissible answers: "not important" and "it is not known". The interested reader will find the details on the setup of this research in Urbański et al. (2016b). Data obtained from *Mind Maze* gameplays form one of the three subcorpora of the Erotetic Reasoning Corpus (Łupkowski et al., 2017). *Mind Maze* subcorpus currently consists of 38 annotated gameplays (69.434 words), which lasted from 5 to 38 minutes. In order to clarify intuitions underlying our formal tools, before turning to a *Mind Maze* task (section 7), we introduce a toy example based on a *20 questions* gameplay (section 6).

3. THE TOOLS: QUESTIONS, SITUATIONS, AND TOPICS

As we demonstrated elsewhere (Urbański et al., 2016b), solutions to all the *Mind Maze* problems consisted of two general phases. They correspond to Stenning and van Lambalgen's (2008) distinction of reasoning to vs reasoning from an interpretation. In the first phase the subjects established interpretation of a problem. As most of the subjects were not very explicit about underlying reasoning processes, first part of our model is based on limited amount of data and offers a rational reconstruction of this phase rather than its full-fledged descriptive model. We employ here elements of Gabbay and Woods' (2005) formal schema of abductive reasoning and Kubiński's (1980) logical theory of numerical questions. In the second phase, which consisted of the actual dialogues with the experimenter (gamemaster), the subjects' information processing can be adequately modelled by means of some extensions of the situational semantics (Wiśniewski, 2013b), incorporating the concepts of topic and question relevance. In our preliminary report (Urbański et el., 2016b) we employed to this end Inferential Erotetic Logic (Wiśniewski, 1995). However, it turned out that our present formalism is better suited for these purposes.

3.1 Setting the cognitive goal: a formal schema of abduction

In *Mind Maze* the players face an abductive problem of making sense of a puzzling fact (Thagard & Shelley, 1997) given in a story and expressed in the initial question. The task is to find pieces of information forming the explanation (as there is the correct solution to each problem in the game) of this puzzling fact. The players are supposed to rely on their general knowledge as well as on their abilities to reason with questions as premises and conclusions.

In order to account for this abductive nature of the tasks we employ some elements of Gabbay and Woods (2005, p. 47) formal schema of abductive reasoning. The authors use the symbol ! to represent the fact that a certain piece of information (represented by T) is a cognitive target of a reasoning subject. Interpretation of this symbol, as well as of some others included in the schema, is not univocal: they may be sentences, theories or rules, or anything forming an abductive hypothesis in a certain context. We shall use the T symbol as a sentential metavariable. Thus in our analysis $T!$ will stand for propositionally construed cognitive target of a subject.

3.2 Interpreting the problem: logic of questions

In representing questions we adopt some essential elements of Kubiński's (1980) analysis of simple numerical questions and also of propositional questions. We shall use simple numerical questions in order to account for the reasoning to an interpretation phase of a problem solution (Stenning & van Lambalgen, 2008, pp. 19-25), in which the subjects need to establish what key pieces of information they need to gather in order to obtain the solution. Propositional questions we shall be using are exclusively simple yes-no questions; a question with A and $\neg A$ as the only direct answers will be represented by $?A$ (which is a slight simplification of the original Kubiński's formalism). However, we need to extend the original Kubiński's concept in order to accommodate into our framework not only direct answers to such questions but admissible ones as well. The interested reader may find a concise summary of Kubiński's formalism in Wiśniewski (1995, pp. 52-62).

On Kubiński's account a *simple numerical question* is an expression of the form Ox_iPx_i, where Px_i (a *desideratum* of a question, as we shall call it after Ajdukiewicz (1974)) is a sentential function with x_i as the only free variable and Ox_i is a simple numerical operator containing x_i as the only variable. Thus the formula $k < x_i\, Px_i$ stands for a question: "for which [more than k] x_i, Px_i?", and the formula $(k)x_iPx_i$ stands for a question: "which are all [exactly k] x_i such that Px_i?". We shall follow the general idea of this formalism. However, it needs to be modified in order to model the cognitive tasks the subjects were solving in *Mind Maze*.

Firstly, on Kubinski's account numerical questions ask for lists of objects exhibiting certain properties. Thus direct answers (defined syntactically) to numerical questions are build by means of first-order sentences in which free variables indicated in a question are either replaced by closed terms or quantified. In *Mind Maze* questions concern pieces of information needed to attain the subject's target and expressed in sentences. As a result, instead of first-order representation, we will be employing propositional one.

Secondly, there is often more than one piece of information needed in order to attain the subject's target, and those pieces are not easily combined in a way analogous to the one offered by compound numerical questions. Thus we allow for multiple question-forming operators to be put in front of desideratum of a question.

We will be using questions analogous to the second example (the question $(k)x_iPx_i$, concerning a complete list of exactly k objects exhibiting certain property), of the form: $(k)AB$, where both A and B are formulas of some assumed language and A occurs in B; we are not using the notion of subformula beacuse we want to allow the "assumed language" to be not only some object-level formal language (the language of Classical Propositional Calculus, CPC, in particular) but a metalanguage as well (incorporating elements of situational semantics). An example of such a question is an expression of the form: *(1) $H_1\, X \models H_1$*, which can be read as: "Which is the only formula H_1 entailed by the set X?". Arguably, the resulting way of representing questions does not meet the standards of the definition of a formalized language. Nevertheless, at the present stage of development of these instruments we consider their flexibility to be the main virtue to be pursued.

3.3 Information processing: situational semantics

In making sense of the concept of situation we shall follow Keith Devlin's (1991, p. 70) claim that "situations are just that: situations", considering it as a primitive concept. We will employ Wiśniewski's (2013b) situational semantics, thus sharing his intuitions concerning the notion, of which the basic is that each atomic sentence refers to a set of situations: "If the relevant set is non-empty, then the set comprises all these situations in which (the claim made by) the atomic sentence holds" (Wiśniewski, 2013b, p. 33). Notice, that this has nothing to do with truth, as yet. However, "the relevant sets of situations are neither supposed to be non-empty nor have to be singleton sets" (Wiśniewski, 2013b, p. 34).

A situational model of the CPC language is defined, after Wiśniewski (2013b, p. 39) as follows ($Form_{CPC}$ stands for the set consisting of all and only CPC formulas).

> **Definition 1.** A *situational model* of the CPC language is an ordered pair **M** = <U, v>, U is a non-empty set (the universe of **M**) and v is a function from $Form_{CPC}$ to 2^U such that:
> i. 1. for each propositional variable p_i, $v(p_i) \subseteq U$;
> ii. 2. for each $A, B \in Form_{CPC}$:
>   (a) $v(\neg A) = U - v(A)$,
>   (b) $v(A \land B) = v(A) \cap v(B)$,
>   (c) $v(A \lor B) = v(A) \cup v(B)$
>   (d) $v(A \to B) = v(\neg A) \cup v(B)$,
>   (e) $v(A \leftrightarrow B) = (v(\neg A) \cup v(B)) \cap (v(\neg B) \cup v(A))$.
>
> **Definition 2.** A formula A is a *situational tautology* iff for every situational model **M**: $v(A) = U$.

Notice, that it is possible, that one atomic sentence subsumes another one, as in:

i. p: Snoopy is an animal,
ii. q: Snoopy is a dog.

In this case $v(q) \subset v(p)$ (for a certain, rather commonsense model **M**). This relation of subsumption may be interpreted as a kind of non-logical entailment. However, it is not tantamount to the intuitive natural-language concept of entailment, as, if A is a situational tautology, for any non-tautological formula B, $v(B) \subset v(A)$.

3.4 Topics

Interpreting the concept of topics in terms of situational semantics we shall follow some general lines proposed by Van Kuppevelt, according to whom "[t]he term topic (...) refer[s] to a topic notion which concerns the 'aboutness' of (sets of) utterances" (van Kuppevelt, 1995, p. 111). As the author claims:

> The notion presupposes that a discourse unit U - a sentence or a larger part of a discourse - has the property of being, in some sense, directed at a selected set of discourse entities (a set of persons, objects, places, times, reasons, consequences, actions, events or some other set), and not diffusely at all discourse entities that are introduced or implied by U. This selected set of entities in focus of attention is what U is about and is called the topic of U. (van Kuppevelt, 1995, p. 112)

4. RELEVANCE

4.1 Topic relevance

We shall be considering topics w.r.t. (or in) some situational model. On this account, topic O may be interpreted as a subset of the model's universe ($O \subseteq U$). One distinguished class of topics, with which we will not be dealing here, are tautological topics, that is, topics which cover all the model's universe (O is a tautological topic iff $O = U$).

Our notion of relevance is relative to a situational model and a topic in it.

> **Definition 3.** Let $\mathbf{M} = \langle U, v \rangle$ be a situational model and $O \subseteq U$ be a topic in **M**. A *situational relevance model* of O w.r.t. **M** is an ordered triple $\mathbf{N} = \langle O, w, \mathbf{M} \rangle$, where w is a partial function from $Form_{CPC}$ to 2^O such that: if $w(A) \subseteq O$, then $w(A) = v(A)$.

Further on we shall refer to such models simply as to relevance models.

> **Definition 4.** A topic O' is *relevant* to a topic O w.r.t. **N** iff $O' \subseteq O$.

Thus a formula A is relevant to a topic O w.r.t. **N** iff $w(A) \subseteq O$. The condition imposed on w in definition 3 may be expressed as follows: If A is relevant w.r.t. **N**, then $w(A) = v(A)$.

Definition 5. A topic O subsumes a topic O' w.r.t. **N** iff O' ⊆ O.

As a result, O' is relevant to O iff O subsumes O' (w.r.t. **N**). Both of these relations hold between two topics simultaneously iff the topics are identical.

In our analyses we found useful also a slightly weaker notion of relevance, which is given in definition 6. However, we shall not be using this weaker notion in the present paper.

Definition 6. A topic O' is *somewhat relevant* to a topic O w.r.t. **N** iff O' ∩ O is not empty.

4.2 Question relevance

The idea underlying the notion of question relevance is that for a question to be relevant w.r.t. certain relevance model, at least one of its direct answers must be relevant w.r.t. that model.

Definition 7. Let Q be a question and let $dQ = \{A_1, ..., A_n\}$ be the set of all the direct answers to Q. Q is *relevant* w.r.t. **N** iff there exists A_i ($1 \leq i \leq n$) which is relevant w.r.t. **N**, that is, such that $w(A_i) \subseteq O$.

Thus Q is not relevant w.r.t. **N** iff none of its direct answers is relevant w.r.t. **N**.

Two relations by means of which we shall be modelling solutions to *Mind Maze* tasks, sifting and funnelling, are in fact special cases of questions relevance. As only simple yes-no questions are allowed by the rules of the game we shall define these relations for such type of questions only.

Definition 8. $?A_1, ..., ?A_n$ are *sifting* questions w.r.t. a topic O of a certain relevance model **N** iff for every i and j ($1 \leq i, j \leq n$): $v(A_i)$ is non-empty, and $v(A_1) \subset O, ..., v(A_n) \subset O$, and if A_i and A_j are distinct, then $v(A_i) \cap v(A_j)$ is empty.

Thus $?A_1, ..., ?A_n$ are sifting w.r.t. a topic O of a certain relevance model **N** iff sets of situations assigned to affirmative answers to these questions are pairwise disjoint and all are subsets of O; in other words, the sets $v(A_1), ..., v(A_n)$ are partitioning O, albeit these partitioning need not to be exhaustive.

Definition 9. A question $?A$ *funnels* a topic O of a certain relevance model **N** iff $v(A) \subset O$, and both are non-empty. A question $?A_1$ funnels a question $?A_2$ w.r.t. a certain relevance model **N** iff $v(A_1) \subset v(A_2)$, and both are non-empty.

Thus a question $?A$ funnels a topic O w.r.t. a certain relevance model **N** iff an affirmative answer to $?A$ narrows down O, and analogously in the case of funnelling holding between questions. This relation is similar to the relation of cognitive usefulness, which holds between implied and implying questions in the case of erotetic implication (Wiśniewski, 2013, p. 72).

Examples of both of these relations are provided in the sections 6 and 7.

4.3 Truth values

In situational semantics it is quite natural to construe truth values in terms of partitions of a universe of situations: Partition of a universe U is an ordered pair $\mathbf{P} = \langle T_\mathbf{P}, F_\mathbf{P}\rangle$, such that:

i. $T_\mathbf{P} \cap F_\mathbf{P}$ is empty,
ii. 2. $T_\mathbf{P} \cup F_\mathbf{P} = U$.

Intuitively: $T_\mathbf{P}$ is the set of situations that hold and $F_\mathbf{P}$ is the set of situations that do not. Notice, that on this account both $T_\mathbf{P}$ and $F_\mathbf{P}$ are topics in U; this coincides with Frege's (1892) ideas on reference of sentences being truth values. Certainly, a suitably defined standard partition is needed in order to align assignment of truth-values to the concept of a situational model. The choice of the meaning of 'standard' depends on the choice of the underlying logic. What we found useful is the following definition of truth value of a formula A in a partition **P** of U, which nicely translates into Kleene's weak three-valued logic (T stands for truth, F for falsehood and N for the third value):

i. $V(A, \mathbf{P}) = T$ iff $v(A)$ is not empty and $v(A) \subseteq T_\mathbf{P}$;
ii. $V(A, \mathbf{P}) = F$ iff $\subseteq F_\mathbf{P}$ (this covers the case of impossibilities, when $v(A)$ is empty);
iii. 3. otherwise $V(A, \mathbf{P}) = N$ (that is, neither $v(A) \cap T_\mathbf{P}$ nor $v(A) \cap F_\mathbf{P}$ are empty).

5. THE MEANING OF ADMISSIBLE ANSWERS

Recall, that besides direct answers to a question ?A we allow for other admissible answers to Q: these are expressions '?A is not relevant' and 'Answer to ?A is not known'.

Consider a relevance model $\mathbf{N} = <O, w, \mathbf{M}>$, such that O is a topic which is not tautological. Suppose that a question of the form ?A is relevant w.r.t. N. Thus one of the following holds:

i. $w(A) \subseteq O$ and $w(A) = v(A)$, and, as O is not tautological, $v(\neg A) \not\subseteq O$; or
ii. $w(\neg A) \subseteq O$ and $w(A) = v(A)$; thus $v(A)' \subseteq O$, and, as O is not tautological, $v(A) \not\subseteq O$.

As a result, at most one answer to a simple yes-no question is relevant w.r.t. a non-tautological topic. For a justification of this, see the next section.

Next, suppose that ?A is not relevant w.r.t. **N**. Then neither $w(A)$ nor $w(\neg A)$ are subsets of O. This may cover two separate cases:

i. the value of $w(A)$ is determined, but neither $w(A)$ nor $w(\neg A)$ are subsets of O, or
ii. the value of $w(A)$ is not determined, as w is just a partial function on $Form_{CPC}$.

There are further more fine-tuned distinctions possible if the weaker concept of relevance will be employed (see definition 6).

Finally, consider the statement 'Answer to ?A is not known'. This in principle does not convey any information concerning relevance. One can argue that, in the case of *Mind Maze*, if ?A were relevant to a certain topic, enough information would be provided for an answer to ?A to be known, but in general this need not to be the case. We shall interpret this answer in terms of truth values as a claim that in a certain standard partition **P** being considered, the value of both answers to ?A is N. To formally represent such a claim we will be using the Łukasiewicz's (1920) operator I: $V(IA, \mathbf{P}) = T$ iff $V(A, \mathbf{P}) = N$.

One may notice that the fact that answer to ?A is not known may be interpreted also as the claim that information conveyed by answers to ?A is not needed in order to solve the task in question. As a result, another possible line of interpretation of this kind of answer is something like 'epistemic irrelevance', which can be expressed in terms of answers to ?A not being epistemic targets for an agent. However, we are not going to pursue this interpretation in the case of *Mind Maze*.

6. A TOY EXAMPLE

Let us exemplify our ideas on a simple gameplay of a *20 questions* game. It should be noted that, however a toy example, it is not an artificial one, as there are some 7-years olds who consider this game rather useful in overcoming the ordeals of prolonged car travels. There are two parties in the game: an Answerer, who chooses an object, and a Questioner, who has 20 chances to guess what it is (there may be many Questioners involved). Only simple yes-no questions are allowed, but our admissible answers ('not relevant' and 'not known') are permitted as well.

> Q1: Is it an animal?
> A1: Yes.
> Q2: Is it a mammal?
> A2: Yes.
> Q3: Is it a rodent?
> A3: Yes.
> Q4: Is it a rat?
> A4: No.
> Q5: Is it a pet?
> A5: Yes.
> Q6: Is it a guinea pig?
> A6: No.
> Q7: Is it a hamster?
> A7: Yes.

Let us describe this gameplay in terms of sifting and funnelling. The first question Q1 funnels the initial topic, which is the universe of objects. Then we have a sequence of funnelling questions (Q2 funnels Q1, Q3 funnels Q2, etc.). An unsuccessful guess in Q4 may be interpreted as lack of relevance of 'being a rat' to the solution. It is then followed by another sequence of funnelling questions (as Q5 funnels Q4 and Q6 funnels Q5). Finally, Q6 and Q7 are sifting questions w.r.t. the topic 'A rodent which is a pet'. Notice, that also Q4, Q6 and Q7 may be interpreted as sifting questions, however w.r.t. a more general topic 'A rodent'.

7. A CASE STUDY

Now let us turn to slightly more complicated example. Our case study is based on a gameplay involving the story *The Traveller* (subject B4). The story goes as follows:

A man, without a single visa, in one day visited eight different countries. Authorities of none of these countries tried to throw him out. What was his profession and how did he manage to do this?

The solution to the considered problem is that the man in question was a courier delivering post to the embassies. Thus there are two key pieces of information which the players needed to identify in order to reach the solution. The first one falls into the topic of the profession of the protagonist, the second one into the topic of visiting many countries in a day. Admittedly, the solution that the protagonist visited embassies and not the countries themselves draws somewhat on a popular belief that all the embassies are sovereign territories of the represented state. Although in fact most of them do not enjoy full extraterritorial status, this did not raise any issues in this research.

In modelling the subject's solution to the task we use the following symbols to represent information involved (s's were put forward by the subject while g's were gamemaster's hints):

$T!$ - cognitive target of the subject (solution to the task);
O_1 - topic: profession;
O_2 - topic: visiting many countries in a day;
H_1 - the first key piece of information, such that $v(H_1) \subset O_1$;
H_2 - the second key piece of information, such that $v(H_2) \subset O_2$;
s_1 - the protagonist's activities were legal;
s_2 - the protagonist travelled in a professional capacity;
s_3 - the protagonist travelled during a global war;
s_4 - the protagonist was an important public figure;
s_5 - the protagonist visited 8 embassies;
s_6 - the protagonist was an ambassador;
s_7 - the protagonist was a caterer;
s_8 - the protagonist was a security officer;
s_9 - the protagonist was a cleaner;
s_{10} - the protagonist's professional duties were performed at the embassies;
s_{11} - the protagonist was a postman;
g_1 - the protagonist's profession is useful for an embassy employees;
g_2 - the protagonist's profession was a common one.

The actual dialogue between the experimenter and the subject B4 can be found in Żyluk (2016). A formal reconstruction of question processing in this gameplay runs as follows:

1. $T!$
2. $H_1 \wedge H_2 \to T$
3. $(H_1 \wedge H_2)!$
4. $(1)H_1, (1)H_2\ (v(H_1) \subset O_1 \wedge v(H_2) \subset O_2 \wedge (H_1 \wedge H_2 \to T))$
5. $v(s_1) \subset O_2$
6. $?s_1$
7. s_1
8. $v(s_2) \subset O_1$
9. $v(s_2) \subset O_2$
10. $?s_2$
11. s_2
12. $v(s_3) \subset v(s_2)$
13. $?s_3$
14. $\neg s_3$
15. $v(s_3) \not\subset v(H_1) \wedge v(s_3) \not\subset v(H_2)$
16. $v(s_4) \subset O_1$
17. $v(s_4) \subset O_2$
18. $?s_4$
19. $\neg s_4$
20. $v(s_5) \subset O_2$
21. $?s_5$
22. s_5
23. $v(s_5) = H_2$
24. H_2
25. $v(s_6) \subset v(s_1) \cap v(s_2)$
26. $?s_6$
27. $\neg s_6$
28. $v(s_7) \subset (v(s_1) \cap v(s_2)) - v(s_6)$
29. $?s_7$
30. $\neg s_7$
31. $v(s_8) \subset (v(s_1) \cap v(s_2)) - (v(s_6) \cup v(s_7))$
32. $?s_8$
33. $\neg s_8$
34. $v(H_1) \subset v(g_1)$
35. $v(H_1) \subset v(g_2)$
36. $v(s_9) \subset (v(s_1) \cap v(s_2) \cap v(g_1) \cap v(g_2)) - (v(s_6) \cup v(s_7) \cup v(s_8))$
37. $?s_9$
38. $\neg s_9$
39. $v(s_{10}) \subset (v(s_1) \cap v(s_2) \cap v(g_1) \cap v(g_2)) - (v(s_6) \cup v(s_7) \cup v(s_8) \cup v(s_9))$
40. $?s_{10}$
41. $\neg s_{10}$
42. $v(s_{11}) \subset (v(s_1) \cap v(s_2) \cap v(g_1) \cap v(g_2)) - (v(s_6) \cup v(s_7) \cup v(s_8) \cup v(s_9) \cup v(s_{10}))$
43. $?s_{11}$

44. s_{11}
45. $v(s_{11}) = v(H_1)$
46. H_1
47. T

Let us now focus on some elements of this reconstruction. The subject starts with setting the cognitive goal (lines 1st-4th) by identifying 'search areas' for two key pieces of information needed in order to solve the task. In other words, at this stage she identifies topics within which the elements of solutions are to be found, as expressed in the question in the 4th line. In the 5th line a topic of legality of the travel is identified as a subset of the topic O_2. On this basis the appropriate question is asked and an affirmative answer to it (6th line) confirms its relevance.

Now, let us turn to the lines 12th-15th. In the 12th line the topic of travelling during a global war is identified as a subtopic of travelling in a professional capacity. Negative answer to the appropriate question (lines 13th and 14th) leads to the conclusion that the 'global war' topic is not relevant w.r.t. the solution been sought and this line of inquiry is abandoned. Both 6th and 13th lines contain funnelling questions: $?s_1$ funnels the topic O_2, whereas $?s_3$ funnels $?s_2$. The operation here is analogous to the one performed in the Q1 - Q4 sequence of our toy example: search areas within which solutions can be found are consecutively narrowed down.

In the lines 26th, 29th and 32nd we have three sifting questions w.r.t. the topic established as an intersection of $v(s_1)$ and $v(s_2)$ (that is, travelling legally in a professional capacity). The subject is trying to guess possible professions of the protagonist, thus she considers different topics at the very same level of generality; they are pairwise disjoint subsets of $v(s_1) \cap v(s_2)$. Another sequence of sifting questions is this: 36th, 39th, 42nd.

In the lines 23rd and 45th the key pieces of information are identified, which lead to the claim that the subject attained her cognitive target (line 47th).

8. CONCLUSION AND FURTHER RESEARCH

In this paper we outlined some formal tools for analyses of question processing involved in solutions to a specific class of abductive problems. The relations of sifting and funnelling, defined with respect to questions and topics, account well for our empirical data. Further studies shall focus on comparative analysis of the ways in which different subjects employ these relations in search for a solution. Also of interest are relationships which hold between sifting and funnelling and other relations aimed at modelling questions dependency, in particular different versions of erotetic implication (Wiśniewski, 1995; Urbański et al., 2016a).

ACKNOWLEDGEMENTS: Research reported in this paper were supported by the National Science Centre, Poland (DEC-2013/10/E/HS1/00172)

REFERENCES

Ajdukiewicz, K. (1974). *Pragmatic logic.* Dordrecht: D. Reidel P. C.; Warsaw: Polish Scientific Publishers.
Devlin, K (1991). *Logic and information.* Cambridge: Cambridge University Press.
Frege, G. (1892). Über Sinn und Bedeutung. *Zeitschrift für Philosophie und philosophische Kritik*, *100*(1), 25–50.
Gabbay, D., & Woods, J. (2005). *The reach of abduction.* London: Elsevier.
Kubiński, T. (1980). *An outline of the logical theory of questions.* Berlin: Akademie-Verlag.
van Kuppevelt, J. (1995). Discourse structure, topicality and questioning. *Journal of Linguistics*, *31*(1), 109–147.
Łukasiewicz, J. (1920). O logice trójwartościowej (in Polish). *Ruch Filozoficzny*, *5*, 170–171. English translation: On three-valued logic. In L. Borkowski (Ed.), *Selected works by Jan Łukasiewicz*, (pp. 87–88) Amsterdam: North-Holland.
Łupkowski, P., Urbański, M., Wiśniewski, A., Błądek, W., Juska, A., Kostrzewa, A., Pankow, D., Paluszkiewicz, K., Ignaszak, O., Urbańska, J., Żyluk, N., Gajda, A. & Marctiniak, B. (2017). Erotetic Reasoning Corpus. A data set for research on natural question processing. *Journal of Language Modelling* Vol 5, No 3, pp. 607–631.
Stenning, K., & van Lambalgen, M. (2008), *Human reasoning and cognitive science.* Cambridge: The MIT Press.
Thagard, P., & Shelley, C. P. (1997). Abductive reasoning: logic, visual thinking and coherence. In M.-L. Dalla Chiara, K. Doets, D. Mundici, & J. van Benthem (Eds.), *Logic and scientific methods*, (pp. 413–427). Dordrecht: Kluwer Academic Publishers.
Urbański, M., Paluszkiewicz, K., & Urbańska, J. (2016a). Erotetic problem solving: from real data to formal models: An analysis of solutions to Erotetic Reasoning Test task. In F. Paglieri, L. Bonelli & S. Felletti (Eds.) *Psychology of argument* (pp. 33–46). London: College Publications.

Urbański, M., Żyluk, N., Paluszkiewicz, K., & Urbańska, J. (2016b). A formal model of erotetic reasoning in solving somewhat ill-defined problems. In D. Mohammed & M. Lewinski (Eds.), *Argumentation and reasoned action: Proceedings of the 1st European Conference on Argumentation, Lisbon 2015* (Vol. II, pp. 973-983). London: College Publications.

Wiśniewski, A. (1995). The *posing of questions: Logical foundations of erotetic inferences*. Dordrecht: Kluwer Academic Publishers.

Wiśniewski, A. (2013a). *Questions, inferences, and scenarios*. London: College Publications.

Wiśniewski, A. (2013b). Logic and sets of situations. In *Essays in logical philosophy* (pp. 33–46). Berlin: LiT Verlag.

Żyluk, N. (2016). *Semantyka sytuacyjna i logika pytań w analizie rozwiązań zadań abdukcyjnych na przykładzie gry 'Takie Życie'* [*Situational semantics and logic of questions in analysis of solutions to abductive tasks on the example of 'Mind Maze' game*]. (MA thesis). Poznań: AMU Institute of Psychology.

59

Strategic Maneuvering with Presentational Devices: A Systematic Approach

TON VAN HAAFTEN
Leiden University Centre for Linguistics (LUCL)
t.van.haaften@hum.leidenuniv.nl

MAARTEN VAN LEEUWEN
Leiden University Centre for Linguistics
m.van.leeuwen@hum.leidenuniv.nl

The aim of our paper is to show how a systematic stylistic analysis of presentational devices can be integrated in a pragma-dialectical analysis of strategic maneuvering in argumentative discourse. It will be argued that the key to such a systematic stylistic analysis is the use of a linguistic checklist. This approach and its added value is illustrated by applying it to a case study: the pleadings that were delivered in a Dutch civil law case.

Argumentative discourse, strategic maneuvering, presentational devices, stylistic analysis, stylistic choices;

1. INTRODUCTION[1]

In the extended pragma-dialectical approach to argumentation (Van Eemeren, 2010), 'presentational devices' are distinguished as one of the three aspects that arguers always employ to maneuver strategically in argumentative discourse. However, the identification and the analyses of the strategic functions of presentational devices or 'stylistic choices' in actual argumentative discourse are often ad hoc: in most cases, a *systematic* analysis thereof is lacking (cf. Fahnestock, 2009; Van Leeuwen, 2014).

---

[1] Both researchers contributed equally to this paper.

The aim of our paper is to show how a systematic identification and analysis of stylistic choices can be integrated in a pragma-dialectical analysis of argumentative discourse. More specifically, we will argue that the key to a more systematic analysis of presentational devices is the use of a linguistic checklist. Although the use of such a checklist has been suggested various times in linguistic and rhetorical approaches to style (cf. Van Leeuwen, 2015, pp. 26-28 for an overview), this suggestion has hardly been taken up in practice.

On the basis of a systematic argumentative-stylistic analysis with a checklist of the oral pleadings that were delivered in a Dutch civil law case, we will illustrate how plaintiff and defendant maneuvered strategically with a variety of stylistic choices to convince the judge – including subtle, 'hidden' stylistic choices that would probably had not come to light with a less systematic approach.

The structure of our paper is as follows. First we briefly sketch some relevant key concepts of the pragma-dialectical argumentation theory. Second, we will elaborate on the case study by providing relevant background information. Third, we will clarify the linguistic-stylistic checklist method we used, followed by some results of our systematic stylistic analysis of the case study.

2. THE IDENTIFICATION AND THE ANALYSIS OF THE STRATEGIC FUNCTION OF ARGUMENTATIVE MOVES

The framework adopted here, the extended pragma-dialectical argumentation theory (Van Eemeren, 2010), assumes that people engaged in argumentative discourse maneuver strategically. 'Strategic Maneuvering' refers to the efforts arguers make in argumentative discourse to reconcile rhetorical effectiveness with the maintenance of dialectical standards of reasonableness. In order not to let one objective prevail over the other, the parties try to strike a balance between them at every stage of resolving their differences of opinion. Strategic maneuvering manifests itself in argumentative discourse in (a) the choices that are made from the topical potential, (b) audience-directed framing of argumentative moves and (c) the use of presentational devices. Although these three aspects of strategic maneuvering can be distinguished analytically, in actual argumentative practice they will usually be hard to disentangle (Van Eemeren, 2010, p. 93-127).

The choice made from the topical potential has to do with the perspective from which the arguer selects his argumentative moves (Van Eemeren, 2010, pp. 96-108). Adaption to the audience covers the tuning of the argumentative moves to audience demand (Van Eemeren,

2010, pp. 108-118). The exploitation of presentational devices pertains to the communicative means that are used in presenting the argumentative moves. (Van Eemeren, 2010, pp. 118-127). In this paper we focus on the third aspect of strategic maneuvering.

When maneuvering strategically speakers or writers make an effort to present their moves in a specific way. In all stages of the critical discussion – confrontation, opening, argumentation and concluding stage – their formulation of the moves may be assumed to be systematically attuned to achieving the strategic purposes they are aiming for. This means that that in analysing strategic maneuvering it is worthwhile to take due account of the stylistic choices that have been made (Van Eemeren, 2010, pp. 118-119).

In analysing strategic maneuvering it is very important to take into account – among others – the following two factors.[2] First, strategic maneuvering may be presumed to be aimed at achieving *a specific result*, so it is worth to take this factor into account in determining the strategic function of a particular argumentative move made at a particular stage of the critical discussion. Another factor to be taken into account consists of *the institutional constraints* imposed on the argumentative discourse in the context in which it is carried out. Taking these constraints into account is necessary in order to track the institutional preconditions that the strategic maneuvering carried out must meet in the specific type of communicative activity (Van Eemeren, 2010, pp. 163).

3. THE YURI-CASE

The case study that we will use to illustrate what the use of a checklist can yield for a pragma-dialectical analysis of strategic maneuvering with stylistic choices, consists of the oral pleadings that were delivered in the so-called 'Yuri case', in which the Dutch gymnast Yuri van Gelder played a leading role. During the Olympics in Rio de Janeiro in 2016, Yuri van Gelder qualified for the final of the rings competition, after a decade full of ups and downs in which he for example became world champion in 2005 but was also several times suspended and withdrawn from the

---

[2] According to Van Eemeren (2010: 163-165), the two other factors that should be considered in every case of strategic maneuvering for determining the strategic function of argumentative moves, are (a) the routes that can be taken to achieve a specific result and (b) the commitments of the parties defining the argumentative situation. For reasons of lack of space we do not discuss these two factors in our analysis of the case study.

team because of the use of cocaine. But during the Olympics in 2016, he could finally compete at the highest level, and he qualified for the final.

In order to celebrate this qualification, Van Gelder went to the 'Holland Heineken House', the meeting place for the Dutch during the Olympic Games. Van Gelder, who had to train the next morning with the Dutch gymnast team, had be warned not to drink alcohol, and he promised his coach via WhatsApp to be back in his hotel around midnight. However, after visiting the Holland Heineken House, where he drunk several beers, he left the Olympic village – which was not allowed because of safety reasons. After going out he returned to his hotel in the early morning; according to his teammates he caused commotion and was drunk. He went to bed and woke up around 3 PM, which meant that he had missed the training with his team. The Dutch Olympic Committee suspended Van Gelder because of his behaviour, and sent him home for violating the team's code of conduct. Back in the Netherlands, Van Gelder – the plaintiff – took the Dutch Olympic Committee – the defendant - to court in a so called "civil summary judgement procedure", and demanded to get reinstated in the Olympic team, denying that he had broken the team rules.[3]

The communicative activity type "civil summary judgement" belongs to the genre of adjudication in the domain of legal communication. Adjudication aims for the settlement of a dispute by an authorized third party rather than by the parties themselves (Van Eemeren, 2010, pp. 147). In the case of a civil summary judgement procedure two parties take a difference of opinion to a public civil court, where a judge makes a reasoned decision in favour of one of the parties.

Like other civil law procedures in the Netherlands, the summary judgement procedure aims for the termination of a well-defined dispute by a judge. The decision is sustained by argumentation that is based on an understanding of the relevant facts and concessions, formulated in terms of conditions for the application of a legal rule, a quasi-legal rule or a contract. Unlike other civil law case procedures, the summary judgement procedure is a very fast procedure meant to deal with urgent cases and there are very few formal procedural legal rules involved.

---

[3] During the trial Van Gelder's case was defended by his lawyer, Cor Hellingman; the Dutch Olympic Committee was represented by its lawyer, Haro Knijff.

4. THE SYSTEMATIC IDENTIFICATION AND ANALYSIS OF THE STRATEGIC FUNCTION OF THE STYLISTIC DESIGN OF ARGUMENTATIVE MOVES

4.1 Method

In order to integrate a systematic identification and analysis of the strategic function of the stylistic design of argumentative moves in the oral pleadings of both plaintiff and defendant, we adopted the following methodological steps.

First we reconstructed the argumentative discourse of both plaintiff and defendant resulting in an analytic overview of the four discussion stages with respect to each oral pleading: the confrontation stage, the opening stage, the argumentation stage and the concluding stage. As part of this reconstruction we also determined the specific dialectical and rhetorical aims of each discussion stage in the speech event.

Next, we searched systematically in both pleadings for stylistic choices that are used to maneuver strategically in each discussion stage. We did this by making use of the Dutch version of the checklist by Leech and Short (2007). The English version of the checklist can be found in the Appendix.[4] It lists numerous linguistic phenomena that can be relevant for the stylistic analysis of a particular text. As such, the checklist can be used as a heuristic tool to find stylistic choices systematically.[5] The result of this analysis was a long list of stylistic choices for each of the pleadings that are assumed to be used somehow to maneuver strategically.

Finally, in order to determine and analyse the strategic function(s) of the identified stylistic choices we took into account the result(s) the plaintiff and defendant aimed for in formulating their strategic maneuvers and the institutional constraints the activity type imposes on the argumentative discourse in this specific case.

In the rest of this paper we will limit ourselves to discussing a set of stylistic choices in the opening stage of the pleadings of the plaintiff and the defendant, and how both parties built on this set of stylistic choices in the argumentation stage. More specifically, we will focus on a series of choices that can be related to the dialectical aim and

---

[4] See Van Leeuwen (2015: 29-32) for the Dutch version of the checklist that was actually used.

[5] See Van Leeuwen (2015: 36-39 and chapter 3) for a more elaborate discussion on the methodology of using a linguistic checklist in stylistic analysis.

the rhetorical aim that both parties pursue in the opening stage of a summary judgment procedure, namely depicting the facts in this case in such a way that it will lead to the most favourable and effective outcome for the party involved while reasonableness is maintained. Applied to the Yuri case: it can be expected that the plaintiff would try to *downplay* the seriousness of Van Gelder's actions, and it can be expected that the defendant would try to *stress* or *underline* the seriousness of Van Gelder's behaviour. These expectations were confirmed on a cursory reading of the pleadings; in the rest of our talk we will highlight stylistic choices contributing to the realization of downplaying / stressing the seriousness of Van Gelder's actions. In other words, we will try to show how the strategic function of stylistic choices can be analysed in the light of the intended results of both parties and the characteristics of the activity type. While discussing the various stylistic choices, we will refer between brackets to which category of the checklist is involved.

4.2 Results

Which stylistic choices did the plaintiff make in formulating his proposals for starting points for the discussion in order to downplay the seriousness of Van Gelder's actions? One of the points of interests mentioned under category A1 in the checklist is the use of words with particular suffixes. It is striking that the plaintiff characterizes Van Gelders' night out as an 'avond*je*' (literally: 'small evening') twice. By using the diminutive 'je', the proportion of the incident is depicted as relatively unimportant. (A1) A further look at the plaintiff's pleading reveals other instances of euphemistic lexical choices as well (A1): for instance, Van Gelder's apparent alcohol abuse is characterized neutrally as 'consumption of alcohol' and 'relaxation'. In addition, the plaintiff does not state that Van Gelder missed the training next morning because he was 'sleeping off his hangover', but because he was 'sleeping' and because he was 'recovering sleep' – which is both more neutral.

Further, the plaintiff does not state that Van Gelder's behaviour caused 'anger' in the gymnastics team, but '*perhaps* annoyance'. In other words, the plaintiff uses a noun that functions euphemistically, combined with an adverb that serves a hedging function (A5) and that suggests even more that the inconvenience caused by Van Gelder's behaviour was not very big. The plaintiff also implies this by using rhetorical questions (B1/C): 'Did the other athletes look pale, then? Was Epke (i.e. Epke Zonderland, one of Van Gelder's team members) not able anymore to perform gymnastics?' From the context it is evident that the implied answer to both questions is 'no' – steering again towards the conclusion that Van Gelder's behaviour did not really affect his teammates negatively. Finally, the plaintiff's statement that Van Gelder 'slipped inside' when he came back from his night out is relevant. The verb (A4/A1) 'to slip inside' suggests, more than possible alternative verbs like 'to go inside' or 'to enter' that no noise was made.

On the other hand, in the defendant's oral pleading, the checklist analysis reveals a series of choices in his formulations of the proposals for starting points for the discussion *stressing* the seriousness of Van Gelder's actions. An interesting, representative passage is the following:

> Van Gelder did not follow instructions from the team leaders on multiple occasions. He was untraceable twice with all the risks this entails. He brought himself in a position in which he could not be present at a training, due to excessive consumption of alcohol and nightly escapades. Time and again he gave contradictive explanations about his behaviours. Contrary to all agreements he drank alcohol, while he still was in competition and had a chance of winning a medal. Van Gelder has a bad influence on the team. His team members have taken offence of his confused and disorientated behaviour after his nightly adventures, and (…) his attitude of acting in defiance of the rules of the team (…).

In this passage, multiple relevant stylistic choices can be observed that can be found elsewhere in the defendant's pleading as well. First of all, a striking lexical choice (A1) is 'nightly escapades': the word 'escapade' suggests that Van Gelder's evening out was quite a bacchanal. In addition, the defendant talks about the incident by using nouns in the plural form (A1). He not only speaks about 'nightly escapades', but also about 'behaviours' and 'adventures'. 'Instructions' were not followed on multiple 'occasions', which lead to certain risks. Further, according to the defendant, Van Gelder gave inconsistent 'explanations', and drinking alcohol was against 'agreements and the rules'. This use of the plural form stresses the magnitude of Van Gelder's offence: it is suggested that he committed various offences and broke several rules.

The seriousness of the incident is further stressed by the defendant's use of intensifiers (A5), i.e. words that stress the intensity of the message (cf. Burgers & De Graaf, 2013, p. 171). For instance, the defendant states that drinking alcohol was against 'all' agreements, contradictive explanations were given 'time and again', there was 'excessive' consumption of alcohol, and Van Gelder had not followed instructions on 'multiple' occasions, indicating an attitude of acting 'in defiance' of the rules. Moreover, the verb tenses (B6) in the fragment are striking: most of the sentences in the passage are formulated in the past tense. However, in the sentence 'Van Gelder has a bad influence on the team' a switch is made to the present imperfective tense. By switching to the present imperfective, it is stressed that the incident had consequences that are still continuing. Finally, it is interesting that the defendant states that Van Gelder '*could* not be present at a training'. The defendant could have said that Van Gelder *was* not present at a training, but by using the modal auxiliary (B6) 'could' here, it is stressed that Van Gelder was not *able* to be on time, which stresses the seriousness of the situation as well.

5. CONCLUSION

As we said in the Introduction, our claim is that the use of a linguistic checklist can be of help to make the identification and analysis of the strategic function of presentational devices more systematic. The use of a checklist has a few times been proposed in stylistics but as stated these proposals have barely been acted upon in analytical practice.

The added value of using a linguistic checklist lies in its heuristic function: the list helps finding stylistic devices that could otherwise easily be overlooked. It more or less 'forces' the analyst to take into consideration a wide variation of stylistic devices and as such, reduces the chance that pertinent stylistic choices are missed (cf. Van Leeuwen, 2014, pp. 237-238). In the Yuri case, for instance, we probably would have overlooked the plaintiff's strategic use of verb tenses and modal auxiliaries: these subtle, more or less 'hidden' grammatical presentational devices would probably not have come to light when the analysis had been carried out without using a checklist. In this way the checklist can be of help also to identify more precisely not only individual strategic maneuvers, but also *discussion strategies*: coordinated modes of strategic maneuvering designed to influence the result of a particular stage of the resolution process, or the discussion as a whole (Van Eemeren, 2010, pp. 47).

Even so, the list is not a panacea. Going through the checklist systematically *reduces* the risk of overlooking pertinent stylistic devices and their systematic relation with other devices but it cannot remove this risk completely (cf. Van Leeuwen, 2014, p. 238). For one thing, the checklist is not exhaustive: a complete list would result in an instrument whose length would make it unmanageable in analytical practice. In addition, the checklist we used here presupposes a lot of linguistic background knowledge: behind every category mentioned, a whole 'world' of linguistics is hidden. This may hamper the use of a checklist, because the categories mentioned in the checklist are often not directly applicable to discourse analysis. For instance, category A1 in the checklist proposed by Leech and Short (2007) steers the analyst among other things to search for 'particular suffixes'. This helped us to identify the defendant's strategic use of plural forms ('escapade**s**', 'behaviou**rs**', 'adventure**s**', etc.). This 'translation' from an abstract category like 'suffixes' to a concrete stylistic phenomenon 'plurals') is something that the analyst needs to do himself – based on linguistic knowledge. However, this is a 'problem' of stylistic research as such: without the use of a checklist, the problem would also exist.

All in all, the use of a linguistic checklist does not provide a panacea for identifying and analysing the strategic function of the verbal presentation of arguments: it does not reduce this to a relatively uncomplicated activity. However, a linguistic checklist is a valuable *starting point* for stylistic analysis (cf. Van Leeuwen, 2014, p. 238): it's heuristic function helps the analyst to identify and analyse the strategic function of presentational devices used in argumentative discourse to manoeuvre strategically in a systematic way.

ACKNOWLEDGEMENTS: We would like to thank lawyers Cor Hellingman and Harro Knijff for providing us with the written versions of the pleas.

REFERENCES

Burgers, C. & de Graaf, A. (2013). Language intensity as a sensationalistic news feature: The influence of style on sensationalism perceptions and effects. *Communications, 38*(2), 167-188.
Eemeren, F. H. van (2010). *Strategic maneuvering in argumentative discourse.* Amsterdam: John Benjamins.
Fahnestock, J. (2009). Quid pro nobis: Rhetorical stylistics for argument analysis. In F. H. van Eemeren (Ed.), *Examining argumentation in context: Fifteen studies on strategic maneuvering* (pp. 191-220). Amsterdam: John Benjamins.
Leech, G. & Short, M. (2007). *Style in fiction: A linguistic introduction to English fictional prose* (2nd ed.). Harlow: Pearson Longman.
Leeuwen, M. van (2014). Systematic stylistic analysis: The use of a checklist. In B. Kaal, I. Maks & A. van Elfrinkhof (Eds.), *From text to political positions. Text analysis across disciplines* (pp.225-244). Amsterdam: John Benjamins.
Leeuwen, M. van (2015). *Stijl en politiek: Een taalkundig-stilistische benadering van Nederlandse parlementaire toespraken.* Utrecht: LOT.

APPENDIX

A checklist of linguistic and stylistic categories (Leech and Short 2007)

A: Lexical categories

1. GENERAL. Is the vocabulary simple or complex? Formal or colloquial? Descriptive or evaluative? General or specific? How far does the writer make use of the emotive and other associations of words, as opposed to their referential meaning? Does the text contain idiomatic phrases or notable collocations, and if so, with what kind of dialect or register are these idioms or collocations associated? Is there any use of rare or specialized vocabulary? Are any particular morphological categories noteworthy (e.g. compound words, words with particular suffixes)? To what semantic fields do words belong?
2. NOUNS. Are the nouns abstract or concrete? What kinds of abstract nouns occur (e.g. nouns referring to events, perceptions, processes, moral qualities, social qualities)? What use is made of proper names? Collective nouns?
3. ADJECTIVES. Are the adjectives frequent? To what kinds of attribute do adjectives refer? Physical? Psychological? Visual? Auditory? Colour? Referential? Emotive? Evaluative? etc. Are adjectives restrictive or non-restrictive? Gradable or non-gradable? Attributive or predicative?
4. VERBS. Do the verbs carry an important part of the meaning? Are they stative (referring to states) or dynamic (referring to actions, events, etc.)? Do they 'refer' to movements, physical acts, speech acts, psychological states or activities, perceptions, etc.? Are they transitive, intransitive, linking (intensive), etc.? Are they factive or non-factive?
5. ADVERBS. Are adverbs frequent? What semantic functions do they perform (manner, place, direction, time, degree, etc.)? Is there any significant use of sentence adverbs (conjuncts such as *so, therefore, however*; disjuncts such as *certainly, obviously, frankly*)?

B: Grammatical categories

1. SENTENCE TYPES. Does the author use only statements (declarative sentences), or do questions, commands, exclamations or minor sentence types (such as sentences with no verb) also occur in the text? If these other types appear, what is their function?
2. SENTENCE COMPLEXITY. Do sentences on the whole have a simple or complex structure? What is the average sentence length (in number of words)? What is the ratio of dependent to independent clauses? Does complexity vary strikingly from one sentence to another? Is complexity mainly due to (i) coordination, (ii) subordination, or (iii) parataxis (juxtaposition of clauses or other equivalent structures)? In what parts of a sentence does complexity tend to occur? For instance, is there any notable

occurrence of anticipatory structure (e.g. of complex subjects preceding the verbs, of dependent clauses preceding the subject of a main clause)?
3. CLAUSE TYPES. What types of dependent clause are favored: relative clauses, adverbial clauses, different types of nominal clauses (*that*-clauses, *wh*-clauses, etc.)? Are reduced or non-finite clauses commonly used and, if so, of what type are they (infinitive clauses, *-ing* clauses, *-ed* clauses, verbless clauses)?
4. CLAUSE STRUCTURE. Is there anything significant about clause elements (e.g. frequency of objects, complements, adverbials; of transitive or intransitive verb constructions)? Are there any unusual orderings (initial adverbials, fronting of object of complement, etc.)? Do special kinds of clause construction occur (such as those with preparatory *it* or *there*)?
5. NOUN PHRASES. Are they relatively simple or complex? Where does the complexity lie (in premodification by adjectives, nouns, etc., or in postmodification by prepositional phrases, relative clauses, etc.)? Note occurrence of listings (e.g. sequences of adjectives), coordination or apposition.
6. VERB PHRASES. Are there any significant departures from the use of the simple past tense? For example, notice occurrences and functions of the present tense; of the progressive aspect (e.g. *was lying*); of the perfective aspect (e.g. *has/had appeared*); of modal auxiliaries (e.g. *can, must, would*, etc.) Look out for phrasal verbs and how they are used.
7. OTHER PHRASE TYPES. Is there anything to be said about other phrase types: prepositional phrases, adverb phrases, adjective phrases?
8. WORD CLASSES. Having already considered major or lexical word classes, we may here consider minor word classes ('function words'): prepositions, conjunctions, pronouns, determiners, auxiliaries, interjections. Are particular words of these types used for particular effect (e.g. the definite or indefinite article; first person pronouns *I, we*, etc.; demonstratives such as *this* and *that*; negative words such as *not, nothing, no*)?
9. GENERAL. Note here whether any general types of grammatical construction are used to special effect; e.g. comparative or superlative constructions; coordinative or listing constructions; parenthetical constructions; appended or interpolated structures such as occur in casual speech. Do lists and coordinations (e.g. lists of nouns) tend to occur with two, three or more than three members? Do the coordinations, unlike the standard construction with one conjunction (*sun, moon and stars*), tend to omit conjunctions (*sun, moon, stars*) or have more than one conjunction (*sun and moon and stars*)?

C: Figures of speech, etc.

Here we consider the incidence of features which are foregrounded by virtue of departing in some way from general norms of communication by means of the language code; for example, exploitation code. For identifying such features, the traditional figures of speech (schemes and tropes) are often useful categories.

1. GRAMMATICAL AND LEXICAL. Are there any cases of formal and structural repetition (anaphora, parallelism, etc.) or of mirror-image patterns (chiasmus)? Is the rhetorical effect of these one of antithesis, reinforcement, climax, anticlimax, etc.?
2. PHONOLOGICAL SCHEMES. Are there any phonological patterns of thyme, alliteration, assonance, etc.? Are there any salient rhythmical patterns? Do vowel and consonant sounds pattern or cluster in particular ways? How do these phonological features interact with meaning?
3. TROPES. Are there any obvious violations of, or departures from, the linguistic code? For example, are there any neologisms (such as *Americanly*)? Deviant lexical collocations (such as *portentous infants*)? Semantic, syntactic, phonological, or graphological deviations? Such deviations (although they can occur in everyday speech and writing) will often be the clue to special interpretations associated with traditional poetic figures of speech such as metaphor, metonymy, synecdoche, paradox and irony. If such tropes occur, what kind of special interpretation is involved (e.g. metaphors can be classified as personifying animising, concretising, synaesthetic, etc.)? Because of its close connection with metaphor, simile may also be considered here. Does the text contain any similes, or similar constructions (e.g. 'as if' constructions)? What dissimilar semantic fields are related through simile?

D: Context and cohesion

- Cohesion: ways in which one part of a text is linked to another (the internal organisation of the text).
- Context: the external relations of a text or a part of a text, seeing it as a discourse presupposing a social relation between its participants (author and reader; character and character, etc.), and a sharing by participants of knowledge and assumptions.

1. COHESION. Does the text contain logical or other links between sentences (e.g. coordinating conjunctions, or linking adverbials)? Or does it tend to rely on implicit connections of meaning? What sort of use is made of cross-reference by pronouns (*she, it, they,* etc.)? By substitute forms (*do, so,* etc.), or ellipsis? Alternatively, is any use made of elegant variation – the avoidance of repetition by the substitution of a descriptive phrase (as, for example, 'the old lawyer' or 'her uncle' may substitute for the repetition of an earlier 'Mr Jones')? Are meaning connections reinforced by repetition of words and phrases, or by repeatedly using words from the same semantic field?
2. CONTEXT. Does the writer address the reader directly, or through the words or thoughts of some fictional character? What linguistic clues (e.g. first person pronouns *I, me, my, mine*) are there of the addresser-addressee subject? If a character's words or thoughts are represented, is this done by

direct quotation (direct speech), or by some other method (e.g. indirect speech)? Are there significant changes of style according to who is supposedly speaking or thinking the words on the page?

60

Criticism and Justification of Negotiated Compromises

JAN ALBERT VAN LAAR
University of Groningen
j.a.van.laar@rug.nl

ERIK C. W. KRABBE
University of Groningen
e.c.w.krabbe@rug.nl

We deal with conflicts of opinion about an already negotiated compromise, taking as our example a debate in Dutch parliament about the approval of the Paris Agreement on climate change of 2015. What worries can we expect from opponents of approval, and what arguments may answer them? We develop a profile of dialogue providing reasonable options for those who test a negotiated compromise critically, and those who attempt to meet the corresponding burden of proof.

KEYWORDS: argument schemes, compromise, criticisms, Paris Agreement, profile of dialogue, stock issues

1. INTRODUCTION: JUSTIFYING A SECOND-BEST OUTCOME

Argumentation and negotiation are intimately connected. But how? That depends on the setting. In advance of negotiation, argumentation may be used to convince others to participate. As negotiations proceed, it may be used to convince others to accept an offer. These two situations we discussed in an earlier paper (van Laar and Krabbe, 2016). But after negotiation, there are again occasions for the use of argument.

In this paper, we attend to some of the arguments that may follow upon a negotiation process. We focus on those settings where negotiators, at an earlier occasion, succeeded in closing a deal but where this agreement meets with critical questions, and possibly also harsher forms of criticism, from those who are to approve or ratify the result.

Such conflicts about an already negotiated compromise, or *conflicts about compromise* for short, are part and parcel of democratic life, and more generally of settings where we need to defend to our friends or partners some kind of collaboration with mutual enemies or competitors.

The ways of criticizing a negotiated compromise are manifold, and the task of justifying it is correspondingly demanding. As compromise brings sacrifice, critics typically scrutinize them closely so as to get clear on how the promised gains balance against the expected losses. To accede to a compromise sometimes requires giving up on realizing one's moral or other deeply engrained ideals, and such an accommodation to the adverse demands of competing parties may threaten or reduce one's integrity and credibility. Conflicts about compromise typically invite arduous discussion. How, if ever, can compromise be justified?[1]

An example of an argumentative probing of the virtues and vices of a negotiated compromise is the plenary debate within the Dutch parliament[2] on May 19, 2016, in which the Minister for Infrastructure and the Environment, Mrs. Dijksma (Labour Party – *PvdA*), presents the Paris Agreement on climate change of December 12, 2015. Notwithstanding the appreciation for the minister's role and the enthusiasm about the agreement of all but one of the parties, the members of parliament question the minister critically. To meet these criticisms, the Minister defends the compromise in various ways.

We shall develop a list of critical questions with which to evaluate a negotiated compromise and to make up one's mind about whether or not to approve of or ratify it. Some of these critical questions we connect to plausible argumentative responses. In this way, we work towards a normative theory of arguments in conflicts about compromise. We illustrate the various dialectical options with fragments from the aforementioned parliamentary debate about the Paris Agreement. To the extent that our dialectical approach sheds light

---

[1] In this paper we deal with the justification of compromise from the normative stance of resolution oriented persuasion dialogue. A different perspective on justification is taken by Fabian Wendt (2016), according to whom public justifiability – understood as justifiability to persons with different beliefs and evaluative standards (p. 119) – is a moral value that provides a moral reason in favour of a compromise (even if the agreed upon arrangement is not fully just), because public justifiability adds to a stable and peaceful society, is respectful towards people, and fosters mutual moral accountability.

[2] More specifically, in the *Tweede Kamer* (House of Commons).

on these fragments, our extended case study supports the utility and appropriateness of our framework.

In Section 2, we elaborate on our concept of *compromise* and introduce the debate about the Paris Agreement that figures as our example of a *conflict about compromise*, the kind of setting targeted in this paper. In Section 3 we deal with possible criticism of the use of negotiation as a means to deal with a particular conflict, such as the many-sided antagonisms between the 196 countries and the EU at the Paris Conference, rather than using some other kind of means. In Section 4, we examine a variety of worries that opponents may advance in response to a particular negotiated compromise, and the variety of arguments in its defence thus invited. We conclude, in Section 5, with a profile of dialogue providing reasonable options for those involved in a conflict about compromise. We also note the role of threats and pressure, but end with a positive view on the role of argumentation.

2. CONFLICTS ABOUT COMPROMISE

In a conflict about compromise, a proponent is committing herself to argue in support of a compromise that has been negotiated at an earlier occasion. Before we investigate, in the next two sections, the various points of criticism with which to assess the acceptability of the compromise it is useful to discuss the concept of a compromise and the kind of conflict a successful compromise may give rise to if it needs the approval of others.

About the concept of compromise, Weinstock writes:

> At a first approximation, a compromise is a position that, with respect to the issue at hand, is from the point of view of parties locked in debate or negotiation inferior to the positions that both (or all) bring to a decision making process (a negotiation, an election, or more trivially a decision-oriented discussion among friends), but which both have reason to accept instead of the position they favour. They may favour X, when only the issue at hand is in view, but favour Y when all things are duly considered (Weinstock, 2013, p. 539).

At an earlier occasion we adopted this definition, and added five clarifying remarks. First, the parties to a compromise have *not resolved* their differences of opinion about policies; second, a compromise is *not imposed*, even though some degree of pressure typically has preceded it; third, a compromise results from some kind of *trade*, and if it settles a difference of opinion some kind of "commodification" has made it possible; fourth, next to action commitments, a compromise generates a *propositional commitment* to explain and defend that the agreement at hand merits acceptance and implementation given that there exists an irresolvable disagreement with peers about what policy to pursue; fifth, a compromise, though not resolving the initial difference of opinion, does typically bring about the resolution of a special, *second-order*, difference of opinion, to wit the difference of opinion about what policy to adopt in setting where the initial, *first-order*, difference of opinion happens to be irresolvable (van Laar & Krabbe, 2016).

The following definition matches with Weinstock's definition and with our conceptual clarifications:

> A compromise for competing parties $P_1,...,P_n$ is an expressed agreement[3] C among the parties about the endorsement of an arrangement[4] A in response to a particular problem or issue, such that:
> - the competing parties have acknowledged that a failure to come to a shared agreement about the issue forms a negative outcome A_0,[5] and
> - each competing party P_i has proposed an arrangement A_i as its preferred solution to the problem or issue, such that these arrangements are not (fully) compatible with one another,[6] and
> - each competing party P_i prefers A_i to A,[7] yet it prefers A to some other A_j ($j \neq i$), as well as to A_0, and
> - A is conceived of by each competing party as involving unforced[8] concessions[9] to other parties.

Underlying the possibility of a compromise is the willingness of each competing party to evaluate arrangements on two levels (cf. Wendt 2016, pp. 21-34). On the first level, a party evaluates arrangements while either disregarding other parties' preferences or supposing itself

---

[3] Thus, a compromise generates commitments, among them propositional commitments.

[4] We follow Wendt in characterizing the substance of a compromise as an "arrangement" (2016).

[5] The reasons for assessing A_0 negatively may, or may not, differ between the competing parties.

[6] This refers to the first-order difference of opinion.

[7] Thus, the first-order difference of opinion has remained unresolved.

[8] Thus, a compromise is not imposed.

[9] Thus, a compromise is based on a kind of trade.

capable of convincing the other parties of what it regards as the superior arrangement. On the second level, the party evaluates arrangements while taking into account that the competing parties' consent is required to make any progress on the issue whereas not all other parties can be convinced to accept what this party regards as the superior arrangement. In a compromise, the competing parties are and remain locked in disagreement on the first level, yet they manage to find agreement on the second level.[10]

The Paris Agreement is an example of compromise in this sense, and a very intricate one. The competing parties, 196 countries and the EU, collectively aimed at reducing climate change and its effects, and managed to find a middle ground among many divergent preferences. This becomes clear from observations by Radoslav S. Dimitrov, who participated in some of the negotiations rounds, and summarized the key sacrifices and gains:

> China failed to obtain legally binding actions in the North and had to concede global stocktaking and stronger international transparency than they liked. Yet, they largely won the battle over differentiation [among developing countries] in both finance and mitigation. The simple binary division between developing and developed countries is now gone but a subtle (and more ambiguous) differentiation remains between developed and "other" countries, "in light of national circumstances". The US managed to weaken the legally binding character of national actions but lost on their [opposition to] mandatory and progressive evolution as well as on [their opposition to a binary] financial differentiation [between developing and developed countries]. The EU won on transparency, finance, and loss and damage - but failed to win quantitative global emission targets and restrictions on bunker fuels from international aviation and shipping. Island nations lost on adaptation and loss and damage but their delegates chanted a song outside the plenary hall, celebrating their success in obtaining a strong reference to a 1.5 degree limit as an aspirational goal of the treaty. (Dimitrov, 2016, p. 5)

---

[10] Thus, they do not resolve their *first-order disagreement*. Yet in those situations where they disagreed about how to accommodate an irresolvable first-order disagreement, a compromise implies a resolution of that *second order disagreement*.

The problem solved by means of the Paris Agreement is not, or at least not directly, the problem of climate change itself. Rather the problem solved is a multifaceted policy problem: China wants to respond to climate change without too much transparency measures, while the EU sets great store by transparency – and so forth for the many other components of the policy problem.

Some main ingredients of the Paris Agreement are: a shared goal of reducing temperature rise "well below 2 degrees Celsius above pre-industrial levels, pursuing efforts to limit the temperature increase even further to 1.5 degrees Celsius" (UNFCCC, 2016); the decision to realize that goal by means of "nationally determined contributions", so that each individual nation itself decides on the required policy measures and reports on its progress every five years to the other signatories; the decision to support developing countries financially to adapt to a changed climate and to reduce emissions.

The chances of realizing any agreement were small, because of the complexity of the issue, the high stakes involved, and the sheer number of negotiating parties. Dimitrov explains the success of Paris Agreement: a) by reference to a change of mind regarding the seriousness of climate change and the economic benefits of limiting such change; b) as enabled by a previous bilateral agreement in 2014 between China and the US regarding climate change and clean energy, and c) as a result of a skilful orchestration by the French hosts. He writes: "Secrecy is common in diplomacy, but the French finesse it to a new level" (p. 6), by step-wise crafting the deal in closed sessions with only key players, keeping all others in the dark by minimizing paper traces, so as to be able to present the final text on the last day as a take-it-or-leave-it deal (p. 6). On December 12, 2015, the parties signed the agreement, which was to become operational upon ratification by at least 55 countries together responsible for at least 55% of global emissions.

On May 19, 2016, the Dutch Minister for Infrastructure and the Environment,[11] Mrs. Dijksma (Labour Party), attended a meeting of the Dutch Parliament (see Note 2), in order to answer parliamentarians' questions about the Paris Agreement, which she herself had helped to negotiate in Paris. We approach this parliamentary setting as including a conflict about compromise, with Mrs. Dijksma (sometimes assisted by other members of parliament) in the role of proponent, committing herself to the defence of the Paris Agreement as meriting endorsement and ratification, and the members of parliament as opponents, critically

---

[11] In Dutch: the *Staatssecretaris van Infrastructuur en Milieu*.

testing and probing the various parts of Dijksma's position. We focus on those fragments of the debate that play a role in the exchange of criticism and argument revolving around this position, leaving other fragments aside, such as those that deal with future policy measures for meeting the objectives of the Paris Agreement.

3. SHOULD WE SETTLE FOR A COMPROMISE?

In a conflict about a compromise, there is – prior to all critical questions that can be asked about specific features of the compromise – the question whether the problem that the compromise is supposed to solve concerns matters that are at all suitable for negotiation and compromise with competing parties. Should one not rather deal with the problem in some other way? For instance, shouldn't one keep trying to convince the competing parties of the superiority of one's own favoured solution? Or would it not be better to bring the case to court or to some other adjudicator? Might it not be wrong to cooperate with the competing parties? Briefly formulated: "Should we settle for compromise?" This critical question expresses what we shall call *general compromise criticism*. It does not target the specific compromise the proponent defends, but targets the fact that the proponent defends some compromise, regardless of the details of the deal at hand.

Opponents can be sceptical about, or even rebellious against, the choice for one or other compromise as the proper kind of response to the policy problem. The Republican Platform 2016, for example, doesn't so much reject the Paris Agreement, but the very idea of crafting a compromise under the guidance of the U.N.'s Framework Convention on Climate Change.

> Example 1. *Rejecting agendas*
>
> We reject the agendas of both the Kyoto Protocol and the Paris Agreement, which represent only the personal commitments of their signatories (…) We firmly believe environmental problems are best solved by giving incentives for human ingenuity and the development of new technologies, not through top-down, command-and-control regulations that stifle economic growth and cost thousands of jobs (The Republican Platform, 2016, p. 22).

There are various ways to respond to such criticism. First, it can be pointed out that there is no good alternative to compromise, for example because the competing parties need to change their policies yet cannot be convinced with persuasive argumentation to act in line with one's cravings. Thus, one may frame the choice for compromise as being instrumental to one's own interests. But one may also stress the moral or ideological reasons for making the best of various suboptimal options. Second, the proponent can support the choice for compromise by favouring compromise as an outcome that is a more democratic than a one-sided solution, that fosters community better than a less cooperative outcome, or that is more epistemically virtuous than an outcome that has not taken competing interests and opinions into account (Weinstock, 2013; van Laar & Krabbe, 2016).

In the debate in the Dutch parliament, hardly any doubts are being expressed about the desirability of collective action based on a global compromise. Mrs. Dik-Faber (Christian Union - *CU*), for example, explains in her contribution that "the urgency is enormous" for the reason that sea levels are rising and that postponing collective action brings great economic risks (Tweede Kamer, 2017, p. 1). However, Example 2 (*Fiddling*) suggests that Mr. Madlener, representing the one climate sceptical party, Party for Freedom (*PVV*), does not consider the situation suitable for any compromise at all.

> Example 2. *Fiddling*
>
> MP Madlener (Party for Freedom – *PVV*): 'Further it has come out that China already has been fiddling with figures about emissions. Of the, I suppose, 190 countries that joined in with this, a gigantic number are corrupt. Of course they love to cash those 100 billion dollars. How can it be that the Minister trusts all those countries? They are all bound to fiddle' (Tweede Kamer, 2017, p. 16).

Mrs. Dijksma in her answers emphasized that the deal is of historic importance by legally binding the ratifying countries to reduce temperature rise, and to compensate developing countries financially, thus making the needed progress towards a climate neutrality (p. 15). Her response to Madlener's skepticism was to underline the obligation of the signatories to report every 5 years about their progress.

It can be expected that considerations for rejecting any compromise can also be used for rejecting a specific compromise. This will be confirmed in the next section, where we turn to criticizing specific compromises.

4. SHOULD WE SETTLE FOR THIS COMPROMISE?

Prompted by a conflict about compromise, opponents can be expected to critically examine the issue whether they should endorse or ratify this specific negotiated compromise. We distinguish between more basic critical reactions regarding that issue and more evolved ones. We start by giving a general sketch of the dialectic that results from advancing and responding to more basic stock issues (critical reactions), and then examine successive, more detailed stock issues (critical reactions) and the kinds of subordinate discussion thus triggered. Thus, we stepwise arrive at a general scheme with which negotiated compromises can be defended. We will reconstruct the various fragments from the Dutch parliamentary debate as raising stock issues, dealing with e.g. with trust, moral acceptability, or legal admissibility, on this more detailed level.

When faced with a compromise that requires endorsement or ratification, the obvious issue to look into is whether the situation that results from endorsing (or ratifying) the negotiated compromise is preferable to the situation that results from not endorsing it. This basic criticism invites the proponent to advance an argument along the lines of the following pattern of reasoning, which we label *the first argument scheme for the Defence of a Negotiated Compromise*:

> (Standpoint) You should endorse (ratify) this compromise, because (Reason 1) by doing so we achieve X at the expense of Y, and (Reason 2) although we sacrifice Y, this arrangement of achieving X at the expense of Y is preferable to not accepting this compromise.

Further criticism of Reason 1 will be discussed in Section 4.1, below. When it comes to Reason 2 we expect opponents to have possibly two worries on their minds. First, they may feel that the status quo is not so bad after all, and possibly better than what results from accepting the deal, for instance because they may expect that a rejection of the deal will leave them with alternative opportunities foreclosed by a ratification of the compromise in hand. If the opponents advance such criticism, the proponent will be inclined to argue that the negotiated compromise at hand outcompetes the status quo, or, more precisely, outcompetes the situation that results from rejecting the deal. Secondly, opponents may feel that a better compromise could have been realized, still can be crafted, or at a future moment will become realizable, so that ratifying the deal now deprives them of a more optimal outcome enabled by reopening the negotiation process. If the opponents advance such criticism, the proponent will be inclined to argue that the negotiated compromise at hand outcompetes other feasible compromise outcomes. These basic criticisms, targeting Reason 2, invite the proponent to advance an argument along the lines of the following pattern of reasoning, which we label *the second argument scheme for the Defence of a Negotiated Compromise*:

> (Reason 2) Although we sacrifice Y, achieving X at the expense of Y is preferable to not accepting this compromise, because (Reason 2.1) an arrangement of achieving X at the expense of Y is preferable to a situation in which the issue is not settled by means of a compromise, (Reason 2.2) as well as to alternative arrangements that happen to have been, still are, or will become feasible outcomes of a (possibly: reopened) negotiation dialogue.

The argument scheme made up from the first and the second argument scheme for the defence of a negotiated compromise we label simply *the argument scheme for the Defence of a Negotiated Compromise*, with basic Reason 1 as its schematic *inventory premise*, basic Reason 2.1 as its schematic *progress premise*, and basic Reason 2.2 as its schematic *optimality premise*.

As arguments for the Defence of Negotiated Compromise are used to convince supporters, clients or allies to accept a deal, they need to be distinguished from *expediency arguments*, which form the kind of argument in which competing parties are to be convinced to accept an offer within a negotiation dialogue. Whereas the latter will try to exploit values and facts as competing parties have accepted them, the former will try to exploit values and facts as they have been accepted along one's own ranks. Both expedience arguments and arguments for the defence of negotiated compromise can be seen as specific kinds of "practical arguments" (cf. Fairclough & Fairclough, 2012).[12]

We will understand the stock issues raised in this parliamentary debate as more detailed critical probes into the acceptability of these three basic premises of a defence of a negotiated compromise.

---

[12] See Lewinski and Mohammed for an analysis of some opening speeches of the Paris conference from the stance of the scheme for practical argument proposed by Fairclough and Fairclough (Lewinski and Mohammed, 2017).

4.1 Examining the inventory premise

First, opponents can question the acceptability of the inventory premise by raising *issues regarding the gains*: How reliable is the proponent's estimation that, indeed, the promised gains X will be realized? Possibly, the proponent overlooks obstacles, relies on an overly optimistic estimation, or has even fallen prey to wishful thinking.

For example, opponents can raise *trust issues*, casting doubt on the reliability of a competing party and the probability that it will live up to a particular promise that is part of the agreed upon arrangement. In the opponents' view, then, this reduces the likelihood that all of the promised gains X will in the end be acquired. In response, the proponent may try and argue in support of the competing party's reliability, either by referring to its track record or by referring to its interest in complying with the agreed upon arrangement, or she may point out that trustworthiness is not that relevant an issue because of the existence of a sufficient measure of implementation guarantees as part of the arrangement. In Example 3 (*They don't give a damn*), it is plausible that the trust issue targets the details of the compromise in hand (in addition to targeting the very idea of settling for compromise).

> Example 3. *They don't give a damn*
>
> MP Madlener (Party for Freedom – PVV): 'How then do you feel about all the other signatories of this agreement, all those corrupt regimes, all those countries that love to take over our industry? Yes, they are joining in with this agreement, but in the end they will simply conclude: just give that employment up to us. The Netherlands are standing up again like a reverend wagging his finger, while our country gets ever poorer. In the Netherlands, too, there will no doubt always be some green oddball (...) whereas in China or wherever on earth people don't give a damn about this agreement and will simply resign from it when it suits them. We want to see this agreement going straight into the waste-paper basket' (Tweede Kamer, 2017, p. 6)
>
> Minister Dijksma (Labour Party - PvdA) points out that all signatories have committed themselves wittingly: '"Wittingly," I would say. We shall also see to it that they observe it. For the beauty of this agreement is namely that for the first time in history all those 195 countries must by a so-called "intended contribution", i.e. a national contribution, show how they are delivering. Next we are going to check again, every five years,

whether each country keeps to its promises. That will result in political pressure all over the world" (Tweede Kamer, 2017, p. 16).

Opponents can also have other issues regarding the gains, regardless of trust issues, and suspect that the positive consequences have been presented overly optimistically on other grounds. Mr. Madlener, in Example 4 (*Burning Canadian woods*), doesn't expect that all positive effects on climate will materialize.

Example 4. *Burning Canadian woods*

MP Madlener (Party for Freedom – PVV): 'A number of measures The Netherlands are about to take are, surprisingly, bad for the environment. Yes, you hear me well. Our government, for instance, supports together with a number of environmentalist groups the proposal to use biomass as supplementary fuel. As a consequence, billions' worth of wood will be cut. Canadian woods will be shipped to The Netherlands and they are going to be burned here in our coal-fired power stations. It is really true!' (Tweede Kamer, 2017, p. 6).

MP Vos (Labour Party – PvdA): "... it is also clear that [biofuels and adding of biomass] actually do have a positive effect on the environment. (…) The beauty of biomass is that one may create negative CO_2-emissions; for one can burn up biomass, in which CO_2 has been stored. You look puzzled, Mr, Madlener, but just listen to this. The CO_2 you may then store, in the North Sea for instance. Consequently, you are drawing CO_2 from the atmosphere. The problem of global warming, about which all present agree except the Party for Freedom, can in that way be solved" (Tweede Kamer, 2017, p. 9).

Second, opponents can question the inventory premise by raising *issues regarding the losses*. For instance, they may query the completeness of the list of sacrifices and other downsides that the deal brings with it, or the transparency of the way in which they have been presented: Are the bad consequences Y all the bad consequences that need to be taken into account? Does the proponent give a fair and frank presentation of the disadvantages, or did she provide a one-sided account by covering up or even neglecting possible problems or flaws of the deal? Raising such issues targets specifically the negative part of the inventory premise.

There are a number of special cases of querying the negative aspects of the deal. One is to point towards the interests and rights of particular stakeholders, such as the citizens in poor countries, and raise doubts about whether their interests have been taken into account to a sufficient measure, or whether their rights will be heeded. If we accept the deal, will we treat them with due respect? Opponents can put emphasis on these specific stakeholders' interests and rights, or on the moral, legal or political principles that have been in place to protect interests and rights like these. Alternatively, opponents can stress their own integrity that gets damaged by endorsing an arrangement that fails to pay due respect to stakeholders or that violates basic principles. Human rights, for one, are often presented as non-negotiable, so that any deal that puts these into jeopardy will be dismissed. If the emphasis is on stakeholders' interests and rights, we speak of opponents who raise *morality issues*, and if it is on one's own integrity we speak of *integrity issues*. Example 5 (*Those countries that are hardest hit*) provides an illustration of a morality issue. Another special group of downside issue regards alleged legal problems of the deal, such as when it violates national laws or international agreements or when its legal foundation falters: *legality issues*, as illustrated in Example 6 (*Legally binding*). Further, opponents can point out that the compromise at hand goes against the present current of popular opinion, so that there is reason to think that it lacks democratic support and legitimacy: *support issues*. Of course, this list of stock issues is not complete, and opponents may point at still other adverse consequences of the deal, as illustrated in Example 7 (*Silly Billies*).

Example 5. *Those countries that are hardest hit*

MP. Wassenberg (Party for the Animals – PvdD): 'I finish by expressing my great concern about the failure to provide adequate climate funding for those countries that are hardest hit, while having contributed least to global warming. They need to be supported as they suffer the disastrous consequences of droughts, floods, and the rising sea level' (Tweede Kamer, 2017, p. 3).

Minister Dijksma (Labour Party - *PvdA*): "I agree with those saying that we should all over the world do collectively more to combat poverty and to install climate policies. In fact, that has always been the position of the Dutch government. We did indeed bring this forward at numerous occasions" (Tweede Kamer, 2017, p. 17).

Example 6. *Legally binding*

MP Madlener (Party for Freedom – PVV): 'The Minister talks about a historical agreement. She also says that the agreement is legally binding. In article 28 of that agreement, however, I read the following. "At any time after three years from the date on which this agreement has entered into force for a party, that party may withdraw from this agreement by giving written notification." So one may withdraw from the agreement. How can I reconcile this with the Minister's story about "legally binding"?'

Minister Dijksma (Labour Party - PvdA): 'If one does not withdraw from it, it is in force'.

MP Madlener (Party for Freedom – PVV): 'Yes. Now Donald Trump did already say that as far as he is concerned this agreement will be brushed aside. He is going to renegotiate and it is to be tackled quite differently. Each party may of course quite easily appeal to article 28 and say: you may very well say that it is legally binding, but if we want, we withdraw from the whole agreement'.

Chair: 'Then, of course, first Trump should happen to be elected' (Tweede Kamer, 2017, p. 16).

Example 7. *Silly Billies*

MP Madlener (Party for Freedom – PVV): 'Big industrial nations such as China need to do much less. The result is a relocation of our manufacturing to other parts of the world. Our businesses will be simply strangled by competition' (Tweede Kamer, 2017, p. 5).

Minister Dijksma (Labour Party - PvdA): 'My last visit to China was in my capacity as Minister of Economic Affairs. You see that the citizens of Chinese cities constitute the upward pressure for reforms in the environmental policies. If you can walk on the street only with a mask because the air is no longer clean, there is a problem. I am grateful and appreciate that the Chinese government takes this very seriously and now in as many as seven regions experiments with an Emission Trading System. There are actually developments that give me positive feelings, though they are not all positive, I am also aware of that. There we must find the point of departure to prevent—Mr. Madlener is right in pointing that out — that we get a water bed effect: production that disappears on this side, but is resumed elsewhere in worse circumstances. Of course we'll watch out for that. I'll have more to say about this in my contribution. We are no silly Billies' (Tweede Kamer, 2017, p. 16).

4.2 Examining the progress premise

Opponents can raise doubts about the positive assessment of the compromise as compared to what plausibly results from not endorsing and ratifying the deal. The opponents may think that the "best alternative to a negotiated agreement" (Ury and Fisher, pp. 99-108) (or even a second- or third best such alternative) is to be preferred to implementation of the compromise, and that the chances of realizing such an alternative outcome are sufficiently great. These *issues regarding progress* come in two kinds. The proponent can be questioned about his portrayal of the no-deal scenario. Doesn't she provide an overly gloomy sketch, possibly even engage in fear mongering? Climate change skeptics or deniers can be characterized as challenging or rejecting, what they regard as, pessimistic and fear-mongering accounts of the consequences of refraining from adopting stricter emission and other "green" policies, and thereby challenging or denying the progress premise.

Alternatively, the opponents may inquire into whether the proponent provides an overly upbeat sketch of the compromise's gains, and possibly even engages in wishful or utopian thinking. In that case, the issue raised is that the scenario based on endorsement of the compromise is not all that good.

Here we must note that the stock issues discussed above in the context of examining the inventory premise may also underlie the opponents' critical assessment of how the negotiated compromise compares to the no deal. After all, the more you think the estimates of gains and losses must be revised so as to reduce the gains or to increase the losses, the less likely you would be convinced that the deal is to be preferred above the no deal option. Thus in examining the progress premise, opponents may raise some of the same questions as in the case of the inventory premise: They may think that the competing party's lack of trustworthiness would be wrecking the deal; or that endorsement amounts to selling one's soul; or that a lack of legal or popular support undermines the deal's legitimacy; or, more generally, that balancing the real gains against the real losses leads one to reject the deal. In Example 8 (*Saving the chance of saving*), Mr. Wassener gives his positive view on how to balance the pros and cons, whereas in example 9 (*100 billion dollars*), Mr. Madlener gives his negative view based on his climate change skepticism.

> Example 8. *Saving the chance of saving*
>
> MP. Wassenberg (Party for the Animals – PvdD): 'The Paris climate summit did not result in rock solid arrangements, but even so something was done. Admittedly, Paris did not save the planet, but perhaps Paris saved the chance to save the planet' (Tweede Kamer, 2017, p. 2.)

> Example 9. *100 billion dollars*
>
> MP Madlener (Party for Freedom – PVV): 'The Party for Freedom is against this climate agreement. We think it is a bad agreement for The Netherlands. This agreement is not going to provide a solution of the so-called climate problem, if there happens to be such a problem. First, it is an agreement that will relocate industries and therefore emissions to other countries, (…)' (Tweede Kamer, 2017, p. 5).
> 'One of the items of the climate agreement, of which the Party for Freedom and Democracy is such a staunch advocate, is the distribution of 100 billion dollars among corrupt countries' (Tweede Kamer, 2017, p. 12).

4.3 Examining the optimality premise

Finally, opponents can raise doubts about the positive assessment of the compromise as compared to more optimal and yet feasible outcomes of the negotiation process, also in light of the possibility of returning to the negotiation table and reopening the negotiation process to try to obtain a better deal. Possibly, it's still feasible to obtain a better result. These, we label *issues regarding optimality*.

It only makes sense to reject a compromise as suboptimal if there still is an opportunity for reopening the negotiation process. The opponents in a conflict about compromise may be in a position to use their influence and reject the current text of the compromise in an attempt to pressure the competing parties into returning to the negotiation table, if only the competing parties' desire for an agreement is sufficiently strong. A reason for the opponents to consider a better outcome feasible can be their belief that the negotiators did not adopt an good negotiation strategy, and failed to seize opportunities.[13] Alternatively, they may have reason to believe that, since the negotiations took place, the situation and the opportunities for bargaining have changed for the better so that a reopened negotiation offers new perspectives.

In Example 10 (*Future generations*) and 11 (*She walked her legs off*), two members of parliament convey the message that, indeed, the outcome is not optimal, yet without criticizing the Minister herself for a flawed performance[14] and without trying to send her back to the negotiation table, which in this specific case would have been pointless. The minister, in example 12 (*Sturm-und-Drang*) shares their assessment.

Example 10. *Future generations*

MP Van Tongeren (GreenLeft – GroenLinks): 'The interests of future generations in the climate issue are of course of great importance but not very conscientiously embedded." (Tweede Kamer, 2017, p. 1)

Example 11. *She walked her legs off*

Van Veldhoven (Democrates – D66): "What solution does the Minister envisage in order to achieve that goal of 1.5° centigrade? In Paris, she walked her legs off for this. It was a great sight; nothing but praise in that regard' (Tweede Kamer, 2017, p. 4).

Example 12. *Sturm-und-Drang*

Minister Dijksma (Labour Party - PvdA): "I understand that part of this House is in Sturm-und-Drang to do more and

---

[13] Elsewhere we explain that the exchange of offer and counteroffer in a negotiation dialogue can be reconstructed as an exchange of *expediency arguments*. The opponents' negative assessment of their negotiator's strategy implies the assessment that the negotiator has been convinced of an expediency argument in support of the final, winning offer, but incorrectly so.

[14] In addition to a critical exchange about the merits of the compromise in hand, members of parliament are expected to hold the minister accountable, and to evaluate her performance as a negotiator at the negotiation table. We refrain from dealing with the accountability issue in this paper, but acknowledge that some optimality issues prompted by the dispute about compromise also adds to the dispute about the accountability issue.

proceed beyond this. I understand that very well. However, I want also for once ask you to keep in mind all that we together — let us share the success — did achieve. That's a tremendous lot. We'll not sit back and watch, as I just clearly stated." (Tweede Kamer, 2017, p. 19)

The Minister's fiercest critic does not even regard the agreement as bringing progress, but in Example 13 (*Relocation*), his criticism focuses on the outcome as being suboptimal in a specific respect, suggesting that the negotiators failed to assign sufficient weight to a key consideration.

Example 13. *Relocation*

MP Madlener (Party for Freedom – PVV): 'The Netherlands are already very efficient in matters of consumption of energy. The country has also fewer opportunities to reduce its emissions in a cost-effective way. Big industrial nations such as China can do with a much lesser effort. It leads to a relocation of our manufacturing to other parts of the world' (Tweede Kamer, 2017, p. 5).

5. CONCLUSIONS

Above, we have dealt with the stock issues (critical reactions) that a conflict about a negotiated compromise gives rise to. We conclude this paper with an overview of the structure of attempts at justifying a negotiated compromise in light of these critical reactions, and we do so by means of a profile of dialogue that sketches the various reasonable moves for the proponent and the opponent in a typical conflict about a negotiated compromise.

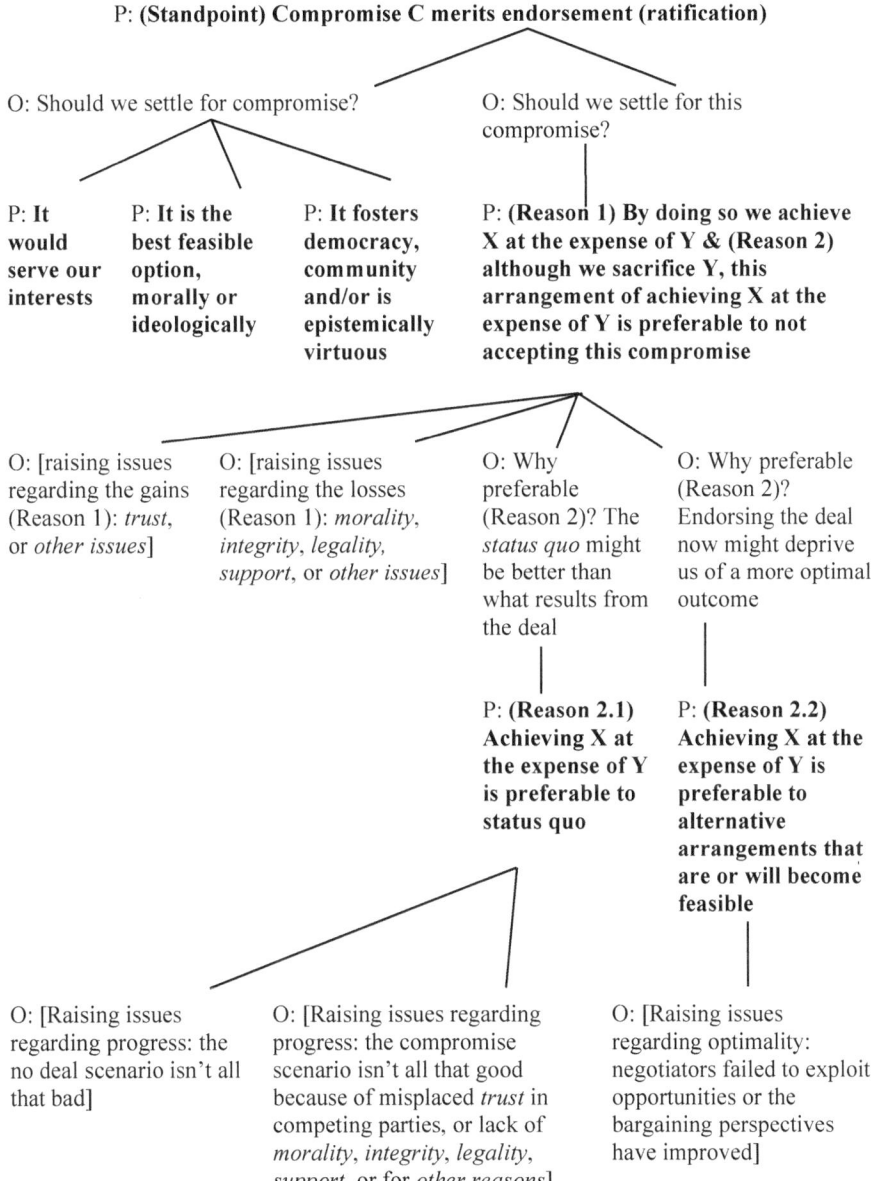

Figure 1 – A profile of dialogue that provides reasonable options for the proponent and the opponent(s) in a conflict about compromise

The justification of a negotiated compromise vis-à-vis those who are to endorse or ratify the compromise gives rise to a specific series of stock issues, and in our dialogue profile we showed how these trigger a myriad of ways in which discussion may reasonably develop. The Dutch parliamentary debate on the Paris Agreement contains many illustrations of criticisms focused at the inventory premise and the progress premise. In the debate, also criticism against the optimality premise as well as general compromise criticism has a role to play, albeit to a lesser extent, and we identified some fragments to illustrate also these types of criticism.

It looks as if proponents of a compromise need to keep many different frogs in a wheel-barrow. First of all, they have to get a process of negotiation going. For this their supporters as well as the competing parties must be persuaded to turn to negotiation and compromise. Second, in the course of the negotiation dialogue itself, the final compromise proposal must be convincing to each and all of the competing parties (signatories). Third, if the compromise requires the endorsement or ratification of the supporters or clients of the negotiators, there must be argumentation available that convinces these supporters or clients of the acceptability of the negotiated compromise. Elsewhere, we discussed arguments for turning to negotiation and arguments used in the negotiation dialogue (van Laar & Krabbe, 2016). Among the latter were *expediency arguments* by which a negotiator can try to convince competing parties to accept an offer based upon their preferences and ideas. In this paper, we have shown how a negotiated compromise may need to be justified towards one's supporters by arguments that are more likely to be based upon one's own preferences and ideas.

If needed, the proponents of a negotiated compromise can intensify the pressure on those to be convinced by increasing the gains of ratification by offering compensations, or by increasing the losses of rejection, for instance by political threats or by mobilizing public support in favour of ratification (which moves would initiate new kinds of negotiation). At various stages of this delicate process, there is room for manipulation, blackmail and threat, and it would require another paper to study how these relate to the argumentation that the process requires.

Here we end with a positive note. In the course of examining how a compromise can be justified in a conflict about a negotiated compromise, we have found that there are cases where the required burden of proof can plausibly be met. There can be good reasons for turning to negotiation and compromise, for accepting at the negotiation table a particular compromise proposal, and also for endorsing or ratifying a particular negotiated compromise arrangement. A compromise, though second-best, can be a most respectable outcome.

REFERENCES

Dimitrov, R. S. (2016). The Paris agreement on climate change: Behind closed doors. *Global Environmental Politics*, *16*(3), doi: 10.1162?GLEP_a_00361.
Fairclough, I., & Fairclough, N. (2012). *Political discourse analysis: A method for advanced students.* London: Routledge.
van Laar, J. A., & Krabbe, E. C. W. (2016). Splitting a difference of opinion. OSSA Conference Archive. 128. Retrieved from https://scholar.uwindsor.ca/ossaarchive/OSSA11/papersandcommentaries/128.
Lewinski, M., & Mohammed, D. (2017). Argumentation towards multilateral consensus on climate change: COP21 in Paris. Manuscript.
Republican Platform (2016). Retrieved from https://prod-cdn-static.gop.com/media/documents/DRAFT_12_FINAL[1]-ben_1468872234.pdf
UNFCCC (2017). *Climate: Get the big picture.* Retrieved, March 20, 2017 from http://bigpicture.unfccc.int/#content-the-paris-agreement.
Tweede Kamer [Dutch House of Commons] (2017). In: *Handelingen 2015 – 2016* [Proceedings, *2015 – 2016*], number 85, item 17, May 19 2016. Retrieved on March 20 2017 from: https://zoek.officielebekendmakingen.nl/h-tk-20152016-85-17.
Weinstock, D. (2013). On the possibility of principled moral compromise. *Critical Review of International Social and Political Philosophy*, *16*(4), 537-556.
Wendt, Fabian (2016). *Compromise, peace and public justification: Political morality beyond justice.* Switzerland: Palgrave Macmillan.

61

Argumentative Functions of Metaphors: How can Metaphors Trigger Resistance?

LOTTE VAN POPPEL
University of Amsterdam
l.vanpoppel@uva.nl

This paper investigates what functions metaphors can have in argumentation and how they may provoke resistance. The pragma-dialectical theory and Steen's (2008) 3D-model of metaphor are integrated to determine the possible uses of 'deliberate' metaphor in argumentation. Based on case studies, it is argued that deliberate metaphors can not only function as single analogy arguments, but also as complex argumentation consisting of both analogy and e.g. causal or symptomatic arguments, that each may counter resistance.

KEYWORDS: analogy, argumentation analysis, criticism, metaphor, pragma-dialectics, resistance

1. INTRODUCTION

Metaphors in language have the power to influence our thoughts or even our behaviour by presenting one concept in terms of another concept. An example is the use of the popular train journey metaphor in the 1990s debate on further European integration (Musolff, 2004, p. 30). The metaphor was employed in an argument supporting the standpoint that Britain should sign the Maastricht Treaty: they should sign, because otherwise *the European train was leaving the station without Britain.* In this metaphor, the target domain, the EU integration process, is compared to the source domain of a train journey in order to convince (British) politicians and the general public of the desirability of ratifying the Treaty.[1]

---

[1] The Maastricht Treaty, or the Treaty on European Union, formed the basis for further economic integration and the introduction of a single European

Although the use of such metaphors may be effective in reaching particular communicative goals, linguists have also pointed at the potential negative effects of metaphor use, e.g. their potential to force undesirable views on the addressee and to hide problematic presuppositions (Chilton, 1996; Lakoff, 1996; Musolff, 2004; Goatly, 2007). When a metaphor is somehow unfitting or undesirable, people may resist the metaphor, i.e. they may not accept it or even react critically to it.

The train journey metaphor appeared to be so attractive that both politicians and media supporting and opposing EU integration made use of it (Musolff, 2004, pp. 60-61), but it also provoked resistance. Former prime minister Margaret Thatcher called it misleading and opposed the conclusions drawn from it by saying that "If that train is going in the wrong direction it is better not to be on it at all" (*The Times*, 31 October 1992, cited in Musolff, 2004, p. 31).

Now that this train journey is indeed about to come to an end for Britain as current Prime Minister Theresa May started negotiations about withdrawing the UK from the European Union or 'Brexit', another metaphor has become popular recently. This metaphor is to refer to the Brexit, as a "divorce". Theresa May has resisted this metaphor as well. At an EU summit on 14 March 2017 she stated: "I prefer not to use the term of divorce from the European Union because very often when people get divorced they don't have a very good relationship afterwards".

Both May and Thatcher reacted critically to the metaphor used, but in different ways. Whereas May disagrees with comparing the Brexit to a divorce, Thatcher seems to accept some parallels between the domains. These reactions may be triggered by certain properties of the metaphors. This raises the question what these properties are. For both speakers and addressees it is important to know what kind of resistance one may encounter or may give oneself to prevent criticism or improve one's critical abilities. To be able to systematically distinguish between different kinds of resistance, this paper sets out to determine what properties of metaphor that are used in an argumentative exchange may provoke resistance. The current study is part of a larger project on such resistance to metaphor carried out at the University of Amsterdam.

---

currency, the euro. The treaty also established policies on other than economic areas, such as security, social issues and European citizenship. (https://www.ecb.europa.eu/explainers/tell-me-more/html/25_years_maastricht.en.html)

Since the intended meaning of metaphors needs to be inferred from underlying conceptual metaphor and not all propositions relevant for the discussion are explicit, the role metaphors have in an argumentative discussion, as (part of) a standpoint or as (part of) an argument, is not directly apparent. Therefore, a theoretical framework is needed that helps to describe the properties of metaphor and that enables an analysis of metaphor use in argumentative discourse. To this end, Steen's (2008, 2017) 3-dimensional model of metaphor is combined with van Eemeren and Grootendorst's (1982, 2004) pragma-dialectical theory of argumentation. I will first shortly describe both approaches. Then I will present some examples of metaphors in argumentation to show how they can function as single or complex argumentation based on different argument schemes. Finally, I will demonstrate how this integrated approach sheds more light on resistance to metaphor.

2. A 3D-MODEL OF METAPHOR

Since the work of Lakoff and Johnson (1980), metaphor is generally seen as a conceptual phenomenon that connects different abstract domains with more concrete ones. According to Lakoff and Johnson's Conceptual Metaphor Theory, our thinking is full of metaphor, for example when considering time in terms of location ('tomorrow' is 'further away'), relationships in terms of a journey and soccer in terms of war. The connections, or 'mappings' that are made between these domains also seep through to our language use and can be employed to clarify complicated matters or change someone's perspective.

To distinguish between the different dimensions of metaphor, Steen (2008, 2011) introduces a 3-dimensional approach to metaphor in which he distinguishes metaphor in thought, in language and in communication. In this approach, metaphors are not only seen as linguistic phenomena and not only as the underlying conceptual mappings, but also as utterances expressed in an interaction between language users with particular communicative goals.[2] According to Steen (2008), not all metaphors in language will purposefully invite language users to view the target domain in terms of the source domain. He therefore distinguishes between *deliberate* and *non-deliberate* metaphors. He describes this distinction as follows:

---

[2] These dimensions are connected to the way psycholinguists, such as van Dijk and Kintsch (1983), look at text comprehension.

> Deliberate metaphors are those metaphors that draw attention to their source domain as a separate detail for attention in working memory, whereas non-deliberate metaphors do not. [...] This means that these source domain details function as separate referents, as for 'summer's day' in Shakespeare's "Shall I compare thee to a summer's day". (Steen, 2017, p. 7).

Thus, the idea is that deliberate metaphors result in an image of the discourse, a situation model (van Dijk & Kintsch, 1983), which includes referents of both the target and the source domain. In non-deliberate metaphor this is not the case, because the intended meaning is inferred directly from the metaphor-related words without forming an image of the source domain (Steen, 2011, p. 102).

A deliberate metaphor is intended to use the source domain to change the perspective on the target domain. The train metaphor in the introduction is clearly an example of deliberate metaphor: this metaphor was used to convince people of the idea that Britain should sign the Maastricht Treaty by making the addressees look at the integration process in terms of train traveling. Thatcher in fact did this and opposed the conclusion on the basis of another mapping drawn from the train journey metaphor.[3]

Deliberate metaphors are thus meant to bring about rhetorical effects, whereas non-deliberate metaphors are not (Steen, 2008, p. 223), making them particularly relevant for argumentative discourse. This does not mean that non-deliberate metaphor could not play any role. To be able to clarify the role of deliberate and non-deliberate metaphor in argumentative discourse, I will introduce the pragma-dialectical theory, which provides a framework to study these phenomena.

3. THE PRAGMA-DIALECTICAL APPROACH TO ARGUMENTATION

The pragma-dialectical theory of argumentation (van Eemeren & Grootendorst, 1984, 2004) helps to determine the specific function metaphors can have in argumentative discourse. The theory provides a framework in which argumentation is regarded as a social and rational activity between multiple parties that are aimed (at least partly) at solving a difference of opinion by advancing (a constellation of) propositions in an exchange of moves and countermoves. This

---

[3] See Steen et al. (2010) on MIPVU, a procedure to identify metaphors in discourse.

framework provides rules for a critical discussion that help to evaluate the reasonableness of discussion moves.

The ideal model of a critical discussion describes the moves in a discussion which can contribute to the resolution of the difference of opinion, e.g. expressing a standpoint, expressing doubt, proposing and accepting starting points, advancing (counter)argumentation, accepting arguments/standpoints (van Eemeren, Grootendorst, Jackson & Jacobs, 1993). The moves, performed by different types of speech acts, are considered argumentative if they are relevant for resolving the dispute. Since metaphors can be found in all the major word classes (nouns, verbs, adjectives, etc.) and in various units of discourse (Goatly, 1997, pp. 82-83, pp. 109-110), there does not seem to be a restriction on the kind of argumentative moves they can play a role in. For a complete picture of metaphor in argumentative discourse, their underlying argumentation structure needs to be reconstructed. In pragma-dialectics, argumentative discourse is analysed by determining which elements in the discourse may be relevant for the resolution process (van Eemeren, Grootendorst, Jackson & Jacobs, 1993, pp. 60-62). The argumentation in the discussion is presented in an argumentation structure which expresses the arguments as separate propositions and shows their interrelation. The structure consists of all propositions to which the speaker can be held committed. Irrelevant elements are deleted from the discourse and implicit elements that are needed to make sense of the argumentation are added.

The simplest argumentation is single argumentation: a standpoint supported by a single argument which is connected to the standpoint by a connection premise, guaranteeing the step from argument to standpoint. Each single argument is based on a particular argument scheme which represents the inference rule on the basis of which the acceptability of the premise is transferred to the standpoint in a particular type of argumentation. In the pragma-dialectical typology of argument schemes, three main types are distinguished, namely causal, symptomatic and comparison argumentation, and several subtypes, such as pragmatic argumentation, argument from example, and argument by authority. More complex argumentation would consist of a combination of multiple of these single argumentations based on various or the same argument schemes (van Eemeren & Grootendorst, 1992).

4. METAPHORS IN ARGUMENTATION

Metaphors in argumentation are often considered as argumentation by analogy or comparison (cf. Hastings, 1962; Schellens, 1985; Perelman & Olbrechts-Tyteca, 1969; Garssen & Kienpointner, 2011). In analogy argumentation, two things, persons or situations are compared on the basis of some similarity or correspondence. This similarity can be that what is mentioned in the argument and what is mentioned in the standpoint share characteristics or that both should be treated in the same way. Indeed, when one would consider a metaphor as an analogy drawn between two domains, it can be used to express the connection premise in argumentation based on the argument scheme of analogy. It thus supports the justificatory force of the argument. In a similar vein, Santibañez (2010), using the Toulmin model, analyses the underlying conceptual metaphors as backing for a metaphorical statement that functions as warrant.

Due to the fact that metaphors in analogy arguments compare concepts from very distinct domains, they are generally seen as a specific type of analogy argumentation, namely figurative analogies (Garssen & Kienpointner, 2011). Garssen (2009) states that "figurative analogy does not involve comparison argumentation at all" (p. 137). He argues that the critical questions belonging to the scheme could not be answered satisfactorily because there would only be one similarity between the things compared. Based on the example Garssen uses to make this claim, I would argue that when using a metaphor there can be many similarities, even between concepts from distinct domains. The example is the much quoted line from Lincoln who supposedly said "I should not resign at this moment, because one should not swap horses while crossing a stream". The two situations compared share multiple characteristics: they both involve someone in charge, they both involve difficult circumstances, they both involve a route that needs to be taken, etc.[4] I would also question the assumption that if the analogy does not pass the testing procedure, there is no comparison argumentation involved. It could simply be seen as unreasonable argumentation.

---

[4] Also, in the same chapter, Garssen (2009) differentiates between 'regular' comparison argumentation and argumentation based on a principle of consistency. This latter category differs from the first in that it "does not involve an extrapolation of characteristics" but "whether the two elements (persons, groups etc.) really belong to the same category and whether this category is really relevant to the claim made in the standpoint" (p.136). The Lincoln example falls in the latter category, so there need not be multiple similarities between the situations compared.

In this article, I will thus consider metaphor as possibly figurative analogy. Yet, not necessarily. Wagemans (2016) shows that metaphors can occur as (part of) a standpoint and as (part of) an argument, in different types of propositions. He therefore argues that metaphors in argumentation do not necessarily constitute argumentation by analogy. The relation between (metaphorical) standpoint and (metaphorical) argument may be a relation of sign or another kind of relation. The type of argumentation metaphors are used in is relevant for determining what kind of resistance they can evoke.

5. RESISTANCE TO METAPHOR IN ARGUMENTATION

According to van Eemeren and Grootendorst (1984, p. 86), criticism against an argument can concern either the acceptability of the propositional content of the argument or the justificatory (or refutatory) force of the argument (see also Snoeck Henkemans, 1997, p. 86). Criticism with respect to the propositional content of the argument represents doubt concerning the content of the argument. Criticism regarding the justificatory force of the argumentation concerns the sufficiency of the argument to justify the standpoint.[5] If metaphors are considered part of the argumentation and were to evoke criticism, this would thus entail criticism with respect to the material premise, the connection premise or both.

Criticism towards the material premise is the same for each type of argument, as it concerns the acceptability of the content of this premise. Possible criticism towards the connection premise is reflected in the critical questions associated with each argument scheme (van Eemeren & Grootendorst, 1984, 1992; Garssen, 1997). Depending on the type of argument the metaphor is used in, resistance to metaphor in argumentation can therefore be manifested in various ways. A clear overview of possible resistance to metaphor in argumentation thus demands a reconstruction of metaphor in terms of argumentation structure and argument scheme.

What makes it so difficult to determine the specific function of metaphor in argumentative discourse is that (1) the intended meaning of metaphors may be hard to infer (so how should argumentation with a metaphor be formulated in the argumentation structure?); and (2) that metaphors involve all kinds of implicit propositions/ presuppositions

---

[5] According to Snoeck Henkemans, criticism concerning the justificatory force includes doubt concerning the relevance of the provided argument (1997, p. 86).

(So how do you add the relevant information to complete the argumentation structure, providing an accurate overview of the commitments of the discussants?). In the following, I will show how these difficulties affect the analysis of metaphors in argumentation and how they may be dealt with.

First, I will discuss a simple, constructed example of an argument that is based on a metaphor in the implicit connection premise. The metaphor is used to link the source domain of cars to the target domain of relationships and thus invites an analogy between what is said in the standpoint about the target domain and what is said in the material premise about the source domain:

(1) Every relationship will end, because every car will break down.

The standpoint 'every relationship will end' is supported with the argument 'every car will break down'. The connection premise can be reconstructed as 'relationships are like cars', constituting an analogy between the two domains. This argumentation can be reconstructed as follows[6]:

1	Every relationship will end.	TARGET DOMAIN
1.1	Every car will break down.	SOURCE DOMAIN
(1.1')	(Relationships are like cars.)	METAPHOR

The argumentation in (1) is based on the argument scheme of analogy and thus faces criticism regarding the acceptability of the material premise (Will every car indeed break down?) and the critical questions pertaining to this scheme (e.g. Are relationships and cars indeed comparable on this point?).

When the metaphor is used in another type of argumentation, different critical questions apply. In the following example, the argumentation is based on the symptomatic argument scheme. According to van Eemeren and Grootendorst, in symptomatic or sign argumentation "the argumentation is presented as if it is an expression, a phenomenon, a sign or some other kind of symptom of what is stated in the standpoint" (1992, p. 97). In this example, the metaphor is expressed in the material premise:

---

[6] In principle, the same proposition used here in the connection premise can also be used in the material premise in another argumentation, but this is not as natural as the argument in example (1): 'Every relationship will end, because relationships are like cars and being like a car is a sign for going to end'.

(2) Relationships need maintenance, because relationships are like cars.

This argumentation can be reconstructed as follows:

1.   Relationships need maintenance TARGET DOMAIN
1.1   Relationships are like cars. METAPHOR
(1.1')   (Being a car characteristically goes together with needing maintenance.) SOURCE DOMAIN

Criticism towards this argumentation may address the material premise (Are relationships like cars in a relevant way?) and the connection premise, in the form of the critical questions belonging to symptomatic argumentation (Is needing maintenance indeed typical of cars? Is needing maintenance not also typical of something else? Are there any other characteristics that relationships need to have in order to ascribe to them a similarity with cars?) (Van Eemeren, Snoeck Henkemans & Houtlosser, 2007, p. 155).

Examples (1) and (2) show that metaphors can be used in analogy argumentation but also in other types, such as symptomatic argumentation. However, these examples are quite straightforward: metaphors are not always used as directly as in examples (1) and (2), which makes it harder to determine which properties may evoke resistance. Consider the following example:

(3) Susan will break up with John soon, because they have had so many bumps in the road (and when there are too many bumps, every car will break down).

In this example, the standpoint that Susan will break up with John soon is supported with the argument that they have had so many bumps in the road. This argument is connected to the standpoint with the premise that when there are too many bumps in the road, every car will break down. In this argumentation, a causal relation is drawn between elements in the source domain, namely between bumps in the road and cars breaking down. On the basis of this relation in the source domain, a claim is made about elements in the target domain, namely that Susan and John will break up soon. The addressee is invited to consider the target domain in terms of the source domain, but without a premise stating that relationships are like cars, as was the case in the previous examples.[7]

In example (3), we are thus confronted with the problem that only some elements from the two domains are expressed in the premises and the link between the two domains is merely presupposed. An added difficulty here is that the metaphorical statements in (3) express specific relations between the elements *within* the domain. This means that the argumentation at the language level expresses a causal or symptomatic or even analogy relation between elements in the source domain, while simultaneously presupposing an analogy relation at a lower level.

What is striking here is that someone who would criticize this argumentation could have doubts regarding three aspects of the argumentation. The critic could question the material premise in (3): did Susan and John indeed have so many bumps in the road, i.e. so many problems? Next to that, the critic could question the connection premise in two respects: namely, with respect to the causal relation that is drawn between bumps and cars breaking down (Will all cars break down because of bumps?), and also regarding the parallel that is drawn between the two domains of cars and relationships (Does what counts for cars also count for relationships?). The argumentation can be disentangled to make visible all relevant underlying presuppositions to which the speaker in (3) can be held committed:

> 1 Susan will break up with John soon. TARGET DOMAIN
> 1.1 Susan and John have had so many problems.
> TARGET DOMAIN
> (1.1') (Having problems in a relationship leads to breakups)
> TARGET DOMAIN
> (1.1').1 When there are too many bumps, every car will break down. SOURCE DOMAIN
> ((1.1').1') (Problems in relationships are like bumps to cars.)
> METAPHOR

The structure above shows that the metaphor here is a metaphor in communication, used to draw an analogy between problems in relationships and bumps to cars. This analogy is used as a justification for a causal argument (1.1') about problems leading to breakups, which

---

[7] One could even imagine a situation in which even the standpoint only refers to the source domain, if the context would enable the addressee to include the target domain in the situation model. For example when one would say, in a discussion about Susan and John: 'that car is going to break down'.

is used to justify the standpoint about Susan ending her relationship with John.

The complexity of such metaphor use in argumentation is partly related to the three dimensions of metaphor and the different types of metaphor distinguished by Steen (2008, 2011). A metaphor in language, such as saying 'divorce' in a discussion about Brexit, is not necessarily deliberate and thus not used *as* metaphor per se. If it is indeed intended to let the addressee make a connection between two domains, and to infer a statement about the target domain, it can be seen as deliberate and be reconstructed as argumentation. I would argue that when the metaphor is used directly, i.e., that the proposition(s) in the argumentation refer to the conceptual domains that are linked in the metaphor, as in (1) ('relationships are like cars'), the metaphor is used deliberately and can be reconstructed as single argumentation. But also when the domains are referred to in the propositions but the cross-domain mapping is not explicitly expressed, as in (3), the metaphor can be deliberate and can be used in the argumentation. When the deliberate metaphor is used indirectly, i.e. it refers to elements from the conceptual domains, the argumentation contains in fact several implicit propositions that are relevant and need to be made explicit, resulting in a complex argumentation structure. So a metaphorical utterance in argumentative discourse can express both analogy argumentation and another type of argumentation at the same time. This results in a complex structure in which the analogy guarantees the step from one (material or connection) premise (about the source domain) to another premise (about the target domain). The place of the metaphor (in material or connection premise) and the type of is thus relevant for determining whether the metaphor fulfils an argumentative function or not and for knowing what kind of criticism the metaphor can evoke.

6. ANALYSIS OF RESISTANCE TO TRAIN METAPHOR

Now having another look at the example of Thatcher referred to in the introduction, we can show more precisely what properties of the used metaphors were criticized. In the example in which Thatcher criticized the train metaphor, the standpoint that the UK should sign the Maastricht Treaty was defended with the argument that if they would not, "the European train was leaving the station without Britain". According to Musolff:

> Thatcher wants to deny the conclusiveness of the TRAIN metaphor [...] by spelling out the   conditions   under

which the presupposed analogy 'DESIRABILITY OF BEING ON THE TRAIN BEFORE IT LEAVES THE STATION equals DESIRABILITY OF JOINING THE EU INTEGRATION PROCESS IN TIME' does not hold – namely, when the train is going in the wrong direction. (Musolff, 2004, p. 38).

To show more clearly what aspects of the metaphor use can be criticized, a reconstruction of the argumentation is needed. This argumentation contains an indirect metaphor in the material premise and can be reconstructed as follows:

1	The UK should sign the Maastricht Treaty.
(1.1a)	(If the UK does not sign the Treaty, the UK will not be part of further EU integration.)
(1.1b)	(Not being part of further EU integration would be undesirable.)
(1.1a-b')	(If something has undesirable consequences, one should not do it.)
(1.1b).1	Not being on the train when it leaves the station (is undesirable).
((1.1b).1')	(being part of EU integration process is like taking a train.)

Several propositions need to be added to make clear how the proposition that 'if Britain does not sign the Treaty, the European train was leaving the station without Britain' is connected to the standpoint. The connection is based on a pragmatic argument which says that acts should or should not be done if they have desirable or undesirable consequences. The metaphorical utterance in this argumentation expresses the (undesirable) consequence of not signing the Treaty while simultaneously making the analogy between being part of the EU integration process and taking a train.

Using this argumentation structure, Thatcher's response can clearly be linked to one of the propositions that is reconstructed from the metaphor, namely (1.1b) that 'not being on the train when it leaves the station is undesirable'. Thatcher argues that "If that train is going in the wrong direction it is better not to be on it at all" (*The Times*, 31 October 1992). She thereby denies that not being on the train when it leaves the station is undesirable and even gives an argument for that claim, namely that it is undesirable to be on the train when it goes in the wrong direction. She thereby implicitly argues that the EU integration process goes in the wrong direction and it is better for Britain not to be part of it. Her criticism is thus directed at the material premise than can be inferred from the metaphor and this criticism is connected to one of the critical questions pertaining to pragmatic argumentation, namely the question whether the proposed or discouraged act is indeed desirable or undesirable, respectively (Garssen, 1997, p. 22; van Eemeren, Houtlosser & Snoeck Henkemans, 2007, p. 166).

7. CONCLUSION

In this paper I proposed to combine Steen's 3-dimensional model of metaphor and van Eemeren and Grootendorst's pragma-dialectical theory to determine what properties of deliberate metaphors that are used in an argumentative exchange may provoke resistance. This approach enables me to incorporate the three dimensions of metaphor, namely language, thought and communication, in the analysis. In the analysis of argumentative discourse, we can thus do justice to the fact that not all metaphors in language have a communicative, let alone an argumentative purpose. In addition, it shows that also indirect metaphors can invite a cross-domain mapping that supports statements about the target domain.

I have argued that metaphorical expressions can in principle be employed in the material and the connection premise and that they can express different types of relations between the elements within a domain, such as a causal or a pragmatic relation. When the metaphor is indirect, the argumentation can be reconstructed as complex argumentation, with subordinative arguments, to make explicit the underlying mappings that justify the steps from main argument to standpoint. By reconstructing the underlying structure in this way, it can be better explained what parts of metaphors can be criticized and how criticism relates to a particular use of metaphor.

ACKNOWLEDGEMENTS: This work is part of the research programme Resistance to metaphor with project number 360-80-060, which is financed by the Netherlands Organisation for Scientific Research (NWO). I thank all members of the ARGA group and Resistance to metaphor research group for our useful discussions.

REFERENCES

Chilton, P. A. (1996) *Security metaphors, cold war discourse from containment to common house.* New York: Peter Lang.
Dijk, T. A. van, & Kintsch, W. (1983). *Strategies of discourse comprehension.* Orlando: Academic Press.
Eemeren, F. H. van (2010). *Strategic maneuvering in argumentative discourse: Extending the pragma-dialectical theory of argumentation.* Amsterdam: John Benjamins.
Eemeren, F.H. van, & Grootendorst, R. (1984). *Speech acts in argumentative discussions: A theoretical model for the analysis of discussions directed towards solving conflicts of opinion.* Berlin / Dordrecht: De Gruyter / Foris.
Eemeren, F. H. van, & Grootendorst, R. (1992). *Argumentation, communication, and fallacies: A pragma-dialectical perspective.* Hillsdale, NJ: Lawrence Erlbaum Associates, Inc.
Eemeren, F. H. van, & Grootendorst, R. (2004). *A systematic theory of argumentation: The pragma-dialectical approach.* Cambridge: Cambridge UP.
Eemeren, F.H. van, Grootendorst, R., Jackson, S., & Jacobs, S. (1993). *Reconstructing argumentative discourse.* Tuscaloosa-London: The University of Alabama Press.
Eemeren, F. H. van, & Houtlosser, P. (2002). Strategic manoeuvring: Maintaining a delicate balance. In F. H. van Eemeren & P. Houtlosser (Eds.). *Dialectic and rhetoric: The warp and woof of argumentation analysis* (pp. 131-160). Dordrecht: Kluwer Academic Publishers.
Eemeren, F. H. van, & Houtlosser, P. (2005). Theoretical construction and argumentative reality: An analytic model of critical discussion and conventionalised types of argumentative activity. In D. Hitchcock (Ed.), *The uses of argument: Proceedings of a conference at McMaster University* (pp. 75-84). Hamilton, Ontario: Ontario Society for the Study of Argumentation.
Eemeren, F. H. van, & Houtlosser, P. (2006). Strategic maneuvering: A synthetic recapitulation. *Argumentation, 20*(4), 381-392.
Eemeren, F. H. van, Houtlosser, P. & Snoeck Henkemans, A. F. (2007). *Argumentative indicators in discourse: A pragma-dialectical study.* Dordrecht: Springer.
Garssen, B. (2009). Comparing the incomparable: Figurative analogies in a dialectical testing procedure. In F. H. van Eemeren & B. Garssen (Eds.), *Pondering on problems of argumentation: Twenty essays on theoretical issues*, (pp 133-140). New York: Springer.
Garssen, B. & Kienpointner, M. (2011). Figurative analogy in political argumentation. In E. Feteris, B. Garssen & F. Snoeck Henkemans (Eds.), *Keeping in touch with Pragma-Dialectics* (pp. 39-58). Amsterdam: Benjamins.

Goatly, A. (2007). *Washing the brain: Metaphor and hidden ideology*. Amsterdam: John Benjamins.
Hastings, A. C. (1962). *A reformulation of the modes of reasoning in argumentation* (unpublished doctoral dissertation). Evinston, IL: Northwestern University.
Lakoff, G. (1996). *Moral politics: How liberals and conservatives think*. Chicago: The University of Chicago Press.
Lakoff, G., & Johnson, M. (1980). *Metaphors we live by*. Chicago: University of Chicago Press.
Musolff, A. (2004). *Metaphor and political discourse. Analogical reasoning in debates about Europe.* London: Palgrave Macmillan.
Perelman, C., & Olbrechts-Tyteca, L. (1969). *The New Rhetoric: A treatise on argumentation.* (Translation of *La Nouvelle Rhétorique. Traité de l'argumentation.* Paris: Presses Universitaires de France, 1958). Notre Dame / London: University of Notre Dame Press.
Santibáñez, C. (2010). Metaphors and argumentation: The case of Chilean parliamentarian media participation. *Journal of Pragmatics*, *42*(4), 973-989.
Schellens, P. J. (1985). *Redelijke argumenten. Een onderzoek naar normen voor kritische lezers.* Dordrecht: Foris.
Steen, G. J. (2008). The paradox of metaphor: Why we need a three-dimensional model of metaphor. *Metaphor and Symbol*, *23*(4), 213-241. DOI: 10.1080/10926480802426753
Steen, G. J. (2011). From three dimensions to five steps: The value of deliberate metaphor. *Metaphorik.de*, *21*, 83–110.
Steen, G. J. (2017). Deliberate metaphor theory: Basic assumptions, main tenets, urgent issues. *Intercultural Pragmatics*, *14*(1), 1-24. doi: 10.1515/ip-2017-0001
Steen, G. J., Dorst, A. G., Herrmann, J. B., Kaal, A., Krennmayr, T. & Pasma, T. (2010). *A method for linguistic metaphor identification: From MIP to MIPVU.* Amsterdam: John Benjamins.
Wagemans, J. H. M. (2016). Analyzing metaphor in argumentative discourse. *Rivista Italiana di Filosofia del Linguaggio*, *10*(2), 79-94.

62

Argument Structures and Frame Theory as Tools of Linguistic Discourse Analysis

SIMON VARGA
Johannes Gutenberg-Universität Mainz, Germany
Université de Bourgogne Franche-Comté, France
varga@uni-mainz.de

> Frame semantics has been a popular field of linguistic research for almost five decades. Yet, so far only little attention has been paid to the frame structures underlying argumentation in discourse. Given their quintessential role in our making sense of and dealing with the world that surrounds us, however, the importance of frames for argumentation is obvious, providing an ideal starting point for working towards a wider integration of both disciplines.

> KEYWORDS: frame semantics, cognitive frames, frame relations, Barsalou, Fillmore, conceptual metaphor, argumentation analysis, discourse analysis, discourse linguistics

1. INTRODUCTION

Since its advent in the mid-1970s, the term *frame* is used in a variety of different disciplines. Originally developed, more or less simultaneously, in artificial intelligence (Minsky, 1975), linguistics (Fillmore, 1975), and sociology (Goffman, 1974), the frame concept has long since established itself in fields as diverse as ethnology, economics, media analysis and argumentation studies. As is so often the case, however, researchers in these fields, while using the same term, are sometimes really talking of quite different things. And although there have been close and mutually beneficial relations between some of these fields—especially between linguistics and the cognitive sciences—the majority of current frame-based approaches seem not overly concerned with questions of interdisciplinarity. However, as suggested by the present article, there is much to gain from bringing some of these approaches closer to the linguistic and cognitive origins of frame theory.

At first glance, argumentation studies already seem to have come a long way in this regard. Recent frame-based approaches, such as Greco Morasso, 2012, and Fairclough & Mădroane, 2016, explicitly refer to the linguistic origins of frame theory in general and to the groundbreaking works of Charles Fillmore in particular. However, as the brief overview in section 2 reveals, they only make limited use of the possibilities offered by frame theory.

This article outlines a more integrated approach allowing for the analysis of argumentative phenomena within the framework of linguistic frame theory (as outlined by Ziem, 2008). This approach is specifically designed to suit the needs of discourse analysts.[1] Lexical analysis and argumentation analysis are two of the major approaches commonly used in this field (metaphor analysis being the third and last). Since the mid 1990s, frame semantics has become a widely used approach to lexical analysis. However, except for some isolated attempts to combine lexical frame analysis and argumentation analysis, both categories have so far remained mostly independent from one another.

2. CURRENT FRAME-BASED APPROACHES IN ARGUMENTATION STUDIES: FRAMES VS. FRAMING

In argumentation studies, the notions of *frame* and *framing* are often used interchangeably to express the idea that semantic choices, e.g. in news media discourse, are used to present the events described from a certain perspective (see Fairclough & Mădroane, 2016, p. 5). This idea of perspectivation is central to current frame-based approaches in argumentation studies, while other aspects of frame theory are still largely ignored. For instance, van Eemeren defines framing as a process that "always involves an interpretation of reality that puts the facts or events referred to in a certain perspective" (van Eemeren, 2010, p. 125). This is achieved by making certain aspects of the events described stand out so as to give them more weight. Framing, in this sense,

> essentially involves selection and salience. To frame is to select some aspects of a perceived reality and make them more salient in a communicating text, in such a way as to

---

[1] Following the definition given by Busse & Teubert in their classic 1994 paper, *discourse* is here understood as a virtual text corpus built around a specific subject (see Busse & Teubert, 1994, p. 16).

> promote a particular problem definition, causal interpretation, moral evaluation, and/or treatment recommendation for the item described. (Entman, 1993, p.52)

This interpretation of frames as representations of particular perspectives on things, states and events has been part of frame theory since the very beginning. Fillmore's frame semantics, for instance, "sees the set of interpretive frames provided by a language as offering alternative 'ways of seeing things'" (Fillmore, 1985, p. 230). And Marvin Minsky, the most influential founder of cognitive frame theory, uses the concept of perspective in a very literal sense when he states that

> the different frames of a system describe [a] scene from different viewpoints, and the transformations between one frame and another represent the effects of moving from place to place. (Minsky, 1975, p. 212)

While developing his frame concept mainly with regard to visual perception, Minsky explicitly extends it to other modalities as well, albeit in less detail. In nonvisual kinds of frames, different frames can represent, inter alia, cause-effect relations as well as changes in metaphorical viewpoint (see ibid.).

Cognitive frame theories such as developed by Minsky, 1975, or Barsalou, 1992, have found little reception in argumentation studies so far. The situation is somewhat different with regard to the purely linguistic frame concept advocated by Charles Fillmore's frame semantics, which is regularly cited in frame-based approaches to argumentation analysis. In contrast to the broad use of the term frame in cognitive science, Fillmore's frame semantics is primarily linguistically-based and has its origins in his early work on case grammar from the 1960s and 1970s. Fillmore's FrameNet project reflects its founder's early interest in case grammar, as most of the frames that are referenced are derived from the syntactic structure of verbs (e.g. Fillmore's famous COMMERCIAL TRANSACTION frame).[2]

One such approach that draws heavily on Fillmorian frame semantics is the concept of contextual frame introduced by Greco Morasso, 2012, who states that

---

[2] Fillmore nevertheless frequently stressed the importance of underlying conceptual structure, even though it was not his main interest (see Barsalou, 1992, p. 28).

putting a certain event into one or the other contextual box (frame) significantly changes its interpretation, even when the reconstruction of the events is seemingly neutral. (Greco Morasso, 2012, p. 198)

Frames, in this sense, represent the background or context against which news items are presented by the media. In her case study on the media coverage of Italian intelligence officer Nicola Calipari's death in Iraq, Greco Morasso identifies three mayor contextual frames that conceptualize the events in different ways by focussing on: "a) Calipari's life in the Italian intelligence services and his missions; b) the war in Iraq; c) the American way of dealing with (presumed) enemies" (ibid., p. 203). Each of these contextual frames makes specific details more salient than others, e.g. the fact that death is one possible "side effect" of an intelligence officer's job and that the Italian government is thus not to blame for what happened (contextual frame a).

Fairclough & Mădroane, 2016, give a very similar definition that describes framing "as a process of offering an audience salient and potentially overriding premises that they are expected to use in deliberation leading to decision and action" (Fairclough & Mădroane, 2016, p. 1). "To frame an issue is to offer the audience a salient and thus potentially overriding premise in a deliberative process that can ground decision and action" (ibid., p. 6). In their analysis of Romanian newspaper articles on the Roşia Montană case, Fairclough & Mădroane find, for example, that the gold mining project is repeatedly conceptualized as ROBBERY, a metaphor that supports only one possible conclusion, namely that it must be stopped.

Frames, in the sense outlined above, are not understood as structures of knowledge representation but rather as a means of presenting knowledge in a certain way in order to impose one's view on others.[3] In this respect they are closer to Lakoff & Johnson's theory of conceptual metaphor (Lakoff & Johnson, 1980) than to the frame concept currently used in discourse linguistics.[4] While frame theory clearly is an effective tool for analysing the way in which information is

---

[3] Interestingly, Fairclough & Mădroane are very well aware that the interpretation of frame theory currently prevalent in both political communication and media studies (and, one might add, in argumentation studies) "seems to be underlain primarily by a notion of the framing process, rather than of 'frames' as systems of inter-related concepts" (ibid., p. 5).

[4] George Lakoff was also the first to introduce the notion of *framing* for taking different views on one and the same situation (see Fillmore & Andos, 2010, p. 163).

presented (i.e. metaphorically conceptualized) in discourse, this is neither the only property nor the only possible application of frames. Indeed, there is much to gain from broadening the scope of frame theory in argumentation analysis. This is especially true with regard to the possibilities offered by frame theory to represent semantic and argumentative structures within the same theoretical framework.

3. TOWARDS A WIDER DEFINITION OF FRAMES IN ARGUMENTATION ANALYSIS

An integrated approach allowing for the analysis of argumentation structures in terms of frame theory requires, as a first step, a more comprehensive definition of frames than those usually given in current frame-based approaches to argumentation analysis. More specifically, what is needed is a definition that sees frames not only as a means of presenting knowledge in specific ways but one that sees them, first and foremost, as a theory of knowledge representation. Such an understanding of frames is at the heart of all major cognitive approaches to frame theory, such as Rumelhart, 1980,[5] who states that:

> A schema theory is basically a theory about knowledge. It is a theory about tow knowledge is represented and about bow that representation facilitates he use of the knowledge in particular ways. According to schema theories, all knowledge is packaged into units. These units are the schemata. Embedded in these packets of knowledge is, in addition to the knowledge itself, information about how this knowledge is to be used.
> A schema, then, is a data structure for representing the generic concepts stored in memory. There are schemata representing our knowledge about all concepts: those underlying objects, situations, events, sequences of events, actions and sequences of actions. A schema contains, as part of its specification, the network of interrelations that is believed to normally hold among the constituents of the concept in question. (Rumelhart, 1980, p. 34)

Frames structure our knowledge at all levels of abstraction, from the very basic to the highly abstract. Evoked frames constitute the backdrop against which we understand spoken or written language, and provide

---

[5] The term *schema* used by Rumelhart is synonymous with the term *frame* as it is used here.

us with rich conceptual knowledge about the objects, situations and events we experience. As such, they are central to our ability to interact with the world that surrounds us in a meaningful way. In accordance with Barsalou's definition of frames as "dynamic relational structures whose form is flexible and context dependent" (Barsalou, 1992, p. 21), frames are here understood as networks of information that can be dynamically rearranged in order to help us satisfy our needs and goals in any given situation.

It is further assumed that the structure of frames comprises four basic elements:

i. *Slots*, that is conceptual blanks that can be filled by specific instances or data. Slots "serve to represent the questions most likely to arise in a situation" (Minsky, 1975, p. 246).[6]
ii. *Fillers*, i.e. the values that are assigned to the slots of a frame in a given situation. Frames being recursive, the fillers instantiated in their slots are themselves frames (see Barsalou, 1992).
iii. *Default values*, that is "initial guesses" for slots whose fillers have we have not yet observed (see Rumelhart, 1980, p. 36).
iv. Knowledge about the relations between elements 1) to 3).[7]

All of these basic elements of frames are derived from experience and present strong prototypicality effects.

In accordance with the definitions of Barsalou and Rumelhart, the properties of frames can be resumed as follows:

a) Frames are structures that represent all of our conceptual knowledge in long-term memory;
b) are derived from our interaction with our environment;

---

[6] Fillmore, for instance, states that "if I tell you that I bought a new pair of shoes, you do not know where I bought them or how much they cost, but you know, by virtue of the frame I have introduced into our discourse, that there have got to be answers to those questions" (Fillmore, 1976, p. 29).

[7] Frame relations being of vital importance to the approach outlined here, they are treated as a separate category. Typically, current frame-based approaches to discourse analysis only name 1) to 3) as basic elements. Frame theorists such as Rumelhart and Barsalou, however, give considerable importance not only to elements 1) to 3) but also to the ways in which they are systematically interrelated.

c) enable us to interact with our environment in meaningful ways;
d) contain knowledge about prototypical relations between the constituents of our concepts; and
e) are dynamic structures that allow us to creatively rearrange their constituent parts to suit our current needs.

4. BRIDGING THE GAP BETWEEN FRAME ANALYSIS AND ARGUMENTATION ANALYSIS

Besides a broader definition of the frame concept itself, an integrated approach allowing for the analysis of argumentation structures in terms of frame theory requires answers to two central questions:

a) What are the relations between cognitive frames and argumentation?
b) How can these relations be analysed with the tools offered by linguistic frame analysis?

4.1 On the relation between frames, reasoning and argumentation

Although argumentation is not a widely discussed topic in frame research, the basic premises of frame theory allow for a relatively straightforward answer to the first question: Given our understanding of frames as "building blocks of cognition [...] upon which all information processing depends" (Rumelhart, 1980, p. 33), cognitive frame theory should be able to account for argumentative phenomena as well; otherwise its most basic assumption—i.e. that frames are central to *all* human cognitive processes—would be fallacious.

From the beginning, frame research in AI and cognitive science has typically focused on seemingly less complex phenomena. Typical questions that have been addressed in this field are: How is the human conceptual system structured? How is it derived from our perception of and interaction with our environment? How do we understand spoken or written language? While argumentation itself has so far received only little attention from frame theorists, a closely related phenomenon—reasoning—has gained broader attention in the field.

According to Trudy Govier, the relation between argumentation and reasoning will here be defined as follows:

> An argument is a publicly expressed tool of persuasion. [...] Reasoning is distinguished from arguing along these lines:

reasoning is what you may do before you argue, and your argument expresses some of your (best) reasoning. But much reasoning is done before and outside the context of argument (Govier, 1989, p. 117).

Following this definition, argumentation is here considered a special case of reasoning, the difference being that an argument is necessarily expressed in public to achieve a specific objective while conscious or unconscious reasoning is also used in a wide variety of other contexts. Argumentation being based on reasoning it seems safe to assume that the cognitive processes underlying both reasoning and argumentation are basically the same.

The main modes of reasoning addressed by frame theorists are reasoning from cause to effect, category-based reasoning and analogical reasoning/reasoning by example. Especially the first category has been widely discussed in frame research: In FrameNet, for instance, there are several frames expressing cause-effect relations, such as CAUSE EMOTION, CAUSE IMPACT, CAUSE TO BE WET. They are all derived from an abstract CAUSATION SCENARIO frame whose slot structure consists of at least three elements: [Cause], [Effect] and [Averted situation]. Rumelhart & Ortony, 1977, assume that our knowledge about all cause-effect relations is ultimately derived from an atomic CAUSE schema that is part of a set of primitives underlying our entire knowledge system (Rumelhart & Ortony, 1977, p. 106). These underlying concepts of causation emerge through "embodied inferences concerning literal force [that] are preserved in the abstract domain of general causation" and give rise to the metaphors we use to conceptualize more abstract cause-effect relations in reasoning and argumentation (see Feldman, 2006, p. 204). In discourse analysis, the importance of argumentations from cause to effect has been demonstrated by Wengeler, 2003, who found that a large part of the *topoi* used in German news media texts on immigration are based on cause-effect relations.

Reasoning by example, too, is usually set against a backdrop of shared knowledge as represented by the default values of frames (see Minsky, 1975, p. 213). The only possible exception are cases in which a completely new example that was previously unknown to the target audience is introduced to provide backing for an argument (and even in this case, frames provide a rich base of shared knowledge necessary for understanding these specific new examples). The same goes for reasoning by analogy. Rumelhart & Ortony even assume that one can regard the entire comprehension process in schema theory as itself being a case of analogical reasoning. When we determine that a situation fits a certain schema we are in a sense determining that the current situation is analogous to those situations from which the schema was originally derived. Moreover, when we make inferences about unobserved aspects of the situations we are, in effect, assuming their existence by analogy from the situations from which the schemata were derived. (Rumelhart & Ortony, 1977, p. 120)

Frames provide us with the conceptual backdrop necessary for categorization. By enabling us to categorize the events, states, objects and entities we experience, they allow us to draw conclusions about an exemplar's features based on the features shared by prototypical members of the category we assign it to (as in the case of the classic syllogism "All men are mortal; Socrates is a man; therefore Socrates is mortal").

This account of how our ability to argue rests upon the knowledge provided by our frames is by no means complete and much work is still to be done. The basic idea, however, should be sufficiently clear. To resume, we can say that argumentation necessarily calls on existing knowledge structures in at least two ways. To successfully engage in an argument, we need

a) knowledge about the phenomena under discussion, and
b) knowledge about the ways in which individual bits of knowledge can be systematically combined in order to build valid argumentations.

All of this knowledge is derived from experience and presents prototypicality effects: Not only do we have knowledge about the more concrete concepts involved in any argumentation but also knowledge about the possibilities of making valid arguments by establishing relations between different bits of knowledge provided by our frames. These relations can be static or dynamic, that is, in real-life situations we can either use remembered arguments stored in long-time memory or build new ones. Ultimately, our ability to make valid arguments rests upon this possibility to creatively combine the knowledge provided by our frames. Which elements are thus combined depends on our current goals and needs as well as on the context in which an argument occurs.

4.2 Describing arguments as frame relations

Following the terminology established by Fillmore, the basic frame-evoking elements in discourse are here called *lexical units*, i.e. words, phrases or grammatical constructions that call on shared underlying knowledge structures. These knowledge structures are represented by propositions (as defined by Searle, 1969) that involve the frame-evoking lexical units. Propositions, in this sense, consist of the acts of *referring* to something and *predicating* things about it. The reference is the frame-evoking lexical unit under analysis; the predications are the fillers that are instantiated in its slots. Following the methodology established by Ziem, 2008, frames are analysed by extracting the predications that are made about frame-evoking lexical units. These predications are then assigned to the frame's slots, which represent the questions that are most likely to arise with respect to the frame under analysis. The questions representing the slots of the CAUSATION SCENARIO frame described in section 4.1, for example, would be *What causes event/state X? What are the effects? How would the situation be different if X had not occurred?* Reasoning being "primarily from propositions to propositions" (Walton, 1990, p. 402), this propositional analysis not only allows us to extract frame structures from text corpora. Since argumentations consist of at least two propositions and a connector, a propositional analysis also allows us to describe the ways in which these structures are used in argumentation.

Within the framework of frame theory, Toulmin's classic example can thus be analysed as follows:

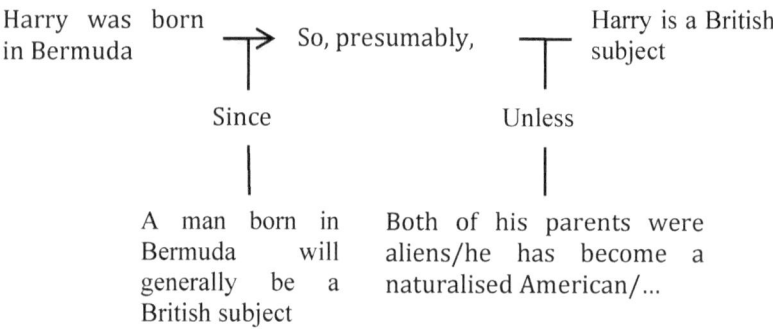

Figure 1 – Toulmin's classic example (Toulmin, 2003, p. 94)

Let us first have a look at the two key propositions in Toulmin's example:

(1) Harry was born in Bermuda.
(2) Harry is a British subject.

In proposition (1) the filler *Bermuda* is instantiated in the slot [Birthplace] of the frame HARRY. Proposition 2) represents an inference about possible fillers for the slot [Nationality], based on the explicit information given in (1).

HARRY [Birthplace: Bermuda] > HARRY [Nationality: British]

Barsalou, 1992, calls such relations between the slots and/or fillers of frames *constraints*. Constraints provide information about the ways in which bits of knowledge are systematically interrelated, e.g. the relation between speed of travel and travel duration or, in Toulmin's example, the relation between a person's birthplace and nationality.

> The central assumption underlying constraints is that [fillers] of [slots] are not independent of one another. Instead, [fillers] constrain each other in powerful and complex manners. [...] Constraints need neither be logical nor empirical truths. [...] Instead, [they] often represent statistical patterns or personal preferences, which may be contradicted on occasion. (Barsalou, 1992, p. 37)

Toulmin's example nicely depicts the way in which frames enable us to reason based on our knowledge about prototypical relations between different aspects of a concept: Our experience tells us that the place where a person is born usually correlates with that person's nationality. The fillers instantiated in the slot [Birthplace] of a PERSON frame thus enable us to infer likely fillers for the slot [Nationality] (and vice versa), this correlation being true of most persons we meet in daily life. This is equally true for a whole range of other information that remains implicit in figure 1. For instance, we can automatically attach default values to fillers such as [Gender] and [Mother tongue] because of our knowledge that a) Harry is typically a male name and that b) a person's birthplace usually correlates not only with their nationality but also with their mother tongue.

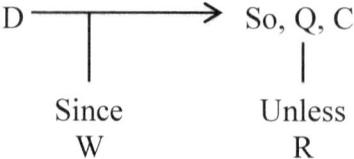

Figure 2 – The elements of Toulmin's model (Toulmin, 2003, p. 94)

4.3. Argumentation and default values

In Toulmin's example, the warrant (W) is based on a) the knowledge that Bermuda is a British Overseas Territory and b) knowledge about the correlation between birthplace and nationality, the latter being a statistical pattern as described by Barsalou (see section 4.2). The conclusion is a default value that is automatically assigned to the appropriate slots of the frame. Default values being only loosely attached to their respective slots, the default value can easily be replaced by new items should the need arise (see Minsky, 1975, p. 213). Possible reasons to do so correspond to the rebuttals (R) in Toulmin's model: While a person's birthplace and nationality usually correlate, we know that one can change their nationality by way of naturalisation. This knowledge allows us to easily update the fillers assigned to the frame's slots when presented with such information.

By analysing arguments with the tools offered by frame theory, it becomes (at least to a certain degree) possible to identify the default values of frames, as shown in the following examples:

(3) He is your friend, yet he betrayed you.
(4) She is his mother but she does not take proper care of him.

The predications involved in premises (3) and (4) are:

(3)' He is your friend
(3)" He betrayed you.
(4)' She is his mother.
(4)" She does not take proper care of him.

The predicational analysis allows us to identify the fillers involved in the propositions as well the slots they are attached to. The problem is that these fillers alone provide less information than the propositions they were extracted from. What is missing in (3)'/(3)" and (4)'/(4)" are two central default values of the frames FRIEND and MOTHER that are activated in (3) and (4): a FRIEND is typically considered loyal, a MOTHER is expected to take good care of her child. As in Toulmin's example discussed above, the frames' default values serve as warrants that support claims such as *You should never trust him again* (3) and *She should by stripped of custody* (4).

In a previous case study (Varga, 2017), this integrated approach was used to study the German and French governments' discourse in the wake of the Fukushima nuclear disaster. The central arguments chancellor Angela Merkel used to justify Germany's faster phasing out of nuclear energy were based on argumentative relations between the slots [Place] and [Consequences] of the German FUKUSHIMA frame: Germany and Japan both being conceptualized as *high-technology countries*, the fact that Japan could not prevent a nuclear disaster serves as data for a series of claims warranted by analogy, as shown in figure 3:

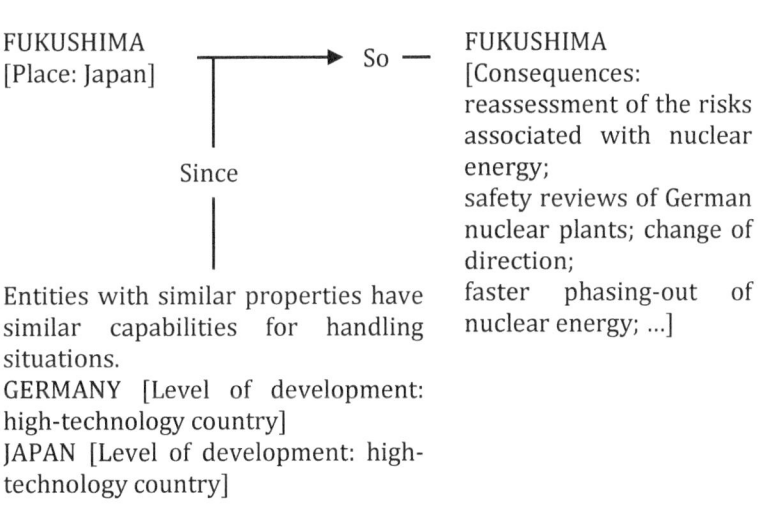

Figure 3 – Argumentative relations in the German FUKUSHIMA frame (Varga, 2017)

By including argumentative relations of the type outlined above in a frame analysis, fillers and slots cannot only be assessed quantitatively but also according to their place in the overall argumentative structure of a discourse. By expressing the arguments identified in a text corpus in terms of frame relations—both at the level of fillers and at the level of slots—lexical analysis and argumentation analysis can be systematically intertwined.

5. CONCLUSION

Much work still needs to be done to deepen the integration of frame analysis and argumentation analysis in discourse linguistics. From a theoretical and methodological standpoint, there is little doubt that this is possible. The issues addressed here—especially the relation between conceptual frames and reasoning/argumentation—provide a good starting point for developing a more fine-grained theory of how argumentation draws on underlying knowledge structures and how an enlarged linguistic frame theory allows us to access these structures.

REFERENCES

Barsalou, L. W. (1992). Frames, concepts and conceptual fields. In A. Lehrer, & E. Feder Kittay (Eds.), *Frames, fields and contrasts: New essays in semantic and lexical organization* (pp. 21–74). Hillsdale, NJ: Lawrence Erlbaum.

Busse, D., & Teubert, W. (1994). Ist Diskurs ein sprachwissenschaftliches Objekt? Zur Methodenfrage der historischen Semantik. In D. Busse, F. Hermanns, & W. Teubert (Eds.), *Begriffsgeschichte und Diskursgeschichte. Methodenfragen und Forschungsergebnisse der historischen Semantik.* (pp. 10–28). Opladen: Westdeutscher Verlag.

Entman, R. (1993). Framing: Toward clarification of a fractured paradigm. *Journal of Communication*, *43*(4), 51–58.

Fairclough, I., & Mădroane, I. D. (2016). An argumentative approach to policy 'framing': Competing 'frames' and policy conflict in the Roşia Montană case. *Rozenberg Quarterly*. Retrieved from http://rozenbergquarterly.com/issa-proceedings-2014-an-argumentative-approach-to-policy-framing-competing-frames-and-policy-conflict-in-the-rosia-montana-case/.

Feldman, J. A. (2006). *From molecule to metaphor: A neural theory of language.* Cambridge, London: The MIT Press.

Fillmore, C. J. (1985). Frames and the semantics of understanding. *Quaderni di semantica*, *6*(2), 222–254.

Fillmore, C. J. (1975). An alternative to checklist theories of meaning. In C. Cogen (Ed.), *Proceedings of the First Annual Meeting of the Berkeley Linguistics Society*, (pp. 123–131).Berkeley, CA: Linguistic Department.

Fillmore, C. J., & Andos, J. (2010). Discussing frame semantics: The state of the Art. *Review of Cognitive Linguistics*, *8*(1), 157–176.

Goffman, E. (1974). *Frame analysis: An essay on the organization of experience.* New York: Harper & Row.
Govier, T. (1989). Critical thinking as argument analysis? *Argumentation, 3*(2), 115-126.
Greco Morasso, S. (2012). Contextual frames and their argumentative implications: A case study in media argumentation. *Discourse Studies, 14*(2), 197–216.
Lakoff, G., & Johnson, M. (1980). *Metaphors we live by.* Chicago: The University of Chicago Press.
Minsky, M. (1975). A framework for representing knowledge. In P. H. Winston (Ed.), *The psychology of computer vision* (pp. 211–277). New York: McGraw-Hill.
Rumelhart, D. E. (1980). Schemata: The building blocks of cognition. In R. J. Spiro, B. C. Bruce & W. F. Brewer (Eds.), *Theoretical issues in reading comprehension: Perspectives from cognitive psychology, linguistics, artificial intelligence, and education.* (pp. 33–58). Hillsdale, NJ: Lawrence Erlbaum.
Rumelhart, D. E., & Ortony, A. (1977). The representation of knowledge in memory. In R. C. Anderson, R. J. Spiro & W. E. Montague (Eds), *Schooling and the acquisition of knowledge* (pp. 99–135). Hillsdale, NJ: Lawrence Erlbaum.
Searle, J. (1969). *Speech acts: An essay in the philosophy of language.* Cambridge: Cambridge University Press.
Toulmin, S. E. (2003). *The uses of argument: Updated edition.* Cambridge: Cambridge University Press.
van Eemeren, F. H. (2010). *Strategic maneuvering in argumentative discourse: Extending the pragma-dialectical theory of argumentation.* Amsterdam: John Benjamins.
Varga, S. (2017). Kernkraft in der Krise? Der Fukushima-Diskurs in Deutschland und Frankreich. *Cahiers d'études germaniques, 73,* 171–184.
Walton, D. N. (1990). What is reasoning? What is an argument? *The Journal of Philosophy, 87*(8), 399–419.
Wengeler, M. (2003). *Topos und Diskurs: Begründung einer argumentationsanalytischen Methode und ihre Anwendung auf den Migrationsdiskurs (1960-1985).* Tübingen: Max Niemeyer.
Ziem, A. (2008a). *Frames und sprachliches Wissen. Kognitive Aspekte der semantischen Kompetenz.* Berlin: Walter de Gruyter.

63

Straw Man as Misuse of Rephrase

JACKY VISSER
Centre for Argument Technology / University of Dundee, United Kingdom
j.visser@dundee.ac.uk

MARCIN KOSZOWY
Faculty of Law / University of Białystok, Poland & Centre for Argument Technology / Polish Academy of Sciences, Poland
koszowy@uwb.edu.pl

BARBARA KONAT
Centre for Argument Technology / Polish Academy of Sciences, Poland
bkonat@gmail.com

KASIA BUDZYNSKA
Centre for Argument Technology / Polish Academy of Sciences, Poland & University of Dundee, United Kingdom
k.budzynska@dundee.ac.uk

CHRIS REED
Centre for Argument Technology / University of Dundee, United Kingdom & Polish Academy of Sciences, Poland
c.a.reed@dundee.ac.uk

The 'rephrase' relation between propositions is introduced in Inference Anchoring Theory to facilitate argument mining (the automated analysis of argumentative discourse). Examining an example from the candidates' debates leading up to the 2016 presidential elections in the United States, we explore the relation between such rephrases and the *straw man* fallacy. Our aim with the structural characterisation of the fallacy is to work towards a foundation for the automated identification of rephrase and inference patterns as a tool in computationally identifying instances of straw man and similar fallacies (e.g. *ignoratio elenchi*).

KEYWORDS: fallacies, Inference Anchoring Theory, rephrase, straw man, television debate, US elections

1. INTRODUCTION

The fallacies have traditionally been a central subject in the field of argumentation studies. Despite continuing progress on the development of computational methods in this field, little work has been done on the computational evaluation of argumentation (Walton, 2016) – that is, the evaluative task, not in terms of persuasive effectiveness or logical validity, but rather in terms of the violation of some conventional (often dialectical) norms for reasonable argumentative conduct. To explore a possible route towards the automated indication of potential fallacies, we propose to characterise some fallacies in structural terms as a pattern of various argumentative relations between propositions. In particular, we focus on the misuse of the rephrase relation by means of which speakers can reformulate their claims and arguments without affecting the inferential structure. The idea is that these patterns are amenable to machine recognition in annotated text corpora or as part of systems for argument mining, and as such can advance the computational evaluation task.

Of course, we do not intend to claim that all instances of the pattern point to fallacies: there are legitimate reasons for, e.g., restating one's argument or claim. However, if these patterns can be identified automatically, then a normative criterion could be brought to bear on these cases to determine their argumentative legitimacy – like other argumentative moves, rephrases can be thought of as operating on a continuum ranging from reasonable to fallacious counterparts (van Eemeren, 2010). In the present paper, we do not concern ourselves with drawing the boundary between reasonable and fallacious, rather we focus primarily on the structural characterisation of a fallacy that we consider to involve rephrase: straw man.

In the next section, 2, we first turn to the notion of 'argumentative rephrase'; how it relates to similar concepts, and its role in our theoretical framework of Inference Anchoring Theory (IAT). Section 3 covers some relevant existing perspectives on the straw man fallacy, and our structural characterisation of straw man in terms of IAT. In Section 4, we illustrate how this structure is realised in argumentative practice, by discussing an example from a television debate in the lead-up to the 2016 US presidential elections. Before turning to the Conclusion, we describe two directions for future work in more detail in Section 5: extension of the approach to the *ignoratio elenchi* fallacy, and linguistic cues for the calling-out of rephrase- related fallacies.

2. REPHRASE IN ARGUMENTATIVE DISCOURSE

2.1 The backgrounds of argumentative rephrase

Restating one's own words can be a powerful argumentative device – just to name a few benefits: it can help the audience to memorise or to understand what the main claim is; it can give the impression that the speaker has more arguments that he or she in fact has; it can protect the speaker's position from being attacked by starting with a strong claim and then gradually withdrawing to "safe ground" by weakening the standpoint by means of a rephrase.

The notion of 'rephrase' bears some resemblance to other communicative phenomena – e.g. paraphrase, textual entailment and summarisation – but is distinct from those in several important respects. Rephrase is intimately connected with the intentions of the speaker: it could be possible for example, that a nuanced or prolix point is rephrased aggressively, succinctly or discourteously, with a very great distance between the two sentences in lexical and semantic terms – yet both points being clearly recognisable as related through a rephrase. On the other hand, just because one span of discourse (even in a single interaction) entails or paraphrases another, this does not mean that the speaker is, in fact, intentionally using the one to rephrase the other. Thus whilst the surface lexicalisation from which paraphrase and entailment have previously been recognised is indeed also useful in recognising rephrase, more is required.

The *text mining* community has long recognised the importance of paraphrases, i.e. two units expressing similar meaning (see e.g. the Microsoft Research Paraphrase corpus (Dolan, Brockett, & Quirk, 2005)). Hirst (Hirst, 2003) defines paraphrase as "talking about the same situation in a different way" with changes in the wording or syntactic structure. Bhagat and Hovy (Bhagat & Hovy, 2013) propose the term "quasi paraphrases" to describe text units "that convey approximately the same meaning using different words". In some cases, the existence of paraphrase can be used as an indicator of the existence of a rephrase relation – but it is neither a sufficient nor a necessary condition: some sentence pairs that happen to be paraphrases might not be used as rephrases, and some rephrases might be far to distant lexically and semantically to count as paraphrase.

Textual entailment (Dagan, Glickman, & Magnini, 2006; Berant, Dagan, & Goldberger, 2012; Zanzotto, Pazienza, & Pennacchiotti, 2005; Dagan & Magnini, 2013; Mirkin, Berant, Dagan, & Shnarch, 2010) is also closely related to rephrase inasmuch as some examples of rephrase that are not paraphrase might involve textual entailment relations. Once again, though, the very fact that textual entailments are (according to the leading approaches in that area) definable on the basis of solely the lexicalisation of the two involved text spans, suggests that it is functioning differently to rephrase, where the contextual embedding of the spans provides essential information about whether or not they constitute a rephrase.

2.2 Rephrase in Inference Anchoring Theory

Our characterisation of straw man as an infelicitous use of the rephrase relation is based on Inference Anchoring Theory (IAT) (Budzynska & Reed, 2011). Building on insights from discourse analysis and argumentation studies, IAT explains argumentative conduct in terms of the anchoring of argumentative reasoning in dialogical interaction, by means of the 'illocutionary connection' between the two. Elaborating on traditional speech act theory (Austin, 1962; Searle, 1969), illocutionary forces are reinterpreted as relations connecting locutions to propositional content. The reasoning appealed to in the argumentation involves three types of relations between propositions. An inference relation holds between a proposition that functions as a premise in an argument and the contested proposition that it supports. A conflict relation indicates that a proposition is incompatible with another. A rephrase relation holds between two propositions when one proposition is used to rephrase, restate or reformulate another proposition. Whether there is a rephrase relation between the propositional content of two text spans depends on the speaker's intention to modify the wording of apremise or a conclusion, by means of, e.g., specialisation, generalisation or instantiation.

Characteristic of IAT is its orientation towards computational linguistic methods and software implementation. To facilitate the required machine-readability, IAT adheres to the extended Argument Interchange Format (AIF+) standard (Chesñevar, McGinnis, Modgil, Rahwan, Reed, Simari, South, Vreeswijk, & Willmott, 2006; Reed, Wells, Rowe, & Devereux, 2008). AIF+ is a graph-based ontology that facilitates the representation of the intertwined locutionary, illocutionary, and propositional structures, resulting from the analysis of argumentative discourse.

The computational orientation comes to the fore in the availability of software to support the use of IAT. For example, OVA+ (Janier, Lawrence, & Reed, 2014) is an online tool for argument analysis facilitating the representation of the structure of argumentative discourse.[1]. Analyses produced with OVA+ (and a variety of other programs) can be saved as 'argument maps' in AIFdb (Lawrence, Bex, Reed, & Snaith, 2012), an online searchable repository of analysed arguments freely available at aifdb.org. The argument maps stored in AIFdb can subsequently be collected in corpora at corpora.aifdb.org (Lawrence & Reed, 2014).

Previous work on the introduction of the rephrase relation into IAT started with an intuitive concept of rephrase used in mediation sessions (Janier & Reed, 2017), leading to a preliminary theoretical model of rephrase (Konat, Budzynska, & Saint-Dizier, 2016). Although more complex constellations are possible, in the simplest case of rephrase two text spans serve (almost) the same function in the argumentation, as is the case in Example (1) from the first television debate for the 2016 US presidential elections (Peters & Woolley, 2016b).

(1)   a.   WALLACE: [...] You support a national right to carry law. Why, sir?
       b.   TRUMP: [...] In Chicago, which has the toughest gun laws in the United States, [...] they have more gun violence than any other city. So we have the toughest laws, and you have tremendous gun violence.

---

[1] OVA+ is freely available at the website ova.arg.tech and has been used to produce all figures in this paper.

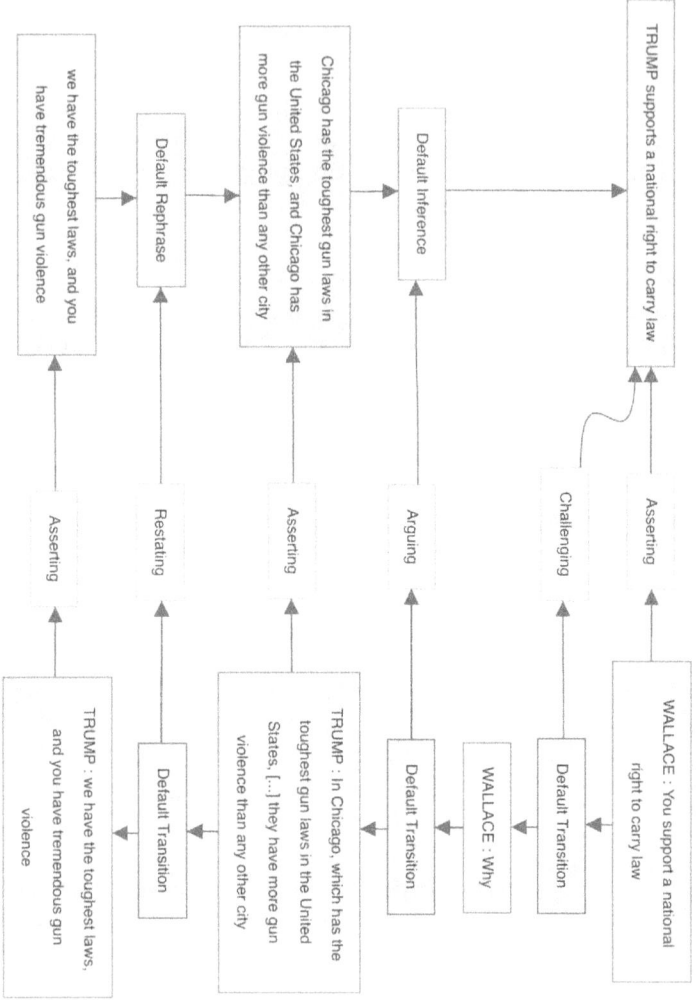

Figure 1 – Diagramming argumentative discourse with Inference Anchoring Theory

The rephrase relation in (then candidate for the Republican party) Donald Trump's response to a challenge from the debate's moderator Chris Wallace constitutes an integral part of the argument's structure but without introducing a new line of argument, as presented in Figure 1. The right-hand side of 1 represents the dialogical dimension of the argumentation, the left-hand side shows the propositional dimension, and the middle row contains the illocutionary connections (see Section 4.1 for more on this style of annotation with IAT). The right side of Figure 1 shows how Example (1) is segmented into four locutions, interconnected with discourse transitions. On the left of 1, three propositional contents are reconstructed (on the basis of the respective illocutionary connections). The top two propositions function as respectively a conclusion (or claim) and a premise (or argument). The bottom proposition is a rephrase of Trump's argument, not constituting a new line of argument, but rather a restatement or reformulation of the original point.

3. REPHRASE IN STRAW MAN FALLACIES

The aim of this section is to answer the question: how the rephrase relation is structurally related to the straw man fallacy? The answer will be given by employing the general method of modelling rephrase in IAT, as discussed in section 2.2 to capture instances of *Default Rephrase* and *Default Conflict* in the straw man technique. This task would require determining how distinctive structural features of straw man can be depicted using IAT structures represented with OVA+.

3.1 The straw man fallacy

In this section we will expose those key features of the straw man fallacy that serve as a source of inspiration for modelling it as misuse of rephrase. Although straw man is a dialogical strategy aimed at pursuing different goals (Aikin & Casey, 2011; Macagno & Walton, 2017) and thus there is a variety of types of straw man, the common way of defining this technique is to associate it with the misrepresentation of someone's position in order to easily refute that position (Walton, 1996; Talisse & Aikin, 2006; Lewiński, 2011; Lewiński & Oswald, 2013; Aikin & Casey, 2016). This general idea of misrepresentation is related to rephrasing because a rephrase might constitute means to modifying original speaker's standpoint in such a way that it is made easier to attack. According to Macagno and Walton, dialogical purposes of employing the straw man technique consist of "attacks to the argument or the claim of the interlocutor in a dialogue (real or fictitious) in order to reject it, and possibly thus supporting the opposing one" (Macagno & Walton, 2017, p. 110). Diagrammatic representation of straw man proposed in this paper allows us to show a functional role of rephrase relation in fallacious attacks on other party's claim or argument.

As Oswald and Lewiński point out, the often overlooked feature of the straw man technique is its meta-discursive role, meaning that "it operates on someone else's discourse that serves as a material for linguistic maneuvering" (Oswald & Lewiński, 2014). Putting an emphasis on this meta-discursive feature of straw man leads to conceiving this technique as "a certain unreasonable, sometimes manipulative, re-interpretation of another position". Our approach relies on emphasising this observation in the graphical representations of instances of straw man fallacy. For the meta-discursive function of straw man is shown in OVA+ diagrams by introducing the rephrase relation in terms of the difference between the propositional content of original speaker's locution and the rephrased statement that has been attacked. This way of representing the meta-discursive feature of straw-man will be shown in Section 3.2.

3.2 The IAT-rephrase approach to straw man

The rephrase approach to the straw man technique can be described in following steps: firstly, the proponent claims C, next, the opponent restates the claim as C' making it easier to attack and finally, the opponent attacks C'. When focusing on dialogical aspects of straw man fallacy, we characterise it as a sequence of two dialogue moves, the first of which is an instance of restating the other party's claims, and the second an instance of attacking the rephrased content. This characteristics allows us to expose the dynamics of the dialogue involving instances of straw man.

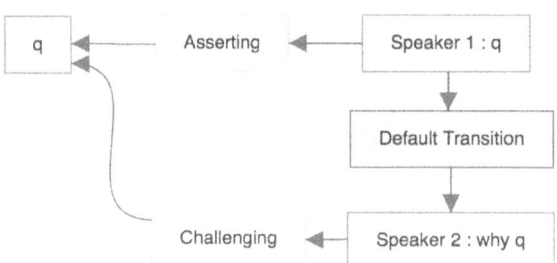

Figure 2 – Non-fallacious challenging

Figure 2 represents an instance of non-fallacious challenging, leading to a non-mixed difference of opinion, where a standpoint is met with a position of doubt. Of course, there is also potential for a straw man to be committed in in explicit disagreement. In this case, instead of a mere expression of doubt, a stronger position is taken by expressing a contrary standpoint. A non-fallacious version of this pattern is shown in Figure 3

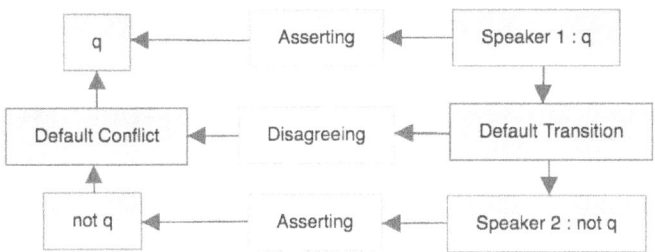

Figure 3 – Non-fallacious disagreeing

Whereas an interlocutor can reasonably challenge or disagree with the propositional content of an original locution, a requirement for this is that the propositional content that is challenged or disagreed with is represented fairly. This is where the rephrase relation comes in. Outwith formal dialogue systems, it would be too strong to expect ordinary language users to express their doubt or disagreement with respect to an exact repetition of the proposition as it was expressed by their interlocutor. Rather, we might expect that the propositional content is restated to some degree. Our suggestion is that there is a continuum of rephrase ranging from literal repetition of a proposition to expressing an entirely different proposition, with the extremes not entailing rephrase and everything in between being an instance of rephrase to some degree. Now, different approaches to rephrase (and its related notions, such as paraphrase and fuzzy quantifiers in logic) will define this continuum in different ways. With respect to the reasonableness of rephrases in the challenging and supporting of standpoints, and their potential to lead to the straw man fallacy, it requires a normative theory of argumentation to determine where on this continuum the boundary may be between reasonable and fallacious rephrase use. Independent of where the boundary of reasonableness is drawn, the structural constellation of interacting propositions, illocutionary connections and locutions would instantiate a common pattern. In Figures 4 and 5 we show two such patterns.

The first pattern, as shown in Figure 4, represents a potential straw man in challenging. It differs from the pattern of non-fallacious challenging (as represented in Figure 2) with respect to the *Default*

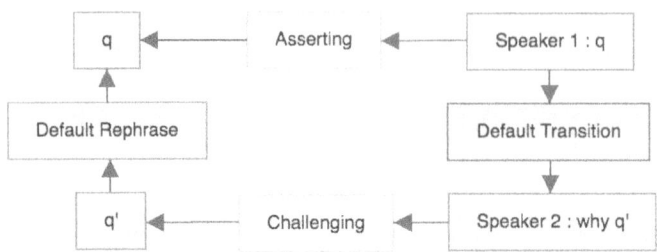

Figure 4 – Potential straw man in challenging

Rephrase node on the left hand side of the diagram and the fact that the illocutionary connection of *Challenging* does not target the content q of the original speaker's locution, but instead the rephrased content q'. These two elements may constitute the structural cue indicating that the straw man technique might have been employed.

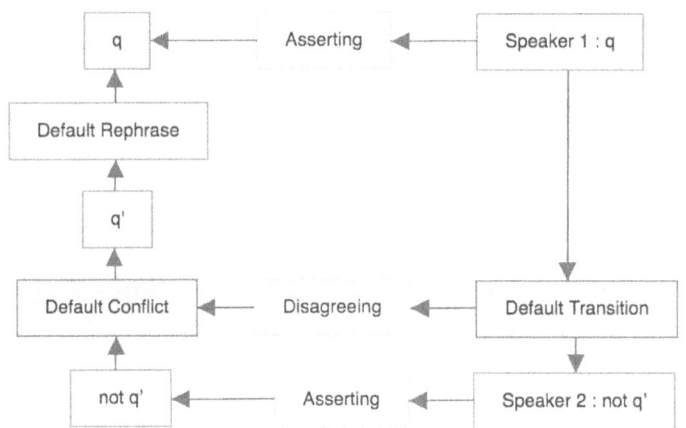

Figure 5 – Potential straw man in disagreeing

The second pattern represented by Figure 5 shows how the straw man technique may be structurally combined with disagreeing. This pattern differs from non-fallacious disagreeing, as shown in Figure 3 with respect to the fact that q has been first rephrased as q' and next that the illocutionary connection of *Disagreeing* targets the *Default Conflict* relation between q' and *not* q' on the left hand side of the diagram. In case of non-fallacious disagreeing there is no *Default Rephrase* relation and the *Default Conflict* relation is between q and *not – q*.

4. EMPLOYING REPHRASE STRUCTURES TO IDENTIFY POTENTIAL FALLACIES IN PRACTICE: THE 2016 US PRESIDENTIAL ELECTIONS

4.1 Annotation of the 2016 US presidential election television debates

To explore how the proposed structural characterisation of straw man can be applied in practice, we will consider an example from the television debates leading up to the 2016 US presidential elections. A selection of three of the televised election debates have been fully annotated on the basis of IAT. The annotated transcripts of the first general election debate, and of the two preceding first television debates for the primaries of the Republicans and of the Democrats, are collected in the US2016tv corpus[2],[3] The transcripts together amount to 58,900 words, which are segmented into a total of 4,671 locutions. The annotation contains 4,277 propositions, connected through 1,551 inference relations, 194 conflict relations, and 333 rephrase relations.

The US2016tv corpus is annotated by four annotators. The annotators are extensively trained to analyse the television debates on the basis of IAT (see Section 2.2), resulting in an inter-annotator agreement on a 10.5% (word count) sample of a Combined Argument Similarity Score κ (Duthie, Lawrence, Budzynska, & Reed, 2016) of 0.679 (the usual Cohen's κ (Cohen, 1960) is less suitable for this kind of complex multi-layered annotation task, yielding a value of 0.516). Below, we summarise the annotation guidelines – the full version of which dealswith, among others: anaphoric references, epistemic modalities, repetitions, punctuation, discourse indicators, interposed text, reported speech, and how to deal with context-specific peculiarities.

Segmentation divides the (transcribed) text into locutions. A locution consists of a speaker designation and an 'argumentative discourse unit' – a text span with discrete argumentative function (Peldszus & Stede, 2013).

Transitions capture the functional relationships between locutions, reflecting the dialogue protocol – a high level specification of

---

[2] The US2016tv corpus is available online at http://corpora.aifdb.org/US2016tv.

[3] While all of the examples we present in this paper are drawn from the 2016 presidential debates, some come from debates that are not included in the US2016tv corpus. They are, however, all annotated in accordance to the guidelines we summarise here. The two larger examples, (2) and (3), are available in a separate corpus at http://corpora.aifdb.org/FallaciousRephrase.

the set of transition types that are available in a particular communicative activity.

Illocutionary connections embody the intended communicative functions of locutions or transitions, such as: *Agreeing, Arguing, Asserting, Challenging, Disagreeing, Questioning, Restating,* and *Default Illocuting* (when none of the other types suffice). Some types of illocutionary connection lead to the reconstruction of a propositional content. Illocutionary connections can anchor in locutions or in transitions, depending on their type and on possible indexicality (*viz.* the different anchoring of *Challenging* in Figures 1 and 2).

Inferences are directed relations between propositions, reflecting that a proposition is meant to supply a reason for accepting another proposition. An argument scheme (e.g., *Argument from Example* or *Argument from Expert Opinion*) can be specified, failing that, it is labelled as *Default Inference*.

Conflicts are directed relations between propositions, reflecting that a proposition is meant to be incompatible with another proposition or relation. Such incompatibility may depend on, e.g., logical contradiction or pragmatic contrariness, or the annotated relation may default to *Default Conflict*.

Rephrases are directed relations between propositions, reflecting that a proposition is meant to be a reformulation of another proposition. Such reformulation may involve, e.g., *Specialisation, Generalisation* or *Instantiation*, or the relation defaults to *Default Rephrase*.

4.2 The radical Islam example

Example (2) illustrates the relation between rephrase structures and straw man (Peters & Woolley, 2015). In this example, the debate's moderator, John Dickerson, asks (then candidate for the Democrat presidential nomination) Hilary Clinton to reflect on what Marco Rubio (then candidate for the Republican nomination) has said at an earlier occasion about being at war with radical Islam. In her response, Clinton first restates Rubio's claim that has been cited by Dickerson and next she disagrees with this rephrased content. These dialogue moves are visualised diagrammatically in Figure 6 with the use of the OVA+ tool mentioned in Section 2.2.

(2) a. DICKERSON: Secretary Clinton, you mentioned radical jihadists. Marco Rubio, also running for president, said that this attack showed [...] that

we are at war with radical Islam. Do you agree with that characterization, radical Islam?

b. CLINTON: I don't think we're at war with Islam. I don't think we're at war with all Muslims. [...]

c. DICKERSON: Just to interrupt. He didn't say all Muslims. He just said radical Islam.

Dickerson, in his move (2-a) reports on what Rubio has said about radical Islam. Reported speech is represented on the top of the diagram presented in Figure 6 with two locutions and one propositional content. Within the same move, Dickerson asks Clinton to express her opinion on Rubio's claim that has just been reported. In Clinton's response (2-b), the original phrase "radical Islam" has been replaced first with "Islam" and then with "all Muslims". Clinton disagrees here with Rubio by saying "I don't think we're at war with Islam". Figure 6 represents this move with the illocutionary force of *Disagreeing* which is anchored in the transition between the two locutions. Then Clinton restates her own statement by saying "I don't think we're at war with all Muslims", which is represented in Figure 6 with a *Default Rephrase* node on the left-hand side of the diagram. The illocutionary force of *Restating* links together the transition between the two locutions with *Default Rephrase*.

Employing OVA+ to the analysis of this example allows us to pinpoint the structural overlap between the argumentative function of straw man and the dialogical structure of rephrase. Clinton, by saying "I don't think we're at war with Islam" (see utterance (2-b)) introduces the rephrased propositional content to the dialogue, namely "this attack showed that we are at war with Islam" (see the second top propositional content on the left hand side of the diagram). As this is the content assigned to Rubio that Clinton disagrees with, in the diagram there is a *Disagreement* node that is linked to the *Default Conflict* node between two contradictory statements, namely: "this attack showed that we are at war with radical Islam" and "we are not at war with Islam". Structurally speaking, if the straw man has not been committed, there would be a *Default Conflict* relation between Rubio's claim "this attack showed that we are at war with Islam" and its negation, e.g. "we are not at war with Islam". This analysis also shows that the dialogical complexity of the straw man technique may consist of the use of the two different instances of the rephrase relation. The second instance of rephrasing is represented in Figure 6 at the bottom left of the diagram. The statement "we are not at war with Islam" has been rephrased again as "we are not at war with all Muslims". The use of two rephrasing dialogue moves related to one instance of the straw man fallacy shows how this technique may be further developed by rephrasing the content that is already an instance of rephrasing.

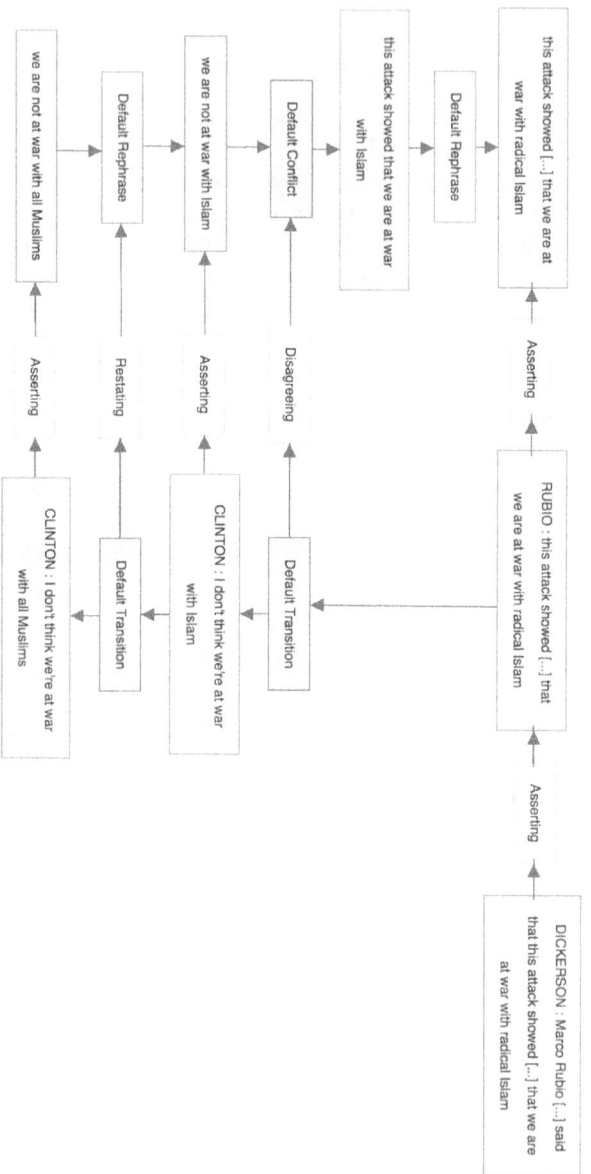

Figure 6 – Diagrammatic visualisation of Example (2)

To sum up, the OVA+ diagram helps to structurally pinpoint the straw man technique as an illocutionary force of disagreeing with the rephrased propositional content instead of disagreeing with the original content. In terms of the conflicts between the propositional contents, Figure 6 helps to identify an instance of a straw man by showing that the *Default Conflict* relation instead of targeting the original propositional content ("we are not at war with radical Islam") targets the content that has been rephrased ("we are not at war with Islam"). This structural difference allows us to emphasise the meta-discursive feature of a straw-man technique that has been discussed in Section 3.1. As this main structural feature of a straw man fallacy is unique in terms of OVA+ structures, the proposed method of structurally representing instances of a straw man fallacy may be a point of departure for an automatic extraction of these particular structures from large repositories of natural language texts.

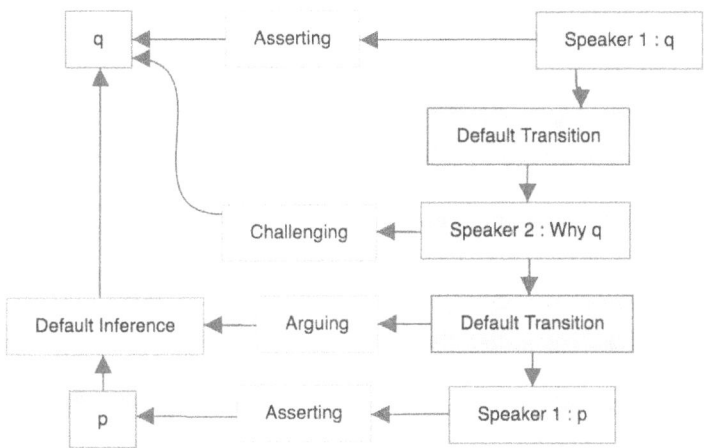

Figure 7 – Non-fallacious supporting

5. PERSPECTIVES ON FUTURE WORK

5.1 Extension to ignoratio elenchi

The rephrase approach to straw man can be extended to other fallacies that rely upon a misuse of the rephrase relation. As a case in point, we can consider *ignoratio elenchi*. While our (narrow) interpretation of straw man involves the restating of an interlocutor's claims, ignoratio elenchi can be narrowly interpreted as the fallacious rephrase of the speaker's own claim. Ignoratio elenchi is a fallacy with various (and rather varying) interpretations. We propose to presently consider a version of ignoratio elenchi that can be conceived of as the mirror image of the straw man fallacy as we understand it in the current paper: a rephrase of one's own standpoint in order to make it easier to defend (*viz.* the straw man fallacy involving the rephrase of an opponent's standpoint to make it easier to attack). A proponent commits an ignoratio elenchi when making a claim C, and then argumentatively defending C', where C' is a rephrase of C making it easier to defend.

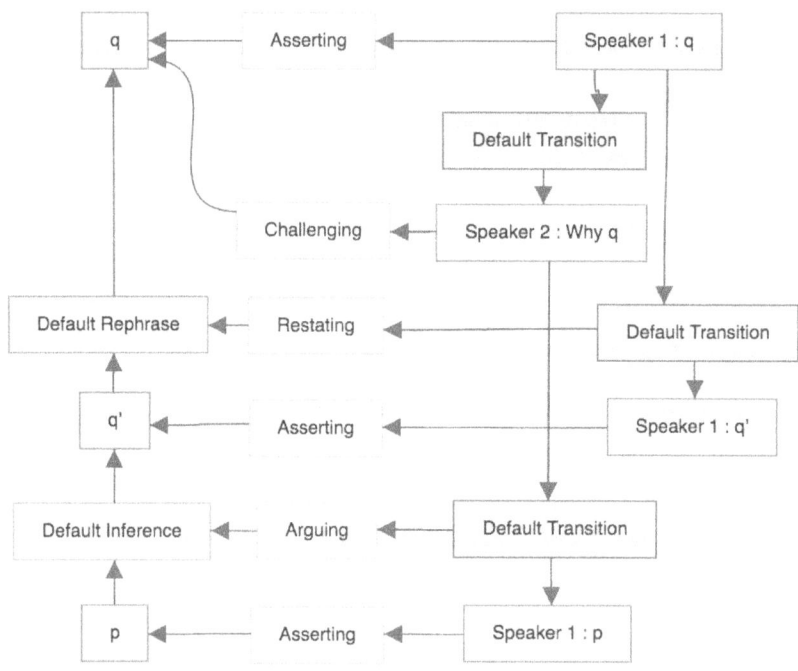

Figure 8 – Potential ignoratio elenchi in supporting

In contrast to the diagrammatic visualisation of a reasonable dialogical counterpart of ignoratio elenchi in Figure 7, Figure 8 shows how in our conception of ignoratio elenchi a proposition q' is introduced standing in a rephrase relation to the original propositional content of the standpoint q. This rephrase relation between q' and q is anchored via an illocutionary connection of *restating*, but this is not part of the ongoing dialogue transitions. The argument p is a dialogical continuation of the request for argumentative defence (i.e. the challenge of q), while the rephrase as q' is on a separate dialogical track; a dead end as it is unconnected to the provision of p as an argumentative defence instigated by the challenge.

In Example (3), we can see this pattern of interacting rephrase and inference relations occurring, as visualised in Figure 9 (Peters &

Woolley, 2016a). During the first head to head debate between presidential candidates Trump and Clinton, the moderator Lester Holt, challenges Trump on his earlier claim that Clinton does not look presidential. Trump quickly restates this into a rephrased claim about Clinton not having the required stamina, and arguing why this is of such importance for a president.

(3) a. HOLT: Mr. Trump, this year Secretary Clinton became the first woman nominated for president by a major party. Earlier this month, you said she doesn't have, quote, "a presidential look." [...] What did you mean by that?
b. TRUMP: She doesn't have the look. She doesn't have the stamina. I said she doesn't have the stamina. And I don't believe she does have the stamina. To be president of this country, you need tremendous stamina.
c. HOLT: The quote was, "I just don't think she has the presidential look."
d. TRUMP: [...] You have so many different things you have to be able to do, and I don't believe that Hillary has the stamina.
e. [...]
f. CLINTON: You know, he tried to switch from looks to stamina.

5.2 Calling out rephrase fallacies

Both examples discussed in sections 4.2 and 5.1 contain dialogue moves that do not belong to the rephrasing structure, but which may serve as discourse cues indicating that rephrase-related fallacies might have been committed. The common feature of these dialogue moves is what we label 'calling out rephrase fallacies' which consists of attempts at signalling that a misuse of rephrase has just occurred in one of the previous dialogue moves.

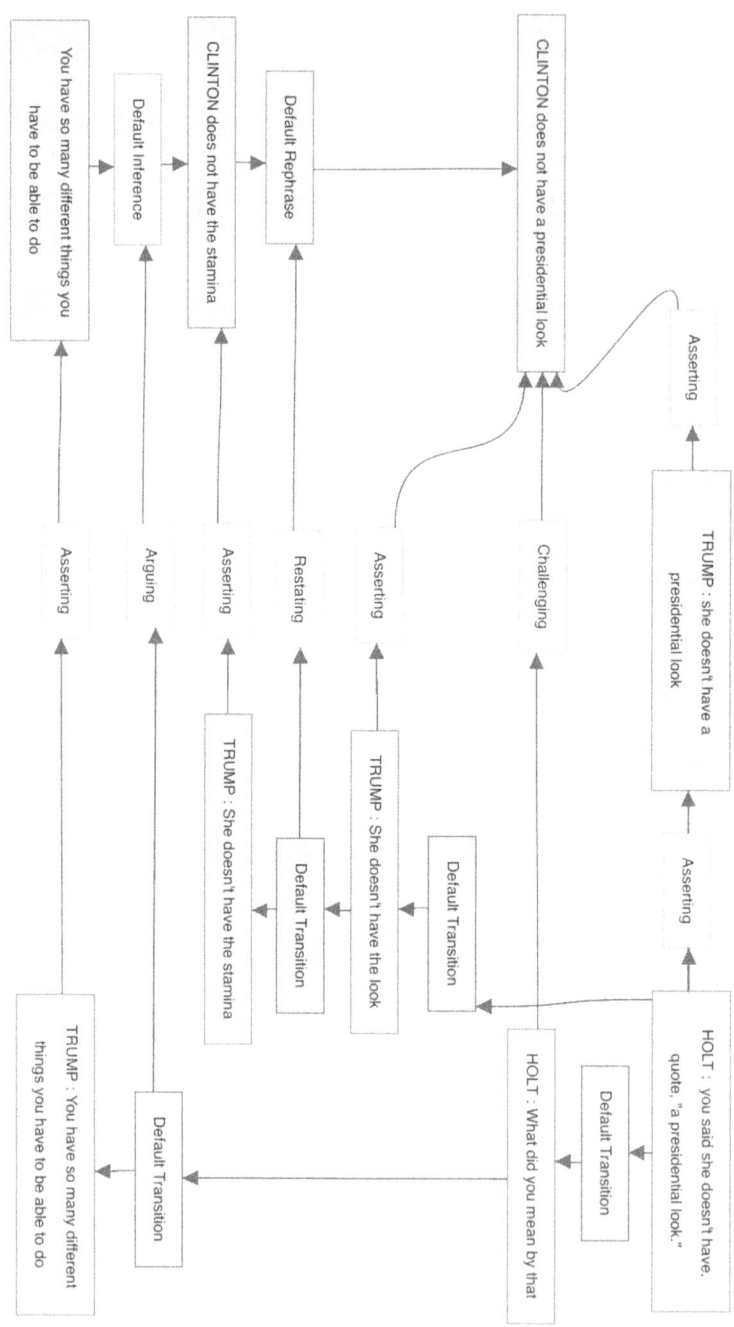

Figure 9 – Diagrammatic visualisation of Example (3)

In example (2) discussed in Section 4.2, Dickerson, by saying "He didn't say all Muslims. He just said radical Islam" (see dialogue move (2-c)), is putting an emphasis on the fact that there was a switch from Rubio's reported claim containing the phrase "radical Islam" to Clinton's view about the rephrased "all Muslims" (see dialogue move (2-c)). Although this content does not strictly belong to the rephrase structure, it shows what type of dialogue moves can be performed when a speaker's aim is to identify an instance of a straw man fallacy. A similar function has Clinton's dialogue move (3-f) shown in example 9 in Section 5.1. By saying "You know, he tried to switch from looks to stamina", Clinton indicates that an attempt at committing a rephrase-related fallacy has been made. Analysing these and other examples of calling out may serve as a preparatory study aimed at exploring linguistic cues for identifying rephrase-related fallacies. For instance, utterances such as "he didn't say..." or "he tried to switch..." may be helpful in finding instances of potential misuses of rephrase in large text repositories.

6. CONCLUSION

Exploring the foundations of automated fallacy detection, we have presented a structural rephrase-based characterisation of the straw man fallacy. By narrowly conceiving of straw man in terms of a misuse of a rephrase of the proposition being challenged, the resulting patterns can facilitate automated techniques for their recognition. Using Inference Anchoring Theory as the framework for our characterisation awards our present theoretical work a close connection to computational implementation. By means of examples from the 2016 presidential election debates in the US we have shown that the patterns we present for the straw man and ignoratio elenchi fallacies do indeed occur in annotated corpora of argumentative discourse. While our structural characterisation does not provide a normative criterion for evaluating any occurrences as reasonable or fallacious, normative theories of argumentation could tell that the examples we discussed can be evaluated as fallacious.

Our exploratory work gives rise to new questions that can be addressed in further research (in addition to that described in Section 5). Some of these questions are of a theoretical or analytical nature, such as: Where are the boundaries of reasonable and fallacious rephrase? How can implicitness of argumentative moves camouflage straw man andother rephrase fallacies? Does rephrase systematically play a part in other fallacies, e.g. circular reasoning? Other questions relate to the computational aspects we lightly touched upon: How can implicitness in natural language be dealt with in argument mining? How can the automated search for structural patterns be implemented? How can we devise computational methods to distinguish between fallacious and reasonable argumentation?

Beyond leading to such questions of an academic nature, working towards the automated recognition of rephrase fallacies, such as straw man, can turn out to have huge societal impact. Amidst ongoing worries about the prevalence of 'fake news', more and more emphasis is put on fact-checking in journalism and the media at large. While the current focus of the fact-checkers is squarely on determining the truthfulness of a specific bit of information, it appears to us that the way that information is phrased is at least of equal importance. This is especially the case since it often involves the checking of a politician's or other public figure's words, and a negative outcome of the fact-checking effectively amounts to an ethotic attack. It goes without saying that it is of the utmost importance in this endeavour that the actual words spoken are not (intentionally or unintentionally) rephrased to such an extend that it becomes a straw man, and neither should a speaker be allowed to weasel their way out of a negative fact-checking outcome by committing an ignoratio elenchi.

ACKNOWLEDGEMENTS: This research was supported in part by the Engineering and Physical Sciences Research Council in the United Kingdom under grant EP/N014871/1, and in part by the Polish National Science Centre under grant 2015/18/M/HS1/00620. We would like to thank Mark Snaith for his technical help in preparing this paper in LaTeX.

REFERENCES

Aikin, S., & Casey, J. (2011). Straw men, weak men, and hollow men. *Argumentation*, *25*(1), 87-105.
Aikin, S., & Casey, J. (2016). Straw men, iron men, and argumentative virtue. *Topoi*, *35*(2), 431-440.
Austin, J. L. (1962). *How to do things with words*. Oxford: Clarendon Press.
Berant, J., Dagan, I., & Goldberger, J. (2012). Learning entailment relations by global graph structure optimization. *Computational Linguistics*, *38*(1), 73-111.

Bhagat, R., & Hovy, E. (2013). What is a paraphrase? *Computational Linguistics*, *39*(3), 463-472.
Budzynska, K., & Reed, C. (2011). *Whence inference* (Tech. rep.). University of Dundee.
Chesñevar, C., McGinnis, J., Modgil, S., Rahwan, I., Reed, C., Simari, G., South, M., Vreeswijk, G., & Willmott, S. (2006). Towards an argument interchange format. *The Knowledge Engineering Review*, *21*(04), 293-316.
Cohen, J. (1960). A coefficient of agreement for nominal scales. *Educational and Psychological Measurement*, *20*(1), 37-46.
Dagan, I., Glickman, O., & Magnini, B. (2006). The PASCAL recognising textual entailment challenge. In *Machine learning challenges: Evaluating predictive uncertainty, visual object classification, and recognising tectual entailment* (pp. 177-190). New York: Springer.
Dagan, I., & Magnini, B. (2013). Entailment graphs for text exploration. In A. Lavelli & O. Popescu (Eds.), *Proceedings of the Joint Symposium on Semantic Processing. Textual Inference and Structures in Corpora*. Trento, IT.
Dolan, B., Brockett, C., & Quirk, C. (2005). Microsoft research paraphrase corpus. *Retrieved March 2005* from http://research.microsoft.com/research/nlp/msr paraphrase.htm.
Duthie, R., Lawrence, J., Budzynska, K., & Reed, C. (2016). The CASS Technique for Evaluating the Performance of Argument Mining. *Proceedings of the 3rd Workshop on Argument Mining* (pp. 40-49). Pennsylvania: Association for Computational Linguistics.
Eemeren, F. H. van (2010). *Strategic maneuvering in argumentative discourse: Extending the pragma-dialectical theory of argumentation*. Amsterdam: John Benjamins.
Hirst, G. (2003). Paraphrasing paraphrased. Paper presented at *Invited talk at the ACL International Workshop on Paraphrase.* Sapporo, Japan.
Janier, M., Lawrence, J., & Reed, C. (2014). OVA+: An argument analysis interface. In S. Parsons, N. Oren, C. Reed, F. & Cerutti, (Eds.), *Proceedings of the Fifth International Conference on Computational Models of Argument (COMMA 2014)* (pp. 463-464). Pitlochry: IOS Press.
Janier, M., & Reed, C. (2017). Towards a theory of close analysis for dispute mediation discourse. *Argumentation*, *31*(1), 45-82.
Konat, B., Budzynska, K., & Saint-Dizier, P. (2016). Rephrase in argument structure. In *Foundations of the Language of Argumentation COMMA 2016 Workshop*. ACM.
Lawrence, J., Bex, F., Reed, C., & Snaith, M. (2012). AIFdb: Infrastructure for the Argument Web. In *Proceedings of the Fourth International Conference on Computational Models of Argument (COMMA 2012)* (pp. 515-516). Vienna, AT: IOS Press.
Lawrence, J., & Reed, C. (2014). AIFdb Corpora. In S. Parsons, N. Oren, C. Reed, & F. Cerutti, (Eds.), *Computational models of argument*, Frontiers in artificial intelligence and applications (pp. 465-466). Netherlands: IOS Press.

Lewiński, M. (2011). Towards a critique-friendly approach to the straw man fallacy evaluation. *Argumentation*, *25*(4), 469-497.

Lewiński, M., & Oswald, S. (2013). When and how do we deal with straw men? A normative and cognitive pragmatic account. *Journal of Pragmatics*, *59*(B), 164-177.

Macagno, F., & Walton, D. (2017). *Interpreting straw man argumentation: The pragmatics of quotation and reporting*. Amsterdam: Springer.

Mirkin, S., Berant, J., Dagan, I., & Shnarch, E. (2010). Recognising entailment within discourse. In *Proceedings of the 23rd International Conference on Computational Linguistics (Coling 2010)* (pp. 770-778). Beijing, China.

Oswald, S., & Lewiński, M. (2014). Pragmatics, cognitive heuristics and the straw man fallacy. In Herman, T., & Oswald, S. (Eds.), *Rhétorique et cognition: Perspectives théoriques et stratégies persuasives* (pp. 313-343). Peter Lang, Bern.

Peldszus, A., & Stede, M. (2013). From argument diagrams to argumentation mining in texts: a survey. *International Journal of Cognitive Informatics and Natural Intelligence (IJCINI)*, *7*(1), 1-31.

Peters, G., & Woolley, J. T. (2015). Democratic candidates debate in Des Moines, Iowa, November 14, 2015. Accessed 11 Aug. 2017.

Peters, G., & Woolley, J. T. (2016a). Presidential debate at Hofstra University in Hempstead, New York, September 26, 2016. Accessed 11 Aug. 2017.

Peters, G., & Woolley, J. T. (2016b). Presidential debate at the University of Nevada in Las Vegas, Nevada, October 19, 2016. Accessed 11 Aug. 2017.

Reed, C., Wells, S., Rowe, G., & Devereux, J. (2008). AIF+: Dialogue in the argument interchange format. In Besnard, P., Doutre, S., & Hunter, A. (Eds.), *Proceedings of the 2nd International Conference on Computational Models of Argument (COMMA 2008)* (pp. 311–323). Amsterdam: IOS Press.

Searle, J. R. (1969). *Speech acts: An essay in the philosophy of language*. Cambridge University Press.

Talisse, R., & Aikin, S. (2006). Two forms of the straw man. *Argumentation*, *20*(3), 345352.

Walton, D. (1996). The straw man fallacy. In J. van Benthem, F. H. van Eemeren, F., R. Grootendorst & F. Veltman (Eds.), *Logic and Argumentation*, pp. 115-128. Amsterdam: Royal Netherlands Academy of Arts and Sciences.

Walton, D. (2016). Some artificial intelligence tools for argument evaluation: An introduction. *Argumentation*, *30*(3), 317-340.

Zanzotto, M. F., Pazienza, T. M., & Pennacchiotti, M. (2005). Discovering entailment relations using "textual entailment patterns". In *Proceedings of the ACL Workshop on Empirical Modeling of Semantic Equivalence and Entailment* (pp. 37-42). Association for Computational Linguistics.

64

Inferences Across Normative Domains

SHELDON WEIN
Saint Mary's University
sheldon.wein@gmail.com

Most societies have several normative social institutions—etiquette, morality, religion(s), and a legal system—which establish and order social cooperation. Argumentation theorists should provide guidance on what arguments are appropriate when norms from different systems give divergent advice. I argue that even if morality is properly characterized as what one should do *all things considered*, one should not accept the argument that morality necessarily overrides other norms. Doubts are raised about using the metaphor of weighting reasons to deal with issues of conflicting norms.

KEYWORDS: addative fallacy, argumentation, conflict, morality, normativity, weighing reasons

1. INTRODUCTION

Most societies have several normative social institutions which serve to establish and order social cooperation. Almost every society has traditions one is expected to respect, a system of rules of etiquette, a morality, religion(s), and a legal system to help ensure that individual actions are sufficiently well coordinated to enable the members of that society to carry on with their lives in ways that permit others to do so also. Of course, through much of our history these normative institutions have not always served all members of society equally well. While it may be that—as conservatives sometimes argue—the survival of these sorts of normative structures shows that on the whole they benefited those societies which had them more than normative free anarchy would have, all of us would grant that some such systems have been defective in various ways. Here I wish to set aside discussion of the fraught matters of the justification for these normative domains and the issues over just what the content and range of such individual systems ought to be. Instead I want to turn to the different, and much ignored, issue of how one should deal with claims from different normative domains when they yield conflicting advice.

Argumentation theorists, logicians, moral philosophers, theologians, scholars of jurisprudence, and etiquette experts have worked both to describe acceptable forms of inference within these domains and to recommend improvements (both by identifying common fallacious inferences and by suggesting new and improved forms of inference). Often—particularly in the case of religion and law—this work is dominated by what methodology to use in interpreting canonical texts. [1] And it is fairly common for the products of inferences within one domain (conclusions reached within that domain) to be accepted within other domains. (For example, people regularly go from the conclusion that they have a religious obligation to do x to thinking they have a moral obligation to do x.)[2] Yet argumentation theorists have done little work on what warrants such inferences, even though some such inferences are clearly unwarranted. There may be a valid, indeed, a sound argument that I have a legal right do y, yet it would be a mistake to infer from that fact that I am morally entitled to act on that legal right or that my religion permits such acts. Special problems occur because the reasons provided by (at least some of) the domains are thought to be exclusionary reasons.

Argumentation theorists and rhetoricians need to devote greater attention to exploring what resources might be used to assist us in deciding when inferences from one normative domain to another are warranted and when they are not. That is to say, we need to work on developing an acceptable set of standards to follow to increase the probability of making only warranted inferences in this area. This, it seems to me, is a huge task which deserves our collective attention.

In this paper I will confine myself to warning against three (very tempting) blind alleys. The first is that we should just weigh reasons from one domain against those from another when such conflict arises. I will first show why this is a tempting strategy and then argue that we should not give into this temptation for the same sorts of reasons that we should not give into thinking that within the domain of morality we should weigh reasons. The second tempting blind alley is that we should think of morality as what one should do all things considered and hence that in determining how to balance considerations from various

---

[1] See Wein 2015 for an illustration of this.

[2] This may be partly because of the halo effect. See Matey 2016.

normative domains we should just do the moral thing. There are (at least) three reasons why this is erroneous:

- Common sense morality (and hence those people who adopt common sense morality) does not hold that what one is morally required is what one is all things considered required to do.
- Even if morality were (in some metaphysical sense) what one should do all things considered, that is not what people take their morality to require of them.
- If morality was nothing but what one should do all things considered and people came to believe this, morality would just vacuously be the thing one should do once one had weighed considerations from competing normative domains. Hence "doing the moral thing" would amount to nothing more than a uninformative label for the very questions we are seeking to answer—we would still need to know everything we now don't know about how to make inferences across normative domains.

Finally, I consider and reject the view that morality is the normative domain that trumps other normative domains. Once we have rejected the mistaken view that morality is what one should do all things considered, it is pretty easy to see that this view is also unsupportable. Of course, one should often do what morality requires, even when doing so conflicts with other norms. Moral demands are strong ones. And typically when there is a conflict between morality and (say) the law, or tradition, or etiquette, the requirements of morality are the ones that should guide us but it does not follow from this that morality trumps other normative domains.

    I conclude with some suggestions about what light these blind alleys throw on what it likely to be the best path forward.

2. THE SORT OF PROBLEM

Argumentation theorists and logicians have been extremely useful to those working within individual normative domains. Moral reasoning, legal reasoning, aesthetic judgments, and even (though to a lesser extent) reasoning in etiquette and the adhering to traditions and customs have all benefited from academic work in these areas. Yet the same is not true when the requirements of different normative domains conflict. I may (correctly) believe that the law requires something of me which my religion looks on unkindly, or that would involve me doing something I think will be rude to my neighbours. Being polite or keeping my promise may require me to do something my society's legal system forbids or my religion condemns. And so forth.

Of course, people frequently decide what they would like to do and then ask themselves whether doing so accords with the various constraints imposed upon them by the normative systems they accept. Dorothea may form the idea of going to the beach on Saturday but then ask herself if any of the normative systems she accepts generate obligations which would conflict with her spending the day at the beach. Has her dermatologist told her to avoid the beach until test results have arrived? Is it her weekend to take care of the children? Has she put in her volunteer hours for the month at the hospice? Has she promised to meet a friend for lunch on Saturday? Does going to the beach conflict with Saturday prayers? And so forth.[3]

In both sorts of cases, argumentation theory and logic are of surprisingly little help. Sometimes, of course, we do get help. If I reason carefully and imaginatively, I may see (what was not obvious before) that there is more than one way to fulfil my obligations within a particular domain—and sometimes this will enable me to avoid an apparent conflict of obligations. But when we cannot avoid conflict—when, say, doing what is polite would violate the law or my religion—we cannot simply turn to argumentation theory or logic for advice. Or if these areas do provide advice, it is so vague or general as to be of no help.

3. WEIGHING REASONS AND THE ADDATIVE FALLACY

It is extremely difficult to resist the temptation to think of reasons as having weight. Values seem to have weight—or something very like it—and since values are what make reasons reasonable (or what make a reason our reason), it seems natural to think of some reasons as weightier than others. But, as I will argue, this is a mistake argumentation theorists need to avoid. I begin with a couple of clarifications. On these matters I have nothing new to say, but I will suppose for the sake of this discussion that things are as favourable as possible to the position that we can weigh reasons. Only then will I turn to the case against that view.

One might think that not all values within a particular domain are commensurable. So, within the domain of painting it might be

---

[3] I thank Michael Gilbert for pointing out the fact that typically this is the way normative obligations enter our thinking.

difficult to say whether Rembrandt's or Vermeer's paintings have greater artistic value. This might be so even if one grants that (say) *The Night Watch* is of more value than *The Little Street*. And one might be reluctant to say whether Vermeer's works (taken as a whole) are more or less valuable than (say) Bach's or Jane Austen's. So even within aesthetics one might hold that values are incommensurable. Moving to other realms might cause even more concerns. Was Galileo more important in the rise of science than Descartes? Was he more important—provide us with more value—than Bach or the Beatles?

For present purposes, however, I am willing to allow that all things of value are commensurable. My pain at reading the President's next tweet is either greater than or less than or equal to the anguish you felt when you twisted your ankle four years ago, which is greater than, less than, or equal to the disturbing fear Jones's nightmare caused her and so on. Similarly for positive values, we can suppose that positive values can be compared with negative ones. So the pain I suffered getting a hepatitis vaccination plus the satisfaction I experienced swimming in the contaminated river was either greater than, less than, or equal to you not getting the vaccination and not swimming. Even granting all this it does not follow that we can weigh reasons (in other than a metaphorical sense).

To illustrate this let me utilize a (slightly modified) version of Peter Singer's famous example of the child in the wading pool (Singer, 1972; Wein, 2001). On my way to work I see a child in a wading pool, and it looks as though she may drown. I have adequate reason to go in to save her even though doing so will muddy my clothes (and cost me a cleaning bill greater than the amount which Oxfam needs in order to save a child starving in some distant land). Now suppose (contrary to Singer's story) that as I dash towards the pool I see another child—a boy—who also appears to be about to drown. I can easily save them both (and doing so will not add to my clothes cleaning costs). Surely I should do so. But we would not say that upon seeing the second child I had twice the reason (or that my reason was twice as strong) for entering the pool. Knowledge of the second child gives me somewhat more reason to enter the pool but not twice the reason even though the value I provide by my action is twice the amount.[4] Reasons just do not work like this. Saving the girl gave me enough reason to enter the pond. Seeing the boy gives me some more reason but not twice much. (And

---

[4] Actually, very slightly more than twice the amount because there is no increase in clothes cleaning costs to weigh against the value of saving the boy; saving the girl already dirtied my clothes.

note that had I seen the boy first, the situation would be parallel. Seeing him would have given me ample reason to enter the pond and then subsequently seeing the girl would have only added to my reason—but it would not have added nearly as much reason as I had when I first saw only her in the pool.)

How reasons function and gain or lose weight or power is a difficult matter. But I take the above to be sufficient to show that we should all avoid what Shelley Kagan has aptly called the addative fallacy (Kagan, 1988). Reasoning about what to do when different normative domains give conflicting advice is necessarily going to be something different than—and more complex than—just weighing different values against one another. How this reasoning properly goes, and what one needs to do to avoid error, is an important and under-explored issue.

4. MORALITY

Some philosophers are inclined to the view that the moral thing to do is that which *all things considered* one should do. There may be some sense in which this is correct. But in general, thinking of morality in this way—at least for the purposes we are concerned with here, that of figuring out how we should react when our different normative systems provide us with conflicting advice—leads to confusion. I will look at a few ways this is so.

First, common sense morality does not hold that morality requires what one should do all things considered. Rather, it requires much simpler things such as not harming others, not killing them, not lying, keeping one's commitments, respecting other people, not stealing, being charitable, and a range of behaviour regarding sexual matters. Typically (and appropriately) common sense morality says little or nothing about what to do in cases of conflict among these things. That is because there usually is not much conflict between the various requirements. And when there is conflict between the various requirements of common sense morality, we have good ways—complex moral theories, well worked out accounts of moral reasoning, veils of ignorance, game theory, techniques such as reflective equilibrium, and so forth—to deal with it. One might characterize doing any of these things as figuring out what all things considered one should do. But obviously that is just to make *all things considered* a garbage category of no intellectual value.

Second, suppose I am wrong in this last claim. (I am not, but just suppose so for the sake of the *reductio*.) Morality is in some deep sense what we should do all things considered. (It is not what Plato or

Aristotle or Aquinas or Bentham or Kant or Mill or Anscombe or Rawls or Hare or Satz or Street or whoever your favourite moral philosopher is thinks it is. It is what all things considered one should do.) Note that this is of no interest for our present purposes. We want to know how to reason when morality as we accept it tells us one thing and some other normative system tells us another thing. Consider an example not from the moral domain. I believe that my legal system tells me I should not do something, but not doing that thing would by the standards of etiquette in my society be rude. Note how foolish it would be for you to say to me.

> Here is how you should resolve this. First, you misunderstand the requirements of the legal system. Your methodology for interpreting the legal system is flawed. You think laws are commands of the sovereign but as Hart and others have shown...

So what? Maybe I do misunderstand the nature of law. Nonetheless I think the law tells me not so do something and that my failure to do that thing would be very rude. I may be wrong about this also; perhaps following the law here is, from the point of view of etiquette, perfectly acceptable. But I think otherwise. And I want to know how to reason when I am in such situations. Telling me I am not in that situation does not help me. It does not answer my question. For, even if I am not now in that situation, I surely may be someday actually be in such a situation since norms from different systems sometimes conflict. What I need from argumentation theorists is advice on how to reason when I am in such situations.

By the way, we should note here that saying that all things considered I should, in the circumstances being considered, do the polite but illegal thing (or that all things considered, I should do the legal but rude thing) would not be, in the ordinary sense of the term, to give moral advice. It would just be to tell one the outcome of reasoning about how one should deal with this case of conflicting advice from different normative systems.[5] And our question is how to properly reach that outcome.

Third, if we did take seriously the idea that moral rules were nothing but what one should do all things considered, morality would then not be a normative system like the others. The term "morality"

---

[5] Of course, the philosopher who holds the all things considered view could bite the bullet here and say this is a case of moral advice. But this trivializes moral reasoning—the very fate for morality I expect they were seeking to avoid!

would just be a term for figuring out what other normative systems tell us to do (Or weighing them, if you are a fan of the addative fallacy!) On this view, those who hold that morality is what one should do all things considered are really moral nihilists or moral eliminativists.[6] They hold that there is no such thing as morality as normally understood. Of course, they want to re-purpose the term "morality" for what one should do when one believes that one is subject to norms from various non-moral domains and one needs to sort out what to do given that these sometimes yield conflicting requirements. But that really amounts to holding that there is no morality but—not wishing to admit this—simply saying that what moral nihilists do when deciding among competing values is moral reasoning.

Furthermore, in all those cases where there is no conflict between different moral domains, the reasoning to follow the dictates of non-moral domains is (magically) going to be converted into moral reasoning. At dinner I consider which knife to use. Etiquette requires using the butter knife. There is no law, tradition, custom, prudential reason, or (in the ordinary sense) moral reason not to use the butter knife. What should I do? On the view that morality is what one should do all things considered (and here the only thing to be considered is that it is good manners to use the butter knife), I should use the butter knife *and* this is a moral duty!

5. MORALITY AS TRUMPS

Ronald Dworkin (1968, 1985) is famous for, among other things, making the idea that rights are trumps over other considerations a staple of moral and legal reasoning. Let us confine our consideration to morality for the moment. Grant *arguendo* that moral rights trump other moral considerations. Thus if you have a moral right to, say, eat the Swiss chocolate in your possession, it would be wrong of me to take it even if I would enjoy the chocolate more than you, or it would harm your health—even lead to your early death—if you eat it. Thus, your right to eat the chocolate is more important than putting the chocolate

---

[6] Amoralists do not care about heeding the demands of morality. Moral nihilists hold that no such demands exist. Of course, moral nihilists (at least typically) acknowledge that folks standardly do think there are such things as moral requirements. They simply hold that the folks are mistaken in this view either because they have not thought about it or because, when they did, they made some sort of philosophic error. See Joyce 2001 for an account of what those errors might be.

to a utility maximizing use or even preventing an unnecessary death. Moral rights trump other moral considerations (at least, so we are now supposing).

Obviously, it does not follow from this that moral considerations always trump other considerations. It does not even follow that moral rights trump other rights. Because a certain type of consideration is a trump *within* a particular normative domain, it does not follow that it trumps considerations *outside* that domain. Suppose that in our system of etiquette, rules governing how guests are to be treated trump rules governing the appropriate placement of cutlery at the dining table. It would not follow that the etiquette of guest-treatment trumped one's religious or legal obligations.

Nor should we think that the requirements of morality always trump those from other domains. (I grant that they are frequently, even usually, more important. I only deny they are always so.) Consider the following example. I have a (admittedly minor) moral obligation to do x for you (perhaps I promised I would do it for you at some point), but my religious leader has just announced a new rule prohibiting x-ing. Or the legislature has just passed a law outlawing x-ing, perhaps going so far to make x-ing a criminal offence. No one—except perhaps those married to the idea that morality always and necessarily overrides or trumps all other considerations—would hold that in some such circumstances the requirements of my religion or my society's legal system do not override a pre-existing moral requirement.

Such examples are, of course, rare. But this only shows the importance we attach to moral requirements. It does not show that, by their very status or nature, they trump all other requirements.

6. CONCLUSION

Different normative domains have different structures and compliance mechanisms and they serve different (though related) social purposes. It is not surprising that most of the time there is minimal conflict between the requirements of norms in one domain and those from another. But just as conflict sometimes occurs within a normative domain, sometimes it occurs between them, and more thought needs to be given to how to make inferences across moral domains.

ACKNOWLEDGEMENTS: I thank Saint Mary's University for generous funding supporting this research. Wm. Barthelemy, Duncan MacIntosh,

Thea E. Smith, and Michael Watkins provided me with helpful comments.

REFERENCES

Dworkin, R. M. (1978). *Taking rights seriously*. London: Duckworth.
Dworkin, R. M. (1985). *A matter of principle*. Cambridge, MA: Harvard University Press.
Griffith, J. (1988). Well-being: Its meaning, measurement, and moral importance. Oxford: Oxford University Press.
Hart, H. L. A. (2012). *The concept of law* (3rd ed., with Postscript replying to R. M. Dworkin). New York: Oxford University Press.
Joyce, R. (2001) *The myth of morality*. New York: Cambridge University Press.
Kagan, S. (1988) The addative fallacy. *Ethics*, 99(1), 5-31.
Matey, J. (2016) Good looking. *Philosophical Issues*, 26(1), 297-313.
Singer, P. (1972) Famine, affluence, and morality *Philosophy & Public Affairs*, 1(3), 229-243.
Wein, S. (2002) Rescuing charitable duties. *International Journal of Social Economics*, 29(1/2), 45-53.
Wein, S. (2015). Mark my words: Vindicating moral and legal arguments. In D. Mohammed & M Lewiński (Eds.), *Argumentation and Reasoned Action, Proceedings of the 1st European Conference of Argumentation, Lisbon 2015*, (Vol II, pp. 1023-1033). London: College Publications.

65

Conspiracy Ideations in Healthcare: A Rhetorical and Argumentative Analysis

ROBERTA MARTINA ZAGARELLA
National Research Council, Institute of Biomedical Technologies
(Rome)
roberta.zagarella@itb.cnr.it

MARCO ANNONI
National Research Council, Institute of Biomedical Technologies
(Rome)
marco.annoni@itb.cnr.it

> This paper investigates the relationship between conspiracy claims and their persuasive effects from a rhetorical perspective. The focus of our research is on the impact and the potential damages of conspiracy-based beliefs for the trust that citizen have in institutions and science, especially with regard to medical issues. In particular, we describe, first, the argumentative structure of conspiracy thinking and, second, we present a case study: the story of the Stamina Foundation.
>
> KEYWORDS: argumentation, biases, conspiracy, health communication, rhetoric, Stamina case.

1. THE CONTEMPORARY RELEVANCE OF CONSPIRACY THEORIES

In *The open society and its enemies,* Popper famously defined the "conspiracy theory of society" as:

> [...] the view that an explanation of a social phenomenon consists in the discovery of the men or groups who are interested in the occurrence of this phenomenon (sometimes it is a hidden interest which has first to be revealed), and who have planned and conspired to bring it about (Popper, 1945, p. 352).

According to Popper, then, a conspiracy theory (henceforth CT) always entails an explanation, and this explanation always entails the public discovery of a secretive plot or agenda.

Today, well-known examples of CTs sharing this pattern include:

i. the belief that a restricted group of powerful people controls the world – be they the *Illuminati*, the *Bilderberg* group or, in disguise, lizard-like aliens;
ii. the belief that tragic events like J. F. Kennedy's assassination and the 9/11 attacks were, in reality, inside jobs;
iii. the belief that governments are in direct contact with alien civilizations;
iv. the belief that the *Shoah* has never happened;
v. the belief that the 1971 moon landing has been staged by the NASA;
vi. the belief that some governmental agency spread chemicals from airplanes to control either the climate or the population – leaving behind "chemtrails";
vii. the belief that the government and/or Big Pharma are actively promoting deadly diseases such as HIV, Cancer and/or withholding a cure for them.

Considering these examples, it would be easy to conclude that all CTs are simply absurd and false; that only a minority of irrational individuals actually believe in CTs; and that CTs are largely innocuous for society. Recent theoretical and empirical researches, however, have proven this simplified view of CTs to be largely mistaken.

First, sometimes CTs turn out to be true, even though they may initially appear false. This is, for instance, the case both of the infamous Tuskegee syphilis experiments and of the "Watergate scandal". Significantly, in both cases, the initial theory was met with scepticism and declared "false", "impossible", "absurd" or "unbelievable", and yet in both cases that initial scepticism proved to be unjustified. In the end, it was true that the American government was conducting unethical experiments using humans as "guinea-pigs"; and it was also true that Richard Nixon's administration attempted to cover-up its involvement in the Watergate scandal. Thus, sometimes a CT may appear unbelievable and nonetheless be true.

Second, contrary to a popular conception, CTs are not the prerogative of a tiny minority of crazy, "tinfoil hat-wearing men who spend most of their time in bunkers" (Miller et al., 2016, p. 824). Rather CTs cut across demographics and political attitudes and are common in all countries across the globe (Byford, 2011). Furthermore CTs are also pervasive; for instance, Oliver & Wood (2014, p. 953) report that across recent nationally representative surveys, "over half of the American population consistently endorse some kind of conspiratorial narrative about a current political event or phenomenon". Therefore, CTs appears to be widespread regardless of which demographic, political and geographical factors are considered. Ironically, one of the few relevant correlations found is the one between a high level of education or knowledge about a specific subject and the tendency to believe in CTs (e.g. Hamilton, 2011).

Third, CTs may directly influence individuals and society far more than we tend to assume (Danblon & Nicolas, 2010; Lewandowsky, Oberauer & Gignac, 2013; Taïeb, 2010; Sunstein & Vermeule, 2009; Taguieff, 2013). In fact, even though a theory may strikes *us* as absurd, *other* people may still be ready to believe and act on it, and this may cause significant public consequences. For instance, Lewandowsky, Oberauer and Gignac (2013) show that the diffusion of CTs about the safety of vaccines has been correlated with a decrease in vaccination rates with negative public-health effects. Indeed, "parents' decisions to forego vaccinating a child, arising from the belief that they are being misled about the safety of vaccines, can have significant negative consequences for public health" (Miller et al., 2016, p. 825). Moreover, evidence suggests that CTs may also affect those who openly discount them (Miller et al., 2016). In the case of vaccination, this entails that CTs about the risks of vaccines may unconsciously influence also the behaviour of those who openly consider such theories false. Finally, conspiracy theories seem to be inducible – at least to some extent. In fact, as showed by Oreskes and Conway (2010) and Swami et al. (2011) it is possible to induce people to believe in new CTs, especially if those people already believe in other CTs. Thus, the beliefs in CTs are to some extent inducible, and people might actually alter their behaviour in response to them, both consciously and unconsciously.

Therefore, the naive view of CTs is wrong: not all CTs are absurd and false; CTs are widespread and pervasive on a planetary scale; and CTs may directly influence individual and collective behaviour with significant implications for society.

This conclusion raises a host of interesting theoretical, practical, and socio-political questions. Clearly, given their implications, understanding what CTs are, and how they originate and propagate, may be important to prevent significant harms at both the individual and the social level. This is especially true whether CTs call into question citizen's trust in science and biomedicine, like in the case of CTs about the risks and effectiveness of vaccines. Indeed, when it comes to health and disease, believing in a CT can sometimes make the difference between life and death.

It is also worth noting that other studies have found "empirical confirmation of previous suggestions that conspiracist ideation contributes to the rejection of science" (Lewandowsky, Oberauer & Gignac, 2013), and "growing indications that rejection of science is suffused by conspiracist ideation, that is the general tendency to endorse conspiracy theories including the specific beliefs that inconvenient scientific findings constitute a 'hoax'" (Lewandowsky, Gignac & Oberauer 2013).

Against this background, the aim of this paper is to investigate the relationship between conspiracy claims and their persuasive effects from a rhetorical and argumentative perspective. What follows is divided in three parts. First, we present our perspective, explaining why, in our view, it is important to study conspiracies from a rhetorical and argumentative point of view. Second, we present our main case-study: the story of the Stamina Foundation. Finally, we explore the argumentative structure of the Stamina conspiracy through the lenses of the three "artistic" or "entechnic" proofs of classical rhetoric – or *logos*, *ethos* and *pathos*.

2. TWO COMPLEMENTARY APPROACHES TO THE STUDY OF CTs

Today, the most popular approaches used to conceptualize and study CTs make extensive use of concepts and methodologies derived from psychology and cognitive sciences. According to these "psycho-cognitive approaches", in order to explain how and why people endorse conspiratorial mindsets we have to look at what these psycho-cognitive disciplines have to say about how we think, reason and, ultimately, decide. The following quote well summarizes the core assumption shared by all these accounts:

> Although conspiracy beliefs can occasionally be based on a rational analysis of the evidence, most of the time they are not. As a species, one of our greatest strengths is our ability to find meaningful patterns in the world around us and to make causal inferences. We sometimes, however, see patterns and causal connections that are not there, especially when we feel that events are beyond our control (Buckley, 2015).[1]

---

[1] Buckley on C. French's approach. See also Brotherton, 2015; Brotherton & French, 2014, 2015; Brotherton, French & Pickering, 2013; French & Stone, 2014; Leman & Cinnirella, 2013; Quattrociocchi & Vicini, 2016.

On this view, the study of CTs equates – entirely or to a large extent – with the study of the cognitive biases[2] and mental processes leading to a belief in such theories. Roughly put, for these approaches, CTs should be explained in terms of the failures of our abilities to think and decide rationally.

Even though we agree that empirical psychology and cognitive sciences provide critical insights about the nature of CTs, we also argue that a purely psycho-cognitive approach illuminates only part of the conspiracy phenomenon. In particular, we maintain that purely psycho-cognitive approaches could incur into two potential shortcomings.

First, framing the phenomenon of CTs only through the lenses of a psycho-cognitive approach may lead to the misleading conclusion that whoever come to believe in a conspiracy is then, *ipso facto*, irrational. As we noted above, it seems too simplistic to label CTs narratives just as fantasy products or to assume that those who believe in conspiracies are merely irrational or overwhelmed by their emotions. On the contrary, as in the case of the two *Washington Post* reporters who uncovered the Watergate scandal, sometimes elaborating, believing and acting on a CT may be a sign of a properly conducted rational inquiry, rather than of fallacious reasoning.

Second, by equating all conspiracies with irrationally held beliefs, one is led to think that, in order to control for or disprove a conspiracy theory, what is needed is simply a higher dosage of rationality. However, integrating the analysis of mental issues with the study of the discursive context complicates the picture. In fact, by focusing primarily on psychological and cognitive variables, psycho-cognitive approaches tend to overshadow other factors that may be important to understand how CTs emerge and evolve. Aside from the various cognitive biases that allow for, facilitate, or reinforce the spread of CTs within a population, what makes a CT more persuasive than another? How can the contents of conspiracies be organized and presented in order to maximize their persuasive impact? What difference it makes, for the success of a CT, the way in which its proponents present themselves to potential believers?

In difference to standard approaches, we argue that these questions are relevant to understand the nature of conspiracies, but require the adoption of a different theoretical and methodological framework, one that draw from the conceptual tradition of classical rhetoric and argumentation theory rather than from psychology. This

---

[2] Such as the confirmation bias or the proportionality bias analysed by Christopher French in his studies (cf. References).

rhetorical perspective builds on the work of other scholars who, in the last decade, have begun to hint and underscore the peculiar argumentative dimension of conspiracies (Angenot, 2008; Danblon & Nicolas, 2010; Dominicy, 2010; Nicolas, 2014; Oswald, 2016)[3]. In synergy with these perspectives, in what follows we propose to consider CTs as a specific form of argumentation and ratiocination endowed with its own internal logic (cf. Angenot, in Danblon & Nicolas, 2010, p. 28).

In sum, in this paper we propose to investigate CTs from a rhetorical point of view as a particularly kind of argumentative strategy. Our claim is that, by making explicit the typical argumentative and inferential structure shared by virtually all CTs, it becomes possible to highlight important aspects of the conspiracy phenomenon that would otherwise remain unthematized by adopting standard cognitive accounts.

3. A RHETORICAL PERSPECTIVE ON CONSPIRACIES

On the surface, each conspiracy is unique. However, as noted also by Popper (c.f. introductory section), we believe that it is possible to uncover a shared argumentative pattern that, like a *fil rouge*, runs through all typical instances of CTs. This patter, we argue, involves at least three elements: (1) the observations of a puzzling phenomenon (someone's death; a pattern or variation in the world-economy; bright lights in the night sky, etc.); (2) an explanation of this puzzling phenomenon in terms of someone's secret agenda; (3) the public exposure of this secret agenda. Our working assumption is that this argumentative structure is common to all CTs.

In order to unpack this claim, in the following sub-sections we shall, first, introduce our case study – the story of the Stamina Foundation –, and then analyse it through the lenses of the three "artistic" or "entechnic" proofs of rhetoric: *logos, ethos,* and *pathos* (Aristotle, *Rh.*, 1356a)[4]. Specifically, we shall analyse, respectively, how the contents of this conspiracy theory have been presented; how the

---

[3] Angenot, 2008, p. 336: «La pensée conspiratoire a retenu l'attention des rhétoriciens dans la mesure où elle est particulièrement argumentative et ratiocinante».

[4] Aristotle distinguished between "entechnic" and "atechnic" proofs. The orator develops the "entechnic" proofs by following a method in a speech; the "atechnic" proof, instead, rely on the provision of physical objects (such as written documents, confessions, etc.). Cf. Aristotle, *Rh.* 1355b-1356a.

orator's credibility has been construed; and how emotions have been recruited by the media in this specific case.

3.1 The Stamina case

A few years ago, the Stamina case spurred in Italy a ferocious debate over issues like the role of science and experts in society, the rights of citizen to receive unproven treatments when all therapeutic options are exhausted, and the role of the State in deciding how "the truth" about the efficacy of a therapeutic regime ought to be established. Given our present constrains, however, here we shall outline only the main aspects of this case[5].

In 2009 two controversial Italian figures – a professor of communication and psychology, Davide Vannoni, and a physician, Marino Andolina, created the Stamina Foundation, a private initiative aimed at administering an innovative "therapy" for incurable neurodegenerative conditions – like the Amyotrophic Lateral Sclerosis (ALS). The treatment consisted in the infusion of mesenchymal stem-cells and had never been tested in controlled trials; all the evidence about its efficacy and safety was therefore purely anecdotal. Yet, by exploiting a series of legal loop-holes, Vannoni and Andolina eventually managed to administer their "method" in a public hospital as a special kind of compassionate and off-label therapy.

Then, in 2013, a popular Italian TV show –"Le Iene Show"–, took up the case in a repeated series of one-sided reports, igniting an animated public debate. On the one hand, the scientific community condemned the Stamina method as being unscientific, unsound, and dangerous. On the other hand, the public opinion, patients' organization, music and TV stars, supported it and demanded it to be available and free for all.

In the resulting epistemic, political, and institutional short-circuit, the Italian Government first decided to run a special clinical experimentation, and then desisted due to the increasing pressure coming from the international scientific community. Eventually, the clinical trial was not run, and Vannoni and Andolina were charged and found guilty of having fabricated their data, deceived the patients, and caused aggravated injuries.

Today, in Italy, the Stamina case evokes the perils of believing in conspiracies about "miraculous cures". For our purpose, however, this story is important because it provides an ideal case to isolate the typical

---

[5] See also Cattaneo & Corbellini, 2014; Abbott, 2016; Capocci & Corbellini, 2014.

features of conspiracies at the level of their rhetorical and argumentative structure.

3.2 The logos of the Stamina case and the role of abduction in conspiracy arguments

Aristotle defined the method to present the contents of an argument as the proof based on "logos" (Aristotle, *Rh.*, 1356a). Indeed, there are many ways in which a logical argument can be construed and presented. In our view, however, the distinct persuasive effect of CTs is often achieved by making use of one particular argumentative strategy based on *abduction*[6]. According to the philosopher C. S. Peirce, *abduction* is an inferential process, different from both deduction and induction, which is characterized by the following structure:

> The surprising fact, C, is observed.
> But if A were true, C would be a matter of course.
> Hence, there is reason to suspect that A is true (Peirce, 1931-1958, 5.189)

Notice that, for Peirce, abductive inferences are triggered in response to a "surprising" or "puzzling" fact that appears inexplicable given our current knowledge. In the Stamina case, this puzzling fact (C) was the existence of a "cure" that had been inexplicably ignored by the scientific community, the media, and the National Health System[7].

Yet, crucially, while for Peirce abduction characterizes the initial stage of *any* rational inquiry – from medical diagnosis to legal case building – in conspiracy argumentations, instead, it tends to become the centre of the whole explanatory strategy. This occurs because, contrary to other kinds of inquiry, CTs base their explanatory and persuasive power on the existence of some *secrets*.

In fact, once the surprising fact (C) is observed and acknowledged, a question arises: How can we explain (C)? Or, in the Stamina case: How is it possible that no one talks about a revolutionary and effective cure?

---

[6] See also Danblon, in Danblon and Nicolas, 2010.

[7] The surprising fact (C) may turn out to be merely anecdotal, indirect and hypothetical – as in the case of the "extraordinary medical progress"; "important results"; "improvement of critical condition", "certified by doctors of several public hospitals" that some patients and the Le Iene Show have hyperbolically reported in the Stamina case. Cf. Le Iene Show (10/03/2013; 17/03/2013), https://www.iene.mediaset.it/.

To provide an answer to this kind of questions, in CTs an abductive inference is formed. This abductive inference serves to explain the surprising fact (C) by linking it to someone's *secret agenda* (A). As a result, the reason why (C) appears *puzzling* to us – i.e. surprising given what we currently know – is explained, in conspiracies, by the appeal to the *secrecy* of (A). After all, a secret is, by definition, something unknown: hence the sense of "surprise". In the Stamina case, this "secret agenda" silencing the cure was attributed to Big Pharma protecting its economic profits, and to the self-interest and envy of the dominant scientific establishment[8]. The presence of this *explanatory secret* is also what motivates some authors to talk about the "self-sealing quality" or "self-insulating nature" of CTs (Sunstein & Vermeule, 2009; Oswald & Herman, 2016; Byford, 2011; Wood & Douglas, 2013). Since (A) was meant to be a secret, for the supporters of a CT, it is often perfectly natural to expect that the evidence about (A) is scarce or non-existing. In this way, the lack of evidence to confirm (or refute) the CT is recruited and cited as evidence in favour of the existence of a secret conspiracy. Thus, a typical CT may sound plausible even though it lacks any supporting evidence (Angenot, 2008, p. 366 ff.).

Moreover, anchoring the explanation of (C) to the secrecy of (A) has the further effect of shifting the burden of proof: rather than verifying the causal relationship between (A) and (C), it challenges others to falsify it. Accordingly, it is not those who propose the CT that have to supply the necessary evidence for its existence but, rather, those who attack it. And even though some evidence is eventually found, it may still not be enough to finally disprove the CT. As other scholars have noted, a conspiracist ideation is difficult to correct because any evidence that contradicts it may be considered as a further "proof" of its existence. For instance, in the Stamina case, the fact that the final clinical experimentation was not conducted by the Government was interpreted by Vannoni's supporters not as the final proof that the Stamina method lacked a scientific experimental protocol, but as the proof that a conspiracy was actually run by politicians, Big Pharma and the scientific establishment in order to silence the Stamina Foundation.

On this issue, it is interesting to note that the argumentations used by Vannoni and his supporters resulted, at least initially, highly persuasive for the public opinion. Those argumentation were meant to cast doubts not just on the issue at hand (in that case, the efficacy of the Stamina method and the secret reasons to boycott it), but also on the institutions involved, as well as on the *bona fide* of the other scientists.

---

[8] Le Iene Show (17/03/2013).

To account for this latter aspect, we need to introduce a second factor in our analysis: the *ethos*.

3.3 Ethos in conspiracy arguments: the Stamina case

As noted by Danblon, one useful criterion to analyse the *ethos* in contemporary societies is to look for the orator's degree of "institutional integration" (Danblon, 2005, pp. 173-176). This criterion, we argue, is especially important in studying CTs, where the orator typically endeavours to ground his credibility by presenting himself as the underdog who bravely denounces the conspiracy for the sake of the common good.

To achieve this effect in a CT, the orator must successfully construct a conflict between two opposing factions: the official and the unofficial, the system and the anti-system, the establishment and the anti-establishment. More precisely, in order to craft a persuasive discourse, in a CT the orator must: (1) identify an "enemy" to be ostracized (the conspirators); (2) adopt for himself a moral role or mission (to denounce and fight the conspiracy); (3) and build for himself a standing as a dissenting voice. In this way, he does acquire – at least in the eyes of those believing in his story – the status of a hero: someone who has the courage to tell the truth and that, in doing so, is ready to take some risks in order to preserve his and others' moral integrity – a *parrhesiastes* in Foucault's sense.

In the Stamina case, the persuasive effect due to the orator's specific *ethos* was achieved through three different strategies. The first strategy was the one used by Vannoni who, being no doctor, made up for his lack of medical expertise by claiming that he had invented the Stamina method after he had been cured of a semi-facial paralysis by a similar procedure. By grounding the credibility of his claims in this personal story, Vannoni successfully managed to construct for himself a very powerful and persuasive *ethos*. First, the now barely visible signs of his paralysis attested that he had directly witnessed the incredible benefits of his "method". Although he was not a doctor, he still managed to convince his audience that he had solid and direct evidence of the "method" efficacy. Second, these signs also attested that he knew what it meant to suffer from an incurable condition. This helped Vannoni to present himself not as an unscrupulous quack selling an unproven therapy to vulnerable patients, but as one of them, a normal citizen and a past patient who, because of sheer destiny, has found himself at the forefront of an epistemic and political battle[9].

The second strategy was deployed by Andolina who, instead, had a medical and scientific background. Like Vannoni, also Andolina portrayed himself as someone standing outside the establishment, and thus uninterested in money and fame, but he also grounded his credibility on his scientific competence about stem cells and on his public humanitarian efforts[10].

The third argumentative strategy was then the one used by the Italian television and other media. "Le Iene Show", especially in the first reportages, represented Vannoni as an independent researcher fighting against the big pharmaceutical corporations for the patients', and truth, sake. They explicitly attempted to draw a parallel between Vannoni and Andolina's situation and the accusations confronted in the past by other scientific anti-heroes. By evoking figures such as Galileo or Copernicus, Stamina's supporters have tried to validate their claims, reminding the audience that sometimes the greatest advance in science must face and overcome harsh resistances.[11] And, of course, the stronger the resistance is, the truer the revolutionary claims must be, thus revealing again the abductive and self-sealing dynamic that is common in most CTs.

In the Stamina case, the combination of the self-sealing *logos* typical of CTs and of the *ethos* built around Vannoni and Andolina was so powerful that, in the end, it led to a paradoxical situation. After Vannoni had been arrested for aggravated fraud, three patients that

---

[9] Le Iene Show (17/03/2013; 13/04/2013).

[10] Le Iene Show (13/04/2013).

[11] Cf. Le Iene Show (15/10/2013, from minute 12.50 to minute 13.30). The reporter Giulio Golia is interviewing Camillo Ricordi (Director of the Diabetes Research Institute and the Cell Transplant Program at the University of Miami; one of the world's leading scientists in the stem cells field):
- Golia (Le Iene Show): "Professor Ricordi, secondo lei un metodo contro cui si schiera compatta quasi tutta la comunità scientifica italiana può mai avere qualcosa di buono?"
- Ricordi: "Difficilmente il programma molto innovativo o rivoluzionario viene accettato da tutto l'establishment scientifico di quel momento. Un po' come ai tempi di Galileo e Copernico; hanno avuto anche loro i loro problemi a far passare una teoria folle: che non era il Sole ad andare incontro alla Terra ma viceversa. Ci si può trovare davanti a delle nuove ipotesi che non erano state pensate da altri prima e che danno dei risultati clinici ancora prima che uno capisca per quale meccanismo funzionano".

were recognized as *victims* in the civil trial still decided to follow him abroad to continue their "cures".

3.4 Pathos in conspiracy arguments: the Stamina case

The *pathos* proof represents another critical component in the argumentative strategy of CTs. Recall that in CTs the orator's *ethos* is construed in explicit opposition toward some powerful group of people or the dominant establishment. This entails that, structurally, in a CT the orator always represents the underdog, the one who, against all the odds, sets out to challenge the received and official version of "the truth".

Given this structure, in order to be persuasive, in a CT the orator needs to evoke a specific set of emotions in the audience. In this respect, resentment and indignation are the most characteristic emotions aroused by conspiracies (cf. Danblon, in Danblon & Nicolas, 2010, pp. 70-71). Therefore, by arousing such negative emotions against the target power or group, the orator may attempt to stir the audience and win his favour.

However, capitalizing too much on the *pathos* at the expenses of the *logos* or *ethos* may prove to be, in the end, a counterproductive strategy. In this respect, the Stamina case is, again, exemplar. Since its very beginning, the story of the Stamina Foundation was presented by the media not as the story of a newly discovered scientific cure, but as the story of Sofia, Noemi, Federico and Ludovica – young children affected by rare diseases and treated with Vannoni's method – and of their desperate families[12]. Initially, the emotional impact of these personal dramas, coupled with the parallel construction of Vannoni and Andolina's heroic *ethos,* was so powerful that it quickly overshadowed any other consideration about the scientific validity and safety of the advertised therapeutic protocol, opening a political and institutional crisis at the national level. Then, a growing series of scientists and journalists began to question both the epistemic validity of the Stamina method, and Vannoni's real interests and moral integrity. In doing so, they appropriately appealed to another set of emotions, such as the fear of enrolling children in a potentially dangerous and risky clinical trial. In response, the public support toward the Stamina method began to fade away. By the time of the civil trial, only a tiny portion of the public opinion was still in support of Vannoni; and by all accounts the highly

---

[12] Le Iene Show (10/03/2013; 17/03/2013; 23/03/2013; 12/05/2013; 27/10/2013; 12/11/2013, etc.).

personal and one-sided reportages done by Le Iene Show were cited as uncontroversial examples of an unprofessional and even unethical way of doing journalism.

Importantly, here our claim is not that an appeal to emotions in argumentation is something fallacious or morally wrong *per se* but, rather, that in reporting scientific and medical news the media have a precise series of deontological responsibilities, which extend also to the way in which they chose to strategically mobilize the emotions of their audience. Thus, emotions play a key-role in all kinds of discourses, including conspiracies, where they are crucial to facilitate socio-political mobilizations[13]. As in the case of the Tuskegee syphilis experiments, sometimes emotions like indignation may be critical in order to uncover a conspiracy and promote positive political changes. Other times, however, emotions may also be used to diffuse false and self-interested conspiracies that may have detrimental effects by jeopardizing individual and public trust in science. Indeed, a problem arises when emotions and passions, rather than being put at the service of democracy, are instead used to ignite extremist behaviours that hijack the political agenda or, even worse, have an impact on the individual and collective choices with respect to clinical experimentation and therapies.

4. CONCLUSIONS

In this paper we have investigated the relationship between conspiracy claims and their persuasive effects from a rhetorical perspective. Our conclusions can be summarized in three general claims.

First, contrary to a popular belief, not all CTs are absurd and false. They are widespread and pervasive; and they may directly influence individual and collective behaviours with significant implications. Therefore, it is increasingly necessary to provide an account of how and why CTs originate, grow and propagate, especially in reference to the health debate.

Second, standard psycho-cognitive approaches illuminate only a part of the complex phenomenon of CTs. In particular, psycho-cognitive approaches tend to focus on mental issues rather than social and discursive processes, thus leaving unthematized other crucial aspects related to how CTs achieve their distinctive persuasive effects. Thus, we

---

[13] On the relation between passions and politics see the interesting analysis of Serra (2017, p. 143 ff.).

have proposed to complement the analysis of CTs provided by standard psycho-cognitive approaches with another methodological approach derived from the conceptual tradition of classical rhetoric and argumentation theory.

Lastly, we have articulated our view that all CTs share some key-structural features from a rhetorical and argumentative point of view. This analysis, we have argued, can shed significant light on how conspiracies achieve their persuasive effect and, thus, it constitutes a first step toward the elaboration of a more comprehensive model to better address the practical and political implications of conspiracy argumentations.

AUTHOR CONTRIBUTIONS: Both authors contributed equally.

REFERENCES

Abbott, A. (2016). Stem-cell scandal gets fresh scrutiny. *Nature*, *539*(7630), 340.
Angenot, M. (2008). *Le dialogue de sourd: Traité de rhétorique antilogique*. Paris: Mille et une Nuits.
Aristotle (1959). *Ars rhetorica*, (W. D. Ross, Ed.). Oxford: Clarendon Press; 1959.
Brotherton, R. (2015). *Suspicious mind: Why we believe conspiracy theory*. New York: Bloomsbury Sigma.
Brotherton, R., & French, C. C. (2014). Belief in conspiracy theories and susceptibility to the conjunction fallacy. *Applied Cognitive Psychology*, *28*(2), 238–248.
Brotherton, R., French, C. C. & Pickering, A. D. (2013). Measuring belief in conspiracy theories: The generic conspiracist beliefs scale. *Frontiers in Psycology*, *4*, Article 279. doi: 10.3389/fpsyg.2013.00279.
Buckley, T. (2015). Why do some people believe in conspiracy theories? *Scientific American Mind*, 1 July 2015.
Byford, J. (2011). *Conspiracy theories: A critical introduction*. London: Palgrave Macmillan.
Capocci, M. & Corbellini, G. (Ed.) (2014). *Le cellule della speranza: Il caso Stamina tra inganno e scienza*. Torino: Codice Edizioni.
Cattaneo, E. & Corbellini, G. (2014). Taking a stand against pseudoscience. *Nature*, *510*(7505), 333-335.
Danblon, E. (2005). *La fonction persuasive. Anthropologie du discours rhétorique: Origines et actualité*. Paris: Armand Colin.
Danblon, E. & Nicolas, L. (Ed.) (2010). *Les rhétoriques de la conspiration*. Paris: CNRS Éditions.
French, C. C., & Stone, A. (2014). *Anomalistic psychology: Exploring paranormal belief and experience.* London: Palgrave Macmillan.

Hamilton, L. C. (2011). Education, politics and opinions about climate change evidence for interaction effects. *Climatic Change*, *104*(2), 231-242.

Leman, P. J. & Cinnirella, M. (2013). Beliefs in conspiracy theories and the need for cognitive closure. *Frontiers in Psychology*, *4*, 378.

Lewandowsky, S., Gignac, G. E. & Oberauer, K. (2013). The role of conspiracist ideation and worldviews in predicting rejection of science. *PLoS ONE*, 10(8), e0134773. doi: 10.1371/journal.pone.0134773.

Lewandowsky, S., Oberauer, K. & Gignac, G. E. (2013). NASA Faked the moon landing—therefore, (climate) science is a hoax: An anatomy of the motivated rejection of science. *Psychological Science*, *24*(5), 622-633.

Miller, J. M., Saunders, K. L. & Farhart, C. E. (2016). Conspiracy endorsement as motivated reasoning: The moderating roles of political knowledge and trust. *American Journal of Political Science*, *60*(4), 824-844. doi: 10.1111/ajps.12234.

Nicolas, L. (2014). L'évidence du complot: Un défi à l'argumentation. Douter de tout pour ne plus douter du tout. *Argumentation et Analyse du Discours*, *13*. doi: 10.4000/aad.1833.

Oliver, J. E. & Wood T. J. (2014). Conspiracy theories and the paranoid style(s) of mass opinion. *American Journal of Political Science*, *58*(4), 952-966.

Oreskes, N. & Conway, E. M. (2010). *Merchants of doubt*. London: Bloomsbury Publishing.

Oswald, S. (2016). Conspiracy and bias: Argumentative features and persuasiveness of conspiracy theories. *OSSA Conference Archive*, 168. Retrieved from http://scholar.uwindsor.ca/ossaarchive/OSSA11/papersandcommentaries/168.

Oswald, S. & Herman, T. (2016). Argumentation, conspiracy and the moon: A rhetorical-pragmatic analysis. In M. Danesi & S. Greco, (Eds.). *Case Studies in Discourse Analysis* (pp. 295-330). Munich: Lincom Europa.

Peirce, C. S. (1931-1958). *Collected papers* (Vols. 1-6, C. Hartshorne & P. Weiss Eds.), (Vols. 7-8, A. Burks Ed.). Cambridge, MA: Harvard University Press.

Popper, K. R. ([1945]2002). *The open society and its enemies*. London and New York: Routledge.

Quattrociocchi, W. & Vicini, A. (2016). *Misinformation. Guida alla società dell'informazione e della credulità*. Milano: Franco Angeli.

Serra, M. (2017). *Retorica, argomentazione, democrazia. Per una filosofia politica del linguaggio*. Roma: Aracne editrice.

Sunstein, C. R. & Vermeule, A. (2009). Symposium on conspiracy theories. Conspiracy Theories: Causes and Cures. *The Journal of Political Philosophy*, *17*(2), 202-227.

Swami, V., Coles, R., Stieger, S., Pietschnig, J., Furnham, A., Rehim, S., & Voracek, M. (2011). Conspiracist ideation in Britain and Austria: Evidence of a monological belief system and associations between individual psychological differences and real-world and fictitious conspiracy theories. *British Journal of Psychology*, *102*(3), 443-463. doi: 10.1111/j.2044-8295.2010.02004.x

Taguieff, P. A. (2013). *Court traité de complotologie*. Paris: Mille et une nuits.

Taïeb, E. (2010). Logiques politiques du conspirationnisme. *Sociologie et sociétés*, *42*(2), 265-289.

Wood, M. J., & Douglas, K. M. (2013). What about building 7? A social psychological study of online discussion of 9/11 conspiracy theories. *Frontiers in Psychology*, *4*, 409.

66

With the Best Intentions, and the Worst Arguments. The "Fertility Day" Campaign in Italy

MARTA ZAMPA
Zurich University of Applied Sciences
marta.zampa@zhaw.ch

CHIARA POLLAROLI
Università della Svizzera italiana
chiara.pollaroli@usi.ch

ABSTRACT: We analyse the argumentative and rhetorical features of the multimodal "Fertility Day" campaign, instituted by the Italian Ministry of Health in 2016. The campaign did not take into account the actual situation of the country and the reasons for its exceptionally low birth rate. This caused heated reactions, expressed in a multimodal online polylogue where the discrepancy between the inferences activated by the campaign in the public and those planned by its authors becomes evident.

KEYWORDS: institutional health campaigns, multimodality, argumentation, rhetoric, polylogue, frames, fallacious communication

1. INTRODUCTION

"Beauty is ageless. Fertility is not", says a young woman with an ironical pout, holding an hourglass. "Fertility is a common good", water drops from a tap. These are only some of the multimodal advertising postcards launched in August 2016 by the Italian Ministry of Health to promote the "Fertility Day", an event scheduled for September 22, 2016 and aimed at raising awareness on fertility issues from a public health perspective. The motive: the dramatically low birth rate in the country and the tendency to reproduce at an older age. Beyond opinions held on its intent, the campaign itself raised broad dissent and critiques as to the communicative strategies chosen. The slogans and pictures have been perceived as insulting and unacceptable because they tell a story of irresponsible young citizens, careless of their health and their country's future. This narrative disregards the true story, i.e. how economic, social and health issues compel many Italians not to have children.

In this article, we investigate the multimedia and multimodal polylogue made up by the official campaign discourse (the advertising postcards and other documents by the Minister of Health) on one side and by social media discourse on it. In particular, we consider tweets and satirical anti-fertility day postcards reacting to the campaign. Attention is devoted to how the polylogical argumentative discussion is developed (Kerbrat-Orecchioni, 2004; Lewiński & Aakhus, 2014), to the (in)appropriateness of the communicative means applied, and to the interpretation of the campaign, as expressed by the citizens and the Government. We look at the argumentative and rhetorical features of the campaign, against the backdrop of the context for which they were created – a country fighting with youth unemployment, women's discrimination and low salaries. We trace out the frames (Fillmore, 1982; Goffman, 1975; Greco Morasso, 2012; Rocci, 2009) and inferences activated in the audience and those thought of by the Minister of Health, as reconstructed and justified in their reactions on the media. To this aim, we make the contextual/cultural premises and the factual premises of the various reasoning lines explicit, relying on the theoretical framework of the Argumentum Model of Topics (Rigotti, 2006; Rigotti & Greco Morasso, 2009, 2010).

2. THE FERTILITY DAY CAMPAIGN

In 2016, the Italian Ministry of Health decided to launch a "National Fertility Plan" (NFP). Its main objectives were to inform the citizens on fertility and provide them with adequate health care, as well as to change the widespread views of fertility. Furthermore, it wished to "celebrate this cultural revolution" (NFP, p. 1) with an event – the "Fertility Day" – whose motto would be "discovering the prestige of maternity" (NFP, p. 1). We here analyse the multimodal campaign advertising the "Fertility Day", taking into account the intertextual chain of discourses (Fairclough, 1995) that it provoked.

The NFP was launched as a reaction to worrying figures concerning the reproductive habits of Italians. As of 2015, the birth-rate in Italy was 1.35 children per woman (Istat, 2015): not only one of the lowest in Europe (European birth-rate: 1.58, Eurostat 2014), but the lowest in the history of the country since its creation in 1861 (Il Fatto

Quotidiano, 19.2.2016). Furthermore, Italian women give birth at 31.67 (Istat, 2015) against a European average of 30.4 (Eurostat, 2014).

Nonetheless, this is neither a recent phenomenon nor a purely health-related one. Socio-political factors such as instable work perspectives and lack of support institutions for childcare have been curbing childbearing for the past decades. After the 2008 crisis, unemployment rates grew in the whole country (from 5.7 % in 2007 to 12 % in 2016, Eurostat on Google Public Data) and nowadays a great number of young employees only get temporary contracts (Mattoni & Vogiatzoglou, 2014). Women face an even more hostile environment: they are employed less often than men (46.6 % in 2007, against 70.6 % of Italian men and 58.6 of the women in the European average), get paid less and occupy only few high positions (Zizza, 2013).

Despite these factual premises, Beatrice Lorenzin, Minister of Health at the time of the campaign, declared in an interview on *La Repubblica* (September 01, 2016) that the "Fertility Day" aims at bringing awareness on fertility from the medical viewpoint only. It does not address socio-economic factors that play a role in deciding not to have babies, for this would lie outside of the authority and expertise of the Ministry of Health. This at least should mean that the campaign uses explicit and well-grounded health-related communication and argumentation. As the data analysis will show (Section 4), it is not the case.

3. THEORETICAL FRAMEWORK

Health campaigns stem from a rhetorical situation, which they aim at adjusting (Bitzer, 1968, 1980). People create discourses because a faulty situation – an *exigence* – requires communicative intervention to get back to the desired state. itself. In our case, the exigence corresponds to the low birth rate in Italy, a situation that the Ministry of Health does not perceive as desirable. Thus, a communicative intervention is required in order to make people aware of the issue and perform actions to fix it.

Many people are involved in communicative interventions. People who are engaged in the communication itself belong to different typologies of stakeholders, each one with their own cultural knowledge, desires, and system of values (Mazzali-Lurati & Pollaroli, 2013; Palmieri & Mazzali-Lurati, 2016). The group of addressers is usually composed by a principal (or group of principals) who is the first to perceive the faulty situation and the main responsible for the communication designed to fix it. Communication (especially mass) is usually designed by authors, on the basis of information provided by the principals. Animators, then, provide the platform where the communication is distributed (e.g., social media or online newspapers). As for mass campaigns, communicators must be aware that the audience might not even be aware of the exigence, or might perceive a different exigence. Needless to say, most mistakes in communication practices are due to a distorted account of the rhetorical situation and the stakeholders.

Social media, online newspapers and online versions of newspapers allow for a polylogical intervention in mass campaigns. As Aakhus and Lewiński (2016, p. 1) point out, "argument in contemporary controversies and deliberation involves many players, many positions, and many places" (see also 2016; 2011; 2015; Lewiński & Aakhus, 2014). Indeed, Web 2.0 is designed to enables stakeholders to participate in argumentative discussions. It is easy to share contents on social media, to comment on them and to reply to other comments as well as to add commentary texts to online newspapers or online versions of newspapers. Users are also technologically well-equipped to design new multimodal contents by manipulating existing ones.

The messages created represent the world according to a certain perspective, a certain *frame* (Fillmore, 1982; Goffman, 1975; Rocci, 2009), that is representations of situations made of entities, attributes, and events. The choice of a frame invokes different systems of values and knowledge. These cultural blocks work as sources where premises of argumentative reasoning are taken from. Following the Argumentum Model of Topics (Rigotti & Greco Morasso, 2009, 2010; Rocci, 2017) an argumentative intervention is usually enthymematic (Bitzer, 1959), that is with an unexpressed premise and/or conclusion. The major premise, for example, is usually easily recovered because it is a proposition based on knowledge or opinion already present in the stakeholders' common ground to which an argumentative communication is addressed and which, thus, can be exploited as the starting point of a critical discussion. This (usually tacit) premise is known as *endoxon* (Rocci, 2006).

Endoxa might generated misunderstanding and conflict (Greco, Mehmeti, & Perret-Clermont, 2017) when an "endoxical discrepancy" is at work (Saviori, 2009, p. 138). An endoxical discrepancy corresponds to a "clash" of knowledge and opinion at the level of the major premise. In such a case, the argumentative discussion is hindered since its very starting point. In newsrooms, for example, lack of clarity regarding *endoxa* can lead journalists to decisions not conform to the mandate Zampa (2015). This results in the production of non-adequate communication, as it is the case for the Fertility Day campaign.

It is indisputable that frames can be represented also by images or by multimodal integrations of different semiotic modalities (e.g., Fauconnier & Turner, 2002; Pollaroli & Rocci, 2015). Metaphorical and metonymical framing, for instance, are widely employed in advertising to (re)conceptualize a product or service through the exploitation of the inferences generated by merging the target frame with a situation belonging to a different frame (the source frame). As recent research on multimodal rhetoric and argumentation has shown (e.g., Kjeldsen, 2012; Pollaroli & Rocci, 2015), propositions deriving from multimodal and visual framing can contribute or correspond to endoxical and/or factual premises in multimodal enthymemes.

4. DATA ANALYSIS

In this section, we analyse the thirteen postcards created for distribution on social media together with the meme reactions to some of them, organized according to the line of reasoning they express and defend (4.1). The issue at stake is whether one should join the Fertility Day or not. The postcards all support the positive standpoint *You should join the Fertility Day*, and can be classified along three argumentative lines: *hurry up and reproduce, care for a common good, care for your fertility.* Then, we consider the reactions to the campaign, including the Minister's replies to critiques (4.2).

4.1 The Fertility Day postcards

We shall start from postcards expressing arguments of the group *hurry up and reproduce* (4.1.1), then move to those related to *care for a common good* (4.1.2), and eventually to *care for your fertility* (4.1.3).

4.1.1 You should hurry up and reproduce

Let's start with the hourglass postcard (Figure 1).

Figure 1 – The hourglass.

The verbal states "Beauty is ageless. Fertility is not". The aim of the postcard is to convince to have kids as soon as possible, because fertility does not last forever. The headline is based on denying an analogy between the frames "beauty" and "fertility". In order to work properly, analogies between frames must show balanced correspondences between the elements of the source and of the target frame. In other words, the relationship between the elements and the events in one frame must mirror the elements and the events in the other (Fauconnier & Turner, 1998, 2002). The two frames must make the audience discover an abstract frame which is true for both the target and the source; this abstract frame is a generic space which might serve as major premise of an argumentation by analogy (Fauconnier & Turner, 1998; Pollaroli & Rocci, 2015; Rocci, Mazzali-Lurati, & Pollaroli, 2018). Here, the Ministry of Health relies on the well-known saying that beauty is ageless, meaning that beauty is a prototypical subjective category, which is not subject to time. This is the case for entities that belong to the category of gifts of nature. But young Italian women should be warned! They might think that fertility is not subject to time as beauty is, because both are a gift of nature. This is not the case: at some point, a woman is not fertile anymore. Therefore, the campaign warns that the analogy is only partially working, because the two entities are not comparable.

This line of reasoning concludes that an endoxon that is attributed to the audience is not valid. The conclusion "Fertility is not a gift of nature" works as a factual premise for a tacit line of reasoning in support of the claim "You should hurry up".

Among the reactions found on the internet, for this postcard (as well as the two analysed right after) there is a meme (Figure 2) by the social activists' network "Act".[1] It features a worried woman holding a glass hour and her thought: "my pregnancy lasts much longer than my contract". Indeed, the glass hour many Italian women feel emptying is the one of their employment status. This is expressed by the sentence in the small font ("anything but fertilityday") which counters the strangling concern of job insecurity with that of expiring fertility, giving priority to the former. In argumentative terms, the small sentence could be translated into the standpoint "I cannot afford to worry about fertility", with the rest of the verbal working as the datum, reinforced by the visual. The visual helps also to activate the standpoint; the girl's expression is undoubtedly worried and the reduced size of the glass hour (compared to the original postcard) puts the topic of fertility in the background in favour of something that is perceived as more important. The endoxical unexpressed premise supporting this line of reasoning is "without a contract, pregnancy cannot be afforded". This shows that the cultural and social situation which is perceived as a faulty situation by the people designing this postcard is different from that of the Ministry of Health.

Figure 2 – The hourglass meme.

The next postcard (Figure 3) also refers to a commonplace: storks bring babies.

---

[1] http://www.act-agire.it

Figure 3 – The stork.

The headline ("Get a move on! Do not wait for the stork!") exhorts the audience to hurry up in having babies, for no fabled intervention will take over that task from them. Even less will it be the stork in this visual, who is standing in an empty nest and does not seem to have any bundle of joy to distribute. Although there is no explicit mention of female fertility, the use of the imperative in the singular (*non aspettare*) and the injunction to hurry up relate to women's expiring fertility. Furthermore, it is already dusk in the visual, very little time is left to conceive. Having babies is thus framed as an urgent responsibility.

The urgency of childbearing for women is again the focus of our third postcard ("Postponing maternity leads to an only child. If ever", Figure 4). The whole responsibility is on them, as the use of the word "maternity" underlines. The only child – so the visual – will feel lonely and unsafe, haunted by the ghost of his unborn brother, who reproaches him for some non-transparent reason.

Figure 4 – The only child.

But this is not all: the long-postponed motherhood could possibly be a non-reachable goal. Therefore, this postcard is based on a twofold reasoning from negative consequences (having only one child, or none) that are presented as inevitable to who postpones maternity, and that in a threatening tone.

The last "hurry up" postcard eventually addresses also men: "Young parents. The best way to be creative" (Figure 5).

Figure 5 – The ball.

Having kids is described as a creative activity, and presupposing that young people value creativity, they should choose this specific creative expression. The contribution of the visual is dubious: as an article on *La Stampa* online[2] notices, "the engineering of the fertile intercourse is difficult to understand" from the picture and the contribution of the 'smiley' is unclear.

4.1.2 You should care for a common good

The postcard in Figure 6 is based on an analogy between water and fertility.

---

[2] http://www.lastampa.it/2016/08/31/societa/la-procreazione-creativa-uno-spermatozoino-con-sguardo-rapace-e-lovolina-rosa-HnOJeyyNGs9Qa8PB3OIoSL/pagina.html.

Figure 6 – The faucet.

"Fertility is a common good", says the headline, whereas the visual depicts a dripping faucet. Both fertility and water belong to the functional genus of common goods, of which citizens are expected do take care. Moreover, the dripping recalls the scarcity of water supplies – it is a common good we have to care for because it is exhaustible, a trait shared with fertility. The visual depicts the source frame of the multimodal metaphor 'fertility is water' and contributes to the identification of the argumentatively relevant generic space 'exhaustible common goods should be taken care of'.

The caring aspect is explicit in the verbal of the next postcard (Figure 7): "The Constitution protects conscious and responsible procreation".

Figure 7 – The little shoes.

Citizens must care for conscious and responsible procreation because the Constitution does so, and what is safeguarded by the Constitution has to be cared for by citizens. Mentioning the Constitution serves as an argument from authority. The visual underlines the institutional involvement by wrapping an Italian-flag ribbon around baby shoes – an image that infelicitously reminds fascist propaganda that exhorted Italians to bear children for their homeland.

4.1.3 You should care for your fertility

The third set of postcards address general health issues, being thus more pertinent to a campaign launched by the Ministry of Health.
A purely female-targeted postcard depicts a woman embracing her womb. "Prepare a cradle for the future", says the headline (Figure 8).

Figure 8 – The womb.

The responsibility of the future mother is in focus here: she is in charge of taking care of her body in the perspective of her future pregnancy and of the future of the country. The choice of 'cradle' as the entity of the frame source reminds of the overused metaphors 'the cradle of culture' and 'the cradle of democracy'. Here it is the source frame for the target 'womb' as well as a component of the metonymical chain which connects, on the one hand, conceiving a baby and letting it grow in a womb and, on the other, a cradle as a place where the baby rests. This care has to start early enough, even before one really thinks of having children (the campaign is in fact addressed also to teenagers, see NFP). In argumentative terms, having children is the final cause towards which a responsible woman should work from early age on. In reply to this piece of advice, a meme was created that adds "and blank resignation letters"[3] to the list of what should be prepared when contemplating a future pregnancy (Figure 9).

---

[3] Often women are requested to sign – illegal – blanc resignation letters (*dimissioni in bianco*), used in case of pregnancy to protect the employer from paying maternity leave (Melchiorre & Rocca, 2013).

Figure 9 – The womb meme.

General health prevention is recommended in a postcard built as the previous one, but with a doctor "embracing" the message "Prevention guarantees your future. In all senses" (Figure 10).

Figure 10 – The doctor.

In this case, reasoning from termination and setting up is activated: prevention is good because it guarantees your future, what is good should be cared for, therefore one should care for one's health (and fertility, of course). This reasoning path is reinforced by the authority of the doctor. It is not clear what "in all senses" should mean. The audience though interpreted it in terms of absolute control of sexual freedom, as the meme in Figure 11 testifies ("We will control your sexual freedom. In all senses").

Figure 11 – The doctor meme.

Again, both sexes are addressed in a postcard regarding sexually transmitted diseases (Figure 12).

Figure 12 – Condoms.[4]

This postcard explicitly invites to protect one's fertility: "Sexually transmitted infections? Rather not. Defend your fertility every day". A series of condoms labelled by day is the means to daily protection – and contraception. This apparent contradiction, stressed by some comments from the audience[5], is resolved in the context of the information and prevention campaign. One should protect his/her sexual health because STDs compromise fertility.

The next threat to fertility is alcohol – dangerous for both sexes, but placed in female hands (Figure 13).

---

[4] Unfortunately it was not possible to find this postcard with better image quality.

[5] https://gitementali.wordpress.com/tag/ipocrisia/.

Figure 13 – Drinks.

In particular, it is said to cut in half the potential to conceive ("Cheers! Alcohol cuts in half your fertility."). The verbal visualizes this reduction by crossing through half of the Italian exclamation for "cheers" (*cin cin*). The habit of drinking should therefore be terminated, as should drug use (Figure 14). Is 'fertile' a category of the frame to which a 'busted', 'doped' and 'stoned' young man belongs? Answering this rhetorical question is easy.

Figure 14 – Drugs.

A termination of smoking is recommended as well, in this campaign especially to men, as its effects on spermatozoa are in the spotlight (Figure 15, "Do not let your spermatozoa go up in smoke"). The visual underlines decrease of potency by showing a bended cigarette, standing for needless to say what. The multimodal metonymy shrinks the cause-effect chain SMOKING LEADS TO DECREASE IN POTENCY WHICH LEADS TO A LACK OF SPERMATOZOA to presents an extremely undesirable situation.

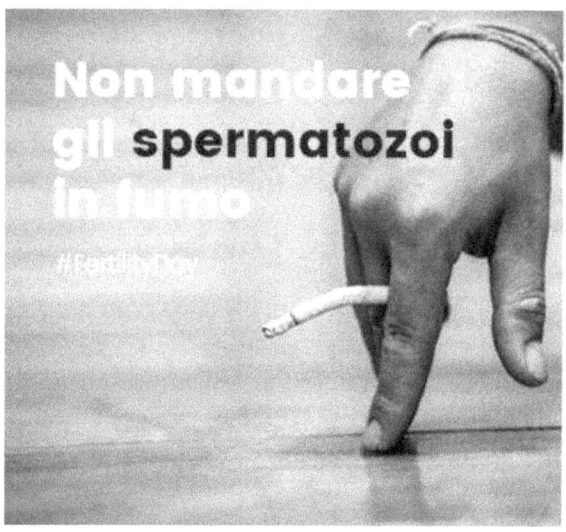

Figure 15 – Smoke.

Male fertility is indeed much in need of protection, it is "much more vulnerable than it seems" (Figure 16). If we were to judge by the banana skin in the visual though, there would not be many chances to save the situation.

Figure 16 – The banana.

4.2 Reactions to the postcards

Reactions bloomed on the web almost immediately after the launch of the campaign, creating a lively multimodal polylogue. The participants to the polylogue are Italian citizens (attacking either the campaign overall or a specific postcard) and the Minister of Health.

The critiques by the general public support the standpoint *you should not join the Fertility Day*. In some cases, the campaign is explicitly attacked, while the Fertility Day is legitimated as an initiative (e.g., Figure 20). In general, though, the critique is directed at both, or no distinction is made between the two levels.

Focusing on Twitter, the arguments proposed can be categorized along the following lines:[6]

    a) the Government should not interfere with the citizens' reproductive behaviour (Figure 17);
    b) socio-economical limitations to childbearing in Italy are neglected (Figure 18);
    c) having children is a free choice (Figure 19):
    d) the campaign is counterproductive (Figure 20);

---

[6] Due to space limitations, we insert only one example for each line of argumentation.

e) the campaign reminds of Fascist propaganda (Figure 21).

Figure 17 – It's my uterus and I manage it.

Figure 18 – A year after the Fertility Day, do not miss the "Look for a Nursery" Day.

Figure 19 – Motherhood is just an option.

L'intento d'una campagna informativa sulla fertilità non è male, ma per come è promossa fa venire voglia di chiudersi le tube #fertilityday

Figure 20 – So badly promoted that it makes one think of having one's tubes tied.

Figure 21 – Fertility Day, what an innovative idea.

Other comments address specific postcards. For example, as a reaction to the glass hour postcard, the pioneer in gender statistics Linda Laura Sabbadini wrote in an opinion article on *La Stampa* online[7]:

---

[7] http://www.lastampa.it/2016/09/01/cultura/opinioni/editoriali/come-togliere-gli-ostacoli-alle-mamme-1mQqxhD0aM6SfmoH586f1I/pagina.html.
Original Italian: "La clessidra tenuta in mano dalla donna vestita di rosso suona

> The glass hour held by the woman in red feels like an interference and an accusation for who delays childbearing and maybe would not want to do so, or simply decides to have a child later because so she prefers.

The visual, instead of being recognized as a playful reminder, is felt as judgmental towards free choices as well as impeding circumstances to reproductive habits, covering thus two of the lines of argumentation mentioned.

In addition, a petition was launched to ask the government to cancel the event, without success. On September 22nd, protests where held with the slogan (and hashtag) "#NoFertilityDay". Besides boycotting the initiative, the protesters requested services in support of families, such as kindergarten, and job security.

The Minister Beatrice Lorenzin replied[8] that it is not any specific picture or slogan, nor the campaign overall what counts, but Italians' reproductive health. The campaign was aimed at attracting attention and provoking a reaction, but not at offending. It can and will therefore be modified without touching the event itself. Lorenzin reasons from means to end: if the means (the campaign) do not serve the end (rising awareness on fertility), as it was the case, the means should be changed – not the end. This shows that she does not consider the importance of communication in itself, for she sees the campaign as a mere instrument. Moreover, Lorenzin argues that she cannot be criticized for having spoken only of health and not of other factors influencing low childbirth. As Minister of Health, she is an authority on health only. Furthermore, she presents health as a necessary precondition for increasing birthrates, without which all socio-economic measures would be useless: "you can build kindergarten, but if we are sterile and cannot have kids, we do not have kids to put there".

---

come una intromissione e una colpevolizzazione per chi ritarda la nascita di un figlio e magari non vorrebbe farlo, oppure semplicemente decide di farlo più tardi perché così preferisce".

[8] http://www.repubblica.it/politica/2016/09/01/news/fertility_day_lorenzin_pronta_a_modificare_campagna_messaggio_va_rimodulato_-147020546/)

5. CONCLUSION

Our analysis of the "Fertility Day" postcards and reactions to them has revealed that the Ministry of Health wrongly evaluated the rhetorical situation of its audience. The campaign ascribes endoxa to the audience which it does not share – even worse, which it rejects as insulting.

Overall, in the postcards fertility is framed as a good Italian citizens should take care of and put to profit early, a good that actually is not their property but belongs to the country. Italian citizens oppose this framing of their reproductive rights and faculty, and advance radically different reasons for why natality is so low. The endoxical discrepancy leads to a failure of the campaign.

Interestingly, the then Prime Minister Matteo Renzi declared his non-involvement with the campaign ("I did not know about this campaign, I had not seen it") and his support to the outraged audience.[9] As Prime Minister he was supposed to be the principal stakeholder of any initiative by his government, but he displays little knowledge of the hierarchy of stakeholders common to any practice of communication. He also attacks the purpose of a campaign for fertility: "I know nobody who has a kid because he sees a billboard". Apparently, such clash of *endoxa* can be found inside the cabinet too.

In general, both the communication by the Ministry of Health and the replies and comments by the audience operate on the common opinion that 'where there is a will, there is a way'. It is peculiar that this meta-endoxon is shared and promoted by a Ministry that should be well aware – if not of the socio-political impediments to building a family – at least of the diseases (e.g., endometriosis) that can result in unwanted infertility or difficulty in conceiving.

The argumentation of this polylogue is clearly multimodal. Pictorial and verbal elements interact in order to construct meaning and we cannot limit the argumentative or rhetorical contribution of pictures to one or the other function. Apart from being sometimes only decorative devices, pictures are able to present either factual premises or endoxical premises, reinforce or specify standpoints, offer a situation for the design of a metaphorical or metonymical framing, activate a line of reasoning, or fix an ambiguous verbal meaning. Pictures do this alone or in combination with verbal cues.

9

http://www.repubblica.it/politica/2016/09/01/news/fertility_day_lorenzin_pronta_a_modificare_campagna_messaggio_va_rimodulato_-147020546/)

REFERENCES

Aakhus, M., & Lewinski, M. (2016). Advancing polylogical analysis of large-scale argumentation: Disagreement management in the fracking controversy. *Argumentation*, *31*(1), 1-29.

Aakhus, M., & Lewiński, M. (2011). Argument analysis in large-scale deliberation. In E. Feteris, B. Garssen, & A. F. Snoeck Henkemans (Eds.), *Keeping in touch with pragma-dialectics* (pp. 165-183). Amsterdam: John Benjamins.

Aakhus, M., & Lewiński, M. (2015). *Toward polylogical analysis of argumentation: Disagreement space in the public controversy about fracking*. Paper presented at the 8th International Society for the Study of Argumentation Conference, University of Amsterdam.

Bitzer, L. F. (1959). Aristotle's enthymeme revisited. *The Quarterly Journal of Speech 45*, 399-408.

Bitzer, L. F. (1968). The rhetorical situation. *Philosophy and Rhetoric*, *1*(1), 1-14.

Bitzer, L. F. (1980). Functional communication: A situational perspective. In E. E. White (Ed.), *Rhetoric in transition: Studies in the nature and uses of rhetoric* (pp. 21-38). Pennsylvania: Pennsylvanian State University Press.

Eurostat. (2014). Fertility. Retrieved from http://ec.europa.eu/eurostat/web/population-demography-migration-projections/births-fertility-data/main-tables.

Fairclough, N. (1995). *Critical discourse analysis: The critical study of language*. London: Routledge.

Fauconnier, G., & Turner, M. (1998). Conceptual integration networks. *Cognitive Science*, *22*(2), 133-187.

Fauconnier, G., & Turner, M. (2002). *The way we think*. New York: Basic Books.

Fillmore, C. J. (Ed.) (1982). *Frame semantics*. Seoul: Hanshin Publishing Co.

Goffman, E. (1975). *Frame analysis: An essay on the organization of experience*. New York: Harper.

Greco Morasso, S. (2012). Contextual frames and their argumentative implications: A case study in media argumentation. *Discourse Studies*, *14*(2), 197-216.

Greco, S., Mehmeti, T., & Perret-Clermont, A.-N. (2017). Do adult-children dialogical interactions leave space for a full development of argumentation? A case study. *Journal of Argumentation in Context*, *6*(2), 193-219.

Istat. (2015). Indicatori di fecondità. Retrieved from http://dati.istat.it/Index.aspx?DataSetCode=DCIS_FECONDITA1.

Kerbrat-Orecchioni, C. (2004). Introducing polylogue. *Journal of Pragmatics*, *36*(1), 1-24.

Kjeldsen, J. E. (2012). Pictorial argumentation in advertising: Visual tropes and figures as a way of creating visual argumentation. In F. H. van Eemeren & B. Garssen (Eds.), *Topical themes in argumentation theory*. Dordrecht: Springer.

Lewiński, M., & Aakhus, M. (2014). Argumentative polylogues in a dialectical framework: A methodological inquiry. *Argumentation*, *28*(2), 161-185.

Mattoni, A., & Vogiatzoglou, M. (2014). Prima e dopo la crisi. L'evoluzione delle mobilitazioni dei lavoratori precari in Italia e Grecia. *Sociologia del lavoro*, *136*(4), 260-275.

Mazzali-Lurati, S., & Pollaroli, C. (2013). Stakeholders in promotional genres. A rhetorical perspective on marketing communication. In G. Kišiček & I. Kišiček (Eds.), *What do we know about the world? Rhetorical and argumentative perspectives* (pp. 365-389). Ljubljana: Digital Library of Slovenia & Windsor Studies in Argumentation.

Palmieri, R., & Mazzali-Lurati, S. (2016). Multiple audiences as text stakeholders: A conceptual framework for analyzing complex rhetorical situations. *Argumentation*, *30*(4), 467-499.

Pollaroli, C., & Rocci, A. (2015). The argumentative relevance of pictorial and multimodal metaphor in advertising. *Journal of Argumentation in Context*, *4*(2), 158-200.

Rigotti, E. (2006). Relevance of context-bound loci to topical potential in the argumentation stage. *Argumentation*, *20*(4), 519-540.

Rigotti, E., & Greco Morasso, S. (2009). Argumentation as an object of interest and as a social and cultural resource. In N. Muller Mirza & A.-N. Perret-Clermont (Eds.), *Argumentation and education* (pp. 9-66). Dordrecht: Springer.

Rigotti, E., & Greco Morasso, S. (2010). Comparing the argumentum model of topics to other contemporary approaches to argument schemes: The procedural and material components. *Argumentation*, *24*(4), 489-512.

Rocci, A. (2006). Pragmatic inference and argumentation in intercultural communication. *Intercultural Pragmatics*, *3*(4), 409-442.

Rocci, A. (2009). Manoeuvring with voices: The polyphonic framing of arguments in an institutional advertisement. In F. H. van Eemeren (Ed.), *Examining argumentation in context: Fifteen studies on strategic maneuvering* (pp. 257-283). Amsterdam: John Benjamins.

Rocci, A. (2017). *Modality in argumentation: A Semantic investigation of the role of modalities in the structure of arguments with an application to italian modal expressions.* Dordrecht: Springer.

Rocci, A., Mazzali-Lurati, S., & Pollaroli, C. (2018). The argumentative and rhetorical function of multimodal metonymy. *Semiotica*, *220*, 123–153.

Saviori, R. (2009). *Argumentation in benevolent interaction: A case study of the United Nations' activity in disaster risk reduction*. Università della Svizzera italiana, Lugano.

Zampa, M. (2015). *News values as endoxa of newsmaking. An investigation of argumentative practices in the newsroom* (doctoral dissertation). Università della Svizzera italiana.

Zizza, Roberta. (2013). The gender wage gap in Italy. In *Questioni di Economia e Finanza* (occasional papers). Rome: Bank of Italy, Economic Research and International Relations.

67

Of Inference and Argumentation in Financial Discourse: The Crisis of 2007-2008

GRISELDA ZÁRATE
Universidad de Monterrey
griselda.zarate@udem.edu

HOMERO ZAMBRANO
Tecnológico de Monterrey, Campus Monterrey
hzambranom@itesm.mx

Financial markets are particularly sensitive to information, in which good or bad news have a strong impact, as well as declarations by key economic, financial or political figures, in the form of bull markets or bear markets. This research paper approaches the role of inference and argumentation in financial discourse in the crisis of 2007-2008, and specifically to the concept of jumps in newspaper articles of specific dates published in *The Wall Street Journal*.

KEYWORDS: Inference, financial discourse, argumentation, conceptual metaphor, Lakoff, Toulmin, Gilbert, jumps.

1. INTRODUCTION

Perhaps the most popular item in finance is the stock market, and more specifically, the corresponding (main) index. In the case of the United States, the leading such indicator is the S&P500, which accounts for roughly 80% of market capitalization in the US, according to the firm behind that index, S&P Dow Jones Indices, and has existed since the early 1940's. That said, the S&P can be said to gauge the health of a very sizable part of the world's finance, and has already witnessed and measured several "crashes". Among the most noteworthy of these are the 1987 crash, the dot-com burst, and the subprime crisis, which led to the 2008 downfall.

That is precisely what we want to contribute to clarify, as least as how those seemingly abnormal movements are related to the surrounding discourse by the relevant economic agents, such as government officials, finance institutions spokespersons, and members of the c-suite in the firms and conglomerates in the market. The market cannot be expected to grow smoothly, or as economist mathematicians say, to be "monotonically increasing". It would be sort of an anomaly to witness a given week with no downward movement in the index, as small as it may be. As an example, from 1996 to 2013, less than 4% of weeks exhibited no negative returns in the index. A return, or more exactly, a rate of return, is defined as the proportional change of the index from one day to the next one.

A traditional assumption is that returns are stationary, which can be translated as saying that an average or mean makes sense over time, as opposed to stock prices, or the index levels, for which an average does not have a meaning. Stationary does not mean constant; returns have some dispersion around the mean, and that dispersion is measured with the standard deviation. It is commonplace for returns measured on a daily basis to be non-conforming to the Gaussian or normal distribution. Such returns, along with weekly returns, and more so the more frequent the measurement as in intraday values, are leptokurtic, which implies that the distribution has "fat tails", or in more practical terms, that the occurrence of extreme events, such as "crashes" is more common than what the normal distribution would prescribe.

This paper is part of a larger interdisciplinary research project which aims to identify the correlation between financial discourse and market volatility in the crisis of 2007-2008, based on the concept of jumps in finance. Previous work has approached the role of metaphorical argumentative processes in financial discourse (Zárate & Zambrano, 2017). The corpus is formed by newspaper articles of specific dates of 2007-2008 published in *The Wall Street Journal*, related to the movements of the Standard & Poor's 500 market index (SP500) of the United States, as well as declarations of key figures in politics and finance. For purposes of this paper the text "Statement of Richard S. Fuld Jr. before the United States House of Representatives, on October 6, 2008" is analysed.

2. METHODOLOGY

The theoretical framework is based on an integrated Operative Model of Argumentation (OMA) (Zárate, 2012; 2015). OMA draws from: Toulmin, Rieke and Janik (1979), includes logical, emotional, visceral and kisceral modes of argumentation from Gilbert (1997), and Lakoff and Johnson's notion of conceptual metaphor (1980). From a financial point of view, the research incorporates Eugene Fama's views as stated in the Efficient Market Hypothesis (1970; 2008), and Robert Merton's jumps concept (1976), as well as Lee & Mykland's methodology to identify jumps (2008).

2.1 Operative model of argumentation (OMA)

The operative model of argumentation (OMA) articulates the 1979 version of Toulmin's model (Toulmin, Rieke and Janik), which stresses the logical mode of argumentation, with the emotional, visceral and kisceral modes of argumentation proposed by Gilbert in the right column, as shown in Figure 1. Emphasis is placed on the interdependence, or close connection, of the elements in the operating model. For example, warrants (W) are supported by backing (B), that is, legal documents such as laws, statutes that transmit the cultural values of a given society, which contain inferences and assumptions.

The emotional mode of argumentation conveys the degree of commitment, resistance, depth and feelings in the discourse, all of which express more than words may seem at the beginning. On the other hand, the visceral mode of argumentation implies a physical demonstration of the arguments (that is, it can be a quick glance, a touch of the shoulder, hitting a bag or hitting a door), all of which can have a broad spectrum of answers. The kisceral mode of argumentation, a term derived from the Japanese *ki*, which means energy or vital force, emphasizes the intuitive, imaginative, spiritual, mystical or religious aspects of discourse, for example, "we are all children of God" (Gilbert, 1997).

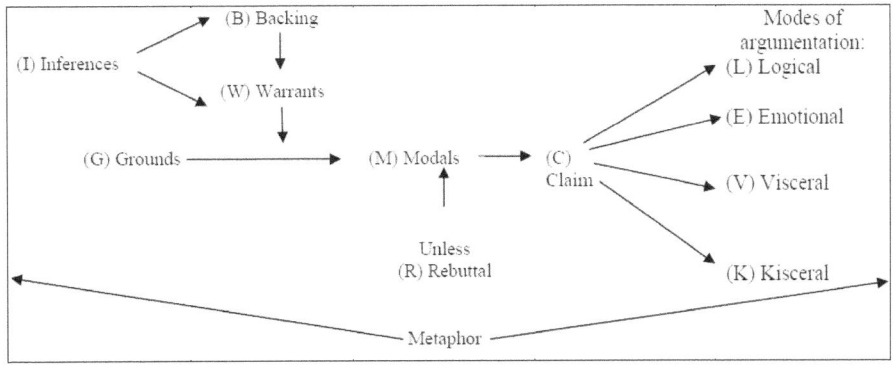

Figure 1 – Operative model of argumentation (OMA) (Zárate, 2012; 2015)

OMA also includes Lakoff and Johnson's cognitive metaphor concept (1980). Metaphors may be found in all elements of this model, except for modals (M). Lakoff and Johnson state that "If we are right in suggesting that our conceptual system is largely metaphorical, then the way we think, what we experience, and what we do every day is very much a matter of metaphor" (3).

3. JUMPS IN FINANCE AND MARKET HYPOTHESIS

In 1976, Robert Merton proposed a concept of jumps in regards to financial processes:

> "...an option pricing formula is derived for the more-general cast when the underlying stock returns are generated by a mixture of both continuous and jump processes" (p.125).

This is, returns (variation in prices of a stock), may undergo petty movements due to usual trading, and significant movements, due to the arrival of relevant information. Merton devised a theoretical model, the jump-diffusion model, in order to simulate future movements. On the other hand, Lee & Mykland (2008) proposed a methodology to detect jump processes in historical (past) series of returns of a stock, and statistically tell them apart from the usual movements.

Cont & Tankov (2003) have an excellent discussion on why the behaviour of market returns cannot be properly modelled after a probability distribution on itself, even when and if it considers fatter tails than the Gaussian distribution. They argue that probably the best way to accommodate the reality of financial markets is superimposing extreme movements in isolation to a conventional diffusion, or probability distribution. As mentioned, Merton (1976) presented the jump-diffusion model, which joins a Poisson process to model the extreme movements as random "jumps", and a "conventional" diffusion as a Gaussian probability distribution.

Modelling a phenomenon (or "process") is relatively easy. The modeller assigns the parameters based on historical values, and loads them in the model. If we talk about a jump-diffusion model, we need a couple of parameters for the diffusive part; for the jumps, we need the average number of jumps in a certain window of time (that will lead to the "how many", through the Poisson process), as well as the parameter that defines the "when" of the jump (e.g., a uniform distribution), as well as the "how much", (or magnitude, as we use herein), that in turn can be itself a Gaussian process (add two more parameters). As a note to understand jump processes a little more, modellers call the "how many" jumps in a given time span "intensity". In short, the modeller can exactly pinpoint which returns are exactly jumps, and which values come from the continuous probability distribution, even if for some statistical whim the latter could be larger than the former, event that cannot be ruled out entirely. Remember that most continuous distributions have at least one tail unbounded, i.e., that can take values up to infinite.

The opposite procedure, telling apart jumps from extreme values from a diffusive process, is less simple. A couple of methodologies for doing so are set forth in Aït-Sahalia (2004), and in Lee and Mykland (2008). Not only are the procedures somewhat complicated, but not as clear cut, in the sense that for model-based back-testing, some actual jumps can be left undetected ("false negative"), and some non-jumps may be wrongly classified as jumps ("false positive"). However, the state of the art in this area does not seem to have made much more progress lately, and such tools are still among the best for our purpose. Those methodologies to identify jumps are not by far the only ones; there are at least two surveys or reviews on the methods; one of them is the one by Schwert (2009), and the other is by Hanousek, Kocenda and Novotný (2012).

The idea is to associate the jumps with comparable statements coming from the sources aforementioned. Stronger statements (arguments) in financial discourse should be positively correlated with stronger movements; if the statements are in some way negative, the index movement should be downward, that is, the return would be negative. Likewise, if the argumentative content shows a positive bias, the movement would be upwards. There are two points here to clarify: comparability of jumps and statements, and the relation between firm-level statements and index movement. At least as a hypothesis, the magnitude of a jump would be correlated with the "strength" of the statement (argument); as an example, if some firm releases posts some unexpectedly high profits, an upward jump of a certain magnitude may be expected. If those profits come along with previously unreleased news about the launch of a very promising product, the associated jump should be higher than the one with the profits news only. If telling apart jumps from outliers in a diffusive process is somehow difficult, classifying statements so that they can be assigned a value, which in turn can be correlated to the magnitude of a jump in the level of an index or the price of a stock, poses an additional challenge.

Generally, downward jumps are the ones that come to mind, partly because of the risk-averse nature of most investors; however, positive jumps do exist, and can be rather considerable, such as the one of Facebook (FB), on July 25, 2013, when the price of said stock climbed from 26.51 to 34.26, or nearly 30% in a single day. To make a point of the magnitude of that jump, let us compare that figure to the long term average of the S&P500, which was around 7.55% at that time, in annual terms. That is, the index would take more than three years just to achieve what FB did in one day.

In the previous example we compared a single stock with a broad index, which is not entirely correct. As Martin (2007) points out, there are firm specific jumps, industry/sector specific jumps, and market jumps. Even if a firm with very high market capitalization exhibits a big jump, the market may barely notice. This is why firm specific jumps may have a bigger magnitude in absolute value than the market. That said, some market numbers are in order: The S&P500 fell more than 20% in a single day on October 19, 1987, and rose around 11% on October 13, 2008, days after the crisis. These are the largest figures in recent history. It must be pointed out that that with respect to the 2008 hike, there was a severe negative return just two days before, in the order of nearly -8%. Two weeks before that, on September 29, the index fell almost 9%, just to grow over 5% the very next day. This might as well be an example of what is called "volatility clustering". The lack of this stylized fact is one of the shortcomings of a number of jump-diffusion models, according to Kou (2008).

Some authors, such as De Bondt and Thaler (1985), regard the phenomenon of opposite price movements following extreme variations as a violation of market efficiency, as set forth by Fama (1970). However, De Bondt and Thaler use a time frame which is probably not applicable nowadays. Fama's original idea is that "security prices at any time 'fully reflect' all available information" (p. 383). Our take in this paper is that, given some abnormal information, an abnormal movement in prices is to be expected. De Bondt and Thaler themselves refer that "Research in experimental psychology suggests that, in violation of Bayes' rule, most people tend to 'overreact' to unexpected and dramatic news events" (p. 793). Thus, an adjustment in the price is warranted, and it should not be interpreted as a violation of market efficiency, unless we made an extreme interpretation of Fama's concept, time-wise.

On this matter, we should bear in mind that financial information is available on a quarterly basis; price movements in-between reports are due mainly to trading, in many cases for liquidity reasons. Of course, there are always random pieces of information that impinge on stock prices, such as, for example, that the CEO passed away.

In the particular case of AAPL, the demise of Steve Jobs was barely noticed by the market, at least on the following (business) day; maybe the fact that he passed away on a Saturday dampened the effect of the news somehow, at least for the stock price. The news needs time for the market to incorporate them into prices, correctly.

Another example would be the 9/11 attacks: the market took a whole month to recover the previous level, which considering the magnitude and significance of the information, could be a relatively short span of time, and this is because American infrastructure and productivity resulted almost intact. As a comparison, the S&P500 did not recover its 09/2008 levels but only over two years, due to the consequences of the subprime crisis. The corresponding change in the U. S. Gross Domestic Product for those years (2001, and 2008 and 2009 combined) was 3.3% and -0.4%, respectively. The damage to the economy was much worse in 2008 than in 2001. In both cases there was not a clear immediate rebound. Upward surges can escape immediate large adjustments too, as was the case of FB on July 2013, mentioned earlier.

3.1 Statement of Richard S. Fuld Jr.

On October 6, 2008, Richard S. Fuld Jr. appeared before the United States House of Representatives to make a statement regarding the Lehman Brothers bankruptcy the previous month. The text shows interesting levels of analysis, including metaphors, assumptions and inferences regarding the financial crisis:

>   (1)   (G) Grounds: "We are in the midst of unprecedented turmoil in our capital markets. The problems that most believed would be contained to the mortgage markets have spread to our credit markets, our banking system, and every area of our financial system. As incredibly painful as this is for all of those connected to or affected by Lehman Brothers –this financial tsunami is much bigger than anyone firm or industry. Violent market reactions to a number of factors affected all of the financial system." (C) Claim: "These problems are not limited to Wall Street or even Main Street. This is a crisis for the entire global economy".

In particular, one can note the (C) claim that the financial crisis has worldwide proportions, as shown in Figure 2. Following the operative model of argumentation (OMA), the (G) grounds for this claim contain metaphors such as "unprecedented turmoil" and "this financial tsunami" which infer and conceptualise the markets as a natural force. They also show an emotional mode of argumentation, which is also present in the discursive expression "As incredibly painful as this is".

Granting that argumentation is the verbal and social manifestation of inferential processes, in particular in the form of warrants and backing in the operative model, one can identify in the example:

(2) (W) Warrants: [Based on systems of values. Context field dependent - Financial crisis conceptualised as a monumental force].
(B) Backing: Laws, statutes, previous similar experiences back warrants as inferences.

The financial crisis is conceptualised as a phenomenon of monumental force, implied in the train of reasoning, which is the logical mode of argumentation. This warrant is backed by similar experiences of financial crashes that have happened in the past, (for example, the 1987 crash, the dot-com burst). This (B) Backing is also implied in the argument.

Two questions arise during this analysis: What role do warrants play as inferences in the operative model of argumentation (OMA) in financial discourse during the crisis of 2007-2008? And, what role do metaphors play as inferences in financial discourse in this operative model of argumentation (OMA)? Warrants, as implied inferences in financial discourse shape the argumentative process in relation to the crisis of 2007-2008. On the other hand, metaphors convey emotional, visceral, kisceral modes of argumentation:

(3) "A storm of fear enveloping the entire investment banking field" (p.8)
"Violent market reactions" (p. 1)
"A litany of destabilizing factors" (p. 8)

For example, "litany" as a metaphor of religious language is taken in this text of financial discourse within a kisceral mode of argumentation. As inferences metaphors frame the financial market as a sick person, or as capable of movement, i.e. jumps, (metaphor of personalization).They make the financial crisis of 2007-2008 as a monumental force (metaphors of natural forces), i.e. "financial tsunami", "storm of fear". These metaphors imply an emotional mode of argumentation.

In his opening statement during the hearings *The causes and effects of the Lehman Brothers bankruptcy*, on October 6, 2008, Tom Davis, Ranking Minority Member for the State of Virginia in the US House of Representatives, mentioned that:

> (4) "So today we start with the case of Lehman Brothers, a venerable investment house that sank into insolvency while others were being thrown Federal lifelines. One lesson from Lehman's demise: Words matter. Rumors and speculative leaks fed the panic and accelerated a flight of confidence in capital from that company. Words matter here as well. Look at the TV monitors. As we watch them, the markets are watching us. In this volatile environment, unsupported allegations, irresponsible disclosures can inflame fears and trigger market stampedes. As these hearings proceed, we should watch the pulse of Wall Street and choose our words with great care".

The quote emphasizes the importance of the use of language with discursive expressions such as "Words matter" and "choose our words with great care". While metaphors such as "the markets are watching us" and "we should watch the pulse of Wall Street" conceptualise and infer financial markets as a living creature, a metaphor of personalization. An emotional mode of argumentation is contained in the metaphor "trigger market stampedes".

4. CONCLUSION

This paper is a first approach to the understanding of the role of inference and argumentation in financial discourse during the crisis of 2007-2008. Inferences shape the argumentative process in financial discourse in relation to the crisis of 2007-2008, and specifically to the concept of jumps as warrants (W). Metaphors are present in financial discourse. They express emotional, visceral, kisceral modes of argumentation. As inferences metaphors make evident conceptualizations of financial market as a sick person, or as capable of movement, i.e. jumps, (metaphor of personalization), show financial crisis of 2007-2008 as a monumental force (metaphors of natural forces).

Research in this area can have progress in several fronts. First of all, we can hope for even better ways to discern jumps from ordinary diffusive processes. Second, also on the front of jump-diffusion models, we can expect models that incorporate more simply volatility clustering and the not so sporadic presence of big opposite movements shortly after a jump. Since this is a multidisciplinary topic, further research in overreaction would be very interesting, especially from the psychological point of view, because nowadays information flows much faster than in 1985. Does that speed increase overreaction, or dampens it? On other front, interest in market efficiency does not seem to ebb, and the topic of this writing could benefit from further discussion around it: Will market efficiency depend not that much on information quantity, or for that purpose, timeliness, but on quality? That is, currently we have often times what has been dubbed "information overload", and investors are not immune to that "disease". Research is needed on how to 1) synthesize financial, economic, and political information, 2) filter and expunge data that is misleading, exaggerated, or plainly fake, and 3) prioritize information sources.

Special attention should be paid to the balance between freedom of speech and control of informational damage. This concept can be interpreted as the consequences of the wording of a fact on financial indicators beyond those warranted by the fact itself. As an example, the effect of just releasing numbers about profit of a firm, may be quite different as those to adding that the results are "extremely disappointing" or "dismal". The converse may also apply. While some steps are already considered, such as in the Code of Ethics & Standards of Professional Conduct of the CFA Institute, perhaps a stronger and wider application of control may be in order to further reduce avoidable volatility in markets.

REFERENCES

De Bondt, W. F. M. & Thaler, R. (1985). Does the stock market overreact? *The Journal of Finance*, *40*(3), 793-805.
Fama, E. F. (1970). Efficient capital markets: A review of theory and empirical work. *The Journal of Finance*, *25*(2), 383-417. Retrieved from http://www.jstor.org/stable/2325486.
Gilbert, A. (1997). *Coalescent argumentation*. Mahwah: Lawrence Erlbaum Associates, Publishers.
Hanousek, J., Kocenda, E. and J. Novotný (2012). The identification of price jumps, *Monte Carlo Methods and Applications*, 18(1), 53–77.
Kou, S. G. (2008), Jump-Diffusion Models for Asset Pricing in financial Engineering (Ch. 2). In J. R. Birge & V. Linetsky (Eds.), *Handbooks in OR & MS* (Vol. 15, pp. 73-116). Amsterdam: North Holland.

Lakoff, G. & Johnson, M. (1980). *Metaphors we live by*. Chicago: The University of Chicago Press.
Lee, S. S., & Mykland, P. A. (2008). Jumps in financial markets: A new nonparametric test and jump dynamic. *The Review of Financial Studies*, *21*(6), 2535-2563.
Martin, M. S. (2007). A two-asset jump diffusion model with correlation (master's thesis). University of Oxford.
Merton, R. C. (1976). Option pricing when underlying stock returns are discontinuous. *Journal of Financial Economics*, *3*(1-2), 125-144. doi:10.1016/0304-405X(76)90022-2
Schwert, Michael W. (2009). Hop, skip and jump – what are modern "jump" tests finding in stock returns? (doctoral dissertation). Duke University.
"Statement of Richard S. Fuld Jr. Before the United States House of Representatives," (October 6, 2008). *The causes and effects of the Lehman Brothers bankruptcy*. Retrieved from http://online.wsj.com/public/resources/documents/fuldtestimony20081006.pdf.
"Statement of Tom Davis, Ranking Member of the United States House of Representatives," (October 6, 2008). *The causes and effects of the Lehman Brothers bankruptcy*. Retrieved from https://www.gpo.gov/fdsys/pkg/CHRG-110hhrg55766/html/CHRG-110hhrg55766.htm.
Toulmin, S., Rieke, R., and Janik, A. (1979). *An Introduction to Reasoning*. New York: Macmillan Publishing.
Toulmin, S. (1995). *The uses of argument*. Cambridge: Cambridge University Press.
Zárate, G. (2015). Argumentación en los textos de Andrea Villarreal, (1907-1910). *Lenguas en Contexto*, *12*, 173-184.
Zárate, G. (2012). El exilio del ningún lugar: Las voces utópicas de la familia Villarreal González (doctoral dissertation). Monterrey: ITESM (220-231).
Zárate, G., & Zambrano, H. (2017). Metáfora y argumentación en el discurso de la crisis financiera de 2007-2008. *Estudios del discurso en México: nuevas prácticas, nuevos enfoques*. Eva Salgado Andrade y Laura Hernández Ruiz, eds. Mérida: CEPHCIS / UNAM, 231-244.

APPENDIX

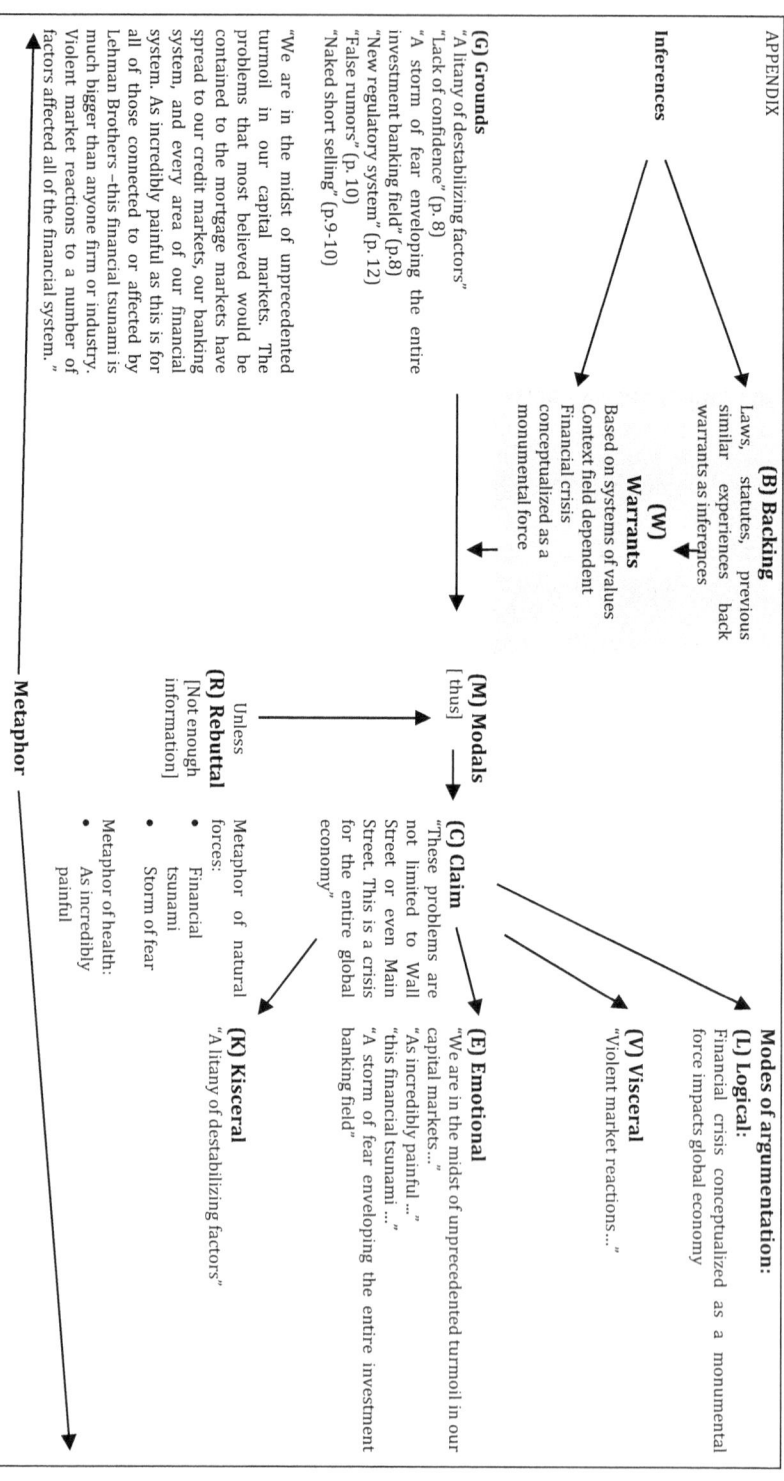

Figure 2. "Statement of Richard S. Fuld Jr. Before the United States House of Representatives," October 6, 2008. Retrieved from http://online.wsj.com/public/resources/documents/fuldtestimony20081006.pdf

What Warrants the Warrant?

DAVID ZAREFSKY
Northwestern University, USA
d-zarefsky@northwestern.edu

> Inferences not entailed must be authorized as acceptable despite uncertainty. Toulmin calls the authorizing agent the Warrant, but what authorized the Warrant? It is the Claim of a supporting argument, and its own Warrant is what Toulmin calls Backing. Disputes would continue in infinite regress were there not a point when a Warrant is accepted as given – a gift to the arguers from their culture. Disputes turn on which arguer can make best use of the gift.

KEYWORDS: audience, backing, claim, consensus, inference, main argument, subsidiary argument, Toulmin, warrant

1. INTRODUCTION

An inference is a mental movement that takes us from the conceptual starting point of an argument to its conclusion. The starting point is something agreed upon – in Perelman and Olbrechts-Tyteca's (1958/1969) categories, facts, truths, presumptions, values, hierarchies, and loci. Toulmin (1958) refers to the starting point variously as data, grounds, and evidence (these terms tend to be used interchangeably).

If in fact there is not agreement on the starting points, then it is necessary to do additional work, preliminary to the main argument, until there is. This can take the form either of providing reasons to agree to the proffered starting point or of providing additional grounds, beyond those originally put forward, in an attempt to secure agreement. In either case, what takes place is a kind of subsidiary argument, in which the new material is the starting point and the original grounds constitute the conclusion. We might think of this as an argument(a). (I am using lettered subscripts to avoid confusion with O'Keefe's [1977] distinction between argument(1) and argument(2).) Implicit in the unchallenged acceptance of grounds is the acceptance by the parties of this subsidiary argument(a).

The conclusion of the argument in valid formal deductive reasoning follows necessarily from the grounds. In a formal deductive argument, the nature of the inference that takes us to the conclusion is of little importance. After all, the conclusion is just a rearrangement of content already present implicitly in the premises. The inference is purely a matter of form. It is either valid or invalid – not an uncertain matter – and this is true regardless of what anyone thinks about it.

2. NON-DEDUCTIVE INFERENCES

Outside the realm of deduction, however, the inference is a matter of critical importance. This is true precisely because the inference is always defeasible. No matter how often it might prove to be correct, it always is rebuttable. That is to say, one always can imagine circumstances in which it does not hold up. This, of course, is why we say that outside of deduction, the conclusion follows only with some degree of probability. (I am bracketing the question of how many patterns of non-deductive inference there are. Some writers define everything not deductive as inductive, whereas others, such as Govier (1987), imagine other reasoning patterns besides deduction and induction, such as conductive reasoning and case-by-case reasoning.)

In this realm outside deduction, the conclusion of the argument is widely referred to as a "claim". This signifies that it is a statement that the interlocutor or audience is asked to accept, a claim on belief or judgment. Whether it concludes the argument or not depends not on its form but on the assent of the audience. The question of how the arguer proceeded from starting point to claim is not trivial, but even more significant is the question of how the *audience* got there, and even more consequential is whether the audience got there in the way that it *should*.

So far, I have covered familiar ground. But doing so makes clear how heavily, outside the realm of formal deduction, the strength of an argument depends upon the quality of its inference. Unfortunately, for all that we are learning about cognitive science, the inferential process is still largely a black box. So far as I can tell, we do not know exactly how the mind works to travel from grounds to claim. We know only that acceptance of the claim is evidence that one has done so.

Theorists have addressed this problem by baptizing and then classifying argument schemes. These are patterns of argumentation in which the grounds and claim appear to stand in the same relationship to one another. On the basis of this similarity, the theorist concludes that the inferences contained in the argument are of the same type. And if inferences of that type generally lead to reliable claims, we will accept it in the situation at hand. Thus we have inference to the best explanation, inference from circumstance to cause, etc. But we still do not know how the inference was made – how a given person determined that a candidate explanation is the best.

One productive approach to this problem is to focus not on the empirical question of how an inference gets made but on the normative ground of whether it is *justified*. On this view, if the inference is authorized as legitimate, what difference does it make how a person made it? The empirical question is subsumed by the normative.

3. WARRANTS

In his layout of arguments, Toulmin (1958) introduces the term *Warrant* as the part of the argument that serves this purpose. Diagrammatically, he represents the Warrant as supporting the inference, holding up the arrow that goes from Grounds to Claim. As with many of his terms, Toulmin's characterization is inexact. He regards warrants as "rules, principles, inference-licences or what you will, instead of additional items of information", and says that they are "general, hypothetical statements, which can act as bridges, and authorise the sort of step to which our particular argument commits us" (Toulmin, 1958, p. 98). The task of the Warrant, according to Toulmin (1958), is "simply to register explicitly the legitimacy of the step involved and to refer it back to the larger class of steps whose legitimacy is being presupposed" (p. 100).

Although some of Toulmin's critics held that the Warrant is no different from the major premise of a syllogism, and although they sometimes look the same, the function of the Warrant is different. It is not a statement about any content. Rather, it is a rule. It provides guidance or direction. It tells us that, given evidence of such-and-such a type, we are authorized to make a certain kind of claim, because evidence of that type generally supports a claim of that type. The inferences usually work out.

Of the various metaphorical terms that Toulmin uses, "licence" is probably the most apt. Just as a driver's license authorizes us to operate a car on the public roads, so a Warrant authorizes us to make an inference of a particular type in putting forward a claim. Conversely, withdrawing the license would undercut our ability to move from Grounds to Claim.

4. JUSTIFYING THE WARRANT

What happens, though, when the Warrant itself is challenged? Or to put it another way, what warrants the Warrant? Toulmin envisions two different kinds of possible challenge. One challenges the Warrant frontally, denying it. The other qualifies the Warrant by suggesting that an exception may be applicable. The first tries to eliminate the Warrant while the second takes advantage of its defeasible nature. For Toulmin, the construct of Backing deals with the first sort of challenge and the construct of Rebuttal deals with the second.

4.1 Direct challenges

"Standing behind our warrants", Toulmin (1958) points out, "there will normally be other assurances, without which the warrants themselves would possess neither authority nor currency – these other things we may refer to as the *backing* of the warrants" (p. 103; italics in original). These "other things", Toulmin implies, gain their function from their authoritative nature. So, for example, "the following statutes and other legal provisions" (p. 105) back up the Warrant that a man born in Bermuda will generally be a British subject. Likewise, "the proportion of Roman Catholic Swedes is less than 2%" backs up the Warrant that "a Swede can be taken to be almost certainly not a Roman Catholic" (pp. 109-110).

If we look closely at what is happening here, we can see that the contested Warrant is functioning as the Claim for a subsidiary argument and that the Backing offers the Grounds for that argument. This subsidiary argument can be identified as argument(b), and it does for the Warrant what argument(a) did for the Grounds. It identifies that the Warrant is legitimate for the argument in which it is used. If the Backing is not acceptable, then there must ensue another subsidiary argument(b) to establish that the Backing can indeed serve as Grounds for the Warrant (serving as the Claim of the subsidiary argument). This process could continue for a seemingly infinite regress unless it reaches a point at which the arguers stumble upon or otherwise find agreement. They take up agreement as a gift from their culture, what my late colleague Thomas Farrell (1976) called the culture's social knowledge.

And this is the key point. The Backing warrants the Warrant not because of intrinsically authoritative statutes and documents, as Toulmin's examples might suggest, but *because the audience accepts it.* Consensus, not external authority, is the important criterion. If one were to deny that the statutes and acts of Parliament settle the matter, for instance, the proponent would need either to convince the interlocutor that the laws and statutes *should* be considered authoritative, or else to provide some other evidence for which authoritativeness could be claimed.

This example illustrates what Toulmin (1958) calls the "warrant-establishing" argument (p. 120). The presence of the Warrant in the main argument, without any explicit Backing, implies that this subsidiary argument(b) is uncontested and hence already has been settled. On the other hand, the call for Backing indicates that the warrant-establishing argument(b) is needed, and in that case the main argument cannot proceed until the subsidiary argument(b) is satisfactorily resolved.

If the argument(b) is called for (by the interlocutor's request for Backing) but cannot be satisfactorily established despite successive attempts to do so, then the situation resembles what philosophers label "deep disagreement" (Fogelin, 1985; Zarefsky, 2016). In that case, rhetorical techniques of various kinds (such as subsumption, frame-shifting, or appealing to urgency) may be able to jolt the arguers into resolving the argument(b) at another level, so that the main argument can go on. Alternatively, the arguers may acknowledge their inability to settle the argument(b) and hence their inability to make progress on the main argument. In such a case, they may abandon the effort, ideally acknowledging that at least the dispute clarified the nature of their disagreement and perhaps resolving to try again another day.

4.2 Indirect challenges

I have referred to a situation in which the Warrant is challenged frontally. But there also is the case in which the Warrant in general is not denied, but it is deemed not to apply in a particular case. This is the suggestion that since the Warrant is not absolute, there may be exceptions that would override it, and that the case at hand is such an exception. The task then is to determine on a balance of considerations whether the Warrant applies in the case at hand or whether the exceptions outweigh it.

Toulmin accounts for exceptions with the concept of Rebuttal in his layout. This is a potentially confusing term because it also is used to refer to the active process of refutation, the attack and defence of arguments. But here it is used to refer to exceptions that would set aside an otherwise generally acceptable Warrant. Ralph Johnson (2000) has held that arguers are responsible for including a "dialectical tier" in their presentation of arguments. This means at minimum that the arguer must identify the most prominent possible exceptions to the Warrant and maintain that they are not operative in the given case. It is more common, however, to expect the interlocutor rather than the protagonist to be responsible for presenting exceptions that *are* claimed to outweigh the Warrant, although the protagonist ultimately has the burden of proof to rule them out.

In such a case, there would ensue a subsidiary argument(c) in which the original advocate would hold that the Rebuttals should *not* be considered because they do not apply to the case at hand. If the interlocutor were not convinced, he or she would offer additional evidence in behalf of the Rebuttal. In such an argument(c), the Rebuttal functions as the Claim. (To be precise, the *absence* of Rebuttals functions as the Claim, since the burden of proof ultimately is on the proponent). Arguers would present Grounds for concluding that Rebuttals should or should not be taken into consideration.

5. CONCLUSION

I have come up against a point that needs to be made explicit. In each of the possible subsidiary arguments(a), arguments(b), and arguments(c) the goal is to secure the assent of the arguers themselves, or of any relevant third-party judge or audience. Their commitment, not any external assumption of authority, is what determines the acceptability of the subsidiary arguments so that the principal dispute can proceed. But we must remember that the consensus that warrants the Warrant is not just any agreement obtained by any means. The consensus is on something that the arguers' culture regards as knowledge (Farrell, 1976). It is offered, as it were, as a gift to people who have committed themselves to engage in argumentation, with the attendant risks that they may be proved wrong, may need to acknowledge their error, and may lose face (Johnstone, 1959).

What can we conclude from this theorizing about the Toulmin model and the specific question of what warrants the Warrant?

First, argumentation consists not only of the explicit arguments but also of subsidiary arguments(a), (b), and (c), which serve respectively to establish the Grounds of the main argument, the Warrant of the main argument, and the absence of Rebuttals in the main argument. They are taken for granted unless contested.

Second, the Grounds, Warrant, and Rebuttals in the main argument do double duty, serving also as Claims respectively of subsidiary arguments(a), (b), and (c). Backing, which applies to Grounds and Rebuttals as well as Warrants (though not shown that way in Toulmin's layout) also serves as Grounds in the various subsidiary arguments.

Third, the subsidiary arguments, as noted, often are not stated explicitly. They are the culture's gift which the advocates try to use to their own advantage. Their seeming omission is a signal that they are implicitly agreed to as support for the main argument. This implicit agreement can be revoked at any time during consideration of the main argument, by request of any arguer for the production of any of the normally missing elements. Such a request signals a devolution of the main argument to one or more of the subsidiary arguments.

Fourth, while this perspective simplifies understanding of Toulmin's *system*, since complex arguments are at root merely the repeated instances of Grounds, Warrant, and Claim, it complicates the schematic *layout* that usually is referred to as the "Toulmin model". This is true particularly because the different components have more than one function at the same time and because a complete schematic reconstruction must reconstruct what is left unsaid. A two-dimensional layout may be especially unsuited to this task.

Fifth, the acceptability of an argument (main or subsidiary) is determined by the audience whose assent is being sought. These are people who, by their choice to participate in argumentation, have *de facto* accepted its norms.

Sixth, and finally, the primacy of the audience in the evaluation of argumentation makes clear the importance of rhetoric as a perspective on argumentation, not just in the obvious case of a speaker addressing a mass audience in an attempt to persuade them, but even in the interpersonal situation in which arguers and judges are the same people.

REFERENCES

Farrell, T. B. (1976). Knowledge, consensus, and rhetorical theory. *Quarterly Journal of Speech*, 62(1), 1-14.
Fogelin, R. J. (1985). The logic of deep disagreement. *Informal Logic*, 7(1), 1-8.
Govier, T. (1985). *A practical study of argument.* Belmont, CA: Wadsworth.
Johnson, R. H. (2000). *Manifest rationality: A pragmatic theory of argument.* Mahwah, NJ: Lawrence Erlbaum.
Johnstone, H. W. Jr. (1959). *Philosophy and argument.* University Park: Pennsylvania State University Press.
O'Keefe, D. J. (1977). Two concepts of argument. *Argumentation and Advocacy*, 13(3), 121-128.

Perelman, Ch. & Olbrechts-Tyteca, L. (1958/1969). *The new rhetoric: A treatise on argumentation* (J. Wilkinson & P. Weaver, Trans.). Notre Dame, IN: University of Notre Dame Press.
Toulmin, S. (1958). *The uses of argument.* Cambridge: Cambridge University Press.
Zarefsky, D. (2016). On deep disagreement. In R. Von Burg (Ed.), *Dialogues in argumentation* (pp. 13-33). Windsor, ON: Windsor Studies in Argumentation (E-book).

www.ingramcontent.com/pod-product-compliance
Lightning Source LLC
Chambersburg PA
CBHW071147230426
43668CB00009B/863